STUDIES AND TEXTS

49

Plate 1. Saint Albert the Great by Tommaso da Modena (1352)
in the Chapter Room of the Dominicans (now the Seminary)
in Treviso, Italy

Beatus frater Albertus Coloniensis de
provincia Theotonie ordinis fratrum
predicatorum episcopus Ratisponensis
clarissimus magister in sacra theologia
Excellentis doctrine sicut eius
scripta multa in omni scientia declarant
multa miracula fecit.

ALBERTUS MAGNUS AND THE SCIENCES
Commemorative Essays 1980

EDITED BY

James A. Weisheipl, OP
Pontifical Institute of Mediaeval Studies

PONTIFICAL INSTITUTE OF MEDIAEVAL STUDIES
TORONTO, 1980

Canadian Cataloguing in Publication Data

Main entry under title:

Albertus Magnus and the sciences

(Studies and texts — Pontifical Institute of
Mediaeval Studies ; 49   ISSN 0082-5328)

Bibliography: p.
Includes index.

ISBN 0-88844-049-9

1. Albertus Magnus, Saint, 1193?-1280 — Addresses,
essays, lectures.   2. Science, Medieval — Addresses,
essays, lectures.   I. Weisheipl, James A., 1928-
II. Series: Pontifical Institute of Mediaeval Studies.
Studies and texts — Pontifical Institute of Mediaeval
Studies ; 49.

Q153.A53   509′.022   C79-094535-5

Sancto Alberto Magno Doctori
Universali ac Patrono
Cultoribus
Scientiarum Naturalium
atque
Quaerentibus
Sapientiam Divinam

Contents

Preface

As I prepared my mind and soul over the past five years to produce a major work on the life and thought of Friar Albert the Great, the year 1980, marking the seven-hundredth anniversary of his death on 15 November, drew ever closer with increasing urgency. The inner compulsion to mark that unique occasion in some memorable way swelled within me like some forceful fountain of living water nourished by three distinct sources.

First, as professor of the history and philosophy of medieval science in the University of Toronto, I was fully aware of the dreadful dearth of serious studies, particularly in English, about Albert, the most influential scientist of the Middle Ages. To most moderns he is known simply as the teacher of St. Thomas Aquinas, if he is known at all. For some reason, contemporary medievalists west of the Rhine have bypassed his unsuspected influence not only on the thirteenth century, but on at least four subsequent centuries. I therefore felt constrained to do something constructive to fill this lacuna in the history of medieval science.

Second, as a grateful and devoted member of the Dominican Province of St. Albert the Great in the United States and a medievalist, I felt a special need to make St. Albert better known to English readers in the Dominican family. It just so happens that there is a special reason for grateful rejoicing at the present time. On 22 December 1979, the Province of St. Albert the Great celebrates the fortieth anniversary of its founding as a separate province in the Dominican Order. This volume, esoteric though it will seem to many, is a small token of prayerful gratitude to St. Albert for the Dominican province established under his Patronage in the United States.

Third, and perhaps the most compelling force, was a profound concern for modern scientists in all fields, who, whether they fully realize it or not, are in need of a sympathetic and saintly "Patron before God" in the new age we have created, perhaps without full realization of the consequences, some forty to fifty years ago.

When on 16 December 1941 Pope Pius XII proclaimed St. Albert the Great "forever the PATRON before God of students of the natural sciences with the supplemental privileges and honours which belong, of its nature, to this heavenly patronage," he may very well have been inspired by divine providence. But he was also fully aware of what the whole scientific community already knew to be a certainty, namely that an atomic bomb, the like of which had never been seen, could in fact be produced. One can truthfully say that St. Albert the Great was proclaimed Patron of natural scientists at the very conception of the atomic age, a phrase that had no meaning whatever to most people until 6 August 1945.

The theoretical possibilities of producing nuclear fission from certain elements, such as uranium, travelling at tremendous speed were long entertained by nuclear physicists throughout the world. By August 1939 Albert Einstein was induced by Leo Szilard and Eugene Wigner to write directly to his personal friend, President F. D. Roosevelt, warning him of the real likelihood of uranium being used to create an entirely new type of bomb, the need for particular vigilance over German activities in this regard, and suggesting immediate government action to coordinate scientific research in nuclear physics. Nothing really came of this or the two subsequent letters from Einstein early in 1940. By July 1941, however, British scientists had succeeded in producing nuclear fission and definitively demonstrating that an atomic bomb could in fact be produced. News from Niels Bohr that the Germans were transporting large quantities of "heavy water" from the Scandinavian countries and the illuminating visit to England by George Pegram and Harold Urey in the autumn of 1941 played the crucial role in the American decision on 6 December 1941 to create a concentrated and expanded program of unified research into nuclear weapons — a few hours before the Japanese attack on Pearl Harbor. In 1942 the Manhattan Project was firmly established with the full support of the United States government and military forces.

When Pope Pius XII proclaimed St. Albert the Great, Patron of natural scientists on 16 December, the whole world was at war and far too busy to be concerned about a medieval saint and his outmoded science. It was not until the world was stunned by the actuality of the first atomic bomb dropped on Hiroshima on 6 August 1945 and on Nagasaki three days later, that pained scientists, philosophers, churchmen, humanitarians of all kinds, and mankind itself were prepared to think about the need of some heavenly Patron in the new, dangerous, and uncharted age of atomic energy. But this is as far as things went. Atheists like Albert Einstein and Sir Bertrand Russell could cry, "Never again," and hold court, denouncing individuals for crimes against humanity. But humanity without God is not human. Mankind cannot "go it alone"; it never could.

The atomic bombings of 1945 ended World War II, but it did not bring peace. It merely ended one era and issued in an atomic age, triggered an arms race between super-powers, and pushed all areas of the technological sciences beyond their imagined limits. In 1952 the British succeeded in exploding the first hydrogen device in history at their proving grounds in the Pacific; two years later the U.S. launched the first atomic-powered submarine. Rocket technology, which had been developing for some decades, brought not only the jet airliner, but also inter-continental ballistic missiles, and successful rocket launchings from the earth, under the sea, and in the air. The successful Soviet launching of Sputnik I as the first man-made satellite in October 1957, jolted Americans out of their technological slumber; but the U.S. managed to put the first man on the moon in July 1969. At the same time, the harnessing of nuclear energy for peaceful purposes was sufficiently developed so that the first commercial nuclear reactors could be built in 1970 and sold throughout the world. It is only recently, however, that both the necessity and dangers of nuclear power even for peaceful purposes are being more fully realized. Neither war nor peace can ever again be as it was before 1941. In this new age, every man of responsibility, whatever his profession, needs caution, sobriety, prudent weighing of risks that cannot be escaped, and some assurance that he is with God. Pope John Paul II notes this need in *Redemptor hominis* when he says, "The development of technology and the development of contemporary civilization, which is marked by the ascendancy of technology, demand a proportional development of morals and ethics" (n.15).

The more recent developments in medical practice, surgery, transplants, the whole range of pharmacology and addiction research, biochemistry, genetic engineering, electronics, and technology of every sort cannot be divorced from moral and human values. Today, more than ever, mankind cannot afford the dangerous illusion of self-sufficiency in human affairs. Not just mankind, but every person who has to make decisions needs divine guidance to bring the greatest benefits, scientific and human, to men of all nations in their quest for God. The "wisdom of this world" may very well, on its own, be its own undoing; it may very well lead to the end of this world and all mankind. One does not have to be a Roman Catholic or even a Christian to see the wisdom of prayer. Prayer does not change God, but it certainly changes the one who prays.

Perhaps it was thoughts such as these that moved Pope John Paul II to write in his very first encyclical: "Theologians and all men of learning in the Church are today called to unite faith with learning and wisdom, in order to help them combine with each other, as we read in the prayer in the liturgy of the feast of Saint Albert, Doctor of the Church" (*Redemptor hominis*, n.19).

It certainly was thoughts such as these that compelled me to devote four strenuous years to the production of this modest commemorative volume exclusively on Albertus Magnus and the sciences. Without the generous enthusiastic collaboration of all the scholars represented here — many of them young, but all with special knowledge of Albert's scientific writings — this volume could not have been compiled. To them, naturally, go my deepest gratitude and encouragement. But scholarly books cannot be published today without substantial subsidies. I am very happy to acknowledge that the entire subsidy for the publication of this volume has been generously supplied by my own Province of St. Albert the Great in the United States. A very special debt of gratitude is due to the Very Reverend Damian C. Fandal, OP, and his Provincial Council for their generosity, encouragement, and trust in a blind venture. While too many individuals could be singled out in gratitude for invaluable assistance in reading, correcting, and preparing manuscripts submitted, I cannot fail to mention the outstanding labours of Fr. Lawrence Dewan, OP, of Ottawa, the valuable suggestions of Dr. William E. Carroll, of Cornell College, Iowa, as well as the careful compilation of the bibliography by Fr. Bartholomew de la Torre, OP, and the final organization of the various indexes by Betsey Barker Price and Steven E. Baldner. Yet without the highly competent staff of the Department of Publications of the Pontifical Institute of Mediaeval Studies, the entire production would not be as presentable a tribute to St. Albert the Great as it seems to be, nor would it have met the scheduled deadline. To all of these individuals and to many more, my deepest gratitude, appreciation, and satisfaction.

James A. Weisheipl, OP
Editor
Feast of St. Albert the Great
15 November 1979

Abbreviations

ed. Colon.   Editio Coloniensis, Alberti Magni *Opera Omnia*, ad fidem codicum manuscriptorum edenda apparatu critico notis prolegomenis indicibus instruenda curavit Institutum Alberti Magni Coloniense (Münster i. Westf.: Aschendorff, 1951 seqq.). Volume, part, page, and line indicated in that order when required; also date of publication when helpful. Still in process of publication; 40 folio volumes projected.

ed. Borgnet   B. Alberti Magni, *Opera Omnia* (revision of 21 folio volumes published by Pierre Jammy, OP, Lyons, 1651), ed. Auguste Borgnet, dioc. of Reims, 38 quarto vols. (Paris: Vives 1890-99). Indicated by volume, followed by page and column *a* or *b* when necessary.

ed. Stadler   Albertus Magnus, *De animalibus libri XXVI*, nach der Cölner Urschrift, ed. Hermann Stadler. *Beiträge*, Bd. 15 (1916): Bk I-XII; Bd. 16 (1920): Bk XIII-XXVI. Pagination consecutive. Referred to by §§ within Books or line on page.

ed. Meyer and Jessen   Albertus Magnus, *De vegetabilibus libri VII*. Kritische Ausgabe von Ernst Meyer und Carl Jessen (Berlin: Georg Reimer, 1867). Pagination consecutive; referred to by page and section.

Albert   Title of work followed by book in roman numerals, *tractatus* (tr.) and *caput* (c.); usual abbreviation indicated within each article.

Aristotle   Title of work, book, chapter with Bekker numbers.

AL   *Aristotles Latinus* (Corpus Philosophorum Medii Aevi). Union Académique Internationale (Bruxelles-Paris: Desclée de Brouwer, 1939-; then Leiden: Brill, 1972 seqq.). Still in process. Indicated by its own series of roman numerals, parts, and sometimes fascicles, usually with editor and date when useful.

Beiträge   *Beiträge zur Geschichte der Philosophie [und Theologie,* from 1928] *des Mittelalters* (Münster i. Westf.: Aschendorff, 1891 seqq.).

MOPH *Monumenta ordinis Fratrum Praedicatorum historica*, ed. B. M. Reichert and subsequent editors (Rome 1896-1904; Paris 1931-33; Rome 1935 seqq.).

Introduction

Albertus Magnus and the Sciences

Edward A. Synan
Pontifical Institute of Mediaeval Studies

Seen across seven centuries, Albert the Great inevitably offers us a whole cluster of puzzles. Nor is the lapse of time the only source of our consternation. How many traits and qualities must be predicated of that astonishing personality! Son of a rich and knightly family, a Dominican friar, a bishop who resigned his see, a canonized saint, an archetypal German professor, Albert poses problems that are by no means of uniform difficulty.

To uncover and to evaluate his achievements in the sciences of nature demand historical techniques of a high order and an exceptional degree of sensitivity, but we know that the thing is possible because our contributors have managed to do it in this volume. Despite many a revolution, families of rank are familiar to this day; German universities still produce prodigies of scholarship who help us to feel at home with this thirteenth-century harbinger of their guild. We have Dominicans among us, to be sure, but not every Christian nor even every Catholic has necessarily comprehended the Dominican mystique of voluntary poverty and study, or knows that it is incorrect (and still less why it is incorrect) to call them "monks." In our world bishops have somehow lost the caste they enjoyed or tolerated in Albert's time when "prince-bishops" shared with kings and dukes and emperors the ambiguous blessings of temporal power.

Sanctity, of course, is neither absent from our world nor is it always unnoticed. Not to speak of saints who have been "canonized" — officially listed by papal pronouncement as worthy of veneration after an adversary process in which even the devil has his advocate

— there are those others whose lives are of such a quality as to persuade us that "saint" is not too strong a term for them: Mother Theresa and Albert Schweitzer, Dag Hammarskjöld, the Pope John whom all the world called "Good" — the list can be lengthened easily. With due allowance for the joyous creativity that marks medieval hagiography and dismays our historians, we may concede that Albert's *Vita*[1] conveys a convincing impression of continuity with those to whom our time spontaneously grants high rank in the order of the spirit.

Not Albert himself, but Pope Piux XII in 1941 provided an enigma of another order when he named Albert "patron of those who cultivate the sciences of nature."[2] Scientists, we dare to think, feel no compelling need for a medieval patron. Besides, what conceivable role can be played by a "patron" who has been seven hundred years in his tomb? Albert's seminal work in a number of empiric disciplines moved the pope to esteem his attitude towards science in the service of peace — "May he stir up hearts and minds to the right and peaceful use of natural realities" — for Pope Pius wrote at a moment when, for the second time in our century, science was serving global war: "Owing to our most doleful condition . . . today's scientific advances are used wretchedly now to carry the disasters of war to civilian areas and cities . . . ."[3] One can hardly deny that it is better to use science for life than to use science for death, but what use can there be in pronouncing a dead medieval theologian the "patron" of scientists, not all of whom share his faith?

Outside the Roman Catholic Church and, indeed within her, official steps with respect to Albert can provoke an understandable perplexity. What can have motivated Church authorities to concern themselves with canonizing a thirteenth-century professor, with naming him not only a "saint," but also a "doctor of the Church"? Why, above all, this preoccupation with labels so little intelligible in our world at a time when our race faces threats of the most ominous sort?

---

[1] *Legenda beati Alberti*, auctore Rudolpho de Novimagio, editor H. Chr. Scheeben, editio altera (Cologne, 1928).

[2] ". . .Cultorum Scientiarum naturalium coelestem apud Deum *Patronum* declaramus et constituimus. . . ": Litterae apostolicae, 16 December 1941, *Acta apostolicae sedis*, XXXIV, (13 April 1942), p. 91.

[3] ". . .Excitet corda mentesque ad pacificum rectumque naturae rerum usum. . . ob tristissimam quoque nostrorum dierum condicionem. . . ad belli calamitates civilibus regionibus urbibusque inferendas nunc hodierni scientiarum progressus misere adhibeantur": ibid., p. 90.

The perplexity may be eased by a most sympathetic pre-Vatican II novelist who thought it right to develop the mind-set that allows churchmen to be fascinated in the most baleful of times with what could well be counted the inconsequential. This is Walter M. Miller, author of *A Canticle for Leibowitz*,[4] a novel that purports to recount centuries of future human history after an atomic war. The Church survives; all political entities have disappeared and new ones have sprung up; genetic mutants (popularly called "pope's children" because the Church spoke for their right to live) bewilder the survivors; material culture has regressed to approximately the level that marked western Europe in the sixth century. Almost everything had to be relearned, rediscovered. Every trace of science or of culture that had escaped the wave of militant anti-intellectualism, "The Simplification," was doggedly preserved by monastic "Brother Memorizers" after copies of what they had memorized had been taken into the deserts and hidden in barrels by other monks, these last the "Brother Bookleggers." Often no monk had any notion of what those relics from before the deluge of fire might mean; blueprints were painstakingly copied and illuminated against the day when their meaning might be recovered; Church business continued in jog-trot Latin.

In the midst of all this, Dominicans are represented as contending with "some theologians of other Orders" on a subtle dispute over the status of the Virgin Mary, to the great disadvantage of the case for canonization of Leibowitz. For the eponymous hero behind the novel was an atomic scientist; he had turned to the Church after the disaster and then fell victim to the mobs who wrought The Simplification. Followers of Leibowitz, dedicated to advancing the mission of the Church by recovering the lost sciences whose abuse had led to so much tragedy were the "Albertian Order of Leibowitz," AOL. Told in the idiom of the future, the tale is a restatement of the past and not of the medieval past only. For it is not by accident that a novel published in 1959 has responded to the 1931 canonization of Albert the Great by Pope Pius XI and to the 1941 initiative of Pius XII in naming him "patron of those who cultivate the sciences of nature."

No doubt in the summer months of 1931 as Pius XI locked horns with Mussolini over the survival of Catholic Action in Italy, the former librarian faced as pope a lesser trial than those of the fictional

---

[4] Walter M. Miller, Jr., *A Canticle for Leibowitz* (Philadelphia, 1959).

Leibowitz. Still, the best historian of those days was, and has remained, convinced that the pope's victory over the Duce masked the magnitude of the peril through which Pius XI had manoeuvred the Church.[5] It was during those very months of struggle that the canonization of Albert went forward; in December the pope proclaimed him a saint and a doctor of the Church, that is, both a man of heroic virtue and one who had constituted himself an effective teacher of what the Church holds ought to be taught. Here it may be remarked that on Albert's own criteria this threefold accolade by the Church he served — saint, doctor of the Church, and patron for his earthbound colleagues — must outweigh all his merely academic or scientific achievements. For us, however, those more pedestrian accomplishments are the more accessible; Albert will pardon our interest in the methods he used to advance the study of his world and ours.

The present volume is not our first evidence that Saint Albert's scientific concerns have intrigued his posterity. Not all of us have been interested in his work on grounds that Albert would have counted "the right reasons." Like Boethius[6] centuries before, Albert's investigations of the properties and interaction of natural items — stars and stones, minerals and herbs — to say nothing of his mathematics, astrology, and alchemy, earned him a reputation, if not as a wizard, at least as a "magician" who, in medieval terminology, might be nothing more sinister than a practitioner of applied science. Such was surely the benign meaning intended by his respectable student, Ulrich of Strasburg, OP, who recorded one of Albert's frequent dissents from the views of predecessors along with the qualities that gave weight to his master's opinion:

My Master (the Lord Albert, sometime bishop of Ratisbon, a man so

---

[5] "In 1931, two years after the Concordat was signed, the smouldering resentment of Fascism against Catholic Action burst into flame. There ensued a period of open warfare. . . . The brief duration of the dispute and its sudden disappearance have induced some foreign observers to dismiss it as 'much ado about nothing,' and they have been reinforced in this belief by the discreet silence on all matters concerning it which has generally obtained in both camps up to quite recently [NB: written in 1941]. But in my judgement at least, this is a distorted view. . .": D.A. Binchy, *Church and State in Fascist Italy* (London, 1941; new impression and preface, 1970), pp. 506, 530, 531.

[6] Boethius, *Philosophiae consolatio* 1, prosa 4; The Loeb Classical Library, p. 152, lines 133-145; PL 63: 628-629 for the accusation of witchcraft. For an idiosyncratic view that Boethius was guilty according to Ostrogothic law see R. Bonnaud, "Note," *Speculum* 4 (1929), 198-206; *sacrilegium* is construed in this context as "magic."

god-like, *divinus*, in all science that he could be called with propriety "the marvel and the miracle of our time," experienced too in the magical arts on which knowledge of this material greatly depends) thought differently from all the aforesaid.[7]

Even today bookshops given to the occult may stock on occasion an "Albertus Magnus Dreambook," yet another good reason for welcoming this serious appraisal of a sober thirteenth-century figure who did pioneer work in a whole range of scientific disciplines. Indeed, one of Albert's major themes was the necessity to discriminate between the merely superstitious and legitimate scientific interest, a discrimination that would inhibit summary condemnations.

If the scientific story ought to be told, it is not the whole story. Albert was both a many-sided scholar and one who had a highly developed sense of hierarchy; a grasp of each aspect of his scholarship demands that it be put into the context of his total work and that we know how Albert himself assessed each major segment of his multiform activity. Let us say it with candor: he did not locate scientific investigation of the natural world at the summit of intellectual endeavor. To appreciate his concern with natural science requires that we remember what he held to be superior to the scientific effort that consumed so much of his energy.

What we term "science" had been pursued since the golden age of Greek philosophizing[8] under the rubric of "physical knowledge," that is, the reliable, cogent grasp of "the natures," $\alpha\iota\ \phi\acute{\upsilon}\sigma\epsilon\iota\varsigma$, encountered in the world of experience. The editor of this volume, J. A. Weisheipl, OP, has more than once examined the medieval assimilation of that Greek tradition.[9] Without calling into question the

---

[7] "Aliter autem ab omnibus praemissis sentit doctor meus dominus Albertus episcopus quondam Ratisponensis, vir in omni scientia adeo divinus ut nostri temporis stupor et miraculum congrue vocari possit et in magicis expertus ex quibus multum dependet huius materiae scientia": J. Daguillon, *Ulrich de Strasbourg o.p. La "summa de bono," Livre 1* (Paris, 1930), p. 139.

[8] Both the "Platonic-Stoic" and the Aristotelian schemes of classification had $\phi\upsilon\sigma\iota\kappa\acute{\eta}$ as one of three species of knowledge, with respectively $\acute{\eta}\theta\iota\kappa\acute{\eta}$ and $\lambda o\gamma\iota\kappa\acute{\eta}$, or with $\mu\alpha\theta\eta\mu\alpha\tau\iota\kappa\acute{\eta}$ and $\theta\epsilon o\lambda o\gamma\iota\kappa\acute{\eta}$; Hegel complained in 1830 that the English were still using "philosophical" to qualify what ought to be termed "scientific." "Logic," *Encyclopaedia of the Philosophical Sciences*, tr. W. Wallace, part 1 (Oxford, 1975), p. 11; cf. remarks by W. James, *Pragmatism* (Cambridge, Mass., 1975) pp. 92-94 (1906 lecture) and J. Dewey, "A Recovery of Philosophy" in *Creative Intelligence* (New York, 1917), pp. 3-69, especially pp. 5, 6.

[9] J. A. Weisheipl, "The Nature, Scope, and Classification of the Sciences," *Studia Mediewistyczne* 18 (1977), 85-101; idem, "Classification of the Sciences in Mediaeval Thought," *Mediaeval Studies* 27 (1965), 54-90.

intelligence with which men of the Middle Ages handled the theoretical classification of whatever sciences were available to them, we must concede that their *physica* and *mathematica* were woefully short on content. No one saw this more clearly than did Albert. As a good Aristotelian he recognized where the trouble lay. The Philosopher has observed that the reason one ought to discuss geometry with geometricians only is that they alone can be counted on to catch an unsound argument, an observation he made in a logical work.[10] Although the caution can be given general application, it bears particularly upon the incapacity of the logician as such to increase the content of any discipline other than logic itself. It is well-known that Aristotle set a splendid example of research for concrete data, not only when dealing with problems in biology or physiology,[11] but in political science as well where his method led him to collect 158 city-state constitutions before attempting an essay on constitutional theory.[12] This lesson, largely lost on medieval academics, was not lost on Albert and the essays that follow establish the point abundantly. Here it will suffice to note a few of Albert's explicit remarks on the issue.

Albert argued that the logician, armed with his syllogisms only, is out of his depth in sciences that bear on nature. Only experience, one's own or that of others, laboriously discovered — *ex dictis eorum quos comperimus non de facili*[13] — holds the key to the scientific city. The reason for this resistance to syllogism as panacea is one that any Aristotelian logican ought to have understood: "In natures so particular a syllogism cannot be had"[14] — syllogistic science is necessarily expressed in universal propositions, whereas the investigator in a particular science must deal with instances that, by definition, fail of universality. The competence of the logician as logician extends no farther than the delimitation of a field for investigation. Sound

---

[10] *Posterior analytics* I, c.12 (77b6-15); cf. ibidem I, c.13 (79a2-6).

[11] *On the parts of animals* I, c.5 (644b22).

[12] W. Jaeger, *Aristotle, Fundamentals of the History of his Development*, tr. R. Robinson (Oxford, 1955), p. 329.

[13] *De vegetabilibus* VI, tr.1, c.1 (ed. Borgnet 10: 159b, 160a). Not only in physics, but also in metaphysics, Albert was chary of the logicians: "Sunt autem QUIDAM Latinorum logice persuasi . . . et huiusmodi multa ponunt secundum logicas et communes convenientias, et hi more Latinorum, qui omnem distinctionem solutionem esse reputant. . . . Sed ego tales logicas convenientias in scientiis de rebus abhorreo, eo quod ad multos deducunt errores" (ed. Colon. 16/1: 5.34-49).

[14] ". . . de tam particularibus naturis syllogismus haberi non potest": *De vegetabilibus* VI, tr.1, c.1 (ed. Borgnet 10: 160a).

method, to be sure, requires that sciences be classified and logicians can classify them with precision and even with elegance, but the masters of syllogism have nothing to offer with regard to the content of any scientific discipline.

In Albert's judgment, a conclusion in physical science that contradicts sensation is at least suspect and a "principle" discovered to be out of harmony with experiential knowledge can only be a pseudo-principle.[15] There is more than a hint that Albert's "experience" (he seems to have used *experientia* and *experimentum* interchangeably) shades from brute observation toward a methodical, systematic "experimentation." Often he recalls remarkable phenomena on which he had stumbled in his travels:

I say, then, that when I was at Venice, as a young man, marble was being cut with saws to decorate the walls of a church. And it happened that when one [piece of] marble had been cut in two . . . there appeared a most beautiful picture of a king's head with a crown and a long beard. . . . A long time afterwards, when I was at Paris, in the number and company of scholars, it happened that the son of the king of Castille came to study there. And when the cooks of this nobleman wanted to buy fish his servants bought a fish which in Latin is called *peccet*, and in the vernacular, plaice. . . . And when they gutted it, they discovered in its belly the shell of a large oyster, which this same nobleman kindly caused to be presented to me. The shell, on its concave side, which was smooth and shining, had the figures of three serpents with their mouths uplifted, so perfectly represented that not even the eyes were missing . . . on the convex outer side, which was rough, it had the figures of many — ten or more — serpents similarly represented in all details. . . . This shell I kept for a long time, and I showed it to many people, and later I sent it as a gift to someone in Teutonia.[16]

---

[15] *Physica* VIII, tr.2, c.2 (ed. Borgnet 3: 564b).

[16] Translation by D. Wyckoff (*Albertus Magnus. Book of Minerals*, [Oxford: Clarendon, 1967], pp. 128-129) of *De mineralibus* II, tr.3, c.1 (ed. Borgnet 5: 48b-49b): "Dico igitur, me essente Venetiis, cum essem juvenis, incidebantur marmora per serras ad parietes templi ornandos: contigit autem in uno marmore iam inciso . . . apparere depictum caput pulcherrimum regis cum corona et longa barba: . . . Post hoc autem longo tempore cum essem Parisiis de numero doctorum et grege, contigit advenire ad studium filium regis Castellae, cuius coqui cum pisces emerent, praenominati nobilis famuli piscem emerunt, qui Latine *peccet*, vulgariter *pleis* vocabatur. . . cum autem exenteraretur piscis, in ventre eius apparuit concha ostrei maximi, quam ad me memoratus nobilis fecit causa dilectionis adaptari: concha ergo illa concavo sui quod est planum et politum, habebat figuram trium serpentum ore elevato optime factorum, ita quod nec figura defuit oculorum. . . exterius autem in convexo quod erat asperum, habebat figuras multorum, decem videlicet et amplius serpentum simili modo per omnia opere factorum. . . Hanc autem concham ego multo tempore habui, et multis ostendi, et postea eam misi pro munere in Teutoniam cuidam."

On the other hand, Albert realized that more might be required if what has been observed is to function as the ground of secure inferences in the sciences of nature:

> For it is necessary to probe experience, not in one way only, but according to all circumstances, in order that it be certainly and correctly a principle for operation.[17]

This sifting inevitably consumes a good deal of time; Hippocrates, Albert knew, had coined the aphorism, "Life is short whereas art is long, experience fallacious, judgment difficult," and it is a view that Albert was willing to concede.[18] In mathematical disciplines (*doctrinalibus*) this time-consuming process is not necessary: enough to consider one triangle and the truth that all of them enclose angles equal to two right angles is established; or to calibrate the time required for a star or stars to pass through one degree of a celestial arc and the rate of all celestial motion will be known.[19]

Naturally he could not see everything for himself and part of the difficulty of the scientist, as Albert saw it, was to check and to evaluate the reliability of witnesses. The draconopodes, for instance, had never come under his observation, but one of those great serpents (they belong in the third order of dragons) had been killed in a German forest where trustworthy witnesses had reported that the carcass had been offered for inspection until it rotted.[20] Sometimes he could not accept the tales that were told. Dragons flying through the air and breathing forth glittering fire he counted impossible, unless the stories referred to certain cases of vapor (described in the Book of *Meteors*) which the uninstructed might mistake for flying and fire-breathing animals.[21]

More than once Albert was impelled to notice critics of his methods and conclusions. True enough, he did not think that every carping critic deserved attention, especially when the criticism proceeded neither from a careful reading of what he had written nor from a comparison with the work of others:

> If one who has not read and compared should register a complaint,

---

[17] "Oportet enim experimentum non in uno modo, sed secundum omnes circumstantias probare, ut certe et recte principium sit operis": *Ethica* VI, tr.2, c.25 (ed. Borgnet 7: 443a).

[18] Ibid., 7: 442b-443a.

[19] *Metaphysics* I, tr.1, c.8 (ed. Colon. 16/1: 11.69-72 and 87-90).

[20] *De animalibus* XXV, c.29 (ed. Stadler, 1567.21-26).

[21] Ibidem, XXV, c.27 (ed. Stadler, 1567.7-16).

then it is clear that he complains out of dislike or out of ignorance and I have small concern for complaints from men of that type.[22]

In one extended passage he proposed an image of lazy and malicious critics drawn from physiology: they function as does the liver in the body and just as the "humor of gall" embitters the whole body, so there are extremely embittered, gall-like men in the academic world who transform all others into bitterness and are unwilling to let those others seek the truth in a pleasant society of scholars.[23] In some cases Albert's fellow Dominicans were given to ignorant protests against the use of "philosophy" and no one in the order offered any opposition to them: "Like brute animals they blasphemed in matters of which they were ignorant."[24] All of this is familiar to scholars of every generation: the innovator whose work meets resistance from lesser men grumbles, but continues.

By no means blind to some faults on the part of some Dominicans, Albert was more than conscious of the intellectual vitality that marked Dominican houses in his time. No doubt some allowance must be made for the *topos* of the teacher whose disciples entreat him for instruction,[25] but if we have given weight to Albert's strictures against some confreres, his benign references to others ought to receive equal attention. He opened his exposition of the eight books of Aristotle's *Physics* with just such a reference:

Our intention in the science of nature is to satisfy (in accord with our capacity) brothers of our order, begging us for the past several years now[26] that we might compose a book on physics for them of such a sort that in it they would have a complete science of nature and that from it they might be able to understand in a competent way the books of Aristotle. Although we consider ourselves insufficient for this task, nev-

---

[22] ". . . Si autem non legens et comparans reprehenderit, tunc constat ex odio eum reprehendere, vel ex ignorantia: et ego talium hominum parum curo reprehensiones": *De animalibus* XXVI, in fine (ed. Stadler, 1598.13-15).

[23] ". . . pro talibus, qui in communicatione studii sunt quod hepar in corpore; in omni autem corpore humor fellis est qui evaporando totum amaricat corpus, ita in studio semper sunt quidam amarissimi et fellei viri, qui omnes alios convertunt in amaritudinem, nec sinunt eos in dulcedine societatis quaerere veritatem": *Politica* VIII, 6 (ed. Borgnet 8: 804).

[24] ". . . et maximi in praedicatoribus, ubi nullus eis resistit, tamquam bruta animalia blasphemantes in iis quae ignorant": *In Epistolam* VII *Dionysii*, 2 (ed. Borgnet 14: 910a).

[25] Cf. *De causis et processu universitatis* II, tr.5, c.25 (ed. Borgnet 10: 619b), as well as the text cited below, note 27.

[26] P. Hossfeld has calculated that those years of unfulfilled requests most likely began with 1248 when Albert was transferred from Paris to Cologne in order to found a *studium generale* of his order there. *De caelo et mundo*, Prolegomena (ed. Colon. 5/1: v).

ertheless, since the requests of our brothers would not cease, finally we have undertaken what we had often refused, vanquished by the pleas of certain ones among them.[27]

In one frequently cited passage Albert associated an apologia of this sort with the explanation that his personal point of view must be sought in his theological writing rather than in his works on natural science:

> . . . for if, perchance, we should have any opinion of our own, this would be proffered by us (God willing) in theological works rather than in those on physics.[28]

This classic expression of his arm's length posture with regard to the Aristotelian physical treatises is also a statement of the way he discriminated between natural science and theology for he prefaced the remark just quoted with these words:

> There is, however, another sort of vision and prophecy according to extremely profound theologians, who speak of divine inspirations and concerning these we say nothing at all for the present on the ground that this sort of thing can in no way be known by means of arguments derived from nature. Pursuing what we have in mind, we take what must be termed "physics" more as what accords with the opinion of Peripatetics than as anything we might wish to introduce from our own knowledge.[29]

This juxtaposition of the "extremely profound theologians" (with whom, of course, Albert dealt in his theological treatises) and the "Peripatetics" invites us to advert to what this theologian thought about the authority of the Peripatetic par excellence, Aristotle himself. Surely it is not necessary to expand on the theme that he

---

[27] "Intentio nostra in scientia naturali est satisfacere pro nostra possibilitate fratribus Ordinis nostri, nos rogantibus ex pluribus iam praecedentibus annis, ut talem librum de physicis eis componeremus, in quo et scientiam naturalem perfectam haberent et ex quo libros Aristotelis competenter intelligere possent. Ad quod opus licet nos insufficientes reputemus, tamen precibus fratrum deesse non valentibus, quod multoties abnuimus, tandem suscepimus, devicti precibus aliquorum": *Physica* I, tr.1, c.1 (ed. Borgnet 3: 1a-b).

[28] ". . . Si quid enim forte propriae opinionis haberemus, in theologicis magis quam in physicis, Deo volente, a nobis proferetur": *De somno et vigilia* III, tr.1, c.12 (ed. Borgnet 9: 195b).

[29] "Est autem et aliud genus visionis et prophetiae secundum altissimos theologos qui de divinis loquuntur inspirationibus, de quibus ad praesens nihil dicimus omnino: eo quod hoc ex physicis rationibus nullo modo potest cognosci: physica enim tantum suscepimus dicenda plus secundum Peripateticorum sententiam persequentes ea quae intendimus, quam etiam ex nostra scientia aliquid velimus inducere. . .": Ibid.

thought Aristotle worth reading and worth explaining; indeed he went so far as to expend considerable effort in filling what he conceived to be gaps in the Aristotelian corpus. In a celebrated passage on his own methodology in expounding the Aristotelian heritage he undertook to specify precisely what that task entailed:

> And we shall also add, in certain places, parts of unfinished books, and in others, books passed over or omitted, ones which Aristotle did not produce or, if perhaps he did produce them, they have not reached us.[30]

But to study is not to worship; Albert did not think Aristotle "Nature's best effort" and a "canon of truth" as did the Peripatetics although, to be sure, not even the Peripatetic school functioned without a certain freedom of interpretation:

> All the Peripatetics, however, agree on this: that Aristotle spoke the truth, for they say that Nature set up this man as if he were a rule of truth in which she demonstrated the highest development of the human intellect — but they expound this man in diverse ways, as suits the intention of each one of them.[31]

Albert defended his own independence too, but he did not feel himself reduced to "twisting the nose of Authority"[32] in order to do so:

> Perhaps some will say that we have not understood Aristotle and that on this account we have not agreed with what he said or that (from their certain knowledge) we contradict him in point of truth on some matter. To him we say that whoever believes that Aristotle was a god ought to believe that he never erred; if, however, one believes him to be but a man, then without doubt he could err just as we can too.[33]

---

[30] "Et addemus etiam alicubi partes librorum imperfectorum, et alicubi libros intermissos vel omissos, quos Aristoteles non fecit, et forte si fecit, ad nos non pervenerunt. . .": *Physica* I, tr.1, c.1 (ed. Borgnet 3: 2a).

[31] "Conveniunt autem omnes Peripatetici in hoc quod Aristoteles verum dixit: quia dicunt quod natura hunc hominem posuit quasi regulam veritatis, in quo summam intellectus humani perfectionem demonstravit: sed exponunt eum diversimode prout congruit unicuique intentioni": *De anima* III, tr.2, c.3 (ed. Colon. 7/1: 182.8-14).

[32] Alan of Lille, *De fide catholica* 1.30, made the often cited joke: "Auctoritas cereum habet nasum, id est, in diversum potest flecti sensum" (PL 210: 333).

[33] "Dicet autem fortasse aliquis nos Aristotelem non intellexisse, et ideo non consentire verbis eius, vel quod forte ex certa scientia contradicamus ei quantum ad rei veritatem. Et ad illum dicimus quod qui credit Aristotelem fuisse Deum, ille debet credere quod numquam erravit, si autem credit ipsum esse hominem, tunc procul dubio errare potuit sicut et nos": *Physica* VIII, tr.1, c.14 (ed. Borgnet 3: 553b).

In the end, theology remained Albert's principal academic interest. Sciences of nature were seen from his perspective as so many examinations of the effects wrought in and with space and time by what metaphysicians call the "First Cause" and what theologians call God. Each science had a degree of autonomy and even so dubious an art as palmistry ought to be given the benefit of every doubt in its claim to scientific status. Should we grant that Albert is the author of the *Speculum astronomiae*, palmistry might be a part of physiognomy, not that moral characteristics are "caused" by exterior corporeal configurations, but both may have a common cause. "I am unwilling," the author wrote, "to make a precipitous decision" on the question.[34] Whatever the specific status of a scientific discipline might be, whatever the causal connections that might bind its materials to those of other arts or sciences, Albert approached them all with a theologically grounded conviction that they proceed without exception from a single cause: the God of Abraham, Isaac, Jacob, and Jesus.

---

[34] "De chiromantia vero nolo determinationem praecipitem ad praesens facere, quia forte pars est physiognomiae, quae collecta videtur ex significationibus magisterii astrorum super corpus et super animam, dum mores animi conicit ex exteriori figura corporis; non quia sit unum causa alterius sed quia ambo inveniuntur ab eodem causata": *Speculum astronomiae*, c.17, ed. S. Caroti et al. (Pisa: Domus Galilaeana, 1977), p. 48, lines 17-21.

1

The Life and Works of St. Albert the Great

James A. Weisheipl, OP
Pontifical Institute of Mediaeval Studies

The "eighty and more" years of St. Albert's life are intertwined with three major movements that characterize the High Middle Ages: (i) urbanization of European society, especially in Germany and Eastern Europe; (ii) reevangelization of Christian Europe, mainly through the mendicant orders founded by St. Dominic de Guzman in 1215 and St. Francis of Assisi in 1223; and (iii) intensive growth and formulation of "scholastic" philosophy and theology in the university centres of Christendom, notably the University of Paris and its spin-offs, such as Oxford, Cologne, Cambridge, Toulouse, and Montpellier.

Although the urbanization of France, Italy, and parts of England had begun vigorously early in the twelfth century, Germany (apart from the Rhine Valley) was a backward country in 1200 "even by medieval standards."[1] "The thirteenth century," as Freed has shown,[2] "was the high point in the urbanization of medieval Germany," and the history of Germany in that century is in large part the history of the Dominican and Franciscan Orders beyond the Rhine, the Elbe, the Oder, and even the Vistula. The same, no doubt, can be said of Bohemia, Poland, and Hungary. When Albert joined

---

[1] John B. Freed, *The Friars and German Society in the Thirteenth Century* (Cambridge, Mass.: Mediaeval Academy of America, 1977), p. 24.

[2] Ibid., p. 43.

the Dominican Order as a young man at Padua in 1223, there was no German Dominican Province, and only two small priories had been established in the territory. But when he died in 1280, there were sixty-two flourishing priories in the Dominican Province of *Teutonia*, which stretched from Vienna, Austria (ca. 1225), to Bern, Switzerland (1269), in the south, and from Stralsund on the Baltic (1251) to Utrecht, Holland (1232), in the north. The German Dominicans even established a missionary house in Riga, Latvia, in 1234, which was incorporated into the Province of *Teutonia* in 1244, well before Albert himself became Prior Provincial of this vast territory. Finally, Friar Albert was the first German Dominican to become a master in theology (1245) from the University of Paris; and he himself established the first centre of higher studies (a *studium generale*) in Germany at Cologne in 1248. Although Albert was already a mature *lector* of theology in his home Province of Germany and well-trained in the "scholastic method" before he encountered the seductive "new learning" that inundated Paris from Greek and Arabic sources translated in the south, he eventually "rewrote" the whole of Aristotelian philosophy in the Latin language, restating, expounding, correcting, expanding, and even adding whole new areas of scientific thought, as we shall see. His younger contemporary Roger Bacon enviously complained in 1267-68 that philosophy was now considered by the bulk of students (a *vulgo studentium*) and some men of repute (*sapientes*) "to be already transmitted to the Latins, and completed, and composed in the Latin language" (*quod philosophia iam data sit Latinis, et completa, et composita in lingua Latina*). Bacon goes on to complain that all of this was done during his own days (*in tempore meo*) at Paris, roughly between 1237 and 1257, and that the author of this philosophy is considered an authority (*auctoritas*) on the same level as Aristotle, Avicenna, and Averroës, even though he is still alive — an unprecedented indecency![3] Albert's reputation as "the Great," even while he was still living, became not only legendary, but was grossly exaggerated, so much so that in the fourteenth and fifteenth centuries it was utterly fantastic.

Our task here is to separate and eliminate the myth from the real man of science, who sought only the discovery of the truth of nature and the vision of the Triune Creator of all things real and beautiful.

---

[3] Rogeri Bacon, *Opus tertium*, c.9, ed. J. S. Brewer, *Opera quaedam hactenus inedita* (London: RS 15, 1859), p. 30.

This is no easy task, not even in our own age of sophisticated historical methods and techniques. Despite the admirable scholarship of recent French and German historians concentrating on Albert's life and works, there remain many uncertainties and contradictions. Certainly there is nothing written in English that can serve as a reliable guide.

A. ALBERT'S BIRTH, YOUTH, AND ENTRY INTO THE DOMINICAN ORDER

Albert, a Swabian by birth, was commonly known to his European contemporaries as Friar Albert the German (*Frater Albertus Teutonicus*) or Albert of Cologne (*Frater Albertus de Colonia*). But to his countrymen and confreres in the German Province, he was more properly and accurately known as Friar Albert of Lauingen, as is indicated on the signet ring (*sigillum*) he received on becoming a master in theology at the University of Paris in 1245: *S. Fr. Alberti de Lavging O. Pr.*[4] Lauingen is a small town in Schwaben situated on the Danube between Ulm a few miles above (south-west) and Dillingen a few miles below (north-east), in the diocese of Augsburg. At that time Schwaben was part of Bavaria, and Austria was "lower Bavaria."

It is certain that Albert came from a military family (*ex militaribus*) of lesser nobility (knights) in the service of the counts of Bollstadt, whose castle, now in ruins, was less than 19 miles (30 km) from Lauingen.[5] But it would seem that the whole of Albert's family, including his younger brother Friar Henry of Lauingen, was not related to those who eventually took Bollestat or some variant as a family name.[6] Albert's knightly family, like all German soldiers, had long been in the service of Frederick Barbarossa of Schwaben, who had been king of Germany since 1152 and Holy Roman Emperor from 1155 until his death in 1190. It is uncertain how many brothers

---

[4] Paulus von Loë, "De vita et scriptis B. Alberti Magni," *Analecta Bollandiana* 19 (1900), 272-84; 20 (1901), 273-316; 21 (1902), 361-71. See Pt. 2, p. 276. All references to Loë are to this fundamental work and to Part 2, the chronology of Albert's life with sources (1901), unless otherwise noted. See also H. C. Scheeben, *Albert der Grosse: Zur Chronologie seines Lebens*, Quellen und Forschungen zur Geschichte des Dominikanerordens in Deutschland, 27 (Vechta: Albertus-Magnus-Verlag, 1931), p. 5. All references to Scheeben are to this basic work, cited simply as *Chronologie*, unless otherwise noted.

[5] Loë, p. 276, n.1.

[6] Scheeben, *Chronologie*, pp. 5-7.

and sisters Albert had; but it seems likely, considering the opposition to his becoming a Dominican, that a military career had been expected of him.

Notwithstanding the contradictory dates usually given for Albert's birth, all that can really be said is that he was born around 1200 or a little before.[7] Certainly the date 1206/07 insisted on by Mandonnet, Glorieux, Van Steenberghen, and others is too late; Mandonnet's disputable evidence is based on the statement of Henry of Herford's *Chronicon* (ca. 1355) that Albert was "a boy of sixteen years" when he entered the Order (in 1223).[8] At the same time, the date 1193 given by Franz Pelster, H.C. Scheeben, and most older authors is much too early; it is based on the statement of Luis of Valladolid (Paris, 1414) that Albert died in 1280, "having completed about (*circiter*) 87 years of his life."[9] Surprisingly, one of the first authors[10] to give both dates without apparently seeing their inconsistency was Peter of Prussia, Albert's first really critical biographer, writing in 1486-87. Peter rightly eliminated many of the myths accumulated by his predecessors, especially by Thomas of Cantimpré and Luis of Valladolid, preserved in the *Legenda Coloniensis*. But somehow he failed to see that Albert could not have been "a boy of sixteen years" in 1223 and a man "having completed about 87 years of his life" in 1280, as reported by Luis of Valladolid.

The only contemporary evidence of Albert's age when he died on 15 November 1280 is that of Tolomeo of Lucca, who says Albert was *plus quam octogenarius* in one passage and *octogenarius et amplius* in

---

[7] A separate paper will be published elsewhere devoted to examining the arguments proposed by P. Mandonnet, "La Date de Naissance d'Albert le Grand," *Revue Thomiste* 36 (1931), 233-56, already assumed in 1912 in *Dict. d'Hist. et de Géog. Ecclés.*, s.v. "Albert le Grand," 1: 1515.

[8] Henrici de Herford, *Chronica seu Liber de rebus memorabilibus*, ed. A. Potthast (Göttingen, 1859), p. 201. See Loë, p. 277, n. 5; Mandonnet, *Revue Thomiste* 36 (1931), 233-56; P. Glorieux, *Répertoire des Maîtres en Théologie de Paris* (Paris: Vrin, 1933), 1: 62; F. Van Steenberghen, *Siger de Brabant*, 2 [Les Philosophes Belges 13], (Louvain, 1942), p. 439; E. Gilson, *History of Christian Philosophy in the Middle Ages* (New York: Random House, 1955), p. 277 and notes.

[9] Luis of Valladolid, *Brevis historia de vita et doctrina Alberti Magni*, c.1, ed. in *Catalogus Codicum Hagiographicorum Bibl. Regiae Bruxellensis* (Bruxellis, 1889), 2: 96; for the history of this work and a better text of the catalogue of writings, see H. C. Scheeben, "Die Tabulae Ludwigs von Valladolid im Chor der Praedigerbrüder von St. Jakob in Paris," *Archivum FFr. Praed.* 1 (1930), 223-63. See Franz Pelster, *Kritische Studien zum Leben und zu den Schriften Alberts des Grossen* (Freiburg-im-Breisgau: Herder, 1920), pp. 34-52; Scheeben, *Chronologie*, pp. 4-5; Quétif-Échard, *Scriptores Ord. Praed.* (Paris, 1719), 1: 169b-170a. All references to Pelster are to his basic work, *Kritische Studien*, unless otherwise noted.

[10] The first seems to have been Luis of Valladolid in his *Brevis historia*, pp. 95-105.

another.[11] Tolomeo was a retired bishop and an octogenarian himself when he inserted the brief lives of Albert and Thomas into his monumental *Historia ecclesiastica* sometime after he had completed it in September 1317. His contemporary Bernard Gui merely copied Tolomeo's phrase in his chronicle: "Hic obiit in conventu Coloniensi anno Domini MCCLXXX, *octogenarius et amplius*."[12] Whatever may be said about the meaning of *et amplius* is sheer guesswork. The only reasonable birthdate consistent with the rest of Albert's chronology is ca. 1200.

As a young man, Albert was sent to Padua under the care of his uncle to study the liberal arts at the incipient university.[13] Many years later, he himself described "Patavia, which is now called Padua, in which a *studium litterarum* flourished for many years,"[14] and he recalled two memorable natural phenomena he had witnessed when he was a youth in Padua and in Venice,[15] as well as his earlier experiences with falcons as a boy.[16] Almost nothing is known about the *studium* at Padua in those early days. Albert says absolutely nothing about the teachers of law who migrated from Bologna to Padua early in the century.[17] He speaks only of a *studium litterarum*, which at Padua was always associated with medicine. Albert's interests at that time were certainly in natural phenomena. It is most likely that he had studied some of Aristotle's works that had been translated by James of Venice ca. 1150-70. But its unlikely that he absorbed much of Aristotle at that time.

Early in the summer of 1223, Jordan of Saxony, the immediate successor to St. Dominic (d. 1221) as master general of the Order of Preachers, came to Padua in the hope of bringing young students into the order by his preaching. There may have been a Dominican

---

[11] Ptolomaei Lucensis, *Historia Ecclesiastica*, XXII, c.19 and XXIII, c.36, ed. in L. A. Muratori, *Rerum Italicarum Scriptores* (Milan, 1724), 11: 1151 and 1184.

[12] Bernard Gui, additions to Stephen of Salanhac's *De Quatuor in Quibus Deus Praedicatorum Ordinem Insignivit*, ed. T. Kaeppeli, MOPH 22 (Rome: S. Sabina, 1949), p. 125; see Loë, p. 309, n. 217.

[13] Gerardi de Fracheto, *Vitae fratrum Ord. Praed.*, P.IV, c.13, § 9, ed. B. Reichert, MOPH 1 (Louvain, 1896), pp. 187-88.

[14] *De nat. loc.*, tr.3, c.2 (ed. Borgnet 9: 570b-71a).

[15] *Metheor.* III, tr.2, c.12 (ed. Borgnet 4: 629a); *De mineral.* II, tr.3, c.1 (ed. Borgnet 5: 48b-49a).

[16] *De animal.* VIII, tr.2, c.4, n.69 (ed. Stadler, 599.25-32); XXIII, tr.1, cc.5-9. On the long list of Albert's personal observations, see ed. Stadler, Index, s.v. "Albertus ego," 1599a.

[17] H. Denifle, *Die Entstehung der Universitäten des Mittelalters bis 1400* (Berlin, 1885), p. 277; Nancy G. Siraisi, *Arts and Sciences at Padua* (Toronto: PIMS, 1973), pp. 16-19.

priory in that city,[18] and Jordan was on his way to Bologna for the annual General Chapter of the order opening on Pentecost (June 11). At first he found "the students of Padua extremely cold," but ten of them soon sought admission, "among them two sons of two great German lords; one was a provost-marshal, loaded with many honors and possessed of great riches; the other has resigned rich benefices and is truly noble in mind and body."[19] The latter of these two has always been identified as Albert of Lauingen. In the *Vitae fratrum*, compiled by Gerard of Frachet between 1254 and 1258 from stories sent to him by brethren throughout the order, the story of Albert's "conversion" to the order by Jordan while he was a student at Padua is narrated at some length, supposedly in Albert's own words: "Hec autem ipse frater narravit sepius."[20] Not only have many untenable legends grown up about Albert's "conversion" by Jordan, but Mandonnet has completely misunderstood the passage in using it to support his thesis that Albert was sixteen years old when he entered the Dominican Order in 1223.

The Dominican Constitutions in vogue at the time of Jordan (1222-37) and clearly formulated in 1228 explicitly stated, "No one under the age of eighteen is to be received [into the Order]." This age was deliberately later than that specified in the Constitutions of Premontré, after which the Dominican Constitutions were patterned and adapted.[21] In the early decades of the order, applicants were considerably older when they received the habit and made solemn vows binding for life; even eighteen was considered "extremely young" until the second half of the thirteenth century. It is untenable that Albert could have been sixteen when he received the Dominican habit from Jordan in the early summer of 1223. Luis of Valladolid, who accepts Henry of Herford's date for Albert's entry as sixteen, also claims that Peter of Tarantaise, later Pope Innocent v (1276), was nine years old when he entered![22]

---

[18] *Vitae fratrum*, pp. 187-188, clearly implies the existence of a Dominican house in Padua at that time, but see Quétif-Échard, 1: vii, where Padua is listed *after* Venice, which was founded in 1234.

[19] *Beati Iordani de Saxonia Epistulae*, Ep. 20, ed. A. Walz, MOPH 23 (Rome: S. Sabina, 1951), p. 24.

[20] *Vitae fratrum*, p. 188.

[21] *Constitutiones Antique Ord. FFr. Predicatorum*, Dist.I, c.14, ed. A. H. Thomas, *De Oudste Constituties van de Dominicanen* (Louvain, 1965), p. 325 line 23; see H. Denifle, "Die Constitutionen des Prediger-Ordens vom Jahre 1228," ALKGMA 1: 202 and n. 3.

[22] Ed. H. C. Scheeben, "Die Tabulae Ludwigs von Valladolid," p. 253; noted by Quétif-Échard, *Scriptores* 1: 352b.

Albert was certainly a "young man" (*iuvenculus* or *adolescens*) when he entered the order, but he was not a "boy" (*puer*) in the canonical sense, nor even a "youth" (*iuvenis*) in the strict sense of being under twenty-one. Relatively speaking, Albert at the age of sixty could refer to his "youth" at Padua; and Roger Bacon, who joined the Franciscans when he was around forty, could speak with an air of disdain of Albert and Thomas as *pueri* when they entered their order.[23] In any case, it is certain that Albert joined the Dominican Order when he was a student at Padua, receiving the habit from Jordan of Saxony around Easter of 1223, despite many personal and domestic difficulties as narrated (with embellishments) in the *Vitae fratrum*.

B. YOUNG FRIAR ALBERT IN GERMANY (1223-CA. 1243/44)

Albert would have been received into the order for some specific province. But when he joined in 1223 there was no Province of Germany (*Teutonia*). There were only two priories: Friesach (1219) in Austria, which was in a critical state,[24] and Cologne (ca. 1220) on the lower Rhine, where Friar Henry, Jordan's classmate and companion at Paris, was then prior. It is most probable that Albert was sent to Cologne to make his novitiate and study theology from the local *lector*. Cologne, then, would henceforth be considered his native priory, even though he could be sent by his superiors anywhere in the known world. There are no grounds whatever for suggesting that Albert remained in the *studium* at Padua or was sent to Bologna or Paris for the study of philosophy. All religious were equally forbidden by church law to study philosophy in any secular *studium*, and at that time the Dominican Order had no *studium artium* of its own.

At Cologne, Albert would not only have completed his one-year novitiate, but would also have attended all the theological lectures of the official *lector* of that priory. No Dominican priory could be established by the General Chapter unless there were at least twelve friars, a prior, and a *lector* of theology.[25] The task of the *lector*, whose importance in the house was second only to that of the prior, was to give theological lectures on some book of the Bible to the entire com-

---

[23] Rogeri Bacon, *Compendium studii philosophiae*, ed. Brewer, pp. 425-26.

[24] See Freed, p. 32.

[25] *Constitutiones Antique*, Dist.II, c.23, ed. Thomas, p. 358 lines 2-3; ed. Denifle, p. 221.

munity, including the prior.[26] According to the primitive Constitutions of 1228, no one could be appointed *lector* unless he had studied theology at least four years.[27] By 1228, Albert himself had become a *lector*. According to Henry of Herford, Albert "lectured twice on the *Sentences* at Cologne," but we do not know when. Henry goes on to say, "[Albert] was at first *lector* in Hildesheim [in Saxony, founded in 1233], then Freiberg [in Saxony, founded in 1236], then Regensburg [in Bavaria] for two years, then Strassburg [in Alsace], then he went to Paris."[28] Albert certainly was in Saxony when he and many others observed the comet of 1240 passing near, as it were, the north pole of the ecliptic;[29] and he travelled great distances to examine various metals in "mining districts" (*loca metallica*), among which he mentions as particularly important Freiberg and Gosslar in the Harz Mountains in Lower Saxony; he also described surface mining of gold, which he observed in the Elbe and Rhine rivers.[30]

During these twenty or so obscure years of Albert's early Dominican life in Germany, he wrote his earliest known treatise, *De natura boni*, based on the whole of Scripture and standard glosses.[31] It has been described as "primarily a devotional work."[32] Rather anachronistically, he quoted only pre-thirteenth-century authors. Nevertheless, in this somewhat "devotional work," he cites explicitly ten works of Aristotle, including six of the *libri naturales* (still proscribed in Paris) in their older versions. Apparently he also composed some hymns and a Marian sequence.[33] Consequently, during these many years, Albert was certainly interested in natural phenomena and went out of his way to see for himself novelties that had reached him by rumor. He even had some knowledge of the Aristotelian *libri naturales* that had been condemned at Paris in 1210 and 1215 and were still prohibited in 1231. But he was completely out of touch

---

[26] See J. A. Weisheipl, *Friar Thomas d'Aquino: His Life, Thought and Works* (Garden City: Doubleday, 1974), p. 153.

[27] *Constitutiones Antique*, Dist.II, c.30, ed. Thomas, p. 363 lines 2-3.

[28] Henry of Herford, p. 201; Loë, p. 277, n. 5.

[29] *Meteor.* I, tr.3, c.5: "Ego autem cum multis aliis anno ab incarnatione Domini MCCXL in Saxonia vidi cometem quasi iuxta polum septentrionalem, et proiecit radios suos inter Orientem et Meridiem, magis dirigendo eos ad Orientem; et constat quod ibi non fuit via alicuius planetae" (ed. Borgnet 4: 504a).

[30] *Mineral.* III, tr.1, c.1 and c.10 (ed. Borgnet 5: 59b-60a and 72a-b).

[31] *De natura boni* (ed. Colon. 25/1 [1974]).

[32] A. Fries, *Die deutsche Literatur des Mittelalters Verfasserlexicon* (Berlin, 1977), s.v. "Albertus Magnus" I, n. 3.

[33] See Fries, ibid., and his forthcoming book, *Marienkult bei Albertus Magnus*.

with the novelties, excitement, vibrant problems, and stimulus of the "new learning" that made Paris the foremost intellectual centre of Christendom. The point is that although Albert knew much about Aristotle and Aristotelian science before he went to Paris, he still had a great deal to learn about the forefront of Christian thought when he got to Paris. Outstanding in his own province, he was the first German Dominican selected by the fourth master general of the order, John "Teutonicus" of Wildeshausen (1241-52), for special studies in Paris. The idea of sending Albert to Paris most certainly originated with John of Wildeshausen, but it would have been with the support of Hugh of St.-Cher, who himself was Dominican Regent Master in Theology at Paris (1230-35), provincial of France (1227-30; 1236-44) and later cardinal (1244-62). Although John of Wildeshausen (*Teutonicus*) was himself a German, and must have known Albert at least by reputation, he spent most of his Dominican life outside Germany and was fluent in French, Italian, and Latin, as well as in his native German. At that time the master general of the order alone had the authority to send individuals to the only *studium generale* in the order where Dominicans held two chairs of theology, the University of Paris. Consequently it was by authority of the master general that Albert was sent to Paris to become a master in theology.

C. ALBERT AT PARIS (CA. 1243/44-1248)

It is by no means certain that Albert was teaching at Cologne when word arrived for him to set off for Paris, although many biographers make that assumption. What is certain is that Albert was eventually sent to Paris to read the *Sentences*: "demum missus Parisius ad legendum *Sententias*."[34] It is difficult to determine exactly when this was, but it was certainly *before* he became a master of theology at Paris in the spring of 1245. The great majority of Albert scholars, even Mandonnet, are of little or no help here because they misunderstand the university system in the Middle Ages, particularly in thirteenth-century Paris.

At Paris, certainly after 1235, no master in theology lectured on the *Sentences* of Peter Lombard; that was the exclusive task of the "bachelor of the *Sentences*" (*baccalaureus Sententiarum*).[35] It would

---

[34] *Brevis historia*, c.1, ed. *Catal. Codd. Hagiograph.*, p. 96; Loë, pp. 278-79, n. 16.
[35] See Weisheipl, *Friar Thomas*, pp. 53-70.

seem that in the 1240s no fixed number of years was determined for the bachelor of the *Sentences*; the years seem to have ranged from one to three or four, and we do not know how many years Albert actually lectured on the *Sentences* before becoming a master. He had, however, to lecture orally on the *Sentences* for at least one year before incepting as a master. Prior to becoming a "bachelor of the *Sentences*," secular clerics had to lecture cursorily (*cursorie*) on the Bible as *cursores biblici* for at least two years, as Mandonnet rightly realized. But there is no indication that this requirement was applied to religious who had lectured on the Bible elsewhere, as Albert had as *lector*.

The seven-volume commentary of Albert (in the Borgnet edition) on the four books of *Sentences* is clearly an *ordinatio*, that is, an edited version prepared for the stationers. It is certain that Albert completed his definitive commentary on Book IV at Cologne after 25 March in the year 1249.[36] Thus the date 1246 given in one of the arguments in Book II of the *Sentences*, d.6, H. art.9 [B. 27, 139a] refers not to the oral presentation in class or preparation for class, but to the final writing *after* he incepted as master in the spring of 1245. By the time Albert was writing Book II, he had also written a substantial part of another *Summa* to which he constantly refers.[37] This *Summa* Pelster identified with a huge *Summa de creaturis*, which is sometimes called the *Summa Parisiensis*.[38] This *Summa*, whatever its name, originated in Albert's public disputations as master in the University of Paris, and has the following order: (1) *De sacramentis*, (2) *De incarnatione*, (3) *De resurrectione*, (4) *De IV coaequavis*, (5) *De homine*, and (6) *De bono*.[39] Parts Four and Five circulated independently for centuries as the *Summa de creaturis*, but today they are known to be only parts of a larger *Summa* completed in Paris. All of these parts were completed by the time Albert was composing Book II of the *Sentences* in 1246, and perhaps even before Book I was begun.[40] How this *Summa* relates to Albert's *Quaestiones disputatae*, which have not yet been published, remains to be seen. It is not known how, when, or where Albert prepared himself for these gigan-

---

[36] *In IV Sent.*, dist.35, E, art.7 ad 2: "inscriptio facienda est sic: 'Anno ab incarnatione Domini nostri Jesu Christi MCCXLIX, presidente Domino N.' etc." (ed. Borgnet 30: 354a).

[37] See Pelster, *Kritische Studien*, pp. 114-28.

[38] P. Glorieux, *Répertoire des Maîtres en Théologie de Paris*, 1: 63.

[39] *De bono*, Proleg. § 1 (ed. Colon. [1951] 28: ixb).

[40] Ibid., Proleg. § 2 (ed. Colon. 28: xiib-xiiia).

tic theological productions. But throughout this whole Parisian period Albert was particuarly concerned with assimilating the "new learning" that so fascinated him.

In any case, when Albert arrived in Paris in 1243 or 1244 (if not earlier), he came to lecture on the *Sentences* under the Dominican Master Guéric of St.-Quentin, who had been Regent Master at the Priory of Saint-Jacques in the Dominican chair for "externs" (Dominicans not from the Province of France) since 1233 — the longest any Dominican had ever held a chair in Paris. Previous to becoming a master, every student in every medieval university had to be enrolled under a specific master and eventually to lecture as a bachelor and to "respond" under a particular master in the university community (not necessarily the same as the one under whom he had enrolled). In Albert's case, his master was Guéric of St.-Quentin; in this matter he had no choice, since he was a "foreigner." When Albert became a master in the spring of 1245, he succeeded Guéric as the "third Dominican Regent Master" in that chair. Both Dominican chairs were at the Priory of St.-Jacques, just as the Franciscan chair was at their Great Convent of the Cordeliers. The most certain fact we know about Albert in this period is that he taught as Regent Master at Paris for three consecutive years. A list of successive masters in the three mendicant chairs may help to put Albert into a fuller context:

	Dominican Chair for France	Dominican Chair for Foreigners	Franciscan Chair
1229-30	Roland of Cremona		
1230-31	Hugh of St.-Cher	John of St. Giles	
1231-32	" "	" "	
1232-33	" "	" "	
1233-34	" "	Guéric of St.-Quentin	
1234-35	" "	" "	
1235-36	Godfrey of Bléneau	" "	Alexander of Hales
1236-37	" "	" "	" "
1237-38	" "	" "	" "
1238-39	" "	" "	Jean de la Rochelle
1239-40	" "	" "	" "
1240-41	" "	" "	" "
1241-42	" "	" "	" "
1242-43	Étienne of Venizy	" "	" "
1243-44	Laurent of Fougères	" "	" "
1244-45	Guillaume d'Étempes	" "	" "
1245-46	" "	ALBERT THE GREAT	Eudes Rigauld
1246-47	" "	" "	" "
1247-48	Jean Pointlasne	" "	William of Melitona
1248-49	Bonhomme Brito	Elias Brunet	" "
1249-50	" "	" "	" "
1250-51	" "	" "	" "
1251-52	" "	" "	" "
1252-53	" "	" "	" "
1253-54	" "	" "	" "
1254-55	" "	" "	Bonaventura da Bagnoregio
1255-56	Florent de Hesdin	" "	" "
1256-57	" "	Thomas d'Aquino	" "
1257-58	Hugh of Metz	" "	Guibert de Tournai
1258-59	Barthélemy de Tours	" "	" "
1259-60	Pierre de Tarantaise	William of Alton	" "
1260-61	" "	Annibaldo d'Annibaldi	Eudes de Rosny
1261-62	" "	" "	" "
1262-63	" "	William of Alton	" "
1263-64	" "	" "	Eustache de Arras

As Regent Master (*magister actu regens*), Albert had clearly defined duties to perform: to lecture as master on some approved text (*legere*), to preside at public disputations and to resolve "questions" he himself had raised (*disputare*), and to preach to the academic community on certain days (*praedicare*). But Albert's main efforts, it would seem, were devoted to writing two massive theological works, namely the *Sentences* and a *Summa Parisiensis* (besides the still unpublished *Quaestiones disputatae*) and in assimilating the "new learning" that had excited the Parisians for at least the preceding decade. The Faculty of Arts had not yet incorporated the forbidden Aristotle into its curriculum, at least not officially, although Parisian masters in arts, such as Roger Bacon and Robert Kilwardby, were in fact lecturing on the *libri naturales* and *Metaphysics* of Aristotle at that time. The earliest mention of these Aristotelian books as "set books" is found in the statutes of the Arts Faculty drawn up on 19 March 1255, almost seven years after Albert had returned to Cologne to open his own *studium*.[41]

Thomas Aquinas arrived in Paris at the direction of John of Wildeshausen in the fall of 1245.[42] Albert was just then beginning his first year as Regent Master in the chair to which Thomas would succeed eleven years later. Those modern scholars who insist that Thomas was sent immediately to Cologne[43] rely on the questionable testimony of Thomas of Cantimpré, the Flemish Dominican from Louvain (d. 1263), who, after painting a vicious picture of the d'Aquino family and Thomas's "ferocious brothers," states that Thomas was sent to the illustrious Friar Albert, lector of Cologne, before the latter obtained the chair at Paris for his incomparable knowledge of theology.[44] The reasons alleged by Thomas of Cantimpré for sending Thomas to Cologne were to get him further away from his family and the Roman Curia (!) and to study under the "praeclarus lector." But the "praeclarus lector" at that time was not in Cologne; he was lecturing as master in Paris. The strongest proof against Cantimpré's allegation is provided by Naples, Bibl. Naz. MS I. B. 54, a collection of Albert's lectures begun at Paris, continued at Cologne, completed

---

[41] See J. A. Weisheipl, "The Parisian Faculty of Arts in Mid-Thirteenth Century: 1240-1270," *American Benedictine Review* 25 (1974), 200-17.

[42] See idem, *Friar Thomas*, pp. 35-37.

[43] Ibid., pp. 36-38.

[44] Thomae Cantimpratani, *Bonum universale de apibus* I, c.20 (Duaci: Baltazar Bellerus, 1627), p. 83.

later and retained by Thomas d'Aquino throughout his lifetime.[45] This famous Naples manuscript proves that Thomas d'Aquino was in Paris when Master Albert wrote or lectured on pseudo-Dionysius's *De caelesti hierarchia* (ff. 1-41v). This first item in the codex is divided into "pecias," and apparently was an apograph of the original Parisian exemplar (prior to certain additions and changes) of one manuscript tradition; the exemplar of the other manuscript tradition, containing the additions and changes, is later. The second item in that manuscript is Albert's lectures on *De ecclesiastica hierarchia*; the third is Albert's *De divinis nominibus* in Thomas's own "unintelligible hand" (ff. 64-130vb); the fourth is Albert's lectures on *De mystica theologia* and the *Ten Letters*, thus completing the Dionysian corpus as known in Paris. Therefore there is every reason to think that Thomas remained in Paris from the time of his arrival in the fall of 1245, possibly studying under Albert, until he accompanied the master to Cologne in the summer of 1248.

The earliest documentry evidence relating to Friar Albertus Teutonicus consists of his signature and seal attached to a university document drawn up by Odo, bishop of Tusculum and papal legate, before the bishop of Paris, William of Auvergne, all officials of the university, masters in theology and law, and "other good men," concerning the Jewish Talmud, dated 15 May 1248.[46] Though regrettable in modern eyes, it would seem (according to Thomas of Cantimpré, *Bonum universale de apibus*, I, c.1) that at the instigation of a certain Dominican Friar Henry of Cologne, the Jewish books called the Talmud were confiscated by papal authority from French rabbis, carefully examined "by men discrete and expert in such things," and found to contain innumerable errors, accusations, blasphemies, and abominable allegations, such that they could not be tolerated in a Christian society, and consequently were officially condemned. Thomas of Cantimpré asserts that the books were then burned, but the document says nothing about that. This document was signed by "Friar Albert the German, Master in Theology," and forty other persons including many "other good men," such as non-regent masters and friars. The other Dominican master in theology who attached his name and seal to the document was Jean Pointlasne (*Johannes*

[45] See Paul Simon in *Super Dionysium De divinis nominibus*, Proleg. § 2 (ed. Colon. [1972] 37/1: vi-ix).
[46] *Chartularium Univ. Paris.* 1: 209-11, n. 178.

Pungensasinum) of Paris. From 15 May 1248 on, the main facts of Albert's public life are clearly documented, but the chronology of his writings is far from certain.

The most perplexing and controverted problem in Albertinian scholarship concerns the chronology of Albert's so-called Aristotelian paraphrases. These works, totaling almost half of his entire writings, are really a "reworking" of all the Aristotelian and pseudo-Aristotelian books with many additions and innovations of his own, each work retaining more or less its original Aristotelian title. Thus Albert's paraphrase of the pseudo-Aristotelian *De vegetabilibus* is simply entitled Albert's *De vegetabilibus* (or *De plantis*). It is ironic that scholars who, like Pierre Mandonnet, insist on a late dating for Albert's birth, that is, 1206/07, are the very ones who insist on an early dating of all the Aristotelian paraphrases. Mandonnet claimed that all of Albert's Aristotelian paraphrases were written between 1245 and 1256.[47] Only one scholar, as far as I know, would wish to date these works earlier. Lynn Thorndike says, "I should be inclined to push these dates back ten or twenty years."[48] At the other extreme, Franz Pelster, insisting on an early date for Albert's birth, that is, 1193, strongly maintained that all of Albert's paraphrases were written between 1256 and 1275.[49] Other scholars today, such as Gilson, Nardi, Vignaux, Maurer, and David Knowles, state unhesitatingly that Albert was born in 1206, but they refuse to devote serious attention to Albert's "reworking" of Peripatetic philosophy, and so refrain from either dating them or taking them as expressive of his own convictions.

The suggestion I propose here is that Albert was born around 1200 and that he wrote all of his Aristotelian paraphrases, including the logical and ethical works and *De causis*, between 1250 and 1270, while he was engaged in a very busy, public life. Certainly by April 1271 all of those paraphrases were finished.

The Dominican General Chapter meeting in Paris at Pentecost, 7

---

[47] P. Mandonnet, "Polémique Averroiste de Siger de Brabant," *Revue Thomiste* 5 (1897), 95-105; *Dict. d'Hist. et de Géog. Ecclés.* (1912), s.v. "Albert le Grand." This view is generally followed by F. Van Steenberghen, *Siger de Brabant* (Louvain, 1942), 2: 470-79.

[48] L. Thorndike, *A History of Magic and Experimental Science* (New York: Columbia, 1923), 2: 525.

[49] F. Pelster, *Kritische Studien*, pp. 156-161; "Zur Datierung der Aristotelesparaphrase des hl. Albert des Grossen," *Zeitschrift f. kath. Theologie* 56 (1932), 423-36; "Die beiden ersten Kapitel der Erklärung Alberts des Grossen zu De animalibus in ihrer ursprünglichen Fassung," *Scholastik* 28 (1953), 229-40.

June 1248, definitively decreed that *studia generalia* be established in four more provinces, namely, Provence, Lombardy, Germany, and England, to which students of all provinces could be sent for higher studies.[50] These "open" *studia* were immediately established in Montpellier, Bologna, Cologne, and Oxford, where there already existed important priories and, in three cases, a secular *studium* of some importance.[51] Thus Henry of Herford (d. 1370) states that after teaching three years as Master in Paris, Albert was sent to Cologne: *post tres annos magisterii sui Coloniam mittitur ad legendum.*[52]

D. ALBERT AS REGENT MASTER IN COLOGNE (1248-54)

In the summer of 1248, Albert, accompanied by Friar Thomas d'Aquino and probably other Dominican friars, traveled by foot to Cologne to establish the first *studium generale* in Germany. Since the proposed establishment of the four new *studia* had successfully passed the two preceding chapters, preparations had been underway in all four places for the expected decision of 1248. The Dominikanerkloster of Heilige Kreuz in Cologne was established on the Stolkgasse (*vicus Stolkorum*) by Friar Henry, the companion and friend of Jordan of Saxony, working continuously from his arrival in 1221/22 until his death in 1229.[53] Around the time of Albert's arrival, the cornerstone of the new cathedral was laid on 14 August 1248, and Albert took a keen interest in the excavations involved in laying the foundations of the new edifice.[54] The Dominican priory on the Stolkgasse (where the main postoffice now stands) extended as far as An den Dominikanern. By 1250, the palatial residence of Duke Walram IV of Limburg adjoining the Dominican property on the Stolkgasse was purchased for the comparatively low price of 150 marks.[55] The total complex was quite ample for the new *studium.*

---

[50] *Chartularium Univ. Paris.* 1: 211, n. 179.

[51] The order of places named in the chapter document relates to seniority of Dominican provinces, not to the importance of the cities intended. At that date, Germany and England ranked seventh and eighth respectively in seniority. After the Chapter of 1248, only three older provinces lacked a *studium generale*: Spain, Tuscany (Roman), and Hungary. The provinces of *Dacia* (Scandinavia) and Poland were already constituted as provinces by 1228, as were Greece and the Holy Land, but none of these ever possessed a Dominican *studium generale.*

[52] Herford, *Chronica*, p. 201; Loë, p. 277, n. 7. Cf. Loë, pp. 279-81, nn. 23-31.

[53] See Freed, pp. 81-85.

[54] *De prop. element.* I, tr.2, c.3 (ed. Borgnet 9: 605a-b).

[55] Freed, p. 93.

During Thomas's student days under Albert, 1248-52, two authentic works of Albert can be dated with some certainty, since they were transcribed by Thomas himself.[56] The first was his lectures on *De divinis nominibus* of pseudo-Dionysius, preserved, as we have said, in Thomas's *littera inintelligibilis* (Naples, Bibl. Naz. MS I. B. 54) and nine other manuscripts. This exposition, however, presupposes Albert's lectures on *De ecclesiastica hierarchia*, which refers to his commentary on Book IV of the *Sentences*, completed at Cologne in 1249. At the same time, this Dionysian commentary presupposes Albert's exposition of *De caelesti hierarchia*, which Albert had delivered in Paris. Thus it would seem that Albert wrote or lectured on *De caelesti hierarchia* while he was in Paris before the summer of 1248, lectured on *De ecclesiastica hierarchia* at Cologne in 1248-49, and on *De divinis nominibus* in 1249-50 (which we have in Thomas's own hand). It is most likely that Albert finished his exposition of the Dionysian corpus by the time Thomas left for Paris in the fall of 1252.

The second authenticated work of Albert at this time is his public lectures in a *studium solemne* on the *Ethics* of Aristotle, "cum questionibus." Only a master in theology as independent and remarkable as Albert would have had the audacity to teach a course in philosophy in such a *studium*. I doubt that he could have gotten away with it at either Paris or Oxford, where there already existed flourishing universities with long-standing traditions. All that had been known of Aristotle's *Ethics* previously were Books II-III (*Ethica vetus*) and Book I (*Ethica nova*). But around 1246/47, Robert Grosseteste, bishop of Lincoln (1235-53), translated for the first time all ten books of the *Ethics*.[57] The temptation to lecture on this treasure was too much for Albert, and he did so at Cologne ca. 1250-52, despite the fact that he was a master in theology charged with directing a theological *studium generale* of the Dominican Order open to all clerics. Although this commentary-with-questions (*per modum commenti*) was taken down by Friar Thomas, his copy no longer exists; but nine other manuscripts of this text still exist, and the work has now been published for the first time.[58] This commentary is quite distinct from the better known commentary *per modum scripti* found in all the old

---

[56] See Weisheipl, *Friar Thomas*, pp. 46-47.

[57] AL, 21/1-3, fasc. 3 (Trans. Grosseteste: *Textus purus*), ed. R. A. Gauthier, 1972.

[58] *Super Ethica: Commentum et quaestiones* (ed. Colon. [1968] 14/1). See ibid. Proleg. § 1; Weisheipl, *Friar Thomas*, p. 46.

printed editions. Albert's audacity in lecturing on Aristotle's philosophy in a theological *studium* is another example not only of his independence, but also of his conviction that philosophy and science are indispensable for theological studies. It was this same conviction and audacity that prompted him to "rewrite" the whole of Peripatetic philosophy. This conviction he would help embody in the first *ratio studiorum* for the Dominicans in 1259.

Albert was an indefatigable student not only of nature, but of everything the ancients, particularly the "Peripatetics," had to say about philosophy, which for him was the totality of human, natural knowledge. He applied himself so sedulously to the natural sciences, which for him included not only natural philosophy but also moral philosophy and metaphysics, that Henry of Ghent (d. 1293), who should have known better, accused Albert of neglecting the sacred sciences. Although Henry, a secular master of theology at Paris, admitted having seen only the first part of Albert's postill (*postilla* or comment) on St. Luke, he complains, "as some people say, while [Albert] intemperately pursues the subtlety of secular philosophy, he tarnishes somewhat the splendor of theological brilliance."[59] Such an accusation is not only unfair; it is false.

In the *Physica*, Albert explains that his Dominican confreres had implored him for a good number of years (*ex pluribus iam praecedentibus annis*) to compose a book on physical science in such a way that they could attain the *whole* of natural knowledge and thereby understand competently the works of Aristotle.[60] This plea of the brethren was current in 1248 when he returned to Cologne, if not long before that. Finally, by the end of 1249 or early 1250, he acceded to their wishes, but his plan was far more ambitious than his brethren could have imagined. Not only would he explain the fundamentals of natural science with all the aids at his disposal, but he hoped to explain systematically the whole of human learning embracing all the natural sciences (inanimate and animate), logic, rhetoric, mathematics, astronomy, ethics, economics, politics, and metaphysics (including its "natural complement" the *Liber de causis*). "Our intention," he said, "is to make all the aforesaid parts [of knowledge] intelligible to the Latins (*Latinis intelligibiles*)."[61] That the plan was deliberate, sys-

---

[59] *De script. eccles.*, c.43, ed. Fabricius, *Bibliotheca Ecclesiastica* (Hamburg, 1718), 2: 125.

[60] *Physica* I, tr.1, c.1 (ed. Borgnet 3: 1a).

[61] Ibid. (ed. Borgnet 3: 2a).

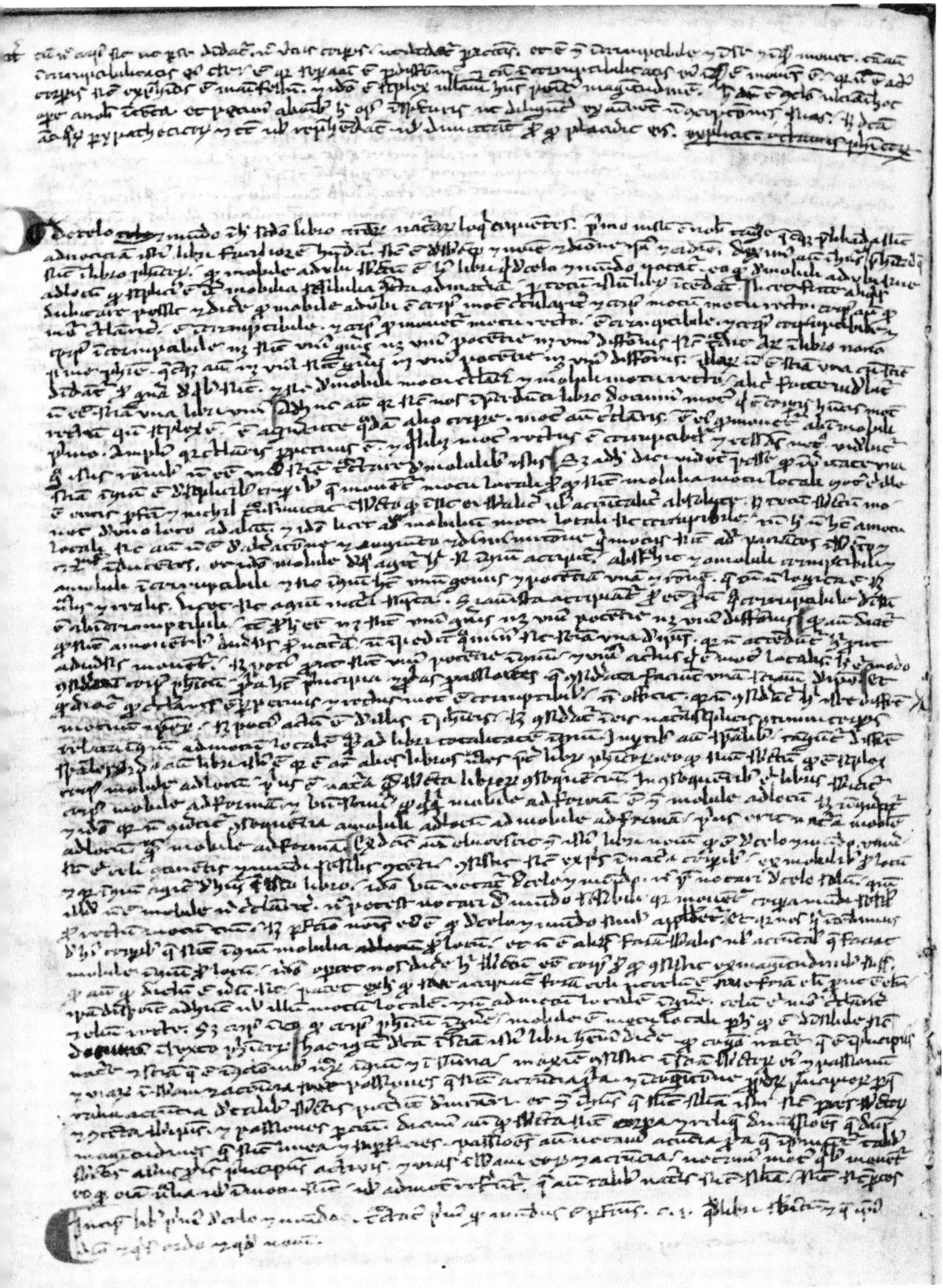

Plate 2. Albert's autograph in Vienna, Oesterreichische Nationalbibliothek,
Cod. misc. lat. 273, fol. 72v:
end of *Physica* and beginning of *De caelo et mundo*

Plate 3. Albert's autograph in Vienna, Oesterreichische Nationalbibliothek, Cod. misc. lat. 273, fol. 142r: end of *De caelo et mundo* and beginning of *De natura locorum*

tematic, and consecutive with the Aristotelian corpus can be seen from the extant autograph copy in Vienna, Oesterreichische Nationalbibliothek, Cod. misc. lat. 273, which contains the last five lines of the *Physics* (fol. 72v) and continues from the same folio to *De caelo, De natura locorum,* and *De causis proprietatum elementorum*[62] (see Plates 2 and 3). The chronological order of the rest of the Albertinian corpus of natural science seems to be the following: *De generatione et corruptione, Meteora, De mineralibus et lapidibus, De anima, Parva naturalia* (eleven distinct works), *De vegetabilibus,* and *De animalibus libri XXVI.*

At the very outset, Albert explained that his procedure would be to follow the order and opinion of Aristotle, presenting whatever seemed necessary to explain and demonstrate his views, but in such a way that no mention is actually made of Aristotle's text. Further, he would make digressions, clarifying difficulties, and supplementing whatever might be wanting in the view of Aristotle. Furthermore, he would add in various places material and sometimes whole books that Aristotle omitted or left incomplete either because Aristotle had not written about such things or, if he had, because these writings have not come down to us.[63] The doctrine Albert presented was systematically and deliberately "Peripatetic," that is, Aristotelian, although he never failed to correct Aristotle when he was in error regarding facts of experience or the teaching of faith. "Whoever believes that Aristotle was a god must also believe that he never erred; but if one believes that Aristotle was a man, then doubtless he was liable to error, just as we are" — that was his reply to the integralists who insisted on the eternity of the world.[64] On countless occasions, Albert rejected a supposed observation of the Stagirite, saying that it is contrary to his own observations.[65] In practice as well as in theory, Albert recognized that "the aim of natural science is not simply to accept the statements of others, but to investigate the causes that are at work in nature."[66]

By personal conviction, Albert was basically an Aristotelian, insisting (i) on the autonomy of the natural sciences in their own field, (ii) on the impossibility of discovering the "real causes" of nat-

---

[62] See *De caelo et mundo,* Proleg. § 2 (ed. Colon. [1971] 5: viii. 1-22).

[63] *Physica* I, tr.1, c.1 (ed. Borgnet 3: 1b-2a).

[64] *Physica* VIII, tr.1, c.14 (ed. Borgnet 3: 553b).

[65] See, e.g., *Metheor.* III, tr.4, c.11; *De animal.* XXIII, tr.1, c.1, etc.

[66] *De mineral.* II, tr.2, c.1 (ed. Borgnet 5: 30a).

ural things *qua* natural via mathematics, and (iii) on establishing the foundations of ethics and metaphysics in the nature of things in the real world, that is, in natural philosophy. This is not to say that Albert was an integralist or literalist in accepting everything Aristotle said, or that he excluded any truth from any source fundamentally compatible with his Christian convictions. It simply means that Albert was very much a realist and accepted the autonomy of human reason in its own field, since nothing truly known by reason can possibly contradict a truth of revelation. The extraordinary point to notice is the amazing number of times Albert rejects the "errors of Plato" or, more commonly of the Stoics (*Stoici*), under which pejorative label he includes Plato, Socrates, Pythagoras, Avicenna, Dionysius, sometimes Augustine, and their followers on certain points. His attacks on the Stoics or Platonists are most frequent in his *Physica, De natura et origine animae, Metaphysica*, and *Liber de causis*.[67]

In March 1252, Albert had his first experience of a role that he would be called upon to play innumerable times throughout his long career: the role of arbiter and peacemaker. The first experience involved a dispute between Conrad von Hochstaden, archbishop of Cologne, and the burgers of the city. On 25 March, Cardinal Hugh of St.-Cher, legate of the Holy See, and Friar Albert, *lector* of the Dominikanerkloster in Cologne, were called upon to arbitrate the litigation; they gave their decision in April, and it was confirmed by Pope Innocent IV on 12 December 1252. Since the long list of litigations that Albert was called upon to arbitrate is well documented, there is no need to mention them further here.[68]

Some time earlier, Albert had made Thomas d'Aquino his official "bachelor," whose duty it was to respond in academic disputations and to read the Bible *cursorie*. Although Albert was particularly fond of Ulrich of Strassburg, whom he frequently chose as a walking companion (and conversed with in German), he was fully aware of the abilities of Friar Thomas of Sicily, the "dumb ox," who was relatively large and knew no German. It would seem that some time in 1252 the master general, John of Wildeshausen, sounded Albert out on suitable candidates to send to Paris, since the position of the men-

---

[67] See, e.g., *Metaphysica* (ed. Colon. [1964] 16/2, index s.v. Plato, Phythagoras, Stoici; also *Probl. determ.* q.17 (ed. Colon. [1975] 17/1: 55.11 and note); J. A. Weisheipl, "Albertus Magnus and the Oxford Platonists," *Proceedings of the Am. Cath. Phil. Assoc.* (1958), 124-39.

[68] See Loë, pp. 281-310; T. M. Schwertner, *St. Albert the Great* (Milwaukee: Bruce, 1932), pp. 120-50.

dicants was becoming precarious. Albert immediately suggested Thomas, but this suggestion was immediately dismissed for many reasons, among them Thomas's youth and John's preference for a German. Albert, however, enlisted the support of Cardinal Hugh of St.-Cher, who was then acting legate in Belgium. When Hugh and John of Wildeshausen met at Constance in August of that same year, the question of the Paris assignment came up and Hugh strongly supported Albert's choice of Thomas as bachelor and eventual master in the chair for externs at Paris. Thereupon Thomas was ordered to prepare himself "for reading the *Sentences*" at Paris; and he began his work at Paris in September.

E. Albert as Provincial of Teutonia (1254-57) and Resignation

At the Provincial Chapter held at Worms in June 1254, following the General Chapter, Friar Albert was elected prior provincial of the Province of *Teutonia*. By 1254, the Province of *Teutonia* numbered thirty-six priories for men (Dominikanerklosters) and more than twenty cloisters of nuns (Schwesterklosters, or *"claustra Sororum*, as the Germans call them"[69]). It was a vast area and the priories were numerous, as can be seen from the accompanying map and list of priories. During his three full years as provincial, Albert made formal visitations of all the houses of his Province on foot, including, apparently, the mission house in Riga, Latvia (*Livonia*).[70] He also established three new priories (Strausberg in the mark of Brandenberg in 1254; Seehausen in the Altmark in 1255; and Rostock on the Baltic [*Slavia*] in 1256). These three priories came to belong to the Province of *Saxonia* when the huge area was divided in 1303. During his provincialate, Albert established at least two cloisters of nuns, the more famous being the Paradisus near Soest in Westfalia. As provincial, he not only presided at three Provincial Chapters, but attended the General Chapter of 1255 in Milan, followed by the Provincial

---

[69] Quétif-Échard, *Scriptores*, 1: i.

[70] *De animal.* XXIII, tr. un., n. 15 (ed. Stadler, 1437.24-27): "iam *expertus sum* esse falsissimum: quoniam *in Livonia* ubi aquilae sunt aquilonares et feroces valde et magnae, nichil penitus talium experimur." *Livonia* is also mentioned among the areas Albert visited as provincial according to *Legenda Coloniensis*, c.4, (ed. Loë, Part 1, p. 274), but the anonymous author of this work (ca. 1483) makes Albert to be provincial of Teutonia before becoming master in theology of Paris, which is inaccurate.

Chapter at Regensburg; the General Chapter of 1256 in Paris (where Thomas had just become a master in theology), followed by the Provincial Chapter at Erfurt.

The most significant event of Albert's provincialate was his summons to the papal curia at Anagni, where he represented the Dominican Order with Humbert of Romans, the master general (1254-63), in its struggle against the attacks of William of Saint-Amour and his colleagues from Paris.[71] St. Bonaventure, minister general of the Franciscans, played a most decisive role in this issue both in his writings and in his public debates. The anti-mendicant controversy was temporarily resolved in favour of the mendicants with the condemnation of William's *De periculis novissimorum temporum* on 5 October 1256, and insistence on the earlier bull *Quasi lignum vitae* of 14 April 1255.

According to Thomas of Cantimpré, Albert, at the request of Pope Alexander IV and all the cardinals, expounded the whole Gospel of St. John and all the canonical Epistles "in such a wonderful and unheard-of manner (*miro et inaudito modo*) that the whole affair of the Preachers and Minors was terminated and concluded, so that their enemies were overcome and stupified."[72] Albert himself states that at the papal curia (in 1256-57) he publicly debated against the Averroist doctrine of one intellect for all men; the material for this debate was later (ca. 1263) turned into a little book called *De unitate intellectus contra Averroistas.*[73]

It was Albert's custom while travelling — always by foot — first to visit the chapel of the religious house where he intended to stay the night, to thank God for the safe journey, then immediately to visit the library to see whether there were any books there that he had not yet seen. Often his candle burned late into the night as he copied long passages of interest to him that could be used later. Hence, Albert frequently cites titles of books and gives direct quotations from works now lost. At the very Chapter in which Albert was elec-

---

[71] Albert, according to Loë and his sources, is supposed to have arrived at the curia on 4 October 1256 and had the decisive debate before the solemn consistory of cardinals on October 6 (Loë, Part 1, p. 284, n. 52). But the papal bull condemning the writings of William of St.-Amour is clearly dated 5 October 1256 (*Chartularium Univ. Paris.* 1: 331-33, n. 288).

[72] Th. Cantipratani, *Bonum universale* II, c.54 § 24, (ed. 1627) p. 176.

[73] *Sum. theol.* P.II, tr.13, q.77, membr.3 (ed. Borgnet 33: 100b): "Haec omnia aliquando collegi in curia existens ad praeceptum Domini Alexandri Papae; et factus fuit inde libellus quem multi habent, et intitulatur contra errores Averrois." See *Libellus de unitate intellectus contra Averroistas* (ed. Colon. [1975] 17/1, Proleg. §§ 1-2).

ted provincial (1254), he explicitly decreed that every friar "is absolutely forbidden the use of vehicles on his journeys," allowing only rare exceptions to this rule. At the Chapter of Augsburg in 1257, the capitular fathers, not Albert, imposed severe penances on the prior of Worms "for having used a carriage and clothed two laybrothers without permission" and on the prior of Minden "for having come to the Chapter on horseback," as well as on all friars who had "come to the Chapter that year in carriages or on horseback."[74]

While Albert was provincial of *Teutonia* (1254-57), he wrote his paraphrase of Aristotle's *De anima*, as clearly shown by Basel, Univ. Bibl. MS F. IV. 34, fol. 50ra: "fratris alberti provincialis fratrum predicatorum per theutoniam liber de anima." This paraphrase could have been written early in his provincialate, and, although he considered it a very important work, it would not have taken much time to compose. For this paraphrase, Albert apparently used both existing translations, namely, the "vetus translatio" from the Greek made by James of Venice (ca. 1160) and the "nova" from the Arabic with the commentary of Averroës made by Michael Scot (ca. 1220), comparing them throughout. In his *De anima*, Albert refers to all his writings on inanimate nature, such as *De caelo*, *Meteora*, and *De mineralibus*,[75] as already completed. *De anima* was immediately followed by a series of eleven works known as the *Parva naturalia* (*De nutrimento, De sensu, De intellectu, De natura et origine animae*, etc.), culminating in the large collection of "Twenty-Six Books," *De animalibus*, sometime in the early 1260s. Later works make constant references to the *De anima*, which Albert obviously considered an

---

[74] English text in T. M. Schwertner, pp. 70-71; Latin in Scheeben, *Chronologie*, pp. 160-61. The source is Peter of Prussia, *Legenda Alberti Magni*, c.26. Albert, however, did not preside at this Chapter, having submitted his resignation as provincial at the General Chapter of Florence a few months earlier.

[75] In Kraków, Bibl. Jagiellońska MS 6392 (s. xv), fol. 7ra-46va, the colophon at the end of Bk V, c. ult. of *De mineralibus et lapidibus libri I-V* (ed. Borgnet 5: 102b) reads on fol. 46va: "Explicit liber mineralium editus a fratre Alberto quo$<$n$>$dam Ratisponense nacione theutonico professore de ordine fratrum predicatorum precipuo philosopho, editus anno domini M°CC°L in civitate Colonia Agrippina, presidente dicto Cunrado archiepiscopo civitatis / (fol. 46vb) memorate. Amen. etc." However, this date seems to be unreasonably early. Could it be that an earlier exemplar had the reading "anno domini M°CC°L iv civitate Colonia Agrippina"? Such a dating would more conveniently fit the known chronology of Albert's other works, and could have been "edited" shortly after 25 March 1254, since the New Year began at Cologne on March 25. Thus the whole of *De mineralibus* could have been written in the early part of 1251. In this MS. Albert's *De mineralibus* is followed immediately by St. Thomas's *De mixtione elementorum* (fol. 46vb-47va), here called *Questio de simplicitate elementorum*, inc.: "Solet esse dubium aput multos quomodo elementa sint in mixta. Videtur enim quibusdam quod quodlibet. . . ."

important contribution to psychology; he did indeed recognize immediately the dangers of Averroës' one intellect for all men, as we have already seen. In the autograph copy of his *Postilla super Matthaeum*, Albert explicitly refers the reader to his fuller explanation in *De anima*.[76]

Finally, after three nomadic years as provincial, Albert was allowed to resign at the General Chapter of Florence in June after Pentecost 1257.[77]

Being in Italy for the first time, Albert remained there from his arrival at Anagni in October 1256 until the General Chapter of Florence in June 1257, that is, for nine months. There is no intrinsic reason why Albert could not have commented on St. John's Gospel and the canonical Epistles during this period, but more evidence and a more likely setting are needed than the picturesque declaration of Thomas of Cantimpré. At this time, however, when he was in Italy (*in Campania iuxta Graeciam*), Albert chanced upon a previously unknown work by Aristotle entitled *De motibus animalium*, which he proceeded to call *De principiis motus processivi* and to comment upon.[78] He noted that although he had already composed a work entitled *De motibus animalium* out of his own ingenuity, he wished to see how closely he had come to Aristotle's own thought. He refers to the basic truth that all movement needs a first unmoved mover as having been proved by him "a long time ago in the Eighth Book of the *Physics*"; he refers to his own *De anima* frequently and to almost all of the *Parva naturalia*; he might even have begun commenting on the huge *De animalibus* XXVI. Clearly he had not yet commented on the *Ethics* ("per modum scripti") nor on the *Metaphysics*.

Immediately after the General Chapter in Florence, the German Provincial Chapter was held in Strassburg for the election of a new provincial.[79] From there Albert returned to Cologne to resume his position as *lector* at the Dominikanerkloster of Heilige Kreuz. In 1258, Albert held a series of scholastic disputations on Aristotle's *De animalibus*, a *reportatio* of which exists in eight manuscripts: "Expliciunt questiones super de animalibus, quas disputavit frater albertus repetendo librum animalium fratribus colonie, quas reportavit quidam frater et collegit ab eo audiens dictum librum nomine cunradus

---

[76] See *De anima*, Proleg. § 2 (ed. Colon. [1968] 7/1: v, note to line 9).

[77] *Acta capitulorum generalium Ord. Praed.*, ed. B. M. Reichert, MOPH 3: 89.

[78] *De prin. motus proces.* tr.1, c.1 (ed. Colon. [1955] 12: 48. 66-74).

[79] Loë, p. 285, n. 59.

de austria. Hoc actum est anno domini 1258."[80] Albert taught, wrote, and dictated almost uninterruptedly from September 1257 until June 1259, interrupted only by the various litigations which required his arbitration.

In 1259, Humbert of Romans, master general of the order, summoned Albert and four other masters in theology to form a special commission for the General Chapter of Valenciennes in northern France, meeting early in June. The commission was to concern itself exclusively with the state of studies in the order. The other members of that commission were Bonhomme Brito, Florent de Hesdin, Thomas d'Aquino, and Peter of Tarentaise.[81] The commission drew up what might be called the first *ratio studiorum* for the Dominican Order in twenty-two clear statements, dealing with the behaviour of *lectors* and students, the importance of philosophy for theology, requisites not only in *studia solemnia* (where there should be bachelors teaching under masters, "repetitions" of lectures by students, and instruction in philosophy), but also in ordinary priories, where there should always be instruction in Sacred Scripture, salvation history, cases of conscience, and the like, "lest the brethren become lazy."[82]

F. ALBERT AS BISHOP OF REGENSBURG AND PREACHER OF THE CRUSADES (1260-64)

Albert returned to Cologne to begin the new academic year in the fall of 1259, while Thomas eventually returned to Naples. But in January, Albert received Pope Alexander IV's letter of 5 January 1260, appointing him bishop of Regensburg in the ecclesiastical province of Salzburg which happened to be in a deplorable state financially and spiritually. On that same date, Alexander IV wrote to the dean and chapter of Regensburg to receive Albert as their bishop and to obey him in all things.[83] Humbert of Romans immediately penned a most fervent plea to Albert, begging him not to accept the dignity that would set a lamentable precedent in the order, and would be a dishonour to his well-known nobility of mind and religious fervour.

---

[80] Milano, Bibl. Ambrosiana MS H 44 inf. fol. 87vb; see *Quaestiones super de animalibus* (ed. Colon. [1955] 12: xxxv. 50-55).

[81] See Weisheipl, *Friar Thomas*, pp. 138-39.

[82] *Chartularium Univ. Paris.* 1: 385-86, n. 335.

[83] Loë, p. 288, n. 72.

But apparently Albert had no choice. He was consecrated in the cathedral at Cologne during March and was invested as a secular prince by a delegate of the Holy Roman Emperor. It is said that Albert entered Regensburg unobtrusively after sunset on 29 March and stayed with the Dominican friars at St. Blasius, where as a younger man he had been *lector*. On the following morning, Tuesday of Holy Week, he entered the ancient cathedral for his enthronment and Solemn Mass, during which all the clergy present promised obedience. It is also said that he found the cupboards of the adjoining episcopal castle bare of all food, the wine cellars completely empty, and the diocese bankrupt.[84] Albert, known to the local Bavarians as "Boots the Bishop"[85] (*episcopus cum bottis*, or *calceatus*), devoted almost two full years to covering the whole of his large diocese on foot, reforming everywhere. The reforms introduced by Albert in his diocese through his own initiative and through the synodal decrees of Salzburg are well known.[86]

During his episcopate, Albert was undoubtedly writing his commentary on *De animalibus*. The phrase "in my villa above the Danube" in Book VII (ed. Stadler, p. 523, v.1) can refer only to the episcopal castle of Donaustauff, about three miles from the city, on the Danube. The entire work must have been completed not much after 1261.[87] At that time he probably also worked on some of his logical commentaries.

Already by the end of 1261, Albert was ready to seek release from this unwanted burden. By the end of December, Albert, having set the diocese in order, left Regensburg for Rome to submit his resignation to Alexander IV. He placed the diocese in charge of Henry as vicar, Leo Torndorf as dean of the chapter, and Ulrich as pastor of the cathedral church. Going by way of Vienna, Albert passed through the Tyrol and arrived at the papal curia at Viterbo in July 1261, only to find that Alexander had died at the end of the preceding May. A new pope was elected on August 29 and consecrated at Viterbo on September 4, taking the name of Urban IV. By that time, Thomas d'Aquino had been *lector* in the Dominican Priory of San

---

[84] Loë, p. 289, n. 77.

[85] Loë, p. 291, n. 94. Clearly the expression was a nickname; see T. M. Schwertner, p. 110. But Scheeben notes, "This nickname is difficult to translate" (*Chronologie*, 63) — even in German. He suggests only that *ligatus calceus* might be translated *Bundschuh*.

[86] See Loë, pp. 280-92; Scheeben, *Chronologie*, pp. 54-64; Schwertner, pp. 101-119.

[87] See *Metaphysics* Proleg. § 1 (ed. Colon. [1960] 16/1: viii. 19 ff.).

Domenico in Viterbo for some time, and old friendships were renewed.[88] Albert's resignation was finally accepted around November, elections were ordered at Regensburg, and Leo Torndorf, dean of the chapter, was elected. But it was not until 11 May 1262 that Leo's election as successor to Albert was confirmed by Urban.[89]

Much to his surprise, Albert learned from Thomas that William of Moerbeke had just finished a new translation of Aristotle's *De motu animalium*, different from the one he had found on his previous visit to Italy. Interestingly enough, Albert had to write a new commentary on it in the same style as he had been using in his other works.[90] Albert himself seems to have remained in the illustrious circle around Urban IV at Viterbo (August 1261 to autumn 1262) and at Orvieto (autumn 1262 to February 1263) by request of the pope; he was even allowed to draw up a last will and testament, depositing a copy in the papal archives.

It is most likely that during these leisurely nineteen months Albert worked on his paraphrases of the *Ethics* ("per modum scripti") and the *Posterior Analytics*. A. Fries suggests that the paraphrase of Aristotle's *Politics* was also written around 1262/63.[91]

In February 1263, however, Pope Urban IV ordered Bishop Albert to preach the crusade in Germany, Bohemia, and all lands that spoke the German language, and he conferred extraordinary powers on him for the successful prosecution of his new mission. The official letter to all the bishops of Germany, Bohemia, and other German-speaking lands was sent by Urban IV on 8 March 1263. On 21 March, Friar Berthold, a German Dominican, was assigned to assist Albert in preaching the crusade and to help him in every way. Albert's movements between March 1263 and the death of Urban IV on 2 October 1264, can be traced easily as he traveled on foot throughout German-speaking countries, preaching a new crusade to the Holy Land.[92] Perhaps many of Albert's extant German sermons belong to this period. With Urban's death, Albert's commission to preach the crusade came to an end.

---

[88] See Weisheipl, *Friar Thomas*, pp. 147-53.

[89] Loë, p. 292, n. 104.

[90] *De prin. motus proces.*, Proleg. § 2 (ed. Colon. [1955] 12: xxix-xxv); Weisheipl, *Friar Thomas*, p. 149. The earliest reference to *De prin. motus proces.* as a distinct work seems to be in Bk XXI of *De animal.* tr.1, c.8, n. 46 (ed. Stadler, 1345.25), which was finished by 1264. On Albert at Orvieto, see Weisheipl, *Friar Thomas*, pp. 147-49.

[91] Fries, *Verfasserlexicon* 1: 8, col. 128.

[92] Loë, pp. 294-98; this was not a crusade against the Albigenses, as some have asserted. For the itinerary, see esp. Scheeben, *Chronologie*, pp. 72-77.

G. Activities and Writings of a Retired Bishop Until His Death (1264-80)

From the end of 1264 to 1267, Albert lived in the Dominican Kloster in Würzburg, where his brother Henry resided. It was probably there, after preaching the crusades, that Albert commented on the *Metaphysics* of Aristotle in the *media* version that was in use only between 1250 and 1270. In this paraphrase, Albert refers to almost all of his earlier works, at least implicitly. It is clear that his *Metaphysics* was written after the *Ethics, De animalibus, Poetics,* and the very important *Posterior Analytics.*[93] The logical paraphrases of Albert are difficult to date because they seem to constitute a concomitant series with the other Aristotelian paraphrases (*Physics* to *De causis*). It is clear that all the books of the logic in their usual order up to and including the *Posterior Analytics* were written before the *Metaphysics*; the *Topics* and *Elenchi* are later. In the *Metaphysics,* Albert refers three times to a work of his own called *Geometry,* but it is still uncertain what this work is or of what it is a paraphrase or commentary.[94] At present it is impossible to determine when Albert began his extensive logical writings (three volumes projected in the Cologne edition); but it would be safe to say that the *Posterior Analytics* and all earlier books were finished before Albert began his *Metaphysics* around 1264.

Albert's primary duty as a master in theology and a bishop was, of course, to lecture on the Bible and to preach. I have tried to show that the vast corpus of Aristotelian paraphrases was not — except in the cases already explicitly indicated — taught in the classrooms or the result of his teachings in any *studium.* They were "written or dictated"[95] for his confreres as an extracurricular avocation and were meant to be *read* by students in order to understand Aristotle better and to acquire the fullest possible range of human wisdom (philosophy) as a necessary preliminary to theology. Albert's biblical commentaries, on the other hand, apparently were the product of actual lectures; but they are extremely difficult to date, because Albert lec-

---

[93] *Metaphysica* I, tr.5, c.14 (ed. Colon. 16/1: 88.45-46); V, tr.5, c.5 (ed. Colon. 16/1: 280.50 ff.). See *Index* of Albert's own works cited by himself in ed. Colon. 16/2: 600-602, including the *Ethics* and Bk. XXVI of *De animalibus.*

[94] See B. Geyer in *Metaphysica* Proleg. § 7 (ed. Colon. 16/1: xix.80-85).

[95] Luis of Valladolid, *Brevis historia,* c.17, in ed. of Scheeben, "Die Tabulae," pp. 243 and 245.

tured on various books of the Bible many times and eventually revised many of his commentaries. All we can be certain of now is that his *Postilla super Isaiam*, already published, seems to have been composed after 1250, since he quotes Aristotle's *Metaphysics* in the *media* version that was not known before that date.[96] But the postills on Matthew and Luke both explicitly refer to his own *De animalibus*. Albert's autograph copy on Matthew 3: 7, explicitly says: "We have said much about this in the book *De animalibus*," referring to *De animal*. Bk. 25, c.2.[97] Likewise in the postill on Luke 3: 7, Albert says in passing, "as explained by us *in libris Animalium*." Therefore these postills on Matthew, Luke, and probably Mark are to be dated after 1262, even though Albert may have lectured on the Gospels on numerous prior occasions. No further attempt will be made here to date his numerous biblical commentaries or other theological works.

In 1268, Albert was in Strassburg at the request of Clement IV to resolve a dispute between the bishop and the burgers of Strassburg.[98] In any case, around 1269, John of Vercelli, then master of the order (1264-83), asked Albert to reside in Cologne as *lector emeritus*.[99] By that date, Albert was already engaged in writing his extraordinary paraphrase of the pseudo-Aristotelian *Liber de causis*, which he knew perfectly well was not by Aristotle.[100] To these late years probably belong his *Elenchi*, which refers to all of his previous works on logic. It would seem that from 1269 until his death in 1280, Albert resided in the Dominican Kloster of Heilige Kreuz, writing new works, revising old ones, while being constantly imposed upon to consecrate churches, altars, choir stalls, and nunneries, and to arbitrate litigations. For example, on 12 September 1276, he consecrated the Dominican church of St. Paulus in Antwerp, which he had caused to be built in 1256 when he was provincial. Albert frequently complained that such duties of a retired bishop left him little time for study and prayer.

One very important fixed date in the chronology of Albert's writings is April 1271 when he received a questionnaire that John of Vercelli sent also to Thomas d'Aquino in Paris and Robert Kilwardby,

---

[96] See *Postilla super Isaiam* Proleg. § 6 (ed. Colon. [1952] 19: xx); also B. Geyer in *Metaphysics* Proleg. § 3 (ed. Colon. [1960] 16/1: x).

[97] *Postilla super Isaiam* Proleg. § 6 (ed. Colon. 19: xx).

[98] Loë, p. 301, n. 163.

[99] Loë, p. 302, n. 171.

[100] *De causis* II, tr.1, c.1 (ed. Borgnet 10: 433b and 435b).

then provincial of England. In Albert's reply to the forty-three questions, he explicitly refers to his earlier *De causis, Metaphysics, Ethics,* and *De animalibus* for further explanation. This reply to the master general, written in April 1271, shows Albert's clear irritation over the "fatuous," "stupid," "fantastic," and "inquisitive" questions sent to him. In conclusion Albert notes that he has taken pains to answer the questionnaire only out of love and reverence for His Paternity, although he is "going blind due to old age" (*caecutientes iam prae senectute*) and "would rather spend the rest of his days in prayer than in answering silly questions."[101]

But this was not the last of Albert's writings. We know, for example, that he composed his commentary on Job in 1272 (Casanatense MS 445) or in 1274 (Munich, Univ. MS 50) and that he prepared a revised version of his commentaries on Matthew, Mark, and Luke between 1270 and 1275. It is generally admitted that *De sacrificio missae* and *De sacramento* (if authentic) are very late compositions, perhaps his last.

It is probably safe to say that, apart from episcopal and para-episcopal duties that took him away briefly from Cologne, Albert resided at Heilige Kreuz from 1269 until his death there on 15 November 1280. There is no evidence whatever that Albert attended the Council of Lyons in 1274; his name does not occur in the list of bishops who attended. In fact, when news reached him of the unexpected death of his beloved Friar Thomas on March 7 at Fossanova, he was definitely in Cologne; it is even said that Albert broke into tears when, through some mysterious vision, he perceived the very moment of Thomas' death.[102]

The statement of Tolomeo of Lucca concerning the last three years of Albert's life most likely cannot be taken seriously, considering its questionable source. Tolomeo concluded his brief account of Albert saying, "Although, as a lesson to others, his memory failed badly in intellectual matters (*multum desipuerit . . . quantum ad memorativam*) about three years before his death — for he far surpassed all others by a most singular grace — nevertheless the vigour

---

[101] *Problemata determinata* (ed. Colon. [1975] 17/1: xxvii-64); see J. A. Weisheipl, "The *Problemata Determinata XLIII* Ascribed to Albertus Magnus (1271)," *Mediaeval Studies* 22 (1960), 303-54.

[102] Loë, p. 304, n. 185. Consequently the legend of Albert's moving sermon at the council (ibid. n. 186) must be dismissed as pure fantasy.

of his devotion to God was in no way wanting, as befitted his religious state."[103]

Legend also has it that Albert went to Paris early in 1277 to protest the rumoured condemnation of some of Thomas' teachings (and his own). The source of this legend, as for the above account of Tolomeo, was undoubtedly Friar Hugo of Lucca, onetime provincial of Tuscany, by way of Bartholomew of Capua, a layman who gave testimony at the canonization process of Thomas, held in Naples from 21 July to 18 September 1319. Don Bartolomeo, who asserted that he had been a friend of Friar Hugo at Anagni and saw him again at Lucca as he journeyed to Provence, testified that he had heard the story from Hugo himself. Since Bartholomew's testimony at the canonization process is the only source for the much-beloved legend of Albert's going to Paris in 1277, it should be related in full:

[Hugo] said that when the aforesaid Friar Thomas died, Friar Albert, who was his teacher, wept profusely on hearing of his death, and thereafter, whenever he was reminded of him, he sobbed, saying that he was the flower and splendor of the world. Indeed the brethren were disturbed by so much sorrow in Albert and thought the tears stemmed from a weakness of mind (*ex levitate capitis provenirent*). Later it was rumoured that the writings of Friar Thomas were being attacked at Paris, the aforesaid Friar Albert said that he wished to rise to the defence of these writings. But the Friars Preachers, fearing the decrepitude of his age and the length of the journey, dissuaded him for a time, particularly because the aforesaid Friar Albert was a man of great authority and reputation (*auctoritatis et reputationis*) at Paris, and they feared that he might become befuddled in memory and awareness of what was going on around him because of age (*ne propter etatem declinaret in memoria et intellectu communi*). But finally Albert, who was an archbishop or bishop, absolutely insisted on going to Paris to defend such noble writings; and he went to Paris, in whose retinue was the aforesaid Friar Hugo, as he asserted to the witness [Bartholomew] himself. But when the aforesaid Friar Albert arrived in Paris and the members of the *studium generale* of Paris were convoked, he ascended the Dominican podium at Paris, taking as his text, "What praise is it for a living man if he is praised by the dead ?" making this to mean that it was the aforesaid Friar Thomas who was alive and the others who were

---

[103] *Historia eccles.* XXII, c.19, ed. Muratori 11: 1151; see testimony of Bartholomew of Capua in *Processus canonizationis sancti Thomae Aquinatis, Neapoli,* n. 82, in *Thomae Aquinatis Vitae Fontes Praecipuae,* ed. A. Ferrua (Alba: Ed. Domenicane, 1968), pp. 324-25; also Mandonnet, "La date . . .," p. 251.

dead; and he proceeded to praise and glorify Thomas in the highest terms, declaring that he was personally prepared to defend the writings of the aforesaid Friar Thomas as the splendor of truth and sanctity before the most competent critics.

After a lengthy panegyric in praise of God and in approbation of those writings, the same Friar Albert returned to Cologne, accompanied by the aforesaid Friar Hugo, as he told the same witness [Bartholomew]. After his return, the aforesaid Friar Albert caused all the writings of the aforesaid Friar Thomas to be read to him in a definite order. Then at a solemn convocation convened by him, he put forward an exceedingly great and glorious commendation, concluding with the assertion that the latter's writings had put an end to the labors of all other men till the end of time, and that henceforth they would all labour in vain." And as the same Friar Hugo related to the witness [Bartholomew] the name of that Friar Thomas could never be mentioned in the presence of Friar Albert without him breaking into tears (*prorumperet ad lacrimas*).[104]

This is the end of Bartholomew's sworn testimony concerning this incident, although he had much else to say about Thomas in the canonization process, and even presented a list of Thomas' known writings.

Obviously, the whole purpose of Bartholomew's testimony about the incident was to show that even Albert the German (an archbishop or bishop), Thomas' own teacher (*doctor eius*), had the highest regard for the sanctity and truth of Thomas' writings. What was at issue in the canonical inquiry at Naples in 1319 was the sanctity and worthiness of Thomas for canonization.

Although it is conceivable that Albert actually went to Paris and acted as reported, it is most unlikely. First of all, both Tolomeo of Lucca and Bartholomew of Capua, relying on Friar Hugo's narration, present Albert as already "senile" before 7 March 1277, a man whose memory was already failing. But, as we shall see, all other contemporary evidence weighs against this view. Second, it was not Thomas' writings that were mainly in question at that time, but certain masters in arts who were recklessly utilizing pagan philosophers, notably Aristotle and Averroës, to the detriment of the Catholic

---

[104] Full Latin text of Bartholomew's testimony in *Proc. canoniz. S. Thomae*, n. 82, referred to in n. 103 above. A freer translation is given by K. Foster in his *The Life of Saint Thomas Aquinas: Biographical Documents* (London: Longmans, 1959), pp. 112-13; on Friar Hugo Borgognoni, see Foster, pp. 124-125 n. 80.

faith. Pope John XXI wrote to Stephen Tempier, bishop of Paris, on 18 January 1277 to ascertain the source and nature of these errors "prejudicial to the faith." There is no way of knowing what form those "rumours" took when they reached Albert in February (presuming they did). But they could not have involved the whole of Thomas' writings. The crucial issue concerning the unicity of substantial form in each material being was then *sub judice* at the papal curia, as Pecham himself informs us.[105] As for the list of 219 propositions actually condemned on 7 March 1277, Gilson rightly observes, "the list of Thomistic propositions involved in the condemnation is longer or shorter, according as it is compiled by a Franciscan or by a Dominican."[106] Even if Bishop Albert had not intended to dissuade Bishop Tempier from such a foolhearty and impetuous condemnation, he could hardly have thought that Thomas' writings were the object of the rumoured condemnation. Third, there is no other evidence whatever of Albert's supposed journey to Paris in defence of Thomas' writings. While every argument from silence is historically weak in itself, it is indeed astounding that there should be no hint of such a momentous event in an independent German tradition of Albert's life or in Parisian circles most concerned with the condemnation of 1277. Until some other evidence is found, it is most unlikely that Albert went to Paris in 1277 or that his memory became much befuddled almost three years before his death.

Up to fifteen months before his death, Albert was clearly competent to negotiate various delicate transactions, including intricate arbitrations, as is evident from six documents dating from 26 September 1277 to 18 August 1279.[107] When Albert made out his last will and testament in January 1279, making his brother Friar Henry of Lauingen, prior of Würzburg, executor of the will, he stated that he was of sound mind and body (*sanus et incolumis*).[108] There is no doubt that when Albert heard of the proposed condemnation, if indeed he did before 7 March 1277, he might have thought of doing something. But apart from Don Bartolomeo's recollection of Friar

---

[105] *Registrum Epistolarum Fratris Johannis Peckham*, ed. C. T. Martin (London: RS, 1885) 3: 866. This letter, dated 7 December 1284, is addressed to the chancellor and University of Oxford.

[106] E. Gilson, *History of Christian Philosophy in the Middle Ages* (New York: Random House, 1955), p. 728, n. 52.

[107] See Loë, pp. 307-09.

[108] Scheeben, *Chronologie*, pp. 123-27.

Hugo's story, there is no evidence whatever that Albert undertook a long journey to Paris, one which could have been nothing but futile.

That Albert was getting senile toward the end of 1279 can hardly be gainsaid. It is quite possible, considering his age and the incredible energy he had put into everything he did, that Albert's memory did begin to fail and that at times he might have been completely befuddled. At that time and for some years before, Friar Gottfried of Duisburg was Albert's *socius et minister*, serving in much the same capacity as Reginald of Piperno for Friar Thomas. It is uncertain when Gottfried became Albert's *socius*, but it would seem that his services were nowhere as long nor as essential as those of Reginald in the life of Thomas. Henry of Herford gives three explicit indications of Albert's growing senility, all of which seem to have taken place during the last fifteen months of his life. For example, once a certain Archbishop Sigfried wanted to visit the aging bishop, but when he was announced, Albert replied, "Albert is not here."[109] There are no official documents involving Bishop Albert after 18 August 1279. He seems to have declined steadily and prayerfully. All that can be said is that neither Albert nor his confreres were unprepared when death came for the saintly bishop on Friday, the feast of St. Geltrud, 15 November 1280. He was then *octogenarius et amplius*; no one knows just how much *amplius*.

Albert spent a very full and active life probing all truth, human and divine, in the service of God, giving to others the fruits of his contemplation in writings, lectures, sermons, counsel, and example. The heroic zeal he showed in his vast apostolate was thoroughly Dominican, as he saw it. In his own day he was with good reason known as *Doctor universalis* and *Doctor expertus*, for his knowledge was truly "universal" and he knew much from personal "experience." In his own lifetime, though late, he was even known as Lord and Friar Albert "the Great."[110] In the words of Ulrich of Strassburg, a disciple and intimate friend, Albert was "a man so superior in every science, that he can fittingly be called the wonder and miracle of our time."[111]

Scholarly respect for Albert the "universal doctor" and popular

---

[109] Herford, *Chronica*, p. 202; Loë, p. 309, n. 216.

[110] *Annales Basileenses*, anno 1277; "Albertus Magnus, lector Coloniae." *Monumenta Germaniae Historica*, Scriptores, 17: 202.10-11

[111] *Summa de bono* IV, tr.3, c.9. See J. Daguillon, *Ulrich de Strasbourg, La "Summa de bono,"* Livre I (Paris: Vrin, 1930), 139.

devotion to Blessed Albert the saintly friar flourished for centuries, especially in his native Germany. Serious study of Albert's authentic writings, however, fluctuated greatly throughout the centuries, although spurious writings continued to grow in number, fertile fantasy, and popular appeal, particularly in times of crises. Around the time of the First World War various academic, scientific, religious, and theological efforts coalesced to bring about the canonization of Albertus Magnus by Pope Pius XI on 16 December 1931 with the additional title of *Doctor Ecclesiae*.[112] As clouds were gathering for World War II, scientists, academicians, philosophers, theologians, and concerned humanitarians throughout the entire world appealed to the Holy See for a special Patron before God for our critical times of scientific advancement and political decision-making that utilize scientific discoveries affecting the whole of mankind. In the midst of a most devastating war Pope Pius XII acceded to this fervent plea of scientists throughout the world, when, on 16 December 1941, he used the fullness of his Apostolic authority to "declare and constitute Saint Albert the Great, Bishop, Confessor, and Doctor of the Church, forever the PATRON before God of students of the natural sciences (*Cultorum Scientiarum naturalium coelestem apud Deum PATRONUM*) with the supplemental privileges and honours which belong, of its nature, to this heavenly patronage."[113]

Two main features are highlighted for "our time" in Albert's patronage of scientific endeavour: his example as a scientist and his intercessory power as a saint. By his example, he devoted his tireless energies (even as bishop) to the pursuit of scientific truth in nature, and he insisted on the indispensability of scientific truth (and indeed of all philosophical truth) for sound theology, "the science of God's special revelation to man." By his actual attainment of eternal life with God, he is in a favoured position to help scientists, individually and collectively, to pursue their special goal with human prudence and dignity. Above all, he is in a unique position to guide scientists, agencies, and government officials to make correct moral decisions — for which they are responsible before God and man — that affect

---

[112] For documents leading to this canonization, see the "Positio pro canonizatione ac doctoratu B. Alberti Magni" collected in *Extensionis seu Concessionis Officii et Missae Addito Doctoris Titulo ad Universam Ecclesiam in Honorem B. Alberti Magni*, and published by the Sacred Congregation of Rites, Rome 1931.

[113] Apostolic Letter *Ad Deum*, AAS, 34 (1942), p. 91. The events leading to this declaration are listed earlier in the Apostolic Letter, pp. 89-91.

the whole of mankind today and tomorrow. Recognition of his Patronage today would be a "miracle of our time."

Addenda: Dominican Houses in Teutonia
(During Albert's Lifetime)

Before Albert became provincial in 1254

1. Friesach, Austria (*Carinthia*), ca. 1220
2. Cologne, Nordrhein-Westfalen (*Lotharingia inferior*), ca. 1220
3. Strasbourg, France (*Alsatia*), 1224
4. Magdeburg, Saxony (*Saxonia inferior*), 1224
   (First Provincial Chapter May or June 1225 elected Conrad von Höxter as provincial, 1221-33)
5. Trier, Rheinland-Pfalz (*Alsatia*), by 1225
6. Bremen, Bremen (*Saxonia inferior*), 1225
7. Vienna, Austria (*Austria inferior*), 1225
8. Worms, Rheinland-Pfalz (*Franconia*), 1226
   (Albert elected prior provincial in June 1254)
9. Würzburg, Bavaria (*Franconia*), ca. 1229
10. Regensburg, Bavaria (*Bavaria*), 1229
11. Louvain, Belgium (*Brabantia*), ca. 1228/29
12. Lübeck, Schleswig-Holstein (*Slavia*), 1229
13. Erfurt, Thuringia (*Thuringia*), 1229
14. Leipzig, Saxony (*Misnia*), 1229
15. Zürich, Switzerland (*Alsatia*), ca. 1230
16. Ptuj (formerly Pettau), Yugoslavia (*Styria*), 1230/31
17. Koblenz, Rheinland-Pfalz (*Franconia*), by 1230
18. Esslingen, Baden-Württemberg (*Suevia*), 1230 (?)
19. Halberstadt, Saxony (*Saxonia inferior*), 1232
20. Utrecht, Netherlands (*Hollandia*), 1232
21. Basel, Switzerland (*Alsatia in Suevia*), 1233
22. Freiburg-im-Breisgau, Baden-Württemberg (*Suevia*), 1233
23. Hildesheim, Niedersachsen (*Saxonia inferior*), 1233
24. Constance, Switzerland (*Suevia*), by 1233
25. Frankfurt-am-Main, Hessen (*Franconia*), 1233 (?)
26. Freiberg, Saxony, (*Saxonia inferior*), 1236
27. Minden, Nordrhein-Westfalen (*Westfalia*), 1236
28. Krems, Austria (*Austria inferior*) 1236 (?)
29. Hamburg, Hamburg (*Slavia*), 1240
30. Soest, Nordrhein-Westfalen (*Westfalia*), 1241

31. Riga, Latvia (*Livonia*), 1234
    (Incorporated into Province of Teutonia in 1244)
32. Antwerp, Belgium (*Brabantia*), 1245
33. Leewarden, Netherlands (*Frisia*), 1245
34. Neu Ruppin, Mecklenburg (*Marchia Brandenburgensis*), ca. 1246
35. Augsburg, Bavaria (*Bavaria*), 1251
36. Stralsund, Rostock (*Slavia*), 1251

During provincialate of Albert, June 1254 to June 1257

37. Strausberg, Brandenburg (*Marchia Brandenburgensis*), 1254
38. Seehausen, Altmark (*Marchia Brandenburgensis*), 1255
39. Rostock, Rostock (*Slavia*), 1256

From Albert's resignation to the General Chapter of May 1277

40. Mainz, Rheinland-Pfalz (*Franconia*), 1257
41. Speyer, Rheinland-Pfalz (*Franconia*), 1260/61
42. Maastricht, Netherlands (*Brabantia*), 1261
    (Began ca. 1232/33)
43. Norden, Friesland (*Frisia*), 1266
44. Plauen, Saxony (*Misnia*), 1266
45. Rottweil, Baden-Württemberg (*Suevia*), by 1269
46. Bern, Switerland (*Suevia*), 1269
47. Wimpfen, Baden-Württemberg (*Suevia*), by 1269
48. Halle, Saxony (*Saxonia inferior*), 1271
49. Wiener Neustadt, Austria (*Austria inferior*), 1275-77
50. Nürnberg, Bavaria (*Bavaria*), ca. 1275
51. Myślibórz (formerly Soldin), Poland (*Marchia Brandenburgensis*), 1275
52. Penzlau, Uckermark (*Marchia Brandenburgensis*), 1275
53. Ziericksee, Netherland (*Zeelandia*), 1276

(At the General Chapter of May 1277 *Teutonia* had 53 priories of men and 40 cloisters of nuns [*Claustra Sororum*]. QE I, i.)

From the General Chapter of 1277 to Albert's death, 15 November 1280

54. Pforzheim, Baden-Württemberg (*Suevia*), 1278
55. Chur, Switzerland (*Suevia*), 1278
56. Eichstätt, Bavaria (*Bavaria*), 1278
57. Colmar, France (*Alsatia*), 1278
58. Leoben, Austria (*Styria*), by 1280

59. Winsum, Netherlands (*Frisia*), 1280
60. Tulln, Austria (*Austria inferior*), 1280
61. Ulm, Baden-Württemberg (*Suevia*), ca. 1280
62. Landshut, Bavaria (*Bavaria*), ca. 1280

Note: When the German Dominican Province was divided at the General Chapter of 1303, there were 49 priories of men and 65 monasteries of nuns belonging to the Province of *Teutonia*, while 47 priories of men and 6 monasteries of nuns constituted the new Province of *Saxonia. Teutonia* retained the territories of *Austria* with adjacent priories, *Bavaria, Suevia, Franconia, Alsatia* all the way to the area around Cologne, and *Brabantia. Saxonia* embraced *Misnia, Thuringia, Saxonia, Slavia, Marchia Brandenburgensis, Frisia, Zeelandia, Hollandia,* and *Livonia.* QE I, ix-x, xiv-xv; A. Walz, *Compendium Hist. Ord. Praed.* (Rome, 1948), pp. 126-27.

2

The Attitude of Roger Bacon to the *Scientia* of Albertus Magnus

Jeremiah M. G. Hackett
Pontifical Institute of Mediaeval Studies

Since the rediscovery of the works of Roger Bacon in the nineteenth century, it has been customary to see the *Doctor mirabilis* as a controversialist, early scientist, philosopher, and theologian. Many scholarly judgments have been passed on the merits of his work. Some would see him as a schoolman who never quite reached the stature of an Aquinas or a Bonaventure. Others would see him as a very significant representative of an important stage in the history of science and philosophy. The life and work of Roger Bacon span the whole educational background of the thirteenth century. Like his contemporary Albertus Magnus, whom he may have known at Paris during the years 1245-1248, he was a *savant* with an enormous encyclopaedic mind. The breadth and depth of their understanding of the whole tradition of learning in their time was very great. To take one example, the reception of the new translations of Aristotle in the university milieu of the first half of the thirteenth century found two diverse interpreters in Roger Bacon and Albertus Magnus. It is well known that Albertus Magnus was renowned as a commentator on Aristotle; it is not so well known, even though the point has been made forcefully in some modern Bacon scholarship, that Roger Bacon stands out as a great example of one who had mastered the new translations of Aristotle in the early years of the thirteenth century. Furthermore, Roger Bacon had completed his *Questiones* on Aristotle's books well before Albertus Magnus came to Paris to take his doctorate in theology.

Perhaps the popular image of Bacon today and the view of his work as mere magic or mere alchemy has been due in no small way to attitudes towards Bacon during the Renaissance. However, the "scientific" work of Bacon was not without its defenders during that period; the *apologia* of John Dee is a case in point. An examination of all of Bacon's writings shows that his works on magic and alchemy form a small though significant part of his work. His criticism of magic stands out as a clear-headed attempt to distinguish magical practice from the art and science of nature. The greater part of Bacon's philosophical work is concerned with the interpretation of Aristotle. There is scarcely a page in his scientific work which does not owe something to the logic, ethics, metaphysics, and natural philosophy of Aristotle. In speaking about the philosophy of Roger Bacon, it is best to avoid traditional labels and to seek out just what he said.

We know very little about the career of Roger Bacon. The only materials which we have for a biography are his own writings. Scholars dispute the date of his birth and the date of his death. The former is usually placed in either 1214 or 1220; the latter is placed at various times from 1284 to 1294. The only definite date from one of his last works is 1292, which is given in Bacon's *Compendium studii theologiae*. It would seem that Bacon was a master of arts at Paris from 1240-1247. We do not know where Bacon was between 1247 and 1250/51. During 1250/51, however, he was in Paris. He seems to have begun a ten-year period of private research after relinquishing his teaching at Paris. It is quite probable that he became a Franciscan friar around the year 1257. Bacon might have remained a forgotten master of arts were it not for the fact that in 1266 Pope Clement IV (1265-1268) requested him to send a copy of his writings on the reform of education and society. Bacon's reply was an embarrassed apology for the non-existence of the work. Still, Bacon wrote with great haste and against serious impediments, and within a year and a half, produced the works for the Pope which are nowadays known as the *Opus maius* (1266-1267), *Opus minus* (1266-1267), and *Opus tertium* (1266-1267). These writings contain Bacon's plans for the reform of education and society. They include an uneven mixture of philosophical comment, polemic, and some scientific work. These works have been seen as the inept ravings of a tired old man. Rather, however, they seem more like a conscious effort to study the state of learning in the universities of the mid-thirteenth century, and to suggest positive means for reform in education.

Bacon believed that two men were responsible for the failures in the educational system of his time. He names one of them in regard to the teaching of theology as Alexander of Hales. The second master, who bears the brunt of the most sarcastic remarks to come from Bacon, is not personally named; he has come to be known in modern scholarship as "the unnamed master." According to Bacon, he is the one who made himself an authority and a writer of many books on the topic of natural philosophy (*ille, qui fecit se auctorem*, and *ille qui composuit tot et tam magna volumina de naturalibus* etc.). Although the "unnamed master" is often thought to be Albertus Magnus, at least in some contexts, there is no universal agreement about the identity of the person intended by Bacon, and almost no appreciation of the reason for Bacon's ire.

Stewart C. Easton,[1] one of the more recent scholars to review the problem, has argued that Albertus Magnus is the only contemporary of Bacon who fits the description in Bacon's polemic, but the argument is made at the expense of Bacon:

> He [Bacon] appears ignorant of the philosophical questions involved; he himself begs most of the questions — not because he was necessarily incompetent in philosophy, but because he did not recognise the competence of philosophy in the sphere of religion.[2]

But this is not the whole picture. In many of his works, Bacon is primarily concerned with the role and place of philosophy and method in the study of theology. Bacon devoted an entire section of his *Opus maius* (Part II) to a discussion of this question, and the plan of the work itself was set out so as to show the regions of knowledge in the light of their final goal in theology. It is rather unfortunate that Easton allows a personal preference for the work of theologians like Aquinas and Bonaventure to blind him to the actual content of Bacon's work. He accuses Bacon of abysmal ignorance of the material studied in the theological faculties of his day, and he rejects Vanderwalle's judgment:

> It is certainly untrue to say, as Vanderwalle does, that Bacon was familiar with the writings of Peter Lombard, Alexander of Hales, Albertus Magnus, and Thomas Aquinas. A close examination of his references to these men shows precisely the opposite.[3]

---

[1] Stewart C. Easton, *Roger Bacon and His Search for a Universal Science* (Oxford, 1952), pp. 210-31.

[2] Ibid., pp. 220-21.

[3] Ibid., p. 20.

Easton devotes an "appendix" of some length to the question of "the unnamed master." He marshalls the various texts in Bacon which refer to him, and insists that he can be none other than Albertus Magnus. While not fully appreciating Albert's notion of *scientia*, Easton sees the opposition of Bacon and Albert:

> Bacon's objections to the science of Albert are more quickly dealt with. As shown in the text of this study, Bacon believed in a universal science which must be complete. . . . This grandiose conception was alien to Albert, even though he made contributions to many sciences. The relation between the sciences for him was by subalternation, a dependence of one science on another, as optics on geometry, the lower dependent on the higher in the scale (degrees of abstraction). But Bacon wanted more than this. He wanted a whole self-contained and beautiful building (his own analogy). Moreover, Albert had omitted optics, and was deficient in mathematics, and knew no languages but Latin and the vernacular.[4]

Bacon's objections to the *scientia* of the master whom the whole world (*vulgi*) followed were, in fact, always very specific: that master was ignorant of ancient languages and mathematics, specifically, perspective. Yet, despite his identification of "the unnamed master" as Albert, Easton seems to share some of the misgivings of earlier scholars about this identification.

The renowned A. G. Little changed his view a number of times.[5] In his last pronouncement (1928) Little thought that it was St. Thomas Aquinas whom Bacon chiefly had in mind: "The chief object of his attacks is Friar Thomas Aquinas (though he does not mention him by name), whom he denounces as the greatest corrupter of philosophy that has ever been among the Latins."[6] Dorothy E. Sharp (1930) argued that it could be either Albert or Thomas.[7] Neither Richard Rufus nor Vincent of Beauvais, whom some have suggested, are likely candidates.[8] Although Vincent produced the most famous encyclopaedia of learning in the thirteenth century

---

[4] Ibid., pp. 230-31.

[5] A. G. Little, ed., *Roger Bacon Essays* (Oxford, 1914), p. 8 n. 9: "Op. Tertium (Brewer), pp. 30, 37-42; Op. Min., pp. 327-8: these passages probably refer to Albert rather than to Aquinas" etc.

[6] A. G. Little, "Roger Bacon," *Proceedings of the British Academy*, 14 (1928), 18.

[7] D. E. Sharp, *Franciscan Philosophy at Oxford* (London, 1930), p. 118.

[8] See Ludwig Lieser, *Vincenz von Beauvais als Kompilator und Philosoph. Eine Untersuchung Seiner Seelenlehre im Speculum Maius* (Leipzig, 1928), pp. 8, 61-69.

(*Speculum maius*), he did not compose works for the schools nor commentaries on Aristotle; and he did not attempt to rewrite the whole of philosophy in the Latin language. Theodore Crowley (1950) thought that the "two men" in question were St. Albert the Great and Alexander of Hales.[9] However, the most cautious, critical, and formidable evaluation of the attempts to identify the "unnamed master" was clearly presented by Lynn Thorndike as early as 1929 in his pioneering *History of Magic and Experimental Science:*

> It seems incongruous for Bacon to speak of his probable senior, Albert, as a boy. Other passages in Bacon's works which have been taken to apply to Albert, though he is not expressly named, seem to me not to apply to him at all closely; and if meant for him, they show that Bacon was an incompetent and unfair critic. Not only was Albert for a short time in Paris; he does not seem to have been in sympathy with the conditions there which Bacon attacks.[10]

Thorndike felt that it was rather incongruous that Bacon should attack a man such as Albert the Great, who manifested even more than he "unmistakable signs of the scientific spirit."[11]

Thus, there is no agreement among modern scholars as to the "unnamed master" who happened to be the target of Bacon's caustic statements. Further, there seems to be no appreciation of the *reason* for Bacon's attack. Bacon may have been a disappointed, disgruntled, and even envious old man in the 1260s and 1270s, but he had a clearly stated point of view, which he defended against the currents of "the past forty some years." He clearly blamed certain individuals for the decline of learning in the Church, and he had a definite program in mind that could remedy a deplorable situation — if only he could be heard.

The purpose of this paper is twofold: to determine as unmistakably as possible the "unnamed master" who has perverted the whole of philosophy in "the past forty some years," and to spell out the precise reasons for the "innumerable errors" that have resulted from the "authority" of this one man, as Bacon evaluated the situation. The procedure will be to examine the works of Bacon, beginning (in

---

[9] Theodore Crowley, *Roger Bacon. The Problem of the Soul in His Philosophical Commentaries* (Louvain-Dublin, 1950), p. 25 n. 42: ". . . From parallel passages in the *Opus tertium* (ed. Brewer, p. 30) and in the *Compendium Philosophiae* (ed. Brewer, p. 425 f.), it can be inferred that the two men in question were St. Albert the Great and Alexander of Hales."

[10] Lynn Thorndike, *History of Magic and Experimental Science* (New York, 1929), 2: 639.

[11] Ibid., p. 535.

reverse order of composition) with the works wherein Albert is mentioned by name to the earlier works in which "that man who has made himself an authority" is, in fact, unnamed. Thus the order of works to be considered are the *Compendium studii philosophiae* (ca. 1271-1272), the *Opus tertium* (1266-1267) to Pope Clement IV, the *Opus minus* (ca. 1266-1267) to the same pope, and finally the first part of *Communia naturalium*, a work begun early in the 1260s and completed at a later date. From this examination not only should the identity of the "unnamed master" be clear, but also the reason for Roger Bacon's objections against the "science."

A. *COMPENDIUM STUDII PHILOSOPHIAE* (CA. 1271-1272)

We may begin with those passages which explicitly name Albert and Thomas together. These are found in a late work entitled *Compendium studii philosophiae*.[12] We know from internal evidence that this was written circa 1271-1272. In this work Bacon mentions that he had sent a tract on these matters to "the predecessor of the present pope"; thus he wrote it in the reign of Pope Gregory X (1271-1276). The *Compendium* shows a remarkable development in tone from that of the three works which he wrote specifically for Pope Clement IV. The topic remains the same, viz. the new "boy theologians" who read the *Sentences* have ruined the traditions of study, which were characteristic of the faculty of arts, and of great prelates such as Robert of Lincoln (Grosseteste). The emphasis has become explicit in identifying the crux in the decline of studies. In the *Opus maius, Opus minus*, and *Opus tertium*, Bacon had not placed the decline in study in the actual context of the conflict between the regular and secular clergy. Bacon shows in the *Compendium studii philosophiae* that he is writing his new ideas as a propagandist who reflects on the current state of university affairs. Here, he makes an effort to situate the problem of studies in historical pattern. Thorndike has rightly drawn attention to the historical awareness of Bacon, who is somewhat unique in discussing the question of the reception of Aristotle in the west in the thirteenth century.

From the opening pages of the *Compendium studii philosophiae* on, Bacon sets out this problem in detail. The work is concerned with the ways and means of achieving speculative and practical wisdom. The

---

[12] Roger Bacon, *Compendium studii philosophiae*, ed. J. S. Brewer (London, 1859).

schools are the ideal place for this endeavour. Again, Bacon is presenting the utility of the sciences for theology. Thus, his overall purpose in this work is unmistakably practical. One of the goals of science for Bacon is the proper direction and reform of society. Very
systematic about learning, he thinks that studies should be based on
a definite method and not on arbitrary decision. In the opening part
of the *Compendium*, he states that method has to do with the study of
things through the discovery of rational causes.[13] It is a search which
is based on authority, reason, and experience. Bacon once again
mentions the normal impediments to learning. The worst impediment, the fourth fault which he gives in the first part of the *Opus
maius*, is the false appearance of knowledge. According to Bacon, the
schools at that time (ca. 1271-1272) are rife with this error. He names
those responsible for this condition — the young students of the two
orders. This deficient state of studies is reflected in the corruption of
society as a whole: the papacy has been vacant for a number of
years; religious have lapsed into a decadent state. The Italian civil
lawyers are responsible for drawing students from the schools, and
for being mechanical in their approach to the study of philosophy.
On the whole, he reserves his wrath for the "boy theologians" of the
two orders; in chapter v he begins a long tirade against them. It is
here that the explicit reference to Albert and Thomas occurs.

He speaks of the "boys" as the embodiment of all the error in
studies. These young men have arisen in the *studia* and have made
themselves into masters and doctors of theology and philosophy.
However, they have learned nothing of value on account of their
state of life. They neglect the arts; they do not know all the parts of
science and philosophy; they presume to know theology even though
they lack the human knowledge which is needed for that task.

> These are the boys among the students of the two orders like Albert
> and Thomas, and others, who enter the orders when for the most part
> they are twenty years of age and less.[14]

He adds that these boys were put to read theology after their profession even though they did not have any formal training in reading
the Psalter or in reading Priscian. And this has been the case since
the establishment of the many *studia*.

---

[13] Ibid., p. 397.
[14] Ibid., p. 426: "Hi sunt pueri duorum ordinum studentium, ut Albertus et Thomas, et alii,
qui ut in pluribus ingrediuntur ordines, quum sunt viginti annorum et infra."

Is this reference to Albert and Thomas as boys quite as self-contradictory as Thorndike would have us think? Is it not clear from the context that the author merely takes them as examples of the many boys who did enter the orders at an early age? In his comments on the "boy" theologians, Bacon is, at least, consistent. He returns to the same criticism in his anonymous texts against the "unnamed master." According to Bacon the latter did not teach (*legit*) in *artes* before becoming a theologian. Therefore, even though he was self-taught, he must necessarily be ignorant of the sciences. Bacon continues in the *Compendium*:

> And so it was proper that they should not profit in any way, especially since they did not seek to have themselves instructed by others in philosophy after they had entered [the orders], and especially since they presumed to investigate philosophy by themselves without a teacher. Thus, they became masters in theology and philosophy before they were students (*discipuli*). Therefore, infinite error reigns among them. . . .[15]

This accusation could be levelled against Albert, and indeed against many of the friars who went out to preach.

Bacon further places blame for the mere "appearance of wisdom" on the apparent sanctity of the two orders. However, the real reason for the lack of wisdom in the centres of Christian culture was the manifest neglect of studies by the secular masters. According to Bacon, contemporary secular masters had betrayed the great tradition of study associated with Robert Grosseteste, Adam Marsh, and William of Shyreswood. The new secular masters do not teach the *Sentences*, or incept in theology, or lecture, preach, or dispute except by means of the *quaternos puerorum in dictis ordinibus*, as is evident at Paris and elsewhere. Bacon consciously discusses the important conflict between the seculars and the mendicant orders, and sees therein the reason for the decline in study.

> Therefore, truly it has already been brought to the notice of the public at Paris for the past twenty years that an unspeakable conflict has arisen among the religious because the seculars revolted against the regulars and the religious revolted against the seculars. And they called each other heretics and disciples of antichrist. . . . And they have not ceased up to this time.[16]

---

[15] Ibid., p. 426.

[16] Ibid., p. 429: "Certum, igitur et jam per viginti annos deductum publice Parisius, quod ineffabilis contentio orta est inter religiosos, ita quod seculares insurrexerunt contra ordines et a converso; et se mutuis assertionibus vocaverunt hereticos et discipulos antichristi. . . . Et adhuc non cessant.

Bacon is not altogether detached in regard to this conflict. It can be seen from Bacon's own words that he was engaged in discussion with the "boy" theologians.[17] He allows that though they are not immune from the corruption of the study of wisdom, they are guarded from the accusation of heresy and the name of the antichrist in that they are members of a holy order.

At this point in the *Compendium*, Bacon reveals a personal conviction which may well shed some light on his motive for joining the Franciscan Order. He says that God has punished the secular masters who blasphemed against the grace of God which is now given to the religious, including his own Franciscan Order. He holds strongly to the view that the whole Church regards the religious state in life as higher than the secular, although he readily admits that the Parisian masters disagree with him:

> But the masters at Paris teach what is plainly contrary and they confirm it with many sophisms.[18]

Bacon is irritated that the seculars solicit the power, support, and authority of pope and prelates to defend themselves. He answers that there are two kinds of authority; the authority of office and the authority of spiritual perfection. For Bacon, the latter is the more perfect kind. Bacon, thus, manifests a tension in his own teaching concern. On the one hand, he has great respect for the tradition of the earlier secular masters. On the other hand, he now despises the position which the new secular masters have taken in the university. Thus, one can say that Bacon's becoming a Franciscan was a much more intense experience than it is generally held to be.

B. *OPUS TERTIUM* (1266-1267)

Bacon is writing in the context of the anti-mendicant controversy. Thus, there is an evident absence of names from all three works which were written for Clement IV. Yet it is certain that anyone acquainted with the problems which Bacon criticises would know

---

[17] Ibid., pp. 429-30: "Multotiens et audiendo et docendo vel dico veritatem fratribus istorum ordinum, et scribo quod respondeant mihi huic argumento: Discipuli sunt heretici et praecursores Antichristi, ut vos dicitis, et verum; igitur vos estis haeretici et discipuli Antichristi. Sed nullum invenio qui dissolvat argumentum, licet omnes negent conclusionem. Pro certo igitur sequitur conclusio ex praemissis, nisi quia status sanctus est, et innocentia juvenum intrantium hos ordines salvarent eos ab haeresi et a titulo antichristi."

[18] Ibid., p. 430: "Sed magistri Parisius docent de plano contrarium, et multiplici sophismate confirmant."

immediately the names which Bacon held responsible for the decline in study. In this respect, it is to be expected that Bacon would refrain from an explicit mention of Albert in the three works for the pope. He does make an explicit reference to Albert in the *Opus tertium*. Apart from the above-mentioned reference from *Compendium studii philosophiae*, this is the only explicit reference to Albert in Bacon's later works. Bacon does praise Albert in the *Opus tertium*, but such praise is merely a stage in an argument intended to show that the science of Albert does not measure up to Bacon's strict standards. At first it would seem that Bacon is showing respect for a great mind. As the argument develops, however, it is evident that Bacon is drawing a contrast between the wisdom of this man and the system of knowledge based on his own theory of perspective. Wishing to explain his delay in writing to the pope, Bacon says that the works which His Holiness requested had not yet been written as the clerk of the pope had believed. Bacon blames his failure in writing on the difficult nature of the subject matter:

> . . . which you can certify through the better known *sapientes* among the Christians, one of whom is brother Albert of the Order of Preachers, while the other is Master William of Shyreswood, the treasurer of the church of Lincoln in England, a far wiser man than Albert. For no one is greater than he in *philosophia communi*.
>
> If Your Wisdom were to write to them concerning the matter of the works which I sent to you, and concerning which I will touch on in this third writing, you will see that ten years will pass before they will send to you those very things I have already written. You will certainly find a hundred places [among my writings] to which they would never attain through those things which they know now, even up to the end of their life. For I know their science well (*Cognosco enim eorum scientiam optime*). And I know at least that neither of the aforementioned, neither the first nor the last, would be able to send to you the works which I have written within the amount of time that has elapsed since your mandate. One should not wonder therefore at my delay in this area of philosophy. For the wisdom of perspective alone, which I will write, could not be written by anyone within a year. But why do I hide truth in this matter? I assert, therefore, that you will find no one among the Latins who would render this area of wisdom within one year or indeed in ten years.[19]

This text poses a problem. Bacon usually contrasts the terms

---

[19] Roger Bacon, *Opus tertium*, ed. J. S. Brewer (London, 1859), p. 14.

sapientes and *vulgus*. Here he calls Albert one of the *sapientes*, yet in another text Bacon refers to the "unnamed master" as the best teacher among the *vulgus*. And he adds that he was the most studious among them. He uses the term *sapientes* again in the *Opus tertium* in those texts which treat of the "unnamed master." A reading of these texts will show that Bacon's words of praise for Albert are severely qualified. In a reference which is placed soon after the above passage from the *Opus tertium*, Bacon takes up the topic of the completion of philosophy in the Latin language.

> The fifth objection is the strongest and the gravest for me; but it is solved through the fourth. It states that it is already thought by the *vulgus studentium*, and by many who are sincere scholars, although they are deceived, that philosophy was given to the Latins, and was completed and composed in the Latin language. And they hold that it was accomplished in my time and spread about at Paris, and the composer of it was held to be an authority (*in tempore meo et vulgata Parisius et pro auctore allegatur compositor eius*). For just as Aristotle, Avicenna, and Averroës are held [as authorities] in the schools, so too is he. And he still lives and has great authority in his lifetime, such as no man ever had in teaching.[20]

Various allegations which Bacon brings against this man will be noted later. The key issue for Bacon is that he (the unnamed master), or rather his followers at Paris, claim that he has rewritten the whole of philosophy in Latin, and that it is now final and complete. Bacon's fourth objection, alluded to in the above quotation, deserves consideration:

> . . . the fourth [objection] is that the author of these works omitted those parts of philosophy of great utility and immense beauty and without which it is not possible to know those things which are commonly taught, concerning which I will write to Your Glory.
>
> And so there is nothing of use in his writings. But there is much in them which is of the greatest detriment to learning. One should not be surprised that his writings have been justly neglected since he heard no part of philosophy and was not taught by anyone. And he was not educated in the *studium* in Paris, nor in any place where a *studium* of philosophy flourished. He did not teach nor dispute nor exercise himself in conferring and disputing with others. Nor did he have a revelation, since living otherwise, he did not prepare himself for this. And gathering false, vain, and superfluous things, he put aside the practical but

---

[20] Ibid., p. 30.

necessary things, which things do not indicate a revelation. But through himself he presumed to treat of those things he did not know.

I have not said these things about this aforementioned author without cause, since not only is it of service to my proposal, but it is to be mourned that the study of philosophy has been corrupted through him more than through all who ever existed among the Latins. For while others failed, they did not presume authority. But this one wrote his books *per modum authenticum.* And so the whole mob at Paris refers to him as to Aristotle, or Avicenna, or Averroës, and other *auctores.* And this man gave great injury not only to the study of philosophy but to theology, as I show in the *Opus minus,* where I speak of the seven sins in the study of theology. And the third sin is especially against him, as I discuss it openly because of him. For I remark on two people there, but he is the principal one in this matter. The other one, who however has died, has a greater reputation. And these things follow clearly from the *Opus majus* and the *Opus minus,* since in respect to matters both human and divine, concerning which he is accustomed to adjudicate, I show that all vain, false, and superfluous things are multiplied, while singuarly renowned, great, and useful things are left aside. And these things will be apparent with sufficient clarity from this third writing.[21]

Bacon is most explicit here. No one has ever composed philosophy in Latin. It was originally given to the Hebrews and renewed through the Greeks, especially through Aristotle, and was renewed in the Arabic language through Avicenna:

It [philosophy] was never composed in Latin, but was only translated from foreign languages, and the better parts were not translated. And nothing is perfect of those sciences which have been translated. The translations are perverse and are not understandable in many sciences, especially in the books of Aristotle.[22]

At this point, Bacon moves to a favourite theme: "only one [of the Latins] knew sciences, that is the bishop of Lincoln [Grosseteste]; only Boethius knew all languages. . . ."[23] He says that there are not even four Latins who know Hebrew, Greek, and Arabic grammar. He does not deny that many translators were at work on the Arabic and Greek texts. Later, in the *Opus tertium,* he refers to them by

---

[21] Ibid., pp. 30-31.

[22] Ibid., p. 32: "Sed nunquam in Latina fuit composita, sed solum translata de linguis alienis, et meliora non sunt translata. Et de his scientiis, quae translatae sunt, nihil est perfectum; et translationes sunt perversae, nec intelligibiles in multis scientiis, maxime in libris Aristotelis."

[23] Ibid., p. 33.

name, and he claims friendship with some of them. But he criticises them for not knowing the grammar of these languages in the manner in which they know their Priscian in Latin.

Bacon then proceeds to talk about the importance of mathematics and Perspective (*Perspectiva*). He is particularly emphatic on the central importance of Perspective. All things are known through mathematics; and the laws of the multiplication of species are known through Perspective. Perspective, then, provides the key to a universal science. It, and not a purely philosophical treatment of physics, is the way to a knowledge of generated things. In Bacon's precise language, Perspective is not just the means of knowing those things which are common elements in a theory of vision, but it is also the key to all sensible things and to "the whole machine of the world, both in the heavens and in inferior things" (*totam mundi machinam, et in coelestibus et in inferioribus*).

Bacon continues the argument:

> However, this science is not yet taught at Paris, nor among the Latins, except twice at Oxford in England, and there are not three people who know the power of this science. Whence that one, who made himself an authority, concerning whom I have spoken above, knew nothing of the power of this science, as appears in his books, because he did not write a book about this science, and he would have done it if he had known it. Nor did he say anything about this science in the other books. However, it ought to be the case that the exercise of this science would be fulfilled in all the others, since all things are known through its power. And so he was not able to know anything about the wisdom of philosophy. But those who know these things are few, just like those who know mathematics, and they cannot be had without great expense. Similarly the instruments of this science, which are very inaccessible and of greater expense than the instruments of mathematics, cannot be had without great expense.[24]

The remarks about money make some sense as Bacon argues strongly in his later works that scientific endeavour is impossible unless some great power such as a king or the pope will support it. And to a great extent, the polemic of the works to the pope is a persuasive attempt to get the pope to finance and set up an organised study in the natural sciences. Bacon evidently knew the public pres-

---

[24] Ibid., pp. 37-38. ". . . Unde ille, qui fecit se auctorem, de quo superius dixi, nihil novit de hujus scientiae potestate, sicut apparet in libris suis, quia nec fecit librum de hac scientia, et fecisset si scivisset, nec in libris aliis aliquid de hac scientia recitavit etc."

tige of this competitor, the "unnamed master." Hence, he needed to argue that his own science was much better for the good of society and the Church. The centrepiece of any such study of the natural sciences would concern itself with Perspective and the multiplication of species, which for Bacon is the *summa et principalis radix sapientiae*.

> But he who multiplied volumes ignores these roots. For he touches on no aspect of them. And so it is evident that he ignores the natural things, and all things which are concerned with philosophy. And he not only himself is ignorant [of these things] but the whole *vulgus philosophantium*, which errs through him, is ignorant of these matters. For if you will write to him about what he would write concerning these roots, you will find him unqualified in these matters.[25]

Thus, no authority, ancient or modern, wrote about these things; but he, Bacon, worked for ten years before he was able to speak to some people about them. And he notes here that he is putting the fruits of his labours into writing on the occasion of the pope's mandate. Bacon's main concern with the multiplication of species and with Perspective meant a widening of the treatment of *Perspectiva* from that of the normal school text of the time. One can recognise this great difference by comparing the brief account of *Perspectiva* in the *De ortu scientiarum* of Robert Kilwardby with the extended mathematical and physical explanation by Roger Bacon.[26] It is obvious from the later scientific tracts of Bacon that his ten-year search for new forms of knowledge outside of the common study of the faculty of arts concerned itself with the study of mathematics, perspective, and *scientia experimentalis*. Bacon, then, claims to have found a new foundation for the sciences. This claim is, perhaps, best stated by Bacon in his *Communia naturalium*.

Bacon repeats the same claim in the *Opus tertium* about *scientia experimentalis* and about alchemy. All of this is significant in that it points to the question of astronomy as the real issue which brought about Bacon's conflict with his order. This aspect has been briefly presented by Theodore Crowley in his study of the problem of the

---

[25] Ibid., p. 38: "Unde ille, qui multiplicavit volumina, ignorat has radices, nam nihil de eis tangit; et ideo certum est ipsum ignorare res naturales, et omnia quae de philosophia sunt; et non solum ipse sed totum vulgus philosophantium, quod errat per ipsum. Scribatis enim ei quod pertractet de his radicibus, et invenietis ipsum impossibilem ad eas."

[26] Robert Kilwardby, O.P., *De ortu scientiarum*, ed. Albert G. Judy (Toronto, 1976), pp. 48-50. See also Roger Bacon, *Opus majus*, ed. John Henry Bridges, vol. 2 (London, 1897), pp. 1-166.

soul in Bacon's philosophical commentaries.[27] In general, Bacon accuses the "unnamed master" of ignoring the basis of all of these sciences:

> But he indeed who composed so many great volumes about the natural things, concerning whom I have spoken above, ignores these basic matters (*fundamenta*), and so his building is not able to stand.[28]

Only one further reference to this "unnamed master" need be mentioned from the *Opus tertium*. It has not, to my knowledge, been used before in regard to the identification of the "unnamed master," but it is important in that it comments on the notion of method. Speaking about his pupil John, whom he is sending to the pope with his works, Bacon claims that John alone of all the Latins knows this *Perspectiva* and mathematics of which he speaks. The others do not know it because they do not know Bacon's method:

> . . . nor that one great master (*magister magnus*), nor any of those whom I have mentioned above, because they do not know my method.[29]

His criticism of the great master is due to the fact that both masters use a different philosophical method.

C. *OPUS MINUS* (1266-1267)

Since Bacon refers to the *Opus minus* in regard to the "unnamed master" in the *Opus tertium*, it is important to examine the context of his argument in the former work. The first sin against theology which displeases Bacon has to do with the place of philosophy in the study of theology. He claims that philosophy has taken on a dominant role in theology. The latter science should be a *scientia dei*. But in the books on the *Sentences* theologians do not generally consider theology and prophecy since most of the questions in the *Sentences* have to do with purely philosophical matters. The theologians thereby are led to neglect the text of scripture.

---

[27] Theodore Crowley, *The Problem of the Soul*, p. 63.

[28] Roger Bacon, *Opus tertium*, ed. J. S. Brewer, p. 42: "Ille vero, qui composuit tot et tam magna volumina de naturalibus, (de quo superius locutus sum), ignorat haec fundamenta, et ideo suum aedificium stare non potest."

[29] Ibid., p. 61: ". . . nec ille magister magnus, nec aliquis eorum de quibus superius feci mentionem, quia nesciunt modum meum — sicut ipse [his pupil, John] qui ore meo didicit, et qui consilio meo est instructus."

The second sin is that theology omits the greater sciences and is quite content with the vulgar sciences. The latter included the grammar of the Latins, logic, natural philosophy according to its worst part, and a certain part of metaphysics. Bacon's point is that these sciences have no practical application. They treat of pure knowledge, and they are without any practical purpose. The greater sciences, which Bacon proposes here and which include mathematics, perspective, moral science, and experimental science, have to do with the practical good of the body, and the soul, and fortune. For that reason, they are more actual and effective.

The third sin, which Bacon relates to the fault of the "unnamed master," is concerned with the same questions as the second. He says that the theologians even neglect the four common sciences which were in general use in the schools at the time he wrote the three works for the pope. According to Bacon, those who wrote *summae* in theology did not know either the natural philosophy or metaphysics in which they now glory. This remark is significant and the reference is unmistakable. He says that all the error of study arose because of these two men, Alexander of Hales and the "unnamed master." Bacon states that he saw these two men who made *summae* with his own eyes, and thus he knew that they never saw or heard the sciences in which they now glory. He argues that they never had a chance to hear or teach the natural sciences. One of them is now dead, the other is still alive. The one who is dead was a good man, a great archdeacon, and a master of theology in his time, and because of this, he was made a great friar when he entered the Order of Friars Minor. This order was new in the world and it had neglected studies, but it still gave to this man authority over all its study. They also ascribed to him a great *summa* which is more than the weight of one horse. Bacon's argument against this man, which is also a point he holds against the second master, is that he had not lectured on metaphysics or on natural philosophy. The books on these subjects, according to Bacon, had been excluded from the arts faculty when Alexander was a student. And soon after that, these books were condemned and forbidden at Paris. Then, when the university, which had been dispersed in 1229, had reassembled in 1231, Alexander had become an old man:

> Whence, as I will say briefly, he ignored these sciences which are now in common use, that is, natural philosophy and metaphysics, in which is found all the glory of the study of the moderns.[30]

---

[30] Roger Bacon, *Opus minus*, ed. J. S. Brewer, p. 326.

Bacon does link Alexander with the "unnamed master." Both are responsible for the decline in study. Alexander is responsible for the fall in the study of theology; the "unnamed master" is the one responsible for the decline in the study of philosophy. In the end, Bacon says that the *studium* at Paris lacked these sciences. He then introduces his remarks on the "unnamed master":

The other one who lives (*Aliter qui vivit*) entered an Order of Friars as a boy. He never taught (*legit*) philosophy anywhere, nor did he hear it in the schools, nor was he in a *studium solemne* before he was a theologian, nor was he capable of being taught in his own order, as he was the first master of philosophy among them. And he taught others; whence from his own study he had what he knows. And truly I praise him more than all of the common students, because he is a most studious man, and he saw many things, and had money (*habuit expensum*). And so he was able to collect many useful things in the infinite sea of authors (*auctorum*). But since he did not have a foundation (*fundamentum*), for he was not instructed or exercised in hearing, reading, or disputing, it was inevitable, therefore, that he did not know the common sciences (*scientias vulgatas*). And again, since he did not know the languages, it is not possible that he would know anything great, on account of the reasons which I write concerning the knowledge of languages. And again, since he ignores perspective, just as others of the common students do not know it, it is impossible that he should know anything of worth about philosophy. And he is not able to glory in the tract which I have composed concerning *scientia experimentalis*, alchemy, and mathematics, since these [sciences] are greater than the others. And if he does not know the lesser he cannot know the greater. God, however, knows that I have only exposed the ignorance of these men on account of the truth of study. For the *vulgus* believes that they [Alexander of Hales and the "unnamed master"] know everything and it adheres to them like angels. For these ones are quoted (referred to) in disputations and lectures as *auctores*. And especially that one who lives; he has the name of *doctor Parisius*, and he is referred to in the studium as *auctor*; which cannot be done without the confusion and destruction of wisdom, since his writings are filled with falsehoods and infinite vanities. Never before did such abuse appear in this world.[31]

Bacon then lists some of the other faults which he later discusses in the *Opus tertium*, and which we have already examined. One should note that Bacon himself intended the cross-reference in these works.

---

[31] Ibid., pp. 327-38.

D. *Communia naturalium*

That Bacon develops his natural philosophy in contradistinction to the thought of the "unnamed master" is clearly seen from a very important introduction to the *Communia naturalium*. This is a purely theoretical work which Bacon probably began in the early 1260s and which he probably completed towards the end of the decade. Like many of Bacon's writings, the text received many revisions. This work lacks the polemic and persuasive character of much of the work which Bacon wrote for Pope Clement iv. He intended the work as a tight *compendium* of the common features of the different areas of natural philosophy. He explicitly leaves it to later times and to others to develop work in the individual special sciences. He leaves us in no doubt that perspective is first among the special sciences. He argues that concision and precision are more important in natural philosophy than are all the volumes of Aristotle and much of the research of the thirteenth century. Thus, he says that there is more value in one book of Aristotle, namely the *De celo et mundo*, than there is in all the other volumes of the *naturalia*. In this respect, he adds:

> And so some moderns are in error beyond measure who exceed the quantity of the volumes of Aristotle and give a greater quantity to one of their own books than Aristotle deigned to present in all [his] books. Truly, therefore, they are convicted of great ignorance on account of which they do not know how to stand in regard to necessary things, although they not only gather most vain things but multiply infinite errors. The root cause of this is that they have not examined the sciences on which they write nor did they teach them in a *studio solemni*, nor did they even hear them. Whence they were made masters before they were disciples, so that they err in all things on account of themselves and they multiply the errors among the *vulgus*.
>
> Again, they are not able to know the *libri naturales* and the common books without knowledge of the seven other special sciences, or even without mathematics. But two glorious moderns (*Sed duo moderni gloriosi*) [have tried to do so], just as they have not heard the sciences about which they speak, nor have they read them or are they exercised in the other ones, as appears from their writings, therefore, it is evident that they are confounded everywhere by errors and vanities. Indeed, their error is multiplied in the natural sciences and in the other common sciences since the translations which they use are perverse and nothing of value can be said from them.[32]

---

[32] Roger Bacon, *Communia naturalium, liber primus*, ed. Robert Steele, Fasc. 2 (Oxford, n.d.), pp. 11-12.

Some lines after this, Bacon makes the following remark which shows that he was consciously defending a particular school of thought, namely that one which is associated with Robert Grosseteste. He argues that these latter sought the sources of natural science in places other than the works of Aristotle and in the practice of mathematics:

> But the other men mentioned, namely, those who heard these sciences, and read and examined them, seeing that through the text of Aristotle and his commentators they were not able to know natural philosophy, turned themselves to the seven other natural sciences, and to mathematics, and to other authors of natural philosophy as to the books of Pliny and Seneca and of many others. And so they came to the knowledge of natural things, concerning which Aristotle in his common books and his expositor were not able to satisfy by [their] study of the natural things.[33]

Whom does Bacon include among the "other men"? He includes Grosseteste, Adam Marsh, Peter of Maricourt, John of London. He also includes Campanus de Novara and Master Nicholas, the teacher of Lord Aumury de Montfort. These names are not given in the *Communia naturalium*, but they are the ones he praises in the *Opus tertium* for their knowledge of mathematics and science. One should also include William of Shyreswood in this group, for he is mentioned together with Grosseteste and Adam Marsh throughout these later works of Bacon.

E. Conclusion

It will be evident from the present review of the texts in Bacon which refer to Albertus Magnus and to a certain "unnamed master" that there is an unmistakable coincidence in these texts. Of the two explicit references to Albert, one is openly critical; the other is a statement of praise which soon changes into critical contrast of Albert's method with that of Bacon. The implicit texts, directed against the "unnamed master," continue the very same argument. The "unnamed master" does not know perspective, and therefore his science is without a foundation. And that charge, according to Bacon, applies to every aspect of his *scientia*. His science lacks a foundation for *scientia experimentalis*, and alchemy. Thus, Bacon, in accordance with his method of experience (*scientia experimentalis*) and by reason of his mathematicization of reality, makes a funda-

---

[33] Ibid., pp. 12-13.

mental methodic objection to the science of Albert. The crux of the question is whether or not mathematics is universal and all encompassing. Does it give the principles of explanation to each area of investigation or are the different areas of knowledge specific in that each of them has its own principles of explanation? Albert's own concern for and criticism of the thirteenth-century *amici platonis* in many places in his works shows that he was involved in controversy with contemporaries who favoured mathematics as the key to a proper understanding of natural science.[34] Bacon's criticism of Albertus Magnus, which is a central part of his *persuasiones* to the pope, should not be dismissed lightly as the pedantry of an old crank, but should be seen for what it is. It is the polemic and persuasive side of a very important debate on the principles of philosophic and scientific method in the schools of thirteenth-century Europe.

One question remains unanswered. Did Bacon include other well-known scholars in his criticism? Who, for example, is the second of the "two glorious moderns" (*duo moderni gloriosi*) mentioned in the *Communia naturalium*? This would not seem to be a reference to Alexander of Hales. It would appear to be a reference to a master who has commented on the works of Aristotle, and especially on his metaphysics and natural philosophy. Since his name is linked in the text to that of the "unnamed master," it could well be a reference to Thomas Aquinas. He, indeed, was the head of the *vulgus studentium* at Paris, who spread (*vulgata*) the newly published works of Albert during this time. His fame as a representative of the standpoint of Albert was well established by the time Bacon had written the *Communia naturalium*. One has to grant that there are problems in regard to scribal changes of personal names in the works of Bacon. Still, one ought not to dismiss the linking of the names of Albert and Thomas, as given in the *Compendium studii philosophiae*, as merely the result of scribal error. For Bacon, they both represented a method of philosophizing which differed from his own.

Finally, it must be said that Bacon quite consciously avoided direct condemnation of Albert by name in his later work, especially in his work to Pope Clement IV. The presence of internal reference in the *Opera* to Pope Clement IV is Bacon's way of criticising the *scientia* of Albertus Magnus without engaging in direct personal name-calling. By means of a system of cross-references Bacon builds up a portrait which would be recognizable to any thirteenth-century reader acquainted with school debates.

---

[34] James A. Weisheipl, "Albertus Magnus and the Oxford Platonists," *Proceedings of the American Catholic Philosophical Association* (1958), pp. 124-39.

3

St. Albert and the Nature of Natural Science

Benedict M. Ashley, OP
Aquinas Institute of Theology

Historians of science tend to emphasize the ways in which the medievals anticipated modern science.[1] Thus modern science is assumed to be the model approach to nature toward which all previous ages were groping. Other modes of understanding nature are viewed as dead-ends of scientific evolution. Recently, however, the "crises of limits" besetting our technological culture is raising doubts about this assumption.[2] Perhaps the Galilean-Newtonian model of science is only one way to understand nature, very successful in terms of its own goals, but incapable of solving all the questions which we need to ask about human nature and its environment. Therefore, in the following I will present another model, first made fully available to the Middle Ages by Albert the Great — the Aristotelian. It will be viewed not as a foreshadowing of the Galilean model with whose great successes we are so familiar, but as a contrasting approach to nature which had modest successes in terms of its own goals. This contrast may suggest a more pluralistic approach to nature in our post-industrial future.

---

[1] E.g. such standard works as Lynn Thorndike, *History of Magic and Experimental Science* (New York, 1923); René Taton, *Histoire générale des sciences* (Paris, 1951); A. C. Crombie, *Augustine to Galileo* (London, 1952); Charles J. Singer, *From Magic to Science* (New York, 1958).

[2] For a broader perspective on "models" see Mary B. Hesse, *Models and Analogies in Science* (Notre Dame, Indiana, 1966) and *Science and the Human Imagination* (New York, 1953); and for the theological roots of such models see Richard Olsen, *Science as Metaphor* (Belmont, Calif., 1971) and Stanley L. Jaki, *Science and Creation: From Eternal Cycles to an Oscillating Universe* (Edinburgh, 1974).

A. Two Kinds of Platonic Models

In the eleventh and twelfth centuries, when the complete works of Aristotle were still unavailable, the predominant approach to nature, exemplified by Thierry of Chartres, might be called the Timaean model because it was so greatly influenced by Chalcidius' (incomplete) translation of Plato's *Timaeus*.[3] This Timaean approach to nature harmonized with Augustinian theology, and was characterized by its *mystical* purpose — to discover in the order of the visible cosmos the vestiges and images of the invisible God, so as to lead the human mind and heart to rest in Him alone. Its outcome was wonder and the praise of the beauty of God manifested in nature. Its mode was esthetic and impressionistic, so that it gave little stimulus to the detailed investigation of natural phenomena.[4] This Timaean model, I would suggest, is still with us in the works of Teilhard de Chardin, not to his discredit, but rather as witness to its perennial fruitfulness.[5] The Timaean approach to nature can be scrupulously faithful to the known facts, yet, as Plato well understood, it goes beyond these bare facts in its use of myth and symbol to explore nature as a sacramental *mystery*.

By the middle of the twelfth century a flood of new astronomical, alchemical and other scientific works, translated from Arabic and from Greek, revealed another side of Platonism — its mathematical or Pythagorean[6] method of investigating natural phenomena. In England especially the interest in this Pythagorean model seems to have been reinforced by purposes no longer so much contemplative or mystical as practical and technological.[7] Thus the Middle Ages

---

[3] *Timaeus a Calcidio translatus commentarioque instructus*, ed. J. H. Waszink, in *Plato Latinus*, ed. R. Klibansky (Leiden, 1962); J. M. Parent, *La doctrine de la création dans l'école de Chartres: études et textes* (Ottawa, 1938).

[4] See Hans Urs von Balthasar, *Herrlichkeit: Eine theologische Ästhetik* (Einsiedeln, 1961), 1: 285-353.

[5] Teilhard was scrupulously faithful to the findings of science, but he goes beyond accepted scientific categories by his introduction of concepts like "radial energy" and the "omega point," nor is it clear how these can be used in scientific research. Yet Teilhard does not attempt a philosophical justification of these views, but presents them as the fruits of contemplative intuition. Cf. *The Phenomenon of Man* (New York, 1961).

[6] See Charles H. Haskins, *Studies in the History of Medieval Science* (Cambridge, 1924), pp. 20-42 and 113-129. I do not mean to deny that the *Timaeus* derives also from Pythagoreanism, but only to say that it does not elaborate the mathematical approach which is the special contribution of that tradition.

[7] "In the last decades of the twelfth century Roger of Hereford, Daniel of Morely, Alexander Nequam, and Alfred Sarashel, following the tradition set up by Adelard of Bath, the chief

were supplied with two Platonic approaches to nature which in many respects were opposite extremes: the one contemplative, mythic, impressionistic; the other practical, mathematical, intensely concerned with research, experimentation, and technological application.[8]

However, along with these works reflecting the Platonic tradition there was also transmitted a vast Aristotelian and pseudo-Aristotelian corpus whose very different orientation was not immediately perceived, especially because the Arabian commentators Alfarabi, Avicenna and Averroës, the chief guides to the exegesis of this difficult literature, were strongly inclined to harmonize Aristotle with Plato.[9] The outstanding figure in the first attempts at the beginning of the thirteenth century to assimilate this new material was Robert Grosseteste whose model of science was not Aristotelian but essentially mathematical and Pythagorean.[10] Between 1214 and 1235, when he became bishop of Lincoln, Grosseteste contributed immensely to the Aristotelian trend by his translations from Greek and his innovative commentaries on the *Posterior Analytics*[11] and the *Physics*,[12] but his own original efforts to explain natural phenomena conform to the Pythagorean model.

---

pioneer in this movement, by Walcher, prior of Malvern (d. 1135), by Robert of Ketene, and by other English mathematicians, astronomers, and scientists, introduced and popularized the new learning in England." Daniel A. Callus, "Introduction of Aristotelian Learning to Oxford," *Proceedings of the British Academy*, 29 (1943), 229-81, see esp. pp. 233-34; A. C. Crombie, *Robert Grosseteste and the Origins of Experimental Science, 1100-1700* (Oxford, 1953), pp. 16-43; James A. Weisheipl, "Albertus Magnus and the Oxford Platonists," *Proceedings of the American Catholic Philosophical Association* (1958), pp. 124-129. Note, however, that those named by Callus did not for the most part teach in England; see Stuart C. Easton, *Roger Bacon and His Search for a Universal Science* (New York, 1952), p. 22.

[8] Of course the "Platonism" of the Middle Ages was highly complex; see M. D. Chenu, "The Platonisms of the Twelfth Century" in his *Nature, Man and Society in the Twelfth Century* (Chicago, 1968), pp. 49-98.

[9] Alfarabi, one of the first Arab philosophers writes: "The philosophy that answers to this description [of true wisdom] was handed down to us by the Greeks from Plato and Aristotle only. Both have given us an account of philosophy, but not without giving us also an account of the ways to it and of the ways to reestablish it when it becomes confused or extinct. . . .So let it be clear to you that, in what they presented, their purpose is the same, and that they intended to offer one and the same philosophy." *Alfarabi's Philosophy of Plato and Aristotle*, trans. by Muhsin Mahdi (Glencoe, Ill., 1962), pp. 49-50.

[10] Crombie, *Grosseteste*; also see Daniel A. Callus, "Robert Grosseteste as a Scholar" in *Robert Grosseteste: Scholar and Bishop*, ed. Callus (Oxford, 1955), pp. 1-69. and S. Harrison Thomson, *The Writings of Robert Grosseteste* (New York, 1940).

[11] *Aristotelis Posteriorum opus cum duplici traductione: antiqua scilicet et Argiropyli; ac eius luculentissimum interpretem Lincolniensem Burleumque* (Venice, 1521).

[12] *Roberti Grosseteste Episcopi Lincolniensis Commentarius in VIII Libros Physicorum Aristotelis*, ed. Richard C. Dales (Boulder, Colorado, 1963).

Grosseteste strongly influenced Roger Bacon in this same Pythagorean direction when Bacon, become a Franciscan, returned from Paris to Oxford in about 1247,[13] although Bacon had been one of the first to lecture on Aristotle's natural science at Paris and always spoke of Aristotle as the greatest of philosophers.[14] A similar tendency appears in the Dominican Robert Kilwardby's *De ortu scientiarum* written about 1250.[15] In Bacon's mature works the Pythagorean model is in plain evidence since he stresses (1) the mathematical understanding of nature, with optics as the fundamental natural science, (2) the practical, technological values of science, and (3) the verification of scientific conclusions by *experimentum*.[16] We might think that this emphasis on experience is Aristotelian, until we notice that for Bacon the paradigm for empirical verification is Ptolemaic astronomy with its "saving the phenomena."[17]

In fact for Bacon *experimentum* means not only sense experience but also interior experience. For him both kinds of experience are *intuitive* as contrasted to abstractive and rational. Bacon distrusts reasoning in abstract terms which can never reach the existent individual.[18] He thinks that the human intelligence was originally illuminated by the divine Agent Intellect so as to have an innate intuition of the whole cosmic order, an intuition which it will perfectly recover only in the next life.[19] However, in the soul's present

---

[13] Easton, p. 87, but Bacon was at least briefly in Paris again in 1250 or 1251 (p. 67).

[14] Easton, pp. 35-66; Theodore Crowley, *Roger Bacon: The Problem of the Soul in His Philosophical Commentaries* (Dublin, 1950), pp. 22-29; F. Van Steenberghen, *Aristotle in the West: The Origins of Latin Aristotelianism* (Louvain, 1955), pp. 108-114.

[15] Robert Kilwardby, *De Ortu Scientiarum*, ed. Albert G. Judy, (London, Toronto, 1976). For the date see introduction p. xvi. Kilwardby also taught on the new texts at Paris at about the same time as Bacon; see Van Steenberghen, p. 114.

[16] Easton, pp. 167-184.

[17] "Set est alia sciencia que considerat futuras alteraciones hujus mundi inferioris, que vocatur Sciencia Experimentalis a Ptolomeo in libro predicto [*De disposicione sphere*]": *Secretum secretorum* in *Opera hactenus inedita Rogeri Baconi*, ed. Robert Steele vol. 5 (Oxford, 1920), p. 9. "But there is another science which considers future alterations of this lower world, which is called 'experimental science' by Ptolemy in his work *On the Globe*." According to Easton, p. 85, this is where Bacon first got the term. On the Ptolemaic "saving the phenomena" see Pierre Duhem, *Le Système du monde*, (Paris, 1913), 1: 484-496.

[18] *Opus majus*, ed. John H. Bridges, 2: 165. See also A. G. Little, "Roger Bacon" in *Franciscan Papers, Letters and Documents* (Manchester, 1943), pp. 72-97. Little remarks, "It is noteworthy that the philosophers of the next generation identified *scientia experimentalis* with *scientia intuitiva*" (p. 95), an essential feature of the nominalist epistemology.

[19] On Bacon's identification of Agent Intellect with God see Crowley, pp. 82-88. *Opus majus*, Part I proves the liability to error of the human intellect in its present state; Part II establishes that even philosophical truth is from divine illumination and that (chapter ix) the plenitude of light was given from the beginning to wise men.

state this original vision has become unconscious or at least confused by its union with the sinful body.[20] Hence reasoning based on such dimly perceived principles (even mathematical reasoning) cannot give certitude until such reasoning is confirmed both by sense experience and also by the tradition of a primitive revelation passed down through the Scriptures and the writings of the saints and philosophers.[21] Moreover, moral purity is the prerequisite as well as the goal for this recovery of the lost vision.[22] This curious version of Platonism is suggestive of ways in which the Pythagorean model might be modified in the direction of the Kantian model which dominates modern science.[23] While Bacon himself did little to give his own model effective application to actual scientific problems, Thomas Bradwardine and the Mertonians at Oxford in the next century were to do so.[24]

B. The Aristotelian Model

Albert the Great was older than Bacon and (as Bacon was fond of pointing out) largely self-taught.[25] Both began to study natural science at Paris in the 1240s where Bacon taught in the arts faculty and Albert in the theological faculty with Thomas Aquinas already his

---

[20] *Questiones super libros octo physicorum Aristotelis*, ed. F. M. Delorme in *Opera hactenus inedita*, ed. Steele, fasc. 13 (Oxford, 1935), pp. 11-12.

[21] Although Bacon, *Opus majus*, Part I insists on the fallibility of authority, in Part II he is no less emphatic on the necessity of authority both in theological and philosophical matters. His concern is to find the most authentic tradition of the wisdom originally revealed to mankind.

[22] *Opus majus* II, c.19, Bridges, 3: 76-79; *Compendium philosophiae* in *Opera Fr. Baconis hactenus inedita*, ed. J. S. Brewer, Rolls Series (London, 1859), pp. 398-413.

[23] Hume destroyed confidence in our power to discover natural causes, leaving only sense data. Kant then rescued the scientific method by his defense of *synthetic a priori* propositions rooted in necessary thought forms which took the place of Bacon's innate experience. Today these *a priori* principles are replaced by hypothetical axioms. Structuralism now again raises the question of innate thought forms. In all these variations of the Platonic model the fundamental dualism between empirical data and mental forms imposed on the data persists and differentiates these models from an Aristotelian one in which the ontological order of the data (as distinguished from various possible logical orderings) exists in the data itself prior to any ordering by the mind.

[24] James A. Weisheipl, *The Development of Physical Theory in the Middle Ages* (New York, 1959) pp. 72-81.

[25] Easton, pp. 210-231, makes a very good case in arguing that Bacon is referring to Albert when he says, "He never heard the parts of philosophy, nor did he learn from anyone, nor was he nourished in the University of Paris, nor anywhere the study of philosophy flourishes" (*Opus tertium*, ed. Brewer, p. 31). See the preceeding papers by Jeremiah Hackett and James A. Weisheipl.

pupil.[26] Van Steenberghen has carefully traced the vicissitudes of Aristotelian studies at Paris, where the strength of the theology faculty with its fears for the possible implications of the Aristotelian world-view raised difficulties not so keenly felt at Oxford, Toulouse, or the Italian universities.[27] About the same time Bacon left for Oxford, Albert, with Aquinas, removed to Cologne, not to a university, but to establish a *studium generale* for his Dominican Order. As a *regens studiorum* for his brethren he had responsibility not only to provide necessary theological studies, but also the philosophical preparation for such studies which he himself had not received, but which he strongly favored.[28] Perhaps this new responsibility explains his decision to write a complete encyclopedia of philosophical disciplines based chiefly on the *corpus Aristotelicum*. We do not know when the logical or mathematical parts of this encyclopedia were written,[29] but the extensive part on natural science begins as follows:

> Our purpose in natural science is to satisfy as far as we can those brethren of our order who for many years now have begged us to compose for them a book on physics in which they might have a complete exposition of natural science and from which also they might be able to understand correctly the books of Aristotle. Although we do not think we are competent of ourselves to carry out this project, nevertheless, because we do not want to refuse our brethren's request, we have finally accepted this task which we so many times rejected. Overcome by the request of certain of these brethren we have undertaken this work first to the praise of Almighty God, who is the fountain of wisdom and the creator, orderer and governor of nature, and then for the benefit of our brethren, and, finally, for the benefit of all those desirous of learning natural science who may read it.[30]

---

[26] J. A. Weisheipl, *Friar Thomas d'Aquino* (New York, 1974), pp. 36-38. Since Thomas was still in the initial stages of his Dominican training this study must have been somewhat informal.

[27] Van Steenbergen, *Aristotle in the West*, pp. 66-114.

[28] William A. Hinnebusch, *The History of the Dominican Order* (New York, 1973), 2: 25-27.

[29] As we find in Bacon and Kilwardby, the normal medieval order of studies was the *trivium* (language and logic), then the *quadrivium* (mathematics), and finally the Bible theologically interpreted. The seven liberal arts were later supplemented by natural science, moral science, and first philosophy. At this stage, theology could make extensive use of metaphysics. See James A. Weisheipl, "Classification of the Sciences in Medieval Thought," *Mediaeval Studies*, 27 (1965), 54-90; Benedict M. Ashley and Pierre Conway, "The Liberal Arts in St. Thomas Aquinas," *Thomist*, 22 (1959), 460-532.

[30] *Physica* I, tr.1, c.1 (ed. Borgnet 3: 1a-b); see also *Analytica Posteriora* II, tr.5, c.2 (ed. Borgnet 2: 232b).

Albert's purpose to instruct students in natural science as an integral discipline leads him to assure them that (1) he intends to supplement the Aristotelian works with material taken from other writers, and (2) he also intends to fill out gaps in the Aristotelian scheme as represented by the extant works.[31] Nevertheless, Albert disclaims final responsibility for the opinions he expounds.[32]

Throughout his commentaries on the *naturalia* Albert exhibits the same concern for logical method as in his logical commentaries, especially that on the *Posterior Analytics*.[33] What is the precise difference between this Aristotelian method and that of the Platonists? Both make use of systematization by deductive reasoning from axiomatic principles, and both guarantee the relation of theory to fact by inductive, experimental (or at least observational) procedures. The difference is that in the Pythagorean model the facts are analytically reduced to theoretical principles justified by the intuition of innate ideas, while in the Aristotelian model these principles are justified by an intuition based on sense observation.[34] However, for some Platonists like Bacon, as we have seen, confidence in the security of innate ideas is weakened, and replaced by the empirical confirmation of the conclusions deduced from these principles, thus opening the way to the modern view that theoretical principles are simply hypotheses subject to empirical verification (or, rather, non-falsification).[35] It is essential to note, moreover, that for Aristotelians the intuition of principles based on empirical data is by no means easily achieved, but ordinarily requires a long, gradual process of research based on the accumulation, classification, and analysis of regularly observed facts. In this process of the *inventio definitionis*, the hypothetical (top-

---

[31] Albert's method is not detailed commentary in the manner of Averroës, and later of Aquinas, nor glosses and questions like those of Grosseteste and Bacon, but a paraphrase after the manner of Avicenna. "And we also add in some places parts of incomplete books and in other places books neglected or omitted which either Aristotle did not write, or if he wrote them have not come down to us": *Physica* I, tr.1, c.1 (ed. Borgnet 3: 2a).

[32] "For in this as in all my works on natural science, I have not taught anything on my own opinion, but have expounded as faithfully as I could opinions of the Peripatetics": *Politica* VIII, c.6 fin. (ed. Borgnet 8: 803-804); see also *Physica* VIII, tr.4, c.7 (ed. Borgnet 3: 633b); *De somno* III, tr.1, c.12 (ed. Borgnet 9: 195b); *De causis prop.* II, tr.5, c.24 (ed. Borgnet 10: 619b); *De animalibus* XXVI, 36 fin. (ed. Stadler 2: 1598); *Ethica* I, tr.1, c.2 (ed. Borgnet 7: 4a), etc.

[33] *Analytica posteriora* (ed. Borgnet 2). For full discussion see article by William A. Wallace in this volume.

[34] *Analytica post.* II, tr.5, c.1-2 (ed. Borgnet 2: 228a-232b).

[35] See Peter Caws, *The Philosophy of Science* (Princeton, 1965), pp. 222-231, for a discussion of verifiability and falsification of hypotheses.

ical or dialectical) method is useful.[36] Thus the Aristotelian method, except for its rejection of innate ideas, is inclusive of the Pythagorean method, rather than in simple opposition to it. Its goal, however, is not merely to establish facts but to give them a theoretical explanation in terms of their "causes," i.e. by a reduction to the first principles of the science, which were themselves established by an analysis of the most general facts of experience.

This reduction to causes in the Aristotelian method is not achieved by a "deduction," if by that is implied that the conclusions of a science are *actually* precontained in the principles and need only to be explicitated, but by an analysis through which the more restricted principles (definitions) of the science are shown to be intelligible in the light of the more generic principles. In every case, however, these more specific principles must be independently verified by reference to sense experience in order to establish that they are "real" (existential) and not merely "nominal" definitions.[37] For example, for Aristotle and Albert the reason dogs have sharp canines is because they are carnivorous animals, and this fact makes sense when reduced to the general theoretical principle that "form follows function." Nevertheless, they never attempted to deduce from that principle the fact that dogs are carnivorous (that was established by observation), nor that carnivores have sharp teeth (that too was referred to observation). However, once the fact was empirically established that dogs have sharp teeth, this fact was considered to be scientifically explained in terms of the real definition of dogs as carnivores and of the theoretical principle concerning the relation of form and function.[38]

C. GENERAL PRINCIPLES OF NATURAL SCIENCE

For Albert there is no distinction between what today might be called a "philosophy of science" or "natural philosophy" and "natu-

---

[36] Albert agrees with Aristotle that the principal use of "topical" logic is in seeking definitions or principles for philosophy, first of all for natural philosophy: *Topica* I, tr.1, c.5 (ed. Borgnet 2: 246b-247a).

[37] Mathematics, however, reduces immediately not to sense but to imagination, i.e. to the construction of ideal geometrical figures or numbers, but this ultimately is founded in sense experience: *Physica* I, tr.1, c.1 (ed. Borgnet 3: 3a-b). See also *Euclid Commentary* in B. Geyer, "Die mathematischen Schriften des Albertus Magnus," *Angelicum* 35 (1958), 159-170, text 170-175, proemium p. 170.

[38] *De animalibus* XII, tr.3, c.6 (ed. Stadler 1: 883), the example of teeth.

ral science." The study of *corpus mobile*, i.e. "body subject to change" and known to us by intelligent reflection on our sense experience, is rooted in a single set of principles and hence forms a single discipline.[39] Albert says *corpus* not *ens mobile*,[40] overlooking the fact, later pointed out by Aquinas,[41] that Aristotle in *Physics* VI demonstrates that all changeable entities are bodies, so that *corpus mobile* is a conclusion not a principle, i.e. not the subject of physics. For Albert it is sufficient to say "body" in order to distinguish natural science from metaphysics which deals with *ens*, and then, to distinguish natural science from mathematics, to add that the "body" in question is one whose dimensions exist in sensible matter.

The first task of such a science is to establish its basic principles by arriving at a real definition of this subject "changeable body" through an analysis of sensible experience, and then demonstrating its most general properties. This general model of a changeable body can then be used in analyzing every kind of natural body, proceeding from the most common features and descending to the ultimate species of things with their specific properties.[42] What is in question is not just single, isolated bodies, but systems of bodies, including the whole universe as such a system.[43] Such a general model is not taken by Albert as a hypothesis, as "models" are usually understood today, but as a real definition established by analytical insight into sense experiences so general that they are the ones by which we recognize the existence of any body whatsoever, e.g. that it changes in various ways, is extended in space and changes in time, that it is acted upon by other bodies and acts on them, etc.

Thus the principles of natural science are not a set of independent axioms, in the modern sense, but are themselves reducible to a single

---

[39] "Every science deals with some generic subject whose properties it proves, and whose properties and different species it investigates. Now in every natural science this subject undoubtedly is changeable body precisely as it is the subject of change. I mean *common* body, and not this or that kind of body, is the subject of natural science, but by 'common' I mean not simply 'body' but body precisely as it is the subject of change": *Physica* I, tr.1, c.3 (ed. Borgnet 3: 6b), a digression from Aristotle's text.

[40] Ibid. (ed. Borgnet 3: 7a).

[41] "This is the book of the *Physics* whose subject is changeable being (*ens mobile*) as such. I do not say 'changeable body' (*corpus mobile*) because in this same work it is proved that every changeable being is a body; and no science proves its own subject. Hence the very beginning of *De caelo*, which is a sequel to the present work, deals with the topic of 'body' ": Aquinas, *In I Phys.*, lect.1 (ed. Leonina 2: 4b).

[42] *Physica* I, tr.1, c.3 (ed. Borgnet 3: 7a).

[43] Ibid., c.4 (ed. Borgnet 3: 8a-b).

fundamental principle, namely, the definition of the subject *corpus mobile*.[44] In the light of such principles the whole order of bodies is accessible to investigation, but not non-bodies (God, angels, human soul) except as these can be inferred to be causes of the processes of change by which bodies are produced or affected.[45]

To achieve a scientific definition of *corpus mobile* it is necessary analytically to resolve the confused whole which we sensibly observe as a changing body into its defining "causes" by conceptually reversing, as it were, the process of change by which such a body is composed by natural processes. Thus Albert does not seek to understand bodies in a static fashion merely in terms of an abstract classification, but dynamically as they come to be through change.[46] Albert pursues this search for a definition of changeable body through the first two books of the *Physics*. In commenting on the first book, he shows (with many elaborating digressions from Aristotle's text) how the pre-Socratics and Plato speculated on the principles of things and then how Aristotle more adequately analyzed changeable body in terms of three principles: matter, form, and the privation of one form by the introduction of another.[47] However, as Nardi has shown,[48] Albert in his understanding of these principles makes way for a conception of matter which inclines to the Neoplatonic doctrine of the *inchoatio formae* where Aquinas was to insist on its pure potentiality.

In commenting on the second book of the *Physics* Albert shows

---

[44] Aristotle in *Physics* VII proves that "whatever is moved is moved by another," so that motion requires the existence of interacting bodies. Furthermore, in VIII he argues that for these bodies to form a coordinated system there must be a first mover. In *De caelo* he identifies this system with the visible universe because it exhibits a coordinated set of motions.

[45] For Aristotle this is true even of the mathematical sciences, i.e. all the axioms of arithmetic can be reduced to the definition of number, and all those of geometry to the definition of the continuum. This reduction, however, is by analysis, not by demonstration.

[46] *Physica* I, tr.1, c.1 (ed. Borgnet 3: 7a).

[47] Ibid., tr.2-3 (ed. Borgnet 3: 18a-91b). Albert's discussion is much more complex than Aquinas' and more influenced by Averroës, whose explanations Aquinas explicitly rejects; *In Physic.* I, 1 (ed. Leonina 2: 6a).

[48] Bruno Nardi, "La dottrina d'Alberto Magno sull' 'inchoatio formae' " and "Albert Magno e San Tommaso," in his *Studi di filsofia medievale* (Rome, 1960), pp. 69-102 and 103-118. Although Nardi draws some of his evidence from the doubtfully authentic *Summa theologiae* (see A. Hufnagel, "Zur Echtheitsfrage der *Summa Theologiae* Alberts des Grossen," *Theologische Quartalschrift* [Tübingen], 146 [1966], 8-39), he cites many texts in which Albert attempts to retain the concepts of *rationes seminales* and of an active tendency of matter in terms which Aquinas explicitly rejects as non-Aristotelian. It should be noted, however, that Albert is speaking chiefly of secondary matter as it is activated by the heavenly bodies rather than of prime matter as such.

how Aristotle somewhat restricts this definition of *corpus mobile* to limit the field of natural science, in distinction to that of mathematics, to bodies as they undergo *natural* change, i.e. as change results from the internal principles of matter and form, thus excluding artificial or violent changes imposed by human art or mere accident.[49] At the same time he makes clear that the natural scientist in demonstrating the natural properties of changeable bodies, uses as middle terms not only the material and formal causes, but also the extrinsic causes of natural agents acting through naturally predetermined processes, i.e. for final causes. Indeed, natural scientists can achieve satisfactory causal explanations of any phenomenon only by working back analytically from some regularly observed result of change (a final cause) to the natural agent (efficient cause) predetermined to produce such a result, in contact with a body susceptible to such transformation (the material cause), so as to arrive at the completed transformation (formal cause) of this body, which is identical with the observed result of change (final cause) which was the datum to be scientifically explained.[50]

Albert strongly defends against the objections of his colleagues (*socii*, probably other Dominicans) the view that not all observed events are natural, since chance events are real coincidences which have no final cause in the natural order.[51] Nevertheless, he also defends the concept of "fate," i.e. that the course of events in our world is in large measure physically predetermined. Hence for Albert natural science can rationally explain and even predict most events, yet only with probability, thus leaving room for chance, fortune, human freedom, and divine intervention. He argues that the divine Providence manifested in the Scriptures is not inconsistent with the natural determination required for scientific explanation, provided we understand this Providence as working through all these various

---

[49] *Physica* II, tr.1 (ed. Borgnet 3: 93a-116b) Mathematics is treated in c.8 (ed. Borgnet 3: 107a-110a).

[50] Ibid., tr.2, c.4 and c.22 (ed. Borgnet 3: 126b-127b and 158b-161b). See W. A. Wallace in this volume.

[51] "Since, however, some moderns among our brethren also deny chance and fortune. . .": ibid., c.10 (ed. Borgnet 3: 138a). "And what is objected by our brethren that nothing comes to be without being preceded by a sufficient cause is true enough; but a chance or accidental effect does not have a legitimate cause *per se*": ibid., c.21 (ed. Borgnet 3: 157a).

modes of causation.[52] Thus for Albert scientific explanations are not absolute, but hypothetical, explaining what happens for the most part (*in pluribus*), but not without exception.[53]

According to Albert the first two books of the *Physics* go no further than to define the subject and appropriate method of natural science. Only in Book III does Aristotle begin to demonstrate the properties of this subject by the method established. The first problem to be dealt with is the most general question: why do all natural bodies undergo change (*motus*)?[54] To answer this, Albert shows that "change" can be defined in three ways: (1) most formally, but least evidently, as "the act of a thing existing in potency as it is such"; (2) more materially and evidently as "the act of a changeable thing as changeable"; (3) most completely and evidently as "the act both of the agent and the patient." He means that the process of change is difficult to observe in its transiency, while the body undergoing change is quite open to observation, but what is most observable is the interaction of one body with another along with the transformation of at least one of the bodies.[55]

Using such definitions as the middle term, it becomes demonstratively evident why any body (i.e. anything composed of matter and form) is liable to change when in contact with a sufficiently powerful agent. Albert points out that his first theorem of natural science also defines the scope of natural science. From the fact that every body is changeable by some agent, it cannot be inferred that every agent is a body. Hence the way is opened to the possibility that agents of bodily change exist which are not themselves bodies, i.e. that reality is wider than the field proper to natural science.[56] In the last part of this

---

[52] On fate see ibid., cc.19-20 (ed. Borgnet 3: 153a-156b). Albert shows that Providence works through the celestial spheres which move with determinate necessity, and these act upon the sublunar elements and living things with regular, but not infallible effect. Mathematical astronomers, however, sometimes fail to realize that God works through natural agents and attribute the effects of these agents to "fate." See also Albert, *De fato*, a.3 (ed. Colon. 17/1: 71-72).

[53] *Physica* II, tr.2, c.12 (ed. Borgnet 3: 142a-144b). Note also Albert's attitude expressed in the following: "Natural things are not the result of chance or will but are from an efficient cause which produces and determines them; hence in natural science it is not our business to inquire how God the Creator uses what he has created according to his free will to work miracles in order to reveal his power, but rather it is our task to inquire what can be done naturally in natural things according to the natural causes intrinsic to them": *De caelo* I, tr.4, c.10 (ed. Colon. 5/1: 103, v.5-12).

[54] *Physica* III, tr.1, c.8 (ed. Borgnet 3: 197b-202b).

[55] Ibid. (ed. Borgnet 3: 202b). See the paper by E. J. McCullough in this volume.

[56] Ibid. III, tr.1, c.1 (ed. Borgnet 3: 177b).

Book III Albert discusses with great care the notion of "the infinite," because some changes are observed to be continuous and any continuum is infinitely divisible. With Aristotle Albert holds that while the notion of the potentially infinite is required for scientific explanations of change, that of the actually infinite is not. The use of the latter notion as an explanatory principle results only in paradoxes.[57]

In Book IV Albert goes on to demonstrate two further properties of all changeable bodies, namely, that they are related to each other both in place and in time. He then eliminates the notion of the vacuum (so essential to a mechanical view of the world) as an explanatory principle, just as he disposed of the actually infinite.[58] He digresses at length on various difficulties about the definitions of place and time, and adds a special treatment of the problem of angelic time and of eternity. Here he takes the occasion to repudiate Aristotle's teaching on the infinite duration of the universe *a parte ante*. In Book V he shows that there are only three kinds of change requiring detailed study: change in place, quality, and quantity. He then treats of "seven intentions" or terms, such as "to be together," "to be separate," "to touch,"[59] etc., and also of what makes a change one or many, in preparation for Book VI, in which he shows how change, time, and place can be measured and hence are susceptible of mathematical study, and to refute the paradoxes of Zeno which would render such study nugatory.[60] In reading all this one is struck with Albert's careful attention with what today would be considered "linguistic problems" by one school of philosophy and as "phenomenological descriptions" by another. He hopes to remove paradoxes (pseudo-problems) by an analysis of everyday language and the common experiences it expresses. Albert also added a commentary on the pseudo-Aristotelian *Liber de indivisibilibus lineis*,[61] although he regards this as a mathematical work, in order to reinforce his

---

[57] Ibid. tr.3 (ed. Borgnet 3: 203a-238b), especially c.17 (235a-6a). Albert shows that mathematics does not need to posit the infinite in act, because it can imaginatively construct any finite quantity it wishes. Cantor's introduction of the infinite in act led to irresolvable paradoxes.

[58] Ibid. IV, tr.2 (ed. Borgnet 3: 272a-304b). See especially c.9 (299a-304b) where Albert refutes the view of Xuthus that unless in the pores of bodies there is a vacuum there can be no explanation of the contraction or expansion of bodies.

[59] Ibid. V, tr.2 (ed. Borgnet 3: 378a-383b).

[60] Ibid. VI, tr.3 (ed. Borgnet 3: 448a-461b) and also VIII, tr.3, c.5 (ed. Borgnet 3: 609b-610a).

[61] Ed. Borgnet 3: 463a-481b. According to W. D. Ross, *Aristotle* (London, 1956), p. 13, Simplicius attributes *De indivisibilibus lineis* to Theophrastus. Albert finds no inconsistency between this work and Aristotle's doctrine.

analysis of the continuum by refuting the view that a line is composed of indivisibles.

In Books VII and VIII Albert shows that Aristotle did not content himself (as today he is so often accused of doing by "process philosophers") with isolated substances, but is ultimately concerned to study interacting systems. Hence, Aristotle attempts to demonstrate that (1) every body undergoing change is acted upon by an agent other than itself (Albert opposes both the Platonic notion of the soul and Galen's notion of "spirit" — energy — as self-moving principles);[62] (2) bodies can act on each other only when in contact;[63] (3) in any system of movers the motions are commensurable, i.e. have a common time;[64] (4) in every system there is a first unmoved agent, which for the whole universe must move the first moved body in circular motion, if the system is perpetual;[65] and, finally, (5) this absolutely first unmoved agent is not a body, but a real immaterial agent, outside place and time.[66] In commenting on these theorems Albert takes great pains to show that the Aristotelian arguments for the eternity of the world are not probative, but he also admits that the argument he himself proposes for its temporal finitude is only probable.[67] However, he insists (leaving it to metaphy-

---

[62] *Physica* VII, tr.1, c.1 (ed. Borgnet 3: 483a-484a).

[63] Ibid. c.3-4 (ed. Borgnet 3: 489a-495a).

[64] Ibid. VII, tr.2 (ed. Borgnet 3: 505a-519b).

[65] Ibid. VIII, tr.2, c.11 (ed. Borgnet 3: 592b-596b).

[66] Ibid. tr.4, c.7 (ed. Borgnet 3: 632a-633b).

[67] Ibid. tr.1 (ed. Borgnet 3: 521a-557b). Albert argues that since God certainly precedes the world in the duration of eternity, and the duration of the world is measured by the duration of eternity, therefore, the world has only a *measured* i.e. finite duration. But he then says, "Thus, therefore, it is proved that the world is created and that God precedes the world in the duration of eternity. This, then, is the reason for our opinion and if we did not have so strong an argument, we would not express any opinion on this subject, since it is a shame and a disgrace in philosophy to present opinions without reasons. However, it seems to us this argument is better than any presented by Aristotle; nevertheless, we do not claim it is strictly demonstrative, nor do we think either side can be so demonstrated." C.13 *fin.* (ed. Borgnet 3: 552b-553a). He also says, "If, however, the question is raised by someone who asks why, if the foregoing arguments are true, did Aristotle, who so well understood many subtle matters, not say so; I reply that it seems clear to me that Aristotle understood very well that his arguments to prove the eternity of the world were not conclusive. He himself indicates this in many places in the *De caelo et mundo* where he says that he makes his investigations from his desire for philosophical understanding, and he presents arguments against which it is more difficult to object than to other arguments. This is a sign that he knew he did not have a strict demonstration, because what can be demonstrated cannot be in any way contradicted, and if they were in fact contradicted by anyone, they would still be just as conclusive as if uncontradicted. Aristotle, however, was not accustomed in his *Physics* to speak of any but strictly natural matters which can be determined by natural principles. Now the beginning of the world was through creation and not by any natural process, nor can it be proved naturally, and, therefore, Aristotle thought it

sics to supply a full treatment of the question) that the universe, whatever its duration, has been created *ex nihilo* by a free act of God.

D. The Researches of Natural Science

Albert was by no means content, as today a philosopher of science might be, with such general reflections on the nature of natural science and its most general abstract principles. For him the value of such principles was in their application to detailed scientific researches on the actual species of things in which these universals were exemplified and concretized.

> In investigations of nature, however, it is necessary not only to consider the changeable understood universally according to its common features, but it is necessary to get down to details so that the primary agent in each individual case may be ascertained, especially in sensible, animate things, because in investigations of nature we must discover the universal principles through singulars, since in such investigations the particulars are better known than the universals. It is through the singulars that we come to believe that it is convenient and necessary for universals and their principles to exist, since it is only those universals which are exemplified in particulars that we accept, while those which are not so exemplified in particulars, we reject.[68]

It would certainly be too much to say that Albert, any more than Bacon or any of his contemporaries, really grasped the notion of unlimited progressive scientific research as it is now conceived.[69] What Albert and Roger had chiefly in mind was the completion of the world view which Aristotle had already developed in its main lines by filling in the gaps which he had left, or the record of which was lost. The medievals did not yet see any great prospect of being

---

best to be silent about the manner of this origin in the *Physics* and did not deal with it expressly except in the book *De natura deorum* which he is [supposed] to have written." Ibid. c.14 (ed. Borgnet 3: 555a-b).

[68] *De principiis motus processivi*, tr.1, c.1 (ed. Colon. 12: 49 v.21-31). See also *De natura locorum* I, c.1 (ed. Borgnet 9: 529a). However, in *De vegetabilibus* VI, tr.1, c.1 (ed. Borgnet 10: 159) before discussing the species of plants Albert says, "In this sixth book of ours on plants, we aim more to satisfy the curiosity of students than to further philosophy: for there can be no philosophy of particulars." Perhaps by this Albert simply means that the kind of certitude which the philosopher craves can hardly be achieved in understanding the details of nature.

[69] For a sketch of the social context of medieval science see Guy Beaujouan, "Motives and opportunities for science in the medieval universities," in A. C. Crombie, ed., *Scientific Change* (London, 1963), pp. 219-236.

able to push the understanding of nature much beyond what had already been achieved. No one really had this hope until the seventeenth century when the first really new instruments and techniques of research began to yield surprising and technologically profitable results. However, Albert sketched out a vast program for the collection, synthesis, and completion of what was known about nature.

The diagram (pages 90-91) gives the quickest view of this program, which I will explain only briefly. The *Physics* treats of what is common to all bodies, but it is necessary then to discuss what characterizes particular classes of bodies according to their different material composition, beginning first with the simplest bodies or elements.[70] Since in *Physics* v it was proved there can be no change without contact between bodies, all other kinds of change depend on change of place. In *De caelo et mundo*, therefore, the elementary bodies are studied as subject to motion in place with the conclusion that there are only five elements in the universe: one moved in circular orbits (the heavenly bodies, composed of inalterable "ether") and four moved in straight lines (fire moving up from the center of the finite universe, earth moving downward to the center, and air and fire in intermediary positions). Albert believed that these conclusions can be determined with certitude on the basis of the fact attested by all observational astronomers that heavenly bodies are completely regular in their motions and unalterable. Such perpetual motion could be explained only if these bodies had circular orbits, so that irregular, finite motions are possible only in our sublunar region.[71] However, he also believed that the theories about the details of astronomy developed by the mathematical astronomers are only hypothetical, based on a mere "saving the phenomena," not on physical principles.[72]

Albert proposed to deal with such a mathematical astronomy in the mathematical section of his encyclopedia, but either he did not complete these works, or they have not been preserved or recovered, unless we have an example of them in the *Speculum astronomiae* whose authenticity is disputed and which does not seem to be textually connected with the encylopedia.[73] This work carefully distin-

---

[70] *De caelo et mundo*, I, tr.1, c.1 (ed. Colon. 5/1: 1-2).

[71] Ibid. I, tr.1, c.5 (ed. Colon. 5/1: 14-16).

[72] Ibid. II, tr.2, c.9 (ed. Colon. 5/1: 161-162).

[73] Ed. Borgnet 10: 629-651. See for discussion P. G. Meersseman, *Introductio in Opera Omnia B. Alberti Magni*, and *Bibliographie*, M. H. Laurent and Y. Congar, eds., *Revue Thomiste* 36 (1931), 422-468; items 289-298 on p. 442.

guishes between descriptive and judiciary astronomy, the latter being what we today call "astrology." Albert, with all the great scholastics, admits in his other works, as does the *Speculum*, that the heavenly bodies are the source of all change on the earth and hence must affect the human body. Hence he believed it possible to predict historical events by a scientific study of these celestial energies, provided that such predictions were recognized to be only conjectural, leaving room for chance and freedom, as we have already noted above. However, Albert and his contemporaries seem to have been insufficiently critical of the way in which these predictions were supposed to be made on the basis of correlations more mythological than empirical. If Albert did write the *Speculum* it provides clear evidence of his concern (1) to oppose superstition and magic, and (2) to preserve from destruction works which might contain truth along with error, for the sake of truth wherever it might be found.

> As for books on necromancy it seems to me (allowing for the better judgment of others) that they should rather be conserved than destroyed, because the time may be near when (for reasons of which I do not now wish to speak) it may be profitable to consult such books at least occasionally, although their readers should be cautious of their possible misuse. However, there are some books of experiments whose titles are similar to necromancy, such as geomancy, hydromancy, aeromancy, pyromancy and chiromancy, which truly do not deserve to be called science, but rather "nonsensomancy" (*garamantia*).[74]

After discussing simple bodies undergoing local motion, Albert studies them as they are also subject to other kinds of change. He deals with this in a general manner in *De generatione et corruptione*, showing that the four sublunar elements are liable to various transformations, some superficial, but others radical, so that these elements are produced by transmutation out of other elements. Thus, while the heavenly bodies may endure perpetually, in our terrestrial region there is no need to posit the indestructible atoms of mechanistic cosmologies. In fact, in the sublunar region the reason that the constant circulation of matter goes on is because new elementary bodies are constantly being produced outside their natural places, and hence must move to these places to maintain the natural homeostasis, e.g., when water is produced from air by some cooling process it then moves downward to its natural position on the earth.

---

[74] *Speculum astronomiae* c.17 (ed. Borgnet 10: 650).

THE DIVISIONS OF NATURAL SCIENCE

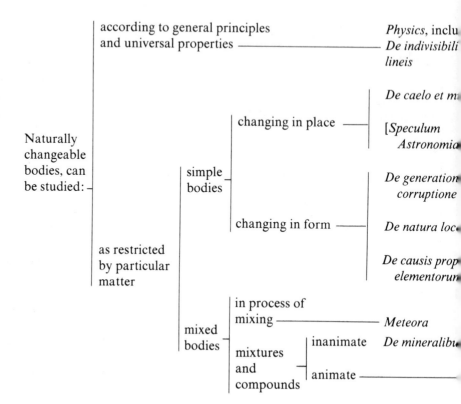

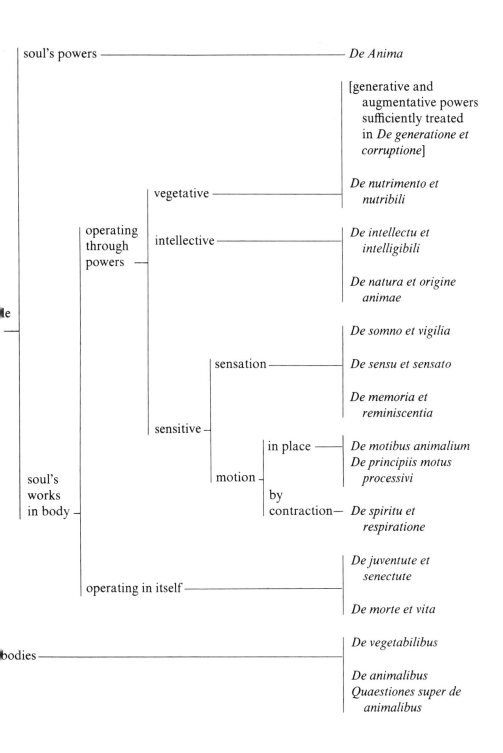

To complete this treatment of the elements Albert added two works based on pseudo-Aristotelian fragments known through Arabic.[75] The *De natura locorum* is a kind of geography discussing the terrestrial sphere, its climactic zones and its water and land masses. The *De causis proprietatum elementorum* treats of how the properties of the elements are related to their positions in the sublunar region.[76] These works comprise the first part of natural science and the consideration of motion in place common to all bodies and characteristic of simple bodies.[77]

The second part of natural science is treated in a single book, the *Meteora*, which deals with the elements in the process of mixture and combination to form compound bodies.[78] In the first three books of this work are discussed the various phenomena produced by this transition, such as meteors, winds, floods, earthquakes and volcanoes, while in the fourth book the nature of compound bodies is considered, dealing with the basic topics of what we today call chemistry.

The third part of natural science, for Albert, treats of compound bodies of various kinds both inanimate and animate. *De mineralibus*, one of Albert's own innovative works, attempts to classify and explain the formation and composition of terrestrial substances.[79] A second group of works deals with compound bodies which are living, and is very extensive not only because of the rich Aristotelian and Arabic medical material available to Albert, but also because of his personal interests.

Albert places *De anima* first because it deals with the principle of life and then treats of the functions which this principle produces in organic bodies.[80] In an original work, *De nutrimento et nutribili*, he discusses vegetative or physiological functioning.[81] Logically, a discussion of sensation would come next, but Albert for pedagogical reasons choses first to discuss the intellective functions in two original works, *De intellectu et intelligibili* and *De natura et origine animae*, the later dealing with the possibility of the survival of the soul after

---

[75] Meerssemann, pp. 34-36.

[76] Ed. Borgnet 9: 585a-653b.

[77] *De generatione et corruptione* I, tr.1, c.1 (ed. Borgnet 4: 345a-347a).

[78] *Meteora* I, tr.1, c.1 (ed. Borgnet 4: 477a-479a).

[79] *De mineralibus*, (ed. Borgnet 5: 1a-103b); translation by Dorothy Wyckoff, *Book of Minerals* (Oxford, 1967).

[80] *De anima* (ed. Colon. 7/1).

[81] *De nutrimento* (ed. Borgnet 9: 323a-341b).

bodily death.[82] After laying this groundwork he then goes back to deal with the related functions of sensation and local motion. Sensation is discussed as regards sleep and waking (*De somno et vigilia*), the sense organs and their objects (*De sensu et sensato*), perception, imagination, instinct, and memory (*De memoria et reminiscentia*).[83] The sense in which living things are self-moving, which might seem to contradict the principle defended in the *Physics* that all moved bodies are moved by an extrinsic agent, is discussed in an original work, *De motibus animalium*, and in a commentary, *De principiis motus processivi* on an Aristotelian work which became available to Albert later.[84] The internal motions of contraction and dilation of the viscera are treated in *De spiritu et respiratione* in which Albert has to confront the Stoic energism of Galenic medicine.[85] This general consideration of what today would be considered physiology and psychology is completed by a consideration of the processes of aging (*De juventute et senectute*) and of death and dying (*De morte et vita*).[86]

Finally, Albert comes to what he considered the real goal of natural science: the study of the specific kinds of living things. Here he applies the general chemical and physiological model developed in the former works to plants (*De vegetabilibus*) and to animals (*De animalibus*)[87] especially the human animal. Throughout this vast synthesis, Albert took pains to follow an Aristotelian methodology, moving always from the more general to the more particular and concrete, and attempting to provide scientific explanations in terms of the model of changeable body developed in the *Physics*. However, especially in the biological works, Albert suffered many hesitations about the exact order to be followed.[88] The most serious departure from Aristotle's own order is Albert's failure to appreciate (as Averroës did)[89] what is perhaps the best example of Aristotle's method of developing a first principle from a very careful analysis of extensive

---

[82] *De intellectu* (ed. Borgnet 9: 477a-521b); *De natura et origine animae* (ed. Colon. 12: 1-75). The reason for the change of order is given in *De intellectu* I, tr.1, c.1 (ed. Borgnet 9: 478a).

[83] Ed. Borgnet 9, *De somno* 121a-207b; *De sensu* 1a-93b; *De memoria* 97a-118b.

[84] *De motibus* (ed. Borgnet 9: 257a-300b); *De principiis motus* (ed. Colon. 12: 48-75).

[85] *De spiritu* (ed. Borgnet 9: 213a-251b).

[86] Ed. Borgnet 9: *De juventute* 305a-319b; *De morte* 345a-371b.

[87] *De vegetabilibus* (ed. Borgnet 10: 1-305). The critical edition of E. Meyer and C. Jensen (Berlin, 1867) was not available to me. *De animalibus*, ed. Hermann Stadler, *Beiträge*, Clemens Baeumker, ed. (Munster 1916) 2 bände. Band 15-16.

[88] Meersseman, pp. 32, 40-41, etc.

[89] Averroës, *Aristotelis opera omnia cum Averrois commentariis* (Venice apud Junctas, 1562-74; Frankfort/Main photo-reprint, 1962), vol. 4: *Meteorologicorum* I c.1, 404a.

empirical data, which is to be found in the way Aristotle moves from description and classification in the *Historia animalium* to theoretical analysis in *De partibus animalium* to the actual process of forming a real definition which is to serve as a principle of demonstration in *De anima*. Here Albert (as Aquinas after him) was misled by the Michael Scot translation of these works of Aristotle (along with the *De generatione animalium*) in which these nineteen books are all lumped together and separated from their proper relation to the *De anima*.[90] Consequently, Albert and Thomas Aquinas both begin their treatment of biology with the definition of the soul in a way that appears *a prioristic*. This was a real misfortune since it deprived the Middle Ages of an adequate understanding of Aristotle's cautious empiricism in the face of the nature of life, "whose existence," Aquinas says, "is most certain, but whose essence is most uncertain."[91]

E. THE RELATIONS OF NATURAL SCIENCE
TO OTHER SCIENCES

Albert's understanding of the general classification of the sciences was much the same as that already established for the scholastics by Dominic Gundissalinus following Alfarabi[92] and represented by Albert's contemporaries Roger Bacon[93] and especially Robert Kil-

---

[90] See A. M. Festugière, "La place du *De Anima* dans le système aristotélicienne d'apres S. Thomas," *Archives d'histoire doctrinale et littéraire du moyen âge* 6 (1931), 25-47.

[91] Aquinas, *In 1 De anima*, 1 (Marietti ed., 1948, nn. 6 and 15, pp. 3 and 5).

[92] *De divisione philosophiae*, ed. Ludwig Baur, *Beiträge* (1903), band 4, heft 2-3; Alfarabi, *De ortu scientiarum*, ed. Clemens Baeumker, *Beiträge* (1916), band 18, heft 3; J. A. Weisheipl, "Classification."

[93] "After I have treated grammar according to different languages as they assist and indeed are necessary for the study of the Latin writers, and also along with these have dealt with the logical arts and after I have treated in a second volume the parts of mathematics, now in this third volume I come to the natural sciences and in a fourth metaphysics will be joined with the moral sciences. For it is evident that grammar and logic are prior in the order of teaching, and the proper place for the natural sciences, as Avicenna says in his *Metaphysics* I, is that they should follow mathematics. And similarly Avicenna teaches that metaphysics follows natural science, since, according to him, the conclusions of the other sciences are principles in metaphysics. And this is certain from Aristotle, since through the conclusions of astronomy Aristotle teaches the unity of the first cause and the plurality of the intelligences, although the metaphysician also has, by another way, to prove the principles of all the sciences, as it should be shown in that science. But moral philosophy is the end of all the other sciences, and, therefore, obtains the last place in philosophical consideration. For all the others are speculations about truth, but this is practical or operative of the good, for which reason it follows in the order of nature, for the knowledge of the truth is ordered to love and good action." Bacon, *Liber primus communium naturalium*, dist.1, c.1, ed. R. Steele, fasc. 2, pp. 1-2.

wardby in his great *De ortu scientiarum*. In this schema the Platonic-Stoic division into logic, ethics, and physics was elaborated by dividing "physics" or theoretical science (as contrasted to morals and technology) into natural science, mathematics, and divinity (metaphysics and sacred theology).[94] Albert, following Aristotle, explains this tripartite division of theoretical science according to three modes (not "degrees" as later manuals were to say) of abstraction. Natural science abstracts only from the peculiarities of *individual* bodies while dealing with all other aspects of concrete, really existing bodies as these exhibit regular, generalizable features. Mathematics, on the other hand, is abstract in a strict sense, since it abstracts from real existence and deals only with one aspect of real things, their quantity, as this can be ideally reconstructed in the imagination. Finally, metaphysics transcends the characteristics proper to physical things altogether, and considers only the aspects of being common to all beings, whether material or immaterial, real, possible, or imaginary, but always with the primary concern of dealing with ultimate reality.[95]

What is more particular to Albert, in sharp contrast to Roger Bacon,[96] is first of all Albert's fidelity to the Aristotelian side of the Platonic-Aristotelian debate over the role of mathematics in the total development of the human intelligence. For Platonists, mathematics, precisely because it is more abstract and thus more removed from matter and motion, is more truly scientific than is natural science and mediates between natural science and the supreme science of metaphysics.

The Aristotelians, on the other hand, regard mathematics as the least of the theoretical sciences, although of greater dignity than logic whose value is purely instrumental to the real sciences. Mathematics is a true science of reality, remarkable for its certitude and clarity, but deficient as regards its subject matter which is merely the quantity of physical objects considered in idealizing abstraction from the existential conditions of such objects. Hence mathematics cannot be considered a step toward metaphysics, because metaphysics as the study of existent being as such, is more directly linked to physics, the science of our sensible world. It is only from the exist-

---

[94] Albert, *De praedicabilibus* I, tr.1, c.2 (ed. Borgnet 1: 2b-5a); *Physica* I, tr.1, c.1 (ed. Borgnet 3: 1a-4a). See also Conway and Ashley, pp. 462-465.

[95] *Metaphysics* I, tr.1, c.1 (ed. Colon. 16: 1-3).

[96] *Opus majus* IV, c.16, (Bridges 1: 175).

ence of visible realities that we can come to know the existence of invisible realities, as from effect to cause.

On the other hand, the role of mathematical-physical sciences, such as optics or astronomy, is that of a *mixed* science in which physical subject matter is open to scientific investigation and demonstration only in restricted terms which usually yield only conjectural (dialectical) solutions. This is why Albert is convinced that many of the mathematical theories of the astronomers are only hypothetical, justified only by the degree to which they "save the phenomena," and incapable of explaining natural change in terms of natural agents. Albert frequently insists on this view of mathematics and mathematical physics. Reproving the "Pythagoreans and Platonists who believed that the nature of things is nothing but mathematical forms,"[97] he goes on to say:

> Since these sciences [natural science and mathematics] have some common ground, so their demonstrations are sometimes mixed, e.g. both prove the earth is a sphere. Thus sometimes the physicist proves this geometrically, and argues that if the earth were not round, the rising and setting of the stars would not be variable in different parts of the earth. But this astronomical proof is only adapted to the purpose of the physicist, and does no more than establish the fact (*quia*); but the reasoned fact (*causa propter quid*) i.e. *why* the earth is round, cannot be established by the astronomer [i.e. mathematically]. However, sometimes the astronomer can give a physical proof of the same fact. Thus he argues that since the earth is a simple body, and a simple body must have a simple figure, and a simple figure does not have one part straight and another part angular, but has no angles [i.e. is a sphere, therefore, the earth is round]. Such a demonstration is [not mathematical, but] physical. Or, [to give another example] when it is argued that since the parts of earth are heavy, and heavy parts fall equally about a center, therefore, the parts of earth fall in a circle [i.e. they form a sphere]. Such demonstrations give the reason (cause) why the earth is a sphere, and they are dealt with in the mathematical sciences [i.e. in the study of the quadrivium] only for the sake of easier instruction. For all the mathematical sciences which deal with questions concerning physical subjects make their inquiry also about the mathematical aspects [of these subjects] through mathematical principles, and, therefore, are subalternated to mathematics rather than to physics.[98]

Albert does not deny the value or importance of mathematics as a

---

[97] *Physica* II, tr.1, c.8 (Borgnet 3: 108b).
[98] Ibid. pp. 109b-110a.

tool of natural science, but he holds that mathematics can only establish a physical fact (*quia*), but it cannot demonstrate the physical reason or cause (*propter quid*) without which scientific understanding is incomplete. Thus, mathematics is related to natural science as a research instrument, but not as a source of its own proper principles.

The relation of natural science to metaphysics is a very different matter. For Albert both disciplines are truly sciences, but only metaphysics deserves the name of wisdom (*sapientia*), because natural science is limited to the study of changeable bodies which are not the ultimate realities, and which themselves cannot be completely understood except in relation to ultimate realities as effects cannot be understood except in relation to their causes. However, this inferior status of natural science does not mean that it is subalternated to metaphysics, as natural science is subalternated to mathematics in astronomy in the foregoing quotation, because natural science never uses metaphysical principles as such in its demonstrations. Albert says that metaphysics "founds" mathematics and natural science, because it stabilizes and defends the validity of the principles of all the special sciences,[99] but the universal principles (*dignitates*, axioms) of metaphysics itself do not enter into the demonstrations of the special sciences, except *dialectically* or in some restricted form.[100] Thus Albert would never suppose that the conclusions proper to natural science (e.g. that the earth is round) could be deduced from abstract metaphysical principles. Such conclusions must stand or fall on the basis of the principles developed in natural science itself — principles rooted directly in sense experience and valid only if confirmed by this experience. Metaphysics defends such principles precisely by defending the validity of sense experience and of intellectual insight based on such experience.[101]

For Albert, also, natural science precedes metaphysics in the order of knowing, since metaphysics is a reflection on insights gathered from the special sciences, and primarily from natural science, since mathematics deals only with idealized objects, not directly with existential realities, while the practical sciences do not aim at theoretical insight.[102] How then is the transition made from natural science to

---

[99] *Metaphysics* I, tr.1, c.1 (ed. Colon. 16/1: 2).

[100] *Analytica post.* I, tr.3, c.3 (ed. Borgnet 2: 74b-75b).

[101] *Metaphysics* IV, tr.3, c.5-6 (ed. Colon. 16/1: 192-195) cf. also I, tr.1, c.6 (pp. 7-10).

[102] *Physica* II, tr.1, c.8 (ed. Borgnet 3: 107a-110a); *Metaphysics* VI, tr.1, c.3 (ed. Colon. 16/2: 305).

metaphysics? For Aristotle the reason that natural science is not "first philosophy" in dignity, although it is first in the order of discovery, is because in the course of the investigation of nature it becomes evident that the cosmic system of changeable bodies cannot be the whole of reality. First, Aristotle in *Physics* VII-VIII attempts to establish that the prime movers or mover on which all the changes in the physical universe absolutely depend cannot be material.[103] Again in *De anima* the investigation of the nature of human intelligence establishes that the agent intellect is something more than the act or form of a body.[104] It too is one of the immaterial prime movers which affect the material world.

Albert, although he does not stress this point, seems to agree with Aristotle that natural science opens the way for metaphysical speculation by establishing the existence of immaterial entities. However, some thinkers in the Aristotelian tradition have preferred to grant to metaphysics a greater independence from natural science than does Aristotle himself. They have argued that since natural science itself assumes the existence (*esse*) of its subject, the human intelligence, independently of any physical proof of the existence of immaterial beings, can come to the intuition that existence as such is not adequately explained by natural science because natural science deals only with the causes of motion but not of existence. Hence the necessity of a discipline of metaphysics is apparent simply from the analysis of the distinction between nature and existence, *essentia* and *esse*. This is not the place to attempt to deal with this question,[105] but it would appear from Albert's proemium to his commentary on the *Metaphysics* that, without explicitly rejecting Aristotle's opinion, he tends to the second position.

> Since the natural scientist supposes the existence (*esse*) of changeable bodies and the mathematician supposes the existence of continuous or discrete quantity, therefore, each posits existence, since they are not able to prove existence from their own proper principles, but existence must be proven from the principles of existence as such. Therefore, this science of [metaphysics] has the task of stabilizing both the subject and principles of all the sciences. Nor can these be established or founded by the particular sciences in which existence (*quia sunt*) or *esse* is left undetermined or supposed. . . . This science is also called divine, because

---

[103] Aristotle, *Physics* VIII, c.10 (266a10-267b25).

[104] Aristotle, *De anima* III, c.5 (430a16-19).

[105] For an excellent treatment of the controversy among Thomists see Thomas C. O'Brien, *Metaphysics and the Existence of God* (Washington, 1960), pp. 61-98.

all such principles are divine, best, and first, furnishing all other things their completion in existence. For existence (*esse*), which this science considers, is not contracted to this or that kind of existence, but rather is considered as it is the first efflux from God and the first creature, before which nothing else is created.[106]

Hence Albert does not hesitate also to quote the Platonist Ptolemy who argues that since natural science deals only with things in temporal flux, therefore, in distinction to metaphysics, natural science is "mixed with opinion and cannot attain to a confirmed, permanent and necessary habit of science."[107] This position is consistent with the fact, (which I cannot fully discuss here) that Albert, while subscribing to the view that all knowledge is rooted in sense experience, could not bring himself to give up completely (as Roger Bacon could not) the notion that it is necessary to suppose some kind of direct illumination by God to guarantee the certitude of the highest principles of human understanding.[108] This last vestige of Platonism (principally via Augustine) is finally shed only by Thomas Aquinas. Even Aquinas in his early *De ente et essentia* still speaks much as Albert does in the above text. Only in his mature works does Aquinas insist that although God is cause not only of motion in the universe but of its total existence, nevertheless, if natural science did not establish the existence of immaterial prime movers (at least God and the

---

[106] *Metaphysics* I, tr.1, c.1 (ed. Colon. 16/1: 2, vv.75-81).

[107] *Metaphysics* I, tr.1, c.1 (ed. Colon. 16/1: 1, vv.24-27). See also *Euclid Commentary*, ed. Geyer, p. 170. Albert seems to intend in these passages to emphasize the relative lack of certitude in natural science. "From this it is clear that metaphysics achieves only a little understanding of what in itself is most clear, and mathematics understands [its objects] very well, while natural science only rarely has a firm and certain understanding. This is so because the divine light overcomes and dazzles the intellect, while mathematical objects are proportionate to our intellect in itself and are blended with the intellect and its light; but because of privation, matter, and motion physical objects fall short of intellectuality. Hence metaphysical objects are said to be superior to the intellect, mathematical objects to exist in the intellect, and physical objects to be below the intellect": *De intellectu* I, tr.3, c.2 (ed. Borgnet 9: 500a). On Albert's attitude toward Ptolemy and natural philosophy, see ed. Colon. 16/1: 1, note on line 27.

[108] "Albert ascribes to the human intellect an active power of its own, but he submits the whole soul of man to the higher illuminating influence of a separate substance, which is God. With respect to God, the human soul, including its intellective powers is "possible," that is, in potency to a higher illumination. The same dual position is to be found in Roger Bacon. The formula (God is the separate Intellect) was so commonly received that even Thomas Aquinas will pay it at least lip service": Etienne Gilson, *History of Christian Philosophy in the Middle Ages* (New York, 1955), p. 670, note 9.

human intelligence), there would be no need or ground for a science of metaphysics distinct from natural science.[109]

What then for Albert is the value of natural science? It is surprising that he does not discuss this question explicitly in the beginning of his commentary on the *Physics* where we might expect it. He never emphasizes its technological applications in the way Roger Bacon does,[110] but realizes their importance in many fields.[111] For Albert, as for Aristotle, the study of nature, as all the theoretical sciences, is valuable in itself, a part of that contemplative life in which human happiness is chiefly to be found. He echoes Aristotle's famous saying that even the viscera of worms are worth our study, because in them is reflected the work of the Divine Artist.[112] Thus, such study contributes to sacred theology, since the creature helps us know something of the Creator by analogy,[113] and in his own theological works Albert seldom misses an opportunity to make use of any relevant scientific findings known to him. He seems to me even more concerned than Aristotle to make use of biological and psychological knowledge in his treatment of moral questions.[114]

However, in one very significant respect, Albert seems to depart from Aristotle in his understanding of the relation of natural science to the moral sciences. For Aristotle not only does natural science open the way to metaphysics and the study of transcendant realities, as I have indicated, but it also raises the question whether such a study is really worth the effort, considering that the human intelligence can only know such transcendant realities mediately and analogically.[115] It is this difficulty which for Aristotle motivates the investigations of the moral sciences which he pursues in his

---

[109] For the texts of St. Thomas in which his view of the relation of natural science to other disciplines are set forth see Conway and Ashley, pp. 520-523 and on its relation to metaphysics see O'Brien, pp. 10-176.

[110] *Opus tertium* c.13 (ed. Brewer, pp. 43-47).

[111] See especially the articles by Nadine F. George in this volume, pp. 235-261, and Pearl Kibre, pp. 187-202.

[112] *De animalibus* XI, tr.2, c.3 (ed. Stadler 1: 793-794).

[113] *Super 1 Sent.* dist.3, D, art.10 (ed. Borgnet 25: 99b). Albert follows the Dionysian notion of the three ways of knowing God: causality, eminence, and negation.

[114] Aristotle in *Ethics* I, c.13 (1102a15-29) emphasizes that the moralist as a man of prudence need not be concerned with the theoretical details of psychology. Albert does not disagree with this, but enlarges on the comparison between the doctor who must thoroughly know the body, and the moralist who must know the soul; *Ethica* I, tr.9, c.2 (ed. Borgnet 7: 141b). Both Albert and Thomas make more extensive use of psychology in their moral writings than does Aristotle.

[115] *Nicomachean Ethics* X, c.7 (1177b26-31).

Nichomachean Ethics and *Politics* and which end in the conclusion that not only does human happiness consist chiefly in the contemplative life devoted to the pursuit of ultimate truth, but that this requires the assistance of well-ordered society. To live according to intellect, although more divine than human, is to live according to what is in us most specifically human.[116] Albert comments on this lofty doctrine as if in full agreement, but he evidently has some hesitations, since in his own proemium to the *Ethics* he adopts the view that moral science, because it alone makes man good in himself, is the highest of all studies, and he seems to accept the view of Avicenna that moral philosophy should be considered not as propadeutic to metaphysics but as its crowning part. Thus Albert joins Roger Bacon in holding that all of human learning is directed toward the study of morality.[117]

Perhaps, this is the reason that when Albert answers the standard scholastic question as to whether sacred theology is a speculative or a practical science, he concludes that it a *scientiam ad pietatem*, an affective or moral science.[118] Aquinas was to go beyond this position, insisting that moral science should be studied before metaphysics[119] and that sacred theology, while it includes moral doctrine, is ultimately and formally speculative and contemplative.[120]

---

[116] Ibid. (1177b32-1178a8).

[117] For Bacon see quote in note 93 above and also *Rogeri Baconis Moralis philosophia*, ed. Eugenio Massa (Turin, 1953), Proemium, I, c.1 6-9. For Avicenna, see *Metaphysica* (*Philosophia prima*), B. Cecilius Fabranensis trans. 1508 (Frankfort/Main, photo-reprint, 1961), tr.10, 107-108. "All other sciences perfect the knower in regard to knowing in some respect, but none of them perfect him in regard to existing as good and worthy, so that he should actually be good and worthy. This is why, as Apuleius says, that the ancient philosophers, after mastering all other science, completed their lives in its study. This is also why Avicenna says this science is the last part of divine science which gives it its ultimate perfection; hence he made the last part of his first philosophy to be moral science": Albert, *Ethica* I, tr.1, c.1 (ed. Borgnet 7: 2a-b).

[118] "It must be said that [the nature of] this science should be determined from its end; and its end is stated in *Titus* 1: 1-2 which says, 'Paul, servant of God, but an apostle of Jesus Christ, according to the acknowledgment of truth as it is according to piety, in hope of life eternal.' And on this the Gloss comments, 'according to piety i.e. the religion of Christ. . .' ": *Super I Sent.* dist.1, A, Proemium, art.4, sol. (Borgnet 25: 18b). On this question see Martin Grabmann, *Die Theologische Erkenntnis- und Einleitsungslehre des Heiligen Thomas von Aquin und Grund seiner Schrift 'In Boethium de Trinitate'* (Freiburg/Schweiz, 1948), pp. 226-236.

[119] Conway and Ashley, pp. 501-53 for references.

[120] *Summa theologiae* 1.1.4 (ed. Leonina 4: 14b).

F. Conclusion

Albert the Great provided his times with a very rich and carefully worked out science of the natural world built on the Aristotelian model. His exposition in some respects, however, lacks the precision and consistency of Aquinas, and is not altogether free of imperfectly assimilated vestiges of Neoplatonism. Neither Albert or Aquinas were able fully to appreciate Aristotle's empirical method in biology, but modern commentators have hardly done better.

Albert understood natural science to be an investigation of the facts and causes of changeable bodies as they undergo regular change through natural processes. Such a science must be based on sense experience. It can and should use mathematics as an instrument of research, but not to provide ultimate explanatory principles. It should be logically systematized by proceeding from the general properties of all naturally changeable bodies, but should extend its researches down to the special properties of every specific kind of body, coming ultimately to a study of the human being. It should seek strict, causal demonstrations when possible, working backward from stable elements, compounds, and organic living things to the processes by which they are regularly produced in nature. Natural science discovers its own limits, leaving room for a metaphysics to consider wider and deeper aspects of reality. It is valuable for its technological application, but more valuable for the contribution it makes to the life of the intellect, and most valuable as leading toward a better understanding of how we should live. For the Christian, however, the light which natural science casts on the world requires support and even correction from the light of revealed truth.

4

Albertus Magnus on Suppositional Necessity in the Natural Sciences

William A. Wallace, OP
The Catholic University of America

Albertus Magnus is commonly recognized as the master who, more than any other, championed the cause of Aristotle's natural science in the University of Paris and thus gave stimulus to the Aristotelianism that was to flourish in the Latin West until the time of Galileo.[1] Considerably senior to his celebrated student, Thomas Aquinas, and thus more subjected to the Neoplatonism and Augustinianism present at the university when he lectured there, Albert is also regarded as more Platonic in his thought than was Aquinas.[2] The latter evaluation of Albert's philosophy poses an interesting problem when juxtaposed with his enthusiastic support for the *scientiae naturales* of Aristotle. It also raises a question as to the meaning of the Aristotelian term *scientia* ($\epsilon\pi\iota\sigma\tau\eta\mu\eta$) when compared with its English equivalent "science," as the latter term has come to be understood in the present day. For Aristotle, *scientia* is the highest form of human knowledge, true and certain because achieved through apodictic demonstrations, and thus yielding conclusions about a subject matter that cannot be otherwise.[3] For Plato,

---

[1] See William A. Wallace, "Albertus Magnus, Saint," *Dictionary of Scientific Biography*, 1 (1970): 99-103.

[2] Some of Albert's teachings that support such an interpretation are given in Etienne Gilson, *History of Christian Philosophy in the Middle Ages* (London, 1955), pp. 277-294, 666-673.

[3] *Posterior Analytics* I.2 (71b8-12).

as is well known, such an ideal could never be realized in the changing world of appearances studied by the naturalist. At best, in the Platonic world view, physics is a "likely story."[4] By and large, philosophers of modern science are no more sanguine in their expectations. The characteristic method of "science," in their view, is not that sketched in Aristotle's *Posterior Analytics*, which claims to yield certain knowledge. Rather, a hypothetico-deductive methodology is generally seen as characteristic of scientific investigation, and this can never yield certitude, but only probability.[5] When evaluating the philosophy of science implicit in the work of Albertus Magnus, therefore, an intriguing question arises. Was this *Doctor universalis* a strict Aristotelian in his commitment to demonstrative methodology, or did he stand midway between the Platonist view of natural science as a likely story and the fallibilist evaluation of modern science as providing a probable but ever revisable account of nature and its operations? An answer to this question based on Albert's commentaries on Aristotle is the burden of this essay.

Aristotle, of course, could know nothing of the direction that recent science and its philosophy were to take, but he surely was aware of the Platonic objection that his ideal of *scientia naturalis* would have to overcome. Nature, for him as for Plato, is particular, variable, and contingent; *scientia*, on the other hand, must be universal, unvarying, and necessary. How, then, can one attain necessary knowledge of a subject matter such as nature, which apparently can always be somewhat otherwise than it is? In the view of the writer Aristotle faced up to this problem and formulated his answer in Chapter 9 of Book II of the *Physics*.[6] He did so in rather cryptic fashion, however, and translations of his Greek text into medieval Latin (as those into English) leave somewhat obscure, if not garbled, the main lines of his solution. Apparently Aristotle also gave hints as to how demonstrative methodology could be applied to the changing world of nature when laying out his general logic of scientific investigation in the *Posterior Analytics*.[7] The connection between these two treatments, i.e., the general logical canons of the *Analytics* and their particular application in the *Physics*, appears not to have been gener-

---

[4] *Timaeus* 29D; also *Republic* 7, 530D-533A; *Philebus* 55D-57E.

[5] For a general description of this methodology, see Carl G. Hempel, *Philosophy of Natural Science* (Englewood Cliffs, N.J., 1966), pp. 19-32.

[6] 199b33-200b9.

[7] I.8 (75b21-36), II.11-12 (94a20-96a19).

ally recognized in the Latin West. To our knowledge the first commentator to become explicitly aware of the connection between them was Thomas Aquinas. In a recent study we have focused on Aquinas' exposition of the technique of demonstrating *ex suppositione finis* as providing Aristotle's basic answer to the problem of how there can be a *scientia naturalis* in the strict sense of *scientia*.[8] What is more, in another study we have argued that this technique as explained by Aquinas was known through the later Middle Ages and was explicitly advocated by Galileo as the basis on which he constructed his *nuova scienza* of motion at the onset of the modern period.[9] Now, since Aquinas was Albert's disciple, and surely would have had access to his commentaries on the *Analytics* and the *Physics* (no less than to his commentary on the *Nicomachean Ethics*), an engaging speculation presents itself — a speculation that, apart from showing a transmission of knowledge from master to disciple, may shed light on the question raised in the opening paragraph of this essay. Could it be that Albertus Magnus had already anticipated the technique of *ex suppositione* reasoning later explained by Aquinas, or at least supplied the basic elements from which Aquinas drew his own, fuller, solution?

This further elaboration of the problem makes it somewhat more complex, but it provides both an historical and a systematic framework within which to work. Our investigation will accordingly proceed in three stages. The first will consider a somewhat enigmatic text of Aristotle in its various translations, and the sense that Aquinas makes of this in terms of the technique of *ex suppositione* demonstration detailed in his commentaries on both the *Physics* and the *Posterior Analytics*. The second stage will then concentrate on corresponding treatments in the Aristotelian commentaries of Albertus Magnus. The third stage, finally, will attempt to situate Albert with respect to Aquinas, as well as briefly to assess the later import of his work for the Scientific Revolution of the seventeenth century and for the solution of contemporary problems in the philosophy of science.

---

[8] "Aquinas on the Temporal Relation Between Cause and Effect," *The Review of Metaphysics* 27 (1974), 569-584.

[9] "Galileo and Reasoning *Ex Suppositione*: The Methodology of the *Two New Sciences*," in *Proceedings of the 1974 Biennial Meeting of the Philosophy of Science Association*, ed. R. S. Cohen et al. (Dordrecht-Boston, 1976), pp. 79-104.

A. Aristotle and Aquinas

The classical locus in which Aristotle treats the method of demonstrating in *scientia naturalis*, as already remarked, occurs in the second book of his *Physics*. Throughout this book Aristotle has been discoursing about nature, how the consideration of the naturalist differs from that of the mathematician, what kinds of causal explanations are available for natural phenomena, how chance and contingency can disrupt the regularity of nature's operation, and the way in which nature acts for an end. The latter consideration leads him to pose his final methodological questions, namely, what kind of necessity characterizes nature's operation, and how this type of necessity is commensurate with the manner of demonstrating that is required of a *scientia naturalis*. The substance of Aristotle's answer to these questions is found in the following passage, given in literal English translation and with the Greek equivalents of the more significant terms indicated in parentheses:

As for that which is of necessity (ἀνάγκης), does it exist by hypothesis (ἐξ ὑποθέσεως) or also simply (ἁπλῶς)? Nowadays it is thought that what exists by necessity does so in generation (ἐν τῇ γενέσει), as if one were to consider the wall as having been constructed by necessity, since what is heavy is carried down by its nature and what is light is carried up by its nature, and so the stones and the foundations are down, then earth right above because it is lighter, and finally wood at the very top since it is lightest. However, although a wall is not constructed without these, still it is constructed not because of these — except in the sense that they are causes as matter (ὕλην) — but for the sake of (ἀλλ' ἕνεκα) sheltering and preserving certain things. Similarly, in all other cases in which there is a final cause (ἕνεκα), although what is generated could not have been generated without the nature (τὴν φύσιν) that is necessary for it, still it is not because of what is necessary — except as a material cause (ὕλην) — but for the sake of something (ἀλλ' ἕνεκά του). For example, why is a saw such-and-such? So that this may come to be or for the sake of this. But this final cause (ἕνεκα) cannot come to be unless the saw is made of iron. So if there is to be a saw capable of doing this work, it is necessary that it be made of iron. What is necessary, then, exists by hypothesis (ἐξ ὑποθέσεως) and not as an end (τέλος); for it exists in the matter (ἐν τῇ ὕλῃ), while the final cause (ἕνεκα) is in the reason (ἐν τῷ λόγῳ).[10]

---

[10] *Physics* II.9 (199b33-200a16).

In this passage Aristotle is distinguishing between two types of necessity, one conditional and the other absolute, as is clear from the first sentence. The remainder of the discussion is apparently making the point that Aristotle's contemporaries believed necessity in nature to be absolute, because a natural thing acts in ways that are determined by its material cause, i.e., by the matter out of which it is made. Aristotle seems to take exception to this view, however, and holds instead that necessity in nature is conditional, and in some way related to the final cause, i.e., to that for the sake of which the natural operation comes about. He concludes his argument, then, with the puzzling statement that what is necessary exists by hypothesis, but "not as an end, for it exists in the matter, while the final cause is in the reason."

The passage as a whole is cryptic, and both translators and commentators have puzzled for centuries over its true meaning. For our purposes it may be sufficient to give the Greek text of the first and the last sentences, and then illustrate the diversity of translations of these two sentences into medieval Latin and contemporary English. The Greek reads as follows:

Τὸ δ' ἐξ ἀνάγκης πότερον ἐξ ὑποθέσεως ὑπάρχει ἢ καὶ ἁπλῶς; . . . ἐξ ὑποθέσεως δὴ τὸ ἀναγκαῖον, ἀλλ' οὐχ ὡς τέλος· ἐν γὰρ τῇ ὕλῃ τὸ ἀναγκαῖον, τὸ δ' οὗ ἕνεκα ἐν τῷ λόγῳ.[11]

In the second half of the twelfth century these sentences were rendered into Latin by James of Venice as follows:

Quod autem ex necessitate est utrum ex conditione sit aut et simpliciter. . . . Et opus ipsius ex suppositione necessarium est, sed non sicut finis; in materia enim necessarium est, quod autem est cuius causa fit, fit in ratione.[12]

Note here that James is not consistent in his rendering of the Greek ἐξ ὑποθέσεως , but translates it the first time as *ex conditione* and the second as *ex suppositione*. In another Latin translation, however, which usually accompanied Averroës' Great Commentary on the *Physics* and is generally attributed to Michael Scot, neither of these two Latin expressions occur, but rather two others, namely, *a*

---

[11] Ibid.

[12] Ibid.; for the Latin text we have used the *editio princeps* of Averroës' commentary on the *Physics* (Padua, ca. 1472-1475), since no critical edition is yet available.

positione and *ex positione*. The latter translation was made from an Arabic text that had previously been translated from the Greek, possibly also via a Syriac version, and reads as follows:

> Et considerandum est de necessitate utrum sit a positione aut simpliciter. . . . Ex positione igitur erit necessitas, non ex fine intenta; necessitas enim est in materia, et illud propter quid est in diffinitione.[13]

Apart from the differences in translating ἐξ ὑποθέσεως , there are other noteworthy changes, such as that from *sicut finis* to *ex fine intenta*, that from *cuius causa* to *propter quid*, and the different ways of rendering λόγος as *ratio* and *diffinitio* respectively.

With such a diversity to work from, it is not surprising that vernacular translators have failed to arrive at a consistent reading. To our knowledge there are now five different English versions, all of which take one or other liberty with the text and none of which is completely clear and unambiguous in its meaning. These five are the following:

(1) Hardie and Gaye:

> As regards what is "of necessity," we must ask whether the necessity is "hypothetical," or "simple" as well. . . .What is necessary then, is necessary *on a hypothesis*; it is not a result necessarily determined by antecedents. Necessity is in the matter, while "that for the sake of which" is in the definition.[14]

(2) Wicksteed and Cornford:

> The phrase "must of necessity" may be used of what is unconditionally necessary or of what is "necessary to this or that.". . . The necessity, then, is conditional, or hypothetical. The purpose, mentally conceived, demands the material as necessary to its accomplishment; but the nature of the material, as already existing, does not "necessarily" lead to the accomplishment of the purpose.[15]

(3) Hope:

> In what sense, then, does anything happen "necessarily"? "Conditionally" [in subjection always to ends]? Or also [without any reference to

---

[13] Ibid., again using the *editio princeps* cited in the previous note.

[14] Aristotle, *Physica*, translated by R. P. Hardie and R. K. Gaye, in *The Works of Aristotle Translated Into English*, ed. W. D. Ross, vol. 2 (Oxford, 1930) 199b.

[15] Aristotle, *The Physics*, translated by Philip H. Wicksteed and Francis M. Cornford, 2 vols., The Loeb Classical Library (Cambridge, Mass., 1957), vol. 1, pp. 179-181.

ends, and thus] unconditionally? . . .Necessity, then, is hypothetical, but not as an end. In other words, necessity is in the *material*; the end is in the "logos."[16]

(4) Apostle:

As for that which is necessary, does it exist by hypothesis or also simply? . . . What is necessary, then, exists by hypothesis and not as an end; for it exists in matter, while final cause is in the formula.[17]

(5) Charlton:

Is that which is of necessity, of necessity only on some hypothesis, or can it also be simply of necessity? . . .The necessary, then, is necessary on some hypothesis, and not as an end; the necessary is in the matter, the "that for which" in the account.[18]

Among these translations, Apostle's is the most literal, and is substantially the text reproduced in its entirety above. Suffice it to add that these five different English versions have literally hundreds of counterparts in medieval and Renaissance Latin commentaries, to say nothing of translations into other vernaculars.

Apart from this classical locus in the *Physics*, Aristotle also touches on the matters that relate to methodology in natural science when elaborating his doctrine of demonstration in the *Posterior Analytics*. Some of these references occur in the first book, where general questions are asked: whether demonstrations must always concern incorruptible and eternal things, and also whether there can be demonstration of things that fall under sense knowledge or of merely fortuitous events. In the second book there are passages that relate to demonstrations made through a number of causes, and the ordering of causal explanations among themselves, which supply a framework in which the discussion at the end of the second book of the *Physics* may be located. Also of methodological interest is Aristotle's treatment of cases involving a temporal interval between cause and effect, and the possibility of demonstrating future events and things that happen only for the most part. Finally, in the first book of *De partibus animalium* Aristotle repeats the canons given for demonstrating in natural science in the *Physics*, and works out in detail the import of such canons for developing a science of zoology.

---

[16] *Aristotle's Physics*, newly translated by Richard Hope (Lincoln, Neb., 1961), p. 39.

[17] *Aristotle's Physics*, translated with commentaries and glossary by Hippocrates G. Apostle (Bloomington, Ind., 1969), p. 40.

[18] *Aristotle's Physics I, II*, translated with introduction and notes by Walter Charlton, Clarendon Aristotle Series (Oxford, 1970), pp. 42-43.

Thomas Aquinas did not comment on the *De partibus animalium*, and thus it is impossible to know what he would have made of the specific instructions laid down there for the study of animals. He did comment, however, on both the *Physics* and the *Posterior Analytics*, and in both of these expositions he worked out a consistent interpretation of Aristotle's methods for demonstrating in the *scientiae naturales*.[19] The central technique, as already mentioned, he identifies as one of demonstrating *ex suppositione finis*. The necessity that characterizes *scientia naturalis*, in Aquinas' understanding, is not so much an absolute necessity (although some types of explanation may involve a necessity of this type), as it is a conditional necessity, which may be understood as the demand for whatever may be required to achieve a certain end. The reason for the latter is that nature is contingent in its operations, or stated otherwise, that natural things come into being through changes that do not always occur invariably. Nature does act for an end, but the agents it employs and the materials with which it works can be defective, and thus it is not completely determined in its operation. Because this is so, one cannot argue from prior causes to the effects they intend to produce; rather, one must proceed in the reverse direction and, on the basis of the effect to be realized, reason back to the causes that will be entailed in its realization. Aquinas illustrates this with the example of the olive tree, for from the fact that one plants an olive seed he cannot be certain that a fully developed tree will be generated from it, whereas, on the supposition of the olive tree's existence, he can reason back to the strict necessity of an olive seed.[20] The example of the house is similar, in the sense that although all the materials that go to make it up dictate necessities by reason of their being different types of matter, the purpose for which the house is being built, which is reflected in the plan of the builder, is the final cause that dictates why the materials come to be arranged in the way in which they ultimately are.

Aquinas' interpretation, which bases the necessity on the end or final cause, may not seem to agree with the last sentence of the passage from Aristotle cited above in its various versions and translations, but it serves to explain rather well how there can be universal

---

[19] See the texts referenced in Wallace, "Aquinas on the Temporal Relation Between Cause and Effect," pp. 572-573.

[20] *In 1 Anal. post.* lect. 42, n. 3; for the English translation, see St. Thomas Aquinas, *Commentary on the Posterior Analytics of Aristotle*, translated by F. R. Larcher (Albany, 1970), p. 148.

and necessary knowledge of the world of nature, even though in the individual case a particular result will not be attained. It also allows the possibility of reasoning from a cause to an effect that is not fully achieved until after some time has elapsed, and from a cause that, because of something that may have happened in the interim, may never produce the intended effect. On the supposition that the effect is to be produced, the prior causes are universally necessitated, and the requirement for their existence could not be otherwise — with the result that all the demands of a strict *scientia* are satisfied. It is further noteworthy that Aquinas, when commenting on the *Posterior Analytics*, explicitly refers to the second book of the *Physics* and the techniques there outlined for demonstrating *ex suppositione finis*, as providing the answers to difficulties involving both defective and time-dependent causality.[21]

This brief sketch of Aristotle's and Aquinas' teaching, truncated though it is, provides a setting in which Albert's teaching on demonstrative methodology in *scientia naturalis* can be explained. A convenient way of doing this is first to take up a question that is not in either Aristotle or Aquinas but is explicitly raised by Albert, namely, whether it is possible to have a science of nature. This will lead to an analysis of Albert's teaching on the finality of nature and the way this entails a conditional necessity in its operation. A third section will then explain how Albert himself envisaged the application of this methodology when elaborating the part of natural science dealing with animals.

B. ALBERT ON THE POSSIBILITY OF A *SCIENTIA NATURALIS*

At the very outset of his exposition of the *Physics* Albert raises some preliminary queries, among which is the question *Utrum sit scientia de physicis, vel non?*[22] The query, he says, is prompted by the objections of Heraclitus and his followers, who bring three difficulties against the possibility of a science of natural things. The first is that natural entities exist in an infinite variety of ways, and so cannot be comprehended by the human intellect. The second argument denies the possibility of obtaining definitions that can serve as middle terms, and thus rejects the possibility of achieving strict demonstrations such as a *scientia* would require. And the third objection rests on the basic instability of natural forms, which are constantly in

---

[21] Ibid.; see also *In I Anal. post.* lect. 16, n. 6 (tr. Larcher, pp. 54-55).

[22] *Phys.* I, tr.1, c.2 (ed. Borgnet 3: 4b).

motion and never remain in the same state, and so cannot be the object of a scientific demonstration, which deals only with things that are unvarying and necessary. Here Albert interjects the remark that Ptolemy was persuaded by the last argument to believe that there could not be a science of nature, but only opinion about it, as witnessed by the diversity of opinions among naturalists, which is far from the agreement found among mathematicians. Albert does not side with Ptolemy, however, but states his own position unequivocally: "We, however, say that there is *scientia* and demonstration of physical things, because they have subjects and properties, and principles through which attributes can be proved of their subjects."[23]

In taking this stand, Albert has effectively rejected the Platonist position as well as the Heraclitean, as becomes apparent from the ways in which he replies to the objections. As to the first, he is quite willing to admit the force of Heraclitus's objection if the work of nature is to produce variety in individuals, but this is not its task, which is rather the production of things complete in natural essentials, and these are nothing more than the species that the natural scientist studies. In Albert's view it is the "complete entity (*ens completum*) that is intended by nature; this is finite, and is made so by its essential causes taken in a real sense, which are form and matter, and [grasped] through the moving cause, which is the efficient agent, and through the cause to which the motion tends, which is the end."[24] Albert handles the second objection in a similar way, holding that univocal definitions can be given of essential species, even though there can be a vast multitude of individual differences that arise from the dispositions of matter, but which are never the principal result intended by nature in its operation.[25] And this, in turn, provides an answer to the third objection, for natural science achieves its necessity through abstraction from individual matter, and this yields a universal concerning which there can be necessary knowledge.[26]

It is noteworthy that throughout his treatment of this question,

---

[23] Ibid. (3: 5b). Elsewhere Albert cites Ptolemy on this point with approval; see *Metaph.* I, tr.1, c.1 (ed. Colon. 16/1: 1 line 27 and note), and below, pp. 122-126. The writer discusses the apparent inconsistency and its relation to Albert's methodology in his "The Scientific Methodology of St. Albert the Great," forthcoming in the *Albertus-Magnus Festschrift* to appear in 1980 under the editorship of Gerbert Meyer, OP.

[24] *Phys.* I, tr.1, c.2 (ed. Borgnet 3: 5b).

[25] Ibid. (3: 6a).

[26] Ibid. (3: 6a-b).

Albert does not explicitly use the expression *ex suppositione* or enter into details of the demonstrative process in a natural science, although his answer to the first Heraclitean objection may be seen as implicitly involving this doctrine. Much the same can be said for Albert's treatment of problems relating to demonstration as these are taken up in his exposition of the *Posterior Analytics*. To our knowledge he does not discuss conditional necessity in that work, or use the expressions *ex conditione* and *ex suppositione* as these are employed in Book 2 of the *Physics*. For example, when discussing whether definition and demonstration must be of "incorruptibles," Albert reviews the opinions of Alfarabi, Themistius, and Alexander of Aphrodisias. He also gives indication of having perused Robert Grosseteste's commentary on the *Analytics* to see how he handled the question, and gives his own solution in terms that are not very different from Grosseteste's.[27] A lunar eclipse, when referred to the moon as its proper attribute, is not always occurring and thus is not a universal and an incorruptible; when referred to the causes that produce it, however, there is always the "universal eclipse," and this is "always necessary since it results from the orderly motion of the sun and the moon. . . ."[28] Similarly, when explaining why there cannot be strict demonstration of fortuitous events and of things that fall directly under sense knowledge, Albert again has recourse to universals as his way out of the difficulty that this poses for developing a *scientia naturalis*. He does attribute a significant role to sense knowledge, however, allowing that it is only because we observe eclipses occurring at different times and under different circumstances that we are able to discover the universal explanation that makes a demonstrative science of eclipses possible.[29]

Albert comes closer to the *ex suppositione* doctrine when discussing in the *Analytics* the cases of multiple causality such as those that concern the naturalist, and also the problem of time interval between cause and effect. Here he stresses that arguments from efficient cause and from final cause are appropriate in both the natural and mechanical sciences, since the end is what is principally intended in these disciplines.[30] Albert also cites Aristotle's example of the freez-

---

[27] See W. A. Wallace, *Causality and Scientific Explanation*, 2 vols. (Ann Arbor, 1972), 1: 66-67.

[28] *Post. Anal.* I, tr.2, c.17 (ed. Borgnet 2: 65a).

[29] Ibid., tr.5, c.7 (2: 143b).

[30] *Post. Anal.* II, tr.3, c.4 (2: 202a-b).

ing of water caused by the absence of internal heat, and acknowledges this as a type of demonstration wherein cause and effect are
simultaneous. If the cause and the thing caused are not simultaneous, however, then Albert admits that this circumstance places a limitation on the way in which one can reason about them. He makes
the statement that "when the cause itself is posited the thing caused
is not posited *de necessitate*, but conversely when the thing caused is
posited then the cause must be posited *de necessitate*."[31] Thus, when
an effect is to occur later in time, the principle of demonstration
must be taken from what is to be achieved later, rather than from the
earlier cause, even though this may rightly be regarded as the
principium essendi of the effect.[32] Throughout these discussions, however, Albert makes no reference to the *Physics*, nor does Aristotle's
text on which he is commenting, although their statements are obviously dictated by the type of problem encountered in the *scientiae
naturales*.

C. ALBERT ON FINALITY AND SUPPOSITIONAL NECESSITY

Albert's explicit treatment of suppositional necessity, not surprisingly, is located in his exposition of the second book of the *Physics*,
where he devotes his third treatise to the problem of nature's acting
for an end and the necessity that this imposes on its operations.[33]
When commenting on the *Physics*, moreover, Albert goes into more
detail than he does in his exposition of the *Posterior Analytics*. His
Physics commentary has more the character of a *postilla*, wherein
effectively he gives a continuous reading of Aristotle's text, while
interjecting his own explanatory phrases and illustrations, and occasionally interpolating an extensive digression on a related subject
matter. In giving the text of Aristotle, moreover, Albert does not pre-
tend to make a literal translation, but rather seems to oscillate
between the alternate readings ascribed to James of Venice and
Michael Scot. This circumstance of his composition enables us to
ascertain the precise sense he gives to the passage from Aristotle
cited above with its various translations. Before coming to that, however, first a few remarks on his understanding of nature's finality.
    Those who deny that nature acts for an end, Albert begins, do so

---

[31] Ibid., c.6 (2: 203a).
[32] Ibid. (2: 203b).
[33] *Phys.* II, tr.3 (ed. Borgnet 3: 162-176).

because they are convinced that the only necessity found in the world of nature is that deriving from the matter of which natural things are composed.[34] This leads them to ascribe everything that happens in nature to chance, and to say that the use of the various parts of animals, for example, follows from the way in which these parts are arranged and not from any utility that guided their formation.[35] Albert rejects this teaching, affirming that those who regard everything in nature as arising from chance are just as much in error as those who wish to eliminate chance events completely and maintain that everything in nature is absolutely determined. For Albert, as for Aristotle, the chance occurrence is a reality in nature that must be taken into account, but it is something of rare and infrequent occurrence, and this enables one to differentiate between it and what is from nature. Natural processes occur regularly and for the most part, whereas chance events do not. Moreover, nature itself can be identified with either the matter or the form, and it is the form that is attained regularly and for the most part in natural processes that is actually the final cause, that for the sake of which natural things come to be.[36]

Albert argues for this interpretation of nature's activity by comparing it with activities that arise from the mechanical arts, themselves also directed to the attainment of the ends intended by the artificer. Following Aristotle, he points out that it is only the existence of a plan or purpose that makes possible a mistake or error in works of art. An analogous case can be made for nature, he says, where the occasional production of monsters or defective organisms is an indication of nature's failure in its purposive effort. Here Albert has an extensive digression on the ways in which monsters originate in the animal kingdom, pointing out the great variety of material and other indispositions that give rise to their occurrence. He observes that there are fewer monstrosities among plants than there are among animals because the seeds of the latter are softer and the mechanisms for their development much more complicated than those that are found in the plant kingdom.[37] Albert also expounds Aristotle's doctrine that purpose can be present in an activity even though there are no signs of the agent's deliberating, contrary to the view of Empedo-

---

[34] Ibid., c.1 (3: 162a).
[35] Ibid. (3: 163a).
[36] Ibid., c.2 (3: 165b).
[37] Ibid., c.3 (3: 169b).

cles, and goes on to dispose of the latter's objections against nature's acting for an end.[38]

This brings Albert to chapter 9 of Book 2 of the *Physics*, and to the passage cited above, which is identified in Albert's exposition as corresponding to Texts 87 and 88 in the Great Commentary of Averroës. As already noted, Albert stays rather close to the text of Aristotle, but interweaves his own remarks and *obiter dicta*, all the while giving a continuous development of Aristotle's doctrine. To enable the reader to discern when Albert is giving Aristotle's text (in the Latin translation made either from the Greek or the Arabic, which he appears to use interchangeably), and when he is interpolating his own clarifications, we shall use a difference of type face when translating his exposition. Thus, passages in italics give the translation of Aristotle, which is usually not literal but can be identified on close comparison with the more literal translation made from the Greek text and indicated above on p. 106. Passages in roman type, on the other hand, indicate Albert's composition. With this understood, Chapter 9, at Text 87, begins as follows:

> We ask therefore first *whether the necessity of physical things is a necessity simply or is a necessity "ex suppositione"* and on the condition of some end that is presupposed. For example, a simple necessity is such that it is necessary that the heavy go down and the light go up, for it is not necessary that anything be presupposed to this for it to be necessary. Necessity "ex conditione," however, is that for whose necessity it is necessary to presuppose something, nor is it in itself necessary except "ex suppositione"; and so it is necessary for you to sit if I see you sitting. For there is a simple necessity in the aptitude and necessity of matter considered alone. But necessity "secundum positionem" is based on some kind of hypothesis, such as it is necessary that you sleep if your sensible powers are to be brought to rest within you.[39]

This, it will be recognized, is the first sentence of the chapter, and one can see that Albert, following the two translators, uses *suppositio, conditio,* and *positio* interchangeably to translate the Greek ὑπόθεσις. He then goes on to describe how some have thought that nature's operation was characterized by an absolute necessity alone:

> *Now indeed it was an opinion* of the ancients who thought that all things

---

[38] Ibid., c.4 (3: 170-171).

[39] Ibid., c.5 (3: 172a); cf. Aristotle, *De interpretatione* IX (18b1-2) and Boethius, *De consolatione philosophiae,* 5.6.

happen by chance *that in nature there is only an absolute necessity* that arises from the demand of matter. Just as if one were to say that it is not "propter suppositionem finis" but on account of the demand and aptitude of matter that a house comes to be, *because one thinks that the wall is made* and erected not so that it will support the roof but *because* the wall is composed of different [materials]. Of these *some are heavy, and so it is necessary that they will go down to the foundation at the bottom*, from the fact that their nature disposes them so to move. *Certain others are light* in intermediate fashion, and so they cohere with the heavier [materials] in the lower parts, *and extend upwards* touching the higher, and so the expanse of the wall comes to be. *For the stones are carried downward and make the foundation, and the woods* being of intermediate lightness go upward, and *the lightest of all are on the top.* Thus they say that there is a motion of components in the composite, and so from the necessity of the matter figures and shapes that are compatible with the motions arise in them. Therefore it is apparent that according to them form follows the necessity of matter, and the matter is not required for the form in the way that matter would not be required by nature except on account of form; and this is absolute necessity, which arises from the demand of matter.[40]

Here Albert is following Aristotle's text, filling out the example of the house, emending it slightly by eliminating the reference to earth but retaining that to stones and wood, and finally presenting the argument in more generalized form. He then continues on, explaining the passage parallel to Text 88 in Averroes' commentary:

But their statement is not true, *for although form does not come to be without the necessary matter, it is not on account of the necessity of the matter that the form comes to be.* For thus matter would not be sought on account of form, but matter could have any form whatever that would result from the necessity of the matter's movements. So also a natural form would be subjected to chance, since it would not be intended by nature. But we say that form is not on account of this, *unless* one wishes to call *the cause of the necessity* that which disposes its subject, which *is matter.* Rather, as in works of art, all of the earlier come to be on account of the later, from the fact that the ends are produced later and the earlier are ordered to the end. Nonetheless the later are not without the earlier, for both are found in natural things, since the end in them is the form that serves as a principle of the entire process, on account of which all other things come to be and are. This indeed is obvious in a house, because that which is *the principle* of all

---

[40] Ibid. (3: 172a-b).

those things that are made in the house is what proposes itself as an effect to the one who makes the house, and this *is as a covering from rains in bad weather and as a storage place of valuables and security* for its contents. And on account of this end the matter of the house is sought and prepared and put together; and everything that comes to be in the house is made on account of this. So we said above that the end which is first in knowledge is last in operation and being, and is the cause of causes, because on account of its being the other causes cause what they do.[41]

This, as is apparent, is the alternate explanation provided by Aristotle, where the causality of the end is introduced. The remainder of the passage thereupon generalizes the explanation, gives further exemplification in terms of the purpose of a saw, and concludes with Albert's exegesis of the puzzling sentence with which Aristotle concludes the passage:

And just as it is in the house, *so it is in all other things that are made on account of something,* which is certainly their end. *For none of these indeed comes to be without the necessary matter,* and this being disposed to receive it, *but at the same time the end is never on account of the matter* or on account of its necessity, but conversely matter and its necessity come to be for the end — *unless* equivocation be allowed in the preposition denoting the cause, such that you say that *the cause is* the "sine qua non," and *the necessity of the matter,* which is the disposing cause in the subject in which the end has existence. *Whence if one inquires why a saw is of this kind,* that is, made of iron, *one would reply* that the form of the saw, which is toothed, requires such a matter, and *on account of this form* it is necessary to seek a matter of this type. And if it be further inquired *why* the teeth should have *irony matter* of this kind, the reply is *on account of* the cause, the ultimate end that is *the work of the saw,* and this is to divide wood. And although this work cannot be performed unless the teeth are of iron, nonetheless the task that is the ultimate end, for which reason the other things come to be, is not on account of the iron and the teeth, whereas conversely they are for it. *If therefore it is necessary that the saw be made of iron* and be toothed, *if it is to complete its work,* this does *not* mean *a necessity of the end* on account of which everything is done that serves to explain a saw, *but* it means *a necessity that is in the matter* and in its dispositions. *For the end is* not in the matter having its necessity, but rather *in the reason,* since this is what moves the artificer and is effectively the principle of the entire operation. Therefore there also flow from it the motion by which

---

[41] Ibid. (3: 172b).

the efficient agent acts and the necessity of the matter whereby it is prepared to achieve that end.[42]

From this account, one can see readily that Albert's interpretation of the passage is the same as Aquinas', and the sense of both is most clearly captured in the English translation of Wicksteed and Cornford. The latter version, unfortunately, is not so much a translation as it is a paraphrase of Aristotle's text. (Hardie and Gaye, as well as Apostle, are accurate in translating, but they leave the text ambiguous in meaning. Hope and Charlton, on the other hand, remove the ambiguity but in so doing they give the wrong sense to the concluding phrases of the last sentence.) Albert, it would appear, is trying to do what Wicksteed and Cornford would later attempt, while at the same time keeping all the words in Aristotle's text and interpolating others that clarify his meaning. The net result is somewhat clumsy, and in the view of the writer would have been more elegant had Albert rendered the last sentence in the passage along the following lines:

> *What is necessary, then, exists "ex suppositione" and not as an* absolutely determined *end; for necessity exists in the matter, while the final cause is in the reason* set out in the "suppositio," which explains why the matter is arranged as it is.

Albert's expression leaves out the explicit reference to necessity *ex suppositione* and is quite involuted, but there can be little doubt that this is the sense he wishes to convey.

The remainder of Albert's exposition of the second book of the *Physics* further substantiates what has just been said. He goes on to contrast the way in which necessity is found in mathematics with the way in which it is found in physics. Albert says that the ancients were thrown off by the fact that there is an absolute necessity in mathematics, which can also be seen as *propter suppositionem finis* in the sense that the *finis* or end of a mathematical demonstration is knowledge of the conclusion.[43] In mathematical reasoning, however,

---

[42] Ibid. (3: 173a).

[43] Ibid., c.6 (3: 173b-174a): "Est autem necessarium in disciplinis demonstrativis quod fuit Antiquis causa erroris, in quo est necessarium simpliciter propter suppositionem finis: licet enim scientia conclusionis finis in demonstrativis, tamen praemissa non sunt necessaria propter conclusionem, sed in se habent necessitatem, et propter necessitatem earum conclusio est necessaria. Et cum ipsa habeat necessitatem rei, non attenditur in ipsis *necessitas consequentiae* tantum, sed potius necessitas rei quae consequitur, quae vocatur a quibusdam *necessitas consequentis.*"

the premises are not necessary only on account of the conclusion but have a necessity in themselves. It is because of their necessity that the conclusion becomes necessary, and thus there is a twofold necessity in mathematical proof, namely, that of the consequence (*consequentiae*) or inference and that of the consequent (*consequentis*) or end result. In proofs in natural science, on the other hand, there is a necessity of consequence only, because the end result is never automatically assured. Yet there is a type of necessity that characterizes its demonstrations, and this is a *necessitas conditionis ex finis suppositione*.[44] This suppositional necessity is one where the end serves as a principle in the same way as the premises serve as a principle in a mathematical demonstration. That is the sense in which natural science exhibits a necessity that is *ex conditione finis*.[45] It serves also to explain why all four causes function in its demonstrations, and particularly the final cause, which is identified with the completed form and its defining characteristics that terminate a natural process.[46]

D. Albert on Demonstrations in Zoology

With this we have effectively answered our question about Albert's conception of the demonstrative methodology to be employed in the *scientiae naturales*. For the sake of completeness, however, it will be well to indicate other passages in his *De animalibus* where he also refers to suppositional necessity, and gives a fuller explanation of the methodological procedures associated with it. These passages reveal that, for Albert, a suppositional type of argument need not be strictly demonstrative but also may be extended into the realm of dialectic — a technique that proves useful in studying animal development, where apodictic certainty is not always attainable.

Book 11 of Albert's *De animalibus* is actually an exposition of Book 1 of Aristotle's *De partibus animalium*, which poses methodological questions relating to the science of zoology. For his exposition Albert made use of the Latin translation of Michael Scot, and

---

[44] Ibid., (3: 174b): "Sic ergo patet quod in disciplinis priora sunt principia sequentium, et ea quae sunt materialia sunt principia finis: sed in his quae fiunt propter aliquem finem, sive in artibus, sive in physicis fiant, e contrario est. Ibi enim (ut diximus) finis movet efficientem, et ab efficiente infunditur materiae motus, et sic finis est principium totius: et ideo est ibi *necessitas conditionis ex finis suppositione*."

[45] Ibid. (3: 174a).

[46] Ibid. (3: 175-176).

expanded it by inserting his own explanations, examples, comments, and digressions in the way we have already seen in his exposition of the *Physics*.[47] Book 11 is divided into two treatises, the first concentrating on the general procedures to be followed when studying animals, viz, those required for suppositional argument, and the second on a discussion of the form that terminates the development of animals and their organs, which should serve as the starting point for demonstrations *ex suppositione*. In what follows we shall concentrate only on the first treatise, and restrict ourselves to contexts where suppositional necessity is explicitly discussed.

The broad methodological questions raised by Aristotle at the outset of *De partibus animalium* include whether there can be a scientific exposition of animals, as opposed to the type of knowledge possessed by a person of general education; if so, whether such scientific knowledge should begin with discussions of particular types; what kinds of causes should be sought in the study of animals; whether the investigation will result in necessary knowledge; and, if so, what kind of necessity will be involved.[48] Albert, like Aristotle, embraces at the outset the ideal of strict scientific knowledge, and even notes a certain parallel between the astronomical and mathematical sciences and those dealing with animals. The former, he says, "first posit those things of which they inquire, such as the eclipse of the moon or of the sun, or that a triangle has three angles equal to two right angles, and afterward they add the causes of these properties, which are the middle terms of demonstrations."[49] Similarly, he goes on, the naturalist should consider the common properties and attributes of animals and proceed from this to an investigation of their causes. The proper procedure, therefore, is "first to recount everything that pertains to the manifest operations and properties of animals, as we have done in all ten of the preceding books, and now we ought to add the causes of those things that we have enumerated and that we said belong to the kinds of animals."[50]

The types of causes to be enumerated, this being a study devoted

---

[47] In what follows we use the Stadler edition, which indicates Aristotle's text and Albert's interpolations by a system of vertical lines inserted into the transcription.

[48] Aristotle, *De partibus animalium* I.1 (639a1-642b4).

[49] *De animalibus* XI, tr.1, c.2 (ed. Stadler, *Beiträge* 15: 765.15-18); the numerals appended to the page number after the decimal point indicate the line numbers in the printed edition.

[50] Ibid. (765.23-27). The first ten books of Albert's *De animalibus* recount the contents of Aristotle's *Historia animalium*; with this descriptive material completed, Albert begins its causal analysis in the eleventh book.

to the generation of natural beings — and here Albert references his prior discussion in Book 2 of the *Physics* — include all four causes, but particularly the final cause, "the cause of causes."[51] In the case of animals, he goes on, the final cause is identical with the form to be generated, and this is a sensible form defined in relation to sensible matter. The zoologist, in this respect, is like the physician who wishes to know the sensible dispositions of the human body so that he can restore it to health. So the naturalist must give his definitive account by looking to the ultimate end that is intended through the process of generation, and this along the lines already presented in the *Physics*.[52]

At this juncture, following the text of Aristotle, Albert broaches anew the different ways in which things are said to be necessary. Here again he enumerates two types of necessity, absolute and suppositional. The first, he states, applies to eternal things, "which exist apart from motion," and exhibit the necessity that is found in the demonstrative disciplines. The second is a conditional necessity, a necessity *per suppositionem finis*, and this is found in all changeable things that come to be by a process of generation. Not only is suppositional necessity found in these, however, but in all things that are ordered to an end, as is seen in the art of carpentry.

Up to this point Albert has been merely repeating what already appears in his exposition of the *Physics*. Now he sounds a new note, for, somewhat surprisingly, he equates knowledge of natural things with probable opinion, and so differentiates it from knowledge that is scientific and grasped through principles. He continues:

> For the necessity is *otherwise in* changeable things that are matters of opinion, such as are natural beings, and in scientific *matters that are intelligible and pertain to the demonstrative sciences. For we have already determined*, in the second book of our *Physics, that in certain things*, such as demonstrables, *there is a first that exists*, and that is the principle of others, as the premises are the cause and principle of the conclusion. *In other things, however*, as in generables and in matters of opinion that could possibly come to be otherwise, *the principle is* not what exists, for this is matter, but *rather* the first and the principle of all others is *what will exist* finally. And therefore everything that is necessary in these matters is necessary according to an intention that is presupposed, and not otherwise. . . .[53]

---

[51] Ibid. (765.28-38); see the text cited above, note 41.

[52] Ibid. (766.1-28).

[53] Ibid. (767.15-26).

Albert likewise stresses that one cannot attribute to an animal the necessity proper to an eternal thing, but can only seek in zoology a necessity of consequence (*consequentiae*) such that, "the last thing being given, it follows necessarily that the earlier exist or have existed, but not conversely. . . ."[54] In the demonstrations of mathematics, on the other hand, there is a necessity of the consequent (*consequentis*), whereby the thing entailed follows necessarily and absolutely as soon as the first thing is posited. And here again Albert refers the reader to the account in the second book of the *Physics*, where all these matters have been generally treated.

Apart from Albert's introducing the notion of opinion and probable reasoning into this discussion of the science of animals, the foregoing agrees substantially with his account of suppositional necessity in the exposition of the *Physics*. In his further development of this treatise, however, a difficulty in the text of Aristotle leads him to go deeper into the teaching on necessity, so as to differentiate two types of suppositional necessity that will be found to characterize the study of animals. The problem arises, Albert notes, because some say that the kinds of necessity noted in the *Physics*, absolute and suppositional, do not seem to be sufficient to account for all zoological phenomena, and so perhaps there is need for yet a third mode of necessity.[55] What requires attention, he goes on, is not so much the manner of seeking "the cause of the generation of animals, but rather [the way to investigate] the cause of the shapes of their members. . . ."[56] What this third mode is, Albert continues, can be discovered among the various modes Aristotle has already enumerated in Book 5 of his *Metaphysics*. Albert thereupon reverts to the text of the *Metaphysics*, remarking that he "will touch briefly on the modes there set forth."[57] These turn out to be four, in Albert's enumeration, and are described by him as follows:

  (1) that which is necessary for a thing's being or coming-to-be, and this is the suppositional necessity found in the generation of natural things;

---

[54] Ibid. (768.7-9).

[55] The English edition of *The Works of Aristotle*, vol. 5, has a confusing note at *De partibus animalium* I.1 (642a6, note 3) which creates the impression that the additional mode of necessity is not contained in Aristotle's *Metaphysics* v.5 (1015a20-1015b8). Albert correctly identifies this source and explains how the two texts can be understood so as mutually to illumine each other, as will now be explained.

[56] *De animal.* XI, tr.1, c.3 (777.7-9).

[57] Ibid. (777.14-17).

(2) that which is not required absolutely for a thing to exist, but which is necessary if it is to be good and perfect in its mode of existence or operation;

(3) that which is necessary because it results from a force that cannot be withstood — and this, Albert says, is needed in the moral sciences but not in those dealing with nature; and

(4) that which is absolutely necessary, which characterizes the mathematical disciplines, but is also required in the natural sciences "if we wish to syllogize" in these matters.[58]

Having listed these types, Albert points out that the person seeking a third mode of necessity should note that the first two modes have something in common, for both involve suppositional reasoning. The second mode, however, is not explicitly mentioned in the *Physics*, for there only the first and fourth modes have been discussed. This suppositional necessity of the first mode, moreover, is treated in the *Physics* only as ordered to the existence of the animal, and not to everything that would prove good or useful for it. Now, apart from such a consideration, one should note that if an animal is to use its members for their proper functions, it is further necessary that these be shaped or structured properly — a necessity that pertains to the second mode. The example Albert adduces is the animal's walking, for which it requires feet that have the right form or shape. To state the matter more generally, since the animal is an organism, i.e., a body made up of parts that serve various functions as instruments of the whole, there must be an end or final cause for each member, and this is nothing more than the operation or function it is to perform for the good of the whole. Possibly an animal could exist without one or other member, or even with a certain amount of malformation, but the necessity of the end that should govern its development as a whole will include the proper formation of all members that contribute to the fullness of its being. This requirement adds a second mode of necessity, above and beyond the first, which is proper to the biological sciences.

These, then, are the main methodological texts in Albert's *De animalibus* where he speaks of suppositional necessity. As noted, he also touches, but briefly, on this type of necessity in his exposition of Book 5 of Aristotle's *Metaphysics*.[59] In addition, Albert seems to have mentioned the topic in his disputed *Quaestiones de animalibus*, Book

---

[58] Ibid. (777.17-36).
[59] Tr.1, c.6 (ed. Colon. 16/1: 220b-222b).

11, for in his *reportatio* Conrad of Austria records the cryptic comment that a *propter quid* demonstration is given through causes, whereas a *quia* demonstration can be given either through effects or *ex suppositione*.[60] The topics mentioned in the *Quaestiones* are fragmentary, however, and add nothing substantial to what is already available in the longer *De animalibus*, or in the treatments of necessity in the expositions of the *Physics* and the *Metaphysics*.

E. ALBERT, AQUINAS, AND MODERN SCIENCE

With this we have provided most of the materials that bear on a solution of the problems posed at the beginning of this essay.

First of all, from the texts of Albert's expositions of Aristotle that have been cited, it now seems most likely that Albert was the proximate source from which Aquinas derived his knowledge of suppositional necessity and the manner of demonstrating in the *scientiae naturales*. Everything that later appears in St. Thomas' commentaries on the *Posterior Analytics* and the *Physics* is already contained in germ in Albert's exposition of the *Physics*, and is developed in yet fuller detail in the methodological canons he elaborates for the study of animals. This circumstance notwithstanding, it could well be that Aquinas' teaching, while not completely original with him, nonetheless played a greater role than did Albert's in the transmission of this methodological doctrine to the later Middle Ages and to the Renaissance. The fact that Albert does not explicitly discuss demonstration *ex suppositione* in his rather brief exposition of the *Posterior Analytics*, whereas Aquinas does, could serve to explain their relative importances for the doctrine's transmission. One can readily conceive that Albert's treatises were better known to biologists, mineralogists, and other investigators concerned with the detailed study of nature. Yet there seems little doubt that Aquinas was the authority who became better known in university circles, for here the *Posterior Analytics* and the *Physics* served as major textbooks for teaching logic and natural philosophy respectively. Since Aquinas discusses reasoning *ex suppositione* in both of these treatises, and connects the expositions in an explicit and meaningful way, he seems the more likely vehicle for its dissemination among later thinkers.

On the matter of Albert's Neoplatonic leanings and the possibility that he may have viewed the study of nature as pertaining more to

---

[60] Q.1 (ed. Colon. 12: 218.51-55).

dialectics than to science in the strict sense, a more nuanced answer would seem to be indicated. St. Thomas' discussion of reasoning *ex suppositione* stresses the requirements for *scientia* and demonstration, if for no other reason than that both contexts in which he discusses this type of reasoning are concerned with demonstrative methodology. On this account Aquinas can be clear and unambiguous in his presentation, and the question of dialectics and its relation to scientific reasoning need not even arise.[61] In Albert's case, on the other hand, his treatments of matters pertaining to the *scientiae naturales* do not remain at the general level of Aristotle's *Physics*, but are pursued in detail down to the *infima species* of animals, plants, and minerals. Surely Albert entertained no doubts that at the general level one could have certain and apodictic demonstrations even when treating of animals, provided the proper norms of *ex suppositione* reasoning were observed. When, on the other hand, he had to broach detailed considerations relating to the shape or structure of the members of animals, and particularly the shapes that might facilitate their characteristic operations, he was aware that it might not be possible to achieve apodictic certainty. In such cases, therefore, scientific conclusions would have to yield gradually to matters of opinion, and there could be a merging of demonstrative and dialectical results. Yet the mere fact that Albert admits "matters of opinion" into his detailed investigation of animals should not be construed as the abandonment, on his part, of the Aristotelian ideal of *scientia*. His own statements in both the *Physics* and the *De animalibus* in favor of this ideal are too straightforward to permit any misinterpretation in this regard.

Precisely how Albert's overall methodology stands in relation to that of modern science is a question that cannot be quickly answered. On the historical side, and particularly when one considers the type of hypothetical reasoning invoked by Galileo, which he referred to as demonstration *ex suppositione*, it seems to the writer that this relates more readily to the teaching of Aquinas than to that of Albert. Having said this, however, we should note that Galileo manifests an acquaintance with Albertus Magnus in his early notebooks, and might even have been exposed to Albert's biological

---

[61] See, however, John A. Oesterle, "The Significance of the Universal *Ut Nunc*," *The Thomist* 24 (1961), 163-174, for a discussion of a text in Aquinas' commentary on the *Posterior Analytics* (*In 1 Anal. Post.*, lect. 9, n. 4) that shows an awareness of dialectical argument and its relation to demonstration.

teachings while a medical student at Pisa.[62] The majority of his citations of both Albert and Aquinas that derive from verifiable sources, however, are traceable to the Jesuits at the Collegio Romano.[63] In this faculty, although the Jesuits did treat the mathematical disciplines and dwelt in considerable detail on the *Physics*, the *De caelo*, the *De generatione*, and the *Meteorology*, there is no evidence that they ever taught the *De animalibus*.[64] Again, since methodological doctrines were treated in the logic course rather than in that on natural philosophy, and Aquinas' commentary on the *Posterior Analytics* entered prominently in the former, it seems that Aquinas, as noted above, is a more likely source of Galileo's knowledge of the techniques of *ex suppositione* reasoning than is Albert the Great.

To come finally to the substantive problems that are being treated in contemporary philosophy of science, it would seem that a rediscovery of Albert's methodological views could contribute substantially to the solution of problems that dominate discussions in present-day literature. Most of these problems arise from an attempt to equate the necessity found in the natural or physical sciences with that of the mathematical disciplines. Profoundly aware that there can be no absolute necessity in nature, but unaware of the possibility of discerning a suppositional necessity in nature's operation, contemporary philosophers have retreated immediately to the dialectics of hypothetico-deductive methodology as the only adequate account of "scientific method." Again, they revere David Hume because he was the first in their eyes to show how readily one can be deceived when postulating necessary connections in nature. Now Albert, like Aristotle before him, never pursued necessary connections of the Humean type; yet neither, on that account, abandoned the search for any necessity whatever in nature's operations — and so they were not tempted to locate such necessity, as Hume was later to do, in men's minds or expectations alone. It was precisely this lack of subtlety on Hume's part that created for him the so-called "problem of induction," which still continues to generate a substantial literature.

---

[62] See W. A. Wallace, *Galileo's Early Notebooks: The Physical Questions* (Notre Dame, Ind., 1977), p. 315, for references to Galileo's sixteen citations of Albert's works.

[63] Ibid., pp. 1-24; for fuller details, see W. A. Wallace, "Galileo's Citations of Albert the Great," forthcoming in a commemorative issue of *The Southwest Journal of Philosophy* to appear in 1980.

[64] Details of the curriculum at the Collegio Romano are given in R. G. Villoslada, *Storia del Collegio Romano dal suo inizio (1551) alla soppressione della Compagnia di Gesù (1773)*, Analecta Gregoriana 66 (Rome, 1954).

For, if there is no necessity of any kind in nature that causes one state of affairs to be entailed by previous states, then certainly a person can never make a valid universal generalization of the type "All crows are black." Yet, by one of those ironies that continue to amuse historians, today medievals are commonly thought to have been very naive in this matter, and to have taught that one could make such a generalization on the basis of a simple enumeration of instances. It is surely enlightening to learn that Albert the Great, who had considerable knowledge of crows, would never make the unqualified statement that "All crows *are* black."[65] He was well aware that exceptions could occur in the generation of crows as in that of other animals, and indeed attempted to explain precisely under what circumstances of egg formation a non-black crow might be produced.[66] But on the supposition that a crow was to be generated with black feathers and the other attributes that characterize its species, he felt confident that he could enumerate the causes that would be required for its generation. In other words, he would see Hume's problem of causation and his problem of induction as pseudo-problems created by a mistaken concept of the necessity to be sought in the *scientiae naturales*. And once a person understood, as Albert did, the complex causality involved in natural operations, and the way in which one has to reason *ex suppositione finis* in order to discern the necessity that makes a science of nature possible, he would be well prepared to shed light on the problems that seem endlessly to perplex philosophers in the present day.

Seven centuries separate us from this *Doctor universalis* who played such an important role in the genesis of the scientific mentality in the Latin West. We honor him in this volume precisely because of his contributions in all areas of research activity. Perhaps it is time that we begin to see him also as a methodologist who had a precise and nuanced knowledge of both the ideals and the limitations of natural science — as one who, to use the modern idiom, merits the title of "philosopher of science" *par excellence.*

---

[65] See Albert's *De praedicabilibus*, tr.7, c.2 (ed. Borgnet 1: 122a-b); tr.8, c.10 (1: 140a).

[66] Ibid., tr.7, c.2 (1: 122).

5

St. Albert on Motion as *Forma fluens* and *Fluxus formae*

Ernest J. McCullough
St. Thomas More College

Etienne Gilson once said that in spite of the fact that Albert the Great is famous he remains little known.[1] The many conflicting interpretations of Albert's works indicate at the very least that a unified understanding of Albert is difficult to achieve. A variety of sources claim that Albert is original, unoriginal, eclectic, a devoted Aristotelian, Avicennian, Averroist, or some combination of these. Historians rate him as both confused and perceptive.[2] Legends obscure an already clouded picture. It is the aim of this paper to examine in detail a rather short but vital section of Albert's *De motu*, which is his commentary on Aristotle's *Physics* III, 1-3. This examination should reveal the precise nature of Albert's doctrine on motion and why the first formulation of a distinction between motion as *fluxus formae* (purportedly an Avicennian view) and motion as *forma fluens* (purportedly an Averroist doctrine) is attributed to Albert. This is the view taken by the eminent historian Anneliese Maier.[3]

---

[1] E. Gilson, "L'âme raisonnable chez Albert le Grand," *Archives d'histoire doctrinale et littéraire du moyen âge*, 14 (1943-1945), 5: "Albert le Grand est beaucoup moins connu qu'il n'est célèbre."

[2] See, for example, George Sarton, *Introduction to the History of Science*, vol. 2, part 1 (Washington: Carnegie Institute, 1931), pp. 935-939.

[3] Anneliese Maier, *Die Vorläufer Galileis im 14. Jahrhundert* (Roma: Edizioni de Storia e Letteratura, 1949), pp. 9-25.

Father W. Wallace says, similarly, that Albert in his *De motu* summarizes the views of Avicenna and Averroës and "makes a number of distinctions which adumbrate the controversy over *forma fluens* and *fluxus formae* that was to arise in the fourteenth century."[4] More hangs on this historical question than the problem of accuracy in tracing historical lineage. To some extent Albert's ability as a perceptive commentator and analyst comes into question, when historians present him as the source of this distinction. Thus, the purpose of this paper is twofold: to consider his conceptual position in a critical way, and to deal with Albert as a possible source of the distinction.

The problem which arises from the position taken by Miss Maier is that close examination of the *De motu* provides little evidence that such a distinction was maintained or argued by Albert himself. Furthermore, it is questionable whether he considered that Avicenna or Averroës held simplified doctrines characterized by the technical phrases *fluxus formae* and *forma fluens*. Why, then, does she locate the source of these phrases in Albert despite the fact that Albert's work seems to provide little textual evidence for the claim?

In order to understand Albert's doctrine we turn briefly to Aristotle's *Physics* and to the commentaries of Avicenna and Averroës. We shall then examine Albert's treatment of three relevant and crucial questions arising from Aristotle, those regarding (i) the genus or the categorical status of motion, (ii) the relation of a general genus of motion to specific instances of motion, and (iii) the causal relationship in movement between the moved body and the mover. Finally, the concluding sections deal with the historical problem of relating the fourteenth-century distinction between *fluxus formae* and *forma fluens* to Albert's doctrine, and with the conceptual difficulties inherent in Miss Maier's analysis.

The issues with which this paper deals are of great philosophical importance, since Albert's analytic power is tested to the utmost in his interpretation and explanation of *Physics*, III, 1-3. This passage in Aristotle is one of the most interesting in the Aristotelian corpus, and one of the most debated. It is a short passage but one which effectively summarizes the entire *Physics*, which itself underpins all Aristotle's physical doctrines. *Physics*, Books III–V develop ideas from chapter 1; Book VI expands on chapter 2; and Books VII and VIII elaborate on ideas in chapter 3. Chapter 1 deals with the historical,

---

[4] W. A. Wallace, *Causality and Scientific Explanation* (Ann Arbor: University of Michigan Press, 1972), p. 68.

logical, and conceptual difficulties which are inherent in a general theory of motion; chapter 2 deals with the dialectical difficulties of relating the species of motion to local motion. Chapter 3 deals with causal relations. The very brevity and succinctness of Aristotle's text challenges commentators and forces them to make considered and supported interpretations based on a careful reading of the text. A detailed account of Aristotle's text is needed before examining Albert's theory.

A. ARISTOTLE'S *PHYSICS*

The *Physics*, in general, provides a sustained argument for a causal source of motion. Books I and II provide physical principles such as matter and form, nature and causal principles (formal, efficient, final, and material). Books IV–V treat key concepts in elucidating motion, such as infinity, place, the void, and time. Book V deals with logical classes, and differentiates change in being or *genesis* from local motion, qualitative change, and quantitative change. Book VI deals with dialectical difficulties, and Books VII and VIII treat causal argumentation and the crucial problems of eternal motion and an eternal First Mover. As we have said, Book III, chapters 1-3, underpins all the subsequent argument of Books IV–VIII.

In chapter 1 of *Physics* III, Aristotle places motion in the genus of a perfection, discusses the connection between the key terms "agent" and "patient" which follows from the active and passive aspects of motion, and relates the genus of perfection to the categories. He then places motion within four species: local motion, qualitative change, quantitative change, and becoming or *genesis* (*genesis* is later removed from motion considered as *kinesis*). Following this logical beginning, he formulates a definition of motion as "the act of that which is potential inasmuch as it is potential" (201a10-11) and defends this definition both inductively and deductively.

In chapter 2 Aristotle notes the difficulty in understanding motion because of its incomplete nature. In order to make the definition clear, motion is placed in the classification of act and given a second, more concrete definition as "the fulfillment of the movable *qua* movable, the cause of the attribute being contact with what can move" (202a7-8). This second definition, called by later commentators a "material definition," is more evident to the observer.

Chapter 3 raises a critical problem implicit in both definitions, whether to place motion in the active source or in the passive recipient. How is the relationship between patient and agent determined? Aristotle's analysis of this problem leads to a causal theory.

Three main difficulties, however, arise from Aristotle's analysis. The most obvious problem is that the division of motion into four categories (substantial change, quantity, quality and local motion) is not in agreement with the position taken in the *Categories* that motion is in the category of a *passio*.[5] In the *Categories*, the apparently incompatible views are held that motion is a post-predicament and hence in several categories,[6] but at the same time in one category, that of a *passio* or a suffering.

The second problem is to relate "motion," an equivocal concept (as its presence in four categories indicates), to different types of motion. How does one relate an imprecise and incomplete perfection to the concrete world and make the study of that relationship scientific?

The third problem is to place motion both within the mobile object and also in the cause of the object's motion. In other words, is motion in the mobile object alone, the moving cause alone, or in both? If motion is in both, how can an agent cause have the requisite of potentiality to motion?

The problems arising from Aristotle's rich account have forced his commentators to present original interpretations of his work. Avicenna and Averroës are the most important commentators, since their commentaries stimulated Albert's thoughts on the crucial issues. The texts which influenced Albert most were Avicenna's *Sufficientia* and Averroës' *Commentary on the Physics*.

B. AVICENNA'S *SUFFICIENTIA*

While Albert has been said to be a committed Aristotelian with Averroist tendencies, his style is much closer to that of Avicenna. Certainly Avicenna's commentaries indicate a very original mind at work on the difficult texts. Miss Maier speculates that Albert's commentary is, at some critical junctures, a paraphrase of Avicenna.[7] This view however, weakens the claim that Albert's theory is philosophically coherent and defensible. The Avicennian work on which

---

[5] Aristotle, *Categories* v (4a10-4b4); ix (11b1-7).

[6] Aristotle, *Categories* xiv (15a14-15b5).

[7] Maier, *Die Vorläufer Galileis im 14. Jahrhundert*, pp. 9-25.

Albert depended was the *Sufficientia*.[8] In the Preface to the *Sufficientia*, Avicenna makes it clear that this will not be a traditional commentary, when he states that he will not spend his life arguing over subtle disagreements in the tradition.[9] In Book I, he deviates from the Aristotelian order considering the causes and principle of motion. Book II deals with motion in general and Book III with the problem of corporeity. Physical studies, as Avicenna sees them, provide a less certain explanation of the key concepts of form and matter than does metaphysics.[10]

The precedence of metaphysical over physical insights is an important feature of Avicenna's doctrine of nature. In spite of this, however, physics provides a more detailed account of a material world characterized by motion and change. Book I of the *Sufficientia* deals with the necessary principles of nature: substance, matter and form, and cause. The subsidiary character of this discussion is made clear when Avicenna notes that efficient causes are either a preparing for or a perfecting of nature. The perfecting of nature, however, lies outside the order of nature. The preparing of matter for the reception of forms is the main function of nature.[11] The removal of an active and perfecting role for motion is crucial in Avicenna's philosophy of nature. The emphasis in natural science is on the receptivity to forms, a more passive role for nature than Aristotle would allow.

Avicenna's designation of motion as a *passio* or a suffering rather than a perfection or an action follows logically from his view that the principles of nature are preparatory for the work of intelligences. As Avicenna sees it, motion could be one of three things: a *passio* or a suffering of a mobile body, a causal force or action, or a metaphysical reality grounded in being and unity, the concept of which is applied univocally. The choice which Avicenna makes concerning motion itself is to regard motion as a *passio* related to various categories. He does not entirely agree with Aristotle's list of the categories involved[12] but still places motion in only four categories.

---

[8] For Avicenna the *Sufficientia* will be cited from the Venice 1508 edition of his works.

[9] *Avicenna, Sufficientia*, I, fol. 13ra.

[10] Ibid., 1.5, fol. 17ra.

[11] Ibid., 1.10, fol. 19ra: "Principium autem motus aut est praeparans aut est perficiens. Sed praeparans est id quod praeparat materiam sicut motus spermatis in permutationibus praeparantibus. Et perficiens est id quod tribuit formam constituentem species naturales et est extra naturalia. . . ." et seq.

[12] Ibid., II.2, fol. 25va.

Avicenna maintains that motion as a *passio* could be conceived in four ways: (i) as the middle between extremes, (ii) as a perfection (a nontraditional view), (iii) as a genus of which there are several species, and finally (iv) as one species which becomes another. Avicenna puts motion as a genus into four species — quantity, quality, location (*ubi*), and position[13] — all generally regarded as sufferings (*passiones*).

The concern in Aristotle's *Physics* III, 3 as to the causal character of motion, whether in the agent or patient, is resolved by Avicenna placing the efficient cause of motion beyond nature itself. The causal source is a source of being; in other words, it is metaphysical.[14]

The three major difficulties in Aristotle's text as to the logical status of motion, its relationship to its kinds, and its causal source, are thus solved by Avicenna. He reduces motion to the category of a *passio*; he places motion in four species within a single class; he locates the efficient cause of motion outside of nature.

C. AVERROËS' *COMMENTARY*

Averroës has a different response to these problems. On Aristotelian grounds, he is unwilling to remove the active and perfecting quality of motion from nature, since motion is defined both as a perfection and as a suffering or *passio*.

In his treatment of Aristotle's doctrine, Averroës notes that there are three ways in which motion can be considered.[15] It can be considered with respect to act and potency, with respect to the categories, and with respect to the relationship between mover and moved. The doctrine of the *Categories* that motion is a *passio* is more widely known. The doctrine that motion is found in the four categories of substance, place, quantity, and quality, determined by the terminal point of the movement, is the more acceptable view of the *Physics*.[16] The definition of motion, however, remains ambiguous unless it is located within a specific category. In order to remove the ambigui-

---

[13] Ibid., II.3, fol. 26va.

[14] Ibid., I.2, fol. 15rb.

[15] For references in Averroës the reprint of the Venice, 1572-1574, edition will be used, the *Aristotelis Opera Cum Averrois Commentariis.*

[16] Averroës, *Physics* III, Text comm. 4: "Via enim ad rem est aliud ab ipsa re, et secundum hoc fuit positum praedicamentum per se, et *iste modus est famosior, ille autem est verior* et ideo Aristotelis induxit illum modum famosum in praedicamentis et istum modum verum in hoc libro." (Italics added.)

ties, Averroës finally argues that motion should be located formally in the category of a *passio*, but materially in the four categories enumerated in the *Physics*.[17] Averroës does not resolve the ultimate classification of motion until he comes to his analysis of Book V and Book VII.[18] There he places motion definitively and formally in the category of a *passio*. The conclusion that motion is a *passio* is more easily understood in the light of the doctrine of act and potency in Averroës.

Being is divided, says Averroës, into what is actual and what is potential. Motion cannot exist outside these two classes. In these classes, every agent may be a mover but not every mover is an agent.[19] Furthermore, motion as actual or potential can exist as a suffering or a *passio* in one of four distinct categories. He favours the doctrine of the *Physics* that motion is in four categories, but the doctrine does not make clear what is the precise relation between acting and suffering. The Averroist solution provided in Book V, text 9 is that motion remains formally a *passio* and is materially in four categories.

D. ALBERT: THE GENUS OF MOTION

Albert, however, regards this solution of Averroës as an unacceptable answer to the problem. He considers that there is a logical difficulty of equivocation and a causal difficulty in the reconciliation among various kinds of movements. Albert's analysis of this question is an attempt to resolve these difficulties.

In the *De motu* (tr. 1, cc. 1-8), Albert follows the Aristotelian order, beginning with definitions, relating the definition to a concrete category, and setting out a causal theory. He adds further elaborations in order to deal with the problems raised by Avicenna and Averroës.

Chapter 1 of Albert's *De motu* opens with a general statement concerning the purpose of the work, in which he discusses the attributes commonly associated with the concept of motion in philosophical discussions: infinity, place, the void, and time. An outline of the general order of the three chapters in Aristotle follows in chapter 2 (of the *De motu*). This provides a formal statement of the three central

---

[17] Ibid., V, 9 Text comm. 9.
[18] Ibid., VII, Text comm. 5, 12, 10-20, 32-34.
[19] Ibid., III, Text comm. 3.

problems of the tractate on motion: the genus of motion, the species of motion, and the relation of mover and moved. Albert then proceeds to the argument itself. First he discusses the genus in which motion falls, namely that of perfection. Second, he treats the notion of potentiality, which is also important in the location of motion within a genus. In chapter 3 he discusses whether motion lies in a single category or in several. In chapter 4 he discusses Aristotle's definition of motion and offers inductive arguments, or arguments from common experience, to support the definition. Chapter 4 argues deductively for the accuracy of the definition. Chapter 6 argues against the alternative definitions of motion given by Plato and Pythagoreans. In chapter 7 he presents Aristotle's second definition, which he calls the "material definition."[20] Chapter 8 resolves the remaining critical problem: whether motion exists in the mobile subject or in the mover.

According to Albert, the three most critical questions in these chapters on motion are dealt with in chapters 2, 3, and 8. These questions are : (i) the genus of motion; (ii) the relation of the genus to the types and species of motion; (iii) the relationship between the mover and moved, and the place of motion in this relationship. In Albert's view, an understanding of these three questions is essential before the concept of motion can be understood properly.[21] The first point constitutes the principal matter of this section, while the two following sections of this paper are related to the two following key questions in Aristotle's text. It is in the handling of these three critical questions that Albert's philosophical capacity will be demonstrated, as will the extent of his divergence from Avicenna and Averroës who provide widely differing interpretations on these matters. The first serious question deals with the genus of motion.

In the introductory remarks to the *De motu*, Albert states that the purpose of the *Physics* is to render the concept of nature intelligible. But, he says, an understanding of nature is impossible unless motion is understood.[22] Towards this end, Albert examines the various shades of meaning of the word "motion."

---

[20] Albert, *Physica* III, 1, 7 (ed. Borgnet 3: 196b): "De secunda diffinitione motus, quae est quasi materialis magis quam prior." The "material" definition is one which is logically consequent upon the formal definition but restricted to the more sensible experience of motion in which there is direct contact between the mover and the moved.

[21] Albert follows Aristotle who discusses three requirements for scientific knowledge: a definition (*quid est*), a discussion of attributes (*qualis est*), and a determination of causal relations (*propter quid est*). Aristotle, *Posterior Analytics* II, 14-19 (98a1-100b18).

[22] Cf. Aristotle, *Physics* III, 1 (200b12-14).

The term "motion" is used in at least three ways. It can be said "to be something"; or "to be of something"; or, finally, "to be in reference to something."[23] From these ways of using the word "motion" the problem of genus, species, and relationship are isolated.

Concerning the genus of motion in the statement "motion is something" the statement means that motion exists or has a reality. This notion is important, since later philosophers, such as Ockham, argue that motion is not a reality over and above the "body" in motion.[24] For Albert, reality is of two kinds. There is either a completely perfected reality or reality having a mixture of imperfection, or potentiality. A third possibility, that reality is pure potentiality, is ruled out by Albert in his analysis of Book I, where he suggests that the term "reality" signifies a perfection or formal determination of some kind. Matter, the passive principle of nature, exists only in substance and does not have a reality independent of act.[25]

What place has motion in this division of reality into the actual or potential? What is its principal characteristic or proper genus? In order to determine this, Albert maintains that the meaning of terms such as "act" and "potency" must be made clearer by an analysis of the uses of the words, and by classification of the ways the words are used.

Neither Avicenna nor Averroës deals with the problem of act and potency in this manner. Avicenna simply states that certain things are in pure act; others are in act in some respect and in potency in some respect; nothing, however, is in pure potency.[26] Similarly, Averroës states that being is divided into actual and potential, but he does not analyse these terms.[27] Albert, in contrast, examines these

---

[23] Albert, *Physica* III, tr.1, c.2 (ed. Borgnet 3: 178b): "Primum autem, sicut diximus, de motu est intendendum, ad cujus notificationem oportet praemittere quaedam. Cum enim motus sit aliquid, et alicujus, et secundum aliquid, oportet nos primum praemittere tres divisiones, ex quarum prima accipiamus quid sit motus in genere, sicut in genere perfectionis et non potentiae: et ex eadem accipiemus secundum quid motus est perfectio, sicut secundum quod genus praedicamenti. Tertiam autem divisionem praemittemus ut sciamus secundum quem modum fluit motus a motivo in mobile, sicut in id quod movetur."

[24] Ockham, *The Tractatus de successivis attributed to William of Ockham*, I, ed. Philotheus Boehner (St. Bonaventure, N.Y.: Franciscan Institute, 1944), pp. 32-69.

[25] Albert, *Physica* I, tr.3, c.13 (ed. Borgnet 3: 78): ". . . materia non habet esse nisi in quantum est dispositio substantiae. . . ."

[26] Avicenna, *Sufficientia* II 1, fol. 23ra: "Debemus ergo prius agere de motu, et dicemus quod eorum quae sunt, quaedam sunt in actu omni modo, quaedam vero sunt in actu secundum aliquid, secundum aliquid in potentia, et impossibile est, ut aliqua res sit in potentia omni modo, quae non habeat esse in actu aliquo modo."

[27] Averroës, *Physics* III, Text comm. 3.

concepts at length in chapter 2, thereby showing his concern for the use of a precise terminology. This concern is characteristic of Albert and of scholastic philosophy in general while it was developing its own scientific methodology.[28]

Albert suggests that the word "perfection" may be understood in different ways. Perfection, in Albert's usage, includes a primary and a secondary sense, both divided three ways. In the primary sense, there is first perfection in those works which are produced gradually from an imperfect to a perfect state; second, there is first perfection in priority of *esse* or reality; and, third, there is first perfection according to priority of formal cause to its effect.[29] Each of these perfections has a related second perfection: to the first corresponds the end or term of motion; to the second correspond accidental forms; and to the third correspond activities (*agere*).[30] Motion is related to all three ways, but diversely.

In the *De motu*, motion is understood to be a first perfection of the mobile and refers primarily to that process from imperfection to perfection resulting from the act of an agent. Thus, while Albert's analysis distinguishes the various meanings of the term "perfection," he limits its meaning in this tractate to the first type, namely to that perfection achieved by the efforts of an agent. This type of perfection is a process which advances through time from an imperfect state.

Potentiality is the second important term in Albert's definition of motion and is understood in three ways: essentially, accidentally, and relationally. The potential in the essential sense is a potentiality to substantial form as in the case of prime matter, which is receptive to form.[31] An example of potentiality in the accidental sense is the potential blackness in a white object. The potential in the relational sense is seen in the notions of place (*ubi*) or local position (*situs*) through which bodies are related to other bodies.

---

[28] I. M. Bochenski, *A History of Formal Logic* (Notre Dame, Ind.: University of Notre Dame Press, 1961), p. 251.

[29] Albert, *Physica* III, tr.1, c.2 (ed. Borgnet 3: 179b): "Dicitur enim perfectio prima secundum prioritatem operis, et dicitur perfectio prima secundum esse, et dicitur perfectio prima secundum prioritatem causae formalis ad actum ejus."

[30] Ibid. (Borgnet 3: 179b-180a): "et omnibus his modis dicitur etiam perfectio secunda respondens primae perfectioni. . . .et secunda perfectio respondens ei est forma secundum quod accipitur in ratione finis et termini motus. . . .et huic respondet secunda perfectio quae est secundum accidens, sive secundum formam accidentalem aliquam. . . .et est perfectum secunda perfectione secundum agere."

[31] Ibid., III, 1, 1 (Borgnet 3: 180a): "est enim aliquid in potentia ad esse, sicut materia."

Having discussed the terms "actuality" and "potentiality" and their unity in motion, Albert now raises the question as to whether these two concepts can be located in a single genus of motion. He argues that it is thoroughly reprehensible to urge, as some do, that there is only one kind of genus of motion, namely local motion.[32] Some philosophers would argue that local motion is the measure of all other kinds of motion or, on the other hand, that motion is one in genus having many species, that it is a kind of logical genus containing subspecies.

In refuting the argument that motion is a single genus, Albert states that motion is present in diverse categories of being. If motion were in a single genus it would be both one genus and many genera. The Aristotelian position, as presented in the *Categories* (c. 14), is that motion, as a post-predicament, relates to several categories. Motion, thus, is a concept used analogously in application to six types of motion. Albert, therefore, concludes that the term "motion" is not used in a single or univocal sense, but in an analogical sense, in the same way that being is analogically said both of perfectly realized being or substance and of dependent being or accidents. Since motion is an analogous concept, it has more than one definition. It cannot, thus, be placed in one genus which can be predicated univocally of each kind of motion.

In order to clarify the meaning of motion, Albert argues for a clear distinction between the concept of perfection or act and that of potentiality. In doing so, he argues that motion must be considered in an analogical way not in a univocal nor in a metaphorical way. Act or perfection is recognized as a denomination of being itself, sharing something of its analogical character. The result of this analogical character of motion is that it can be defined in several ways. Albert suggests three definitions in the *De motu*: a formal definition, a material definition, and a definition expressing the relationship between the mover and the moved object. These definitions will be dealt with in sections E and F.

Albert's treatment of the Aristotelian concept of act and potency goes beyond Avicenna's and Averroës' simple acceptance of the terms. His postulation of an analogous concept of motion goes beyond the simple logical classification of motion within a single genus. Since motion cannot be restricted to a single type which can

---

[32] Avicenna argues for one genus of motion as a *passio* (*Sufficientia*, fol. 25va) which may be beyond the ten categories of Aristotle.

be analysed in detail, it is necessary to consider motion as a type of perfection related to various genera. There are three definitions of motion, expressing formal reality, material reality, and causal relation.

E. ALBERT: THE SPECIES OF MOTION

The second critical problem of the tractate arises from the view that motion cannot be restricted to a simple type. Motion as a perfection must be related to the types of motion. Again, Albert begins with a linguistic question: what does the phrase "of something" (*alicuius*) mean as it applies to "motion"?[33] Motion is a perfection "of something" but is not restricted to a single category or class and for this reason the relation of motion to types of predicates or categories becomes important.

In the *Categories* Aristotle had faced the same problem. There he stated that there are classes of terms which can be set out in a limited number of categories. However, the way in which he clarified the relation of motion to these predicates is ambiguous and leads to a wide variety of doctrinal positions.[34] Aristotle appears to hold at least two doctrines concerning the place of motion in the categories. At the beginning of the *Categories*, motion is considered to be in the category of *passio*, or suffering, which the mobile body undergoes.[35] Later in the same work, he treats motion as a "post-predicament"; this term crosses into several categories:[36] substantial change (which he later withdraws from the list), local motion, qualitative motion, and quantitative motion.[37] This ambiguity has forced the Aristotelian interpreters to clarify his precise meaning.

Averroës states that, although the view that motion is a *passio* is better known, the argument in the *Physics* that motion belongs to several categories is more tenable.[38] However, this solution is ambiguous. In text 4, Book III, he argues that motion is in several catego-

---

[33] Albert, *Physica* III, tr.1, cc.2-3 (ed. Borgnet 3: 178b-181b).

[34] Ibid., III tr.1, c.3 (ed. Borgnet 3:182a): "Est autem in his quae diximus, multa ambiguitas, et multorum Philosophorum, sententiae diversae."

[35] Aristotle, *Categories* IX (11b1-6): "Action and affection both admit of contraries and also of variation of degree. Heating is the contrary of cooling, being heated of being cooled. . . ."

[36] Ibid., XIV (15a14-15b16).

[37] Aristotle, *Physics* III, 1 (200b31-201a8). He later restricts motion as $\kappa\iota\nu\eta\sigma\iota\varsigma$ to the last three. Physics V, 1-2 (224a21-226b18).

[38] Averroës, *Physics* III, comm. 4. See n. 16.

ries and not in the category of *passio*, while in text 9 of Book v[39] he argues that motion is formally a *passio* and materially in various categories. Because of this ambiguity, Albert discards Averroës' solution and approaches the problem in his own way.

Albert begins his discussion of the problem in the following way:

> Because the solution of Averroës is obscure and doubtful, before I discuss it I will consider all the diverse views of the Peripatetics concerning motion; and these, indeed, Avicenna seems to have considered before us in the *Sufficientia*, saying that in general there are three diverse opinions concerning the genus of motion. There are, indeed, certain men who compared motion to the mover because they saw that motion is the act of the mover. . . .[40]

Albert then goes on to classify the views on motion in the following divisions:

1. action (*actio*);
2. a suffering of the mobile body (*passio*);
3. a flow of a being or reality to a terminal determination:
   a. the terminal and the flowing form do not differ in essence but only in way of participation in the substantial form (*esse*). In this interpretation black and blackening are essentially one in definition but differ in realization. This view he ascribes to Averroës and the peripatetics generally;
   b. the term and the flowing form differ in essence:
      i. motion is in neither a genus nor a species but is a process or road (*via*) to a predicamental reality and a principle leading to it. This view he ascribes to Avicenna and later to Averroës as well. The reason why he places Averroës in two classes is explained by Averroës' material and formal view of motion. More will be said about this;
      ii. motion is a predicament in its own right univocally predicated of the species of motion which fall under it.[41]

---

[39] Ibid., v, Text comm. 9.

[40] Albert, *Physica* III, tr.1, c.3, (ed. Borgnet 3: 182b): "Sed quia ista solutio Averrois est obscura et dubia, ideo antequam inquiramus de ea, tangemus omnes diversitates Peripateticorum de genere motus et has quidem ante nos videtur tetigisse Avicenna in *Sufficientia*, dicens quod in generali tres sunt diversitates opinionum de genere motus. Sunt enim quidam qui comparaverunt motum ad movens, quia viderunt quod motus actus est moventis. . . ."

[41] Ibid., III, 1, 3 (Borgnet 3: 182b-184a).

As noted in Albert's classification there are three divisions in the peripatetic tradition. Motion is (1) an action, (2) a passion, or (3) a flowing of some being (*fluxus alicuius entis*). Motion as a flowing being can be said to be (a) related to an end (*fluxus a fine in quo stat*), or (b) different essentially from the end (*fluxus per essentiam . . . differt ab eo a quo fluit*). In differing from the end, motion can be a process to a predicamental reality or a category in its own right, univocally predicated of its species.

This last view, that motion is a category univocally predicated, was rejected by Albert when he argued that motion applies to several categories and is used in an analogous sense.[42] The four remaining choices are that motion is action, passion, a flowing form related essentially to an end, and a flowing form differing essentially from the end. Albert proceeds to examine each of these four alternatives.

The argument for the position that motion is action is based on the following view. Analysis of motion reveals three distinct elements: the action of the mover, the reception by the moved, and the motion itself. Since these three cannot be separated, they are one and the same in essence. Against this position Albert argues that if motion is an action, then it is in the agent as a subject, whereas motion, in fact, is in the mobile object. Again, if motion considered as an action were in the body moved, the object would be a self-mover. Every moving body would then be a self-mover, which is not the case. It appears, thus, that motion is a *passio* or an effect of an agent existing in a patient.

Against this second view that motion is a *passio*, Albert argues that form mixed with potency and pure form do not differ in definition, but in manner of participation in the final substantial form. These final forms are of different types, so that motion lies in several categories. If the motion of a certain type had no relation to its term, then change of color could terminate in change of location. Thus the term must be intrinsic to the changing form. Again, in the analogy of a line, the point flows into the line to its completion. Hence, in agreement with Averroës, Albert states that motion is the generation of one part after another of a perfection to which the motion is directed.[43] Furthermore, the name and the definition fit the things to

---

[42] Ibid., III, 1, 2 (3: 180b-181b).
[43] Averroës, *Physics* III, Text comm. 4.

which they are applied inasmuch as they express something essential to the things. But the name and definition of motion are made to fit the special instances of motion by indicating the *termini* of the motions. Thus, the *termini* are one in essence with the very motions viewed according to their nature. And so, according as the *terminus* is placed in one of the categories, so also will the corresponding motion be reduced to that category.[44] Finally, Albert argues that pure forms and forms mixed with potency are essentially the same. A form mixed with potency is a flowing form (*forma fluens*), and this is motion. This form is the same in essence as the completed form, although differing in degree of perfection, in the various predicamental determinations which it reaches.[45] These last arguments establish the third view, the position of almost all peripatetics and the one which Albert himself supports: that motion is a process (*continuus exitus formae*) which is essentially the same as its term, but differing from it, since it is a flowing form rather than a static form.

However, Albert returns to his role as a dispassionate critic of philosophical theories by now presenting arguments against this position. Two of these are drawn from Avicenna. Arguing for the distinction between motion and terminal form resulting from motion, Avicenna says that motion is not essentially the same as its term, since the form either *is* or *is not*.[46] Furthermore, since the final term and the form in process differ as a form pure and unmixed and a form mixed with its contrary, therefore they differ essentially.[47]

Albert argues in reply generally against an essential distinction between motion and its term,[48] saying there is rather a difference in perfection (*in esse imperfectionis*) in form.[49] This position is not the same as that of Averroës, as we shall see.[50] In Albert's view, motion can be seen (i) from the aspect of the mover, as an action, or (ii) from the aspect of the moved, as a *passio*,[51] or (iii) from the aspect of form

---

[44] Albert, *Physica* III, 1, 3 (Borgnet 3: 185b).

[45] Ibid., III, 1, 3 (3: 185b).

[46] Avicenna, *Sufficientia* II, 2-3, fol. 25ra-25va; see Albert 3: 185b-186a.

[47] Ibid., II, 2, fol. 25ra-b; see Albert 3: 186a.

[48] Albert, *Physica* III, 1, 3 (Borgnet 3: 186b).

[49] Ibid., III, 1, 3 (3: 186b): ". . . et haec est opinio quam credo esse veram."

[50] Ibid., III 1, 3 (3: 189b): "Quod autem dicunt, quod secundum quod est via, est passio: *dicendum* quod hoc dixit Averroes, et forte melius fuisset si dixisset affectionem mobilis esse passionem."

[51] Ibid. (3: 187a).

alone moving to a term.[52] Albert favours the view that motion flows to a term.[53]

In his argument against motion as action or *passio* (against Avicenna), as well as in other replies to objections, Albert is led to formulate his own position more clearly. Motion is not an action (*simpliciter*) because it is not a simple perfection in an agent but a mixed perfection in a patient involving perfection and imperfection. Motion is not a *passio* (*simpliciter*) although a *passio* and motion are both together in the same subject. The motion does not take anything away from the subject whereas the *passio* does. The notion of *passio* is helpful in understanding motion, but motion is not a *passio* as Avicenna and, to some extent, Averroës held.

Against the two arguments of Avicenna previously mentioned, Albert argues first, that love and "to love" are essentially the same while differing in realization (*esse*), and so while black and blackening differ in realization (*esse*) they are also essentially the same: to become black is flowing blackness (*nigrescere est nigredo fluens*). His second point is that the true nature of an intermediate determination as contrasted with a final determination is recognized in the continuity of the intermediate determination with the term of motion.[54]

Averroës had stated that there is a sense in which motion is not a separate genus, nor a *passio*[55] as Avicenna had claimed.[56] Motion, said Averroës, is reducible to the category at which it (motion) terminates. With this view Albert has no disagreement. But he does not identify his own position with that of either Averroës or Avicenna.[57] In contrast with Averroës, who sees motion as both related to its term and essentially different from the term as a *passio*, Albert argues that it is a form flowing to its term and essentially one with it. Averroës view is clearly not the same as Albert's.

Having dealt with the genus and species of motion in chapters 2

---

[52] Ibid.: "quia esse motus absque dubio in mobili est, aut ut forma fluens, aut ut affectio mobilis, sicut diximus." It should be noted that Albert later explicitly rejects the "affectio mobilis" view of motion; cf. Borgnet 3: 187b: "motus aliud est quam affectio mobilis facta in motu et per motum. . . ."

[53] Ibid. (Borgnet 3: 187b).

[54] Ibid. (3: 187b-188b).

[55] Averroës, *Physics* v comm. 9: "Ponentes vero receptionem esse praedicamentum passionis non bene fecerunt, quoniam receptio est potentia ad rem, et iam diximus quod potentia ad aliquam rem est de genere illius rei."

[56] Avicenna, *Sufficientia* II, 2, fol. 25va.

[57] Albert, *Physica* III, 1, 3 (Borgnet 3: 189b).

and 3 of *De motu*, Albert elucidates the definition of motion in chapters 4 to 7. In chapter 4, Albert maintains that form is a perfection of the potential as actual, while motion is a perfection of the potential as potential. In chapter 5, he argues the case deductively and illustrates the presence of the perfecting principle throughout nature with the analogy of copper which is potentially a statue but not in motion unless it is in process of formation. In chapter 6, he compares the Aristotelian definition of motion with the Platonic and Pythagorean notions of difference in equality and non-being. Averroës interprets the Pythagorean view on contraposed lists of contraries to indicate a privation in *habitus* or disposition. Albert rejects this view, and argues that motion involves a privation in *esse* or being. In chapter 7, he elaborates Aristotle's attempt to clarify the definition by using a "material" definition. In this definition the mover is "touching" or in contact with the mobile body. Avicenna's attempt to place motion in the genus of a "flux" fails, since each flux involves a different species of motion. Thus chapters 4 to 7 complete Albert's treatment of the relationship of the genus to the species of motion. Albert certainly disagrees with the Avicennian view of motion as a flux of form (*fluxus formae*). However, he does not posit a clear alternative position that motion is a flowing form (*forma fluens*) related to a term, since motion is basically incomplete and relational. A flowing form is not manifestly relational with respect to its causal source but only with respect to its term. The remaining Aristotelian problem is that of determining the nature of the relation between the causal source in the mover and the change effected in the moved object.

F. Albert: The Cause of Motion

Is motion in the mover or in the mobile body? This is the third major problem of the *De motu* and is the central question of chapter 8 of the tractate. At issue is the nature of the relationship between a mover and a moved object. This question arises from the notion that motion is "according to something" (*motus sit aliquid, et alicujus, et secundum aliquid*).[58] Albert had stated the problem at the beginning of the *De motu* when he asked how motion flows from the mover to the moved.[59] The question may be answered by attempting to determine in what way motion is in the mover and in the moved.

---

[58] Ibid., III, 1, 2 (3: 178b).
[59] Ibid.

Since motion is imperfect, it is not a being (*ens*), properly speaking, but of a being (*entis*)[60] and must, therefore, be seen in reference to beings. The two realities to which it can be related are the mover and the moved. To which of these does motion properly belong, or does it belong to both? If looked at from the side of the mover, it would seem to be an action (*actio*), but, if looked at from the side of the moved, it appears to be a passion (*passio*) or suffering. However, there is a concept of the intermediate between any two relative terms: for example, the unit one exists in the relationship of one between one and two, or space exists between two related cities, Athens and Thebes. What is the relation involved between mover and moved?

At least three dialectical solutions are possible. Both acting and suffering exist in mover and moved; acting exists in the mover, suffering exists in the moved; and finally acting exists in the moved and suffering exists in the mover. Albert rejects all three since they lead to impossible consequences: equivocation on the meaning of "acting" and "being acted upon," a motion which has motion and is not moved; and both passivity and activity existing in the same subject in the same respect.

In these dialectical arguments, Albert follows Aristotle's presentation closely. The problem arises, as Albert sees it, from the ambiguity involved in the word "motion" (*motus*). This word can signify the force of an active source of motion or the motion of a passive recipient. Both agent and recipient can be said to "move" but they "move" in different ways, one as mover, and the other as moved. Once the ambiguity in the use of the word is recognized, the problem of the interpretation of the relationship of mover to moved becomes apparent. Albert presents four possible interpretations. If motion flows from the mover, then the phrase "to move" (*movere*) is correctly applied only to the mover, making motion and "to move" identical and present in the mover. If motion is also that which the moved object receives from the mover, then to move (*movere*) and to be moved (*moveri*) would be identified in the same subject. The third interpretation states that the mover and moved thing differ in physical motion since to move (*movere*) and to be moved (*moveri*) are contrary forms. Finally, the act of the mover and the act of the moved seem to be of different species.

---

[60] Ibid., III, 1, 3 (3: 189b).

Albert's solution to these dialectical problems lies in his image of motion as an uninterrupted flux (*fluxus*) which is from the mover to that which is moved.[61] Albert states that the flux between the mover and the recipient of motion is essentially the same motion, just as the distance from Athens to Thebes and from Thebes to Athens is essentially the same distance. The phrase "to move" can be used in two ways: to signify either the flux of the mover or the flux of the moved. The phrase "to be moved" signifies the flux as conceived *in a subject*, just as the phrase "to move" signifies the flux as caused *by an agent*. The word "motion" signifies the relationship of one to the other, of mover to moved. Motion is a flux,[62] just as the ray of the sun is a flux with a differing source and term. In the same way, motion can be seen in its source and in its subject. In this analysis Albert has tried to bring out the distinction between the phrase "to move" (*movere*) which can be applied to mover as cause and "to be moved" (*moveri*) which applies only to the subject moved.

Where then is motion? The answer is that, as produced *by a cause*, motion belongs to the mover; as present *in a subject*, it is in that which is moved. To move and to be moved are different in meaning. The phrase "to move" signifies motion as coming from the cause; the phrase "to be moved" is one in essence with "to move" but differs in point of reference (*penes esse et esse*).

This doctrine solves the problem of the relationship between agency and patiency. In a case such as teaching and learning, there is one process but a source and a recipient different in possession of the knowledge or of the formal reality. The process is essentially one but

---

[61] Ibid., III, 1, 8 (3:200a): "diximus motum esse sicut fluxum quemdam. . . ." Albert's use of the word *fluxus* here would seem to indicate that he is not using the words *fluens* and *fluxus* with any technical precision. J. H. Randall misses the point when he says (*Aristotle* [New York: Columbia University Press, 1960], pp. 191-192): "It is the peculiarity of Aristotle's usage that he assigns the locus of this cooperation (acting and being acted upon) to the thing being acted upon."

[62] Albert, *Physica* III, 1, 8 (Borgnet 3: 200b): "Sed tamen quia movere non tantum nominat motum illum fluxum, sed cum fluxu nominat esse fluxum a motore: et moveri non tantum dicit fluxum, sed cum fluxu conceptionem ejus in subjecto: motus autem non dicit nisi motum qui est ab uno in aliud: ideo movere non est moveri: et tamen motus est unus fluxus ejus qui est ejus quod est moveri, sicut spatium ab Athenis ad Thebas, et a Thebis ad Athenas. Unde et idem est in essentia, sed secundum tamen quod terminatum est ad Athenas, non est spatium terminatum ad Thebas."

differs in "quiddity."[63] The formal reality of teaching in the teacher differs from its formal reality in the student, although the process constitutes a unity and a single motion: there is no teaching if the student is not learning.

Albert's conclusion to the *De motu* follows Aristotle's text and the Averroist interpretation. Both provide two definitions of motion: a formal definition (i.e., the act of the potential as potential) and a material definition (i.e., the act of the moveable as moveable). The formal definition is more universal than the material definition but is more abstract. The material definition is more particular and is more readily recognizable in particular kinds of movement. Albert, however, goes one step beyond Averroës[64] by finding three definitions of motion in Aristotle's treatise. The third definition states that motion is a fulfillment of the mover *and* of the moved, thus revealing that motion is a causal relationship involving a motive impulse from the mover and a causal connection with the moved. Motion is "the perfection of both the mover and the mobile."[65] This definition is clearest to us since it reveals motion related to both the mover and the moved. While this third definition has achieved a balanced view of motion, it is not a significant advance from Aristotle's position save that it has clarified one of the more difficult problems in Aristotle's text, that of understanding the relationship between mover and moved in nature. The actuality which is in the mover turns out to be the actuality of and in the moved, just as the actuality which is in the building is of and in the house being built.

In his explanation of Aristotle, Albert exhibits philosophical competence through skillful interpretation and soundness in argument. Through his interpretation of the texts, he has brought a deeper

---

[63] Ibid., III, 1, 8 (Borgnet 3: 202a). ". . . hoc enim non est nisi fluxus unus, et quidditas est diversa: quia quidditas eorum est penes esse: et esse ejus secundum quod est a movente, aliud est ab esse ejus secundum quod est in eo quod movetur:. . ." Averroës, considering the same problem, says that ". . .quod sunt idem secundum subjectum, et diversa secundum definitionem." Averroës, *Physics* III, Text comm. 21 (compare also Aristotle, *Physics* III, 3, 202b19-22). Averroës goes on to distinguish between the unity in the subject, motion, and quiddity (*Physics* III, Text comm. 22). In Aristotle, as well, there is a unity in subject and a difference in definition; Averroës simply repeats this view. Albert adds to this the notion of essential unity in subject, a difference in quiddity which is more than a simple difference in definition; it involves the difference in participation in a formal reality. The difference in definition hinges on this realization.

[64] Averroës, *Physics* III, Text. comm. 23.

[65] Albert, *Physica* III, 1, 8 (Borgnet 3: 202b): ". . . est endelechia et moventis et mobilis, quae dicit totum quod est motus, secundum quod est fluxus a movente in mobile. . . ."

understanding of the concept of motion by clarifying the notions of act and potency. Through sound argument, he has identified motion, as a perfection (*actus*), with an imperfect form flowing to a term. His position is logically developed and leads to a dynamic view of forms. Finally, in his answer to the problem of whether motion lies in the mover or the mobile object, he combines careful interpretation with solid argument and presents a distinctive third definition of motion. Albert's theory of motion is the work of a discriminating philosopher.

G. Miss Maier's Interpretation

Having dealt with the Aristotelian doctrine of motion and the interpretations of Avicenna, Averroës, and Albert, two questions remain. First, is the synthesis which Albert presents of significance historically as an influence on his successors, as Miss Maier claims? Second, is Albert's doctrine a significant advance conceptually over the doctrines of Avicenna and Averroës? These are not wholly unrelated questions since Albert's conceptual rigour may be doubted if his work is taken to be either a simple paraphrase of Avicenna or a restatement of Averroës' position.

Albert was well placed historically to consider anew the problems of the genus, species, and causal aspects of motion, since he had the two contrasting Islamic doctrines at hand. Miss Maier says that Albert is the source of fourteenth-century debate on motion as *fluxus formae* and *forma fluens*. The difficulty with this view is, first, that a clear distinction between them is not drawn by Albert in those technical phrases, and second, that if Albert does teach such a simple distinction, then there can be no valid reason why he placed Averroës in two distinct classes with respect to flowing form (*forma fluens*). Closer analysis of the key text in the *De motu* may provide solutions of these difficulties.

Albert classifies views concerning motion into three groups:

(1) as a simple action (*actio*);

(2) as a suffering of the mobile body (*passio*); and

(3) as a flow of form (*fluxus formae*), (a) essentially one with its term, or (b) essentially distinct from its term (and as a word either equivocal or univocal). Albert rejects the view that motion is an action alone, since it is unfulfilled and he also rejects the view that motion is a passion of the body alone. Motion could be either a flow of a being or reality to a term essentially related, or a flow of being or

reality to a term not essentially related. He settles for the first alternative of the third view, which puts him in agreement with one Averroist position but not with the second Averroist position that motion differs materially from its term.

Looking at this interpretation from an historical perspective Miss Maier contends that Albert's analysis is the *locus classicus* for succeeding medieval debates on the nature of motion. She sees the distinction between motion as a form related to its term and motion as a form unrelated to its term as the source of a fundamental distinction made in the fourteenth century between motion as *forma fluens*, which she considers an Averroist position, and *fluxus formae*, which she argues is an Avicennian view.[66] Miss Maier interprets the key text in the *De motu* as a paraphrase of Avicenna.[67] However, it could also be interpreted as a personal account by Albert of the peripatetic tradition, a possibility clearly present in the introduction to the classification of various approaches as we have explained above.

Miss Maier notes that Albert sets out three classifications, as does Avicenna, although the third classification is divided into three parts. She sees the sentence introducing the classification — "There are, indeed, certain men who compared motion to the mover"[68] — as the beginning of a free paraphrase of Avicenna. Serious difficulties are inherent in this interpretation. If Albert's text is regarded as a paraphrase of Avicenna, the question immediately arises whether the views, namely Albert's and Avicenna's, are similar enough to warrant such an assumption.

When one examines Albert's classification of the various interpretations of motion by the peripatetics, it is at once clear that they are almost totally different from those of Avicenna's. This becomes immediately obvious when their classification of views of motion are juxtaposed:

---

[66] Miss Maier discusses this distinction in several key writings in *Zwischen Philosophie und Mechanik* (Roma: Edizioni di Storia e Letteratura, 1958), pp. 61 et seq.; in *Die Vorläufer Galileis im 14. Jahrhundert* (Roma: Edizioni di Storia e Letteratura, 1949), pp. 11 et seq.; in "Die scholastische Wesenbestimmung der Bewegung als forma fluens der fluxus formae und ihre Beziehung zu Albertus Magnus," *Angelicum*, 21 (1944), 97-111. Cf. also E. J. Dijksterhuis, *The Mechanization of the World Picture* (Oxford: Clarendon Press, 1961), pp. 174-175.

[67] Maier, *Zwischen Philosophie und Mechanik*, p. 73: "Er hat in seiner Physik in einer Paraphrase, die sich manche Freiheiten und manche Abweichungen von dem zugrunde gelegten Text erlaubt, das eben betrachtete Kapitel aus Avicennas Sufficientia wiedergegeben, in dem von den verschiedenen Auffassungen der Bewegung berichtet wird."

[68] Albert, *Physica* III, 1, 2 (Borgnet 3: 182b): "Sunt enim quidam qui comparaverunt motum ad movens...."

Albert	*Avicenna*
(1) Action	(1) Passion[69]
(2) Passion	(2) Chance participation in a name[70]
(3) A flux of some being to a term of motion	(3) A genus predicated univocally.[71]

If Albert is paraphrasing Avicenna, then the difference in classifications indicates serious confusion on Albert's part. Miss Maier, in fact, draws this conclusion and indicates that Albert did not understand Avicenna.[72]

Further examination of the texts of Albert shows how serious his confusion would be, given Miss Maier's interpretation. Albert places Avicenna's position in the third classification, i.e. that motion is a flowing form.[73] Avicenna, in his own text, says that motion is a *passio*, the first of his own classifications.[74]

Aside from the difference in classifications and the misplacement of Avicenna, there is also a fundamental difference in philosophical points of view expressed. Albert tries to remain close to the Aristotelian categories in his classification, while Avicenna rejects the limita-

---

[69] Avicenna, *Sufficientia* II, 2, fol. 24vb: "Quidam enim dixerunt quod motus est praedicamentum patiendi."

[70] Ibid., II, 2, fol. 24vb: "Alii vero dixerunt quod hoc nomen motus cadit super maneries quae sunt in illo sola casuali participatione nominis." Avicenna explains that in this theory each type of motion falls within the category in which it is actualized. There are divisions in this group: (1) those who refer motion to action and passion; (2) those who relate motion to the changing form alone, which group is divided into (a) those who see a specific difference between moving form which is being added to and the resting form, or (b) those who argue that motion is a species in which there is no change through addition. Avicenna argues against both these views, first on the basis that the form in question is or is not (25ra) and second on the basis of the specific difference between a flowing and resting form (25ra). This subdivision is close to Albert's classification but not close enough to indicate a paraphrase.

[71] Avicenna, *Sufficientia* II, 2, fol. 25ra: "Est autem adhuc hic tertia sententia quam dicunt quod nomen motus quamvis sit commune sicut dictum est, tamen maneria quae sunt sub ipso non sunt species predicamentorum secundum modum quae dixerunt. . .predicatur motus univoce. Quid perfectio quam accipimus in eius descriptione quasi genus, est de numero verborum scilicet ens et unitas."

[72] Maier, *Zwischen Philosophie und Mechanik*, p. 76: "Wir wollen dahin gestellt sein lassen, ob Albert mit seiner Klassifizierung und Zuschreibung die Ansichten Avicennas und Averroës' wirklich ganz exakt erfasst und in ihrer Eigenart und Gegensätzlichkeit richtig herausgestellt hat. Besonders was Avicenna anbelangt, kann man zweifeln."

[73] Albert, *Physica* III, 1, 3 (Borgnet 3: 183b).

[74] Avicenna, *Sufficientia* II, 2, fol. 25va: "Unde melius est eis: ut praedicamentum passionis et motum ponant unius continentiae sive unius aequipollentiae."

tion of categories to ten.[75] In Miss Maier's reading, Albert is both confused in the classifications which he provides and unmindful of Avicenna's fundamental deviation from the Aristotelian doctrine.

In opposition to Miss Maier's interpretation, one may ask whether the section in Albert under consideration is a personal statement outlining the peripatetic tradition, and not a paraphrase of Avicenna. This second interpretation is the only one which grants Albert any ability as a critical philosopher. Miss Maier's explanation of Albert's text undermines Albert's status as a serious thinker. Her interpretation does not account for the closely reasoned arguments which indicate that he provides much more than a paraphrase of Avicenna. The more detailed examination of the texts indicates the justice of this claim, as we have tried to show in sections D–F.

Miss Maier ultimately suggests that Albert's analysis in c. 3 constitutes a general acceptance of the position of Averroës but, as Albert himself notes prior to his analysis, the position of Averroës is ambiguous. First, Albert locates Averroës in the class of those holding that motion is a form flowing (*forma fluens*) to its term and essentially one with it. Second, Albert locates Averroës among those who hold that motion is a flowing form essentially different from its term, and in the category of a *passio*. Albert's choice of the first of these alternatives for himself constitutes a rejection of what he considered Averroës' final view, viz., that motion is, at least formally, a *passio*.[76] This would indicate that Albert's personal position is that motion is a flowing form essentially one with its term and that he arrived at it independently, thus avoiding conceptual difficulties in Avicenna's and Averroës' positions.

Miss Maier's view is correct that Albert holds an important historical position, since the *De motu* is the *locus classicus* for interpretative discussions of Aristotle's theory of motion. Her view that he exerts an important influence on fourteenth-century thinkers is sound. However, the problem of reconciling Albert, the paraphraser of Avicenna, with Albert, the supposedly Averroist interpreter,[77] is not solved by historical analysis or by tracing historical lineage alone. A closer examination of his conceptual contributions is needed.

---

[75] Ibid., II, 2, fol. 25va.

[76] Albert, *Physica* III, 1, 3 (Borgnet 3: 189b).

[77] Maier, *Die Vorläufer Galileis im 14, Jahrhundert*, p. 12: "Und in dieser Wiedergabe nun fuhrt Albertus jene Begriffe ein, die dann zugleich zur Interpretation der Averroesstelle dienen."

What reasons can be provided for the attribution to Albert of the distinction between motion as *fluxus formae* and *forma fluens*? First, Albert does distinguish carefully between the positions taken with respect to motion by Avicenna and Averroës. The differences can be represented by two quite distinct conceptual options involving emphasis on form itself or on form related to its term. It would appear that the fourteenth-century interpreters oversimplified Albert's position in the interests of clarity, and hence Miss Maier may be justified in placing the historical ground for the distinction in Albert. The conceptual issues are more complicated, however, and Albert does not argue the distinction in the form in which it appears in the fourteenth century. Father Wallace's more careful use of the word "adumbrate" with reference to Albert's relation to future controversies perhaps represents the actual relation between Albert and fourteenth century theorists.[78] Albert's account is perhaps a faint outline or a faint foreshadowing of the future. But overemphasis of this simple distinction and reading it back into Albert's works has led to considerable distortion of Albert's view.

---

[78] See note 4.

6

The Physical Astronomy and Astrology of Albertus Magnus

Betsey Barker Price
University of Toronto

Astronomy and astrology, it is often believed, were one and the same in the Middle Ages. Medieval astronomers and astrologers, it is said, were "the same men, writing indiscriminately on both subjects."[1] There is admittedly some ground for the assumption that the two sciences were conceived as one, or perhaps two but indistinguishable from one another. It was, indeed, a commonplace of medieval terminology to use the words *astronomia* and *astrologia* interchangeably. But this does not mean that the scholastics themselves were unaware of a clear distinction between two different sciences.

There is little reason today to persist in the belief that Albert the Great or any of his thirteenth-century contemporaries did not recognize the two distinct bodies of knowledge (*scientiae*), which the twentieth century calls "astronomy" and "astrology." In theory there was a marked difference between the two. In practice, however, the distinction might not have been so obvious, for both sciences deal with the same subject matter, namely the heavens and the celestial bodies. Each, nevertheless, possesses a particular point of view, a unique perspective from which are posed different questions for investigation. Albert respected the distinction which Ptolemy, a second-century Greek astronomer and astrologer, made between the two:

---

[1] A. J. Meadows, *The High Firmament* (Leicester: Leicester University Press, 1969), p. 44.

> It ought to be stated that there are two parts to astronomy (*astronomiae*), as Ptolemy says: one is about the locations of superior [*heavenly*] bodies, their quantities and their individual phenomena (*passionibus*); and one arrives at the knowledge of this part through demonstration (*demonstrationem*). The other is about the effects of the stars on inferior [*terrestrial*] things, which effects are impermanently assumed by the mutable things; and therefore one arrives at knowledge of this part only by conjecture, and it is necessary that the astronomy of the latter kind exist according to something physical, and that it be conjectured by physical signs.[2]

To designate the first part of Ptolemy's astronomy Albert's "astronomy," and the second part his "astrology" would not be far from the truth. Albert defines astronomy several times as the science which seeks knowledge concerning the number of motions, the duration of motions, and the locations of the mobile heavenly bodies. Any book about astronomy, according to him, should definitely tackle the abstract, mathematical descriptions pertaining to planetary motions, as, for example, bodies moving in eccentric circles. Albert's own descriptions of astronomical systems, however, never lack a consideration of their physical reality. He maintained that the physical aspects of the subject matter of both astronomy and astrology could not be ignored.

Although Albert speaks of *astrologia* most frequently, his use of the word does not reveal his definition of astrology. His descriptions of astrologers down to their particular specialty are, however, generously illustrative of his conception of the whole science. Astrology stood, for Albert, as Ptolemy portrays it, on equal footing with astronomy, worthy of equal study and perhaps of greater value in the daily affairs of men than its associate science. One senses throughout Albert's writings his feeling that perhaps the ultimate purpose for the existence of the heavenly bodies is to exert an influence on the earthly realm. Albert did seem more secure with his understanding of astrology than with his grasp of astronomy, but he left the details of each to the practitioners.

During Albert's lifetime the scope of both astronomy and astrology was to change enormously. Under his influence, after 1255 the University of Paris would include in the curriculum of the faculty of arts some of Aristotle's most important works, which had been repeatedly condemned in the first half of the thirteenth century. The

---

[2] *De fato*, a. 4 (ed. Colon. 17/1: 73, ll. 36-44). Cf. Ptolemy, *Tetrabiblos* I, c.1.

physical features of astronomy would thereby obtain an even more secure footing in the Aristotelian corpus. Vague, non-technical astronomical treatises of earlier curricula would yield to a growing collection of works on mathematical astronomy. Meanwhile both astrological and astronomical writings new to the Latin West, translations from Greek and Arabic, waited to be understood and evaluated. Newly translated treatises of Muslim determinism threatened any role for astrology inside Christian theology. Attempting to embrace all facets of astronomy and astrology, Albertus Magnus played an active part in these thirteenth-century changes.

A. ASTRONOMY

Albert the Great has been identified as one of the students at Padua whom Jordan of Saxony, master general of the Order of Preachers, brought into the Dominican order in 1223 (Jordan, *Epistulae* 20). Born about 1200 Albert would by that time have been following a liberal arts programme in the Paduan "studium." One text used by beginning and intermediate students of astronomy at Padua, as elsewhere, was, most likely *On the Sphere* (*De sphaera*) by John of Sacrobosco.[3] This work written about 1220 may even have been conceived as a textbook and delivered in the form of lectures.

From it Albert would have gleaned only the most rudimentary spherical astronomy. It names the circles used to describe the motions of the five planets, Saturn, Jupiter, Mars, Venus, and Mercury, of the two luminaries, the sun and moon, and of the collection of all the stars which appear fixed in position relative to one another. It treats with equal concern the division of the spherical earth into climes or zones based on the length of the day at a particular latitude. Containing almost no calculations, no diagrams, and no star catalogue, *On the Sphere* is a purely descriptive work. To the student, John presented a popular astronomy simplified and synthesized from his predecessors, in natural philosophy, Aristotle, and in mathematics, Ptolemy.

There is a faint possibility that while in Padua Albert may have encountered astronomical texts of a more practical nature. However, there is no direct evidence that he saw either treatises dealing with

---

[3] Lynn Thorndike, *The Sphere of Sacrobosco and its Commentators* (Chicago: University of Chicago Press, 1949), esp. pp. 14 & 21. Nancy G. Siraisi, *Arts and Sciences at Padua*, Studies and Texts, No. 25 (Toronto: Pontifical Institute of Mediaeval Studies, 1973), p. 94.

instruments for observational astronomy such as the astrolabe, or others which apply astronomy to time-reckoning, such as the *Compotus* also by Sacrobosco. Nor does it seem correct to assume that Albert studied a popular anonymous work, *The Theory of the Planets* (*Theorica Planetarum*) in Padua. This treatise ascribed by some modern authors to Gerard of Cremona (d. 1187) concentrates on the essential features of Ptolemaic planetary theory.[4] Although its origin is unclear, the work does not enter the corpus of educational material on astronomy until the latter half of the thirteenth century, too late for Albert's study at Padua.[5]

This does not mean that *On the Sphere* was the only early source of Albert's learning about astronomy. Some of Aristotle's works on natural philosophy, known as the *libri naturales*, had been translated from Greek by James of Venice (Jacobus Grecus Veneticus) who flourished from 1136 to 1148. These included versions of the *Physics*, the *Metaphysics*, and other works whose contents pertinent to astronomy Albert was to explicate later. It is probably safe to assume that these works were known by Paduan scholars at the *studium* from the first decade of the thirteenth century.[6] Through them Albert may in Padua have first become directly acquainted with some of Aristotle's ideas on astronomy.

However, it was only by 1245 that Albert, nurtured by other writings and later translations, began in his theological writings to reflect his knowledge of astronomy and belief in astrology. Around 1250 he started the monumental tasks of rendering all the works of Aristotle intelligible to his fellow Dominicans. This undertaking, completed some twenty years later, involved the systematic explanation of the whole of human knowledge in all areas of logic, ethics, social order, metaphysics, natural philosophy, and astronomy. Amidst the running commentary and paraphrasing of the Aristotelian texts Albert inserted true "digressions," in which ideas from many sources including his own experience are discussed and weighed. This is the vast stage for his ideas on astronomy and astrology.

---

[4] Olaf Pedersen, trans., "The Theory of the Planets" in *Source Book in Medieval Science*, ed. Edward Grant (Cambridge, Mass.: Harvard University Press, 1974), pp. 451-465. Francis J. Carmody, ed., *Theorica Planetarum Gerardi* (Berkeley: privately printed, 1942).

[5] Olaf Pedersen, "The Corpus Astronomicum and the Traditions of Mediaeval Latin Astronomy" in *Studia Copernicana* 13, Colloquia Copernicana 3 (Wrocław: Polska Akademia Nauk, 1975), pp. 76-79.

[6] Cf. Fernand van Steenberghen, *Aristotle in the West* (Louvain: Nauwelaerts, 1955), esp. pp. 62-66.

The principal expression of Albert's ideas, however, is found in a few of his commentaries. Because of an astronomer's special field of investigation, that of physical yet immutable celestial bodies, he has to rely on the explanations of the "physicist" or natural philosopher and of the mathematician to solve some problems of his own science. It is not surprising, therefore, to find the majority of Albert's statements about astronomy in his own commentries on Aristotle's works of natural philosophy. Information is found in Albert's commentary on Aristotle's *Physics*, a work which Albert described as "a book of the physical realm examined," and in his commentary on the Stagirite's *On the Heavens* (*De caelo et mundo*), "the study of the mobile body with respect to its location."[7] A most important discussion is also provided in Albert's commentary on Aristotle's *Metaphysics* XI. Both Albert's commentary on the *Meteorologica* of Aristotle, in which he included a number of non-Aristotelian explanations of comets, and his commentary on the Philosopher's *On Generation* (*De generatione*), "the study of the mobile body with respect to its form,"[7] contribute to an understanding of Albertus Magnus' astronomy and astrology.

Albert rightfully felt that no work of Aristotle dealt specifically with astronomy, a physical and mathematical science. He clearly rejects that role for Aristotle's *On the Heavens*: "Those discussing nature talk about the shape of the sun and the moon and whether the earth is spherical and whether the universe is spherical or not, as is apparent in the book *De caelo et mundo*. Astronomy also discusses all these things as is revealed in the first book of the *Almagest* of Ptolemy."[8] Instead he designated, the $\mu\epsilon\gamma\acute{\iota}\sigma\tau\eta$ $\sigma\acute{\upsilon}\nu\tau\alpha\xi\iota\varsigma$ or *Almagest*, a remarkable, non-Aristotelian work on planetary motion and theory, as the one which handled those questions via astronomy.

Like other writers of later antiquity and the Middle Ages, Albert considered the *Almagest* to have been written by one of the Ptolemys, the dynasty of Egyptian kings. Although little is known about the real author, Ptolemy probably lived near Alexandria from about AD 100 to 178.[9] His work written in Greek after 141 was translated into Latin from the Arabic "al-majasti" by Gerard of Cremona in

---

[7] *Physica* I, tr.1, c.4 (ed. Borgnet 3: 10a and 8).
[8] *Physica* II, tr.1, c.8 (ed. Borgnet 3: 107b-108a).
[9] Franz Boll, *Studien über Claudius Ptolemäus* (Leipzig: Teubner, 1894), p. 64.

1175.[10] It circulated throughout the Latin West and there is little reason to doubt that Albert had access to a copy in Paris. It is more important, however, to appreciate that Albert saw the *Almagest* as a work designed to describe a physically real astronomy.

In every age since its writings, Ptolemy's major work has been considered by some a strictly mathematical conception of the motions of the celestial bodies. The Euclidian foundation of the *Almagest* and its exact mechanism for calculating any and all planetary positions do consume almost all its thirteen books and cannot be ignored. But neither can one disregard Ptolemy's statement of Aristotelian philosophy and his subsequent comments on the topics Albert noted. For Albert this union of the physical and mathematical was essential to a work of astronomy.

He himself desired to write such a work, perhaps modeled on the *Almagest*. In the *Physica* Albert wrote, "It would be lengthy to demonstrate (*demonstrare*) how a chord is converted (*convertatur*) into an arc such that afterward a line equal to the arc will be obtained (*accipiatur*), but this will be taught (*docebitur*) in Geometry (*geometria*) and in Astronomy (*astronomia*), God willing."[11] Although Ptolemy included a table of the ratio of chord to arc in a circle (*Almagest* I, c.11), even he was not able to "demonstrate" his procedure, for exact values cannot be obtained by geometry, the ancient mathematicians' most sophisticated tool. Whether Albert was unaware of the ultimate empirical source of Ptolemy's coefficients cannot be determined, but neither is it possible to know what Albert had in mind by way of "proof." "Certain stars are first and certain lower ones are last [in order of distance from the earth]. And certain ones are removed from others by a greater or lesser distance in longitude, as is shown in the *Almagest* (*Et elongantur quaedam ab aliis majori vel minori longitudine...*)." "And all these things ought to be stated and determined adequately by mathematical principles in *Astronomy*."[12] Like Ptolemy, Albert would have considered both the mathematical theory and the observable realities of planetary motion.

---

[10] Ptolemy, *The Almagest*, trans. R. Catesby Taliaferro, Great Books of the Western World (Chicago: Encyclopaedia Britannica, 1952), 16: 5-456. The earliest translation of the *Almagest* into Latin was from Greek around 1160 by a Sicilian author whose name and work have been lost.

[11] *Physica* I, tr.2, c.1 (ed. Borgnet 3: 22a).

[12] *De caelo* II, tr.3, c.11 (ed. Colon. 5/1: 167, ll. 80-85). Albert seems here to be referring (1) to the arrangement of the planets in order from the farthest away to the closest to the earth, and (2), to the longitudinal differences between the planets. Cf. *Almagest* IX, c.1; XII, cc.9-10. On *motus longitudinis, v De caelo* II, tr.1, c.6 (ed. Colon. 5/1: 120, l. 28).

He even provided an outline for his proposed work. "Nevertheless with God's consent we shall make a comparison in the *Science of Astronomy* between the way [of explaining celestial motion] which al-Bitrūjī discovered and the way which Ptolemy followed taking it from the Babylonians and Egyptians, whose learning Aristotle says he verified in *On the Heavens*, from which it appears that he agrees, because he also consented to their opinions (*Nos tamen domino concedente collationem faciemus in Scientia Astrologiae inter viam, quam invenit Alpetrauz Abuysac, et viam, quam secutus est Ptolemaeus accipiens eam a Babyloniis et Aegyptiis, quorum scientiam se verificasse dicit Aristoteles in Libro Caeli et Mundi, ex quo videtur innuere, quod et ipse consensit opinionibus eorum*)."[13] Hence although this work was probably never written, the twentieth-century historian of astronomy has every indication as to the goal of Albert's scattered ideas on astronomy. Due to his framework, the Aristotelian corpus, it is not surprising that Albert concentrated on physical astronomy to the unfortunate neglect of anything mathematical. However, from it alone, one can construct his appreciation of the Eudoxan-Callippic homocentric system adapted by Aristotle, Ptolemy's contribution, and al-Bitrūjī's model of the universe. Albert revealed his motive for this comparison in the *Metaphysica*: to answer the all important question, "how many *movers* are required to cause the observed motion of the heavens?"

Apart from their differences, the systems of Aristotle, Ptolemy, and al-Bitrūjī have several features in common. All three are geocentric, positing a spherical universe. Each, following the Pythagorean ideals, posited a point about which a body would move with constant speed in uniform circular motion; this point was, however, neither the same for each planet nor was it always the earth in all three systems. In the *Almagest* Ptolemy did not discuss the material element of the planets or the celestial spheres on which they move. However, his views were derived indirectly from another of his works concerning an analogous but physical system, *Planetary Hypotheses*. Thus the seven planets were generally conceived as luminous ethereal globes affixed to or actually part of ethereal spheres which revolve. Aristotle and al-Bitrūjī envisioned whole spheres, hard and transluscent as if made of crystal; the physical systems of

---

[13] *De caelo* II, tr.3, c.9 (ed. Colon. 5/1: 162, ll. 77-84).

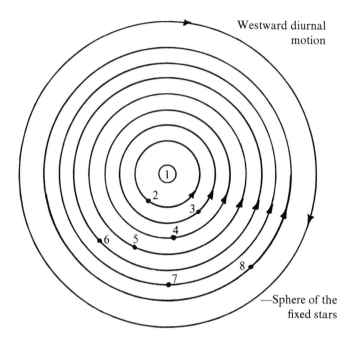

Westward diurnal motion

—Sphere of the fixed stars

Figure 1.   Cross section of the basic homocentric model of the universe

1. Earth	5. Sun
2. Moon	6. Mars
3. Mercury	7. Jupiter
4. Venus	8. Saturn

all three "astronomers" required that the spheres while leaving no empty spaces between them touch but not intersect.[14]

The most primitive model of celestial motion simply allocates one sphere to each celestial body. A total of only eight spheres could result, one for each planet and one to account for the motion of the fixed stars, otherwise an incomprehensible number might be required. Albert notes that the Spanish-Jewish philosopher, Moses Maimonides (1135-1204) "absurdly" held that each individual star like the planets had its own sphere (*De caelo* II, tr.3, c.11). A single sphere carrying a celestial body in a circle of constant motion around the earth cannot, however, account for latitudinal variations periodically different from the revolution, nor for the observed phenomena of stations and retrograde motion whereby a planet appears

---

[14] The values Ptolemy gives for the sizes of some epicycles and eccentrics and for some degrees of eccentricity in the *Almagest* would cause intersections of planetary spheres. These values are altered in his later work *Planetary Hypotheses* to reveal a physically possible system.

to stand still or move backward in its orbit against the backdrop of the constellations. Thus planetary models of greater sophistication employed sets of spheres functioning together to reproduce these and other anomalies of motion for each separate body.

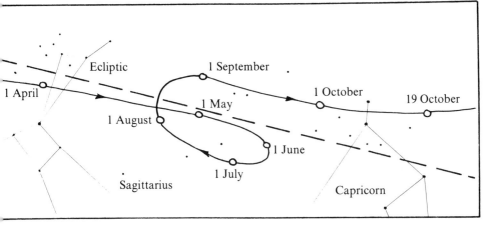

Figure 2. Retrograde motion of Mars against the background of the fixed stars. Mars makes a backward loop in some part of the sky every 780 days.

i. The Eudoxan System of Astronomy

The oldest known kinematic model of this type is that of Eudoxus, a Greek of the first half of the fourth century BC. Albert's exposition of it offers little more detail than his source, Aristotle's account, the earliest extant version. That source (*Metaphysics* XI, c.8, 1073b17-1074a14) provides only a general description of sets of concentric, rotating spheres which Eudoxus assumed for each planet. In each set the outermost sphere is the sphere of the fixed stars (*aplanorum* – "without erring") which performs the daily revolution of all celestial bodies from east to west. A second sphere moves the planet with uniform velocity along the ecliptic or zodiac belt from west to east with the speed of its sidereal mean motion.

In the case of the five planets a third and fourth sphere work together to produce their individual oscillations in latitude and longitude. Albert was not alone in failing to realize how the Eudoxan model could actually represent planetary motions. During the Middle Ages and even until Schiaparelli's work in the nineteenth century the use of an algebraic curve resembling a figure eight on its side, called a "hippopede" or horse-fetter, seems to have gone unrecognized.[15] This pattern could be generated by the motion of the two inner concentric spheres if they were to rotate with constant but opposite angular velocities about two axes inclined to each other. Instead of noting their combined effect, however, Albert followed Aristotle in designating the third sphere as the cause of latitudinal variation with respect to the equator and to the ecliptic for all planets and the sun and moon, and the fourth as responsible for longitudinal planetary motion along the zodiac and for conjunctions with the sun causing eclipses.

Albert did not mention the directions of the motions of the inner spheres but he did include a word about the axis of the third sphere. In each set it is extended from the zodiac belt or middle of the second sphere at the same place for all the planets except in the cases of Venus and Mercury. They both require axes positioned differently, for their orbits are consistently north or south of the ecliptic respectively, unlike the others which are "as much inclined from the zodiac to the north as they are inclined from it to the south."[16] Except for different pole locations, Albert was obviously not concerned with how the third sphere alone could cause varying degrees of planetary latitude. In fact without the cooperation of a fourth sphere, absent in the case of the luminaries, Albert and Aristotle have represented solar and lunar latitudes based on their mean motions.

Albert introduced two definitions essential to his own understanding of the Eudoxan model and its subsequent modifications. "I call a deferent sphere (*deferentem*) one which carries a planet (*astrum*) through its continuous motion. A sphere is called a 'back-turning' sphere (*revolvens*) which turns a planet backwards for the purpose of returning it to the place in the zodiac whence it was previously moved."[17] Only the deferent sphere was used by Eudoxus, Albert wrote. He reminds the reader that it carries the planet, instead of the planet propelling itself by its own proper motion, as Maimonides and others would have celestial bodies move (*Metaphysica* XI, tr.2, c.24).

The sum total of these deferent spheres is finally calculated at

---

[15] Thomas L. Heath, *Aristarchus of Samos* (Oxford: Oxford University Press, 1913), pp. 194-223, esp. the drawing on p. 203.

[16] *Metaphysica* XI, tr.2, c.22 (ed. Colon. 16/2: 512, ll. 8-10). All planets including Mercury and Venus actually follow an orbital path which carries them both north and south of the ecliptic. Cf. *Almagest* XIII, cc.3-5.

[17] *Metaphysica* XI, tr.2, c.22 (ed. Colon. 16/2: 511, ll. 75-79).

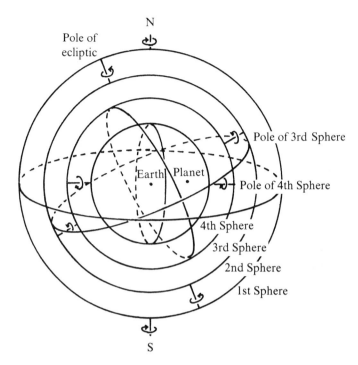

Figure 3. A unit of homocentric spheres for one outer planet in the Eudoxan system. (The planet lies outside the plane of the page.)

twenty-six, three for the sun and moon, four for each of the planets. Albert concluded that these twenty-six spheres would require twenty-six movers, one for each sphere. After Eudoxus two later attempts to modify his spheres increased their number and their movers to fifty-five. One was made by Callippus in Athens about 330 BC, the other, by Aristotle himself who accepted Callippus' increased number of spheres and combined them with more to form a connected mechanical system. Albert attributed all modifications to Callippus and went on to interpret their function.

Albert thought that Eudoxus had accounted for the diurnal rotation, and longitudinal and latitudinal motions for all the planets. He assumed that Callippus introduced modifications needed to reproduce two more phenomena in a planet's course around the earth: (1) a variation in its distance from the earth, observable by changes in brightness for the planets, and by changes in size for the moon, and (2) a perceived variation in its speed. These two anomalies Albert considered to be true, readily observable, and worthy of representation; he took them as Callippus' motive for change. Unfortunately the lack of detail in Aristotle's original description renders it virtually impossible to know the real "improvements." Neugebauer, a prominent historian of ancient mathematical astronomy, suggests, "it would be better to admit our total ignorance of the character of Callippus' modifications of the Eudoxan model."[18]

The Latin translation of the *Metaphysics* which Albert primarily used, *translatio media*, seems, however, to have given Albert cause for his assumptions.[19] *"Callippus autem positionem quidem sphaerarum eandem posuit cum Eudoxo, hoc est absidentiarum ordinem. . ."* ("Callippus moreover posited indeed the same position of the spheres as Eudoxus, that is the [same] order of the apsides" — *Metaphysics* XI, c.8, 1073b32). The apsis (perigee) and aux (apogee) Albert explains, are a celestial body's points of closest and farthest distance from the earth (*"auges summae elevationes et absides infimae depressiones"*).[20] Since there would be no mention of these points in connection with concentric spheres, Albert deduced that Callippus posited eccentric spheres, or spheres whose centre is not the earth, as the deferents of the planets. A celestial body whose deferent is geocentric suffers no apogee or perigee as it is always at a constant distance from the earth. Callippus is indeed generally thought to have added an extra sphere to each of the five planets. Albert's unique assumption, however, is that to account for varying distance this additional sphere was an eccentric deferent, which really came from another source.

To the luminaries Callippus added two concentric "back-turning" spheres. Albert tried to explain why "back-turning" and not deferent spheres were necessary to account for that characteristic of the sun and moon, namely of moving through equal parts of the orbit at unequal speeds. According to Callippus, Albert wrote, the only way to cause such unequal motion is by a composite motion of diverse spheres moving in opposite directions. "However, spheres moving in opposite directions negate themselves if they are on the same poles."[21] But as the slowing down and retrograding, indications of unequal speed, take place in the same signs of the zodiac and on the same arc where the luminaries had previously moved ahead, a sphere

---

[18] O. Neugebauer, *A History of Ancient Mathematical Astronomy*, Studies in the History of Mathematics and Physical Sciences, 1 (New York: Springer-Verlag, 1975), p. 684.

[19] *Metaphysica*, ed. Bernhard Geyer (ed. Colon. 16/1: x).

[20] *Metaphysica* XI, tr.2, c.23 (ed. Colon. 16/2: 512, ll. 44-45).

[21] *Metaphysica* XI, tr.2, c.23 (ed. Colon. 16/2: 513, ll. 11-12).

representing them would have to be on the same poles as the original Eudoxan spheres which carry the body forward.

The strange conclusion is left unstated: "back-turning" spheres are used, for although they are on the same poles as the deferents, they do not negate their motion. Further "back-turning" spheres are also allotted to each of the five planets to reproduce their inconstant motion. The first of a set of four is located inside the innermost deferent. It gives a planet its irregular motion in latitude. The other three follow: one to cause retrograde motion, one to effect the times when a planet comes to a standstill, its orbital stations, and a third to produce simple forward motion (*cursus directo*).

Albert and perhaps Aristotle's translator curiously misunderstood Aristotle's addition of "back-turning" spheres. Aristotle envisioned a system of contiguous homocentric sets of spheres nested within one another like the layers of an onion's skin. This arrangement would allow motion to be mechanically transmitted from an ultimate first mover at the extremity of the universe continuously inward toward the earth through one rotating set of spheres to the next set below it. However, in order for each planet to maintain its own motion with reference to the fixed stars it must not be carried along by the motion of the set of spheres belonging to the planets above it. Thus Aristotle employed "back-turning" spheres designed to function just as Albert suspected counterrevolving, concentric spheres would; they were to negate a superior planet's motions with respect to the planet below it. So when n is the number of spheres a single planet would require to represent its motion, Aristotle needed $n - 1$ "back-turning" spheres to eliminate its effect for the next planet; in the case of each planet five (revolving spheres) minus one, or four "back-turning" spheres were required, in the case of the luminaries, $3 - 1$ or two "back-turning" ones.

Reason will allow this Aristotelian system, Albert conceded, but not necessity. Necessary acceptance would require further inquiry by "those who are stronger in such investigations," by "those who consider the proper principles of the heavens."[22] Albert was aware that this was merely a general description lacking observational detail to confirm it. However, the only faults he found with the system were Aristotle's location of the luminaries as the closest bodies to the earth and his frequent statement that the sphere of the fixed stars is

---

[22] *Metaphysica* XI, tr.2, c.23 (ed. Colon. 16/2: 513, ll. 87-88).

the outermost sphere. Callippus' assumed use of eccentrics, although against Aristotelian principles of regular motion about a physical centre, did not disturb Albert.

ii. The Ptolemaic System of Astronomy

In fact Albert defended the eccentrics and epicycles of Ptolemy against strict Peripatetics like Averroës. Albert denied the necessity for all motion to have one centre, the earth. Epicycles, little planet-bearing spheres with points on other spheres as their centres, could be posited. Eccentrics or "circles with the centres removed [from the earth]" ("circuli egressae cuspidis")[23] could exist with separate centres. Since, contrary to Averroës, the celestial spheres are not all "of one nature, of one species and of one matter,"[24] they do not all have to have one centre.

Although Albert did acknowledge the validity of Ptolemy's collection of "eccentrics, epicycles, and diverse centres," he did not discuss its intricacies. Neither did Albert feel obliged to explain the Ptolemaic system in a simplified version as some of his sources, Sacrobosco and al-Fārghānī, a ninth-century Arab astronomer, had done. Instead it was specific features of the work of Ptolemy "whom almost all the moderns follow" which impressed Albert, for example the Ptolemaic order and the system's strength that any part of its whole could withstand comparison to observation. Most astronomers agreed in assuming that Saturn, Jupiter, and Mars, in that order, were closer to the earth than the fixed stars, yet farther away than the other planets. Differences of opinion existed, however, concerning Venus and Mercury which were placed by the "older" astronomers, as Ptolemy notes (Almagest IX, c.1), between the moon and the sun, whereas the more recent astronomers placed all five planets beyond the solar orbit[25] arguing that the other arrangement would imply the occasional occurrence of eclipses, as Mercury and Venus would pass between the sun and the earth.

Although Ptolemy realized that such phenomena had not been observed he returned to the positioning of Mercury and Venus between the sun and moon. Albert attributed to him three arguments

---

[23] De caelo I, tr.1, c.3, (ed. Colon. 5/1: 10, ll. 81-82).

[24] De caelo I, tr.1, c.3, (ed. Colon. 5/1: 10, l. 87).

[25] Both orders imply what were believed to be real geocentric distances. Specific figures for each are attributed to Archimedes; the "recent" order has also been ascribed to Plato. V. Neugebauer, History, pp. 227, and 690-693.

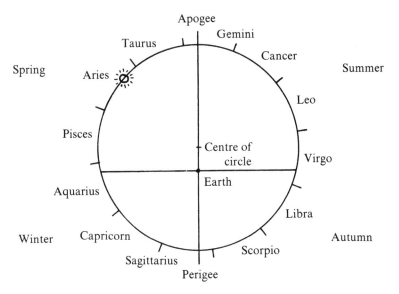

Figure 4.   Solar eccentric model

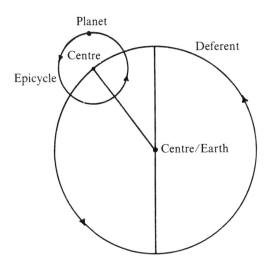

Figure 5.   Basic epicyclic model

denying the necessity of eclipses with this arrangement (*De caelo* II, tr.3, c.11). The first points to motion in latitude which can exclude transits of these planets in front of the sun, for Mercury and Venus never encounter the sun on its ecliptic path. The second and third refutations (not Ptolemy's actually) explain that even if Mercury and Venus were to cross the sun in orbit the resulting superimposition would not be perceptible. Albert makes no mention of Avicenna's claim to have observed transits of Mercury,[26] stating simply that the smallness of both these planets[27] and their proximity to the sun would prevent any observable effect; the translucence of Venus and Mercury would allow the sun's light to penetrate them from behind and thus the sun would constantly be seen unobscured. Unnoted by Albert, Ptolemy did, in fact, find in the brightness of the sun the true cause for the impossibility of naked eye observations of transits of Mercury or Venus.

Albert did not see Ptolemy as the innovator some modern historians of science perceive him to be. Albert felt that his primary achievement was to confirm the appropriateness of the eccentric-epicyclic method proposed by the Chaldeans. Ptolemy adopted the eccentric sphere for the observable unequal progression of the sun through the zodiac signs. He did the complex calculations to determine what number of revolutions of the planet on the epicycle and of the epicycle on the eccentric circle were required to reproduce the observed motions of a planet, as Saturn, for example. To Albert's satisfaction Ptolemy's system with a small modification by the Arab mathematician, Thābit ibn Qurra (826/7-901) accounted for all the motions which observation, astronomical instruments, and reason lead one to believe exist.

iii. Al-Bitrūjī and Other Astronomical Systems

Albert seems to have had a predisposition toward mechanical models, and the descriptive system of al-Bitrūjī, a Spanish-Arab astronomer of the second half of the twelfth century, in particular seemed to haunt him. In the *Metaphysica* Albert wrote, "And thus

---

[26] Bernard R. Goldstein, "Some Medieval Reports of Venus and Mercury Transits," *Centaurus*, 14 (1969), 49-59.

[27] "Mercurius et Venus, quae sunt corpora parva respectu solis. . .": *De caelo* II, tr.3, c.11 (ed. Colon. 5/1: 168, ll. 56-57). The source of Albert's conviction about their size is unknown. It is probably the same source which led him to believe the sun to be 180 times greater than the moon (ibid., ll. 40-41). Cf. *Almagest* V, c.16.

the concept of his imagination is described, but the information of astronomy as far as it concerns an observation of the quantity of motions is not fully supplied."[28] Yet despite this critical assessment of the system's inability to reproduce faithfully all observed motion, Albert describes its general workings without reproach in numerous texts. Perhaps it was the dynamics of this homocentric system that Albert found too philosophically pleasing to ignore it entirely. According to al-Bitrūjī the only true motion of the planets is the diurnal one; their observed passage from west to east is a retardation or lagging behind this twenty-four hour rotation. This idea is based on two philosophical principles: (1) motion implies that the moved object is moved by something else, and (2) a body integral in its nature, as is a planet, cannot suffer two natural motions different in species. Motions which are on different poles about different great circles and which go in opposite directions are specifically different revolutions.

This single east-west motion finds its source in a single mover. The force of this mover is greatest in that part of the heavens immediate to it, less so in those parts joined to it by intermediate spheres. Therefore, the motion of the outermost sphere, the one closest to the first mover, completes a revolution most quickly, in twenty-four hours. The sphere of the zodiac is already diminished from that perfection; it lags behind on a full circle enough to total 1° every one hundred years, Ptolemy's figure for precession.[29] In 36,000 years this retardation results in the completion of a west-to-east circle.

The sphere on which Saturn sits does not complete an east-west rotation either, but lags behind a little each day such that the total of its daily losses amounts to one circle completed in the opposite direction every thirty years. Jupiter is still farther removed from the power of the mover such that its delays grow into one complete circle every twelve years. The east-west motion has even less power for Mars which finishes an opposite circle in two-and-a-half or three years, depending on which of Albert's accounts one follows.[30] The planets

---

[28] *Metaphysica* XI, tr.2, c.24 (ed. Colon. 16/2: 514, ll. 64-66).

[29] "Based on his observations and those of his predecessors, [Ptolemy] noted that the starry sphere moves in the direction opposite to the motion of the universe... moving one degree in a hundred years. Thus it completes a full revolution in 36,000 tropical years." Bernard R. Goldstein, ed. and trans., *Al-Bitrūjī: On the Principles of Astronomy*, Yale Studies in the History of Science and Medicine, 7 (New Haven, Conn.; Yale University Press, 1971), 1: 68-69.

[30] "In duobus annis et dimidio": *Metaphysica* XI, tr.2, c.24 (ed. Colon. 16/2: 514, l. 53); "in tribus annis": *Problemata Determinata*, q. 8 (ed. Colon. 17/1: 52, l. 16); *Summa theologiae*, II, q. 53, m. 2 (ed. Borgnet 32: 568b).

become increasingly more relaxed from the diurnal rotation with the sun completing an opposite motion in one year, Venus, in less than one year, and Mercury, in about nine months. The moon loses as much as $13°$ from every daily rotation, completing its west-east circle in only one month.

The periodic motion of the planets is thus accounted for, while diversity of latitude, distance from the earth, stationary points, retrograde and direct motion, as well as eclipses at various times should all be caused by different positions of the poles of each sphere in the first sphere and by the motions of those poles about the axis of diurnal revolution. Only nine spheres then need be posited, that of the first most simple mover and eight planetary spheres inside it. As each part of this model is moved either as the first mover or by the poles of each sphere, it is necessary to posit only nine other movers or transfers of motion. Although Albert's vague description of this system might reflect a lack of thorough understanding of its complexities, he does accurately perceive its inability actually to reproduce retrograde motion and stationary points. His mention of two specific details about the system, al-Bitrūjī's positioning Venus beyond the sun and Mercury inside it (*De caelo* II, tr.3, c.11) and projected spiral celestial motion ("laulab")[31] indicate that Albert probably read al-Bitrūjī's work *On the Principles of Astronomy* translated into Latin as *De motibus celorum*[32] in 1217 by Michael Scot (ca. 1127-1235).[33]

Many other writings on astronomy influenced Albert. Without doubt he was familiar with his contemporaries' works: *De celestibus* and *Compotus* of Roger Bacon,[34] an English scholastic (ca. 1214-1292), and *De motu supercaelestium* by Robert Grosseteste (ca. 1175-1253), Bacon's teacher. In Paris Albert must have reencountered *On the Sphere* by Sacrobosco with commentaries by Michael Scot and

---

[31] *De caelo* II, tr. 2, c.5 (ed. Colon. 5/1: 137, ll. 24-30). "By its rotation it [a point marked out on the surface of a sphere] will generate a figure (called) a spiral (*lawlab halazūnī*) — a curve (*dā'ira*) which begins at a point and after a complete rotation reaches another point in another plane": Goldstein, *Al-Bitrūjī*, 1: 85.

[32] Francis J. Carmody, ed., *Al-Bitrūjī: De motibus celorum* (Berkeley: University of California Press, 1952).

[33] "Scot or Scott. Scotch philosopher, alchemist, astrologer, translator from Arabic into Latin. . . . Born in Scotland in the last quarter of the twelfth century": George Sarton, *Introduction to the History of Science* (Baltimore: Williams and Wilkins Co., 1931), 2: 579.

[34] Many ideas similar to Albert's are also found in Bacon's *Opus Maius*: cf. *De caelo* II, tr.4, c.11 (ed. Colon. 5/1: 200, ll. 77-78) and *Problemata Determinata*, q. 2 (ed. Colon. 17/1: 49, l. 4). However, as this work was written for Pope Clement IV and there is no evidence of its publication, it is unlikely that Albert ever read it.

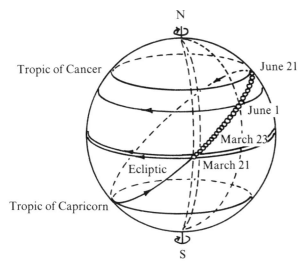

Figure 6.   The motion of the sun is seen as a spiral in the al-Bitrūjian system due to the sun's 4' lag behind each circle completed from east to west by the sphere of the fixed stars in 24 hours.

Robert Anglicus. Albert had access to Chalcidius' fourth-century fragmentary translation and commentary on Plato's *Timaeus*, but he also relied on a fifth-century commentary by Macrobius on Cicero's *Dream of Scipio* for his Platonic theory of the heavens.[35] Both Pliny and Vitruvius, first century AD, provided collections of astronomical tidbits to which Albert alluded.

The technical practices of astronomy were barely touched by Albert. He noted several astronomical instruments which he seemed to consider essential to the science, the armillary sphere or spherical astrolabe, the planispheric astrolabe, the equatory, and one separate instrument for calculating the latitude of stars. Their use was necessary for one mode of astronomical inquiry, "investigation by reason."[36] This is the method of deducing from results of accurate observations recorded by many individuals those heavenly phenomena whose existence can be recognized only after a space of time greater than that afforded by one lifetime, such as the precession of

[35] See Leopold Gaul, *Alberts des Grossen Verhältnis zu Plato*, *Beiträge*: 12/1 (1913), pp. 1-73, for use of Plato's *Timaeus*. Use of Macrobius, *In somnium Scipionis* I, cc.12-15 in *De anima* I, tr.2, cc.3, 6 and 7 (ed. Colon. 7/1: 24, l. 55; 31, l. 15; 33, l. 49); *Summa theologiae* I, q. 4, m. 2, a. 5, p. 1 (ed. Borgnet 32: 966-972).

[36] *Metaphysica* XI, tr.2, c.22 (ed. Colon. 16/2: 519, ll. 52-66).

the equinoxes. Albert did reflect that a knowledge of astronomy could be applied to navigation (*De natura boni*)[37] and the reckoning of geographic distances (*De caelo* II, tr.4, c.11). He displayed, however, no real fascination for numerical values in any practical application.

Since Albert recognized that all the astronomical systems he described were hypotheses, that is, that they were neither exclusively revealed nor demonstrable, his main concern was the dynamics of any system which agreed with the observations of skilled astronomers. The primary source of his ideas was Aristotelianism; no driving force whatsoever is discussed in Ptolemy's *Almagest*. Albert believed that no matter what the arrangement of the spheres, the origin of their motion was one prime mover at the outer periphery of the whole. Beneath this prime separate substance undefined by matter were the material celestial spheres in the required number of units. Each sphere was mobile and contained its own form predicated to it by the prime mover. While Albert was sure that all planetary motion was voluntary, i.e., could not be accounted for by the nature of the celestial body, he did not commit himself as to the ensuing possible causes of motion in each planet, individual intelligences, a motive desire, or even soul-like movers.[38] Albert did, however, adhere rigorously to the Aristotelian law that whatever type of mover actually existed it must be conjoined to the moved object, namely, located "in" the unseen as well as the visible spheres of each planetary unit. These few principles, flexible to the particular system, help to explain Albert's generous embrace of divergent world pictures in astronomy and an unspecified planetary system in astrology.

B. ASTROLOGY

In addition to a work on astronomy Albert also desired to write one on astrology. "With God's favour we shall continue to speak about the stars in *Astronomy* and in the *Science of Elections* and we shall determine those things which are omitted here."[39] While this work does not appear to have been written either, even its contents

---

[37] *De natura boni*, tr.2, p.3, c.2, 2, 3, A, 1, 1 (ed. Colon. 25/1: 56, ll. 65-72). Cf. ibid., "n." 68-70 and Alexander Neckam, *De naturis rerum*, ed. Thomas Wright (London: Longman, Green & Roberts, 1863), p. 183.

[38] Albertus Magnus, *Liber de causis*, I, tr.4, c.7 (ed. Borgnet 10: 423b-427).

[39] *De caelo* II, tr.3, c.6 (ed. Colon. 5/1: 154, ll. 87-89).

were never so clearly defined as those of the projected *Astronomy*. The actual "science" of Albert's astrology, that is, his account of the mechanism and usefulness of celestial influence would certainly have filled a major part of his work. But Albertus Magnus, philosopher and theologian, could not have failed to confront the Stoic, Moslem, and other interpretations of astrology, which were fatalistic, with his own concept of celestial determinism.

The real structure of the universe is of greater importance to Albert's astrology than to his astronomy. Since distance theory did not enter into his understanding of the astronomical systems he discussed, Albert seemed to be aware in astronomy only of the use of the length of a planet's period to determine its relative order.[40] Astrology, on the other hand, presented reasons to him: a planet's position, motion, and qualities produced certain terrestrial phenomena. From simple observation Albert accepts the general premise of astrology, namely, that celestial bodies influence certain events on earth. His belief in the specific effects of certain planets and their arrangements is supported by empirical evidence, reason or written authority, and philosophical principles.

Some effects of the heavens are easily observed. The diurnal rotation of the highest sphere causes day and night every twenty-four hours, for example, and to the moon's motion in phases is linked the ebb and flow of the sea waters. Other effects require more careful scrutiny and an understanding of the principles of astrology to recognize their true celestial cause. Among these are generation, especially spontaneous generation, births of monsters, illness, and "chance" occurrences. The most basic influential factor in all terrestrial events Albert considered to be the nature or qualities of the individual planets.

From various sources the thirteenth century had inherited the Empedoclean doctrine of the four terrestrial elements, fire, air, water, and earth and their corresponding qualities, hot-dry, hot-wet, cold-wet, and cold-dry, etc. But there are two specific works in the Middle Ages which provided a detailed astrology based on the elements: (1)

---

[40] The ancient Greek ordering of the seven planets, the one recognized today as correct, was based upon sidereal periods, an order suggested by Aristotle's statement (*De caelo* II, c.10, 291a30-291b10) that the farther away a planet is the slower it appears to move. Ptolemy, however, while he based his planetary order on considerations including the sidereal periods, realized that the question could only truly be solved by measurements of parallaxes, imperceptible to the naked eye (*Almagest* IX, c.1).

another of Ptolemy's works, *Tetrabiblos* translated into Latin as *Quadripartitum* in the early twelfth century by Plato of Tivoli,[41] and (2) a widely read work attributed to Ptolemy, *Centiloquium*, a compilation of one hundred astrological sentences.[42] In them a relationship is identified between the planets and these pairs of qualities and further between those qualities in terrestrial objects and the planets possessing the same qualities. The premise was that each object on earth is constituted by a mixture or combination of the primary qualities. That object is in turn linked to one or more celestial sources of influence which govern its components; all living things, containing vital heat, for example, are connected to the sun as their source. Albert undoubtedly based his attributions of pairs of qualities to the seven spheres of the seven planets on these sources.

Albert explained that to avoid impossible oppositions there were only four ways of pairing the qualities, cold with dry, cold with moist, hot with dry, and hot with moist. Among these qualities are those which when mixed with matter cause an active, strong bond and a living union; others reject being mixed. Each planet has the properties of one pair of qualities "not inasmuch as they [the planets] are fashioned by them, but inasmuch as they are producing them in matter which is susceptible to contrariety."[43] This explanation of Alexander of Aphrodisias (ca. AD 200), a Greek commentator on Aristotle, adopted by Albert, allows the planets to "have" the qualities and yet remain themselves made of the fifth element, quintessence.

Albert built his astrological arrangement of the planets around the sun located as the fourth celestial body from the earth. From that central position it can inform the planets with influential light, for just as the visible light of all the stars is provided by the sun, so through it their invisible power to affect the terrestrial globe is obtained. The sun's direct effect on the earth, to initiate and sustain life there (*Metaphysica* XI, tr.2, c.25), is caused by its simple motion, the size of the solar body and its qualities of heat and dryness.

Next to the sun is Mars, also a motivator of matter through heat

---

[41] Ptolemy, *Tetrabiblos*, ed. F. E. Robbins (Cambridge, Mass.: Loeb Classical Library, 1940).

[42] Ptolemy, *Karpos*, ed. Ae. Boer in *Opera quae extant omnia*, 3/2 (1952; Leipzig: Teubner, 1961).

[43] *De caelo* II, tr.3, c.2 (ed. Colon. 5/1: 145, ll. 94-98).

and dryness. Mars, however, is so hot that it must be situated at a greater distance than the sun from the earth.

Saturn has the hostile qualities of cold and dryness, those which attempt to cause the dissolution of earthly mixtures. Of all the planets it is located farthest away from the earth and given the slowest[44] motion.

Jupiter is placed just inside Saturn to counter its death-bearing effects. Its qualities are moisture and heat, a "spiritual" heat, "one from which the spirits which are the bearers (*vectores*) of the powers of life are created."[45]

Venus also of life-giving influence is like Mars suitably joined to the sun. Its moisture, balanced to the other quality, cold, is highy subject to combination in living things.

The moon, although it shares the qualities of cold and moisture with Venus, has need of the closest proximity to the earth and powerful motion. In the instance when one of a planet's qualities predominates, as is the case with the moon's moisture, the effect on terrestrial bodies is weak if it is not enhanced by a strong motion of that planet.

Located between Venus and the moon Mercury is the only planet which has an effect upon the actual mixing, not just on the ingredients of a mixture or the already mixed thing. This explains its many intricate motions, for each planet must have as many motions as are necessary to bring about its particular effects.

Thus for Albert instances of planetary influence are effected by the following order of the planets: Saturn, Jupiter, Mars, the sun, Venus, Mercury and the moon. However, the composition of being with matter in generation and destruction is attributed to three other spheres outside the planetary seven.

The outermost sphere, the cause of being in all things, has a diurnal motion from east to west. Its motion is continuous and permanent reflecting divine existence. The second sphere has a motion in the opposite direction allowing the mixture of being and matter. The determination of the quality and shape of that which is composed is the role of the sphere of the fixed stars. It is the third of all ten spheres and the first visible one.

---

[44] "Tardissimi motus": Albert, *Metaphysica* XI, tr.2, c.25 (ed. Colon. 16/2: 515, l. 62). Albert is considering here the west-east periodic motion of the planets and not their al-Bitrūjīan motion, in which case Saturn would be the fastest of all the planets from east to west.

[45] *Metaphysica* XI, tr.2, c.25 (ed. Colon. 16/2: 515, ll. 75-76).

Like the fixed stars the planets must be attached to spheres in order that the influences of them might be transmitted and distributed to individual things. A motion relayed from the prime mover outside all the spheres can then extend over the whole surface of each successive sphere, while the particular motion of the planet on the sphere, up or down, closer or farther from the earth, allows its effect on one terrestrial object to be different from its influence on another. A celestial sphere is the only shape which can give constant attention to its centre, the earth, from all sides and still provide an individuation of effect.

Light is the instrument of influence for each celestial body. Albert used the analogy of the intelligence of an artist to explain the production of the effect (*Metaphysica* XI, tr.2, c.26). It brings about a form, a work of art, by means of the artist's hands and tools. The supreme active intelligence of the heavens (*intellectus unicus universaliter agens*) uses the light of a star as its instrument. Once the intelligence has brought a form to the lighted celestial body, that body transfers this form over into terrestrial matter which it changes from a potential to an actual thing.

The science of astrology is then for Albert the reading of the influence of planetary light on earthly things from the positions and interrelationships of the planets and stars in the described system. Albert reflects his belief in particular celestial configurations, especially the positions of planets relative to the houses and/or signs of the zodiac, as portents of wars, natural disasters, and deformed births. Although he mentioned only the desire to write a work on electional astrology, that which deals with the most propitious moment for initiating a personal effort, he was obviously aware of another branch of astrology popular in the Middle Ages, judicial astrology; it considers the positions of the planets with respect to their influence upon entire populations, countries, or cities. In *De caelo* Albert attributed to Ptolemy two works he consulted, one on electional astrology, *De accidentibus parvis particularibus* (*Concerning Particular, Small Events*) about "events in the life of an individual born under this or that constellation," and one on judicial astrology, *De accidentibus magnis universalibus mundo* (*Concering Great Universal Events in the World*), an eight part work about great social events involving large numbers of people.[46] Another work of eight books,

---

[46] *De caelo* II, tr.3, c.5 (ed. Colon. 5/1: 150, ll. 58-67).

the largest surviving Latin treatise on astrology, *Mathesis* written by the Roman astrologer, Julius Firmicus Maternus, between AD 334 and 337 was known to Albert.[47] He also explicitly notes a major work by Abū Ma'shar, an Arab astrologer (d. 886) *De conjunctionibus planetarum* (*Great Conjunctions*) as a source for interpreting the portent of the arrangements of celestial bodies.

Abū Ma'shar is just one of many authors mentioned in a brief thirteenth-century treatise called the *Speculum Astronomiae*.[48] During years of debate about its authorship the work has been attributed to, among other known figures of the High Middle Ages,[49] Albertus Magnus.[50] It is an extremely valuable work for its bibliographical content; authors' names, titles, and/or incipits of both astronomical and astrological writings probably available to Albert and his contemporaries are listed by an author intent on recommending their value in a Christian context. The unknown author's main concern was judical astrology which he subdivided and discussed in detail. While there is some question as to how representative of the thirteenth century the *Speculum* author's vague understanding of astronomy was, his qualified praise of astrology (and his library) would certainly have been appreciated by Albert.

"Changes of the general state of the elements and of the world" is at the mercy of the effects produced by the planets and their spheres, Albert wrote.[51] To a conjunction of Jupiter and Mars, both occupying the same degree position in the zodiac, with other planets aiding in Gemini, he attributed pestilential winds and the corruption of the air which results in a great plague. Albert also mentioned the opinion of "certain Arabs" that floods are due to the imagination of the intelligence which moves the sphere of the moon.[52] Jupiter and Saturn when either in conjunction or in the same trigon, a set of three signs,

---

[47] Julius Firmicus Maternus, *Matheseos*, ed. W. Kroll and F. Skutsch, Bibliotheca scriptorum graecorum et romanorum Teubneriana (Leipzig: Teubner, 1897-1913).

[48] Albertus Magnus, *Speculum Astronomiae*, ed. Caroti, Pereira, & Zamponi, under direction of Paola Zambelli, Quaderni di storia e critica della scienza, New series, 10 (Pisa: Domus Galilaeana, 1977).

[49] P. Mandonnet, "Roger Bacon et le *Speculum astronomiae*," *Revue néo-scolastique*, 17 (1910), 313-335.

[50] Paola Zambelli, "Da Aristotele a Abū Ma'shar, Da Richard de Fournival a Guglielmo da Pastrengo," *Physis*, 15 (1973), 1-26 (extr.); Lynn Thorndike, "Further Consideration of the *Experimenta, Speculum Astronomiae* and *De secretis mulierum* ascribed to Albertus Magnus," *Speculum*, 30 (1955), 423-427.

[51] *De causis elementorum* I, tr.2, c.9 (ed. Borgnet 9: 620b).

[52] *De causis elementorum* I, tr.2, c.9 (ed. Borgnet 9: 619a).

usually signify major events in the world. Even meteorological phenomena beneath the sphere of the moon can be astrologically significant. From the seventh tract of Abū Ma'shar's *Great Conjunctions* Albert recounts that wars are indicated by comets; Mars, as was generally accepted, was a portent of such disasters, and comets, as are all objects of the sublunary world, are governed by and hence reflect the dominance of Mars.[53]

Individual matters as births and conceptions are subject to planetary design as well. Propagation of most species requires at least the seed of that species which is predisposed to perpetuating its kind, matter, and the heat and light of the sun.[54] In the case of lower creatures, the sun itself can provide almost everything needed for generation; it supplies the heat and light which enables available decaying matter to acquire a new form and generate worms, eels, and some insects spontaneously. But species of greater complication require these conditions as well as a position of the celestial bodies amenable to their generation.[55]

From a work attributed to Ptolemy, *De nativitatibus* Albert learned that abnormal births such as Siamese twins could be ascribed to two causes: (1) a particular cause, the formative power of the seed and the preparation by the mother of the matter to accept a human form, and (2) a general cause, the location and relationship of all the stars at the time at which the seed falls into the mother (*Physica* II, tr.2, c.17). Although it cannot be known which of these causes is responsible for a particular deformed infant, Albert noted that there are several recognized malefic planetary configurations under which conception or birth should be avoided. Albert knew of a warning by Firmicus Maternus among others that children born under a new moon might be defective in sense and discretion.[56] He wrote specifically that normal human generation is not possible when the luminaries meet in Aries near the star Algol (*versus caput Gorgonis*) if Jupiter is not helping and Venus is not visible.[57] (Note

---

[53] *De meteoris* I, tr.3, c.11 (ed. Borgnet 4: 507b-508).

[54] Cf. Isaac Newton, *Mathematical Principles of Natural Philosophy*, ed. Florian Cajori, rev. 2nd. trans. Andrew Motte (Berkeley: University of California Press, 1962), 2: 547.

[55] *Quaestiones super de animalibus* XVII, q. 14 (ed. Colon. 12: 295, ll. 61-80).

[56] *De natura boni*, tr.2, p.3, c.2, 2, 3, A, 1, 1 (ed. Colon. 25/1: 49, ll. 76-79). Cf. Firmicus Maternus, *Matheseos* IV, c.1, n.10.

[57] *Problemata Determinata*, q. 35 (ed. Colon. 17/1: 61-62, ll. 80-13); *De fato*, a. 4 (ed. Colon. 17/1: 73, ll. 20-24).

that a star in a certain arrangement, triplicity or sextile, can also take on the qualities of the planet or planets.)[58]

Empirically Albert knew this to be true, for he claims to have seen twice the results of conceptions so timed, human beings born with truncated legs and arms who "will not have the appearance of a human body."[59] However, another kind of monster, piglets born with human heads, is actually the more conclusive "test case" for stellar influence. Such a phenomenon could not be the result of the sperm's own active quality. Its source cannot be a combination of pig and human sperm, for sperm diverse in species would corrupt each other and nothing would be generated from them. Matter is able to be prepared by the celestial bodies to acquire the shape of a human being, which results in a human head, but the predisposition of the sperm cannot be completely determined by the stars. If that were true, a man could be generated from seed which did not stem from man but rather from a goat or other animal. The planets and stars, Albert reiterates, can only induce specific effects on the seed which have a result with respect to the same species to which the seed is predisposed.[60]

Albert did not consider himself an astrologer but merely one who was aware of the logos behind the science. He recommended the practice of others for guidance with the particulars. A group of astrologers called genethlialogists could prognosticate more specifically the results of conceptions at certain times and offer predictions for the future of an individual based on the configuration of the heavens at the time of birth. Doctors with their knowledge of astrology should be entrusted with the care of one's body, composed as it is of four humours corresponding to the four elements and their qualities.[61] Albert noted several times how through prognostication a wise doctor could deter the celestial inducing of a quartan fever on one's melancholy humour by causing a predominance of the opposing humour, blood, in the body.[62]

Melancholics, he believed, were the best at foretelling future events and predicting fatal ones because they were less worried than

[58] Cf. Ptolemy, *Centiloquium*, v. 73.

[59] *Problemata Determinata*, q. 35 (ed. Colon. 17/1: 62, ll. 10-11).

[60] *Physica* VIII, tr.2, c.10 (ed. Borgnet 3: 591b).

[61] Cf. Aristotle, *De partibus animalium* II, c.1 646a12-24.

[62] *Problemata Determinata*, q. 9 (ed. Colon. 17/1: 52, ll. 53-56); *De fato*, a. 2 (ed. Colon. 17/1: 69-70, ll. 69-4).

other people about possessions and less diverted from the observation of immutable things than were other people (*Physica* II, tr.2, c.21). They were well adapted to being augurs who see events in favourable stellar arrangements or diviners who practice the "science" of interpreting dreams. Dreams are the key to stellar influence. They exemplify for Albert the fine line between the individual free will and the so-called fatalistic determinism of the stars. A dream is fate, celestial power, using the nature of specific bodies to impress on the soul images which are signs of fated events, i.e., events to which the disposition of the heavens is already inclined.[63]

Albert held that dreams do, in fact, incline men's minds to diverse desires, fantasies, and awarenesses of future events but what is dreamt does not necessarily happen.[64] Man is able to divert himself by his own will, or, if not, is able to be diverted from fated things ("cum tamen ab omnibus his averti possit homo").[65] The terrestrial sphere is a realm of contraries, in motion, up and down, in quality, hot and cold, black and white, etc.; by taking a contrary course to that determined by the stars a human being can provide his own impediment to the realization of fate. "Therefore," Albert warned, "often the astrologer speaks the truth and nevertheless what he says does not occur because his statement was most true according to the disposition of the heavens, but this disposition was prevented by the mutability of inferior [*terrestrial*] things."[66] Dreams and other forms of foreknowledge impose no necessity on man; in fact, they allow him to contravene celestial influence.

Albert realized that for many the role of the celestial bodies as omnipotent determinators was not to be so easily dismissed. To treat the entire question of fate, its existence, and its role, Albert wrote a small treatise on fate, *De fato*. He began with summaries of twenty-one attempts to explain fate, collected from the writings of Boethius, Aristotle, Augustine, Macrobius, and others. The discussion, despite Albert's sparse representation of it, had been heated for centuries. The medieval Church was generally opposed to any wholehearted commitment to astrology.

Some Church leaders, Augustine included, attacked the art of astrology for its lack of precision, for example, in calculating

---

[63] *Physica* II, tr.2, c.21 (ed. Borgnet 3: 157b).
[64] *De somno et vigilia* III, tr.2, c.5 (ed. Borgnet 9: 202-203a).
[65] *Physica* II, tr.2, c.21 (ed. Borgnet 3: 157b).
[66] *De fato*, a. 4 (ed. Colon. 17/1: 73, ll. 52-56).

moments of conception, which rendered it unable to predict anything exactly. But opposition was principally based on the limitation the Church felt was imposed upon the individual free will to choose between right and wrong. At least two schools of thought, active in the thirteenth century, did, in fact, believe in such absolute determinism.

The Moslem school, represented by Avicenna, the Persian philosopher (980-1037), underlined the idea that the prime mover, as the universal necessary agent, had excluded free will by ordering every action to a predetermined end. There was also one branch of the Aristotelian school which interpreted Aristotle's dynamic model of celestial motion to imply that since one prime mover regulated everything, there could be no deviation from its referred motion either through the act of a separate intelligence, as the Neoplatonists thought, or by chance and fortune.

However, for Aristotle himself, there was no absolute determinism for both a logical and an ontological reason. Future contingent events are per se indeterminate. "It is not of necessity that everything is or takes place; but in some instances there are real alternatives, in which case the affirmation is no more true and no more false than the denial; while some exhibit a predisposition and general tendency in one direction or the other, and yet can issue in the opposite direction by exception."[67] Aristotle seems to have had the application of this conclusion to determinism, including astrological determinism, in mind when he wrote: "For events will not take place or fail to take place because it was stated that they would or would not take place, nor is this any more the case if the prediction dates back ten thousand years or any other space of time."[68]

Another major objection Aristotle had to absolute celestial determination in the individual man was that man has free will. "For nature, necessity and chance are thought to be causes, and also reason and everything that depends on man."[69] Albert added, "The human soul (*anima*) according to the Philosopher is an image of the world (*imago mundi*). For that reason in that respect in which it is the image of the prime cause and intelligence, it is impossible that it

---

[67] Aristotle, *On interpretation*, ed. and trans. W. D. Ross (London: Oxford University Press, 1942) 1: IX, 19a18-22.

[68] Aristotle, *On interpretation*, ibid., 1: IX, 18b38-40.

[69] Aristotle, *Ethica Nicomachea*, ed. 2nd trans. W. D. Ross (London: Oxford University Press, 1942) 5: III, c.3 1112a31-33.

should be subject to celestial motions."[70] God, the voluntary creator "is the cause and causes through his own virtue and his own will, which are his essences," Albert wrote.[71] Hence, man, whose soul is made in His image, reflects Him in the use of his own will, free from exterior influence.

With the aid of Aristotle and both the *Tetrabiblos* and *Centiloquium* with commentaries on both by Haly ibn Rodan, Albert presented in *De fato* his own explanation of fate: "Fate is the form of the order of being and of the life of inferior things effected in them by the periodic motion of the celestial sphere which surrounds their births with its rays."[72] His continuation explains why he, like Aristotle, believed that celestial forces are not binding on the free will.

> This form is not a form giving being, but rather the form of the universal order of being and life, simple in essence, multiple in power; it has a simple essence for the simplicity of the general rotation of the sphere, it has a multiplicity of power from the multitude of those things which are contained in the sphere. It flows from many stars, locations, spaces, constellations, (*imaginibus*), rays, conjunctions, *praeventionibus*, and multiple angles which are defined by the intersections of the rays of celestial bodies, and by the production of rays around the centre [*the earth*] in which alone, as Ptolemy says, all the powers of those which are in the celestial sphere are gathered and joined together. This form is halfway between necessary and possible; whatsoever is in motion of the celestial sphere is necessary, however, whatsoever is in the matter of the generable and destructible is possible and mutable. That form effected by the celestial sphere and adhering in generable and destructible things is halfway between both.[73]

Fate, for Albert, is not the forced execution of celestial influence at all. It is the disposition of the heavens infused with its whole assembly of possible effects. All things have a relationship to two causes: (1) to causes acting on universals, and (2) to causes acting on particulars. The second kind comprises the celestial motions; the first is the ultimate cause, God. Many confuse the first cause which effects necessary motion in the heavens with the second, but while the latter emanates from God and from the heavens, it effects only motion on earth which has its own contrary and hence is not bound by necessi-

---

[70] *De xv problematibus*, p. 4 (ed. Colon. 17/1: 36, ll. 52-62).

[71] *Summa theologiae* I, tr. 13, q. 55 (ed. Borgnet 31: 557a).

[72] *De fato*, a. 2 (ed. Colon. 17/1: 68, ll. 31-33).

[73] *De fato*, a. 2 (ed. Colon. 17/1: 68, ll. 36-54).

ty. Most often forgotten entirely is a third cause by which human beings are affected, the free motive intelligence of man. It is the most proximate and hence the most decisive cause of all to human action.

But while there is no binding fate, neither is there complete absence of celestial influence. The early Greek atomists, such as Democritus, believed that nothing in the universe was subject to determinate causes. Albert assumed that this error resulted from the fact that a science of the stars had not at that time yet been discovered. Retrograde motion and changes in planetary distance appeared to be caused by "chance" because their place in a system of motion had not yet been conceived. Chance and fortune do happen, but they are the events which happen outside the intention of both celestial motion and the predisposing intention of nature. Their existence is relative to and dependent upon the total, ordered astrological/astronomical structure of a universe. Both "scientiae" comprise the specific natures in the world and the freely moving spheres of the heavens, each created by a free, first universal cause and directed freely by a first universally active intelligence.

7

Albertus Magnus on Alchemy

Pearl Kibre
City University of New York

Albert's interest in alchemy,[1] the art, in his words, that best imitates nature,[2] is revealed in the references to the subject in his authentic writings, particularly the *Book of Minerals* (*Liber mineralium*), his *Commentary on Aristotle's Meteorology*, and other tracts. He had investigated and made a careful study of the subject in the course of his inquiry into the nature of metals, for guidance in which he had sought in vain for the treatise by Aristotle.[3] Without that guide, he was, as he reported, obliged to follow his own devices and to set down what he had learned from philosophers or from his own observations. He had thus at one time become a wanderer, journeying to mining districts to "learn by observation the nature of metals." "And," he stated, "for the same reason I have inquired into the transmutations of metals in alchemy, so as to learn from this, too, something of their nature and accidental properties."[4] Among the

---

[1] For previous accounts of this subject see especially J. R. Partington, "Albertus Magnus," *Ambix* 1 (1937), 3-20; Lynn Thorndike, *History of Magic and Experimental Science* (New York, 1923) 2: 569-573; also my "Alchemical Writings Ascribed to Albertus Magnus," *Speculum* 17 (1942), 499-518; and Albertus Magnus, *Book of Minerals* translated into English by Dorothy Wyckoff (Oxford: Clarendon Press, 1967; henceforth indicated as Wyckoff), introduction pp. xxx-xxxii.

[2] Albertus Magnus, *Liber mineralium* (*Opera omnia*, ed. A. Borgnet, Paris 1890, vol. 5, p. 61; henceforth indicated as *Min.* with number of book, tractate, and chapter): *Min.* III.1,2; Wyckoff, p. 158.

[3] *Min.* I.1,1; III.1.1; Wyckoff, pp. 9, 153, 263, item 14; and 267, item 8.

[4] *Min.* III.1,1; Wyckoff, p. 153; Partington, p. 11.

names of the philosophers to whom Albert had turned were some of the principal authorities on alchemy, current in the twelfth and thirteenth centuries in Latin translation from the Arabic, comprising chiefly Hermes, Callisthenes (that is Khālid ibn Yazīd), Democritus, Gigil, and Avicenna.[5] Of these Albert relied principally upon Avicenna (Abū 'Alī ibn Sīnā, 980-1037), utilizing both the apocryphal and authentic tracts appearing under his name. Albert thus cited a section of the supposititious Avicenna tract *On the Soul in the Art of Alchemy* (*De anima in arte alchemiae*), one of the most influential of the alchemical tracts upholding the possibility of the transmutation of metals.[6] Naming the work, "The Physical [Stone]," Albert referred only to the final section, the "Exposition of the physical stone by Avicenna to his son Aboali (Abū 'Alī)," which circulated as a separate tract.[7] In addition, Albert utilized the so-called "Letter of Avicenna to King Hasen (or Hazen), the philosopher," in which is set forth the view commonly held in the thirteenth century, that quicksilver (mercury or *argentum vivum*) and sulphur are the materials of all metals and hence basic to the alchemical process.[8] This so-called newer theory of the components of metals, Albert contrasted with the older view expressed by Aristotle, that metals are formed from a subtle fatty moisture (*humidum unctuosum subtile*) combined with a subtle earthy tincture.[9]

In addition, in his search for "immediate efficient causes existing in the material and transmuting it,"[10] Albert continued his critical

---

[5] For Hermes and the others named, see below.

[6] *Min.* III.1,4 and 9; Wyckoff, pp. 161 and 177; also 283-284. Ascribed to Avicenna, the work is printed in J. J. Manget, *Bibliotheca chemica curiosa* (Geneva, 1702) 1: 633-636, especially p. 634. Although the tract purported to have been prepared in 1012, it was probably written in Spain about 1140 and appears to have been translated into Latin from the Arabic in 1235 (Partington p. 3).

[7] *Min.* III.1,4; Wyckoff, pp. 161 n.1, 284; Dorothy Waley Singer, *Catalogue of Latin and Vernacular Alchemical Manuscripts in Great Britain*, 3 vols. (Brussels, 1928-1931) 1: 117.

[8] *Min.* III.1,4,9; Wyckoff, pp. 161, 177-178, 284. *The Letter of King Hasen* is printed in L. Zetzner, *Theatrum chemicum*, 6 vols. (Strasbourg, 1659-1661) 4: 863-874. For manuscripts see L. Thorndike and P. Kibre, *A Catalogue of Incipits of Mediaeval Scientific Writings in Latin*, new and augmented edition (Cambridge, Mass., 1963; henceforth indicated as TK with column number), 1036. The authenticity of this tract as by Avicenna has not generally been accepted although Stapleton (H. E. Stapleton, R. F. Azo, M. Hidayat Husain, and G. L. Lewis, "Two alchemical treatises attributed to Avicenna," *Ambix* 10 (1962), 41-83) who discovered an Arabic manuscript of this text suggested it was an early work of Avicenna written before he "abandoned his belief in transmutation": Wyckoff, p. 284.

[9] *Min.* III.1,2; Wyckoff, pp. 155-159; Partington p. 10.

[10] *Min.* I.1,4; Wyckoff, p. 19.

evaluation of so-called authorities and the operations of alchemy. He characterized as erroneous the opinions of Hermes, Empedocles, Democritus, "and some of those in our own time who are practitioners of alchemy," and who are concerned with the generation of stones. These authorities mistakenly suggested, he reported, that all stones are produced by accident without a specific cause for their production,[11] whereas the true or productive cause, Albert asserted, is a "mineralizing power."[12] The making of stones, he concluded, by the operations of the alchemists is even more difficult than the making of metals.[13] Albert had earlier noted that just as metals are formed from water congealed by intense cold and dryness,[14] just so is the work of the alchemists performed, that is by separating and sublimating the humidity of iron.[15] Moreover, in the operations of alchemy, he noted that mercury (*argentum vivum* or quicksilver) that is dried by much burning and mixture with sulphur will be coagulated by heating in a furnace with green wood.[16]

Albert next went on to draw attention to the discussions unfavorable to the transmutation of metals which he attributed also to Avicenna, wrongly citing the "Letter to King Hasen" as the source rather than Avicenna's tract on minerals, *De congelatione et conglutinatione lapidum*.[17] In this treatise which has been shown to be an excerpt from the authentic book *Book of the Healing of the Soul* (Kitāb al-shifā'),[18] Avicenna disparaged the "claims of the alchemists" whose "power to bring about any true change of metallic species," he denied with the words, "Let practitioners of alchemy know that they cannot transmute one form of metal into another, but only make something similar. . . .As to the rest, that is that specific differ-

---

[11] *Min.* I.1,4; Wyckoff, pp. 18-20.

[12] *Min.* I.1,5; Wyckoff, p. 22.

[13] *Min.* I.1,5; Wyckoff, p. 23.

[14] *Min.* III.1,7; *Meteor.* III.5 unicum (Borgnet 4: 701).

[15] *Meteor.* IV.2.9 (Borgnet 4: 761a).

[16] *Meteor.* IV.3,2 (Borgnet 4: 775); Partington p. 10.

[17] The *De congelatione* was translated into Latin ca. 1200 by Alfred of Sareshel. Frequently cited by Latin authors, it was often attributed to Aristotle as Albert noted. *Min.* III.I,9; Wyckoff, pp. 177, 178; Partington p. 4. It was cited with an English translation by E. J. Holmyard and Mandeville, *De congelatione et conglutinatione lapidum* (Paris, 1927), pp. 45-54. For MSS and earlier printed editions, see TK 1565; and Wyckoff, pp. 283-284; also A. C. Crombie, "Avicenna's Influence on the Medieval Scientific Tradition," in *Avicenna: Scientist and Philosopher. A millenary symposium*, ed. G. M. Wickens (London: Luzac and Co., 1952), 87-88; and bibliography in n. 14.

[18] Holmyard and Mandeville, p. 41.

ences between metals may be removed by some clever method, I [Avicenna] do not believe. . .possible."[19] Albert went on also to paraphrase Avicenna's further statement (in the "Letter to King Hasen on Alchemy") that he had examined the books of those upholding "the art of transmutation" and had found them devoid of the reasoning that belongs to every art. He had found most of their content nonsensical. Moreover, an examination of the books of those who opposed the art of transmutation revealed that their arguments and reasoning were too feeble and trivial to destroy belief in the art.[20] Hence it appeared prudent to add that "specific forms are not transmuted, unless perhaps they are first reduced to prime matter (*materia prima*) . . . and then, with the help of art, developed into the specific form of the metal" desired.[21] Albert then added to Avicenna's stricture his own criticism of the alchemical literature: "I have examined many alchemical books, and I have found them lacking in [evidence] and proof." They merely rely "on authorities," and conceal "their meaning in metaphorical language, which has never been the custom in philosophy. Avicenna is the only one who seems to approach a rational [attempt], though a meagre one, towards the solution of the above question, enlightening us a little."[22]

Albert, moreover, was critical of opinions expressed by alchemists which did not coincide with those of Aristotle or Avicenna. He cited the undeniable but nonspecific statement that metals are made up of all the elements, expounded in the *Book of Alchemy*, by Hermes, the mythical or legendary founder of alchemy, who was probably of Greek origin but was known to the west through Latin translations from the Arabic.[23] Albert also characterized as "the strangest and most ridiculous of all opinions, the one that held that '*calx* (quicklime) and *lixivium* (lye) are the material of all metals'," an opinion he attributed to Democritus. Albert asserted that this statement about

---

[19] Text in Holmyard, p. 54; *Min.* III.1,9; Wyckoff, p. 177; Partington, p. 3.

[20] *Min.* III.1,9; Wyckoff, p. 177 n.2.

[21] *Min.* III.1,9; Wyckoff, p. 178. The quotation is from the *De congelatione*, Holmyard and Mandeville, p. 55.

[22] *Min.* III.1,7; Wyckoff, p. 172.

[23] *Min.* III.1,4; Wyckoff, p. 162; also pp. 282-283. Hermes' work referred to here may have been the *Emerald Table* (*Tabula smaragdena*) included in the Pseudo-Aristotle, *Secret of Secrets*, translated from the Arabic into Latin by Plato of Tivoli (fl. 1134-1135). Hermes is credited with the ascription of the names of the planets to the names of the metals. *Min.* III.I,6: Wyckoff, pp. 168-169. He is also credited with the quotation from the *Emerald Table* that "earth is the mother of metals and heaven their father": Wyckoff, p. 169.

the material of metals does not fit the fact, and is incorrect.[24] He further referred to the opinion "that alchemy is the science that confers upon inferior metals the nobility of the superior ones," expressed by Callisthenes;[25] and to the attempt "to prove that fused ash is the material of metals," reported in the book of *Secrets* by "Gigil of Moorish Seville," which Albert noted, "has now been returned to the Spaniards."[26] Albert went on to characterize Gigil's arguments in defense of his assertion as "unconvincing and stupid," and "Gigil himself" as "a mechanic and not a philosopher" who relied too greatly upon "the mechanical operations of alchemy" and was guilty of making incorrect assertions regarding natural science.[27]

Albert next drew attention to the procedures and objectives of the alchemists. "The experience of the alchemists," he asserted, "confronts us with two grave doubts. For they seem to say that the specific form of gold is the sole form of metals and that every other metal is incomplete — that is, it is on the way toward the specific form of gold, just as anything incomplete is on the way toward perfection." Thus metals lacking the form of gold in their material "must be diseased." Hence to cure or remove these diseases the alchemists endeavor "to find a medicine which they call the *elixir*, by means of which they may remove the diseases of metals. . ." and bring "out the specific form of gold."[28] Continuing further with the use of the *elixir* and the transmutation of metals, Albert asserted that since "it is found by experience that by means of the *elixir* copper turns to silver, and lead to gold, and iron likewise to silver," the alchemists erroneously conclude "that the specific form of all metals is one and the same, but the diseases of the material are many," an opinion with

---

[24] *Min.* III.1,4; Wyckoff, p. 162. See also *Min.* I.1,4; Wyckoff, p. 20 n. 8, and p. 281; Partington p. 10. Albert, as Professor Wyckoff suggests, may have thought he was citing Democritus of Abdera (fifth cent. BC), one of the founders of the atomist philosophy, known only through Aristotle who was a severe critic of Democritus' theories. More likely, however, as Professor Wyckoff suggests, Albert may have seen a reference to a Democritus in the *Turba philosophorum* (*The Conference of Philosophers*), edited by J. Ruska (Berlin, 1931) and by Plessner (1954). Albert, however, would have thought them to be one and the same person.

[25] *Min.* III.1,7; Wyckoff, pp. 171-173, 283. Callisthenes, apparently for Khālid ibn Yazīd ibn Mu'āwiya, author of the *Book of the Three Words* (*Liber trium verborum*). The work is printed in J. J. Manget (1702) 2: 189-191.

[26] Gigil (Abū Dā'ud ibn Juljul) a physician of Spain in the tenth century. Wyckoff, p. 163; *Min.* III.1,4; Wyckoff, pp. 161-164, 284. As Professor Wyckoff indicates, the Christians reconquered Seville in 1248.

[27] *Min.* III.1,4; Wyckoff, p. 163.

[28] *Min.* III.1,7; Wyckoff, p. 171. On the *elixir* as a medicine, see *Min.* I.1,1; Wyckoff, p. 10.

which Albert would not concur. He went on rather to discuss the means, that is the use by the alchemists of "calcination, sublimation, distillation, and other operations," to induce the *elixir* to penetrate into the material of metals, and hence possibly "to destroy the specific forms of metals that originally were in the material. The material that remains can then with the help of the alchemical art be reduced to another specific form, just as seeds are helped by ploughing and sowing or [as] nature is helped by the efforts of the physician."[29] This explanation, Albert noted, was not acceptable to "Hermes and Gigil, and Empedocles and almost all that group of alchemists," who appeared to defend the "stranger" principle "that in any metal whatever there are several specific forms and natures, including one that is occult and one that is manifest."[30] Albert himself had earlier expressed his opinion that in the case of the "experiments which the alchemists bring forward," to establish the validity of their conclusions, they do not offer enough proof.[31]

Albert then went on further to compare the procedure of the skillful alchemists with that of the skillful physicians, and also to enlarge upon his theory of nature's role. The skillful physicians, he asserted, "by means of cleansing remedies clear out the corrupt or easily corruptible matter that is preventing good health — the end which physicians have in mind. In doing so, they strengthen and aid the power of nature, directing it to bringing about natural health. [Good] health will [thus] be produced by nature, as the efficient cause; and also by art as the means and instrument." The skillful alchemists also proceed similarly in transmuting metals. They first cleanse thoroughly the mercury or quicksilver and sulphur, the constituents of metals; then, when this is done, "they strengthen the elemental and celestial powers in the material, according to the proportions of the mixture in the metal that they intend to produce." Thus "nature itself performs the work, and not art, except as the instrument, aiding and hastening the process."[32] "The alchemists appear, in this manner, to produce and make real gold and real silver, since the elemental and celestial powers can produce in artificial vessels, if they are formed like those in nature, whatever they produce in natural vessels. Hence "whatever nature produces by the heat of the sun and stars, art also

---

[29] *Min.* III.1,7; Wyckoff, pp. 173-174.
[30] *Min.* III.1,8; Wyckoff, pp. 174-175.
[31] *Min.* III.1,7; Wyckoff, p. 173.
[32] *Min.* III.1,9; Wyckoff, p. 178.

produces by the heat of fire, provided the fire is tempered so as not to be stronger than the self-moving formative power in the metals."[33] The inclusion of the "stars" as an agency influencing chemical operations is further exemplified in Albert's assertion in his *De causis elementorum* that "when skilled alchemists work during the waxing of the moon, they produce purer metals and stones." Albert also referred to the introduction by Hermes of the association of the seven planets with the seven metals so that the substitution of the names of the planets for the metals, such as *sol* for gold, *luna* for silver, and so on, became common practice.[34] In general, Albert noted, "of all the operations of alchemy, the best is that which begins in the same way as nature," that is "with the cleansing of sulphur by boiling and sublimation, and the cleansing of quicksilver, and the thorough mixing of these with the material of metal; for in these by their powers, the specific form of every metal is induced." Moreover, the alchemist proceeds by destroying "one substance by removing its specific form, and with the help of what is in the material producing the specific form of another [substance]."[35]

Although Albert recognized in the above directions the theoretical validity of the alchemical process he was obliged to admit that he had never seen it successfully carried to completion. He thus stated that "we have rarely or never found an alchemist, as we have said, who [could] perform the whole [process]." It is true that "One of them may indeed produce with the yellow *elixir* the color of gold," and with "the white *elixir*, a color similar to silver"; and "may endeavor to make the color remain fast when it is placed in the fire and has penetrated the entire metal just as a spiritual substance is put into the material of a medicine." He may by this operation induce a yellow color, while at the same time "leaving the substance of the metal unchanged."[36] Such operators Albert denounced as "deceivers." Without doubt they are deceivers. . .since they do not make real gold and real silver." And yet most alchemists follow this practice in whole or in part. "For this reason I have had tests made on some alchemical gold, and likewise silver, that came into my possession, and it endured six or seven firings, but then, all at once, on fur-

---

[33] *Min.* III.1,7; Wyckoff, p. 178.
[34] *De causis elementorum* I.2,7 (Borgnet 9: 615); *Min.* III.1,6: Wyckoff, pp. 168-169.
[35] *Min.* III.1,9; Wyckoff, p. 179.
[36] *Min.* III.1,8; Wyckoff, p. 176.

ther firing, it was consumed and lost and reduced to a sort of dross."[37]

In accord, moreover, with his view that the aim of a natural scientist is not merely to accept the statements of others, but rather to make an effort to observe the phenomena at first hand, Albert supplemented the knowledge of alchemy that he had derived from past authorities in his reading of books on medicine and alchemy, with the results of his own observations and experiences. From his visits to mining districts, metal workshops, and alchemical laboratories, he had acquired a practical acquaintance with the nature of metals by a direct observation of processes in nature. "I have learned," he explained, "by what I have seen with my own eyes, that a vein flowing from a single source was in one part pure gold, and in another silver. . . ." And "[from what] miners and smeltermen have told me. . .[that] what artisans have learned by experience is also the practice of alchemists who, if they work with nature, transform the specific form of one metal into another."[38] Furthermore, from visits to laboratories, in all probability in Cologne and Paris, Albert reported on the results of his inquiries into "the transmutation of metals in the art called alchemy" which he had directed to contemporary workers in the field, that is the "alchemists of our time," whose names he does not reveal.[39] He referred to alchemical experiments which showed that watery moisture is easily converted into vapour, and to the use of the alembic.[40] He noted that minerals that seem to be intermediate between stones and metals are important reagents in alchemy since they may be influenced by laboratory treatment. And he added, "On these substances depends most of the science of those who endeavor to convert one [metal] into another."[41] He had further reported on operations similar to those of the alchemists, such as drying of mercury by frequent burning and mixture with sulphur or when placed in a hot furnace with green wood,"[42] or of the forming of metals "from water congealed by intense coldness and dryness."[43] And he also went on to note in his exposition of the alchemical art, which as earlier noted he had

---

[37] *Min.* III.1,9; Wyckoff, p. 179.

[38] *Min.* III.2,6; Wyckoff, pp. 200-201; also *Min.* III.1,1; Wyckoff, pp. 153-154.

[39] Partington, pp. 9-11; *Min.* I.1,4; also III.1,1; Wyckoff, pp. 20, 153-154.

[40] *Min.* III.1,3; Wyckoff, p. 156.

[41] *Min.* V.1; Wyckoff, pp. 239-240, 241 ff.

[42] Partington, p. 10, citing *Meteor.* IV.3,2 (Borgnet 4: 775).

[43] See the preceding note.

termed the best imitator of nature, that of the two major constituents, sulphur is known as the father and quicksilver as the mother, "as the writers on alchemy metaphorically" suggest.[44] Moreover, he drew attention to the fact that since in alchemy there is no better way of proceeding than with the yellow *elixir* made with sulphur, the alchemists have observed that there is an unctuousness in sulphur so intensely active in burning that it burns all metals, and in burning blackens everything on which it is cast. Hence, the alchemists recommend that the sulphur be washed in acid solutions and that it be cooked until no more yellow liquid comes forth. These solutions may then be sublimed until all the unctuousness capable of burning has been removed, and there remains only as much subtle unctuousness as can endure the fire without being reduced to ash. This is, Albert added, "expressly stated by the authorities, Avicenna, Hermes and many others, who are men of great experience in the nature of metals." Albert had also noted unskilled alchemists at work in the digestion or boiling by moist heat of the earthiness in the moisture in metals.[45]

The foregoing details regarding Albert's concern with alchemy provide convincing evidence of his interest in the subject. They also demonstrate Albert's belief in the possibility of the transmutation of metals, although he judged the process to be very difficult and beset with the errors and imitations of imposters. "For [as quoted by Partington] alchemical gold does not gladden the heart like the real gold, and is more [easily] consumed by fire, yet transformation may really be produced by exspoliation of properties by alchemical operations, as Avicenna teaches."[46] Albert's desire to explore the entire matter of the possible transmutation of metals as thoroughly as possible is further exemplified by his study of the principal authorities and direct observation of alchemical procedures in laboratories as well as by association with contemporary alchemists whose names he does not provide. However, there appears to be a dearth of contemporary evidence to attest that Albert himself was considered an adept alchemist or that he engaged in or performed the alchemical processes he describes. He appears rather to have been an acute observer, an onlooker, but not an active participant in the laboratory experiments.

---

[44] *Min.* IV,1; Wyckoff, p. 204.
[45] *Min.* III.1,2; Wyckoff, p. 158.
[46] Partington, p. 13, citing Albertus Magnus, *Sententiae* II.vii, F. 8 (Borgnet 27: 154 f.)

Despite the lack of contemporary and specific evidence of Albert's direct participation in alchemical laboratory procedures, his fame and repute as a skilled alchemist became manifest not long after his death. By the mid-fourteenth century he is mentioned in catalogues as author of an alchemical tract[47] and is credited with having had as a disciple in this art, Roger Bacon, the English schoolman.[48] Nor did this repute diminish in the fifteenth and sixteenth centuries. In a collection of Stowe manuscripts, Hugh of England named Albert as one of the several authorities in the field.[49] This fame, whether deserved or not, appears to have motivated the attachment to Albert's name of some twenty-eight or more tracts on alchemy.[50] They appear in manuscripts dating from the close of the thirteenth century with the major number in the late fourteenth or fifteenth century. How much of this literary output can be attributed to the common practice in the Middle Ages of attaching to such treatises the names of prominent churchmen in order to give them respectability and insure their acceptance cannot be determined. Nor does the explanation that Albert was confused with a chemist who lived at Cologne, provided by Naudé in his "Apology for all Great Personages Who Have Been Falsely Suspected of Magic,"[51] seem adequate. What does appear clear is the fact that for the most part the alchemical tracts bearing Albert's name as author reflect, in keeping with Albert's authentic writings, an active interest in not only the philosophical bases of the alchemical art but also in its practical and experimental aspects.

The individual alchemical treatises that appeared under Albert's name have elsewhere been enumerated and analyzed briefly together with the manuscripts and printed editions in which they are found.[52]

---

[47] See my "Alchemical Writings Ascribed to Albertus Magnus," *Speculum* 17 (1942), 499; also "The *Alkimia Minor* ascribed to Albertus Magnus," *Isis* 32 (1940 [1949]), 267-268 n. 5, for the MS 4 Qq A 10, formerly in the possession of the Speciale family and now in the Communal Library of Palermo.

[48] Kibre, "Alchemical Writings Ascribed to Albertus Magnus," p. 499, and the references there cited; also Paris, Arsenal MS 2872, 14th cent., f. 46, "et le dit Roger estoit disciple de frère Albert. . . ."

[49] London, British Library MS Stowe 1070, 15th cent., ff. 32-37: Hugh of England, *Secreta secretorum artis philosophorum*, inc. "After the mynde of Hermes, Plato, Morien, Geber, Aristotill, Virgill, Albert, Avicen etc. . . ." See also Partington, p. 17, for other citations.

[50] See my "Alchemical Writings Ascribed to Albertus Magnus," pp. 499-518; also my "Further Manuscripts Containing Alchemical Tracts Attributed to Albertus Magnus," *Speculum* 34 (1959), 238-247.

[51] Kibre, "Alchemical Writings," p. 499; and G. Naudé, *Apologie pour tous les grands personnages qui ont esté faussement soupçonnes de magie* (Paris, 1625) chap. xviii.

[52] See the references in note 50 above.

Hence it will perhaps suffice here merely to draw attention first to some general characteristics of the tracts and second, to choose some examples for more specific comparison with the information contained in Albert's authentic works. In general the treatises are free from the mystifying and allegorical features upon which Albert himself in the *Book of Minerals* had cast aspersion as characteristic of alchemical tracts. In this respect the imprint of Albert's didactic method, noteworthy for clarity of expression and systematic arrangement, was strong enough to influence those who professed to write under his name. As in Albert's exposition to his confreres of the Aristotelian scientific corpus, the professed purpose of the authors of the alchemical tracts, in the majority of cases, was to explain to their readers in as simplified a fashion as was necessary for their understanding, the art of alchemy, its theory and practice.

The above features are exemplified specifically in the *Little Book of Alchemy* (*Libellus de alchimia* or *Semita recta*), the tract most consistently attributed to Albert and extant in manuscripts dating from the close of the thirteenth century.[53] The clear, concise, and well ordered account of alchemy contained in this tract resembles Albert's treatment of other topics of natural science in his authentic works. It also provides an excellent introduction to the alchemical art of the late thirteenth and fourteenth centuries. The author is particularly concerned with making known to his confreres, the aims, accoutrements and processes of alchemists and the alchemical art. His instructions are detailed and even repetitious in character. They are, moreover, together with the frequent cautions and admonitions, largely practical in nature. Yet, true to the professed author's ecclesiastical calling, the work opens with the phrase from Ecclesiasticus: "All wisdom is from the Lord God." However, despite the fact that the suggestions contained in the tract are largely practical in nature and might perhaps have emanated from Albert, the author's style, beginning with the introduction, differs pointedly from that of Albert in his authentic works. For example, the introductory phrases con-

---

[53] See my "Alchemical Writings," pp. 500, 511-515; and "Further Manuscripts," pp. 238-244, for example Vatican MS Palatine 978, late 13th or beginning 14th cent., ff. 33r-46v; also British Library Additional MS 41486, end of 13th cent., ff. 7-8, containing selections of the work. This manuscript was brought to my attention by T. C. Skeat, then Deputy Keeper of the Manuscripts at the British Museum. See also *Libellus de alchimia ascribed to Albertus Magnus*, translated [into English] from the Borgnet Latin edition by Sister Virginia Heines, S.C.N. (Berkeley: University of California Press, 1958; henceforth cited as *Libellus de alchimia* [1958]), p. 75.

tain the author's stated conviction that he has found what he was seeking; not, however, (in his words) "by my own knowledge, but by the grace of the Holy Spirit. Therefore, since I discerned and understood what was *beyond nature*, I began to watch more diligently in decoctions and sublimations, in solutions and distillations, in cerations and calcinations and coagulations of alchemy and in many other labors until I found possible the transmutation into gold and silver, which is better than the natural [metal] in every testing and malleation."[54] This does not coincide with Albert's view in the *Book of Minerals*. There he had expressed his belief that while the transmutation of baser metals into gold was theoretically possible, it had not so far been accomplished by the alchemists; also that while the alchemists were able to produce a metal similar to gold, their product was inferior to natural gold or silver and did not stand the accepted test for gold.[55] Moreover, while the author of the *Little Book of Alchemy* looked "beyond nature," Albert had repeatedly stipulated that the observation of nature and natural processes provided the best guide.[56]

Yet despite these essential differences it appears evident that the author of the *Little Book of Alchemy* was acquainted with Albert's work, or at least utilized similar sources. He repeated from Avicenna's *Congelatio* the phrase which he mistakenly attributed to Aristotle, "Let the masters of alchemy know that the species of things cannot be changed," and the accompanying statement, here also attributed to Aristotle, "I do not believe that metals can be transmuted unless they are reduced to prime matter, that is purified of their own corruption by roasting in the fire." Only then is transmutation possible.[57]

The treatise itself also has some interesting features. Among these are the enumeration of possible errors and the list of specific precepts to guide those undertaking the enterprise. For example, attention is drawn to the fact that some were incapable of carrying out certain sublimations "because they failed to grasp the fundamentals"; "others because they used porous vessels." Also, in the precepts, the first provided that "the worker in this art must be silent and secretive

---

[54] *Libellus de alchimia* (1958), pp. 2-3.

[55] See above, and *Min.* III.1,9; Wyckoff, pp. 178-179.

[56] See above, also *Min.* III.2,6; Wyckoff, pp. 199-201.

[57] *Libellus de alchimia* (1958), pp. 9 and 10; and see above; also *Min.* III.1,4; Wyckoff, pp. 161-162.

and reveal his secret to no one";[58] and the eighth "that no one should begin operations without plenty of funds. . .for if he should. . .lack funds for expenses then he will lose the material and everything."[59] Contributing further to the practical nature of the tract are the descriptions of the various utensils, furnaces, ovens, and flasks; then the spirits: quicksilver, sulphur, orpiment, and sal ammoniac; and finally the *elixir* or *fermentum*, the medicine or philosophers' stone capable of transmuting baser metals into gold and silver.[60]

Another even more practical tract, the *Alkimia minor*, is attributed in the manuscripts to "Brother Albert of Cologne of the Order of Preaching Friars."[61] Best described, perhaps, as a laboratory manual, it has directions for the preparation of chemical substances, for the dyeing of metals red or white, that is the transmutation into gold or silver, and for the preparation of the *elixir* or medicine, the transmuting agent. Like the *Little Book of Alchemy* or *Semita recta*, of which it seems to include abstracts, the *Alkimia minor* appears to have been in circulation by the mid-fourteenth century, although no manuscripts of the text earlier than the fifteenth century have been located so far. However, the tract is listed with the same opening words among the books contained in a collection of alchemical treatises of the early fourteenth century.[62] The text itself gives no indication of the date of composition, although the chemical knowledge coincides with that of similar writings of the thirteenth and fourteenth centuries. The tract provides details of laboratory procedure and of apparatus and utensils. The directions are simple and straightforward with no attempt to mystify. Yet, the work does conform to the common practice of assigning names of planets to the minerals, and it also makes use of the term medicine as a synonym for *elixir*, the transmuting agent. Similarly, the use of "to redden" (*ad rubeum*) or "to whiten" (*ad album*) for the gold or silver making recipes can be found. An explanation for the use of these terms is given in the *Book of Minerals*.[63] Many of the chemical substances utilized were already described in such works as the *Book of Minerals* and the *Little Book*

---

[58] *Libellus de alchimia* (1958), pp. 3-4, and 12.

[59] Ibid., p. 14.

[60] Ibid., pp. 12-19.

[61] For the edition and analysis of the text, see Kibre "The *Alkimia Minor* Ascribed to Albertus Magnus," *Isis*, 32 (1940 [1949]), 268-300.

[62] Ibid., p. 268 n 5.

[63] "*Alkimia Minor*," p. 270; *Min.* IV.7; Wyckoff, p. 231.

of Alchemy. In general, the *Alkimia minor* is characterized by a total lack of attention to theory; the emphasis is on actual procedure and practice. Many of the processes listed, such as sublimation and distillation, for example, are common to pharmacy as well;[64] and there is great variety in the laboratory apparatus.[65] Included were furnaces, the baker's oven (*furnus panis*), and the furnace of reverberation, the dung bath, marble slab, alembics, aludel, recipient (*ampulla*), and various kinds of jars, flasks, and vessels; earthen, copper, and glass, closed or open; a descensory, that is a vessel or retort used in distillation by descent, and pestles of iron or wood, as well as a mallet or hammer.

A further tract attributed to Albertus Magnus for which there are no fourteenth-century manuscripts extant but which is named in the same fourteenth-century alchemical miscellany as the *Alkimia minor*,[66] is that entitled "On the hidden things of Nature" (*De occultis naturae*).[67] This treatise provides a survey of the various alchemical doctrines set forth by alchemical authorities chiefly of Arabic origin. The work bears only a slight resemblance to the other alchemical tracts ascribed to Albertus Magnus. It professes to have been written in response to a request by a reverend father,[68] addressed in the course of the work. Unlike either the *Semita recta* or the *Alkimia minor*, with which it is frequently found in the manuscripts, the *De occultis naturae* relates more to the theoretical side of alchemy than to the practical although both aspects are covered. The author has utilized a large number of writers with the intention, he informs us, of making it unnecessary for the reader to consult them further since their principal doctrines will have been transferred in briefer form to the present compendium. In general the treatise appears to resemble more the alchemical writings of the fourteenth rather than those of the thirteenth century in its predilection for alchemical jargon, allegorical devices, and mystical phraseology.[69]

---

[64] "*Alkimia Minor*," pp. 271-272.

[65] Ibid., pp. 272-273.

[66] "Manuale d'alchimia miscellanea membranceo del secolo XIV," Palermo, MS 4 Qq A 10. See "*Alkimia minor*," p. 268 n. 5; also P. Kibre, "The *De Occultis Naturae* attributed to Albertus Magnus," *Osiris*, 11 (1954), 24 n. 7.

[67] Ibid. pp. 23-39; also "Albertus Magnus, *De Occultis Nature*" *Osiris*, 13 (1958), 157-183, for an edition of the text.

[68] Kibre, "*The De Occultis Naturae*," (1954), pp. 24-25.

[69] Ibid., pp. 34-35.

Of the remaining alchemical treatises appearing under Albert's name, it may suffice to note here two further examples. The tract *On Alchemy* (*De alchimia*)[70] which appears not to have been available before the fifteenth century, bears a close relationship with the *Book of Minerals* in several of the arguments presented and in the discussion regarding alchemy. It is distinguishable from the *Little Book on Alchemy* (*Libellus de alchimia* or *Semita recta*) by the opening words: "Callisthenes one of the earlier founders of our art after Hermes. . . ." In addition to Callisthenes, the author names other authorities similar to those included in Albert's discussion in the *Book of Minerals*, such as Hermes, and Avicenna. In addition he names Geber Hispanus[71] instead of Gigil and attributes to him the statement with some verbal changes, that Albert expresses as his own, in the *Book of Minerals*, namely that he has examined certain alchemical books and has found them to be without distinction and with their intention hidden under the guise of allegory. The author here also follows the current practice of using the names of the seven planets as synonyms for the metals, a practice that Albert attributed to Hermes. But he fails to repeat here the references to the influence of the heavens in the alchemical process found in the *Book of Minerals*. On the other hand he follows Albert's emphasis upon the principle that alchemy imitates nature and hence that it is necessary to observe carefully and closely natural processes.[72]

The other tract attributed to Albert that we would note here, namely the *Compound of Compounds* (*Compositum de compositis*) attracted attention in the fifteenth to seventeenth centuries when it was translated into French and English. However, the text in the late manuscripts resembles closely that found without Albert's name as author in a fourteenth-century manuscript at Edinburgh. In that text the work is said to have been collected and promulgated by masters at Paris in the year 1331. However, in the later manuscripts at Paris and the Vatican it is clearly attributed to Albertus Magnus.[73] In the course of the work reference is made to "our *Book on Minerals*" which is suggestive of Albert. There are included in the tract discus-

---

[70] P. Kibre, "An Alchemical Tract Attributed to Albertus Magnus," *Isis*, 35 (1944), 303-316.

[71] Ibid., p. 303.

[72] Ibid., pp. 303-304.

[73] P. Kibre, "Alchemical Writings Ascribed to Albertus Magnus," p. 506, item 8.

sions on the theory of alchemy along with practical recipes for the preparation of vermillion and of white sublimate.[74]

Of the remaining alchemical treatises appearing under Albert's name, none, with the exception of those that are also ascribed to other authors as well, appeared before the fifteenth century. Although the relation of these alchemical texts to Albert is tenuous to say the least, they do attest to his repute as an adept alchemist in the decades and centuries following his death.

---

[74] Partington, pp. 17, 19-20. A copy in French translation is contained in a manuscript at the Beinecke Library (Yale University, New Haven, Conn.), MS Mellon 19, 15th cent., ff. 28v-32r: "Incipit tractatus Alberti qui intitulatus Compositum de compositis." I am preparing an edition of this text.

8

Albert on Stones and Minerals

John M. Riddle and James A. Mulholland
North Carolina State University at Raleigh

In common with a host of nature's observers before him, Albert knew that all things are either animal, vegetable, or mineral. Few before Albert, however, devoted study to minerals. Those that did were either lapidarists, astrologists, magicians, encyclopedists, or medical men interested in their therapeutic effects. No previous writer observed and recorded information on the entire compass of minerals, enough so that one could say to a modern scientist's satisfaction, "this is a pioneer work in mineralogy." To some extent Albert was aware that he was venturing in a new branch of *scientia*, one without a previous tradition, because he could cite no authorities who combined the theory of mineral formations and their properties together with the practical knowledge of the lapidarists, alchemists, pharmacists, miners and other practitioners in stone and metals lore. Aristotle had written on minerals but Albert could find only excerpts, and Avicenna's work, it seemed to Albert, treated the subject too briefly and insufficiently.[1] Albert wrote the *Book of Minerals* in five books. He wrote neither merely to record and synthesize all prior authorities, nor merely to add his own observations. The task was too great, the subject too vast. What he intended, in his own words, was that "on the basis of what has been said [in his work], anything else [relating to minerals] that has not been mentioned here

---

[1] Albert, *De mineralibus* I, tr.1, c.1 (ed. Borgnet 5: 1b; trans. Dorothy Wyckoff [Oxford, 1967], p. 9).

can also be readily understood."[2] This is what a branch of science is all about. Albert provided a theoretical structure for the organization and explanation of data in a category of physical nature and to that extent founded a "scientia de mineralibus."[3]

In 1967, Dorothy Wyckoff published a detailed English translation and commentary to Albert's *Minerals*.[4] Her work is so thorough, so scholarly, and so clear that we can do little more than supplement her contributions. Wyckoff, a mineralogist by profession, a classicist by avocation, saw in Albert's study an important step in the field's foundation, for few before her saw Albert's work as anything but a curious blend of scholasticism, lapidarist folklore, and alchemy.[5]

Albert's *Minerals* has a logic in the design of presentation, as we shall see more fully below. In the first book Albert outlined his procedure. In keeping with the logic of Aristotle and the conventional, Empedoclean explanations of the compositions of minerals, he explained that minerals are not alive and have no souls, but are compositions of earth or of water. Stones are the subject of his first discourse. Even the driest stones formed of earth have water which binds the stone together. Some stones are congelations primarily of water, and this provides an explanation of glass and quartz. He rejected the explanation of alchemists who said that stones "born" in water were necessarily stones formed of water, because they might be solidifications of the earth material in water.[6] The power of the elements is the *material cause*. The *efficient cause* is the production of stones through a *mineralizing* power which is, he observed, a mysterious natural process produced by heavenly powers and difficult to explain except through analogy. Albert rejected previous theories of stone production including those of the alchemists who said that

---

[2] *De min.* V, tr.1, c.9 (ed. Borgnet 5: 102b; trans. Wyckoff, p. 251). Unless otherwise noted all quotations in English from Albert's *De mineralibus* are of Wyckoff's translation.

[3] *De min.* III, tr.1, c.1 (ed. Borgnet 5: 60a): ". . . et complebimus in eius totam istius scientiae de mineralibus intentionem" (Wyckoff, p. 155).

[4] *Albertus Magnus Book of Minerals*, trans. by Dorothy Wyckoff (Oxford, 1967).

[5] As late as 1955 a commentary on a translation of Agricola dismissed Albert as well as all other medieval writers of lapidaries with the observation: "There was no important work on mineralogy from the time of Pliny until Agricola published his *De natura fossilium* in 1546 and the shorter introductory work *Bermannus* in 1530. During the intervening fourteen centuries that spanned the rise and fall of the Roman Empire and the Dark and Middle Ages, writers on mineralogical subjects merely elaborated on the information and misinformation contained in Pliny's *Natural History*." Mark C. Bandy and Jean A. Bandy, trans., *George Agricola De Natura Fossilium*, Geological Society of America Special Paper 63 (Menasha, 1955), p. v.

[6] *De min.* I, tr.1, c.3 (ed. Borgnet 5: 4a-b; Wyckoff, p. 20).

stones were purely an accidental production by dry heat, for example a brick produced from clay by baking. Were this so, Albert said, stones would not differ one from another, and there were obvious differences in properties, appearance and powers. Only a discerning mineralizing power can effect the variety of stones. Further, if all stones were merely dry heat, they could be dissolved by moist cold, and "we do not see this happen" (quod non vidimus contingere).[7]

Albert's preference for Aristotelian theories of stone formation was occasionally an *impedimentum* to his insight into minerals. Favoring Aristotle's explanation over vague hints to the contrary in Avicenna, Albert believed that rocks originated where they were formed, that is, *in situ*.[8] Were he to have expanded on Avicenna he might have observed the corrosive effects of water wearing smooth river pebbles or erosion on sedimentary rock layers or the action of glaciers. River pebbles are formed, Albert said, by the action of the heat of the earth on river bottoms which bakes the mixture of earth permeated by water in the pores so that vapor cannot escape, thereby cooking, as it were, a river stone.[9] More acceptable to modern theory is the following statement:

> From all this it seems impossible to report anything certain about the [kind of] place that produces stones. For [stones occur] not in one element only, but in several, and not in one clime only, but in all. . . .For all things produced must have a certain place of production, and away from this they are destroyed and dispersed.[10] (Wyckoff trans.)

A description of the destruction and dispersion (*corrumpuntur et destruuntur*) of stones would have opened new vistas but, from this, it is unclear whether Albert thought the destruction and dispersion was a natural process or accidental.

Similarly Albert came close to developing a classification scheme for stones. Given the ancient, especially Aristotelian, propensity for systemization and classification, one would have expected earlier attempts at classification. Albert seemed to have had thoughts in this direction, but he aborted them. As an example he said there was a

---

[7] *De min.* I, tr.1, c.4 (ed. Borgnet 5: 7a; Wyckoff, p. 21). On the Aristotelian background for Albert's ideas here, see James A. Weisheipl, *Development of Physical Theory in the Middle Ages* (Ann Arbor, 1971), pp. 37-38.

[8] *De min.* I, tr.1, c.7 (ed. Borgnet 5: 9b-10b). See also Wyckoff's discussion on the problem together with references to Aristotle and Avicenna, pp. 16, 26-27, 36-38.

[9] *De min.* I, tr.1, c.8 (ed. Borgnet 5: 11a-b; Wyckoff, p. 31).

[10] *De min.* I, tr.1, c.7 (ed. Borgnet 5: 10b; Wyckoff, p. 29).

group (*genus*) of marbles which included porphyry, alabaster and so on, and that there were other groups.[11] In discussing the cause of colors in nontransparent, nonprecious stones, he said there were four groups, namely flint, tufa, freestone and marble.[12] And yet, he did not develop a systematic classification scheme, bowing instead to the convenience and custom of relating information about specific stones in alphabetical order. "This method," Albert said elsewhere about alphabetization, "is not suitable in philosophy."[13]

Following Aristotle's method Albert dealt with the *formal* cause, namely whether *forma* and *species* can be applied to stones. In modern times, the same question reformulated is whether stones have individual chemical composition, and how categories of similar physical qualities relate to one another, for example, whether all transparent stones are crystalline. "We find," Albert said, "in stones powers which are not those of any element at all," but powers based on "the particular mixture of their elements."[14] Albert followed the Avicennian pharmaceutical theory, although probably not directly from Avicenna's works. Avicenna held that a compound has as its characteristics not only the sum of its constituent elements but, as a result of a "fermentation," unique, specific qualities as well. Thus one cannot predict the qualities of an object, such as a stone, by analyzing its chemical makeup. Its *forma specifica* can be known through experience in its use.[15] Albert's doctrine is important because, as will be seen later in his discussion of the qualities of individual stones, each has unique powers which are empirically tested.

Albert implicitly rejected a major pharmaceutical theory over which controversy was raging, probably in the 1240s in Paris where Albert lived before writing the *Minerals*.[16] In the *Minerals*, Albert did not seem to accept the Galenic theory on degrees of intensity for the active and passive qualities of simples and compounds. In a few

---

[11] *De min.* I, tr.1, c.6 (ed. Borgnet 5: 9a; Wyckoff, p. 26).

[12] *De min.* I, tr.2, c.3 (ed. Borgnet 5: 17a; Wyckoff, p. 44).

[13] Cited as *Animals* XXII, tr.2, a.1 by Wyckoff, p. 68; Albert explained that the alphabetical order for stones was most convenient inasmuch as medical men followed this custom in describing simples (*De min.* II, tr.2).

[14] *De min.* I, tr.1, c.6 (ed. Borgnet 5: 8b; Wyckoff, p. 24).

[15] Ibid. "Experience" as the means of knowing the specific qualities can be seen in *De min.* II, tr.1, c.1 (ed. Borgnet 5: 24a, 30a; Wyckoff, pp. 56, 68-69). For more precise statement and corroboration, see Avicenna, *Canon.* V, tr. spec., fol. 507v (Venice 1507); *Canon.* I, fen.2, summa 1, cap.15, fol. 33v; Michael R. McVaugh, *Arnaldi de Villanova, Opera Medica Omnia*, vol. 2: *Aphorismi de gradibus* (Granada-Barcelona, 1975), pp. 18-19.

[16] McVaugh, *Arnaldi*, 2: 31-32.

instances throughout the whole of *Minerals* Albert does allude to the active qualities of warmness and coldness and the passive qualities of dryness and wetness. One of each is possessed by everything, be it animal, vegetable or mineral.[17] Although Albert doubtless knew of Roger Bacon's pronouncements on the subject and may have known of Peter of Spain's attempt to place the physical actions of substances on a high theoretical plane,[18] he made little attempt to incorporate the current controversy in the *Minerals*, and only later in *Plants* did he accept the basic rudiments of Galenic theory.[19] On the surface it seems strange that Albert, given his Aristotelian penchant for systematization, was not attracted to Galen's theory which provided an explanation and perhaps predictive indicator for the physiological action of minerals. In fact, Albert must have consciously rejected this theory. He extensively employed Constantine's (d. ca. 1085) *On degrees* as a source for individual stones in Book Two of the *Minerals*, but Albert omitted Constantine's ascription of intensity of action for the minerals even when quoting directly.[20] We can only conclude that Albert was not satisfied by the theory and, while not speaking against it, was unwilling to employ its rationale for the explanation of various minerals' behavior. Probably he did so because the theory did not satisfy his strong empirical bias[21] or his assumptions.

---

[17] *De min.* IV, tr.1, c.1; IV, tr.1, c.2 (ed. Borgnet 5: 84a, 85b; Wyckoff, pp. 204, 207) Albert gave the degrees of intensity of qualities (and elsewhere) when relating qualities, but simply as cold and dry, and without "degrees"; see *De min.* II, tr.1, c.3; V, tr.1, cc.4, 5, 7, and 8 (5: 27a, 100a, 102a, 102b; trans. Wyckoff pp. 62, 245, 249, 250). It is curious that all citations to Albert's intensity of qualities come in the last two books, thereby raising a question whether, if the work were composed over an extended period and Books IV and V were written last, Albert had not by then come to accept to some small extent Galen's drug theory.

[18] McVaugh, *Arnaldi*, 2: 32; Lynn Thorndike, *A History of Magic and Experimental Science* (New York, 1923), 2: 488-510. We know of no definitive dating of Peter's *Summule logicales*, which contains his pharmaceutical theory, but Heinrich Schipperges (*Die Assimilation der arabischen Medizin durch das Lateinische Mittelater*, in *Sudhoffs Archiv*, Beiheft 3 [Wiesbaden, 1964], p. 180) places it as the earliest of his works. L. M. De Rijk places the date of Peter's *Summule* in the early 1230s (Peter of Spain, *Tractatus, or Summule logicales* [Assen, 1972], pp. lvii-lxi).

[19] For example, Albert, *De veg.* III, tr.1, c.6; VI, tr.1, c.2.

[20] For example, Albert, *De min.* II, tr.2, c.7, *granatus* (ed. Borgnet 5: 38a; Wyckoff, p. 96) quoting directly and by name Constantine's citation of Aristotle; whereas in Constantine's work *De gradibus* (*Opera* [Basel, 1536], p. 352), the degrees are given in sections on drugs intensive to the first degree; "Quos omnes Aristoteles cal[ida] et sic[ca] dixit esse." Compare also *De min.* II, tr.2, c.6, *falcones* (ed. Borgnet 5: 37a; Wyckoff, p. 92) with Constantine *De gradibus* [Basel, 1536], p. 383.

[21] *De min.* II, tr.1, c.1 (ed. Borgnet 5: 24a; Wyckoff, p. 56).

Albert looked to Hermes who, "of all the ancients," gave "the most probable reason for the powers of stones."[22] One by one Albert related the various authorities' explanations before summarizing those of Hermes. The heavenly power operated through stars and constellations to impress powers into every specific form of stone.[23] "Nevertheless this statement," Albert cautioned, "is not enough for natural science (*physicis*), although perhaps it may be sufficient for astrology and magic. For natural science discusses the cause that acts upon matter."[24] The implanted powers in stones (and metals) are both indirect and accidental. It is indirect because the power goes through the intermediary of the elements and the fermentation. The power is accidental in so far as not all objects in their various locations receive the same distribution. Thus stones have accidental properties, such as color, transparency, hardness, fissility, porosity and size, according to their mixture.[25] What Albert calls mixture is today called chemical composition. Thus each type of stone is unique. Even individual stones of the same class may differ from one another as, for instance, a *saphirus* which is said to lose its power to cure abscesses once it has cured one. But even here there are variations from the norm because Albert claimed that he had personally seen a *saphirus* cure two abscesses in a four-year interval.[26]

Tractate Two of Book Two is the most familiar section of the *Minerals* because it is the traditional lapidary. Often in many manuscripts this section of the *Minerals* was separated and stood as an independent treatise.[27] The names of some ninety-nine "precious stones" as well as their descriptions and powers are related by Albert as they are known "either by experience or from the writings of authorities."[28]

In this section Albert gave advice on how to conduct successful business deals, win battles, test for virginity, prevent storms, protect

---

[22] *De min.* II, tr.1, c.3 (ed. Borgnet 5: 27b; Wyckoff, p. 63). But Albert often disagreed with Hermes on other matters of detail, e.g., *De min.* III, tr.1, c.3 and 8; IV, tr.1, c.4 (ed. Borgnet 5: 63a, 69b, 87a; Wyckoff, pp. 162, 171-172, 213).

[23] *De min.* II, tr.1, c.2-3 (ed. Borgnet 5: 24a-28a; Wyckoff, pp. 58-64).

[24] *De min.* II, tr.1, c.3 (ed. Borgnet 5: 27a; Wyckoff, p. 63).

[25] *De min.* I, tr.2, c.1-8 (ed. Borgnet 5: 14a-21b; Wyckoff, pp. 36-53).

[26] *De min.* II, tr.2, c.17 (ed. Borgnet 5: 44b; Wyckoff, p. 115).

[27] Inc.: "Supponamus autem nomina praecipuorum lapidum et virtutes secundum . . ." — Cambridge, Univ. Lib. MS Dd III 16, fol. 7v-15v (with Book II, tractate 3, on sigils); London, British Libr. MS Sloane 1009, fol. 68v-72v; Toledo, Bibl. del Cabildo MS 157, fol. 78 ff.; Vienna, Nationalbibliothek MS Pal. 2303 (s. xiv), fol. 62-64v; MS Pal. 12,901 (s. xiii-xiv), fol. 94-125v.

[28] *De min.* II, tr.2, c.1 (ed. Borgnet 5: 30a; Wyckoff, p. 68).

against robbery, reduce or eliminate fever, stop breathing, confer happiness, and cure scabs, dropsy, heart attacks, kidney stones, bladder stones, hemorrhoids, belching, stomach ache, jaundice and diarrhea. In short, stones can control almost any aspect of the environment as well as most physical ailments as diagnosed. But for someone unfamiliar with medieval tradition it comes as a surprise to read of the saint advising the use of stones to thieves for successful robbery,[29] to women to prevent conception or to produce a miscarriage,[30] to men to betray secrets,[31] and to all people to arouse sexual desire.[32] The initial shock is mitigated perhaps when one reads that there are stones which counteract these powers, such as stones which drive away phantoms,[33] keep travellers safe from robbery,[34] help childbirth,[35] restrain sedition,[36] moderate licentiousness[37] and check hot passions and desires.[38] Generally most stones are recommended for qualities of making people happy and alleviating pain and illness. In this section more than any other Albert was relying on previous authorities, including Marbode (1035-1122), bishop of Rennes, and possibly as well the Venerable Bede.[39]

Albert, true to his prefatory remarks, has throughout related his personal experience in attesting to the stones' powers. Albert often differentiated what his authorities say and what he learned through experience. When he said chalcedony is good for fanciful illusions for those afflicted with melancholy and causes and preserves the powers of the body, he added, "The last is a matter of experience" (*hoc ultimum est expertum*).[40] For him it worked. He said about

---

[29] *De min.* II, tr.2, c.13, *ophthalmus* (ed. Borgnet 5: 42b; Wyckoff, p. 110).

[30] *De min.* II, tr.2, c.8, *jaspis*; c.13, *oristes* (ed. Borgnet 5: 39b, 43a; Wyckoff, pp. 100, 110).

[31] *De min.* II, tr.2, c.15, *quiritia* (ed. Borgnet 5: 44a; Wyckoff, p. 114).

[32] *De min.* II, tr.2, c.1, *alectorius* (ed. Borgnet 5: 31b; Wyckoff, p. 73).

[33] *De min.* II, tr.2, c.3, *chrysolitus* (ed. Borgnet 5: 34b; Wyckoff, p. 83).

[34] *De min.* II, tr.2, c.8, *hyacinthus* (ed. Borgnet 5: 38b; Wyckoff, p. 98).

[35] *De min.* II, tr.2, c.6, *galaricides* (ed. Borgnet 5: 38a; Wyckoff, p. 95).

[36] *De min.* II, tr.2, c.5, *epistrites* (ed. Borgnet 5: 36a; Wyckoff, p. 90).

[37] *De min.* II, tr.2, c.7, *gelosia* (ed. Borgnet 5: 37b; Wyckoff, p. 94).

[38] *De min.* II, tr.2, c.17, *sardonyx* (ed. Borgnet 5: 45a; Wyckoff, p. 117). In the case of onyx, a stone which has bad effects (*De min.* II, tr.2, c.13, *onyx, onycha* [ed. Borgnet 5: 42a-b; Wyckoff, pp. 108-110], the stone is specifically counteracted by *sardinus* (II, tr.2, c.17 [ed. Borgnet 5: 45a; Wyckoff, p. 117]).

[39] Wyckoff, *Albertus*, pp. 266-268. Venerable Bede's lapidary is an example of a group of lapidaries, the so-called "Christian Symbolic Lapidary," a discussion always of twelve stones, which may differ. There is less uniqueness in this type of lapidary, thus making it unclear whether Albert ever used Bede on stones.

[40] *De min.* II, tr.2, c.3, *chalcedonius* (ed. Borgnet 5: 33a; Wyckoff, p. 78).

carnelian that "it has been found by experience that it reduces bleeding, especially from menstruation or hemorrhoids."[41] In powder form chrysolite helps one with asthma but in relating information, Albert failed to embrace with his own testimony his source's assertion that it expels stupidity and confers wisdom.[42] We might assume that he omitted his personal verification not because he doubted that chrysolite had these powers but either because he simply had not experimented with it or because of an economy of space. He challenges the credulity of his modern reader by telling of his experience with the stone *ramai*. He said experience gives certain proof that through its powers (". . .virtus pro certo experta est. . .") it overcomes looseness of the bowels and especially the bleeding of dysentery and menstruation.[43] He related first hand experience with powers of the stones rock crystal (*crystallus*, II. 2, 3), dragonstone (*draconites*, 4), jet (*gagates*, 7), amber (*ligurius*, 10), onyx (*onycha*, 13), sapphire (? *saphirus*, 17), emerald, (? *smaragdus*, 17), topaz (*topasion*, 18), and *virites* (?, 19). In no other lapidary does the author attempt to relate personal experience to the testimony of others on stones' powers. In addition, Albert attempted to straighten out descriptions of stones and testify to their locations, for example, dragonstone, perhaps a fossil ammonite (*draconites*, 4), eaglestone (*echite*, 5), pearl (*margarita*, 11), emerald (? *smaragdus*, 17), and *specularis* (?, 17). It is clear from his section on stones, and is even clearer in the section on metals discussed below, that Albert has not simply compiled previous sources but has added observations and judgments of his own which constitute a distinct contribution to the subject.

A modern mineralogist, more concerned with minerals themselves than with methodological problems, would observe that opal has two to four different names in Albert's work, differentiated by coloring patterns, *ophthalmus* (13), *pantherus* (14) and maybe *hiena* (8) and *agates* (1). Each one has different powers. Similarly there are four names, descriptions and powers for amber: *chryselectrum* (4), *succinus* (17), *ligurius* (10), and *kacabre* (9, maybe jet as well). Under *chryselectrum* he said that the story of its being a "solidification of an ignoble substance. . .is not true," but under *ligurius* (10), meaning in Greek "lynx-urine," he accepted the tale that the lynx, jealous of its urine, buries it and it subsequently hardens into amber. A modern expert, reading Wyckoff's translation, would wonder whether Albert

---

[41] *De min.* II, tr.2, c.3, *cornelius* (ed. Borgnet 5: 33b; Wyckoff, pp. 81-82).

[42] *De min.* II, tr.2, c.3, *chrysolitus* (ed. Borgnet 5: 34a; Wyckoff, p. 82).

[43] *De min.* II, tr.2, c.16, *ramai* (ed. Borgnet 5: 44a; Wyckoff, p. 44).

might not have known that *exacolitus* (5), *filacterium* (6), and *gecolitus* (7) were scribal errors in his authorities' texts for stones otherwise already discussed?[44] Why does Albert, he would think, attribute sexes to two stones, *balagius* (2) and *peranites* (14), when there is no hint of stone sexuality in Books One and Two where the theory of stones and minerals are discussed? Again, why would Albert advise and believe that stones have such wondrous powers — preternatural powers, in the modern view?

From a logical and rational point of view, many of Albert's theoretical explanations could be considered scientific even in the twentieth century. The acceptance of the assertion of stones' powers will be better understood by a physican and psychologist than by a mineralogist or even a gemologist. Albert accepted the claims of Costa ben Luca (Qustā ibn Lūqā, fl. late 9th c.)[45] who argued in a treatise called *On physical ligatures, incantations and suspensions around the neck* that, contrary to a fundamental aversion among the Greeks to magic and the occult, many of the body's afflictions are not because of bodily disorders, but are attributable to mental disorders. To some extent the mind controls the body, as Indian medical people have claimed and even some Greeks allowed. A belief in a cure is frequently itself a cure for the body's ill. It is superfluous to argue a stone's power because experience demonstrates the power, the mind believes it, and, therefore, it has the power.[46] It works. Costa ben Luca's work was known to Albert, but Albert seems to have had trouble in citing it because sometimes he calls it by title[47] and only

---

[44] Trans. Wyckoff, pp. 91-92, 95.

[45] See Fuat Sezgin, *Geschichte des arabischen Schrifttums*, 3 vols. (Leiden, 1970), 3: 270-274; Albert Dietrich, *Medicinalia Arabica* (Gottingen, 1966), p. 198; Carl Brockelman, *Geschichte der arabischen Literatur*, 2 vols. (Leiden, 1943) 2: 222-224.

[46] *De physicis ligaturis*, Brit. Libr. MS Add. 22,719 (s. xii), fol. 200v (-202v). The Latin translation is printed in *Opera Constantini* (Basel, 1536), pp. 317-320, and in *Opera Arnaldi de Villa Nova* (Venice, 1505), pp. 344-345. Both Constantine and Arnald are credited in various manuscripts as being Costa ben Luca's translator for *De physicis ligaturis*. The British Library's twelfth-century manuscript of the text, and the use of Costa by Constantine for *De gradibus* and by Marbode for *De lapidibus* makes it certain that the translation, if not by Constantine, was prepared by the eleventh century and certainly available to Albert.

[47] With the information coming from Costa ben Luca's *De physicis ligaturis*, Albert cites the work as "in physicis ligaturis" in *De min.* II, tr.2, c.3, *chrysolitus*. But in three other citations Albert seemingly cites the same work as "in incantationibus autem et physicis ligaturis", but the information does not come from Costa ben Luca's work; cf. *De min.* II, tr.2, c.5, *epistrites*; c.7, *galarcides* ("in libro de ligaturis physicis"); c.13, *oristes*, (ed. Borgnet 5: 36b, 38a, 43a; Wyckoff, pp. 90, 94, 110). Possibly Albert had yet another text (or texts) for *On physical ligatures*, and he merely compressed them into the one title, which was confused with Costa ben Luca's work.

once by name, *Constabulence*, where the quotation makes certain the text is related to Costa ben Luca's.[48] Further, at least two other of Albert's authorities accepted Costa ben Luca's assertions, namely Constantine the African and Marbode.[49] If one reads the lapidary section of Albert's *Minerals*, the tractate on images and sigils (II, tr.3) and to some extent the books on metals (III, IV), with Costa ben Luca's ideas in mind, then the powers inherent in stones, sigils, and metals make sense as psychotherapy. Much will make sense, not all. Thus when one reads that the dragonstone bestows victory[50] (II. 2, 4 *draconites*), then possibly the power of suggestion was operative.[51] Self-confidence derived from the stone's power. What about the second quality of dragonstone that it dispels poisons? Would a twentieth-century psychiatrist be willing to state unequivocally all *toxicity* or a *belief* in toxicity could not be willed away by faith in a cure or prophylactic?[52] So, when Albert says of carnelian (II. 2, 3, *cornelius*), "It has been found by experience that it reduces bleeding, especially from menstruation or hemorrhoids. It is even said to calm anger," one might readily agree to a possible psychotherapeutic effect of the latter and be cautious in challenging the efficacy of the former.[53]

The lines between the physiological and psychological (spiritual to Albert) are not clear, nor is it clear always whether the various stones are pharmaceutically active or inert as a placebo. Hematite, Albert advises, is "a powerful styptic, and therefore experience shows that if crushed, mixed with water, and drunk it is a remedy for a flux of the bladder or bowels, or menstruation; and it also heals a flux of bloody saliva."[54] Hematite is a red oxide of iron containing ferric chloride

---

[48] *De min.* II, tr.3, c.6 (ed. Borgnet 5: 55b-56a; Wyckoff, pp. 146-147).

[49] John M. Riddle, *Marbode of Rennes' (1035-1123) De lapidibus*, in *Sudhoffs Archiv*, Beiheft 20 (Wiesbaden, 1977), pp. 9, 16-20.

[50] *De min.* II, tr.2, c.4, *draconites* (ed. Borgnet 5: 35a; Wyckoff, p. 87).

[51] For an exploration of this phenomenon in a modern context, see Jerome D. Frank, *Persuasion and Healing: A Comparative Study of Psychotherapy* (Baltimore, 1961).

[52] For a recent reappraisal of medieval psychiatry and a revision upward in appreciation of medieval approaches to mental disorders, see Jerome Kroll, "A Reappraisal of Psychiatry in the Middle Ages," *Arch. Gen. Psychiatry*, 29 (1973), 276-283.

[53] *De min.* II, tr.2, c.3, *carneleus* (ed. Borgnet 5: 33b; Wyckoff, pp. 81-82). Arthur K. Shapiro, a physician, ("The Placebo Effect in the History of Medical Treatment: Implications for Psychiatry," *Amer. Jour. of Psychiatry*, 116 (1959), 298-304, esp. 299) defines the placebo effect "as the psychological, physiological or psychophysiological effect of any medication or procedure given with therapeutic intent, which is independent of or minimally related to the pharmacologic effects of the medication or to the specific effects of the procedure, and which operates through a psychological mechanism."

[54] *De min.* II, tr.2, c.5, *ematites* (ed. Borgnet 5: 36a-b; Wyckoff, p. 90).

which acts as an astringent.[55] But in comparing Albert's stones with modern pharmaceutical compounds there is always an element of uncertainty especially when we cannot even know what Albert's stone was, as for instance with *ramai*. Albert said that *ramai* was certain to overcome looseness of the bowels and especially the bleeding of dysentery and menstruation.[56] If we do not know what *ramai* was, we are unable to confirm or reject, however timidly, the alleged physiological actions. In most instances the actions may be those of a placebo but, as is recognized by modern medicine, placebo's have positive results in short-term psychotherapy.[57]

An interesting and largely typical example of Albert's recommendations is coral about which he said:

> And it has been found by experience that it is good against any sort of bleeding. It is even said that, worn around the neck, it is good against epilepsy and the action of menstruation, and against storms, lightning, and hail. And if it is powdered and sprinkled with water on herbs and trees, it is reported to multiply their fruits. They also say that it speeds the beginning and end of any business.[58] (Wyckoff trans.)

Coral is almost entirely calcium carbonate. Modern experts affirm, as any gardener knows, that calcium carbonate helps plants. Probably it would aid in coagulation either as a topical application or, possibly, internally.[59] For epilepsy, on the other hand, it is more difficult to determine how the disease would respond to calcium carbonate. An epileptic has an acidosis condition, which is an abnormally high production of acids or an abnormal decrease of alkalinity. Modern therapy would include a ketogenic diet in treatment because it produces acetone or ketone bodies which are helpful to an epileptic. An alkaline such as coral would be beneficial, but Albert advises its use as a necklace! The examples show that the ancients did know some

---

[55] Ferric oxide has no currently recognized astringent qualities but ferric chloride does. Pliny (*Nat. Hist.* XXXIV. 45. 152-153, ed. H. Rackham [Cambridge, 1952], 9: 238) recommends iron rust to unite, dry, and staunch wounds.

[56] *De min.* II, tr.2, c.16 (ed. Borgnet 5: 44a; Wyckoff, pp. 114).

[57] In five separate studies when patients with various emotional states were given placebos, fifty-five percent showed significant symptomatic improvements: Lester H. Gliedman, Earl H. Nash Jr., Stanley D. Imber, Anthony R. Stone, and Jerome D. Frank, "Reduction of Symptoms by Pharmacologically Inert Substances and by Short-term Psychotherapy," *A.M.A. Archives of Neurology and Psychiatry*, 79 (1958), 345-351.

[58] *De min.* II, tr.2, c.3 (ed. Borgnet 5: 33b; Wyckoff, p. 81).

[59] We are grateful to Prof. Samuel Tove, a biochemist at N. C. State University, in assisting in this judgment.

specific effects of the substances they dealt with and that their experiments tended to further this knowledge although in other respects they were credulous and accepted claims that could not be verified.

In trying to understand his attitude to stones, it is important for the history of science to realize the hermetic origins of much of Albert's thought. For in speaking of the occult (our term) powers of the *liparea* stone, he said, "If this is true, it is very marvellous, and undoubtedly is to be ascribed to the power of the heavens: for, as Hermes says, there are marvellous powers in stones and likewise in plants, by means of which natural magic could accomplish whatever it does, if their powers were well understood."[60] Albert knew of Evax, king of the Arabians, whose letter to Emperor Tiberius (14-37 AD) claimed that God had placed in each variety of stone certain powers beneficial to man.[61] These powers were made known to the Egyptians and almost lost in a fire which burned the library, presumably the Alexandrian library. The secrets were rescued and held in a trust by the Arabians. These secrets are there for man's discovery. Marbode said that for man each herb has certain powers, but even greater than those in herbs are the powers in stones.

Is not this line of thought, stemming from the hermetic tradition accepted by Albert, just as important in the history of modern science as the thinking of the more highly acclaimed natural philosophy which sought logical explanations for natural things, but whose theories were more separated from observation and empiricism? For Albert, God was not in each rock, but he had put certain powers into them through secondary causes, including the celestial bodies. Those powers, whatever they are, can be discovered only by observation of their effects.

Albert saw a division between natural science and the science of magic; the latter he saw as a legitimate field of inquiry but inferior to natural science.[62] When explaining why images of things are formed

---

[60] *De min.* II, tr.2, c.10 (ed. Borgnet 5: 402; Wyckoff, pp. 102-103).

[61] For text of the Evax letters, see Riddle, *Marbode*, pp. 28-31.

[62] When explaining that he was to omit a discussion of a method for the discovery of metal ores, an omission we regret, Albert said that the science dealing with this "depends not upon [scientific] demonstration, but upon experience in the occult and the astrological." He called that science, "the science of magic called treasure findings" (*De min.* III, tr.1, c.1 [ed. Borgnet 5: 60a; Wyckoff, p. 154]. After relating Hermes' reason for powers of stones, Albert stated: "Nevertheless this statement is not enough for natural science, although perhaps it may be sufficient for astrology and magic" (*De min.* II, tr.1, c.3 [ed. Borgnet 5: 27b; Wyckoff, p. 63]). But throughout this work Albert gathers information from magicians, astrologers, and incantators, e.g., *De min.* II, tr.2, c.8, *iaspis*; c.11, *magnes*; II, tr.3, c.5 (ed. Borgnet 5: 39b, 40b, 53b; Wyckoff, pp. 100, 104, 141).

in the patterns of gem stones and not in other things, for example in mineral ores, bones, etc., he theorizes that gemstones are more amenable to heavenly impression during their formation. He said, "These things are not pure science, but because they are good doctrine they are included here."[63]

Why should an image of a king's head appear in a marble slab which Albert saw when he was a young man in Venice? Albert had satisfied the curious spectators to the scene who wondered why the king's forehead was disproportionately large. Young Albert explained that, while the mineralizing power was forming the marble, the vapor at the forehead rose disproportionately higher, like a cloud.[64] He accepted the mysteries of the mineralizing power and the general power of nature. He was only a spectator, an observer, who called witness to the holy mysteries to his fellows. They all need only learn and know the powers God had given them in the secrets of things.

Another significant feature of the *Minerals* is the extended attention which Albert devoted to a consideration of metals and to materials which he classified as "intermediates," possessing the characteristics of both metals and stones. Three of the five books and nearly half of the space in his treatise were assigned to these topics, and in the discussion of metals Albert made many noteworthy observations. More than in the books on stones, he was forced to draw on his own experience since the sources available to him had even less to say about metals than about stones. Albert explained the problem facing him, and his method of solution, at the beginning of Book III.

> In [writing] this as well as the preceding books, I have not seen the treatise of Aristotle, save for some excerpts for which I have inquired assiduously in different parts of the world. Therefore I shall state, in a manner which can be supported by reasoning, either what has been handed down by philosophers or what I have found out by my own observations. For at one time I became a wanderer, making long journeys to mining districts, so that I could learn by observation the nature of metals. And for the same reason I have inquired into the transmutation of metals in alchemy, so as to learn from this, too, something of their nature and accidental properties. For this is the best and surest method of investigation, because then each thing is understood with reference to its own particular cause, and there is very little doubt about its accidental properties.[65] (Wyckoff trans.)

---

[63] *De min.* II, tr.3, c.2 (ed. Borgnet 5: 51a; Wyckoff, p. 134).

[64] *De min.* II, tr.3, c.1 (ed. Borgnet 5: 49a-b; Wyckoff, pp. 128-129).

[65] *De min.* III, tr.1, c.1 (ed. Borgnet 5: 59a-b; Wyckoff, p. 153).

To discuss metals, then, Albert sought to avail himself of the best sources of information he could obtain. From Peripatetic and Arabic philosophers Albert took elements with which to create a comprehensive theory explaining the formation of metals and their distinctive properties. The writings of the philosophers were not sufficient, however. From miners Albert learned much about where metallic ores and intermediates were found and the characteristics of ore bodies in relation to surrounding geological formations. That information, supplemented by personal observation, was cited to support various features of his theory. When miners could tell him little about the properties of metals, Albert turned to the alchemists, who knew of metals through their efforts to effect transmutations. By drawing upon the experience of two contemporary groups whose interest in and constant association with metals made them most knowledgeable on the subject, the miners and alchemists, Albert was able to effect a unique synthesis. Books III through v of the *Minerals* contain information on mineralogy and on metals which helps to bridge the gap in our knowledge of these subjects between the writings of Pliny in the first century and the sixteenth-century works of Biringuccio and Agricola. By focusing attention on metals it also represents a milestone in the literature, one which establishes the content, and to some extent the format, of later studies.[66]

Considering first the topic of mining and mineralogy, it is surprising that today we know as little as we do about the expansion of mining and the use of metals in the twelfth and thirteenth centuries. As John Nef, in his analysis of medieval mining, has noted,

> The increasing curiosity about the material world and the increasing agricultural, commercial, industrial and artistic needs for gold, silver, iron, lead, copper, tin and alloys of these metals made men eager to explore beneath the soil, to examine and to exploit the substances they

---

[66] Following Albert, the first major work devoted to metals was the *Pirotechnia* of Vannoccio Biringuccio (1540). Biringuccio began with a preface describing the location of ores, a book devoted to the ores of metals, and a book discussing "semi-minerals," a number of which, like marchasite, alum, and arsenic, are Albert's "intermediates," before proceeding to the discussion of assaying and smelting (Vannoccio Biringuccio, *Pirotechnia*, trans. by Cyril Stanley Smith and Martha Teach Gnudi [New York: American Institute of Mining and Mineralogical Engineering, 1941]). Agricola's primary work on mineralogy was *De natura fossilium* (1546). While devising a more "modern" classification system for minerals based on natural properties, Agricola's format followed Albert in treating metals in a separate section subsequent to the general discussion on "stones." See Georgius Agricola, *De natura fossilium*, trans. M. C. Bandy and J. A. Bandy, Geological Society of America Special Paper 63 (Menasha, 1955).

found. Not until the eve of the Reformation, when fresh waves of settlers pushed into the same regions, was there another comparable movement of exploration and discovery.[67]

The discovery of gold in the river sands of the Rhine and the Elbe, the opening of silver-bearing ore bodies in the Carpathians and Erzgebirge, in Devon and Alsace, elicited the attention of commercial interests and civil authority, but drew scant notice from the schoolmen. Albert is practically the only major medieval figure to discuss the state of mineralogy before the sixteenth century. Although he had little to say about the technology of mining, which had little bearing on his main thesis, he did wrestle with the difficult problem of ore formation, thereby providing excellent descriptions of many primary ores. His own field observations, made on visits to such famous mining sites as Goslar and Freiberg, constitute an important contribution to our understanding of medieval mineralogy.

A full consideration of Albert's discussion of ores and of mining lore has been made by Dorothy Wyckoff in her article, "Albertus Magnus on Ore Deposits."[68] It suffices here to repeat the major points established by Wyckoff. Of primary concern for Albert was the need to establish, in the scholastic tradition, the causes for the occurrence and properties of metals "in a manner which can be supported by reason."[69]

The basis for the organization of Albert's *Minerals* was a synthesis of Peripatetic concepts of matter, central to which was the doctrine of the four elements, earth, water, air and fire, and Arabic alchemical ideas, which emphasized the importance of quicksilver and sulfur. The order of discussion was dictated by the degree of complexity involved. "First, then, we shall investigate stones, and afterward metals, and finally substances intermediate between these; for in fact the production of stones is simpler and more obvious than that of metals."[70] The material substance of stones, which are infusible, is some form of earth or some form of water. Metals, on the other hand, exhibit properties not possessed by stone, in that they are fusi-

---

[67] John U. Nef, "Mining and Metallurgy in Medieval Civilization" in *The Cambridge Economic History of Europe* (Cambridge, 1952) 2: 437. See also the important paper by Nadine F. George, "Albertus Magnus and Chemical Technology in a Time of Transition," in this volume, 235-261.

[68] Dorothy Wyckoff, "Albertus Magnus on Ore Deposits," *Isis* 49 (1958), 109-122.

[69] *De min.* III, tr.1, c.1 (ed. Borgnet 5: 59a; Wyckoff, p. 153).

[70] *De min.* I, tr.1, c.1 (ed. Borgnet 5: 1a; Wyckoff, p. 9).

ble and malleable, and these properties arise from the admixture of quicksilver and sulfur which, in turn, are combinations of the simple elements. The efficient cause for the production of metals was, according to Albert, heat which digests the unsuitable materials and allows for the combustion of the opposed passive properties of moisture, associated with quicksilver, and dryness, associated with sulfur, which give metals their unique characteristics.[71] A key concept in the genesis of metallic ores was that the nature of the formations in which the ores were "generated" influenced the proportions and the degree of purity with which the simple elements were mixed, thus determining the particular metal to be found:

> In order to know the cause of all the things that are produced, we must understand that real metal is not formed except by the natural sublimation of moisture and earth, such as has been described above. For in such a place, where earthy and watery materials are first mixed together, much that is impure is mixed with the pure, but the impure is of no use in the formation of metal. And from the hollow places containing such a mixture the force of the rising fume opens out pores large or small, many or few, according to the nature of the [surrounding] stone or earth; and in these [pores] the rising fume or vapour spreads out for a long time and is concentrated and reflected; since it contains the more subtle part of the mixed material it hardens in those channels and is mixed together as vapour in the pores, and is converted into metal of the same kind as the vapour.[72] (Wyckoff trans.)

Thus it is that gold and copper, both partaking of the redness imparted by sulfur, differed in their first properties because of the relative purity and admixture of the constituents. Gold had both pure quicksilver and sulfur mixed in the ideal proportions, and, for Albert, the nature of the place was a determining factor in the final product.

> But gold which is formed in sands as a kind of grains, large or small, is formed from a hot and very subtle vapour, concentrated and digested in the midst of the sandy material, and afterwards hardened into gold. For a sandy place is very hot and dry; but water getting in closes the pores so that [the vapour] cannot escape; and thus is concentrated upon itself and converted into gold. And therefore this kind of gold is better. And there are two reasons for this: one is that the best way of purifying sulfur is by repeated washing, and the sulfur in watery

---

[71] *De min.* III, tr.1, c.5 (ed. Borgnet 5: 65b-66a; Wyckoff, p. 167).
[72] *De min.* III, tr.1, c.10 (ed. Borgnet 5: 72b-73a; Wyckoff, pp. 182-183).

places is repeatedly washed and purified; and for the same reason the earthy quicksilver is often washed and purified and rendered more subtle. Another reason is the closing of the pores underneath the water along the banks; and thus the dispersed vapour is well-composed and condensed, and is digested nobly into the substance of gold, and hardens into gold.[73] (Wyckoff trans.)

For copper, however, its occurrence in intrusive veins meant that the vapor could not be concentrated so that heat could digest the unsuitable materials. The resulting effect was a degraded mixture which was similar to gold in appearance, but inferior in form and properties.

Let us assume, then, that the quicksilver is good, not full of dross and dirt, but still not completely cleansed of extraneous moisture; and that the substance of the sulfur is full of dross, burning hot and partly burnt, and in this condition it is mixed with the quicksilver, both in substance and in quality. Then undoubtedly it changes the quicksilver to a red color; and because neither is sufficiently subtle, they cannot be mixed well. And this will make copper, which is not at all well mixed, and much dross is separated from it, and it evaporates greatly in the fire.[74] (Wyckoff trans.)

The reference to an excess of dross and the occurrence of impure sulfur most likely stems from firsthand experience with the copper ores of the Rammelsberg at Goslar. There the sulfide ore of copper, chalcopyrite, appears in places in graded beds containing other heavy metal sulfides, and in a portion of the old bed, intercalated with slate in a network of ores including pyrite, galena and sphalerite.[75] The separation of copper from such an ore body would have involved repeated roastings, with material loss, and the copious release of sul-

---

[73] *De min.* III, tr.1, c.9 (ed. Borgnet 5: 73b; Wyckoff, p. 184). Albert believed that metals were formed where found, and so missed the true significance of alluvial gold deposits. However, nearly three hundred years later Biringuccio repeated many of Albert's observations on stream deposits while only suggesting the possibility of water transport to account for them. It was left to Agricola to suggest the true explanation of the formation of alluvial deposits.

[74] *De min.* IV, tr.1, c.6 (ed. Borgnet 5: 90a; Wyckoff, p. 223).

[75] Pyrite, galena, and sphalerite are the sulfides of iron, lead, and zinc respectively. A full discussion of the constitution of the Rammelsberg ore beds can be found in F. H. Bayschlag, J. H. L. Vogt, and P. Krusch, *The Deposits of Useful Minerals and Rocks: Ore Deposits*, trans. S. J. Truscott (London, 1914-16), 2: 1145-1148.

furous fumes.[76] Albert's theory, then, has been carefully constructed to conform with observed phenomena, graphically demonstrating that he was engaged in a scientific investigation into the nature of minerals, rather than elaborating on the knowledge of metals in the encyclopedist tradition.

The discussion of ores in the *Minerals* was not intended to include their identifying characteristics and properties. For one thing, Albert, like all writers on the subject before the sixteenth century, did not make a clear distinction between metals and their ores as having separate chemical identities. The occurrence of ores was used to illustrate the general theory for the constitution of metals. Yet, the descriptions are sufficiently precise for us to be able to identify those ores of which Albert had firsthand knowledge, and to determine where his information had been derived from other sources. In particular, his descriptions of mercury and tin clearly indicate a lack of familiarity with the ores of those metals. His treatment of tin ore nevertheless is of interest because of the possible light it throws on the date of composition of the *Minerals*.

Albert's discussion on tin ore is very brief: "Two [kinds of] tin are found, namely a harder and drier kind which comes from England or Britain, and a somewhat softer kind which is found more abundantly in parts of Germany."[77] In both regions the tin-rich lodes are associated with granitic intrusions. In England the Cornwall-Devon complex had been mined since pre-Roman times, and by the thirteenth century extensive underground mining of the granitic matrix already was being undertaken. The German mines occur in a broad, north-south zone intersecting the Erzgebirge (Ore Mountains) along the Saxony-Bohemia border not far from Freiberg.[78] Mining for tin had begun near Graupen, in Bohemia, by the end of the twelfth century,

---

[76] Agricola gives a detailed description of the complex process used to reduce copper pyrites to the metal. As a preliminary step, "The cokes of melted pyrites are usually roasted twice over. . . first. . . in a slow fire and afterward in a fierce one." The preliminary roasting drove off some of the sulfur, but additional roasting and refining was necessary to convert the black, brittle matte from the initial treatment to relatively pure blister copper. From Albert's descriptions it is clear that he had witnessed the process. Georgius Agricola, *De re metallica*, trans. by Herbert Clark Hoover and Lou Henry Hoover (London, 1912), p. 349; see also Hoover's footnote on the refining process, p. 407.

[77] *De min.* IV, tr.1, c.4 (ed. Borgnet 5: 88b; Wyckoff, p. 217).

[78] The geomorphology of the Cornwall-Devon complex is described in Charles F. Park, Jr., and Ray A. MacDiarmen, *Ore Deposits*, 3rd ed. (San Francisco: W. H. Freeman & Co., 1975), pp. 163-173; that of the Erzgebirge, in Beyschlag, Vogt and Krusch, *Useful Minerals*, 1: 425-429.

and, according to Albert's contemporary, Matthew Paris, tin ores had been discovered in Germany in 1241.[79] Certainly, this would refer to deposits on the northern slope of the Erzgebirge. From Albert's description, however, it appears that early tin mining in Saxony-Bohemia was confined to the exploitation of alluvial concentrations of the major tin ore, cassiterite. This was still largely true four hundred years later when Agricola described a number of "ancient" methods for working alluvial tin deposits.[80] By comparison to the Cornish hard rock mines the Saxon deposits of ore then would be "soft," as Albert described them.

From Albert's failure to describe cassiterite, the principal tin ore in the deposits of Saxony, one can only conclude that he never saw active tin mining.[81] This omission indirectly tends to support the completion date of 1250 for the *Minerals*. If Albert had visited the silver mines at Freiberg during his youth, as seems probable, there would have been little reason to make a side trip over the mountains into Bohemia to visit the then relatively minor tin works. However, if, as Dorothy Wyckoff has suggested, Albert continued to seek information for the book on his travels as prior provincial in 1254-1256 a visit to the cloister of Andelhausen near Freiberg or to the chapter house of St. Michael at Litomerice in Bohemia, just south of the Erzgebirge, almost certainly would have included examination of nearby tin mines.[82] The absence of a firsthand account of tin ores in the *Minerals* would seem to imply, then, that Albert had completed his work before 1254.[83]

---

[79] *Matthew Paris's English History From the Year 1235 to 1273*, trans. J. A. Giles (London, 1852), 1: 373.

[80] Agricola, *De re metallica*, trans. Hoover, pp. 336-341. Agricola noted that of eight common methods for mining tin, only two were of recent origin. The passage contains an illustration showing a miner digging into the side of a stream bank to tap an alluvial deposit with a mattock, indicative of the softness of the deposits.

[81] At one point Albert claimed to have seen tin incorporated with stone, but no details are given. This, however, was probably a reference to the mixed, metallic-looking pyrites of the Rammelsberg. The earthy brown-black nodules of cassiterite found in alluvial deposits would not fit the description given. *De min.* III, tr.1, c.1 (ed. Borgnet 5: 59a-60b; Wyckoff, p. 154).

[82] V. J. Koudelka gives the date for the founding of the chapter house of St. Michael at Litomerice in Bohemia as 1236, "Zur Geschichte der bohemischen Dominikanerprovinz im Mittelalter," *Archivum Fratrum Praedicatorum*, 25 (1956), 145-146. The founding of the cloister of Arndelhousen at Freiberg in 1234 is indicated in a manuscript footnote quoted in Heibert C. Scheeben, "Handschriften I," *Archiv der deutschen Dominkaner*, 1 (1937), 174-175.

[83] See Wyckoff's arguments on the "Date of Composition of the Book of Minerals" in introd. to her trans., pp. xxxv-xli, particularly p. xxxvi, where she argues that some observations made at Freiberg most probably date from his term as prior provincial.

Although incidental to his main purpose, Albert included several observations which do much to expand our knowledge of the state of thirteenth-century geology. One of these is his discussion of a formation later termed the *gossan* by Cornish miners or *Eisenhut* by the Germans: "If the metal is incorporated with the whole stone, the upper part is full of slag and useless, while the inside is better and more noble."[84] The passage accurately describes the weathered crust of oxides for a sulfide ore body. Whether Albert ever saw a formation is uncertain, but his passage shows that thirteenth-century miners recognized its importance. Finally, Albert recorded his personal observation of a formation peculiar to ore bodies, the pinching out of ore veins passing from one type of rock to another.

> As to natural processes, I have learned by what I have seen with my own eyes that a vein flowing from a single source was in one part pure gold, and in another silver having a stony *calx* mixed with it. And miners and smeltermen have told me that this very frequently happens; and therefore they are sorry when they have found gold, for the gold is near the source, and then the vein fails. Then I myself, making a careful examination, found that the vessel in which the mineral was converted into gold differed from that in which it was converted into silver. For the vessel containing gold was a very hard stone — one of the kind from which fire is struck with steel — and it had the gold pure and not incorporated [with the stone], but enclosed in a hollow within it; and there was a little burned earth between the stony part and the gold. And the stone opened out with a passage into the silver vein, traversing a black stone which was not very hard but earthy. And the black stone was fissile, the kind of stone from which slates are made for building houses. This proves, however, that from a single place which was the vessel of the mineral matter both [gold and silver] evaporated, and a difference in the purification and digestion had been responsible for the difference in the kind of metal.[85] (Wyckoff trans.)

While this passage is significant because it is the first apparent record of what now is recognized to be a common mineralogical phenomenon, it is important for a second reason. No passage in the *Minerals* is more indicative of Albert as scientist. There is the careful examination and accurate description of the formation uncolored by *a priori* assumptions about the nature of stones or ore formation. Yet the example is not one of random observation, for Albert has cited it

---

[84] *De min.* III, tr.1, c.10 (ed. Borgnet 5: 73b; Wyckoff, p. 183).
[85] *De min.* III, tr.2, c.6 (ed. Borgnet 5: 81a-82b; Wyckoff, pp. 200-201).

within the context of his theory of ore formation, thereby, in a sense, providing an explanation for the phenomenon while establishing greater credibility for the theory. Finally, Albert added to his own observation the corroboration of miners and smeltermen, expert testimony indeed. The juxtaposition of observation, hypothesis and authority constitutes the essence of scientific writing as we recognize it today; the same was no less true for Albert.

As the frequent references to "vessels" in the foregoing passage signify, throughout the treatment of metals and ores a second contemporary influence can be noted in the *Minerals*. For much of the discussion of the properties of metals and for the mechanisms of ore genesis Albert had recourse to the growing body of alchemical literature. In the work of the alchemists could be found artful processes analogous to the natural processes by which metals were generated in the earth. Since "art imitates nature," by studying the alchemical efforts to effect the transmutation of baser metals to silver and gold, one could better understand the way in which natural processes functioned. "For whatever the elemental and celestial powers produce in natural vessels they also produce in artificial vessels, provided the artificial [vessels] are formed just like the natural [ones]."[86] Hence, the genesis of ores could be likened to the alchemical operations of washing, boiling, sublimation and condensation, "because, of all the operations of alchemy, the best is that which begins in the same way as nature."[87]

While alchemical operations could be cited to help explain the way in which the natural mechanisms of ore generation functioned, Albert also found much information in the alchemical corpus concerning the properties of metals which could be used to support his theories. For a metal such as tin, of which he had little first hand knowledge, the alchemical corpus provided the primary source of information.

Tin, according to Albert, "has a 'stuttering' constitution;" hence, "it makes all metals with which it is mixed 'stuttering,'" too, and takes away their malleability, as Hermes says; and when it is itself drawn out, it is quickly and easily broken."[88] In one sense Albert is confused here, because pure tin is not, as the passage implies, brittle.

---

[86] *De min.* III, tr.1, c.9 (ed. Borgnet 5: 71a; Wyckoff, p. 178).
[87] *De min.* III, tr.1, c.9 (ed. Borgnet 5: 71b; Wyckoff, p. 179).
[88] *De min.* IV, tr.1, c.4 (ed. Borgnet 5: 87b; Wyckoff, p. 215).

In the twelfth century Theophilus had given detailed directions on the manufacture of tin leaf by hammering.[89] Rather, it is more likely that Albert appears to refer to the embrittling effect on metals alloyed with tin, a fact well known to the alchemists, as indicated in this passage from the works of the Latin Geber:

> Therefore, not omitting to discourse of *Jupiter*, We signifie to the *Sons* of *Learning*, that *Tin* is a *Metallick Body*, white, not pure, livid, sounding little, partaking of little *Earthiness*; possessing in its *Roots Harshness, Softness*, and swiftness of *Liquefaction*, without *Ignition*, and not abiding the *Cupel*, or *Cement*, but Extensible under the *Hammer*. . . .yet its vice is, that it breaks every Body, but *Saturn* and most pure *Sol*.[90] (Russell trans.)

It is not surprising that Albert might confuse the effect of alloying with tin with the properties of the metal itself. Moreover, the source of the confusion between the nature of pure tin and its alloys appears to have originated in the unknown source, Hermes, which may not have been as explicit on the properties of tin as on its effects on other metals.[91] That Albert himself was not in error, but was accurately transcribing the opinions of the alchemists as he found them, may be illustrated by another example. In describing the origin of tutty (zinc oxide) Albert reported that "It is made from the smoke that rises upwards and solidifies by adhering to hard bodies, where copper is being purified from the stones and tin which are in it."[92] German copper ores are more commonly associated with zinc ores than with tin, and the tutty would result from the volatilization of the zinc, its oxidation and subsequent condensation. Yet the Latin Geber, drawing on the same alchemical tradition as Albert, gave a nearly identical explanation for its origin.

---

[89] Theophilus, *On Divers Arts*, trans. John G. Hawthorne and Cyril S. Smith (Chicago, 1963), pp. 180-182. Theophilus described several manufacturing operations using tin, but did not mention ores or tin mining.

[90] Geber, "Of the Sum of Perfection," tr.1, c.31, *The Works of Geber*, trans. Richard Russell (1678), introd. by E. J. Holmyard (London, 1928), p. 66. An early manuscript of *Summa perfecti* is from the thirteenth century (see, Dorothea Singer, *Cat. of Lat. & Vern. Alch. MSS in Gr. Brit. and Ireland*, 2 vols. (Brussels, 1928), 1: 94-96), which means that Albert may have seen the work despite Holmyard's dating the translation later.

[91] Albert made frequent reference to Hermes' *Book of Alchemy*, which neither Wyckoff nor the current writers can identify, although the same source apparently was used by the contemporary authors Arnold of Saxony and Bartholomeus Anglicus. As the quotation from Geber shows, however, Albert's citation accurately reflected the alchemical knowledge of metals such as tin.

[92] *De min.* 1, tr.1, c.8 (ed. Borgnet 5: 102a; Wyckoff, p. 250).

But *Tutia* is the fumes of *White Bodies*; and this is evidenced by manifest *Probation*. For the *Fume* of the *Mixtion* of *Jupiter* [Tin] and *Venus* [Copper], adhering to the *Sides* of the *Forges*, or *Furnaces* of *Artifices Working* in those *Metals*, makes the same impression as it.[93] (Russell trans.)

Throughout the *Minerals* the accuracy of Albert's citation is never in question. One can only regret that the scope of the project was so great as to preclude his firsthand observations of many such phenomena.

By drawing upon the alchemical tradition Albert, as he had done with miners, provided clues to the metallurgical knowledge and skill possessed by the artisans of the thirteenth century. Referring once again to the discussion of tutty, Albert reported that when mixed with copper by the alchemists it changed copper to the color of gold. This is a direct reference to the making of brass using zinc oxide instead of the traditional method of adding calamine (zinc carbonate) to copper, the method Albert had observed at Paris and Cologne.[94] Although it is probable that tutty had been used in this context since the Roman era, Albert may have been the first observer to distinguish clearly between the use of tutty and calamine. This is particularly noteworthy, since the commercial use of zinc oxide did not develop until the sixteenth century.[95]

The references to some metallurgical phenomena are not so easily interpreted, nor can they be assigned to the alchemical tradition with certainty. But they are of interest to historians of chemistry and of metals, because they constitute the earliest record we have on the subjects. One intriguing example is in the case of tin, where Albert wrote, "They say that cast tin quickly decays."[96] One could wish for a fuller explanation of what was occurring, but reference probably was being made to the phenomenon of "tin disease" or "tin pest," an

---

[93] Geber, "Of the Sum of Perfection," II, tr.1, c.4; p. 129.

[94] *De min.* IV, tr.1, c.6 (ed. Borgnet 5: 90b; Wyckoff, p. 224). Albert attributed the information that tutty gave copper the color of gold to Hermes.

[95] After Albert, Biringuccio may have been the next major writer on metals to note the use of tutty to make brass, in a passing reference: "In addition to calamine, copper is also colored yellow by tutty" (Biringuccio, *Pirotechnia*, p. 75). Smith's footnote to this quote gives the date of 1550 for the introduction of brass manufacture from zinc oxide at Rammelsberg.

[96] *De min.* IV, tr.1, c.5 (ed. Borgnet 5: 88a; Wyckoff, p. 216). "They," in this case, would seem to refer to smeltermen from the subsequent comment: "Now it has already been stated that tin is poorly mixed, and this is the reason it is damaged by fire; and if it is removed from the place where it originated, it is destroyed more rapidly than other metals."

allotropic transformation of malleable white tin to a brittle gray, powdery phase which takes place normally below 18°C. Such a transformation also would reinforce the belief that the metal itself was of an inherently brittle nature, as previously noted. The failure of tin plate by "tin disease" also is implied in the discussion of iron. According to Albert, "Tin poured over it [iron] penetrates into its substance. But after this penetration it becomes so brittle that it cannot be worked."[97] The significance of these passages is that the phenomena to which they apparently refer went unnoted in any of the later works on metals and were not explained in the technical literature before the start of the twentieth century.[98]

Some of the more perceptive observations concerning metals in the *Minerals* involve the effect of metals on health. The classification of stones in Tractate 2 followed the lapidary tradition of ascribing medical properties to minerals. Albert did not completely neglect medical properties in his discussion of metals, although, clearly, they were of secondary significance compared to the "accidental" or metallic properties of substances. For lead, Albert reiterated Pliny's claim that lead had a special power over lust and nocturnal emission.[99] "But," he continued, "care must be taken lest the lead, by its coldness contracting the material [below] too forcibly drive it upwards into the head, and cause madness or epilepsy; and care must also be taken lest it cause paralysis of the lower limbs and unconciousness."[100] The later passage is a clear reference to the symptoms of lead poisoning, resulting from the inhalation of lead fumes, as first described by Vitruvius.[101] Albert could not have seen Vitruvius but it is quite likely that Albert had knowledge of the noxi-

---

[97] *De min.* IV, tr.1, c.8 (ed. Borgnet 5: 94b; Wyckoff, p. 234). The plate would decay or become embrittled by tin disease, not the iron.

[98] Mantell, in discussing the allotropic forms, credits the first observation of the effect of extreme temperature change on tin to the pseudo-Aristotelian *De mirabilibus auscultationibus*. Modern observations of tin disease date from 1851 in tin objects, from 1908 for plated objects. The physical basis for tin disease was established in 1899. See C. L. Mantell, *Tin, Its Mining, Production, Technology and Application*, 2nd ed., American Chemical Society Monograph, 51, (New York, 1949), pp. 7-12.

[99] Pliny, *Natural History* XXXIV, c.50, 1. 166 (London, 1952) 9: 247.

[100] *De min.* IV, tr.1, c.3 (ed. Borgnet 5: 86a-b; Wyckoff, p. 210).

[101] Pliny warned of the dangers of breathing the "deadly vapour" of the lead furnace, but without describing symptoms. Vitruvius was more explicit: "For when lead is smelted in casting, the fumes from it settle upon their members, and day after day burn out and take away all the virtues of the blood from their limbs." Vitruvius, *The Ten Books on Architecture*, trans. Morris Hickey Morgan (Cambridge, Mass., 1914), pp. 246-247.

ous effects of lead from his acquaintance with both refineries and alchemists, where exposure to lead fumes in cupelation and assaying and in alchemical procedures might be expected to produce chronic lead poisoning with some regularity. In a similar vein Albert noted that quicksilver "is said to be a kind of poison. It is cold and moist to the second degree, and for this reason it causes loosening of the sinews and paralysis."[102] The passage appears to be the first description in Western literature referring to the affliction known as "hatters' shakes," a form of mercury poisoning characterized by trembling in the extremities resulting from inhalation of mercury vapors.[103]

While not directly affecting health, Albert also noted that metals have peculiar odors and tastes. In particular, he remarked on the ability of copper vessels to taint the taste of most liquids. Other authors, from Pliny onward, had made similar observations, but Albert added the perceptive distinction that the effect was more pronounced for brazen (*aeneus*) vessels.[104] Today, the greater solubility of copper ions from brass alloys in the presence of weak acids and bases is experimentally demonstrable.[105]

Throughout the consideration of metals Albert exhibited a sure instinct for the chemical basis for metallurgical processes. That instinct influenced his own observations and dictated the examples to be drawn from the blend of myth and fact which comprised the store of knowledge possessed by miners and alchemists. These observations together with his classification of minerals and stones gives Albert an important place in the history of the geological sciences.

---

[102] *De min.* IV, tr.1, c.2 (ed. Borgnet 5: 85a-b; Wyckoff, p. 207). There seems no question that Albert's information is drawn verbatim from some unidentified alchemical source. The reference to "degree" of cold and moist so indicates.

[103] Avicenna may have been the first person to describe hatter's shakes in his *Canon of Medicine.* However, this was not one of Albert's sources for compiling the *De mineralibus.* See discussion and citation in Leonard J. Goldwater, *Mercury, A History of Quicksilver* (Baltimore, 1972), p. 211.

[104] *De min.* III, tr.2, c.4 (ed. Borgnet 5: 79b; Wyckoff, pp. 195-196).

[105] Compare Albert's observation with the following quote from a modern metallurgical study of copper alloys: "In the presence of materials such as certain foodstuffs, sufficient copper may sometimes be dissolved, even though in traces, to effect the taste or flavor of the product. In such cases, tin coating of the copper alloys effectively overcomes the situation" (Henry L. Burghoff, "Corrosion of Copper Alloys," *Corrosion of Metals* [Cleveland, 1946], p. 127).

APPENDIX 1: DATE FOR THE COMPOSITION OF *THE BOOK OF MINERALS*

Estimates for the date of the composition of the *Minerals* has varied widely from 1248 to 1263. Dorothy Wyckoff suggested that it was probably not written until 1261-1262, or 1256-1257 at the earliest.[106] She thought that he might have started work on the project before 1254 while he was in Cologne where he was composing four treatises at about the same time, namely, *The Nature of Places, The Properties of the Elements*, the *Meteorology*, and *Minerals*. She argued that Albert delayed the final version of *Minerals* while travelling to find Pseudo-Aristotle's lapidary, about which Albert said he had "inquired assiduously in different parts of the world" to no avail.[107] She raised an objection to Paneth's theory concerning the short anonymous, fourteenth-century tract which bears resemblance to some sections in Albert's Books III and IV of the *Minerals*.[108] She wondered why, if Albert had written this unattributed tract in Bologna during his Italian trip as an early draft of the *Minerals*, a theory suggested by Paneth, did Albert not make reference to locations in Italy and Alpine regions of minerals as he did throughout the *Minerals* to places in Germany and France. Albert's reference to locations of mines is especially frequent in Books III and IV, the same section resembling the Paneth manuscript. Nonetheless, Wyckoff accepted Paneth's thesis because it supported her view of composition of the *Minerals* around 1258. She supposed that Albert's discussion of silver ores at Freiberg[109] and of alluvial gold in Westphalia,[110] a petrified bird's nest at Lubeck,[111] probably dated from Albert's trips when he was prior provincial in 1254-1256. The date 1248, as the earliest date that *Minerals* could have been written, is certain because Albert refers back to his time in Paris which he left in 1248 and mentions the recovery of "Moorish Seville, which is now returned to the

---

[106] Wyckoff, *Albertus*, pp. xxxv-xli.

[107] *De min.* III, tr.1, c.1; cf. also II, tr.3, c.6 (ed. Borgnet 5: 60a and 57; Wyckoff, p. 153, cf. p. 151).

[108] Wyckoff, *Albertus*, pp. xxxviii-xxxix; Fritz Paneth, "Ueber eine alchemistische Handschriften des 14. Jahrunderts und ihr Verhältnis zu Albertus Magnus' Buch 'De Mineralibus'," *Archiv für Geschichte der Mathematik der Naturwissenschaften und der Technik*, n.f. 3, 12 (1929), 35-45; 13 (1930), 408-413; and study of text by Karl Sudhoff, "Codex Fritz Paneth, Eine Untersuchung," *Arch. f. Gesch. der Math.*, n.f. 3, 12 (1929), 2-26.

[109] *De min.* III, tr.1, c.10 and probably IV, tr.1, c.5 (ed. Borgnet 5: 72a-b, 89b; Wyckoff, pp. 181, 220-221); cf. Wyckoff, p. xxxvii.

[110] *De min.* IV, tr.1, c.7 (ed. Borgnet 5: 93a; Wyckoff, pp. 230-231); cf. Wyckoff, p. xxxvii.

[111] *De min.* I, tr.1, c.7 (ed. Borgnet 5: 7a; Wyckoff, p. 28); cf. Wyckoff, intro., p. xxxvii.

Spaniards."[112] The Reconquista of Seville was in 1248. Thus, modern scholars have placed the writing of Albert's *Minerals* as between 1248 and 1263, with Wyckoff hypothesizing a date close to 1262 as most likely.[113]

As cogently argued as Wyckoff's thesis is, her later date seems incorrect in light of an explicit of Albert's *Minerals* in a fifteenth-century manuscript now at Krakow. The colophon states: "Here ends the Mineral Book written by Brother Albertus, of Teutonia at one time from Regensberg, professor of the Order of Preaching Friars, an excellent philosopher, [which was] written in the city of Cologne in the year 1250 of our Lord, under the distinguished guidance of Conrad, archbishop of the aforesaid city."[114] Certainly the Krakow text is copied, but it could hardly have been copied from a manuscript that did not trace back to a manuscript with the same colophon first composed in or near to Albert's lifetime. Albert was teaching in Cologne between 1248 and 1252 when Conrad of Hochstadt was archbishop.[115]

The date of 1250 seems likely when other evidence is considered. As stated above, Albert's knowledge of tin revealed no firsthand experience but, if he had written *Minerals* as late as 1258, he almost certainly would have come into contact with tin mines during this interval when he was travelling in the area. Wyckoff's belief that Albert's travels to Freiberg, Westphalia, and Lubeck were more likely after his Cologne post, is circumstantial when one considers Albert's statement that as a youth he travelled widely to learn of minerals.[116] His visit to Freiberg could have been earlier. Finally Paneth's thesis regarding the text which is connected to Albert's *Minerals* must be rejected out of hand. His thesis that the text is Albert's first draft (and Albert's missing *De alchimia*) written at or

---

[112] *De min.* II, tr.3, c.1; III, tr.1, c.4 (ed. Borgnet 5: 49a, 63b; Wyckoff, pp. 128, 163).

[113] James A. Weisheipl, "Albert the Great, St.," *New Catholic Encyclopedia* (1967), 1: 257b ("before 1263").

[114] Kraków, Biblioteka Jagiellońska MS 6392 III, fol. 7-46v, ending: "Explicit liber mineralium editus a fratre Alberto quodam [*sic*] Ratisponense nacione theutonico, professione [*sic*] de ordine Fratrum Predicatorum precipuo philosopho editus a. D. MCCL in civitate Colonia Agrippina, presidente dicto Cum[ra]do archiepiscopo civitatis memorate. Amen." as reported by Anna Zabrzykowska, Zerzy Zathey, et al., *Inwentarz Rękópisow Biblioteki Jagiellońskie*, 7 vols. (Kraków, 1962), 2: 179. But see a more accurate view above in "Life and Works of St. Albert the Great," p. 35 and n. 75.

[115] Conrad was archbishop of Cologne between 31 May 1238 and 28 September 1261: U. Chevalier, *Répertoire des sources hist.*, 1220.

[116] *De min.* III, tr.1, c.1 (ed. Borgnet 5: 59a-b; Wyckoff, p. 153).

near Bologna is based on no greater evidence than that the text was in a north Italian hand, one of the early fourteenth century.[117] Since it is not an autograph, it is a copy and one made from a text which could have been written almost any place in Europe. Further, the evidence that it was Albert's work is no stronger than a counter hypothesis that it is a modification of sections from the *Minerals*. The possibility is, of course, present that Albert delayed his *Minerals* until he despaired of finding Pseudo-Aristotle's lapidary. However, when his search began, and how long his patience held before he wrote is conjectural. His search quite possibly could have succeeded because one of his source's for the *Minerals*, Arnold of Saxony whom Albert quoted extensively, had a copy of Gerard of Cremona's translation of the Pseudo-Aristotelian lapidary. There are two manuscripts of it.[118] Thus, there were in Albert's time manuscripts of the text. Within the time between 1250 and 1262 the chances of his locating the lapidary would have increased. But he did not know it except through other's works when he wrote the *Minerals*.[119] A date of 1250 for the composition of the *Minerals* keeps the time within the bracketed frame previously suggested but moves this creative interest in natural history closer to his tract on the *Physics*, composed between 1245 and 1248, and at the same time as his teaching of St. Thomas. Even without the evidence afforded by the Krakow manuscript, a 1250 date, or one no later than 1252, seems likely.

APPENDIX 2: NOTES ON SOURCES FOR *THE BOOK OF MINERALS*

Professor Wyckoff identified most of Albert's sources. As was his usual practice, Albert frequently named his authorities. In the lapidary section (II tr.2) Albert relied on other writers, naming some fifteen authors and titles, more than he did, for instance, in Books III–V, where his outside authorities were more restricted. Some new

---

[117] Paneth "Ueber eine alt. Handschrift," *Arch. f. Gesch. der Math.*, n.f. 3, 12 (1929), 45, who based the location on K. Sudhoff's conclusion that the manuscript is *probably* copied in Northern Italy, perhaps Bologna or Padua ("Codex Fritz Paneth" *Arch. f. Gesch. der Math.*, n.f. 3, 12 [1929], 24).

[118] Liège, Bibl. del'Univ. MS 77 (s.xiv), fol. 146v-152v, and Montpellier, Ecole de Med., MS 277 (s.xv), fol. 127-130v. See discussion in Riddle, *Marbode*, pp. 11-12.

[119] See above, n. 118. Albert knew of Pseudo-Aristotle's lapidary through Arnold of Saxony, Marbode of Rennes, and Constantine's *De gradibus*.

evidence, however, has come to light which supplements Wyckoff's study.

Probably the largest unresolved questions regarding Albert's authorities are his use and relationship to the encyclopaedists, principally Thomas of Cantimpré and Arnold of Saxony, and the means of Albert's knowledge of Aristotle's lapidary. The question regarding Aristotle's lapidary was discussed above, p. 230. There can be no question as to the close relation, often word for word, between Albert's lapidary section (and in his section on sigils) and Thomas of Cantimpré's *The Nature of Things*. Since Thomas wrote before Albert, since he was a fellow Dominican, and since Albert normally cited his sources, why did Albert not cite Thomas? Wyckoff suggested either that Albert and Thomas used a mutual source which was anonymous, or perhaps that Albert had a copy of Thomas which lacked attribution.[120] We have confirmed the evidence of Thorndike and Rose that there are many copies of Thomas' encyclopaedia.[121] Thomas' lapidary section was often extracted and stood in manuscripts as an independent work, frequently without attribution.[122] Following the lapidary section of his encyclopaedia, Thomas of Cantimpré lifted Zael's (Thetel) tract on sigils and put it within his work almost intact.[123] This being the case, might not Thomas have bor-

---

[120] Wyckoff, *Albertus*, pp. 99, 269-270; see also, the older studies which noted the relationship between Thomas of Cantimpré and Albert, e.g., H. Stadler, "Albertus Magnus, Thomas von Cantimpré und Vinzenz von Beauvais," *Natur und Kultur*, 4 (1906), 86-90; F. Bormans, "Thomas de Cantimpré indiqué comme une des sources où Albert le Grand et surtout Maerlant ont puisé les materiaux de leurs écrites sur l'histoire naturelle," *Bulletin de l'Academie royales des sciences . . . de Belgique*, 19/1 (1852), 132-159.

[121] Thorndike, *A History of Magic and Experimental Science*, 2: 396-398; Valentin Rose, "Aristoteles *De lapidibus* und Arnoldus Saxo," *Zeitschrift für deutsches Altertums*, 18 (1875), 335-337.

[122] I have notes on the following MSS in addition to those noted in Thorndike and Kibre's *Incipits*, col. 582, and in Thorndike, *A History*, 2: 396-398, with the Inc.: "Generaliter primo dicendum est de lapidibus preciosis. . . ." This is the beginning of the lapidary section in Thomas of Cantimpré's *Liber de natura rerum*, e.g., in Brit. Libr., MS Egerton 1984 (s. xiii), fol. 126. But in many of these MSS the tract stands alone and is without attribution, e.g., Paris, Arsenal MS 1080 (anno 1343), fol. 206v-217; Bibl. Nat. MS lat. 523A, fol. 12; Erlangen, Bibl. Univ. MS 434 (s. xiii-xiv), fol. 152-156; Vatican, MSS Vat. lat. 724 (s. xiv), fol. 67-76; Vat. Pal. lat. 1144, fol. 154-161v; Vienna, Nat. Bibl. MSS lat. 1365 (s. xiv), fol. 81; lat. 2317 (s. xiv).

[123] Wyckoff, *Albertus*, p. 276. Zael's lapidary is in Thomas' *De natura rerum*, Brit. Libr., MS Egerton 1984, fol. 139-140. In addition to the MSS cited by Thorndike (*A History*, 2: 399-400), we have noted Zael's lapidary as a separate work in Milan, Ambrosiana, MS I 65 sup. (1), (s. xv), fol. 1-66 (cited by hand written catalogue); Oxford, Bodl. MS Ash. 1471 (s. xiv), fol. 65v-67v; Florence, Laurentian MS Ashburnham 1520 (s. xiv), fol. 51-55 (*Libellus sigillorum*); Naples, Bibl. Naz. MS XII.E.31 (s. xv), fol. 69v-81.

rowed fairly literally, at least not reworking his material to much extent, from an anonymous source? This unknown source Albert might also have used. The supposition is given support by the fact that manuscript texts without attribution of authorship exist which are parallel to Thomas' lapidary section. Interestingly there are two manuscripts of this tract, found in Thomas' encyclopaedia, which give Albert as the author.[124] An anonymous copy of this tract is found in another manuscript of the thirteenth century, Sloane MS 2428, which contains a text close to Albert's source, but which is not found in the manuscript version of Thomas of Cantimpré, cited as being most reliable, namely Egerton MS 1984; it is not found in the variant text of Thomas, in Bodleian MS Rawl. 545.[125] Specifically this anonymous lapidary manuscript adds the passage for the stone *isciscos* which is not found in Thomas' full encyclopaedia in Egerton MS 1984. Further the Sloane text adds to Thomas' encyclopaedia in the Egerton text the following stones: *karabre, kabrate, kacamon* and *liparia*.[126] They are not found in the variant Bodleian text, but the text on them seems to have been used by Albert for his entries on these stones. Finally Albert has an entry on the Jew Stone, which Wyckoff thought came from Avicenna's *Canon of Medicine*. The text, however, is not related to Avicenna, but it is related to the anonymous lapidary in Sloane MS 2428.[127] Although certainty cannot come until a thorough study of Thomas of Cantimpré's work has been completed the evidence cited here is enough to give Wyckoff support and even probability in her suspicion that Albert and Thomas were using a mutual source.

Albert was not beyond quoting an authority directly from an encyclopaedia without attribution. Albert used Arnold of Saxony but he never cited him by name. A recent study has added to our information of Arnold's souces, and, thereby, has given us a major source for Albert.[128] Wyckoff could not locate a work by "Dioscorides" whom both Arnold and Albert cite. It was not the well-known first-century Greek herbalist. Arnold of Saxony had two frequently

---

[124] Erlangen, Univ. MS 434 (s. xiii-xiv), fol. 152-156, and Vatican, MS Vat. lat. 724 (s. xiv), fol. 67-76.

[125] A Bodleian MS text is published by Joan Evans, *Magical Jewels of the Middle Ages and the Renaissance* (Oxford 1922), pp. 223-234, from MS Rawl. D.35A.

[126] Brit. Libr., MS Sloane 2428, fol. 5r-v.

[127] Ibid.; Wyckoff (*Albertus*, p. 100) gives Albert's reference to Avicenna as *Canon of Medicine* II, tr.2, c.394, but it should read c.404 (fol. 126, ed. Venice 1507).

[128] Riddle, *Marbode*, pp. 11-17.

repeated citations to Aristotle's lapidary, "Aristotle's lapidary translated by Gerard" and "Aristotle's lapidary translated by Dioscorides." There is no doubt but that the translation by Gerard was the text that Albert sought but could not find except through Arnold's fragments as well as fragments in Marbode, Constantine, and Costa ben Luca. On the other hand what Arnold cited as Aristotle's lapidary translated by Dioscorides is the same in content as that cited by Bartholomew the Englishman as being by "Dyascorides" without reference to Aristotle. The study of the context of these fragments revealed that the work allegedly by Dioscorides, whether as author or translator, was the work of Damigeron, a little known lapidarist, probably of the first century. Albert's source "Dioscorides" was Damigeron.[129]

Albert cites among the highest authorities on stones: "Hermes [Evax], king of the Arabs, and Dioscorides, Aaron, and Joseph."[130] In many manuscripts prefacing the lapidaries of both Damigeron and Marbode are two letters written by Evax, king of the Arabs to the Emperor Tiberius (14-37 AD) about the secrets of stones. Albert knew both Damigeron's and Marbode's works.[131] Damigeron's lapidary is in prose, Marbode's in verse. Probably either because the texts Albert used did not attribute correctly the authors or because Albert saw too close similarities between Damigeron and Marbode, he chose not to cite either except as "Evax" since Evax's letter preceded both works.

Albert's "Hermes" is more difficult for the modern researcher to trace. Dorothy Wyckoff noted the "bewildering number or books" ascribed to Hermes.[132] Although Wyckoff identified some of the Hermetic treatises employed by Albert, she was unable to determine all of Albert's Hermetic material, nor have we been able to add to Wyckoff's study.

Aaron's work escapes modern identification. Whereas Albert refers several times to Aaron in association with "Evax" and "Dioscorides," he three times cites Aaron for specific information on stones. In our study of Latin lapidary manuscripts, we did not find any lapidary text attributed to Aaron. Arnold of Saxony and Costa

---

[129] Ibid., pp. 103-105.
[130] De min. I, tr.1, c.1 (ed. Borgnet 5: 2a; Wyckoff, p. 10).
[131] Riddle, Marbode, pp. 28-30.
[132] Wyckoff Albertus, p. 273.

ben Luca cited Aaron, however.[133] On *amethysus*, and *hiena* Albert cites Aaron and the information is found in Arnold but without attribution. On the stone *iscustos*, Albert cites Aaron but the information is not in Arnold[134] but is found in the Sloane lapidary, MS 2428 which is related to Thomas of Cantimpré.[135] It is possible that Albert had a copy of a tract by Aaron, and it is equally possible that his knowledge of Aaron was indirect. For instance, Albert three times cited Isidore, but in the first case Albert's direct source was Thomas who named Isidore as his source[136] and, in the second and third instance, Albert's source may be found in the Sloane MS 2428 which names Isidore for the source.[137] It may be that Albert had no actual text of a lapidary by Aaron but instead employed him indirectly as an authority.

---

[133] Arnold, *De coelo et mundo*, 3 (ed. Emil Stange [Erfurt, 1905], p. 73), who begins the section: "Nam que utiliora, meliora et notabiliora ab Aristotele et Aaron et Evace, rege Arabum et Diascoride sparsim tradita sunt, excepi . . ." (3; ed. Stange, p. 67). Costa, *De physicis ligaturis*, in Brit. Libr., MS Add. 22,719, fol. 201: "Aaron dixit, stercus elefantum cum lacte. . . ." Noteworthy is the fact that Aaron's lapidary is not found cited in the Arabic lapidary tradition, viz. Al-Kitāb al-Muršid, *Über die Steine. . .*, trans. Jutta Schönfeld (Freiburg, 1976).

[134] *De min.* II, tr.2, c.1 and 8 (ed. Borgnet 5: 31b, 38b, 39a-b; Wyckoff, pp. 74, 96-100); Arnold, *De coelo*, 3 (ed. Stange, pp. 70, 73).

[135] Fol. 5.

[136] *De min.* II, tr.2, c.17, *syrus* (ed. Borgnet 5: 46b; Wyckoff, p. 122); cf. Thomas, London, Brit. Libr., MS Egerton 1984, fol. 136.

[137] *De min.* II, tr.2, c.8, *iscustos, judaicus lapis* (ed. Borgnet 5: 39b; Wyckoff, pp. 99-100); cf. Brit. Libr., MS Sloane 2428, fol. 5.

9

Albertus Magnus and Chemical Technology in a Time of Transition

Nadine F. George

Hamilton College

The chemical tradition of the Middle Ages is a complex blend of alchemy, workshop practice and theory derived from Arab and classical sources. Albert's position with respect to alchemy is ably discussed elsewhere in this volume; I will here attempt to describe his role in the transmission of chemical processes less clearly associated with transmutation of base metals into gold. "Chemical technology," for the present purposes, is both broadly and humbly defined to include examples of any effort to modify the characteristics and qualities of material substance in some way that will also modify the utility of that substance. Although Albert's observations on such subjects are widely scattered through his works, the chief source used here will be his *Meteora*, a commentary on Aristotle's *Meteorologica*.[1] The choice may seem peculiar to contemporary scholars, since neither the Aristotelian work nor Albert's commentary has attracted much attention in this century. An extreme example of modern distaste for Aristotle appears in H. D. P. Lee's introduction to his translation of the *Meteorologica*, where he says, "the main interest of the work is to be found. . .in the fact that all his conclusions are so far

---

[1] Albertus Magnus, *Meteororum* (ed. Borgnet, vol. 4). Aristotle, *Meteorologica*, trans. H. D. P. Lee (Cambridge, Mass., 1962). I have used Lee's translation when quoting Aristotle.

wrong and in his lack of a method which could lead him to right ones."[2] As for Albert's commentary, all is silence; there are one or two brief citations in Partington, Thorndike and Sarton,[3] but I have been unable to find that any significant study of the *Meteora* has appeared in the last fifty years.

It may be well to admit at the beginning that present-day opinion of Albert's chemical knowledge is not high.[4] Certainly he did not make discoveries or perform laboratory experiments. His role was that of a scholar-observer who combined extensive learning with an ability to take note of detail that might escape the eyes of others. Even in the sphere of technology, he has left records that deserve further study. One of the more notable examples occurs in his expanded commentary on Aristotle's report of iron and steel processing. Aristotle says,

> Wrought iron indeed will melt and grow soft, and then solidify again. And this is the way in which steel is made. For the dross sinks to the bottom and is removed from below, and by repeated subjection to this treatment the metal is purified and steel produced. They do not repeat the process often, however, because of the great wastage and loss of weight in the iron that is purified. But the better the quality of the iron the smaller the amount of impurity.[5]

Aristotle uses the word $\tau\acute{\eta}\kappa\epsilon\tau\alpha\iota$, "melt" or "dissolve," but this should not be taken to mean that iron liquefies. The furnaces of that day were capable of producing a maximum temperature of about 1,200°C. Iron melts at 1,537°C. Until the development of the blast furnace, iron was smelted but never melted; it became a spongy mass known as the "bloom," while fusible impurities melted and ran down to the bottom of the furnace. The infusible impurities remained mixed in the bloom, which was taken to the forge where they were hammered out. Production of steel depended on carburization of the iron; during repeated firings some of the carbon from the charcoal would diffuse into the iron, making it much harder but

---

[2] Aristotle, *Meteor.*, p. xxvi.

[3] J. R. Partington, "Albertus Magnus on Alchemy," *Ambix* 1 (1937), 9-10. George Sarton, *Introduction to the History of Science* (Baltimore, 1931), 2: 936, 938. Sarton does not refer to *Meteor.* in his discussion of Albert's chemistry, 937. Lynn Thorndike, *A History of Magic and Experimental Science* (New York, 1923), 2: 523, 524, 543, 547, 577, 581, 583. Most references are to curiosities of astrology.

[4] The most balanced account in English is Partington, "Albertus Magnus on Alchemy."

[5] Aristotle, *Meteor.* IV.6 (383a32-b5).

also much more brittle, especially if it was quickly cooled by quenching. The operation known as "tempering" involved reheating the steel to restore some ductility. All of these processes developed slowly, perhaps through accidental discovery. Although Aristotle does not mention quenching, we know from a reference in Homer that it was practiced in the early Greek civilization; there is no reliable evidence of tempering (at least in Europe) in antiquity.[6]

The production of cast iron depends on development of a furnace that can melt iron, and also on improved methods of working the iron itself. Archaeological evidence for the blast furnace does not begin in Europe before the early part of the fourteenth century, although there have been efforts to date it earlier. Recently Cyril Stanley Smith and John G. Hawthorne revived Otto Johannsen's suggestion that the furnace may have been in use as early as the eighth or ninth century.[7] The evidence is literary, and very slight, depending upon the use of the word "running" to describe smelted iron. A conservative view dictates that one interpret this word as Aristotle's "melt" is interpreted: it expresses only the soft, spongy appearance of the bloom.

In the *Meteora* Albert follows his usual practice of keeping to the Aristotelian text but embellishing it with extended comment and many new examples. This is what he made of Aristotle's passage on iron and steel:

> When they take the iron out of the ground it is mixed with stones and earth and its dross (*scoria*) is great. They make the fire exceedingly hot, and then they distill iron in the bottom [of the furnace] and the stones and dross are thrown out on top and it [the iron] becomes moist and flows. And if it is again solidified and again dissolved by powerful fire it will flow, and each time this is done the dross always comes down from it, and steel will be made. For in this way they make steel from iron, because steel is a kind of metal different from iron. However, the smiths (*fabri*) make this with much change of fortune because it causes great loss in the iron, [which] weighs less on account of the great consumption of moisture and burning away of many parts of earthiness, and they make no profit. The better and more noble iron is that whose impurities (*purgamenta*) are few, because this kind is better mixed and

---

[6] Robert Maddin, James D. Muhly and Tamara S. Wheeler, "How the Iron Age Began," *Sci. Am.* 237 (October 1977), 122-131.

[7] *Mappae clavicula*, ed. and tr. by Cyril Stanley Smith and John G. Hawthorne, Transactions of the American Philosophical Society, n.s. 64, pt. 4 (Philadelphia, 1974), p. 62.

made from better mercury and sulphur with less filth. Indeed if iron is more often purified it is made drier and more easily breakable; and therefore they do not liquefy it often, lest it lose (*emittat*) its flexibility (*curvabilitatem*).

But this is open to doubt, because we have said in the end of the third book of this science that iron differs from other metals in that it is softenable only, and not liquefiable; but now we grant that it is liquefiable by more powerful fire and with greater difficulty than the other [metals]. This seems to be opposed to our teaching, for lightness or violence of fire do not transmute species; therefore that which a violent fire liquefies has the nature of that which is dissolved by a slow [fire]. A stronger fire seems to dry out more, and whatever is more dry is less soluble and liquefiable; therefore a strong fire ought rather to keep iron from liquefying than to liquefy it.

Now we must say that iron is certainly soluble with difficulty, because much congealed moisture is introduced into it by cold, and a metal into which less congealed moisture is introduced is more extensible under the hammer (*magis est productibile malleis*), and one into which more congelation is introduced is less extensible. And thus cold iron is not extensible, unless heat first releases some of the congelation of essential coldness in it. And it seems proper to agree that metals do not differ in kind according to the ease or difficulty of their solution; but rather they differ with respect to matter according to the manner of their admixture and retention of miscibles, and they differ with respect to kind in their forms. Therefore what was said in the end of the third [book] was said as evidence of the strength of retention and coagulation of moisture that is exhibited by iron, whereas now we are saying that iron liquefies in some fashion. As for the objection concerning heat and strong fire, we must say that a slow fire opens and dissolves; but it is not excessive enough to gather sufficiently together the homogeneous parts, or separate out the heterogeneous parts, of dissolved [iron]. For this is done by powerful heat, that which dissolves moisture and burns up earthiness and separates each in turn. If moisture is retained, it cannot completely escape, and therefore it begins to distil with subtle earth, and the earth is liquefied, and then the gross earthiness that held it fast is burned up in the dross. And this indicates that greater dryness is brought about, yet not [complete] dryness because moisture is retained and cannot escape. It follows that if in certain ways they [the smiths?] assist fire in the separation of the heterogeneous [parts], that is, the gross earthiness and subtle waters, in iron, they will liquefy the iron straightway. For if iron is filed, and the powder of sulphur and orpiment is projected over [it] and well mingled with the iron filings, and if afterward it is [put] in a strong fire, it straightway liquefies because the sulphur assists in the burning up and dissolution of gross earthiness from the moisture retained in it. And this is how the work of alchemists is performed; they often separate the earthy and subtle moisture of iron, finally sifting out from iron that which is like silver; for it is made flexible by the subtle moisture, and begins to whiten as earthiness separates out. Concerning such things, however, there will be another treatise. Now in the same way steel, warmed by gentle heat, softens; for since there is not enough moisture dissolved in the steel and not drawn out, if the heat is gentle the [remaining] moisture begins to run about through the dry parts of the steel, and softens them. In this way swords and such things are restored and softened by heating, and afterwards by cooling slowly, little by little. Now there is another way of softening, which Nicolaus Peripateticus puts in the *Alchemica*: if steel is hollowed out in the fashion of a hemisphere, and many small holes are made in the steel, and glowing lead is cast over the steel, the lead will be evaporated, and leave only a slight tincture on the steel, and its moisture will attract the steel, where it will drink and be softened. And if this is done very often, the steel will at last be made so soft that it will be squeezable and shapeable in the hands. Glowing irons are hardened whenever they are dipped into cold water, because then the heat is quickly pushed back into the depths of the iron by cold, which draws with it the moisture diffused through the iron. The heat will be gathered into itself by cold, in the center, where it then burns up much moisture, and thus the iron is hardened. And if it is a water of powerful dryness in which the iron is quenched, then it will be quickly consumed by rubbing together with other bodies. And if it is a sword, it will cut other iron easily (*fortiter*), especially (*sicut si*) if it has been quenched often in radish water mixed together with the liquid which is squeezed from the worms called earthworms (*lumbrici terrestres*). Thus skilled men (*ingeniatores*) harden the axles on which wheels revolve, and some soldiers harden the edges of their swords and the points of their lances in this way.[8]

The passage has been given *in extenso* as an outstanding example of Albert's technique. Characteristically, he does not quote directly; he also telescopes an alarming number of observations and allusions into a single unit. Beginning with the smelting process, he goes on to combine this with carburization after the manner of Aristotle (note that neither makes a distinction between smelting and forging), proceeds to a vaguely "alchemical" recipe for preparation of an arsenical iron alloy, adds the curious instance of softening steel with "lead," and describes quenching and tempering in terms reminiscent of a much better known passage from his *Mineralium*:

---

[8] Albert, *Meteor.* IV.2.9 (ed. Borgnet, 4: 760a-761b).

Steel is distilled and repeatedly purified until it has almost the white-
ness of silver; and then engravers' tools are formed of it, with suitable
sharp points. Then the juice is squeezed out of a radish, and mixed with
an equal amount of water extracted from earthworms which have
been crushed and pressed through a cloth. Then the tool, heated white
hot, is quenched in this water two or three or more times, or as many
times as may be necessary. And it becomes so hard that it scratches
gems and cuts any other iron like lead.[9]

The radish-earthworm quenching water has stimulated considerable
scholarly effort; early suggestions that these odd ingredients provided carbon have been rejected, and the theory currently in favor
posits an effect on the color of the hot steel. Unfortunately, modern
metallurgists maintain that there is no color change observable in the
effective range of temperatures.[10] As we shall see later, the tradition
of special quenching waters is long and honorable, at least in "practical" literature. First, however, we must return to the question of
melted iron.

Albert says that the iron "flows" (*fluat*), that the "dross is thrown
out on top" (*scoria ejiciuntur sursum*), and seems to indicate that the
iron melts in the bottom of the furnace. All of this suggests cast iron:
in that process, slag forms a layer over the top of the molten iron,
whereas in the bloomery process melted slag flows down to the bottom. Albert's description of steel-making, on the other hand, is conventional and Aristotelian: the dross "comes down" (*purgabitur deor-
sum scoria*) with each firing. Failure to distinguish between the
smelting and forging operations is crucial here, since ancient steel
was produced by the bloomery process with extensive hammering
and reheating; it was not necessary or even desirable to melt the
iron. Yet Albert's observation that iron melts seems authentic, since
he stresses the effect this fact will have on his theory of solution and
admits that it seems to contradict earlier statements. In the
Mineralium, Albert says of iron that "it cannot be liquefied like wax,
but is liquefiable only in that it can be softened."[11] Both language
and emphasis in the *Meteora* passage suggests a real change of opinion. Albert has seen melted iron and noted the particularly high heat
that such melting requires.

---

[9] Albertus Magnus, *Book of minerals*, trans. Dorothy Wyckoff (Oxford, 1967), p. 133.

[10] Maddin, et al., "Iron Age," p. 131.

[11] Albert, *Book of minerals*, p. 234. On the chronological problems involved, see above, J. A.
Weisheipl, "Life and Works of St. Albert the Great," pp. 30-31, and below, Appendix 1, p. 568.

However, Albert's statements in the *Mineralium* introduce a dating problem, since this work was probably written after the *Meteora*. Vincent of Beauvais uses material from the *Meteora* (he refers to Albert as "Philosophus") but not from the *Mineralium*; this seems conclusive aside from other arguments. Albert may have revised the *Meteora* but not the *Mineralium*, and his efforts to resolve conflicting statements about iron-melting may be inspired in part by his knowledge that the latter work also maintains that iron will not melt. But why did he let the assertion stand in one work and not the other? Perhaps he revised the *Meteora* when he was growing old, or perhaps he saw no necessity to change the *Mineralium*. Aristotle's *Meteorologica*, especially Book Four, represents the closest approximation to an applied theory of chemical change that we have from the Stagyrite; the principles laid down in *De generatione et corruptione* are here applied to an extensive range of substances. If Albert observed melted iron after both the *Mineralium* and the *Meteora* were completed, he might have wished to revise both, but the need to revise the *Meteora* would be more pressing because that work represented fundamental theory whereas the *Mineralium* was primarily descriptive. Such conclusions are speculative, and may be confirmed or altered when the definitive editions of these works become available. At present, one cannot rule out the possibility that the passage is an interpolation added by another author; this makes any firm statements about the dating of cast iron impossible.

If we accept the report as Albert's, we may wish to say that he has given evidence that the blast furnace was operating around the middle of the thirteenth century, somewhat earlier than dates established archaeologically. We must remember, however, that accidental production of cast iron is not the same thing as deliberate preparation; most experts agree that furnaces capable of melting iron appeared at least as early as the thirteenth century, but at that time the melted iron was a catastrophe. There was no technology for handling the melt, which was abandoned as dross. Even in the sixteenth century, according to Biringuccio, cast iron was not a particularly desirable material.[12] More conservatively, we may suggest that Albert's testimony provides literary evidence reinforcing the idea that furnaces with iron-melting capabilities were in operation, though this does not

---

[12] Vannocio Biringuccio, *Pirotechnia*, trans. Cyril Stanley Smith and Martha Teach Gnudi (Cambridge, Mass., 1943; Dover reprint, 1966), p. 66. Note that Biringuccio says that iron melts because of impurity in the ores, suggesting that the blast furnace was still very primitive.

prove that cast iron was a commercial product at that time. This is important, since there has been great uncertainty about the exact date of such furnaces; on the archaeological evidence they seem to have appeared rather suddenly in the early fourteenth century.

If Albert saw melted iron fairly late in his career, the radical implication would be that this dates furnaces with iron-melting capabilities quite precisely in the third quarter of the thirteenth century. But this is unwise, since Albert himself was not a technician, nor was he able to visit foundries regularly or extensively. His observation of the phenomenon confirms that it could happen, but says nothing about the possibility that it had happened much earlier. At best, we may say that he has given testimony to an important transition phase in iron technology.

The alchemical recipes that follow deserve comment in this context. The juxtaposition is unusual; by the Middle Ages at least, the traditions of alchemy and the workshop were completely separate. It is true that laboratory techniques and equipment originated in alchemical practice, but these were not related to the smithy. Indeed, iron in metallic form was not popular with alchemists, as a brief summary may show.

The "seven metals" of alchemy were, in Albert's time, gold, silver, copper, tin, lead, iron and mercury. This list was not established without variation; the liquid character of mercury presented problems, and many early lists use electrum (a gold-silver alloy) or even glass as the seventh metal. Iron itself was also problematical: it rusted easily, and was therefore considered corrupt; it did not amalgamate with mercury in the way other metals did;[13] it did not melt. Some of the earliest Latin documents of alchemy do not even mention iron; examples are the *Book of Morienus* and the *Turba philosophorum*, which utilize various iron compounds such as vitriol, but do not use the pure metal in recipes.[14] A work known to Albert and to his contemporary Vincent of Beauvais, the *De aluminibus et salibus* attributed to Rasis, refers to melting iron by the addition of arsenic or sulphur compounds; oddly enough Vincent, even though

---

[13] On the "artifice" for making an iron amalgam, see Ernst Darmstaedter, *Die Alchemie des Geber* (Berlin, 1922), p. 7.

[14] The *Book of Morienus* recently appeared as *A Testament of Alchemy*, ed. and trans. Lee Stavenhagen (Hanover, N.H., 1974). Waite's translation of the *Turba* is unsatisfactory. I use Julius Ruska, *Turba philosophorum; ein Beitrag zur Geschichte der Alchemie* (Berlin, 1931) for Latin text and German translation.

he uses material from the *De aluminibus* extensively, flatly denies that iron is of any use whatever in alchemy.[15] Not even the Latin Geber, a model of late medieval technical skill, speaks of melting iron without an arsenical additive. Whatever the technical progress of the foundry may have been, alchemists remained separate. Indeed, we may say (with the usual precautions) that alchemical interest in iron was limited both in extent of application and time; the recipes are relatively sparse, and these seem to appear mainly between the ninth and the fourteenth centuries. After Geber, alchemical literature *per se* became more and more exotic, and with the exception of that written by Paracelsus and Basil Valentine, less and less valuable technically. In contrast, the day of workshop literature was beginning to dawn, as the classic works of Biringuccio and Agricola indicate.[16] Albert's reference to both technical and alchemical traditions is therefore notable because such eclecticism is extremely rare. Alchemy was a pursuit of the learned, most of whom knew nothing of foundry practice or any other art of the workshop.

The addition of sulphur to iron produces a fusible sulphide on heating. It melts, but it has no properties that would be desirable in the smithy; the sulphide is much too brittle to serve as material for tools or weapons. The same is true of arsenical alloys of iron. These favored devices of the alchemist might appeal to a jeweler of light morals, since the arsenic alloy resembles silver, but otherwise their practical value is limited. They do "liquefy" iron, but they change it radically in the process. Albert seems to have wished to cover all instances of liquefication known to him as a way of completing Aristotle's example. Further evidence of this desire is the inclusion of information on quenching and tempering, also lacking in the Aristotelian model. Granted the scholarly interests of Albert's day, his use of alchemical literature is not surprising, but the reference to workshop practice is most unusual. Not even the very advanced Geber shows any familiarity with ordinary metalworking traditions. The sixteenth-century manuals of Biringuccio and Agricola are usually cited as the earliest evidence that iron could be melted. Thus Albert's report may be said to span two very different traditions.

---

[15] Vincent de Beauvais, *Speculum quadruplex*; vol. 1: *Speculum naturale* (Graz, Austria, 1964; reprint of 1624 ed.), 8.54: 458. For the Rasis text, see Robert Steele, "Practical Chemistry in the Twelfth Century," *Isis* 12 (1929), 10-46.

[16] Georgius Agricola, *De re metallica*, trans. Herbert Clark Hoover and Lou Henry Hoover (New York, 1950: Dover reprint of 1912 ed.)

Ironically, but perhaps inevitably, neither Biringuccio nor Agricola has a high opinion of Albert's work; both knew him primarily through the *Mineralium* and some spurious works, and both chose to criticize his lapses rather severely.[17]

Recipes purporting to soften steel (*chalybs*) with lead are a curiosity. One appears in the *De aluminibus*, and Vincent of Beauvais also insists that lead can soften steel; some authorities, notably J. M. Stillman, believe that the "lead" is galena, a lead sulphide.[18] Albert's recipe is not the same as that in the *De aluminibus*; I have been unable to identify its source in materials available to me. Both Vincent and Albert stress the plastic quality of the resultant product. Vincent compares it to wax, and as we have seen Albert says that it is "squeezable and shapeable in the hands." Vincent may have taken his information from Albert, both may have used a common source, or Vincent may be relying on the *De aluminibus*. If the lead is galena, the product might be iron sulphide, but Albert's recipe makes that doubtful. Hot lead would have very little effect on steel, at least not when applied in the manner that Albert describes.

Julius Ideler, a great classical scholar of the early nineteenth century, was also unimpressed by Albertus Magnus; his variorum edition of Aristotle's *Meteorologica* cites few significant variations in Albert's *Meteora*, and Ideler's introductory words are unflattering if witty.[19] His shortcomings are basically those of a scholar trained in a single discipline. For example, in his discussion of iron he notes only the alchemical recipes and the radish-earthworm quench. In overlooking the "direct" method also mentioned by Albert which we have been discussing, his classical scholarship was perhaps no less closed to practical technology than was the art of the alchemists which he regarded as fanciful.

Of course, the exotic quenching water could not fail to attract attention. A belief that waters of particular chemical constitution could affect the temper of iron is found in classical sources, particularly Pliny, and repeated by Isidore of Seville, the sixth-century encyclopedist.[20] Oil is also recommended in such sources, but the

---

[17] Biringuccio, pp. 32-33, 36, 115. Most of Agricola's more hostile comments are not in *De re metallica*. See notes to that work, pp. xxx, 609, and Agricola's own remark, p. 76.

[18] John Maxson Stillman, *The Story of Alchemy and Early Chemistry* (New York, 1960: Dover reprint of 1924 ed.), p. 243.

[19] *Aristotelis meteorologicorum libri xx* (Lipsiae, 1836), 2: 536.

[20] Plinius Secundus, *Historia naturalis*, xxxiv, c.41; trans. H. Rackham, Loeb ed. (Cambridge, Mass., 1952), 9: 144-146. Isidore of Seville, *Isidori hispalensis episcopi etymologiarum sive originum libri xx*, ed. by W. M. Lindsay (Oxford, 1911), 2: 14.4.

addition of peculiar organic substances seems to arise during what we may call the "alchemical" period. Albert's recipe may have come from observation, but more probably stems from some literary source; however, he receives credit for it in the *Natural Magick* of Giambattista della Porta, and I have been unable to identify any similar prescription in sources available to me. Like his contemporaries Biringuccio and Agricola, Porta sees nothing of value in Albert: the radish and earthworm water does not harden steel, but *softens* it to the consistency of wax, and indeed, none of Albert's recipes ever work! Porta himself recommends a variety of waters for "tempering" steel (the distinction between quenching and tempering is modern). Some of the ingredients are asafoetida, urine, and even "the foul moysture of the serpent Python," which is said to be particularly effective.[21] As noted earlier, the use of additives in quenching water has been variously explained; one generation of scholars suggested that organic substances provided carbon for the steeling process, but this has been rejected. A later suggestion that the organics might aid in the production of surface colors to indicate the correct temper must be modified since we now know that the range of effective temperature in the process does not affect the color of the steel. Though the steel itself will not change color, it is still possible that the organic mixture might leave a thin surface film which would respond more sensitively to temperature change; only laboratory tests can confirm this suggestion. Porta is supremely confident that color changes are indices of correct temper, but in view of his statement that he tried Albert's recipe and found that it softened steel (which of course it would not), we must be cautious. Albert does not discuss color change in the tempering process, but does refer to the "tincture" left by lead in steel-softening. This is a standard alchemical expression, derived from metal-coloring, and may have no other significance, though it is just possible that Albert was led to use the word because he was thinking of the description of tempering that would follow. In any case, that play of color on the surface of quenched steel seems to have been important to the ironworker, since even Biringuccio discusses it in some detail.[22]

Simply by following Aristotle's text and by trying to provide more complete discussion of relevant examples, Albert has given us some

---

[21] Giovanni Battista della Porta, *Natural Magick* (New York, 1958; reprint of 1658 English language ed.). For Albert's recipe, see p. 309; for the python, p. 310. I have retained the spelling.

[22] Biringuccio, p. 371.

valuable information about the state of iron technology, using both the technical and alchemical traditions. In the case of another metal, much less important to the technician but much more important to the alchemist, Albert left a provocative puzzle. It seems that Aristotle was the first classical author to mention mercury. In the *Meteorologica* he refers to ὁ ἄργυρος ὁ χυτός , which may be translated "liquid silver."[23] Later, a compound word was devised which became the Latin *hydrargyrum*; Pliny uses this term to differentiate between an "artificial" mercury prepared from cinnabar and a "natural" kind found in its pure state, called *argentum vivum*.[24] This confusion in terminology was not completely eliminated by Albert's time, as his passage in the *Meteora* shows:

> Also incoagulable are all things in which water participates, but many such things with the aqueous humor have an airy humor as well, particularly if they are viscous, like oil and mercury (*hydrageros*); i.e., *argentum vivum*. But oil is completely consumed [by fire] before it coagulates, as we demonstrated above. Mercury, however, because of the great admixture of its moisture with its earthiness, is not easily held fast and dried out; since the moisture is proportional to the dryness, it does not stick to what it touches; the dryness keeps the moisture from sticking. And because moisture touches dryness everywhere, it [the mercury] has rapid motion as though moving by itself; that is why it is called *argentum vivum*. Yet in the words of alchemists it is dried by strong burning and by mixing it with sulphur when it is not completely burned. Also it is said that if it is put in a furnace, and green hazel (*corillo*?) branches (*ligna viridia de corillo*) are stirred around in it one after another, it will be hardened and coagulated because hazel attracts moisture powerfully. And the furnace heat takes away the moisture contained in the hazel itself, and this, as it is carried off, attracts that [moisture] coming from (*alienum ex*) the mercury. And by so interchanging the hazel wood for a long time, especially if it has been peeled, they can fix the mercury, and it is coagulated by heat and burned to great dryness. Now, that hazel attracts the [watery] humor more than all other woods is apparent from this: if it is planted next to a vine, it hardens it by attracting its moisture. And know moreover that if mercury is thus dried by heat, it cannot afterwards be dissolved by heat: thus it is more accurate [to say] that *argentum vivum* is dried out than that it is coagulated. For the earthiness which is in it is then dried out and brought to the condition of brick. Therefore this is not an

---

[23] Aristotle, *Meteor.* IV.8 (385b3).
[24] Plinius Secundus, *Hist. Nat.*, XXXIII, c.40 (Loeb ed., 9: 123).

instance of what was brought up in the previous example, referring to the coagulation of things dissolved and the solution of things coagulated; these retain the form itself, as copper is solid and liquid in a single form. Mercury (*hydrageros*) however, fixed in the manner previously mentioned, does not retain a single form in itself or in its component parts. In the same way, if anything is moist with much humidity, that is, very viscous, that thing will not be coagulated as glue is.[25]

This is a remarkable extension of Aristotle's brief reference. For Aristotle, mercury was just another example of highly viscous substances that refused to solidify when heated; for Albert, the liquid metal was an exception to the general theory of coagulation just as liquefied iron was an exception to the general theory of solution. Therefore the processes involved in coagulating mercury had to be discussed, like the processes for liquefication of iron. And again, three processes seem to be described, although there is some obscurity: Albert says that mercury "is dried by strong burning and by mixing it with sulphur when it is not completely burned." This seems to indicate two different processes; the simple "drying" would yield a solid oxide of mercury, whereas the addition of sulphur would produce a sulphide. Most alchemical recipes for the sulphide specify addition of the sulphur before heating begins.[26] Other alchemical methods of "fixing" mercury — that is, of turning it into a solid that would not, at least in theory, dissolve when heated again — involved sublimation with salt or sal ammoniac, which produced chlorides of mercury. The *De aluminibus* includes a wide selection of such recipes, and an early, complex instance from the *Liber sacerdotum* is of interest here:

If you wish to congeal mercury, take pumice and make powder of it, and in the same way, the dung of wood or mountain geese and mountain chicks, and make powder of all these and put [it] in a crucible or some earthenware vessel, and cast the powder down into it and also the mercury, and afterwards [more] powder above, and put it [the vessel] on the coals not far from the hearth; cover it with a cover having a hole in the top so that it can be stirred up with an iron or with wood; dissolve it over a slow fire for a third or half of a day, and look; and if it is not congealed, add a little powder of living sulphur.[27]

---

[25] Albert, *Meteor.* IV.3.2 (ed. Borgnet, 4: 775b-776a).

[26] For example, *Mappae clavicula*, p. 42, c.105.

[27] Marcelin Berthelot, *La chimie au moyen âge* (Paris, 1893), 1: 224; Latin text without translation, number 194.

It is just barely possible that the exotic mixture given above might yield a compound known as "white infusible precipitate," with chemical formula $Hg(NH_2)Cl$. To be sure the process might also result in simple chlorides of mercury, or — as the suggestion to add sulphur indicates — in nothing at all. The ammonia needed for the precipitate could be supplied by the dung; and the pumice would probably contain some chlorine, or be contaminated by sodium chloride (common salt).[28] The recipe is of interest because it is one of the very few that specifies stirring the mercury during heating; either iron or wood stirrers could serve to reduce mercury oxides and thus prevent these from appearing in the final result. Albert's third method for solidification of mercury is a puzzle because it emphasizes use of wood stirring rods in circumstances that appear to make oxidation desirable: if the only ingredient placed in the furnace is liquid mercury, the only way to get a solid by heating is to oxidize the metal; any reduction would be self-defeating.

Albert seems to have had some difficulties with the use of wooden poles in metallurgy; in his *Mineralium* he mentions "pieces of green wood propped up against the copper ore" that are consumed by the sulphureity and fatness of the copper. In a note, Wyckoff suggests that this description indicates imperfect understanding of the poling process, where green wood was forced into the copper melt to reduce oxides still present.[29] But in the case of mercury, an oxide should have been satisfactory. Without further comment on the matter, Partington summarizes this passage from the *Meteora* by saying "mercury is dried by many burnings and mixture with sulphur, or by heating in a furnace with green wood,"[30] but as we have seen, some explanation is required. Perhaps significantly, Albert appears to differentiate between his green wood process and the other two; he seems to be thinking only of the wood process when he says, "if mercury is thus dried out by heat, it cannot afterwards be dissolved by heat." If this is his meaning, some affiliation with the *Liber sacerdotum* recipe may exist; mercury compounds are notoriously unstable to heat, and the "white infusible precipitate" earned its common name because its behavior is unique. It does not melt when heated, though it will dissolve in boiling water and can lose nitrogen

---

[28] On the chemical composition of alkaline pumice, see Robert F. Mueller and Surendra K. Saxena, *Chemical Petrology* (New York, 1977), pp. 334-335, 356-357.

[29] Albert, *Book of minerals*, p. 199.

[30] Partington, "Albertus Magnus on Alchemy," p. 10.

and ammonia under some heating conditions. Other mercury compounds, especially those commonly produced by alchemists, will break down into mercury vapor and other gases when strongly heated.

Of course, Albert may merely be parrotting a common assurance of alchemical texts, where one finds solemn insistence that mercury "fixed" (as an oxide, chloride or sulphide) according to directions will not "flee the fire." It is perhaps rash to suggest that Albert knew the *Liber sacerdotum* recipe and considered the dung and pumice superfluous, though I am tempted. There are other, no doubt, more rational explanations: one might say that at the temperatures used to oxidize mercury (necessarily below its boiling point of approximately 357°C) the wood did not char appreciably and therefore could not act as a reducing agent. It would then be merely a source of moisture, which would speed up the oxidation process. It is also conceivable that some reducing agent would be necessary because of impurities in the mercury itself, or even that Albert has confused the solidification of mercury with its production. Mercury was obtained by roasting cinnabar, the naturally occurring sulphide; one process, still used in the sixteenth century and described by Biringuccio and Agricola, involved condensation of mercury vapor on green leafy branches introduced into the roasting chamber.[31] This does not sound much like Albert's process, to be sure; if he did confuse the congelation with the production of mercury, it seems more likely that he knew of some precise analogue to copper poling, applied during a production phase for mercury, perhaps when the liquid metal began to separate from the cinnabar ore.

Though it is not my purpose to provide a complete summary of Albert's chemical theories, there are some remarks in this passage that may need comment. Albert distinguishes between "coagulation" and drying out; by this he appears to mean that a coagulated thing is simply a more solid state of its liquid phase, whereas the drying actually changes the nature of the substance affected. If so, this is an astute observation with respect to mercury, which is certainly *not* the "same" when it has been solidified by any of the processes discussed. The language is confusing; in this case, drying seems to mean loss of something other than water, and coagulation is what we would mean

---

[31] Biringuccio, pp. 83-84. Agricola copies this description but provides a better illustration (p. 430).

by drying. Significantly, it is the stability of the "dried" mercury to fire that inspires these remarks. This suggests that Albert was fully aware of the general instability of mercury compounds to heat, and that he postulates a fundamental change of state to explain the occurrence of some apparently stable substance derived from mercury.

Such technical details must remain unresolved, but we may say that Albert demonstrates a lively interest in the progress of metals technology and has been eager to include not only alchemical lore familiar to the learned but also examples drawn from the workshop. To that extent he is a precursor of Biringuccio, Agricola, Porta and a host of sixteenth-century writers of manuals which circulated in a less bookish society than that of Renaissance alchemists. Albert also seems to be affiliated with another tradition, that of the incendiary compilations: both Berthelot and Partington have studied the resemblance between the *Liber ignium* and the *De mirabilibus mundi*, a work attributed to Albert and perhaps written by one of his students.[32] We may add to this a passage from Albert's *De causis et proprietatibus elementorum*:

> To bring back the glow of a candle ignited and extinguished, finely powdered sulphur is molded above [it] before the fire is completely dead, and the sulphur provides a flame. In the same way, if a glue of mud is scattered with naphtha and sulphur and stirred up, and lint dipped in it is ignited, it will burn almost inextinguishably. Naphtha is a certain kind of bitumen found in Persia; it is very sticky and has a glutinous and viscous fatness like amurca oil [the dregs of olive oil]. When mixed together with sulphur it becomes inflammable. Its fire clings wonderfully wherever it is thrown, and cannot be extinguished unless completely covered up. Another way of extinguishing it, so it is said, is to throw urine over it. However, water does not extinguish it easily, because water does not enter into it or stick to it because of the fatness.[33]

These snippets of incendiary lore are of small importance in themselves, but serve to indicate Albert's knowledge of another aspect of chemical technology that was only indirectly related to the alchemical tradition. Fire was an elaborate mystery to the alchemist, but to the conjuror it was an endless source of parlor-tricks. Thorndike

---

[32] Berthelot, *Chimie au moyen âge*, 1: 109-120. J. R. Partington, *A History of Greek Fire and Gunpowder* (Cambridge, 1960), pp. 42-90; see especially pp. 81-87.

[33] Albert, *De causis elemen.* II.2.2 (ed. Borgnet, 9: 646a-b).

describes an elaboration of Albert's candle, in which the mouth of an image painted on a wall is smeared with sulphur and turpentine; the glowing but extinguished wick is pressed to the mouth and bursts into flame. This example was taken from a manuscript of the fifteenth or sixteenth century.[34] Porta made an extensive collection of recipes for artificial fires;[35] some of these have a military derivation, but others are of the "home entertainment" variety. The material in *Natural Magick* is taken from classical sources and from those books of "secrets" that form a kind of lower echelon genre of alchemical writing. Albert's composition of mud, naphtha and sulphur can claim kinship with a great variety of references to petroleum, ranging from Livy's account of the torches of the Bacchae to an incendiary preparation attributed to Aristotle by the *Liber ignium*.[36] Works like the *Mappae clavicula, Liber ignium* and *Natural Magick* include instructions for extinguishing such fires; mud, sand, alum, urine and vinegar are variously favored.[37] It is impossible to determine the exact character of Albert's "naphtha" in this passage, but it seems to resemble that described by Strabo, who was one of the first to note the difficulty of extinguishing it once ignited.[38]

The earliest recipes for distilling alcohol appear in "incendiary" sections of chemical manuals. That in the *Mappae clavicula* is now generally considered to be the earliest identified with any certainty; the word "alcohol" was not used to designate this substance until the fourteenth century.[39] Albert mentions distillation of alcohol twice in the *Meteora*; these passages have been cited by Wyckoff[40] but deserve further study. In Book IV tr.3 c.18, Albert says:

> Wine in some ways behaves like oil, in some ways like water. For sweet wine, especially if it is old and dry, evaporates like oil, since it contains much subtle fattiness; and therefore it has many properties in common with oil. For, like oil, it is not solidified by chilling — though it must be admitted that oil is thickened by cold. And like oil it is combustible and disappears completely in burning, and for that reason such wine is famous. But it does not behave at all like wine, since its humor is not

---

[34] Thorndike, 2: 792-793.

[35] Porta, *Natural Magick*, pp. 289-304.

[36] Partington, *Greek Fire*, pp. 6, 46-47.

[37] *Mappae clavicula*, p. 70, c. 279. Partington, *Greek Fire*, p. 48. Porta, p. 298.

[38] Cited by R. J. Forbes, *Studies in Ancient Technology* (Leiden, 1955), 1: 36.

[39] Berthelot, *Chimie au moyen âge*, 1: 136.

[40] Albert, *Book of minerals*, p. 157. Lacunae in Wyckoff's translation supplied from *Meteor.* IV.3.18 (ed. Borgnet, 4: 790a).

vinous, but an oily unctuosity; no such wine intoxicates, but rather it produces gross vapors that often stop up the passages and motive powers of animals. It has, nevertheless, a very subtle vapor. Evidence of this is that it emits a flame; for if it is placed on the fire and hollow reeds are inserted above it [the vapor coming out of them] flames like oil; and what is sublimed from such wine is the nourishment of a subtle flame, as we showed in the preceding tractate of this science, if it is mixed with a little salt and a little powdered sulphur; for the sulphur increases its unctuosity and the salt its warmth.

Wyckoff omitted the sections beginning "for this reason such wine is famous" and ending "motive powers of animals," and beginning "as we showed in the preceding tractate" to the end. The first of these may on casual evaluation seem merely another paraphrase of Aristotle, who says,

Sweet wine fumes, being fat and behaving in the same way as oil, for cold does not solidify it and it will burn. And though called wine, it has not the effect of wine, for it does not taste like wine and does not intoxicate like ordinary wine. It gives off few fumes and so it is inflammable.[41]

Aristotle's wine is apparently only partially fermented; such wine would have a high sugar content with little alcohol, and would therefore be sweet and relatively non-intoxicating. A confirming detail is Albert's evidence that the wine can produce vapors harmful to animals, since continued fermentation would yield carbon dioxide gas. But carbon dioxide gas does not support combustion, and both Aristotle and Albert say that the wine will burn. This apparent contradiction may be resolved by evidence from Porta, whose *Natural Magick* contains several recipes for distillation of *aqua vitae*; one in particular tells how to obtain alcohol without the use of fire, by a process so simple that "it does not require the attendance of a learned artist, but of an ignorant clown, or a woman."[42] The apparatus is a condensing tube set over the fermenting must; alcohol given off as vapor during fermentation condenses in the tube and is drawn off.

There is some difficulty in the assertion that vapor from such wine will ignite readily. Albert says that if one puts the wine on the fire and fixes reeds above it (presumably by inserting them in holes of a cover on the wine-pot) one can ignite the vapor, but this would be

---

[41] Aristotle, *Meteor.* IV.9 (387b9-14).
[42] Porta, p. 247.

true only if alcohol predominated over the carbon dioxide. The problem gains complexity when one considers Albert's other reference to wine, as given by Wyckoff:

> But you may know that when wine is distilled in the same way as rosewater, what is first emitted from it is a watery insipid moisture, and when that has been drawn off, the earthy parts of the wine are left imbued with an oily fat. And if that substance is further distilled over a slow fire, an oil comes off. In this respect one wine differs from another because the stronger the wine, the less water and the more oily liquid is distilled from it; and the weaker and thinner the wine, the more water and the less oily liquid.[43]

On this evidence, Albert's description of the flaming reeds is highly compressed; it would be some time before an alcohol vapor sufficiently free of steam to ignite would be produced, and careful management of the fire would be necessary. Wyckoff was puzzled by the statement that the first distillate was watery, since water has a higher boiling point than alcohol, and rightly suggested that the wine must first have been brought to a boil. The first vapors would then be mixed with steam, and only a later heating at lower temperatures would produce alcohol unadulterated by water. Early distillation seems to have been marked by that difficulty; one of Porta's recipes includes instructions for testing the purity of the alcohol: a rag dipped in the distillate and set on fire will burn up if the *aqua vitae* is sufficiently pure, i.e., sufficiently free from water.[44]

Vincent of Beauvais was so intrigued by Albert's "nourishment of a subtle flame" that he devoted a chapter to it in the *Speculum naturale* and included what appears to be his own theory of the propagation of the flame: sulphur combines its dryness with the subtle moisture of water released by sublimation.[45] It is this subtle moisture that keeps a cloth held in the flame from burning. Vincent gives Albert (called "Philosophus") full credit for this passage, but quotes the description of watery and oily "evaporations" from wine without mentioning a source.[46] He compares the "double evaporation" of wine with the recalcitrance of iron, which does not evaporate at all, and specifies red wine (generally stronger) as the source of the oil.

In his chapter on the *flammae nutrimentum* Vincent does not, as

---

[43] *Meteor*, IV.4.2 (ed. Borgnet 4: 796a); Wyckoff trans., p. 157 n.8.
[44] Porta, p. 257.
[45] Vincent, 6.91: 423.
[46] Vincent, 4.93: 291.

Albert does, suggest that the sulphur is an additive. Vincent apparently accepts the commonly held notion that sulphur is a kind of inflammatory principle inherent in all flammable things; Albert is aware, perhaps due to his knowledge of incendiary recipes, that sulphur was often added to wine before distillation. In some cases, quicklime was also added; according to Partington, this would yield an especially strong alcohol.[47] Berthelot cites an unusual use of the word "wine"; it meant not only the beverage, but in collections of incendiary recipes, it sometimes stood for a mixture of calcium polysulfides and organic matter.[48] This mix would ignite when sprinkled with water, and was understandably fascinating to men of the Middle Ages. Though Berthelot does not speculate about the derivation, it seems possible that the incendiary mix was called "wine" because it developed from the residue left when wine with quicklime and sulphur was distilled; the calcium and sulphur might combine with tartrates to produce a substance that would indeed be "nourishment" of a flame. If Albert means us to take the inflammable vapor and that "nourishment" as separate substances — and while the reference is not clear it is suggestive — he may possibly have the incendiary in mind. In any case, such references to alcohol show that its technology was very new, and that its properties fascinated the careful observer. As Porta illustrates, the making of *aqua vitae* was surrounded by its own mystique even in the sixteenth century.

Albert does not clearly state that the "oily liquid" of his second reference is inflammable, but the term "oily" may imply that property. Oil, like iron and wine, was of great interest to both Albert and Aristotle; one of Albert's interpolations in the *Meteora* is a lengthy discussion of *lateritium* (oil of bricks):

> And therefore whatever things are soft and moist, as many earths are, do not become fat, but lose moisture when evaporated by heat, like a baked brick, which is moist when raw. Stones are solidified in the earth by natural heat: this is the action of heat known as roasting (*optesis*). The reason why it does not fatten is this: that fattening comes from air remaining in water, and from subtle earth containing moisture of water and air, as we say. In softness however there is gross earthy moisture not well mingled, and therefore the action of heat separates one thing from another, for those which are not well mixed will not hold to each

---

[47] Partington, *Greek Fire*, p. 148.
[48] Berthelot, *Chimie au moyen âge*, 1: 116.

other; separated from light things, the earth is hardened and the moisture evaporates.

Now if anyone objects that the oil which is called *lateritium*, extracted from bricks, could not be extracted unless the brick were made fat, we must say that the oil is not to the point, since here we are speaking of fatness as the perfected form of mixture limited by mixable qualities. And certainly brick does not take such form from heat, but rather, so to say, it takes the form of solidity called hardness. Nor do we here deny what alchemists say, that oil and glass [reading *vitrum* for *nitrum*] and gold are drawn out of everything by fire [both] successive and continuous, sharp and powerful or slow, according to the proportion required by the things to be transmuted. For it is not possible that the heat of fire should be extractive of all moisture, since in everything there is some part of moisture inseparable from the dryness it contains; and that dryness contained in moisture is subtle, and its moisture is viscous. And therefore when it begins to stay outside the fire, against the cold of air, it will be fat, like oil. Now moisture collected in this way will be continuous with the body from which it is extracted, and [the body] will be more aqueous, and from it, since it is much burned, glass will be made. But should there be everywhere in it earth of most subtle dryness, and should the subtle earth in it be only a little burned, and most properly combined with the subtlest water by means of heat, then gold will flow out from thence. Indeed these opinions are very difficult, and their principles are not completely possible by artificial means. It is therefore apparent that this objection is not at all contrary to the intended proposition; for the oil which among some physicians is called oil of bricks (*lateritium*) is spread over the top of the brick and is not cognate with it: they take hot brick ignited by strong burning, and quench it by dipping it in oil, or by pouring oil over it, and afterward they prepare the brick by grinding and sublimating as though to make rosewater, extracting from it a very sharp oil both hot and dry. And they use it in medicine.[49]

Both bricks and oil (usually olive oil) play important roles in the roaster of Aristotelian examples, but in most other ways the resemblance between this passage and any revelant segment of the *Meteorologica* is slight. Aristotle discusses some properties of bricks in the sixth chapter of Book IV, and the elusive nature of olive oil in the seventh chapter. The problem of "fattiness" does not arise. Bricks are an example of an earthy substance solidified by heat; olive oil is difficult because it grows more dense when heated or

---

[49] Albert, *Meteor*. IV.2.8 (ed. Borgnet, 4: 757b-758b).

cooled, but does not solidify completely in either case. In his effort to reconcile the Aristotelian teachings with Arab theory (and perhaps with his own observations) Albert greatly sophisticated — and complicated! — the Aristotelian doctrine of four elements and their four "qualities." Fundamentally, one is still dealing with the hot/dry fire, the hot/moist air, the cold/moist water, and cold/dry earth, but there is a congeries of subdivisions. Albert recognizes several kinds of moisture, earth, etc. For present purposes, a complete development of Albert's scheme would be diversionary; we need only notice that in the passage dealing with oil of bricks Albert is trying to differentiate between some innate fattiness that may be brought out by heating and the accidental, external fattiness of the bricks from which *lateritium* was distilled.

Albert's choice of oil of bricks tempts one to psychologize: surely this is an example of free association! Albert seems to have been thinking of Aristotle's brick and oil examples, and to have gone from there directly to the exotic preparation. His digression on alchemical operations might follow naturally, especially since we have evidence that the assertion that alchemists could extract "oil, glass and gold" from all substances was one that stayed in Albert's mind: he used it also in the *Mineralium*.[50] This intriguing statement will be discussed later. For the present, it seems important to note that *lateritium* is more closely allied to the incendiary tradition than Albert's words reveal. A recipe for oil of bricks appears in one manuscript of the *Liber ignium*, and even the earlier versions name it as an ingredient in incendiary mixtures. Another recipe is in the *Liber sacerdotum*,[51] but oddly enough it is the *Liber ignium* prescription that refers to medical uses. The reader is advised to drink the oil with balm of cardamom as a nerve medicine, to use it for the chill of gout, and to enjoy the "marvellous heat" it develops when used as a rubbing oil. The recipe also says that a fisherman anointed with this oil will catch many fish. In his comments, Bertholet called the oil an "empyreumatic," that is, a kind of red hot liniment, analogous to juniper oil (*huile de cade*). A recent edition of the *Petit Larousse* describes juniper oil as "a black, stinking inflammable liquid used to treat sores of horses and skin diseases." None of this will do much to mitigate any grim notions we may hold about medicine in the Middle Ages; indeed,

---

[50] Albert, *Book of minerals*, p. 231.

[51] Berthelot, *Chimie au moyen âge*, 1: 128, 133, 203. Partington, *Greek Fire*, p. 56..

one is tempted to suggest that oil of bricks looks best in its under-played incendiary role.

In practice, the incendiary role may have been the only dependable one. The preparation of oil of bricks is a rudimentary catalytic cracking process, remotely resembling those used in today's petroleum refineries, but such crude methods could not assure a uniform product. Albert has given the essential details: to make oil of bricks, one heated bits of brick or tile to white heat, quenched these in oil, cooled and ground them, put the ground brick in a distillatory and heated it to drive off the oil. In the *Liber ignium* it is called a "philosophical oil," which suggests alchemical ancestry. The term is very imprecise since it was used indiscriminately to designate everything from oils to acids. A footnote in the Borgnet edition refers readers to "Damascenus in the chapter on philosophical oil." This is not much more informative. The "Damascenus" is probably not John of Damascus, an eighth-century churchman and saint. He may be Ibn Māsawaih, more commonly known as Mesuë the Elder—or as Joannes Damascenus. He could even be Nicolaus Damascenus, whose *De plantis* is known to have been familiar to Albert.[52] I have been unable to consult works of these authors, but suggest Mesuë the Elder as a likely candidate since his medical writings seem to have been very popular in the Middle Ages.

Was Albert aware of the incendiary affiliations of *lateritium*? There is reason to believe that he knew both the *Liber ignium* and the *Liber sacerdotum*, or at least works with similar recipes. But since all such compilations are inconsistent in content, we cannot maintain with certainty that Albert read a recipe for oil of bricks in some incendiary manual. More probably, this oil represents evidence of Albert's interest in another branch of chemical technology, the preparation of pharmaceuticals. Indeed, the insertion of the alchemical statement that one can extract oil, glass and gold from any substance, on the face of it a puzzling digression, may reinforce this conclusion. The Hellenistic tradition of alchemy was primarily metallurgical, but the Arabs greatly expanded the pharmacological literature. Though I have been unable to identify a specific source of the statement about oil, glass and gold, there are many suggestive references available. For example, the *Liber de septuaginta* discusses both the general fattiness of stones and the ubiquity of oil, assuring

---

[52] Sarton, 1: 574, 226; 2: 561, 938. Thorndike, 1: 162, 164; 2: 734-735.

readers that they can learn to extract oil from all substances.[53] This work may be a relatively unchanged Latin version of a treatise by Jābir ibn Ḥayyān, the original "Geber" who lived in the eighth century; his fourteenth-century Latin namesake, amusingly enough, displays an ambivalent attitude toward glass: "vitrificatory fusion" can ruin the work of the alchemist.[54] Glass was at least as mysterious to alchemists as iron or mercury; according to Vincent, Rasis (in his *Liber de animalibus*) says that glass softens and liquefies all bodies, and that it is removed from them by fusion. Vincent also says that there is oil in everything, and refers this remark to one J. Damascenus.[55] Though I know of no explicit statement that gold can be extracted from everything, this conclusion would follow naturally on the standard alchemical doctrine that all things are ultimately a unity. Lists of suggestive fragments might be extended indefinitely; though Albert may have taken his remark from a specific source, it would hardly be necessary. Wyckoff's suggestion that the remark is derived from the assay of some gold-containing sulphide is ingenious, but perhaps strained.[56]

The pharmacological-alchemical relation was so close as to be generally indistinguishable during the Middle Ages, and this in itself may explain Albert's curious interpolation. His exclusion of the incendiary affiliations of oil of bricks then becomes mere testimony to the strong bond between alchemical operations and the manufacture of drugs. Certainly Albert's musings carried him far from Aristotle in this instance, and it seems odd that Ideler, who cited the variant passages on liquefication of iron and solidification of mercury with green wood, did not remark this passage. It may have been repulsive to his classical tastes, but it appears to offer a rich lode for the student of medieval psychology.

Some minor features of the chemical information in Albert's *Meteora* deserve notice. The domestic and homely nature of Aristotle's examples provided a paradigm for Albert's emendations, but the zeal he showed in this pursuit is remarkable. There is, for example,

---

[53] Berthelot, *Archéologie et histoire des sciences avec publication nouvelle du papyrus grec chimique de Leyde et impression originale du Liber de septuaginta de Geber* (Paris, 1906), pp. 339-340.

[54] Jābir ibn Ḥayyān, *The Works of Geber Englished by Richard Russell, 1678* (London, 1928), p. 47. Note that this work, though cataloged as that of the Arab Jabir, is by a fourteenth-century author; see Darmstaedter, *Alchemie des Geber*, pp. 3-7.

[55] Vincent, 6.79: 416; 7.95: 485.

[56] Albert, *Book of minerals*, p. 231 n.24.

the matter of Aristotle's comments on the density of salt water. Aristotle says, "If you make water very salt by mixing salt in it eggs will float on it, even when unblown, for the water becomes like mud."[57] Albert adds the information that an egg will sink in sweet water, with the caution that the egg must be fresh, since older ones will float even on sweet water.[58] He acknowledges the subtlety of Aristotle's "unblown" by saying that a fresh egg is full (*plenum*). This observation may be of little significance in itself, except perhaps to signal the comparative newness and wonder of observing such a difference in behavior, but it is one more instance of the way in which Albert strove to make his *Meteora* a true completion of the *Meteorologica*, particularly with respect to examples of chemical interest. Another of these instances occurs when Albert addresses the properties of milk. Aristotle discusses the coagulation of milk with rennet, and the separation of whey from cheese; Albert adds butter:

> Milk from which whey is separated makes fat cheese by a mixture of the substance of butter with the substance of cheese, and finally hardens. But there are some who by artifice hold back the fat from cheese; they separate butter from milk, and afterward separate whey from cheese by coagulation of the milk.[59]

Butter was not ordinarily an item of diet in the classical world; writers like Dioscorides and Galen stress its medical uses, and the word does not appear in Aristotle's *Meteorologica*.[60] Albert's language suggests that it was almost equally rare in the thirteenth century. This ability to find the wonder of simple domestic processes is one of Albert's more appealing traits.

Although this is by no means a complete survey of Albert's remarks on chemical problems, it may be taken as approximately representative. Major categories are: the fusion and combustibility of materials, their various physical states (i.e., liquid, solid or "subtle," which may be taken to mean gaseous in some contexts), and some oddments of medical or domestic interest. These are precisely analogous, in most cases, to problems addressed in the

---

[57] Aristotle, *Meteor.* II.3 (359a13-15).

[58] Albert, *Meteor.* II.3.16 (ed. Borgnet, 4: 579a.)

[59] Albert, *Meteor.* IV.2.8 (ed. Borgnet, 4: 758b-759a.)

[60] Forbes, *Studies in Ancient Technology*, 3: 101. According to Bonitz, Aristotle uses the word "butter" only in *Frag.* 593 (1574a30).

Meteorologica, particularly in Book IV, but both the theoretical structure developed and the examples presented have undergone considerable sea change. In particular, the problem of oils is addressed in ways foreign to Aristotle's text. The notorious wrongness of both works may in fact be a function of the kind of problems each addressed; problems of mixing, of combustion and related phenomena, and above all of fats and oils are notoriously difficult branches even of modern chemistry. In an article on "Aristotelian chemistry" which unaccountably ignores the *Meteorologica* but analyzes Aristotle's more general treatment of the problem of change, R. A. Horne remarks that the very failure of the Greeks in chemistry may have had alchemy as its consequence.[61] The suggestion seems founded on an assumption that lack of Greek rationality produced the irrationality of alchemy; this may be true to a degree, but analysis of so-called "alchemical" elements in writers of Albert's stature does not always bear out claims that all alchemy was irrational. Documents produced from the ninth through fourteenth centuries are classified as alchemical literature without much regard for the content; if the pursuit of gold was a factor in many of these, it was nevertheless often incidental to the practical chemistry of particular recipes. Certainly the alchemists' disregard of quantitative relations and their bootless concern for "qualities" was a deterrent to the progress of chemistry, and here both Aristotle and Albert — not to mention every other writer in between — must share that opprobrium. But C. F. Mayer remarks that the theory of "qualities" was an effort to establish a thermodynamic basis for physical combinations;[62] to the extent that he attempted this, Albert was simply premature. The laboratory apparatus and structural theory required to develop such a premiss was far in the future.

Given this fact, what can we say of Albert's reports on problems of chemistry? He made no original contributions to that field, nor was he a skilled technician; yet he has some claim to historical significance, if only because his miscellaneous observations furnish material that may link the speculation of alchemy with the achievements of medieval and Renaissance technology. In particular, his report on iron-melting seems significant, and his remarks on the coagulation of

---

[61] Horne, "Aristotelian Chemistry," *Chymia*, 11 (1961), 26.

[62] Claudius Franz Meyer, "Die Personallehre in der Naturphilosophie von Albertus Magnus," *Kyklos*, 2 (1929), 201.

mercury are thought provoking. More intensive (and extensive!) comparison of the items of chemical interest he discusses with other literature, not merely to establish sources but also to determine affiliations between particular processes and their results, might serve to confirm that Albert has an important, if not major, position in the chain of evidence we must use to reconstruct developments in chemical technology. He represents a time when technological change was imminent, if not already in progress, and his talent for observation provides some useful clues to the state of the chemical art in that time of transition.

10

Albert on the Psychology of Sense Perception

Nicholas H. Steneck
University of Michigan

By the mid-fourteenth century, when the anonymous *Tractatus ad libros Aristotelis* containing the accompanying diagram of the senses was copied (see Plate 4),[1] most Latin writers in the scholastic tradition held in common a conceptualization of sense perception that served well the needs of natural philosophers, theologians, and physicians alike. While there was debate about the fine details of this conceptualization, its basic outline was clearly understood by all involved. Two centuries earlier, when Adelard of Bath wrote his well-known *Quaestiones naturales*, the situation was quite the reverse. Numerous ancient teachings on sense perception were known in part, but no single theory was available to tie these teachings together and provide a common ground upon which further debate could take place.[2] In the events that transpired between these two stages in the history of psychology one figure that stands out above all others as playing a major role is undoubtedly Albert the Great.

The development of the psychology of sense perception between

---

[1] *Tractatus ad libros Aristotelis introductorius cum commentario interlineari et marginali*, Prague, Universitni Knihovna MS 770 (IV. H. 6), fols. 1r-39r.

[2] For Adelard's discussion of sense perception, see *Quaestiones naturales* 12-14, 17, 18, 21-31, trans. Hermann Gollancz, *Dodi ve-Nechdi* (London, 1920), pp. 102-105, 109-110, 112-124. A brief, general discussion of twelfth-century views on sense perception can be found in Pierre Michaud-Quantin, "La Classification des puissances de l'âme au XIIᵉ siècle," *Rev. Moyen Age Lat.* 5 (1949), 15-34.

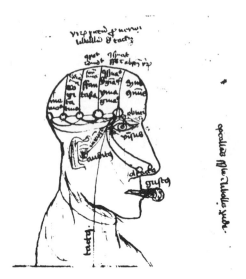

Plate 4. Diagram of the five external senses (*visus, auditus, olfactus, gustus, tactus*) and five internal senses (from right to left: *sensus communis, conservatio imaginationis/imaginatio, formativa/ffantasia, estimativa/cogitativa, memoria*) with their respective connecting nerves, contained in Prague, Universitní Knihovna, MS 770 (IV.H.6), fol. 22v.

the twelfth and fourteenth centuries is evident in a comparison of the writings of Adelard with the assumptions that are implicit in the schematic diagram in the *Tractatus ad libros Aristotelis* (Plate 4). Adelard clearly was working within the general framework of the ancient tradition of sense perception when he wrote:

> whatever operation of this sort the mind performs in the body, it performs with a certain amount of assistance from the body, and this is done one way in the brain, in another in the heart, and in yet another in the other members [i.e., in the senses].[3]

But Adelard was not aware of most of the details of the ancient theories that lay behind this framework. The highly organized, almost mechanistic view of the body that was so familiar to the author of the *Tractatus*, the view that tied the five external senses to the brain by connecting nerves and then localized a number of internal senses (usually four or five) in the cells of the brain, had yet to find its way

---

[3] Adelard, *Quaestiones*, p. 109.

into the Latin West through the writings of Aristotle and his commentators. Since Albert played an important role in bringing these works to the attention of his contemporaries, it should come as no surprise that by virtue of this role he became an important figure in the history of the psychology of sense perception.

That Albert did play an important role in the development of the scholastic theory of sense perception was widely recognized by fourteenth-century contemporaries of the anonymous author of the *Tractatus*. In their discussions of the actions and localization of the senses, Albert's name is the only contemporary one (post 1200) mentioned with any frequency and with an authority equal to that of the ancients.[4] Even though he may not have been the first writer to sort through the newly translated works of Aristotle, Avicenna, Averroës, and others with an eye toward elucidating and systematizing their thoughts, he was clearly the author quoted (and presumably read) when a weighty opinion was needed to settle a dispute among these authorities. The reason for this is not difficult to understand. The comprehensiveness of Albert's discussion of sense perception in the *Summa de creaturis* and later works far exceeded that of other thirteenth-century writers and made them an ideal introduction to the topic at hand. For our purposes they provide as well an entry into the scientific thought of this important scholastic.

The hundreds of folios that Albert devoted to the problem of sense perception make the task of summarizing his thoughts on this subject a difficult one. However, over the course of his lifetime his psychology of sense perception seems to have changed very little; the basic description set out early in his career in the *Summa de creaturis* is followed fairly closely in his commentary on *De anima* and the *Parva naturalia* and is implicit in *De animalibus* and miscellaneous references to the senses in works not devoted strictly to science. As a result, his earliest major treatment of sense perception in the *Summa* is in general a good guide to this aspect of his thought, and it will be focused upon first. Thereafter, the later works will be drawn upon to help round out Albert's views on sense perception and place

---

[4] For examples, see John of Jandun, *Super libros Aristotelis De anima subtilissimae quaestiones*, q. 37 ([Venice, 1589], cols. 213-217), where Albert is quoted in addition to Averroës; John Buridan, *Quaestiones in tres libros De anima* II.27 ([Paris, 1516], fol. 21rb), where he is quoted in addition to Averroës and Avicenna; and Nicole Oresme, *Quaestiones super librum De anima* (Munich, Bayerische Staatsbibliothek CLM 761, fol. 24rb), where he is quoted in addition to Aristotle, Averroës, Avicenna, and Galen.

them very briefly into an overall epistemological and methodological framework. This latter discussion is designed to explain in part how Albert, and other scholastics, could spend so much time discussing how the senses functioned while still falling far short of developing a rigorous psychology of sense perception.[5]

A. *SUMMA DE CREATURIS*

For Albert, as indeed for most scholastics, the topic of sense perception was most commonly broached within discussions of the soul and its powers. The soul, which is one in substance and the active form of the body, moves the body in many ways.[6] It does so not because it consists of several individual souls or a number of substances — one soul is not responsible for sense perception and another for reason — but because the one soul of the body exerts its actions in different ways (called powers of the soul) in the many parts of the body.[7] Or conversely, the parts of the body can be said to pervert or corrupt the activities of the soul in different ways, thereby accounting for its many powers. Just as an old man who

---

[5] The major works that contain discussions of Albert's psychology of sense perception are: Josef Bach, *Des Albertus Magnus Verhältniss zu der Erkenntnislehre der Griechen, Lateiner, Araber und Juden* (Vienna, 1881); Jacob Bonné, *Die Erkenntnislehre Alberts des Grossen mit besonderer Berücksichtigung des arabischen Neoplatonismus* (Bonn, 1935); Ulrich Dähnert, *Die Erkenntnislehre des Albertus Magnus gemessen an den Stufen der 'abstractio'* (Leipzig, 1933); George Klubertanz, *The Discursive Power: Sources and Doctrine of the 'Vis Cogitativa' According to St. Thomas Aquinas* (St. Louis, 1952); C. Mazzantini, "La teoria della conoscenze in Alberto Magno," *Riv. di filos. neo-scholastica* 29 (1937), 329-335; Pierre Michaud-Quantin, "Albert le Grand et les puissances de l'âme," *Rev. Moyen Age Lat.* 11 (1955), 59-86; Henri-D. Noble, "Note pour l'étude de la psychophysiologie d' Albert le Grand et de S. Thomas. Le cerveau et les facultés sensibles," *Rev. Thomiste* 13 (1905), 91-101; S. Ogarek, *Die Sinneserkenntnis Alberts des Grossen verglichen mit derjenigen des Thomas von Aquin* (Fribourg, 1931); Arthur Schneider, *Die Psychologie Alberts des Grossen. Nach den quellen Dargestellt*, in *Beiträge* 4 (Munich, 1903); George M. Stratton, "Brain Localization by Albertus Magnus and Some Earlier Writers," *The American Journal of Psychology* 43 (1931), 128-131; B. Trum, "La dottrina di S. Alberto Magno sui sensi interni," *Angelicum* 21 (1944), 279-298; Karl Werner, *Der Entwicklungsgang der mittelalterlichen Psychologie von Alcuin bis Albertus Magnus* (Vienna, 1876); and Harry Wolfson, "The Internal Senses in Latin, Arabic, and Hebrew Philosophic Texts," *Harvard Theological Review* 28 (1935), 69-133. For a discussion of the strengths and weaknesses of these works as well as my own views on Albert's psychology of the internal senses, see Nicholas Steneck, "Albert the Great on the Classification and Localization of the Internal Senses," *Isis* 65 (1974), 193-211.

[6] Albert *Summa de creaturis* II.7.1 (ed. Borgnet 35: 93b); cf. Aristotle *De anima* II.1 (412a28-412b1).

[7] The attributes of the soul are discussed in *Sum. de creat.* II.4-5 (ed. Borgnet 35: 31-84).

receives (*accipiat*) the eye of a young man will see like a young man (*sicut juvenis*), so too the remaining senses and even the rational soul will act differently depending on the organs of their activity.[8]

Given this general definition of the soul and its powers, the investigation of sense perception very rapidly came to focus on four basic issues: the precise definition of the sense (power of the soul) under consideration, its organ, its mode of action, and the medium or media that are responsible for initiating its activity. Definition clearly established which of the powers of the soul was being discussed; the question of organ established a proper part of the body for each power to act through; the discussion of actions explained how each power actually acts through its organ; and the discussion of medium tied the actions of the senses to external stimuli.[9] Except for a few general questions, Albert's discussion of these issues, which proceeds sense by sense, comprises the psychological portions of the *Summa*.

i. The Definitions of the Senses

Albert's classification of the senses has led to a great deal of controversy among modern scholars regarding his consistency. I have dealt with this issue elsewhere and endeavored to show that the apparent inconsistencies that previous scholars have pointed to in his works involve differences in his use of terms and not in his overall conceptualization of sense perception. Throughout his writings Albert remains faithful to his particular understanding of the Greco-Islamic tradition he received, as described initially in the *Summa*.[10] According to this tradition the actions or powers of the sensitive soul can be divided into two major subgroupings, the external and the internal senses, with the former including the five proper senses — vision, hearing, smell, taste, and touch — and the latter

---

[8] *Sum. de creat.* II.7.1 (ed. Borgnet 35: 94a).

[9] The last issue, the media of perception, is discussed at some length in Lawrence Dewan's article in this volume, "St. Albert, the Sensibles, and Spiritual Being," and therefore will not be covered in the present article.

[10] The main misunderstanding that has arisen with regard to Albert's classification of the senses relates to his use of the term "*deforis*" (from without) to characterize the common sense, a sense normally classified as an internal sense (*Sum. de creat.* II.19. intro. [ed. Borgnet 35: 164]). This misunderstanding is eliminated when it is realized that he does not use the terms "*deforis*" and "*interior*" interchangeably; Steneck, "Classification and Localization," see especially pp. 197-203.

three, four, or five internal powers, depending on the authority followed. (Albert discusses five such powers in the *Summa*: common sense, imagination, phantasy, estimation, and memory.) The distinction being pointed to here is the distinction between those senses (the external senses) that have the capacity to sense only that which is proper to them alone (their proper sensibles) and those (the internal senses) that respond in one way or another to information received from many senses.[11]

The fact that each of the five external senses was assumed to have a proper object or objects that it alone can perceive and that each was assumed to reside in a proper organ provided sufficient information to establish suitable definitions for them. Thus Albert at one point defines vision as the power that has its seat of activity in the eye (*visus sit vis ordinata in oculo*),[12] taste the power that is limited to the nerve that covers the surface of the tongue (*diffinitus a nervo expanso in superficie linguae*),[13] and so on for the other senses. Similarly, vision can be defined as the sense that perceives color, hearing the sense that perceives sound, smell the sense that perceives odor, and so on for taste and touch.[14] This much was fairly evident. Few commentators who dealt with the external senses had trouble defining each of the five commonly assumed ones.

However, the fact that suitable definitions could be established for five of the external senses did not end the problem of definition. If other organs or sensibles exist, beyond those associated with vision, hearing, smell, taste, and touch, then it might be possible to argue that there are other senses that need to be defined. Albert resolves this issue by demonstrating that animals need only five external senses and therefore there are no additional ones. The senses serve two functions: they preserve the being of the living creature (*ad esse*), and they allow it to survive in some semblance of comfort (*ad bene esse*).[15] For preserving the being alone the living creature needs to

---

[11] Perhaps the clearest statement on this major division and the one that Albert most consistently follows is that given by Avicenna, *Liber De anima seu sextus de naturalibus* 1. 5, ed. Simone Van Riet (Louvain-Leiden, 1972), 1: 83-90.

[12] *Sum. de creat.* II.19.1 (ed. Borgnet 35: 165a); cf. Avicenna *De anima* 1.5 (Van Riet, 1: 83.59).

[13] *Sum. de creat.* II.32.1 (ed. Borgnet 35: 273a); cf. Avicenna *De anima* 1.5 (Van Riet, 1: 84.74).

[14] Proper sensibles and their relationship to the five external senses are discussed in *Sum. de creat.* II.34.2 (ed. Borgnet 35: 297a-310a). This means of definition is the one that Aristotle relies on most heavily in *De anima*; see Aristotle *De anima* II.6 (418a7-26).

[15] *Sum. de creat.* II.34.4 (ed. Borgnet 35: 305b); cf. Aristotle *De anima* III.12 (434b21-25).

grow (*ad esse constituendo*), which requires touch, or simply to survive (*ad esse conservando*), which requires taste. To preserve well-being, sense is needed either to regulate eating (*secundum regimen in cibo*), which requires smell, or to pursue progressive motion (*secundum regimen in motu processivo*). The latter encompasses both direct and circular motion, which require vision and hearing respectively. Since these are the only senses that are necessary and since nature provides only that which is necessary, it can safely be argued (*posset probare*) that there are only five external senses.[16]

Having established that there are five and only five external senses Albert turned next to a related problem, their order. When faced with several entities in a single larger grouping, the scholastic mind frequently attempted to arrange these entities in a hierarchy, to establish a first sense and a last sense among the external senses. Since, according to Albert, this order can be established in at least two ways, the first and last among the senses differs. If the senses are judged on the basis of their capacity to contribute to one of the most fundamental properties of an animal, its power to sense, then touch is of primary importance. Touch is the only sense that animals cannot live without. Since "in the destruction of touch the animal is destroyed and this is not so for the other senses, namely for hearing, vision, smell, and taste," touch constitutes the foundation (*fundamentum*) of the other senses.[17] If, however, the senses are ordered not in terms of what actually makes an animal an animal (*in constituendo animal*) but in terms of the primary function of the sensitive soul, cognition (*secundum . . . rationem cognitionis*), then vision is first and touch last.[18] We receive more information through vision than the other senses, followed by hearing, smell, taste, and touch. How the senses are ordered depends, therefore, on the criteria that are used for ordering.

The internal senses, like the external senses, were sometimes defined by reference to either the object or organ of their activity. How-

---

[16] *Sum. de creat.* II.34.4 (ed. Borgnet 35: 305a-306b).

[17] ". . . Et destructo tactu destruitur animal: et sic non est in aliis sensibus, scilicet in auditu, visu, odoratu, et gustu . . . ": *Sum. de creat.* II.33.1 (ed. Borgnet 35: 282a); cf. Avicenna *De anima* 2.3 (Van Riet, 1: 132.96-6).

[18] *Sum. de creat.* II.19.2 (ed. Borgnet 35: 166b-168b); cf. Gregory of Nyssa *De natura hominis* 7 (PG 40: 647-650) and John Damascene *De fide orthodoxa* 2.18 (PG 94: 934). For a discussion of the ordering of the senses, see David Lindberg and Nicholas Steneck, "The Sense of Vision and the Origins of Modern Science," in *Science, Medicine and Society in the Renaissance*, ed. Allen Debus (New York, 1972), 1: 29-45.

ever, since the internal senses all reside in one organ, the brain, and seem not to receive distinct stimuli, definition via object and organ did not produce a very precise understanding of their division. For example, on the basis of the object of perception Albert notes that one internal sense (internal by virtue of the fact that it resides within the brain), the common sense, apprehends through things that are external to the brain (*apprehensiva deforis*), while the remaining four senses (imagination, estimation, phantasy, and memory) apprehend through things that are within the brain (*apprehensiva deintus*).[19] But this distinction does not do justice to the many senses that reside within the brain. As a consequence, Albert tends initially to accept the fivefold classification of Avicenna (common sense, imagination, phantasy, estimation, and memory)[20] and waits to distinguish them more precisely from one another on the basis of their actions, as will be discussed in the next section.

ii. *The Acts of the Senses*

Since sense perception, within the framework of an Aristotelian epistemology, must of necessity be initiated by the actions of external objects, the senses, as recipients of these actions, were commonly understood to be passive powers. "It must be said, in accordance with the pronouncements of all the philosophers, that 'sense' is a passive power and that it is said to be acted upon."[21] However, as passive powers the senses do not undergo physical, form-matter transformations. If they did, the action of the form of light on the eye would change it to light, which clearly does not happen. In sense perception, "there is no physical alteration in the soul" (*in anima nulla est alteratio physica*).[22] Therefore, the act of sensing can be described as the senses being acted upon but not being acted upon by the forms of objects *per se*. Instead, the senses are acted upon by the representatives of the objects, called sensible species, which convey the active intention of the object and not the form of the object itself to

---

[19] *Sum. de creat.* II.19.intro. and 37.intro. (ed. Borgnet 35: 164 and 323); see also *Summa* II.42.2 (ed. Borgnet 35: 360a-361b); cf. Avicenna *De anima* 1.5 (Van Riet, 1: 83.58 and 88-92).

[20] *Sum. de creat.* II.39.1 (ed. Borgnet 35: 337ab); cf. Avicenna *De anima* 1.5 (Van Riet, 1: 87-90) and Algazel *Metaphysica* 2.4, ed. J. T. Muckle (Toronto, 1933), pp. 164-171.

[21] "Dicendum secundum sententiam Philosophorum omnium, quod *sensus* est potentia passiva, et quod *pati* dicitur . . .": *Sum. de creat.* II.34.1 (ed. Borgnet 35: 295b); cf. Aristotle *De anima* II.5 (416b32-417b8).

[22] *Sum. de creat.* II.34.1 (ed. Borgnet 35: 297a).

the senses. When this happens the senses, as passive powers, receive (*accipit*) these intentions and perception takes place.[23]

Although there is no physical alteration of the senses during perception, sense perception is a physical process. Moreover, it is only when the soul and body are in proper harmony with their sensible species that sense perception takes place. Such harmony is established through three successive activities: first, the power of the soul is joined with its organ; thereafter, its disposition is established through the activity of the animal spirit and the natural heat and harmony of the organ (*dispositio fit per spiritum animalem et calorem naturalem et harmoniam organi*); and finally, the sensible *species* are received (*quando vero habet speciem sensibilem*). At this last stage, the passive power of the soul apprehends its proper object and the act of perception is completed (*tunc est potentia completa per actum*).[24]

The grounding of the passive action of each of the senses in one organ not only establishes the mechanism through which sense perception takes place but also the object that each of the senses is able to perceive. Unlike the intellect, which is able to receive all intelligibles (*potest recipere omnia intelligibilia*) because it does not operate through an organ, the senses, whose organs are not made to receive all sensibles (*organum suum non est fabricatum ad naturam omnium sensibilium*), receive only those sensibles that are proper to them, their proper sensibles.[25] This is why the eye receives only color and not sound or smell or some other sensible, and so on for the other senses. In sum, it is the nature of a sense, as determined by the soul acting through and being influenced by the potency of its organs, that determines its actions. The nature of the eye is to perceive color, and vision, which is the proper action that the sensitive soul exerts through the eye, is the action of the eye.[26]

The coincidence of sense and proper sensible serves one additional function besides confining particular pieces of information to particular senses. The coincidence of sense and proper sensible provides the assurance in the chain of cognition from object to intellect that the information that is perceived by the senses is correct. A sense acting in accordance with its proper sensible does not err (*non contingit errare*). It may err with regard to information that is not within

---

[23] *Sum. de creat.* II.34.1 (ed. Borgnet 35: 295b-296a).
[24] Ibid.; the role of the sensitive spirit in sense perception is discussed below, pp. 283-286.
[25] *Sum. de creat.* II.56.3 (ed. Borgnet 35: 482b).
[26] *Sum. de creat.* II.34.2 (ed. Borgnet 35: 297a-299b).

the bounds of its proper sensible, such as when the sense of vision attempts to determine the composition of a particular color (*circa compositionem colorum*), but it does not err when it perceives the information that pertains strictly to its proper sensible — that the colored body is red or green and so on.[27] To this extent, sense cognition can be said to be true and accurate.

Having set out this general framework for understanding the acts of the senses, the specific details that Albert relates for each one, particularly for the external senses, are fairly straightforward. The act of vision is to receive the sensible species of color, the act of hearing to receive the sensible species of sound, and so on for taste, smell, and touch. Nonetheless, there is always room for subsequent debate and questioning.

In his discussion of vision Albert launches into a lengthy recapitulation of past opinions, particularly those of al-Fārābī (Averroës) in *De sensu et sensato* and Avicenna in *De anima*,[28] which discussion prompts him to append a detailed analysis of a problem that had long currency in discussions of vision, the problem of extra- versus intromission theories of vision.[29]

A full assessment of the act of hearing rests on an understanding of the exact location of its organ, as will be discussed in the next section. As to the object of hearing, sound, Albert queries whether or not it comes to the ear very rapidly (*subito*), as does light. He concludes that it does only if that which produces the sound is very close, thus making the time of transmission imperceptibly small. Otherwise, hearing requires time (*percipitur tempore*).[30]

Clarifying the act of smell leads Albert to question whether or not different animals can smell the same odor differently. Since odors initiate the act of smell, it would seem that the same odor would produce the same act in all animals. This obvious inconsistency — some animals can smell better than others — is resolved by pointing out

---

[27] Ibid.

[28] *Sum. de creat.* II.22 (ed. Borgnet 35: 210a-214b); Averroës *Compendium libri Aristotelis De sensu et sensato*, ed. A. L. Shields (Cambridge, Mass., 1949), pp. 5-8 and 13-18, and Avicenna *De anima* 3.1-8 (Van Riet, 1: 147-283). Albert incorrectly attributes Averroës *De sensu* to al-Fārābī, as is demonstrated by R. de Vaux, "La Premiere entreé d'Averroës chez les Latins," *Rev. des sciences philosophiques et théologiques* 22 (1933), 238-40.

[29] *Sum. de creat.* II.22 (ed. Borgnet 35: 215a-228b). Albert's theory of vision is summarized very briefly by David Lindberg in *Theories of Vision from Al-Kindi to Kepler* (Chicago, 1976), pp. 104-107.

[30] *Sum. de creat.* II.27 (ed. Borgnet 35: 253a-254b).

that since the organ of smell is closely associated with the brain and since the brains of animals differ in their dispositions, animals smell differently. The sensible species of smell are able to act upon the dry brain of some animals much more easily than the wet, cold brain of humans, and consequently these animals have better senses of smell.[31]

Finally, the definition of the acts of taste and touch depends upon resolution of the problem of whether they comprise one sense or two. It was sometimes suggested that they do not comprise two separate senses since they seem to have the same objects (hot, cold, wet, and dry) and are at times in the same organ — we both feel and taste with our tongue. However, since the act of taste is to judge flavors (*judicium saporum*) and the act of touch to judge that which can be felt (*tangibilium judicium*), they are, Albert concludes, two separate senses.[32]

The simple data received directly from proper sensibles accounts for only a small portion of the information that is eventually known about an object. Moreover, since each of the senses apprehends only its own proper sensible, none is able to compound this information with information received in other senses and form a composite image of an object: that a particular tree is green, has the smell of pine, and so on. Consequently, Aristotle and most of his commentators argued in favor of the existence of a common sense to receive species from the external senses and form them into a composite image of the object. Having formed this composite image, common sense then has the capacity to compare the composite image of one object with that of another and reach comparative judgments about two objects: that one object is sweeter or whiter than another object.[33] Thereafter the sensibles species received from the five external senses are passed from the common sense to the imagination, the second internal sense, where they are stored for future reference or further transmitted to phantasy, estimation, and memory, the three remaining internal senses.

---

[31] *Sum. de creat.* II.31 (ed. Borgnet 35: 271a-272b).

[32] *Sum. de creat.* II.32.2 (ed. Borgnet 35: 274a) and II.33.1 (ed. Borgnet 35: 282a). Albert's discussion of the senses is drawn primarily from Avicenna *De anima* 2.3-5 and 3.1-5 (Van Riet, 1: 130-234), Gregory of Nyssa *De natura hominis* 7-11 (PG 40: 658-659), and John Damascene *De fide orthodoxa* II.18 (PG 94: 934-938).

[33] *Sum. de creat.* II.35.2 (ed. Borgnet 35: 312a-313b) and II.36.2 (ed. Borgnet 35: 320b-322a); cf. Aristotle *De anima* III.2 (426b8-427a15) and Avicenna *De anima* 4.1 (Van Riet, 2: 1-3).

The fact that imagination stores the images of things in their absence means that an entirely new type of sensitive action begins, the internally apprehensive action (see p. 270 above). From this point on in the cognitive process, additional sense data can be derived even in the absence of any material object simply by reflecting upon (imagining) the images that are stored in the imagination or the memory. But imagination does more than store images. It also prepares them (*praeparat imaginationes quadrupliciter aliis virtutibus operantibus in ipsa*) for the future actions that will follow (1) in phantasy and estimation, (2) in memory, (3) in the intellect, or (4), by a reflowing action, in common sense.[34] In this way, Albert is able to incorporate into this power the comprehensive actions assigned to imagination by John Damascene, Gregory of Nyssa, and Augustine.[35]

The need for additional internal powers, besides common sense and imagination, stems from the fact that animals and humans apprehend certain things that are not sensed by the external senses. The most common example given in support of the contention was a sheep apprehending the hostility of a wolf and fleeing even though it has never seen a wolf before. Obviously some internal power, in this case "estimation," is called for to evaluate the suitableness or unsuitableness of the intent (*intentio*) of an object by a special apprehension and thereupon motivate the apprehending creature to approach or flee from the object (*sit determinare de fugiendo et imitando per apprehensionem convenientis et inconvenientis*).[36] Since this additional information is of great importance in directing the actions of animals, a fourth internal power, phantasy, was added, which has the capacity to compose and divide intentions with the less complex information that is received in the common sense and stored in the imagination.[37] Phantasy also has the power in humans to act under

---

[34] *Sum. de creat.* II.38.4 (ed. Borgnet 35: 329a).

[35] Because John Damascene, Gregory of Nyssa, and Augustine distinguished only three internal senses, they tended to assign to each broader powers than were assigned in either the fourfold classification of Averroës or fivefold classification of Avicenna; see John Damascene *De fide orthodoxa* 17, 19, 20 (PG 94: 934, 938, 939), Gregory of Nyssa *De natura hominis* 6, 12, 13 (PG 40: 631-638, 659-666), and Augustine *De Genesi ad litteram libri duodecim* 7.18 (PL 34: 364). Albert was able to incorporate these broader acts into his own narrow definition of imagination by calling them preparatory acts.

[36] *Sum. de creat.* II.39.4 (ed. Borgnet 35: 339b): "Et videtur secundum Avicennam et Algazelem, quod actus eius sit determinare de fugiendo et imitando per apprehensionem convenientis et inconvenientis in partibus." Cf. Avicenna *De anima* 4.1 (Van Riet, 2: 6-8).

[37] *Sum. de creat.* II.38.4 (ed. Borgnet 35: 333b-334b).

the influence of the intellect, thereby making it in some ways a cogitative power (*vis cogitativa*).[38] And then ultimately all of this information is stored in the memory where it can be recalled by the simple action of the senses (true memory) or by the intervention of reason (reminiscence),[39] thereby completing the acts of the internal senses.

iii. The Organs of the Senses

Interest in the organs of the senses arose to large extent after the time of Aristotle. To be sure, there were numerous references to sense organs scattered through his writings, but the systematic localization of the senses, particularly the internal organs, in specific parts of the body had to wait for the anatomical writings of Galen to be fused with the Aristotelian corpus by later commentators. Few Latin writers in the thirteenth and fourteenth centuries paid more than routine attention to this aspect of sense perception. Writers such as Thomas Aquinas mention the organs of the senses, but seldom in any detail or with an eye toward actual organic placement. Albert does not share this common disinterest. Throughout his writings it is clear that he is localizing the senses in an anatomically real body and not simply repeating descriptions handed down from an earlier tradition. This is especially true of the discussion of the senses in *De animalibus* (as noted below, pp. 285-286), but also to a lesser degree is the conclusion reached from a careful reading of the *Summa de creaturis*.

Of all the sense organs, Albert pays the most attention by far to the organ of vision. Here he is following a long tradition that extended from the writings of Aristotle through those of Avicenna and Averroës (his principal sources) to the works of his own contemporaries.[40] In reviewing this tradition, Albert attempts to steer a middle course between Gregory of Nyssa, who held that vision resides in the forepart of the brain, and Avicenna, who assigned vision to the optic nerve and crystalline humor, by outlining three stages through which the act of vision proceeds. Vision begins in the sensitive portion of the eye (the crystalline humor), where the sensi-

---

[38] *Sum. de creat.* II.38.1 (ed. Borgnet 35: 331a). The history of the *vis cogitativa* is described by George Klubertanz in *The Discursive Power*.

[39] *Sum. de creat.* II.42.2 (ed. Borgnet 35: 360a-361b). For an expanded treatment of memory and reminiscence, see Albert's *Liber de memoria et reminiscentia* (ed. Borgnet 9: 97-118).

[40] Albert's use of past authorities is discussed by Lindberg, *Theories of Vison*, pp. 106-107.

ble species of color is received, is advanced toward greater perfection in the optic nerve, where it is mixed with the sensitive spirit, and is perfected (finally and completely perceived) in the anterior part of the brain.[41] Thus to a certain extent the eye, the optic nerve, and the brain can each be considered to be the origin of vision, depending on the definition of the act of perception being considered.

The appearance of the auditory nerve, as described by Aristotle and his followers, led to some confusion regarding the act of hearing. According to this description the auditory nerve is hollow and contains within it, as it extends from the brain to the tympanum of the ear (*apparet versus tympanum auris*), air that is similar to the air outside the ear (*claudat in se partem connaturalem aeris*). It would seem, accordingly, that the ear and its nerve are capable of receiving sound as it exists materially in the air (*esse soni secundum materiam*), thereby obviating the need for sensible species. However, since apprehensive powers can only perceive immaterial species (*speciem sine materia ejus*), clearly this is not how hearing takes place. Albert concludes, therefore, that air is in contact with the auditory nerve only at its beginning (*aer tangit nervum in principio sui*). Beyond this point, in the spirit that flows outward from the brain to the tympanum, only sensible *species* are impressed (*non imprimitur nisi species soni*). Thereafter, if hearing follows the same path as vision, these species would be carried to the brain where the act of hearing would be completed.[42]

After passing briefly over the organs of smell and taste, which one assumes begin in the nose and tongue respectively, are perfected in the connecting nerves, and completed in the brain,[43] Albert turns to the difficult problem of touch. At first glance touch appears not to have a single organ but to be diffused throughout the body and to perceive objects in and of themselves without the intervention of an external medium. Albert agrees that in one way this is true. As the first of the senses, touch is the form and perfection of the animate

---

[41] *Sum. de creat.* II.19.1 (ed. Borgnet 35: 166a); cf. Gregory of Nyssa *De natura hominis* 7 (PG 40: 642) and a similar view given by John Damascene *De fide orthodoxa* 2.18 (PG 94: 334), as compared to Avicenna *De anima* 1.5 (Van Riet, 1: 83.59-60).

[42] *Sum. de creat.* II.27 (ed. Borgnet 35: 253a-b); c.f. Aristotle *De anima* II.8 (420a3-11) or Averroës *De sensu* (Shields, p. 8).

[43] *Sum. de creat.* II.31 and II.32.3 (ed. Borgnet 35: 271a-272b and 279a-b). That this localization pertains to smell and taste can only be inferred from the discussion in the *Sum. de creat.* Albert's most specific statement is with regard to taste: ". . . sic nervus gustativus principiatur a cerebro, et expanditur in lingua et palato": *Sum. de creat.* II.32.3 (ed. Borgnet 35: 279a).

body (*est forma et perfectio animati corporis*) and therefore has the entire body as its organ (*pro organo*) and senses without a medium. Its capacity to sense in this instance depends on the degree to which a particular part of the body is influenced by the sensible spirit (*participant spiritum sensibilem*). Those parts that are influenced the most, such as nerves, flesh, and skin, sense better than those that are influenced the least, such as bones, brain, and hair. The latter, Albert suggests, sense

> insofar as they are surrounded by membranous nerves, in the break-
> down of which, such as around the brain and around the bones, pain is
> sensed.[44]

As the last of the senses, touch is simply the sense that receives tangibles, and in this case it acts like the other senses; it is completed in the brain, perfected in the sensible spirits that flow out from the brain through the nerves of the body, and begun in the flesh.[45] In this latter way touch senses those qualities that have opposites and are active toward touch (*prout habent contrarietatem et motum ad ipsum*), such as hot, cold, wet, and dry. In the first way, touch senses those things that are harmful and beneficial to life (*quod est dissolvens continuationem talis corporis, et . . . conservans*), such as the pain of a flogging or the pleasure of intercourse.[46]

Since Albert most commonly holds to a fivefold classification of the internal senses, his need with regard to their localization is to find five loci for them within the three Galenic cells of the brain.[47] Avicenna, Albert's primary source for localization, had accomplished this by assigning two powers to the anterior cell (common sense and imagination), two to the middle cell (phantasy and estimation), and memory to the posterior cell.[48] Albert agrees with this description, but adds one qualification. The two-two-one arrangement, Albert argues, applies to the senses of animals. The internal senses of animals are not, however, entirely analogous to those of humans. In

---

[44] " . . . Inquantum circumposita sunt panniculis nervosis, in quorum dissolutione sentitur dolor, sicut circa cerebrum, et circa ossa": *Sum. de creat.* II.33.3 (ed. Borgnet 35: 289b).

[45] *Sum. de creat.* II.33.3 (ed. Borgnet 35: 289b-290a).

[46] *Sum. de creat.* II.33.4 (ed. Borgnet 35: 292b-293a). Aristotle discusses touch in *De anima* II.11 and *De sensu* II.

[47] For Galen's threefold division of the brain, see *De locis affectis* 3.9, ed. D. Carolus Kühn (Leipzig, 1825), 8: 173-175. This division was almost universally accepted throughout the course of the Middle Ages.

[48] Avicenna *De anima* 1.5 (Van Riet, 1: 87-88).

humans the internal senses are in one crucial way decidedly different, and that is insofar as they act as a cogitative power that is influenced by the intellect. Albert assigns this cogitative activity to the middle portion of the brain and to accommodate it moves phantasy and estimation in humans toward the anterior cell. This displacement prompts him to present the unconventional conclusion that in humans the anterior cell of the brain has four powers: common sense, imagination, phantasy, and estimation.[49]

B. LATER WORKS

i. De anima

The years that passed between the writing of the *Summa de creaturis* and the commentary on *De anima* undoubtedly afforded Albert the opportunity to read more widely and reflect more deeply on, among other issues, the problem of sense perception. Certainly this is the impression that is received on turning to the commentary on that portion of *De anima* that begins the treatment of the senses (II, tr.3). The clear delineation of problems and the structured presentation of the opinions of previous authorities give every indication that Albert's thoughts on this subject are now firmly established and that he is himself fully in control of the material being presented. Although his basic psychology of sense perception seems to have changed very little in the dozen or so years that separate the two works, his confidence in and mastery of this psychology seems to have grown and matured greatly.[50]

Albert's increased mastery of the problem of sense perception is evident in more than his style of presentation. By the time he commented on *De anima* he had clearly thought through and was prepared to make explicit a suggested metaphysical unity to sense perception that is implicit but never clearly articulated in the *Summa*. His thoughts on this subject are presented early in the sections on the

---

[49] *Sum. de creat.* II.38.3 and II.40.3 (ed. Borgnet 35: 328a and 349a-b). I have discussed Albert's localization of the internal senses in more detail in "Classification and Localization," pp. 204-209, pointing out in particular how the last, seemingly radical departure fits very nicely with his anatomical description of the brain.

[50] Albert's confidence is reflected in such expressions as "et ideo frustra quaeritur," "videtur mihi stulta quaestio," "opinio . . . omnino ridiculosa est," "opinio . . . est multa probabilior, licet pauci modernorum teneant": *De anima* II.3.6 (ed. Colon. 7/1 [1969], ed. Stroick, 106). Such expressions are few and far between in the *Summa de creaturis*. All references to Stroick are to ed. Colon. 7/1.

senses as "a digression explaining the degrees and manner of abstraction" (*digressio declarans gradus abstractionis et modum*).[51] In this digression Albert looks beyond the common element that had informed his discussion in the *Summa*, the common element of the senses as powers of the sensitive soul, and seeks to uncover the unity of the metaphysical process that underlies all perception. He presents this unity in the form of an explanation of the way in which the senses can be arranged hierarchically through a consideration of their relationship to the objects of perception.

Building on the same basic definition of perception set out in the *Summa*, "to apprehend is to accept the form of that which is apprehended" (*apprehendere est accipere formam apprehensi*),[52] Albert notes that that which is accepted, the representative of the object (*intentio ipsius et species*), is received in four distinct ways.

The first and least abstract way is when the representative of the object is accepted in abstraction "from its matter but not from its presence or its appendices" (*a materia, sed non ab eius praesentia nec ab eius appendiciis*). This is the abstractive level of the externally apprehensive power (*vis apprehensiva deforis*), which, Albert notes, is sense (*quae est sensus*). By sense there can be little doubt that he is including, as he had in the *Summa*, the five external senses and the common sense.[53]

The second level of abstraction contains imagination, which apprehends the form of the object apart from both its presence and matter but not in abstraction from the conditions or appendices of matter (*sed non ab appendiciis materiae sive condicionibus*). By "conditions" Albert is referring to the attributes of the form as it exists or existed in a particular subject.[54]

At the third level of abstraction the intentions (*intentiones*) of the form are known by estimation and the compounding action of, one assumes, phantasy and estimation (*et numquam est sine aestimatione et collatione*).[55]

---

[51] *De anima* II.3.4 (Stroick, 101.48-49). Although Albert does not identify the source for the ideas presented in this digression, he is probably drawing on Avicenna *De anima* 2.2 (Van Riet, 1: 114-121). The broader implications of this theory are discussed by Dähnert, *Die Erkenntnislehre*, pp. 9-26.

[52] *De anima* II.3.4 (Stroick, 101.62-66).

[53] Ibid. (Stroick, 101.68-71).

[54] Ibid. (Stroick, 101.72-78). Albert explains that appendices include "quas habet subiectum formae, secundum quod est in tali vel tali materia" (Stroick, 101.78-80), or, in other words, the senses know things only as particulars and not universals.

[55] Ibid. (Stroick, 101.90-102.10).

And lastly, the form is known simply and separated from all the prior conditions. This is the cognition that is achieved solely in the intellect.[56]

Translating this description into more concrete terms and adding to it the initial advances toward the complete perception of proper sensibles that take place between the senses and the brain,[57] Albert would describe the process of apprehending an object, such as a particular person, as follows. When that person is present before us, its species, or representative is accepted by the organs of the external senses, where perception begins. The species received in the external senses are then transmitted by the sensitive spirits that flow in the connecting nerves to the brain, where a composite image of the object is formed. These steps take place only in the presence of the object. Thereafter, the form of the object is transmitted to the imagination and at the same time abstracted from the presence of the matter. We now know the form of the person even if that person should leave, but we know this form only as it existed in that person and not as it could have existed in another person (*in uno individuo unius speciei, quod non sunt in alio*). That is to say, we as yet do not apprehend the form of "*homo*" apart from a specific set of attributes — a particular placement of limbs, facial color, age, and so on — belonging to the person originally perceived. At the third level of abstraction, certain tangential information, called "intentions" (*intentiones*), is apprehended by phantasy and estimation, such as the potential friendly or unfriendly nature of the person in question. This is information that is received along with the sensible species but not impressed in the senses (*cum sensibilibus accipimus, et tamen eorum nullum sensibus imprimitur*). Finally, at the level of the intellect the form "*homo*" is known as a universal concept, now in abstraction from all specific limiting features associated with the existence of that form in a particular subject.[58]

Having established the physiological background for dealing with sense perception, Albert then continues his discussion of the external senses along the same general lines followed in the *Summa*. Under sight he once again focuses on the proper sensible of vision and the

---

[56] Ibid. (Stroick, 102.11-20).

[57] See pp. 275-276 above. Although Albert does not dwell on this initial stage in the abstractive process in *De anima*, he does refer to it upon occasion; see, for example, *De anima* II.4.11 (Stroick, 163.40-53).

[58] *De anima* II.3.4 (Stroick, 101-102).

manner in which it is transmitted to the eye.[59] His discussions of hearing, smell, and taste entail detailed considerations of the nature of sounds, odors, and flavors and the manner in which they are transmitted from object to sense organ.[60] In fact, the only major difference between these two works, from the point of view of this study, is their corresponding treatments of the organs of perception. In line with Aristotle's own discussion of the senses *De anima* and unlike his earlier interests expressed in the *Summa*, Albert for the most part ignores the problem of localization in *De anima*. The only exception to this generalization is found in his discussion of touch. The lack of any obvious sense organ for touch had prompted Aristotle to discuss its localization in some detail,[61] thus providing Albert with an opportunity to consider at least this one organ in *De anima*.

The reason given for a detailed consideration of the organ of touch is the apparent disagreement between Alexander of Aphrodisias, Themistius, and Avicenna on the one hand and Aristotle on the other over the role of flesh in touch. Aristotle clearly states that flesh is the medium and not the organ of touch, whereas the other three authorities argue that the flesh that is imbued with nerves (*carnem nervosam*) is the organ of touch.[62] Albert's compromise, which is similar to one set out in the *Summa* but now discussed in more detail, rests on an understanding of the types of flesh that exist. True flesh (*id quod vere caro est*) seems to be what Aristotle has in mind by flesh and as such Albert does not object to calling it the medium of sense perception. However, flesh that has sensitive nerves mingled with it or that is situated in the vicinity of other senses does perceive what touches the body and as such can be considered the organ of touch.[63] The manner in which flesh senses in the latter way is explained through mention of two apparent anomalies: flesh-like organs that do not sense and nonflesh-like organs that do sense.

It is clear, Albert argues, that the sensitive power of nerve-imbued flesh extends to parts of the body that are not flesh-like in appearance, such as teeth. Teeth obviously feel pain. The cause of this,

---

[59] *De anima* II.3.7-16 (Stroick, 108-123).

[60] *De anima* II.3.17-29 (Stroick, 123-141).

[61] Aristotle *De anima* II.11 (422b32ff).

[62] *De anima* II.3.31 (Stroick, 143.10-16); cf. Aristotle *De anima* II.11 (423b25-27), Themistius *De anima* 1.4. (ed. G. Verbeke [Louvain-Paris, 1957], p. 174.40-43), where Alexander of Aphrodisias is also quoted, and Avicenna *De anima* 2.3 (Van Riet, 1: 138.2-4).

[63] *De anima* II.3.34 (Stroick, 147.40-47).

Albert conjectures, is the vivifying influence of the surrounding flesh that is carried to the teeth along with the nutriments they receive. That teeth do receive nutriments from the surrounding flesh is evident from the fact that they sometimes regrow after being extracted (*dentes extractos recrescere*) or that they increase in size when a facing tooth is removed (*dentes superiores vel inferiores habet extractos ... dentes illis oppositos super alios prolongari*).[64] This would only happen if the teeth were influenced by the nutriments of the body and explains how they can participate in sense perception. But not all nerve-imbued flesh senses, to turn to the second anomaly. The brain and liver, for example, were believed by Albert to have no sense capacity of their own. The reason for this is that the brain and liver are more influenced by their own qualities than by the surrounding flesh (*quorum complexio ad medietatem carnis non accedit*) and hence do not sense. If we feel pain in these organs it is due to the nerves that surround them (*hec est in panniculis, qui sunt circa substantias eorum*) and not the organs themselves, as was noted above. The same would be true of the nerves of ligaments and sinews (*funes sive ligamentum*) that attach to bones and likewise do not sense.[65] Ultimately, then, whether or not a particular part of the body can be said to be the medium or organ of touch depends on the vivifying and sensitive spirits that are active within it and the degree to which they are active.

Following his discussion of the external senses, Albert goes on to question whether or not they are five in number, as had Aristotle, and then orders the remainder of his discussion of sense perception with "a digression clarifying the five interior powers of the sensitive soul" (*digressio declarans quinque vires animae sensibilis interiores*). These powers are clearly delineated as common sense, imagination, phantasy, estimation, and memory, and localized within the three ventricles of the brain in accordance with the teachings of "the Peripatetics" (*Peripatetici*), in this case, Avicenna.[66] Thereafter, the actions of each is explained in turn and various problems discussed, such as an error of the ancients that maintained "that to know and to sense would be the same" (*quod intelligere et sentire essent idem*).[67] A

---

[64] *De anima* II.3.34 (Stroick, 147.48-61).

[65] *De anima* II.3.34 (Stroick, 147.48-82); cf. Avicenna *De anima* 2.3 (Van Riet, 1: 138.9-139.14).

[66] *De anima* II.4.7 (Stroick, 156-158); cf. Avicenna *De anima* 1.5 (Van Riet, 1: 87-88).

[67] *De anima* III.1.5 (Stroick, 170-171).

great deal of this material simply repeats the discussion of the senses in the *Summa*, although in *De anima* its organization is more straightforward and easier to follow. Throughout, Albert's initial psychology of sense perception remains essentially that of the earlier work. If his views in *De anima* differ at all from those of the *Summa* it is in emphasis and not in content.

ii. Parva naturalia

The *Parva naturalia* turn from thinking about the soul in and of itself (*secundum seipsam considerata*) to a consideration of the soul as it acts through the bodies of animals, which means ultimately to a consideration of the natures of animals, (*considerationem de animalium naturis*).[68] Since one aspect of the nature of animals is their capacity to sense, sense perception appears as an important topic for discussion throughout these shorter works, with the major treatment being found in the two treatises devoted specifically to sensation, *De sensu et sensato* and *De memoria et reminiscentia*, and in a treatise that dealt with an unusual form of sense perception, dreaming, as discussed in *De somno et vigilia*. The latter work adds an important dimension to an understanding of Albert's psychology of sense perception, the role of the heart in sensing, and therefore needs to be considered briefly at this point.[69]

Confusion over the heart's role in sense perception stemmed from yet another apparent disagreement between Aristotle and his commentators; Aristotle specifically states at several points that sense perception begins in the heart whereas most of the commentaries on his works tend to stress the importance of the brain in sensing.[70] Albert's solution to this problem, which becomes standard in later works, eliminates this disagreement by explaining how the heart functions in perception. Just as the sun is the source of all things that are generated in the macrocosm, so too the heart is the source of all

---

[68] *De sensu et sensato* I.1 (ed. Borgnet 9: 1a-b); cf. Aristotle *De sensu et sensibili* I (436a1-5).

[69] *De sensu* and *De memoria* both give indications of the slow refining process that Albert went through over the course of his life as he incorporated new readings into his science. A comparative, developmental study of these works, vis-à-vis the *Summa de creaturis*, should be carried out.

[70] Aristotle *De somno et vigilia* II.(456a1-10) and *De partibus animalium* III.4(666a10-15). I have discussed the history of this issue in more detail in "A Late Medieval Debate concerning the Primary Organ of Perception," *Proceedings of the XIIIth International Congress of the History of Science* (Moscow, 1974), 3.4: 198-204.

vital actions in the body, and like the sun, which is active through light, the heart too has its vehicle, which is called spirit.[71] The spirit that arises in the heart, from its heat, flows and reflows through the body, changing in subtlety in the process and motivating the vital actions of the body, such as sense perception. In the brain, spirit "perfects the animal powers, which are to sense, to imagine, and to understand" (*perficit in cerebo virtutes animales, quae sunt sentire, imaginare, et cogitare*). From here they flow to the organs of the external senses, where seeing, hearing, and the like are carried out.[72] Accordingly, both opinions are in a way correct; the spirit that activates sense perception arises in the heart but is perfected in the brain.[73]

Given this general description of the origin of perception, the reason for the senses ceasing to be active in sleep follows with little difficulty. During sleep the spirit that activates the senses and causes them to sense is withdrawn (*somnus autem est retractio spiritus ab exterioribus organis*). This withdrawal renders the senses ineffective (*impotentia earum ad agendum*), thus explaining why in sleep we are not aware of external stimuli.[74] However, when the animal spirit is withdrawn from the external senses a different type of perception occurs, the sense perception of dreams.

Dreams arise from the mixing and mingling of the images stored in the brain through the action of the vapors that arise during sleep.[75] Such mixing does not take place during the day because the images received from the external senses tend to dominate any internal mixing that may take place. But when these external images are no longer present, due to the withdrawal of the animal spirit during sleep, the internal mixing of forms takes over and produces the images that we see in sleep.[76] Exactly how this happens and the causes of various types of dreams are topics that were of great interest to Albert, as is clear from the discussion in *De somno* and even

---

[71] *De somno* I.1.7 (ed. Borgnet 9: 132a-b).

[72] *De somno* I.1.7 (ed. Borgnet 9: 132b-133a).

[73] This is the distinction that Albert undoubtedly has in mind when he notes in his commentary on Matthew that ". . . sensus communis, qui in corde est, unus est: et ille componendo phantasias sensibilium turbat cordis secretum": *In Evangelium Matthaei* 6:6 (ed. Borgnet 20: 241b).

[74] *De somno* I.1.7 (ed. Borgnet 9: 133a-b).

[75] *De somno* II.2.2 (ed. Borgnet 9: 171a).

[76] *De somno* II.2.1 (ed. Borgnet 9: 169a-b).

more in the lengthy treatment of this subject in the *Summa*.[77] This sidelight to sense perception forms an interesting chapter in the history of medieval psychology that is well worth a detailed study, especially in relation to developments in physiology and anatomy. Unfortunately, to date very little has been written along these lines.

iii. De animalibus

The physiological and anatomical background to Albert's theory of dreams as well as to his entire psychology of sense perception is treated most fully in *De animalibus*. In this mammoth work, which ranges broadly over human and animal anatomy and physiology, he broaches the issue of sense perception on numerous occasions. Most frequently the resulting discussions focus on anatomy and provide descriptions, sometimes in very careful detail, of a particular sense or some aspect of the sensitive process. Less frequently he engaged in speculations on the relative role or placement of the senses in humans and animals or on some other comparative topic. In sum these discussions add very little to an understanding of Albert's psychology of sense perception. Very seldom does he attempt in *De animalibus* to explain the origin of any but the most obvious apprehensive processes. His goal in this work is not to explain *what* the senses know but rather *how* they know.[78]

The anatomy and physiology of sense perception set out in *De animalibus* clearly reaffirms Albert's belief that sensation begins in the heart. The nerves of the body, which some physicians (*secundum multos medicos*) suggest come from the brain, have their place of origin in the heart and are only divided at the brain and base of the neck (*a corde oriantur, et a cerebro et a nucha dividantur*).[79] Even the brain's location is dependent on the heart; since the heart is in the front of the body, the brain, which requires blood for perception, is

---

[77] *Sum. de creat.* II.44-50 (ed. Borgnet 35: 402-441).

[78] Perhaps the best example of Albert's preoccupation with the mechanisms instead of the end products of the apprehensive process is his discussion of the brain and central nervous system, *De animalibus* I.2.17 and XII.2.3-4 (ed. Stadler, 15: 121-125 and 842-851). In this discussion there are only one or two very general allusions to the actual information that the brain acquires, which is quite the reverse of the emphasis in the discussion of the brain in *De anima* and the *Summa*. In the latter works very little is said about the detailed anatomy of the brain and most attention is given to explaining its apprehensive capacities.

[79] *De animalibus* XII.2.5 (ed. Stadler, 1: 851.135); see also *De animalibus* I.2.18 (ed. Stadler, 1: 126.356).

also located in the front part of the body (*cor . . . est in anteriori corporis, et ideo etiam cerebrum*).[80] Therefore, it can be argued that sensation begins when the warm vapors and animal spirit of the heart rise, like water vapors in the macrocosm, to the brain where they enliven the sensitive process (*spiritus enim venit corde ad cerebrum, et digeritur ibi ad operationes animales*) and are cooled.[81] (Cooling at this point is not simply an adventitious action, it is a necessary one. If the warmth of the heart were never overcome, animals would never sleep and, of perhaps even greater consequence, the constant flow of heat from the heart into the external senses would eventually destroy them [*fluerent ex ipso humores calidi in oculos et aures et olfactum, et destruerentur operationes organorum*].)[82] In brief, then, sense perception rests on the same basic heating, cooling, moistening, drying, and enlivening actions that the physicians of this period, working in the Galenic tradition, used to account for the rest of the body's actions.

The loci within which this array of physiological activities are carried out comprise the organs of sense perception. Once at the brain, the spirit that arises in the heart passes through the complex cerebral, neural, and sense anatomy of the body to those places where perception actually takes place. Ultimately, then, it is in the organs of sense perception that body and soul join together and render those who possess the attributes of animals capable of receiving and responding to external stimuli. It is in the organs of perception that the universal principles that lie behind sense perception, the principles of animal soul and animal spirit, are shaped and formed by the substance of the body into the various forms of sense cognition that are classified under the activities of the external and internal senses. As a consequence, at this most basic level, sense perception falls squarely within the Aristotelian metaphysics of form-matter composition and Albert's task, as a medieval psychologist, has been to determine how the form of the animal soul is active through the body. This in essence, and as was stated at the beginning of this article, comprises the psychology of sense perception in the Middle Ages.

---

[80] *De animalibus* XII.3.1 (ed. Stadler, 1: 866.171).

[81] *De animalibus* XII.2.4 (ed. Stadler, 1: 849.132). The macro-microcosm analogy is presented in *De animalibus* XII.2.3 (ed. Stadler, 1: 844.117).

[82] *De animalibus* XII.2.4 (ed. Stadler, 1: 846.124).

C. Miscellaneous References and Conclusions

Having placed Albert's psychology of sense perception squarely into its Aristotelian framework, it should be pointed out that it is not, in my opinion, this framework as such that separates his science from its modern counterpart. The assumption is too often made by historians of science who are not familiar with medieval science that Aristotelian science is synonymous with a preoccupation with final causes and that it is only when this "incorrect" focus is overturned that modern science emerges.[83] A moment's reflection on Albert's psychology of sense perception quickly indicates the fallacy of this assumption. Throughout his writings on the senses Albert is concerned primarily with material and efficient causality. It could not be otherwise within a psychology of the senses that stresses the role of the body in shaping and directing the activities of the soul. Albert's scientific explanations do not rest on final causes. At the most he could be accused of allowing too many issues to be explained in terms of formal causes, but even this criticism misses the point. The search for formal causes, in Aristotelian science, means the search for the most fundamental characteristics of things — their propensity or nature to act in certain ways — which is akin to, although certainly not in any way as sophisticated as, the modern search for the fundamental properties of matter.[84] As a consequence, it is incorrect to look to the Aristotelian framework of Albert's psychology of sense perception *per se* as the ingredient that sets him apart from today's thinking on the same subject. Rather, it is the values that he holds as a scientist that are so alien to our present way of thinking, values that are apparent if one turns to Albert's views on sense perception advanced outside the context of his strictly scientific works.

Albert's fascination with sense perception clearly extended beyond the technical context of his commentaries and *summae*. Just as mention of the ten men in Zacharia provided a ready excuse to

---

[83] A particularly blatant example of this misreading of medieval science would be Hugh Kearney's recent work, *Science and Change: 1500-1700* (New York, 1971); see especially pp. 22-27. The basic sentiments expressed by Kearney are implicit in the many statements on causality in such influential works as E. A. Burtt, *The Metaphysical Foundations of Physical Science* (New York, 1954). For a more balanced analysis of the role of causality in scientific explanation, see William Wallace, *Causality and Scientific Explanation*, 2 Vols. (Ann Arbor, 1972-74).

[84] For Aristotle's statements on "nature" see *Physica* II.1(192b8-23), or Wallace's summary of Aristotle's views in *Causality*, pp. 11-18.

once again remind his audience that ten is also the number of the senses, so too mention of the ten maidens in Matthew 25:1 prompted a similar enumeration, this time with a more detailed description of some of their actions:

> The philosophers assume that there are five internal senses: common sense, which compounds that which is sensed; imagination, which retains and brings back an image in the absence of its object; estimation, which draws forth friendliness and hostility, compatibility and incompatibility to itself from the sensibles; phantasy, which composes and divides that which is drawn forth [by estimation]; and memory, which preserves everything, as in a repository.[85]

Similar partial listings can be found in the commentaries on the *Sentences* and on Dionysius' *De coelesti hierarchia*.[86] There can be little doubt that Albert's world view contained as part of its working vocabulary the psychology of sense perception set out in the *Summa* and elaborated in his many later works.

Within the context of such digressions in works not directly on natural philosophy Albert continued to explain and clarify the more subtle points associated with sense perception. Confusion over Augustine's use of the term memory in *De trinitate*, which use seemed to contradict Avicenna's and Algazel's distinction of memory from imagination, prompts Albert to comment that Augustine was speaking only generally when he noted that memory retains corporeal images and that he did not intend to address himself to a more subtle distinction of powers.[87] In the *Liber topicorum*, judgment regarding truth and error (*rectitudo et peccatum*) is assigned to the common sense, "which composes and divides designated particulars, and by composing and dividing it judges concerning them through the mingling action of the estimative power."[88] Mention of the doubt

---

[85] "Vel, Quinque sensus interiores sunt, quos ponunt Philosophi. Sensus communis, qui sensata componit: imaginatio, quae imaginem absentis rei retinet et repraesentat: aestimatio, quae amicum et inimicum, conveniens et inconveniens sibi ex sensibilibus elicit: et phantasia, quae elicita componit et dividit: et memoria, quae sicut in thesauro omnia reponit": *In Evangelium Matthaei* 25:1 (ed. Borgnet 21:118a). See also *In Zacharium prophetam enarratio* 8.20 (ed. Borgnet 19: 567).

[86] *Comm. in II Sent.*, 24.8 (ed. Borgnet 27: 406b); *De coelesti hierarchia* 15.5 (ed. Borgnet 14: 417a-b).

[87] *Comm. in III Sent.*, 24.G.14 (ed. Borgnet 28: 430a-431b); cf. Augustine, *De trinitate* 13.2 (CCSL 50A: 385.130-136).

[88] ". . . Qui componit et dividit signata particularia, et componendo et dividendo judicat de ipsas per mixtionem aestimativae potentiae": *Liber topicorum* II.2.6 (ed. Borgnet 2: 304b).

of the Apostle Thomas in John 24:25 prompts Albert to suggest that Thomas was seeking first the simple verification of vision, then, since vision can deceive, the reassurance of information gained through the touch of a finger, and finally, to reassure himself that the touch of this single finger was not deceived, the infallible experience (*simul infallibile nuntient tactus experimentum*) of all the fingers.[89] Again, the conclusion is reached that his Aristotelian understanding of sense perception formed a working part of his world view.

However, as important as this understanding was to Albert and despite the amount of time he spent discussing sense perception, his scientific sensitivities in this area of investigation fall far short of being in any way equivalent or even preparatory to later and more modern developments in psychology. To a certain extent this is perhaps obvious. The simplified description of the senses that he received from his predecessors missed the mark on so many points that it is difficult to imagine how it could ever have evolved into our modern description. But it is not at this level that I would set Albert apart nor would I agree that his basic approach to the study of sense perception is particularly "unscientific." Certainly wherever possible he resorts to experience and he is profoundly interested in the physical processes that underlie cognition. Accordingly, it is not his science of sense perception that is so alien to us as moderns as it is the importance he assigns to this science and the fruits he would expect to derive from an exhaustive study of how the senses function.

Albert believes, in harmony with the Aristotelian tradition of his day, that knowledge begins with the senses. It is objects that lie outside the soul that lead to *scientia* (*res extra animam existentes sint causa nostrae scientiae*).[90] But just because the senses are necessary to the intellect it does not follow that the senses themselves attain much knowledge nor is there much pleasure to be gained through the senses alone. In fact, just the opposite is true. Since the purity of a sense, and hence its capacity to give pleasure, is directly proportional to its remoteness from matter (*purior . . . sensus est remotior a materia*) and since in the order of abstraction the external senses are the most closely joined to matter, followed by the internal senses, and then the mind, which is the most remote from matter, it follows that there is

---

[89] *In Evangelia Joannis* 20:25 (ed. Borgnet 24: 689a).

[90] *Metaphysica* x.1.6 (ed. Colon. 16/2 438.61-62); see also *Liber physicorum* VIII.2.1 (ed. Borgnet 3: 559a-560a) and *Liber posteriorum analyticorum* I.5.7 (ed. Borgnet 2: 142a-144a).

much more pleasure to be gained from the activities of the mind than from the activities of the senses.[91] Moreover, it is only insofar as the senses are directed by the mind that Albert finds any real dignity in their actions. When faced with the problem of the frailty of the senses in humans, he makes no effort to defend their strength as senses *per se*, vis-à-vis animals. Since the quality of a sense depends on its organ, if animals have better sense organs they can sense better; dogs have superior senses of smell, wolves and wild boar better hearing, geese better vision. For Albert, the only essential superiority found in the senses of human beings as such is derived from their immediate and proper ordination to reason (*secundum quod coniungitur rationi, et in illa excedit*).[92] Just as matter is ordained to form as to its final cause, so man's senses are ordered to reason, and thereby derive all their nobility and perfection.

---

[91] *Liber ethicorum* IX.1.11 (ed. Borgnet 7: 618b-619b).

[92] *De coelesti hierarchia* 15.4 (ed. Borgnet 14:410a-b). Note especially the references in n. 90 above.

11

St. Albert, the Sensibles, and Spiritual Being

Lawrence Dewan, OP
Collège dominicain de philosophie et de théologie

While the present paper is about St. Albert's doctrine of external sensation, its precise topic is a feature of the theory of sensation having no modern counterpart. Present-day Aristotelians willingly talk about the immateriality or spirituality of intellectual knowledge, but some balk at using this notion for the explanation of sensation itself.[1] Even more difficult to appreciate, then, is the application of the notion of spirituality to the account of sensible things even before they enter the senses.[2] Yet that is the application we will be seeing Albert make. Whether that means that the doctrine here discussed is irrevocably in all respects a thing of the past is more difficult to judge; which theories we were well rid of, in the general rejection of Aristotelian science which ushered in the modern era, has surely not been completely decided.

---

[1] Cf. the controversy between John N. Deely, "The Immateriality of the Intentional as Such," *New Scholasticism* 42 (1968), 293-306, and Mortimer J. Adler, "Sense Cognition: Aristotle vs. Aquinas," ibid., pp. 578-591. Adler (p. 587 n. 8) admits intentionality and spiritual immutation of the organ (though expressing dislike for this use of the word "spiritual"), but denies immateriality.

[2] Sheila O'F. Brennan ("Sense and the Sensitive Mean in Aristotle," *New Scholasticism* 47 [1973], 279-310), having interpreted "*intentio*" as essentially cognitive (p. 304), has difficulty with St. Thomas Aquinas' statement that colour is in the medium outside the sense after the manner of an *intentio* (p. 305 n. 61).

In any case, the history of science and of its advancement is not merely the history of the discovery of what we now take to be truth. As in contract bridge, so in science, under certain conditions the good player should fail. To give the "right" answer for the wrong reason is never a glorious chapter in the history of anything. Accordingly, what I promise the reader is not an account of how Albert discovered this or that truth; rather, I propose to put on display St. Albert's mind as a probing mind, dealing with difficulties, setting out theories, reexamining the difficulties, and revising the theories. I say simply that he exhibits the dispositions which have in fact advanced science.

A remarkable feature of Aristotle's doctrine of sensation is the sort of item called "the medium." Whereas we tend to distinguish merely between knower and known, sense and sensible thing, Aristotle has *three* items in his schema: the sense, the medium, and the sensible thing. If one places the visible thing directly on the eye, nothing is seen; the transparent body, e.g., air or water, must be placed between the visible thing and the eye if vision is to occur. The coloured body is something naturally suited to have its effect on the transparent as such; the transparent, as so affected, is naturally apt to affect the thing which sees. The medium, i.e., in our example the transparent, is thus one of the essential components of the world of sensibles and sense. The medium is essential even for the sense of touch, according to Aristotle. In the case of touch, the medium is not outside the animal but in it; the flesh is not the *organ* of touch but rather its medium, the organ being deep within the flesh.[3]

Theories of the medium thus constitute an integral part of Aristotelian theory of sensation, and over the centuries the various Peripatetic commentators presented the media of the senses in various ways. In the case of Albert the Great, we are fortunate in having two treatises discussing in detail the doctrines presented by Aristotle in his *De anima*. First, there is the *De homine*, i.e., the second part of the so-called *Summa de creaturis*; this part is sometimes referred to by Albert himself as "our treatise *De anima*," and dates from about 1245-1246, being in the form of disputed questions.[4] Secondly, there

---

[3] Aristotle, *De anima* II.11 (423b17-26).

[4] Cf. Odon Lottin, "Problèmes concernant la 'Summa de creaturis' et le Commentaire des Sentences de Saint Albert le Grand," *Recherches de théologie ancienne et médiévale* 17 (1950), 321. For the date, cf. James A. Weisheipl, "Life and Works of St. Albert the Great," in the present volume, p. 22. The treatise is printed in ed. Borgnet, vol. 35; we will refer to it as *De homine*, in order to avoid confusion with the *De anima* (the later Aristotelian paraphrase).

is Albert's "paraphrase" of Aristotle's *De anima*, written between 1254 and 1257; the word "paraphrase" should not lead one to think this is mere repetition, since it contains lengthy supplementary discussion.[5] We propose to examine in turn these two works to see what happens to Aristotle's doctrine of the medium in them. More especially, we will be focussing on how the sensible *exists in* the medium, which in some way "conveys" things to sense.

A. THE *DE HOMINE*

In the particular part of Aristotle's *De anima* which concerns us, there is first a discussion of sensation in general, then treatments of the types of external sensation taken severally, in the order: vision, hearing, olfaction, taste and touch, after which again discussion of sensation in general. This order will be exactly followed in Albert's *De anima*, but in his *De homine* there is no initial general discussion. We begin accordingly with Albert's views on vision.

i. Vision

How, asks Albert, are the *species* of visible things present in the medium of vision and in the organ? The term *"species"* is technical here, and refers to what we might call the thing in the form of a message. More will be said about it later, but for the present that will suffice; we will use the Latin term *"species"* and its synonym *"intentio"* throughout the paper.[6] That the sensible, here the visible, must somehow be present in the medium is clear from the fact that the medium affects the eye in function of the particular visible thing. The medium must somehow be programmed. And yet this seems to

---

[5] Of this work we have a critical edition, viz Alberti Magni *De anima*, ed. Clemens Stroick, OMI, in ed. Colon., 7/1 (1968). For the date of the work, cf. Stroick, p. v.

[6] Albert, *De homine*, q.21, a.5 (ed. Borgnet, 35: 205b). Regarding the word *"species,"* its etymological link with vision should be remembered: thus, the English "specious" means "of good appearance," and we have such words as "spectacle."

"Intentio" was the word selected by the Latin translators of Avicenna to translate the Arabic *ma'nā*; the fundamental Arabic verb involved here, *'anā*, they translated *velle dicere* (cf. French *vouloir dire*), i.e., "to mean" or "to *intend* to say". Thus, *"intentio"* is best rendered by such English words as "meaning" or "notion." In our context of sensibles and sense, it means the message sent from the sensible to the sense. It is misleading to put emphasis on the notion of tendency in the etymology of *"intentio."* Cf. the Arabic-Latin lexicon contained in *Avicenna Latinus. Liber de Anima seu Sextus de naturalibus I-II-III*, ed. Simone Van Riet (Louvain/Leiden, 1972), p. 346, 536.

Concerning *"species"* and *"intentio"* in Albert's *De anima*, cf. below, p. 303.

mean that contrary colours are simultaneously present in the same part of the air. Here is how Albert puts it:

> Let A be an eye placed in an easterly position, but looking towards the west; and B an eye in a westerly position looking east, and let the distance between them be ten cubits. Now, on the east side, let there be C, a white body of intense whiteness affecting the air right to the eye placed opposite on the west side. And on the west side let there be D, a black body of darkest black affecting the air right to the eye placed to the east. And let these bodies and these eyes be set apart in diametrical opposition. And at the same time, one of the eyes will be affected by the white, and the other by the black. Therefore, the two colours necessarily meet in the medium, and simultaneously are actual there; which would be impossible if they were possessed of contrariety and were present in the air as an accident in a subject.[7]

We have, thus, the difficulty presented. The presence must be real, yet are we to say that contraries are present in the same subject, that the same air is both black and white?

Two solutions, Albert tells us, have been proposed, the first describing the presence as one according to "spiritual being" (*esse spirituale*); this follows Averroës in his *De anima*. The second calls the presence "potential" (*sicut in potentia*). These are considered in turn.[8]

Some wish to say that the colours are not in the medium and in the eye according to the colours' own proper and natural being (*secundum suum esse proprium et naturale*), the way they are in something as in a matter and as in a subject, but according to spiritual being (*esse spirituale*), separated from matter and subject; and in this mode of being they have no contrariety, just as neither do their intelligible notions (*rationes intentionum*) in the intellect (we obviously entertain together the *notions* of contraries). But this position seems to run counter to the fact that the colours are present in air and water, i.e., the media are bodies; should not these bodily subjects give bodily being to the colours? Also, colours, if too strong, damage the eyes (Albert mentions glass jars of mercury placed in sunlight); this suggests the mode of being of colours in the medium is material (*secundum esse materiale*), not spiritual. For such reasons as these,

---

[7] Albert, *De homine*, q.21, a.5 (ed. Borgnet 35: 206a).

[8] Ibid. (206b-207b). Cf. Averrois Cordubensis *Commentarium magnum in Aristotelis De anima libros* II, com. 97, ed. F. Stuart Crawford (Cambridge, Mass., 1953), p. 277.

some take "spiritual" here as a special sort of corporeal presence, the word "spirit" meaning a subtle and luminous sort of body.

Those who find this position of Averroës no help say the *species* are present in the medium as "in potency," i.e., the way the form towards which motion tends is already present in the motion. Contraries, they say, can exist together in potentiality. But to this Albert objects that contrary movements of bodies directly colliding ought to bring each other to a halt; in the case of the black and white bodies, neither eye should be affected; and if one says there is deflection of one colour by the other, this is against what sensibly appears.

We come now to Albert's own solution to the problem. While Albert conceives colour as something present in the surface of the coloured thing, nevertheless colour is able to show itself to an onlooker, i.e., able to affect the intervening medium and ultimately the eye, only by itself being activated by light. Light is regarded as an agent which "abstracts," i.e., dissociates from its material subject, "liberates," one might say, the colour. Three points, then, are to be noted if one is to understand how the *species* of visible things are in the medium. First, there is the nature of the agent producing the colours in the medium, i.e., abstracting them from the bodies of which they are the colours. This agent is light, a quality which has no contrary, whose passage through the medium is instantaneous, not in time. Hence, since colours act on the medium and on the eye only in virtue of light, in their action they have no contrariety with each other. The mode of agency is that proper to light itself.

Secondly, there is the nature of the abstracted colour. It is not abstracted along with the causes which generate it in a subject (which causes are the hot, the cold, the wet, and the dry). Rather, what is abstracted is the proper *species* of colour alone, without any part of matter, without any material cause. This is what Averroës, in his *De anima*, calls "spiritual being" (*spirituale esse*). Now, colours, just by themselves, do not bring about either active or passive qualities; this they do rather by virtue of the hot, the cold, etc. Thus, in their abstract condition in the medium they do not act upon each other, do not interfere with one another.

Thirdly, there is the nature of the medium. Air and water are not media for vision precisely as air and water, but according to what they have in common with the celestial bodies, namely transparency (*transparentia*). This is the aptitude to receive the *species* of visible things, but to receive in the way a conveyer receives, not in the way a holder or retainer receives. This is what people mean by the visibles being in the medium as in potency. And so, because of the nature of the transparent, neither from this direction do visibles in the medium have contrariety. Albert concludes that thus, in a way, both of the solutions proposed, spiritual being and potentiality, are true. And he goes on to show that his position offers answers to the objections brought against them.

Averroës' term "spiritual," he tells us, is not to be taken as used of soul in distinction from body. It is taken from "spirit" said of the body with which the transparent has something in common, i.e., the celestial body. He also argues that it is not colour and light which in excess directly harm the eye, but it is rather certain corporeal humours whose flow is overstimulated by the intensifying of visual activity.

At the very end of the discussion, Albert answers the question: if one says that colour is present in the medium and in the eye according to spiritual being, what is the difference between presence in the medium and presence in the eye? One difference is taken from the anatomy of the eye, which as it were collects and retains the *species*; air does not do this. Secondly, in the eye the *species* is present as a thing possessed and as a disposition, and thus as a principle for knowing the entire coloured thing; in which way it is not in the air. Thirdly, in the air it is present as *in via*, while in the eye it is present as an actuality.

This, I would say, brings out the point that both doctrines, spiritual being and potentiality, are needed to explain the medium as a medium. When one says that the *species* is an actuality in the eye, one means an actuality of spiritual being.

To review, the basic problem was the co-existence of contraries in the medium. The solution was the non-contrariety of the celestial body, as found derivatively in light, colour, and transparency. This is the spiritual being of colours in the medium, colours themselves being essentially a type of form susceptible to the influence of light, even though they also have a type of being according to which they are tied in with matter and contrariety.[9]

---

[9] Albert, *De homine*, q.21, a.5 (ed. Borgnet 35: 208b-210b). The point that the heavenly corporeal nature has no contrariety proper to it and its operations (in contrast to the four subcelestial elements whose upward and downward movements, and whose qualities of hot, cold, wet and dry, have contrariety) is fundamental to the Aristotelian theory of the heavens. It was argued on the basis of the naturalness of the observed circular movement of the heavens. Thus Albert, in his paraphrase of Aristotle at *De caelo* I, c.4 (270b33), says: "But we have said also that there is no contrary whatsoever to circular motion, and upon that we have based all our

ii. Hearing

Unlike the visible, sound exists only as a movement, as an event, one might say, in air. Is air, then, a medium here, or is it rather a matter, a subject? Does sound exist only at the initial place of striking, and is it carried to the ears by the medium, or does it actually exist in ever increasing circles? Do contrary sounds in the medium interfere with each other? Do sounds have spiritual being in the medium, the way this is said of colours?

Albert teaches that sound is generated, not in one part of the air only, but rather in every part, right up to the sense of hearing. Every sensible thing, he says, has being only in its own matter, and is as extensive as its own matter is. Thus, however much air is struck by the first striker whence emanates the sound, that struck air strikes more air and generates sound in it, and so on with more air, just as long as the violence of the original striker endures; and according to the weakening of that violence the sound likewise becomes weaker.[10]

Albert also maintains that sounds can very well cancel each other out. The difficulty of maintaining this is that we do not seem to find simple cancellation by opposed sounds. Accordingly, some people attribute spiritual being (*esse spirituale*) to sound in air. Albert refuses to accept this. In order to explain why there is not simple cancellation, he makes some use of wave-theory and echo-theory.[11]

We should note especially one line of argument presented and rejected by Albert. Some say that sound is in air and in water, as in its media, according to spiritual being (*secundum suum esse spirituale*). They argue (a) that a medium is one sort of thing, a matter is another; sensibles are in their matter according to material being (*esse materiale*), in their medium according to spiritual being; and also (b) that since the sense is receptive of the *species* of sensibles without matter, and comes to be such by the action of the medium, it would seem that the sensibles must be in the medium spiritually, i.e., without matter. According to this line of argument, some say that the media for all three senses having external media (in contrast to taste and touch, where the medium is internal to the sensing animal) are

---

proofs concerning the fifth [i.e. celestial] body; and so we must here prove and demonstrate that" (Albert, *De caelo et mundo* I, tr.1, c.9 [ed. Colon. 5/1 (1971): 24.59-63]). The *De homine* discussion of the visible and its diffusion views light, colour, and the transparency of the medium all in function of the "fifth body" and its immunity from contrariety.

[10] Albert, *De homine*, q.24, a.5, part 1 (ed. Borgnet 35: 239b).

[11] Ibid., a.6 (241a-242b).

air and water as agreeing in the nature of the transparent (*diaphani*), i.e., as having something in common with the celestial bodies. They say that light and the transparent confer spiritual being on the sensibles in the medium.

This is to make light the agent producing sound and odour as well as colour. Albert rejects this position, because such diverse effects ought to have diverse agents. He also points out that sound and odour exist actually in the dark, and so it is absurd to propose light as the agent.

Albert's own doctrine of the medium for sound seems designed to explain why air is so much more suitable as a propagator of sound than water is. Air is the medium and the matter in which sound is generated; without air, sound does not have its being. Water is a medium for sound, not its matter; indeed, it is a medium in which and through which sound travels in its own matter, which is forced air. That is why water confuses and interferes with sound, whereas air preserves sound: the conservation of a thing is due to its proper matter. It is by dividing the water that forced air, with its sound, passes through the water. It is not as though the water were altered according to the mere *species* of sound: if water were a medium receiving the pure *species* of sound, the way it receives the *species* of colour, then it would be more retentive of sound than air is.

Coming back to the arguments of those who posit a spiritual being for sound, Albert rejects them. He says the case of sound cannot be compared to that of colour. Colour exists in a medium which is a medium only; sound has its material and natural being (*suum esse materiale et naturale*) in the medium and not outside it. The case of odour will be discussed subsequently. As for the argument that spiritual being in the sense presupposes spiritual being in the medium, this is not so. It is true that every sense, as judge of the sensible, receives only the *species* of the sensible, not its matter. But it is not necessary that the sensible be already in the medium with that sort of being (*secundum esse tale*); otherwise, this would have to happen also in the case of touch and taste, where clearly it does not.

What we see here, then, is a complete rejection of the doctrine of spiritual being for the *species* of sound in the medium. The doctrine of spiritual being applies, thusfar, only to the visible.[12]

---

[12] Ibid., q.26 (251b-253b).

iii. Olfaction

Air and water are media for odour. The common factor which makes them media is a receptiveness of the odoriferous evaporation, so as to retain and yield up again this evaporation. This is more strongly a property of air than of water, and more of warm air than of cool. Is odour present in the medium by mere alteration of the medium, or by the material presence of the evaporation (the latter meaning that small particles of the odoriferous body would be mixed in with the medium)? Albert tells us he follows Avicenna in replying that it is present in both ways. Material presence is not further explained, but as regards the purely qualitative change of the medium it is said that the nature of the air is dominated by the evaporation, so that the evaporation infects air in function of odour, quite apart from the air's receiving any part of the evaporating thing. Water is similarly affected, as when wine in a jar becomes tainted with the odour of things placed near it.

That this qualitative sort of influence takes place is shown by the case of the vultures and "tiger-birds" which, on the occasion of a massacre in Greece, flew to the carcasses from a distance too great to be accounted for by particle-diffusion. Avicenna, Albert acknowledges, does say this might be accounted for by vision from great heights. On the other hand, that particle-diffusion does take place is shown by the fact that apples become lighter in weight after giving off odour for a long time.[13]

It is to be noted that the expression "spiritual being" is not used concerning the transmission of odour by sheer infection or alteration of the medium. Moreover, Albert says he is following Avicenna. Now, Avicenna proposed three possible explanations, viz the two mentioned by Albert, together with transmission after the manner of colours, in which case there would be no affecting the medium at all, but a pure conveying of the odour by the medium. Avicenna rejects only this third position.[14] Thus, it is probable that Albert is here speaking of a qualitative influence which would be according to material being, as opposed, on the one hand, to mere mixture of small parts, and on the other, to a doctrine of spiritual being.

---

[13] Ibid., q.30 (269a-271b).

[14] Cf. Avicenna, *Liber de anima seu Sextus de naturalibus*, pars II, c.4 (ed. Van Riet, pp. 148-152).

iv. Taste

Speaking of saliva as an extrinsic medium for taste (though not a medium essential for the existence of taste), Albert says it is medium and matter, not medium only. It is comparable to air as medium for sound, not to the case of sight where air is medium only and not matter for colour.[15]

v. Touch

Following Avicenna, Albert distinguishes two aspects of touch, one as form, the other as power. As form and perfection of the whole animal body, it has the whole body for organ and has no medium. So considered, it has its primary seat in the heart. However, as power, and judge of tangibles, it has its primary seat in the brain, whence power is communicated via the nerves to flesh and skin, the vehicle of this communication being the subtle body called "spirit." From this point of view, the flesh has the role of medium: the action of the tangible on the flesh and nerve is simultaneous, but comes to the nerve via the flesh. This, says Albert, is what Aristotle considers.

The doctrine of Aristotle, that one cannot sense what is placed directly on the organ, cannot be maintained for the sense of touch, since the nerves are more sensitive than the flesh; or rather, it is true only if one takes as organ the brain itself: the brain is not sensitive to touch, except for its surrounding threadwork.[16]

Here, nothing is said of spiritual being, and we had already been given reason to think nothing would be.

vi. The Senses in General

This section contains some surprising features for the reader who has followed Albert through the senses individually. The question is asked: according to what sort of being does the sensible exist in the object, in the medium, and in the organ? Aristotle's discussion of the issue: whether anything is affected by odour except the sense of smell (and so on with the other senses),[17] is influential here. Aristotle presents reasons for thinking that only the senses are affected by the sensibles (it is not the sound but the air which splits the tree, in the

---

[15] Albert, *De homine*, q.32, a.4 (ed. Borgnet 35: 280a-b).
[16] Ibid., q.33, a.3 (289b-291a).
[17] Aristotle, *De anima* II.12 (424b3-20).

case of thunderbolts). This would incline one to say the sensible is in no way in the medium, since the medium is not the sense. But against this is the fact that the medium does act on the sense in function of the sensible.

Albert proposes a special type of being for the sensible in the medium, for all three senses having a genuine external medium. In the object, it has material being (*esse materiale*); in the sense, it has spiritual being only (*esse spirituale tantum*); but in the medium it has *sensible* being (*in medio vero sensible*). This is explained first with reference to vision: the object of vision is not in the medium as in a matter, as we saw above. Furthermore, though sound has being in air as in a matter, nevertheless it does not act upon the hearing with the action of matter, but with the action of a medium. And the case of smell is similar.

Albert goes on to expand on the two modes of action. The action of matter transforms the matter, or at least disposes the matter according to the quality (e.g. either the thing becomes red, or at least is put into the genuine condition of becoming red). Accordingly, air having colour in it as in a medium does not act with the action of matter in function of the colour: it does not colour anything. Similarly, the air having sound does not act with the action of matter in function of the sound (at least, not when it acts on the sense of hearing): for it does not make the thing acted upon (i.e. the sense of hearing) to sound. But the air acts with the action of matter inasmuch as the thunderbolt splits the tree. It is, then, by the action of a medium, not of matter, that the air acts on the sense: it actualizes the sense by virtue of the sensible which is in it (the air) as in a medium.

By means of this action of a medium, Albert interprets the saying that the sensible "is, potentially" in the medium. As in matter, and as in sense, the sensible "is, actually." At the same time, Albert keeps reminding us that for some of these senses the sensible is in one and the same thing as both matter and medium. The doctrine that the sensible is in the medium only "potentially" and as "in transit" is seen as explaining the Aristotelian position that the sensible affects only the sense: as being only potentially, the sensible in a way neither is nor acts in the medium.

In the case of the senses of taste and touch, which have internal media, Albert tells us they have sensibles which are present as actualities in the object and in the medium and in the organ. Their sensibles, he goes on, acquire "spiritual being" (*esse spirituale*) from the animal spirit (*per spiritum animalem*) drawing and conveying their *intentiones* to the brain: and thus the *intentio* of heat, by virtue of the attracting animal spirit which issues from the origin of touch, does not heat the brain.

Thus, here again, as in the case of the other senses, the *species* are seen as acquiring a special kind of being *before* they enter into the sense proper. The sense of touch gives rise to a subtle body, an animal spirit, which boosts the merely material influence to the level of spiritual being, so that it is the *intentio* alone which acts upon the sense.[18]

To resume, in the *De homine* Albert has insisted on the material being of sound, has indeed refused to attribute "spiritual being" to it. At one point he has refused to accept the argument that spirituality is needed already in the medium, so as to produce spirituality in the sense, saying that this is obviously not the case for touch. Yet here in his synthesis he has posited a special type of being for sensibles in the medium, based on their exhibiting a special type of action. This is a being distinguished from the spirituality found in the sense, and from the materiality found in the object. This doctrine of a middle mode of being reminds one (though Albert makes no mention of this) of Averroës' doctrine in his *Compendium libri Aristotelis De sensu et sensato*. Albert calls it "sensible being", which doubtless means "a being peculiar to the pure sensible, i.e. to that which acts upon the sense," but certainly does not exclude the meaning suggested by ordinary English usage: "a being which can actually be sensed."[19] Also, confusingly enough, Albert uses the expression "spiritual being" concerning the sensible being proper to touch.

---

[18] Albert, *De homine*, q.34, a.2 (ed. Borgnet 35: 298a-b and 300b-301a). The treatment of touch as having the "spirit" come from the sense and "spiritualize" the sensible reminds one of the general position criticized below, p. 304, according to which the soul itself is source of spirituality for sensibles. At *De homine*, q.33, a.3 (ed. Borgnet 35: 290a), Albert refers this doctrine of touch to Algazel. Cf. al-Ghazzālī (Algazel), *Metaphysics*, pars II, tr.4, c.3 (ed. Joseph T. Muckle, CSB [Toronto, 1933], p. 165).

[19] Cf. Averrois Cordubensis *Compendium libri Aristotelis De sensu et sensato*, ed. Aemilia Ledyard Shields (Cambridge, Mass., 1949; volume title: *Compendia librorum Aristotelis qui Parva naturalia vocantur*), p. 31, line 45–p. 32, line 48. That the pure sensible, as communicated by the medium, is what is actually sensed, i.e., is the proper object of the sense (at least for the senses other than taste and touch), is clear from the texts referred to below in nn. 43 and 41 (at any rate, for Albert in his *De anima*).

B. THE *DE ANIMA*

i. The Senses in General

At the outset, let us note a statement by Albert as to what he means by the "*intentio*" or "*species*" of the thing. In our acts of knowledge of corporeal things, Albert distinguishes diverse levels of knowledge: we know corporeal things in sense-knowledge, and also in acts of imagination, and in acts of understanding. The explanation of any act at any level is that the knower possesses a *species* or *intentio* of the thing known, having left aside the matter of the thing. All knowing, even external sensation, leaves matter aside, i.e., "abstracts" from matter. Albert distinguishes carefully between the *form* of the thing and the *intentio* of the thing. The form is in the thing and is a part of the thing (which is a composite of matter and form). The form, by informing, gives actual being to the matter and thus to the composite of form and matter. The *intentio* (which is found in the sense, in the case of sense-knowing, and in the intellect, in the case of intellection) does not confer being on the sense or on the intellect. It rather constitutes a *sign* and a *knowing* of the thing known (*signum facit de re et notitiam*). Albert stresses that whereas the form is a *part* of the thing, the *intentio* is a principle making known the thing *as a whole* (*totam rem notificat*). By the *intentio* of the coloured thing, which is in the eye, we know, not merely colour, but the coloured thing. By the *intentio* in the imagination, we imagine the thing as a whole.[20]

The *intentio* or *species*, at whatever level of abstraction, is thus a principle of specification (or programming) of the act of knowing the thing as a whole: it is the sign of the thing as a whole. It is this, nevertheless, in function of some particular objectivizing principle, such as colour.

Let us return to our own quest. Above, in the *De homine* discussion of the senses generally. Albert first states the general problem, involving external media demanded spiritual being for the sensible in the medium, together with the proposal that the source of this spiritual being for all three senses was light and transparency. Here in the *De anima* we find that this has become a major issue for the discussion of the senses generally. Albert first states the general problem, then arguments for positing the existence of one single source for all

---

[20] Albert, *De anima*, II, tr.3, c.4 (ed. Colon. 7/1: 102.28-53).

spiritual being for sensibles, then the arguments of two schools of thought on the nature of this source (one opting for light, the other for the soul itself), then individual replies to all these arguments, and lastly a careful statement of his own position. Here we can give only the absolutely essential points.

The immediate occasion for the discussion seems to be Aristotle's doctrine of the passivity of sense.[21] It being acknowledged that the sense must possess the *species* in an immaterial way, this demands that the *species* have a certain immateriality or spirituality even while still in the medium, i.e. on the side of the agent. Albert now takes this argument seriously. He no longer rejects the doctrine of spiritual being. What he rather rejects is the notion that this spiritual being is uniform for the various sensibles: and thus he rejects the need for a single factor as source of the spirituality. He will posit a multiplicity of such sources.[22]

Thus, Albert says that spiritual and intentional being (*esse intentionale et spirituale*) does not have the same character in diverse types of sensible. In one case it is more spiritual than in another. In the case of objects of touch, the sensible affects the medium and the organ, acting upon it according to material being (*esse materiale*). And even those sensibles which have the same medium, viz colours, sounds, and odours, do not have that medium according to one and the same medial nature, but according to diverse natures. And the being that the sensibles have in the medium is not of one character: that of colour is more spiritual than that of sound, that of sound more spiritual than that of odour (thus, the wind neither brings nor removes colour, but it does dull sounds, taking them away in part though not altogether; odours it can take away entirely). Albert says it seems one ought to admit that for some of the senses and sometimes the sensible is according to one sort of being in the thing sensed, and according to another sort of being in the medium and in the organ: this is true for senses having an external medium. But still, this being which they have in the medium and in the organ is not the same for any two senses.[23]

---

[21] Aristotle, *De anima* II.5 (416b34 and 418a1-5).

[22] The problem is first posed at *De anima*, II, tr.3, c.3 (ed. Colon. 7/1 101.29-47); the whole of ibid., tr.3, c.6 is devoted to it; uniformity of spiritual being is rejected there (ed. Colon. 7/1: 105.82-87).

[23] Ibid. (pp. 105.79-106.14). Albert, contrasting his own position with those of his adversaries, proclaims that he is concentrating strictly on the natural: "Nos autem simpliciter naturalibus insistentes. . ." (p. 105.79). Notice that here (p. 105.88-90) there is said to be material action on the *organ* of touch; below, pp. 315 and 318, this will not be admitted.

We can see that Albert's position has changed. Sound is now said to have a more spiritual sort of being in the medium than does odour, and generally there is a greater willingness to concede the presence of spiritual being.

As for the question: whence comes this spiritual being for a sensible, Albert considers it a foolish question. He says that he has previously shown that every active power is of itself perfect as regards action, requiring no extrinsic power.[24] And so:

> I say that the form of the thing sensed, by virtue of itself (*per seipsam*), generates itself in the medium of sense according to sensible being (*esse sensibile*). And of this the necessary demonstration, proved by all the philosophers and by the force of truth itself, is that that is sensible by virtue of itself (in the second way of saying "of itself (*per se*)") which through its own essence is cause of its own being sensible (*esse sensibilis*). And so it is a useless question: what confers that [sensible being] upon it; it is as if one were to ask what confers upon light its operation of actual illumination.[25]

In taking this position, that colours and sounds and odours, etc., are by their very own essences causes of being sensible and of actual sensing, Albert affirms that he is following the common doctrine of the Peripatetics.[26]

Albert concedes the argument that there is need of an immaterial agent to confer intentional being (*esse intentionale*) (as he here often calls the spiritual being we are discussing). But as for the source of that immateriality, he tells us that the form, which is in the thing, sometimes acts by means of the qualities of the matter in which it is, and then it acts materially; but sometimes it acts by itself alone, and then it acts immaterially, for it is by itself an immaterial essence. In

---

[24] Albert's reference backward is to *De anima*, II, tr.1, c.1 (ed. Colon. 7/1 64.45-65.26).

[25] Albert, *De anima*, II, tr.3, c.6 (ed. Colon. 7/1: 106.18-26). Where, at lines 22-23, it reads, ". . .per se sensibile esse, *quod* in secundo modo dicendi per se per essentiam suam est causa sui esse sensibilis," I have read, ". . .per se sensibile esse, in secundo modo dicendi per se, *quod* per essentiam suam est causa sui esse sensibilis."

The reference to the "second way of saying 'of itself' (*per se*) is to Aristotle, *Post. Anal.* 1.4 (73a35-b1); there, various ways for one thing to be said of another "of itself" or "through itself" are distinguished. The first way is, e.g., "animal" belonging to "man" of itself: here, the predicate is of the very essence of the subject. Albert takes as second way "visible", e.g., said of "colour," where the predicate "visible" expresses an essential *result* of the nature of the subject, colour. Thus he says, ". . .through its own essence is *cause* of its own being sensible. . . ." Cf. Albert, *In Post. Anal.*, I, tr.2, c.9 (ed. Borgnet, 2: 43b): "Unde omnes praedicationes in quibus effectus praedicatur de causa, reducuntur ad secundum modum." Concerning essential results, cf. the text referred to above, n. 24.

[26] Albert, *De anima*, II, tr.3, c.6 (ed. Colon. 7/1: 107.40-51).

the latter sort of action it requires nothing but itself alone, whereas in the former it requires something other than itself (Albert points out that thus, just the opposite of what the adversaries say obtains: the form by itself can provide spirituality; it is as regards material action that one must seek a "something else" besides the sensible).[27]

Spirituality of the sensible is here no longer being traced to any body as such, whether subtle or celestial. Rather, Albert maintains that the form itself is a simple essence multiplicative of itself; indeed, in this way every form multiplies its own *intentio*; and since a form is an essence simpler than any body, there cannot be found any bodily form capable of conferring intentional being on simple form.[28]

Here, then, we find that the doctrine of two modes of action mentioned in the *De homine* concerning the medium has now been, as it were, "backed up" until it is seen as pertaining to the very form of the thing sensed. Moreover, there is some change in the conception of the spirituality involved. Whereas previously Albert was content to explain the spirituality of the colour in the medium as derived from the "spirit" said of the celestial body, now it is clear that the

---

[27] Ibid. (p. 106.69-78).

[28] Albert says, ". . .aliud esse agens formam in materia et aliud agens formas tantum. . . .Agens autem formas tantum non est agens materiale, sed potius ipsa forma; et sic agit se per hoc quod ipsa est essentia simplex sui ipsius multiplicativa, *et sic omnis forma multiplicat intentionem suam.* . .forma sensibilis multiplicat se in esse spirituali et sufficit sibi ad hoc, sicut omnis forma in propria et essentiali actione sibi sufficit" (ibid. [p. 107.51-73]).

Has one to do here with the doctrine of "*multiplicatio specierum*" such as one finds it in Robert Grosseteste (cf. A. C. Crombie, *Robert Grosseteste and the Origins of Experimental Science 1100-1700* [Oxford, 1953], pp. 109-111)? It is, of course, a doctrine of multiplication of *species*, but is evidently very different from that of Grosseteste. Albert is speaking about a special kind of agency of form, that proper to the order of sensibles and sense. Grosseteste, on the other hand, envisages one sort of emanation, which may affect either ordinary bodies or sense-receptors: "Agens naturale multiplicat virtutem suam a se usque in patiens, sive agat in sensum, sive in materiam. Quae virtus aliquando vocatur species, aliquando similitudo, et idem est, quocunque modo vocetur; et idem immittet in sensum et idem in materiam, sive contrarium, ut calidum idem immittit in tactum et in frigidum. . . .Sed propter diversitatem patientis diversificantur effectus. In sensu enim ista virtus recepta facit operationem spiritualem quodammodo et nobiliorem; in contrario, sive in materia, facit operationem materialem. . ." (Grosseteste, *De lineis*. . ., ed. Ludwig Baur, in *Beiträge* 9 [1912], p. 60.16-27; quoted in part, in English translation, in Crombie, p. 110). This conception is in accord with the fact that Grosseteste has an Augustinian conception of sensation, i.e., sense as an act of vital attention, rather than the Aristotelian conception of sense as the soul's undergoing an influence: cf. his *De generatione sonorum*, ed. L. Baur, in *Beiträge* 9 (1912), p. 7.23, where we find the formula for sensation — "a passion of the body, which is not concealed from the soul" — a formula taken from St. Augustine, *De quantitate animae* XXV.48 (PL 32.1063). Concerning the active character of such sensation, cf. Etienne Gilson, *History of Christian Philosophy in the Middle Ages* (New York, 1955), p. 75.

intentional being of the sensible in the medium cannot be derived from any body, no matter how noble, as a body; it is an effect of form as form.[29]

The change of location for this discussion is worth remarking. In the *De homine*, with Albert rejecting spiritual being even for sound, this issue was merely a moment in the distinction of sound from vision. Now that Albert has admitted spiritual being for, at the very least, all the senses having external media, it is seen as pertaining to the discussion of sense as sense. There is also, here in the *De anima*, and by virtue of this issue, from the very outset a marked new insistence on the naturalness or intrinsic character of sensible things being sensible.

ii. Vision

Surprisingly little pertaining directly to our topic is to be found in the *De anima* discussion of vision. The problem of the contraries simultaneously present in the medium does not occur at all. Perhaps this is because in the *De homine*, where only colour had this spiritual being in the medium, the contraries problem was paramount, whereas now spirituality in the medium has become the common characteristic of the sensible, pertaining to the sensible's being sensible by virtue of itself: every form multiplies itself in intentional being. We shall, however, see something of the contraries problem later in connection with the spiritual being of sound.

Much discussion is devoted to the question: is light needed in vision in order to light up the transparent medium (position of Averroës) or to make colour actually visible (position of Avicenna)? If one says the latter, is not one denying what has just been so strongly affirmed, namely that colours are visible by themselves? Albert holds that there is truth in the two positions. He conceives of light as constituting colour (if "colour" names the formally and actually visible), and also as illuminating the medium. Since "colour" means the surface of the coloured body as rendered visible by light, then one can say that colour, of itself (i.e. as including light), is visible. On the

---

[29] For the explicit rejection of the celestial body as source of intentional being, cf. Albert, *De anima*, II, tr.3, c.6 (ed. Colon. 7/1: 106.35-37). Cf. St. Thomas Aquinas, *De potentia*, q.5, a.8, who presents the type of multiplication or communication of *species* of which we are speaking as an action of *bodies*, but inasmuch as they participate in a mode of action proper to separate substances, i.e., pure subsisting intelligences.

other hand, because colour, taken even in this formal way, is light tied to matter, it cannot have its effect on the *dark* transparent medium. Its exactly fitting medium is the *illuminated* transparent. Hence, Averroës is right in saying that light is needed for the medium. As for the fact that when the surface of the coloured body is illuminated, but the space from there to the viewer is dark, the surface is nevertheless seen with its colour, Albert contends that the medium is in such a case illuminated in the immediate vicinity of the object, and that this is enough to make it receptive of the influence of the object.[30]

Of more direct interest to our topic is the doctrine concerning light. Albert uses the term "*lux, lucis*" for the quality, light, in the originally luminous body, e.g., the sun, whereas he uses "*lumen, luminis*" for light in an illuminated body, such as air. He says:

> The light (*lumen*) which is. . .generated by the luminous body in the transparent is related to the light (*lucem*), which is the form of the luminous body, as the *intentio* of the colour, generated in the transparent, is to the form of the colour, which is in the coloured body.
>
> And therefore some say, fittingly enough, that light (*lumen*) is an *intentio* having spiritual being (*spirituale esse*) in the transparent, just as colour has intentional being (*esse intentionale*) in the medium; but they differ in this, that light (*lux*) is a more noble and simple form than colour is, and so it perfects the transparent and gives it actual being (*esse actu*) and colour cannot do this. Similarly, it gives actual being (*actu esse*) to colours, and this again colour cannot do. . . .[31]

At c. 14, the paraphrase reaches the point where Aristotle says that light is the perfection of the lucid (or luminous transparent: *Perfectio autem lucidi* sive perspicui *lumen est*), and that colour moves the actually lucid. A sign of this essential role of the lucid as intermediary between colour and eye, says Aristotle, is that if something is placed directly on the eye it does not become seen. Albert comments:

> . . .and the cause of this is that the coloured thing does not act on anything according to touch except by physical action, which is by its material principles. But colour is not made to be in sight by physical action but by formal and spiritual action. . .and so it needs a body in which it is previously rendered spiritual, before it is generated in the

---

[30] Albert, *De anima*, II, tr.3, c.7 (ed. Colon. 7/1: 108.39-110.15).

[31] Ibid., c.12 (p. 116.67-80). St. Thomas Aquinas rejects this doctrine of light as having intentional being in air: cf. his *Summa theologiae* I, q.67, a.3.

eye; and this is the necessity for there being a medium for the sense of sight. . . .[32]

It is hard to see this as very satisfactory. Surely the coloured thing does not act on the medium by physical action either. Why would not such reasoning give rise to an infinite regress? Perhaps the point is that once the *fact* of the transparent having this role is observed, one can reasonably enough conceive of its being necessary in some such way as this.[33]

iii. Hearing

Whereas in the *De homine* it was in the section on sound that we saw spiritual being rejected most unqualifiedly, the *De anima* presents a rather different position. Water is now regarded as a medium in which sound exists according to spiritual being, the earlier doctrine of sound with accompanying air forcing its way through the water having been abandoned. But more important is the change regarding air. Air is a medium for sound in two different ways: (a) first it is both matter and medium, and (b) subsequently it is medium alone. Sound is no longer thought to be found in air only as in a matter which simultaneously has the role of medium. It is also found in air merely according to spiritual being. In this latter situation it is comparable to the coloured thing, which produces its *intentio* alone in the air, though we are reminded once more that sound's spiritual being is not as spiritual as that of colour.[34]

No example is given as regards this purely spiritual being of sound in air. However, in the following chapter, echo is explained as reflexion of sound, and all sound is said to involve reflexion (though for the most part it is not perceivable). In the midst of this discussion, the problem of interference of contrary sounds is raised. Why

---

[32] Albert, *De anima*, II, tr.3, c.14 (ed. Colon. 7/1: 119.34-42); cf. Aristotle, *De anima* II.7 (419a10-15). Also to be noted is Albert, *De anima*, II, tr.3, c.16 (ed. Colon. 7/1: 123.20-36), along with Aristotle, *De anima* II.7 (419a26-33). Albert there repeats the doctrine of spiritual being for the first three senses (with explanation of lesser immateriality for sound and odour), while affirming the doctrine of the medium for the last two senses but not that of spiritual being.

[33] This same argument seems to be behind the doctrine of spiritual being throughout the *De anima* treatment; cf. below, pp. 311-312 and 315-316. The facts of vision, and to a lesser extent, of sound and odour, persuade in the direction of spiritual being on the side of the agent (i.e., the sensible). In the *De homine*, it was the facts of touch and taste which were allowed to prevail (cf. above, p. 298).

[34] Albert, *De anima*, II, tr.3, c.18 (ed. Colon. 7/1: 126.11-21).

do not equal and opposing sounds cancel each other out? Albert offers as a solution the behaviour of circular waves radiating from a disturbance in water. Where two such waves radiating from different centres intersect, there is cancellation, but only at a point; in their other parts the two waves do not touch or break each other. And from the places where the sound remains intact, it is diffused again even to those places where it was destroyed by the interference. But this comes about, not by re-sounding, but by reflexion:

> for the generation of sound is easy, because it is in some air only through its *intentio*. . . .[35]

Albert seems to be associating reflexion of sound, i.e. echo, with spiritual being of sound. The general point seems to be that for there to be so little evidence of interference, when yet interference is real, there must be this sort of spiritual diffusion as well as the physical diffusion. Only thus could the gaps be filled in as fully as they are. This is to say that ultimately it is the coexistence of contraries which is lending support to the doctrine of spiritual being.

We have clearly come a long way from the *De homine*, where we were told that sound could not be compared to colour.

iv. Olfaction

Here we find considerable change of doctrine, and all in the direction of making the spiritual being of the sensible in the medium an essential feature of sensation as such. Albert takes his stand explicitly against Plato and Avicenna, and with Aristotle and Averroës.

The doctrine of Plato, Albert tells us, is that odour is the fumy evaporation of the odoriferous thing. Albert presents a sign that this is not so, based on the limits of rarefaction of bodies. Fumes are subtle parts of the fuming body, released from it. The maximum rarity possible for generable and corruptible things is that of fire. Suppose one handful of earth is rarefiable to one thousand handfuls of fire. But the parts of the odoriferous thing which fumes, the parts released into the fumes, do not amount to a handful, nor is the resultant fume as extenuated as the substance of fire. It has been explained, moreover, that sensed things are generated spherically from the source; and the odour of something is diffused in a circle for five hundred leagues at least; thus, the spherical diameters of the odour are one

---

[35] Ibid., c.19 (pp. 127.87-128.1).

thousand leagues all about, and the odour spherically diffused fills all that space. Therefore, if odour were a fumy evaporation, it would be necessary that the fumy part be rarefied by a space of one thousand leagues, not lineally but spherically, in length, breadth, and depth. So much matter does not have so much extension, even under the form of fire, and so the fumy part would be incomparably rarer than fire; which is impossible. And so it is false that odour is a fumy evaporation.

That odour sometimes extends to such great distances is shown by the case of the vultures and tiger-birds flying to the carcasses at the massacre in Greece.[36]

Albert now gives us his own judgment in the matter:

> Therefore, just as we said that every sensible *species* multiplies itself in the medium by spiritual and intentional being (*esse spirituale et intentionale*), so also does odour. And therefore, without any matter of the odoriferous thing, just the quality is diffused in the medium.[37]

We should notice how generally Albert speaks here. This clearly has become the model for the senses. And he goes on with a short series of arguments:

> For if it were diffused in fumes, then it would materially come to the sense; and then the medium would not be necessary for sense, but only for its improved condition, and we disproved this earlier; for the sense, on that account, would not be receptive of the *species* alone, according as it is a *species*, but rather according to the being which it has in matter; and later we will show that to be altogether false.[38]

The reference to earlier disproof seems to be to the places where Aristotle maintains the necessity of a medium for sight and for all the senses.[39] It looks as though the fact that the medium is needed is taken from the experience of placing the sensible directly on the organ, whereas the reason for the fact seems to be the spirituality of sense-knowledge itself, coupled with its passivity: this suggests that the sensible, while still on the side of the agent, already has spiritual

---

[36] Ibid., c.25 (p. 135.12-51). We saw the vultures earlier, at p. 299. Albert's argument is a development of that of Averroës, *In De anima* II, com. 97 (ed. Crawford, pp. 277-278) (Aristotle, 421b8-12).

[37] Albert, *De anima*, II, tr.3, c.25 (ed. Colon. 7/1: 135.55-58).

[38] Ibid. (pp. 135.58-66).

[39] Cf. above, n. 32 and the text cited on pp. 308-309.

being. This interpretation accords with the argument which Albert now adds:

> Further, the sensible would not be of itself (*per se*) the mover of the sense, but rather [would be the mover of the sense] through something of its matter; which is altogether false, as was proved above.[40]

The doctrine that it is the sensible itself which is the mover, the agent, making itself known to sense, demands that the sensible itself be already constituted in spiritual being in the medium. Albert wants to posit the sensible according to some sort of purity of being sensible, and the medium is the means of doing it.[41]

Albert assures us that it is these considerations which moved Aristotle to say that odour is not the fumy evaporation; he seems to mean the arguments concerning sensation generally, as distinct from the experiences concerning vultures, etc.

Albert now criticizes Avicenna, seen as following Plato. Avicenna, he says, conceded that the medium is not necessary for there to be actual sensation, being merely for sensation's improved condition. He argued that if some sensible can be conjoined to the sense according to spiritual being, without the contribution of a medium, then that sensible will not require a medium; and such a sensible, he claimed, is evaporation, because it is spiritual and odour is present in it according to spiritual being.

In this, Albert's interpretation of Avicenna, the expression "spiritual being" (*esse spirituale*) is Albert's. So also, Avicenna is seen as granting the general scheme of sensation, with sensation as passive and spiritual, and so as demanding spiritual being of the sensible outside the sense. The fault of Avicenna seems to consist in making the sensible itself, in its material being, also spiritual. Thus Albert goes on to mention some people who try to agree with both Aristotle and Avicenna. They say that evaporation is twofold, one material and one spiritual, and that Aristotle is denying that odour is material evaporation while Avicenna is affirming that it is spiritual evapora-

---

[40] Albert, *De anima*, II, tr.3, c.25 (ed. Colon. 7/1: 135.67-69).

[41] We saw earlier, at pp. 307-308, that it is *formal* colour, i.e., light on a surface, that is the "of itself, visible." So also Albert, in the *De anima*, carefully distinguishes between sound, the primary sensible of the ear, which is the *simple form* of sound, and what is material to it, namely the commotion of the air; this latter is not, for the most part, sensed, because the simple form of sound affects the ear according to spiritual being. Cf. *De anima*, II, tr.3, c.18 (ed. Colon. 7/1: 126.1-10).

tion. Albert counters by saying that if "spiritual" means a subtle body, then every evaporation is spiritual, but if "spiritual" means the alteration of the medium brought about by the pure and simple quality of the odour, this is certainly against Plato's conception.

Albert's own position, then, is that Plato and Avicenna have not spoken the truth in this matter; that Aristotle and Averroës are rather to be followed here: that odour is the pure and simple quality having spiritual being in the medium. Albert says it is not the experiences (he seems to mean the vultures, etc.) which convince him of this, but rather the arguments (seemingly the general considerations of the nature of sensation, assimilating odour to sound and colour). Albert points out that Avicenna has his own explanation for such phenomena as the vultures.

Lastly, Albert provides answers to some difficulties raised by Avicenna, and in doing so shows how, while not what odour essentially is, evaporation nevertheless frequently accompanies odour. Avicenna had mentioned that some odours are poisonous to animals, which seems to suggest a material invasion of the animal by the odour. Albert concedes that the odour sometimes penetrates in conjunction with evaporations of a noxious body, but he insists that what effects the odour is the alteration of the medium by the simple quality of the odour.

Also, he points out that since the evaporations are spiritual (here, he means subtle and airy), just as the medium (i.e., the air) is, sometimes the evaporation substitutes for the medium and is conveyed right through to the actual organ of smell. In the medium, then, the evaporation can be both matter for odour and also medium, but when it comes to the sense of smell, the evaporation acts on the sense not by material but by intentional being (*non. . .per esse materiale, sed per esse intentionale*), which it has in itself inasmuch as it is the medium of odour, not its subject according to material being.[42]

Thus, Albert is saying that the evaporation may well travel to the organ, but that it brings about sensation according as it has the role of medium, containing the sensible as an *intentio*. This position accounts for the sort of thing Avicenna had objected, and yet maintains the position of Aristotle.

In sum, having followed Avicenna in the *De homine* concerning odour, Albert has now turned quite against him.

---

[42] Ibid., c.26 (pp. 135.72-136.44).

v. Taste

Following Aristotle, it is explained that the sense of taste has as a medial instrument, detached from the body of the taster, the saliva. This moisture is not a medium, in the way we have seen an external body play the role of medium for the already considered three senses. If we lived in water, a sweetness mixed with the water would be experienced as a property of the water, and this is quite different from the way air or water is a medium for colour. Colour is not present in them as mixed with them. It is not present according to material being but according to intentional being only. We receive the sensible "by" or "from" the medium, but not "in" it (*potius a medio. . .quam in medio*).

Explaining further on his own account, Albert says that taste does not receive the *species* of flavour stripped of the flavoursome body, the way we have explained for the previous three senses. Taste receives flavour by the flow of the parts of the flavoursome body to it. Thus, that which is tasted is not the quality or *intentio*, but is the body, which comes to the tongue. For this reason, taste is truly a sort of touch, and it requires an internal medium which is a part of the animal. This medium is not the saliva, but rather the limits of the tongue or mouth, in which the gustatory nerves are distributed.[43]

Here Albert does not, as in the *De homine*, compare flavour to sound, and saliva to air as carrying sound: the conception of sound and its transmission is now quite different than it was there. And what we are speaking about here is the fact that the thing outside the animal affects the animal itself (at the point where it does its "tasting," i.e., the external or surface point) by a material operation. It is truly the flavoursome body, as a body, that operates. The process of spiritualization, which is essential for all sensation, has yet to commence. Thus, it is within the animal that we find the true medium, the body in which the sensible (still not yet in the sense) will acquire spiritual being.

An interesting additional remark by Albert helps to confirm what we have already seen of olfaction. Odour, he adds here, cannot be the fumy evaporation, because, since every fume is a body, if odour

---

[43] Ibid., c.27 (pp. 137.86-138.50). That which is tasted is not an *intentio*; the strong implication is that that which is seen, heard, smelt *is* an *intentio*; cf. also ibid., c.5 (p. 103.38-58). An *intentio*, of course, is a sign of the material thing: cf. above, p. 303. Within the confines of this paper, we could not deal with the general theory of knowledge implied in our texts.

attained the sense only by fumes, it would be a sort of touch, just as taste is. By this we see that Albert clearly means the corporeality of the action to be essential to taste and quite incidental to olfaction.

vi. Touch

Why must the medium for touch be internal to the animal? Albert says:

> The cause of this is that all animated sentient bodies are composed out of tangibles, by the very essence of body, and so it was necessary that, not through their *species* alone, but also through their own essences they [the tangibles] be conjoined to the animated bodies in sensing (*in sensu*); and so it was necessary that the medium, by virtue of which it senses such things, be actually part of the animated thing.[44]

I take the thrust of this to be that, in the case of touch, since the object itself, i.e., the material thing sensed, cannot be separated from the sensing thing, *a fortiori* the medium, which must come between the object and the sense, cannot be so separated.

And why is a medium necessary at all for the sense of touch? Albert says:

> . . .the cause, why the organs do not sense except from the medium, is this, that the sensitive power is in the organ, and it [the power] receives only the *species* in the case of every sense, and therefore it is necessary that in the organ there be only the *species*, and therefore sensation (*sensus*) cannot be brought about by material contact of the body, but it is necessary that the *species* be in the medium and be made to be in the organ by the medium.[45]

It is notable that here the organ is said to have only the *species* (and so does not receive materially), whereas earlier, even here in the *De anima*, both medium and organ of touch were said to be acted upon materially.[46] Also, the force of the argument is that already in the medium there is some kind of liberation of the *species* from its material ties. Indeed, what we have here is the general argument used for the other senses right from the start in the *De anima*, namely the spirituality of sensation coupled with its passivity, as demanding spirituality already on the side of the agent (the sensible thing while still

---

[44] Ibid., c.31 (p. 143.49-56).
[45] Ibid., c.34 (p. 147.8-14).
[46] Cf. above, p. 304.

outside sense). This argument was proposed and rejected in the *De homine*, but it has now become paramount for Albert.[47]

We might also note that here in the *De anima* there is no mention of the spirit issuing from the brain, as in the *De homine*, the doctrine of spiritual being of sensibles having a wholly different source here in the *De anima*, viz the proper operation of form *qua* form.[48]

vii. The Senses in General (Revisited)

Here Albert begins a new treatise or section, the fourth of the second book. In c. 2, he paraphrases the text where Aristotle discusses whether sensibles affect anything but the sense. The general answer, for Albert, is that the sensible, taken formally, affects only the sense, but the sensible, taken materially, affects other things (the air which sounds also splits the tree or sets the teeth on edge). But the problem of the medium is discussed, since the medium is not the sense and yet does seem to be affected by the sensible, taken formally.

It is pointed out, following Aristotle, that the media are indeterminate and that the *intentiones* do not last in the medium as permanent forms. The media are thus not truly and physically affected by the sensible *intentiones*; they are rather affected *in a way*.

But the problem remains: if the air and the other media do undergo this sort of influence from the *intentio*, why do they not sense? Surely, to sense is to undergo the influence of the sensible *species*? Albert answers that to smell, e.g., is not merely to undergo, but "to sense and to judge odour," such judging being the second perfection of sense, and an operating, not merely an undergoing. Thus, the sense's act of judgment is introduced as a means of distinguishing the perfect presence of the *intentio* from the sort of presence it has in a medium. This seems a somewhat more definite, if less material, answer to the question than we got in the *De homine*.[49]

---

[47] For its proposal in the *De homine*, cf. above, p. 297; for its rejection, and precisely because of touch, p. 298.

[48] While Albert does mention Avicenna's doctrine of touch as the form of the whole animal, he does not endorse it, as he seemed to do in the *De homine* (above, p. 300). Moreover, he takes considerable pains to save Aristotle's doctrine of flesh as medium of touch from Avicenna's criticism: cf. *De anima*, II, tr.3, c.34 (ed. Colon. 7/1: 147.24-82).

[49] Ibid., tr.4, c.2 (pp. 150.60-151.7). Cf. Aristotle, *De anima* II.12 (424b3-20). For the *De homine*, cf. above, p. 296. As regards this answer of Albert, nevertheless, one might wonder how merely to sense (as distinguished from judging) differs from the presence of the *intentio* in the medium.

In c. 3, in connection with Aristotle's attempt[50] to show that there are only five senses, Albert explains the media as a sort of system. We will note only a few points. First of all, he says that a medium is required for every sense. Secondly, speaking of the external media, he explains the medium of vision as a nature derived from the superior, i.e., celestial body, but the media of sound and odour are explicitly said to be based on something pertaining to the element (air or water) itself. Thus, the explanation of sound and odour is seen as, in a way, altogether subcelestial.[51]

But Albert concludes his presentation of the media with the question of the modes of being of the sensible in them. He says:

> But it must be realized that there is no local motion of the sensibles through their media, but rather . . . the sensible formally generates (*formaliter generat*) its *intentio* in every part of the medium, circularly and spherically all around. Nor is the sensible *intentio* numerically one for every part whatsoever of the medium, but diverse, as we said above also concerning the generation of sound. And this is more of an alteration than a local motion; nevertheless it is not truly alteration, by the fact that not a thing, but the *intentio* of a thing is generated in the medium, which *intentio* is not a thing firmly established in being, but rather the spiritual likeness of a firmly established thing.[52]

Albert suggests two objections to this, one being that if there is no local motion of *intentiones*, then distance ought not to impede sensation, and we ought equally quickly to sense the near and the far. The other is that if there is alteration in the medium, and every alteration involves some change in the subject, then the media ought to be changed.

In answer to the first, Albert cites the philosopher al-Biṭrūjī (Alpetragius),[53] to the effect that every efficient power is more effective for action close up than for action on things farther off, and says that accordingly sensibles generate their forms in proportion to proximity.

The answer to the second is more satisfying, at least from the viewpoint of our own present inquiry. Albert concedes that the

---

[50] Aristotle, *De anima* III.1 (424b23-25a14).

[51] Albert, *De anima*, II, tr.4, c.3 (ed. Colon.7/1: 151.74-152.31).

[52] Ibid. (p. 152.32-42). The ending in Latin runs: "quae intentio non est res rata in esse, sed potius ratae rei similitudo spiritualis."

[53] Concerning whom cf. al-Biṭrūjī, *De motibus celorum*, ed. Francis J. Carmody (Berkeley and Los Angeles, 1952).

media are changed, in a way, though they are not changed, speaking unqualifiedly. This is because the influences which affect the media are not things but certain *intentiones* of things; to which Albert adds the following precision:

> ...especially in vision; but in hearing, sound has material being in air, and by reflexion it also has spiritual being; odour has intentional being at a distance and up close has also something of material being. Tastables and tangibles are joined to the media by matter, but nevertheless the medium does not actuate in the organ anything but intentional being; and so it appears that as the sensibles are related to the medium for the other senses, so in touch and taste the media are related to the sense.[54]

This presentation sums up the situation we have been exploring, bringing out as much materiality as possible. We should underline that it is "reflexion," i.e., echo, which manifests the spiritual being of sound, and also that the organ of touch is said to receive only intentional being.

C. CONCLUSION

Without repeating the individual differences we have noted along the way, we will here call attention to the main features of Albert's shift in doctrine. The difference in the two accounts is not extreme. In both treatises a special mode of being of certain sensibles is recognized, and, more important, the spirituality of sense being affirmed, the necessity for an immaterial mode of action by the sensible on the sense is asserted.

In the *De homine*, the question of the mode of being of the sensible in the medium seems to lack unity. There is a doctrine of "spiritual being" peculiar to the visible, and in contrast to this everything else is viewed as material. Again, there is a doctrine of "spiritual being" proper to the sensible in the sense, and in contrast to this the being in the medium is an intermediate mode, called "sensible being"; however, in the case of hearing and olfaction, this intermediate mode shows itself only in action (so that these instances of special being are rather weakened). Again, there is the "spiritual being" of the tangible, which seems to mean its sensible being; this also seems to manifest itself in action, and to flow from the soul itself.

---

[54] Albert, *De anima*, II, tr.4, c.3 (ed. Colon. 7/1: 152.57-66).

In the *De anima*, the term "spiritual being" (as well as "intentional" and "sensible" being) is quite definitely established as applicable to the sensible in the medium,[55] and not merely to the sensible in the sense. And while the variety to be found in the spiritual being of the diverse sensibles is stressed, nevertheless the source of spiritual being, in general, is identified as form *qua* form. This makes the doctrine of the intrinsically sensible character of sensible things a main feature of the *De anima* presentation.

Reconsideration of the phenomena of sound and colour has made it possible for Albert to give greater play to the general theory of sense in his account of the particular senses, and so to present a more unified doctrine, though he is constantly careful to bring out what is proper to each sense and sensible.

Symptomatic of the general trend of his thinking in this matter is what we have called the "backing up" (as one backs up a vehicle) of the two modes of action from the sensible in the medium to the form in the thing. What in the *De homine* were accounts of sound and odour apply in the *De anima* more to the case of touch. Thus, in the *De anima* Albert tells us that as the thing is to the medium for sight, hearing, and olfaction, so the medium is to the organ and the sense for taste and touch.

In closing, I would like to indicate something of the perennial interest which I believe the just summarized doctrines of St. Albert have. Their interest does not lie primarily in their being a presentation of the notion of spiritual or intentional being: that notion is probably best met with in the context of the direct contrast between mind or sense, on the one hand, and being or nature, on the other. Rather, the foregoing doctrines have their interest as pertaining to an ecology of knowledge, in particular of sense-knowledge: i.e., the conception of the knower as fitting into a world, a knowable world. How different is Albert's schema of sensation from, e.g., Robert Grosseteste's,[56] where beings or their forms radiate just one sort of influence, and where the difference between mere material reception

---

[55] This use of the term "spiritual being," not only concerning colour (as in the medium) but also for odour (and implicitly for sound) is to be found in Averroës, *In De anima* II, com. 97 (421b8-12) (ed. Crawford, pp. 277-278); one finds there also (Crawford 278.68-72) the doctrine of colour being more spiritual than odour.

[56] At least as regards the texts indicated above in n. 28; we should note the complexity of texts of Grosseteste on sensation, in which, e.g., great play is given to vision as *sending out* power from the eyes. Cf. *De iride* (ed. L. Baur, *Beiträge* 9 [1912], p. 73.5-10).

(e.g., becoming warm) and the reception which is sensation (e.g., feeling warmth) is explained entirely in function of the thing receiving. The knowing being is, as it were, an anomalous atom of awareness in the midst of an otherwise dark or silent (or, more precisely, extra-sensible) world. Moreover, the knowing being's knowledge is entirely action; it is domination of the surrounding milieu. In contrast, Albert's knowing being, i.e., the sensitive animal, is the terminus of influences themselves special, belonging to an order of sense and the sensible right from their roots in the sensible beings themselves. The senses allow a being to partake of a peculiar mode of communicability belonging to the world at large. Indeed, sensation is a passivity, an undergoing the influence of a world already in itself communicable.

Again, the primary interest of these questions of the spiritual being of the sensibles in the medium is not, to my mind, their being a way of explaining corporeal reality precisely where it defies sensible observation (e.g., the coexistence of contraries), but rather their being the occasion to isolate for consideration the peculiar agency (and accordingly the peculiar ontology) of the sensible as such. The medium provides a kind of theatre in which one can consider the extent to which the sensible is a formal order unto itself, a communion of entities somewhat diverse from the merely natural or material communion.

12

Albertus Magnus' Universal Physiology: the Example of Nutrition

Joan Cadden
Kenyon College

Albert's early admirers were guilty of no hyperbole when they called him the Universal Doctor. The meaning of this title refers, no doubt, to the wide range of his intellectual activity, one small part of which this volume celebrates. His investigations of physiological subjects reflect his universal nature: he is concerned with virtually every known aspect of the life processes, and is familiar with virtually all the available sources. Here too we find a different and equally characteristic universality: in writing about living things, one of his goals is to articulate the general, all-encompassing principles of the subject. This chapter will consider Albertus Magnus' views on nutrition as an illustration of the universal character of his treatment of animate creatures. The subject of nutrition offers a fruitful and important example for a number of reasons. First, throughout the period of Albert's greatest productivity in natural philosophy, he concerned himself repeatedly with various aspects of the topic. Thus substantial sections of his commentaries on Aristotle's *On the Soul* and *On Generation and Corruption*, and of his monumental works *On Plants* and *On Animals* are devoted to it. In addition, he wrote a brief original treatise about it, *On Nourishment and What Can Be Nourished*.[1] Sec-

---

[1] In this paper we shall consider mainly the following works of Albertus Magnus: *De anima libri III* (ed. Colon. 7/1 [ed. C. Stroick, 1969]), abbrev. as *De an.; De generatione et corruptione libri II* (ed. Borgnet, [Paris, 1890] 4: 345a-457b), abbrev. as *De gen. et corr.; De vegetabilibus libri VII* (ed. Meyer and Jessen [Berlin, 1867]), abbrev. as *De veg.; De animalibus libri XXVI* (ed. Stadler, *Beiträge*, 15, 16 [1916, 1920]; abbrev. as *De animal.; De nutrimento et nutribili* [part of the *Parva naturalia*] (ed. Borgnet [Paris, 1891] 9: 324-341), abbrev. as *De nutr*. As to the chronological order of these writings of Albert, see J. A. Weisheipl, "Life and Works of St. Albert the Great," above, pp. 31, 35-36, 38, and below, Appendix 1, pp. 567-573.

ond, nutrition plays a crucial role in Albert's view of life. It is the most fundamental of the body's functions, because it is common to all creatures, even the most rudimentary, and so it is central to the definition of life *per se*.[2] It continues throughout the life of the individual, and when the creature's power to assimilate food fails, it dies.[3] Third, nutrition is the primary operation of the vegetative or nutritive soul, in the sense that the other operations, growth and generation, depend upon it. For it is by the assimilation of food in certain quantities that a plant or animal grows, and reproduction requires seed or semen which the soul creates from nutriment.[4] And finally, Albert displays in his treatments of nutrition the full colors of his virtuosity in the use of sources.

Because Albert's accounts of the process by which bread is turned into flesh vary according to the nature of the works in which they appear, I shall attempt to present a general view of the subject, based on all the works mentioned, giving some attention, however, to some important differences among the treatments. Albert considered, among other things, the purpose, the preconditions, and the process of nutrition. We shall treat each in turn, focusing on the process of nutrition, and evaluate the advantages and the difficulties of being a "Universal Doctor."

The purpose or final cause of nutrition is to preserve the individual. Living things are made up of a mixture of the elements, and heat is prominent among the qualities of their bodies. It is in the nature of heat, no matter how benign, to consume the substance in which it resides, so the natural heat is continually using up and drying out the body. If unreversed, this action would result in death: the fluids, on which, as we shall see, life depends, would be used up, and the heat, instrument of the soul, would run out of fuel.[5] But "nature, mindful of her work, lest it perish," thought up nutrition as a means of restoring what had thus been lost.[6] In addition to restoration, its primary

---

[2] *De an.* II, tr.2, c.2 (ed. Colon. 7/1: 84v.15-21).

[3] *De gen. et corr.* I, tr.3, c.15 (ed. Borgnet 4: 387a-b); *De nutr.* tr.1, c.5 (ed. Borgnet 9: 333b-334a).

[4] *De nutr.* tr.2, c.1 (ed. Borgnet 9: 336a-b) and c.2 (339a-b); *De an.* I, tr.2, c.13 (ed. Colon. 7/1: 53 v.55-57).

[5] *De nutr.* tr.1, c.4 (ed. Borgnet 9: 331b-333b); *De an.* II, tr.2, c.6 (ed. Colon. 7/1: 90-91); *De gen. et corr.* I, tr.3, c.8 (ed. Borgnet 4: 383b-384a).

[6] *De nutr.* tr.1, c.4 (ed. Borgnet 9: 332b): "... natura sollicita ne periret opus ejus...."

purpose, the assimilation of food bestows two secondary benefits upon the creature. First, it renovates, for, "just as in animals that breathe, air held from a long time cannot be made useful for the operation of respiration, so too in animate things matter which is kept for a long time becomes hard and is rendered unable to perform the operations of organic life."[7] Animals which shed shells or feathers and then grow new ones attest to the importance of the renewal provided by food. If it did not occur, the same cooling and drying might endanger life. The other boon bestowed by nutrition is a fortification of the body's matter, so that it is not so quickly consumed and threatened by its own heat.

Like natural heat, which is both necessary and dangerous to life, nourishment has limitations, even disadvantages, as well as benefits. However great its contributions to the reparation of bodily parts, which have suffered depletion, desiccation or decay, and to a balanced, humid vehicle in which the natural heat can subsist, nutrition — the incorporation of matter from outside the body — cannot restore or replace the radical fluid (*humidum radicale* or *seminale*), whose functions include the perpetuation of the body's proper form. This moist substance, of which more is said below, is of particular importance in the maintenance and reproduction of complex animals; its gradual weakening is irreversible, and eventually fatal.[8] Indeed, Albert occasionally speaks almost pessimistically: the process is like adding water to the well of a lamp which is running out of oil. (Death, after all, is a natural phenomenon with a natural explanation.)[9] Although he generally portrays the transformation of food as complete and perfect, he does use Aristotle's metaphor of diluting wine with water,[10] and warns us that food, even after conversion, may retain harmful properties: "The blood generated from lettuce is cold, and the flesh will be cold."[11] Closely connected as it is

---

[7] Ibid. (332a): ". . . sicut in respirantibus diu tentus aer non efficitur utilis ad operationem respirationis, ita etiam in animatis materia diu tenta duratur et efficitur inhabilis ad vitae organicae operationes. . . ."

[8] Ibid. (332b-333a); *De gen. et corr.* I, tr.3, c.8 (ed. Borgnet 4: 383b-384a). See below, p. 325-326. For background on the radical fluid, see Michael McVaugh, "The '*Humidum Radicale*' in Thirteenth-Century Medicine," *Traditio*, 30 (1974), 259-283.

[9] *De gen. et corr.* I, tr.3, c.8 (ed. Borgnet 4: 383b-384a). See Peter H. Niebyl, "Old Age, Fever, and the Lamp Metaphor," *Journal of the History of Medicine*, 26 (1971), 351-368.

[10] Aristotle *De gen. et corr.* I.5 (322a31-34). Albert, *De gen. et corr.* I, tr.3, c.6 (ed. Borgnet 4: 380a).

[11] *De gen. et corr.* I, tr.3, c.15 (ed. Borgnet 4: 387b): ". . .sanguis enim de lactuca generatus, frigidus est, et caro erit frigida. . . ."

with the basis of life and with causes of death, nutrition has an important place in Albert's study of nature. He takes great care, therefore, to define the conditions under which it can and cannot occur.

Not everything can be nourished, only those objects which have souls and which have bodies with particular characteristics. In these animate bodies reside the final, formal, and efficient causes of nutrition. A material cause, food, is also required, and Albert argues that this too must be specifically suited to the process: one cannot, after all, feed upon stones. The insistence that the elements involved in nutrition be appropriate to their roles derives in part from general Aristotelian metaphysics, according to which there is a special relationship between the form and the matter of a compound substance (among other things, the parts of the body are made for the sake of their form, the soul), and in part from Albert's general experience of particular conditions under which nutrition is actually observed.

Only objects with soul, that is to say, living creatures, are nourished and grow. Soul is the defining principle of life, and although some inanimate objects, such as air expanding, may appear to be nourished and to grow, they do not in fact do so.[12] Only the most rudimentary form of soul, the vegetative soul, is necessary for these processes to occur, for it is precisely over the functions shared by plants, animals and human beings alike (nutrition, growth, and reproduction) that this aspect of soul presides. Because the vegetative soul is the very essence of life *per se* on the philosophical plane, and because nutrition, growth, and reproduction are fundamental to life on the physical plane, Albert's concern with them reflects his interest in the nature of life itself. While this interest is often manifested in such works as *On Plants* and *On Animals* in terms of variety and with attention to specific differentiation, in his work on nutrition and growth it appears in the more general terms of universal functions. Although Albert does occasionally refer to the differences among plants and animals, he most often emphasizes the identity or equivalence of the processes in all living things. This is precisely because, in terms of their essence with respect to nutrition, that is to

---

[12] Ibid., c.5 (379a-b) and c.11 (385b-386a); *De an.* II, tr.2, c.5 (ed. Colon. 7/1: 89 v.35-56).

say, with respect to their possession of a vegetative soul, all physically living creatures are equal.

Since, in nature, there is no form without matter, there is no vegetative soul without body. And both because of its relation to the soul and because of its relation to the particulars of nutrition as it actually occurs, the body of the animate being must have certain characteristics. From Albert's writings on the subject there emerge a set of specifications for the living body: (1) it must have the proper qualitative balance, with the hot, the moist, and the subtle holding sway over the cold, the dry, and the earthy; (2) it must have certain fluids, a seminal fluid, a nutrimental fluid, and blood or its analogue, which function as the vehicles of both form and matter in nutrition; and (3) it must have specific structural or anatomical properties, such as a stomach (or something that serves as a stomach) and a kind of porosity. The rationale for this composite picture is drawn from a variety of sources, including observations, specific theoretical problems, and more general philosophical concerns.

Because of his concern with the formal aspects of the assimilation of food, he requires that the living body possess some mediator between form (soul) and matter, some vehicle by which the essence of the creature is preserved in the course of the changes it undergoes during nutrition. He invokes the elusive medical constructs, virtues and faculties,[13] but his interest in the material aspects of the natural world as it actually exists leads him away from reliance on such abstract mechanisms, and he adopts the notion of a radical or seminal fluid. This is called "radical" because it is the original and most fundamental fluid in the body, and it is called "seminal" because it contains the germ of the form that determines the actuality of the creature. (It is substantially the same as the seminal fluid involved in reproduction.) Specifically, the radical fluid carries or embodies the formative virtue of the soul, the power which resides in the various members, and enables them to convert the product of the earlier stages of nourishment into their own likeness. This product Albert calls the nutrimental fluid, and it serves both to supplement and to undermine the radical fluid. These two fluids are essential to the assimilation of food: the radical as a characteristic and instrument of the plant or animal, the nutrimental as the last form taken by the food before it is completely converted and also as an imperfectly inte-

---

[13] *De an.* II, tr.2, c.6-9 (ed. Colon. 7/1: 90-95); *De nutr.* tr.1, c.2 (ed. Borgnet 9: 327b).

grated component of the body itself. If, as I suggested above, Albert was looking for a naturalistic response to the question, What mediates between the soul and the body in nutrition and growth? then his particular formulation of the response was undoubtedly guided by his experience of the general moistness of living things, as well as by the apparent role of particular fluids in their survival and development.

The observed connection between life and moistness is one of the many reasons Albert supported traditional notions of the physical properties of living things. All plants and animals can be seen to have some moisture in them — even the hardest old oak tree and the crunchiest insect — and all may be seen or assumed to require the intake of moisture for their survival.[14] Most important for our purposes is the role that moisture plays in digestion, in the distribution of the product of digestion, and in other aspects of the assimilation of food. Related to moistness is the property of subtlety. It is contrasted with earthiness, and shares with the moist the characteristic of fluidity. The fluid has no natural shape or boundary, so that moist and subtle substances can flow easily and can be adapted to the forms of the bodily members which they permeate. (Crass, earthy substances, on the other hand, tend to remain in fixed shapes, and so are neither easily diffused nor able easily to adapt to other forms.) Thus, though the body must have some earthiness for other reasons (such as the permanence and solidity of the parts), from the point of view of nutrition it must have moist and subtle fluids. The notion of subtlety probably also derives from Albert's feeling that the operations of so noble a principle as the soul must be carried out by the most refined instruments possible, and from the commonplace observation that life and breath are essentially linked in higher animals.

The presence of fluids and the properties of moistness and subtlety have reasonable foundation in the observation of *all* living things, except for the implied association of subtlety and breath. The necessity for another crucial quality, heat, is clearly derived from higher animals. Here as elsewhere, Albert, following Aristotle's habit, takes his model from warm blooded animals. But because nutrition is the *shared* operation of all living things, Albert is insistent on applying the model universally, even in the face of the fact that plants and

---

[14] *De animal.* xx, tr.1, c.1 (ed. Stadler, pp. 1273-1275); *De nutr.* tr.1, c.1 (ed. Borgnet 9: 325a-b) and c.5 (334b); *De veg.* iv, tr.1, c.1 (ed. Meyer-Jessen, p. 213, para. 4 and p. 214, para. 6).

lower animals are not warm to the touch ("The heat in plants is moderate").[15] Aside from its empirical association with higher animals, the notion of heat recommended itself to Albert for a number of reasons. The analogy with cooking, which is the basis of the Galenic account of digestion, reinforced the importance of heat. Appropriate because of its association with food and its role in visible substantial change, the idea of coction provided a strong explanatory image for Albert, as it had for his predecessors.[16] Albert's emphasis on heat as a precondition for nutrition is further supported by the importance of heat in other branches of science, such as alchemy and mineralogy, which deal with analogous operations.[17] While never hesitating to compare, even, indeed, to identify the heat requisite for nutrition with the heat that effects changes in the inanimate world,[18] he explicitly puts limits on this apparently mechanical model by distinguishing among the various powers of heat operating in nutrition, some of which are unique to living things: (1) the ordinary fiery power of heat, able to alter things, (2) the power of heat operating in the fluid of the body's members, able to digest by boiling, (3) the power it has from the various members to bestow their forms upon the nutriment, (4) the power from the heavens, especially the sun, which affects the body's humors, and (5) the principal power which it has from the creature's soul.[19] The physical properties of the living body, warmth and moistness, are sustained by and perhaps embodied in the radical and nutrimental fluids, which thus play an important theoretical role in Albert's model of nutrition. Those fluids do not seem to be directly observable: Albert normally identifies them by their operations and effects. They are sometimes associated with a perfectly observable fluid, the blood. Furthermore, according to the standard system of four humors, blood is warm and moist. Thus blood (or its analogue) unifies in its substance several physiological principles of life in general, and nutrition in particular.[20]

---

[15] *De veg.* II, tr.1, c.1 (ed. Meyer-Jessen, p. 108, para. 14: ". . . calor in plantis est modicus. . . ." Plants get helping heat from the sun and the earth: ibid. (para. 13); *De nutr.* tr.1, c.2 (ed. Borgnet 9: 326b-327a).

[16] *De nutr.* tr.1, c.3 (ed. Borgnet 9: 331a); *De animal.* XX, tr.1, c.4 (ed. Stadler, p. 1281 v. 19-22).

[17] *De animal.* XX, tr.1, c.4 (ed. Stadler, p. 1282 v. 5-15).

[18] Ibid. (pp. 1281-1284); *De veg.* IV, tr.1, c.1 (ed. Meyer-Jessen, p. 213, para. 5).

[19] *De nutr.* tr.1, c.3 (ed. Borgnet 9: 330b). See also *De an.* II, tr.2, c.3 and 4 (ed. Colon. 7/1: 86-88).

[20] *De animal.* III, tr.2, c.6 (ed. Stadler p. 339-342) and IX, tr.2, c.3 (p. 717 v.17-p. 718 v.9).

Albert does not, however, give blood a prominent place in his discussions of the assimilation of food.

Albert's system of nutrition requires not only that a body have a soul, certain fluids, and specific physical qualities, but also that it have particular anatomical structures. In general, living creatures have organs, that is, instruments through which the soul carries out its operations. Because nutrition is the most rudimentary process, it requires only the most rudimentary organs: a mouth through which the food can enter, a stomach (or equivalent receptacle) in which the food can be transformed, and pores to allow the digested nutriment to permeate the various parts of the body. As we shall see in Albert's description of the conversion and assimilation of food, creatures of the higher orders have other members, such as the liver, which assist in the conversion; in plants the simplest forms suffice.[21] While something resembling a mouth and something resembling a stomach could easily be identified in most living things, and the necessity of an opening and a vessel for the food needed no elaborate justification, the functions of the pores were less obvious. The existence of tiny channels and spaces in the substance of the body was a matter of general agreement; the exact function of these pores in the assimilation of food was a matter of long and subtle debate. Albert took Averroës' treatment as his point of departure on the subject.[22] The Commentator criticizes Alexander of Aphrodisias' position on the relation of the formal and material aspects of growth; Albert takes up the standard and shifts the battle to a new front. As Albert understands Alexander, one aspect of Alexander's error regarding the material conditions of growth is grounded in a misconstrual of what happens to the nutriment once it has been distributed to the interstices of the body.[23] Yet Albert does not disagree with the anatomical view, which he attributes to Alexander, that "nature gave channels to the body which grows, which are veins and pores and sponginess."[24] For, as he says elsewhere, otherwise the food would have no way to

---

[21] *De nutr.* tr.1, c.2 (ed. Borgnet 9: 327b-328a); *De an.* II, tr.2, c.8 (ed. Colon. 7/1: 93 v.56-p. 94 v.4); *De veg.* II, tr.1, c.1 (ed. Meyer-Jessen, p. 107, para. 11-p. 108, para. 13).

[22] Averroës, *Commentarium medium in Aristotelis De generatione et corruptione libros*, ed. Francis Howard Fobes, Corpus Commentariorum Averrois in Aristotelem, versionum Latinarum, 4.1 (Cambridge, Mass., 1956), pp. 48-55.

[23] *De gen. et corr.* I, tr.3, c.8 (ed. Borgnet 4: 382b-384b). Albert's reading of Alexander goes beyond what Averroës says in the passage referred to.

[24] Ibid. (383a): ". . .dedit natura corpori quod augetur vias quae sunt venae, et poros, et spongiositatem."

be distributed to the various parts for its final conversion.[25] Indeed, like the other structural features of bodies that grow this one is essential, and, like the others, its full definition and meaning reside in its function. For Albert there was a teleological relation between the structure of the living body and its vital operations, to which we shall turn after a brief consideration of the requirements for that by which creatures are nourished: the food.

In spite of the importance of the physical features which Albert attributes to the thing that grows, what distinguishes it from other things is not that it has a body, but that it has a soul, a principle of life; not that it is the subject of change, but that it possesses the motive principle to change itself naturally. In contrast, what is important about food is its passivity and materiality: it is operated upon in the process of nourishment. Nevertheless, according to Albert, if stuff is to serve as food it must meet certain requirements, and, since it does have formal and qualitative characteristics, these may affect the results of nutrition. In the most general terms, food is potentially the thing which is being fed (the creature); it is actually separate and different from — even contrary to — that which is nourished, but it has the possibility of becoming one with and the same as that which is nourished.[26] It must also be potentially a certain quantity, not simply because all body has extension, but more specifically because whether the nourished body diminishes, remains the same size, or grows, depends upon quantitative relations. These general characteristics are, as we shall see, the foundations for Albert's abstract consideration of the nature of nutrition and growth.[27] But there are more concrete properties of food which Albert explores, some of which bear upon the conversion of the nutritive matter into flesh, bone, and other parts, and some of which bear upon the effect of the conversion on the living creature. It follows, for example, from the nutriment's potential similarity to what is nourished, that simple substances cannot serve as food. For, Albert argues, a substance can properly be called "potentially similar" only if the similarity can be actualized by a single motion, that is, if the change can be effected simply. But for an unmixed or elemental material to be transformed into flesh or bone, two motions

---

[25] *De nutr.* tr.1, c.2 (ed. Borgnet 9: 327b-328a).

[26] Ibid., c.1 (324a); *De gen. et corr.* I, tr.3, c.9 (ed. Borgnet 4: 385a); *De an.* II, tr.2, c.5 (ed. Colon. 7/1: 88 v.63-89 v.15).

[27] *De gen. et corr.* I, tr.3, c.12 and 13 (ed. Borgnet 4: 386a-b). See below, p. 331-334.

would be necessary: one from the simple into a compound (brought about by the motions of the heavens), and a second from that mixed body into the thing being nourished (brought about by its soul).[28] Not all compounds, however, are appropriate for the nourishment of living things, and in explaining why rocks and metals cannot serve as food, Albert gives us a sense of some of the ways in which the characteristics of the food are limited by the demands of the nutritive process and of how a living thing may be affected by what it eats. Stones and metals are too close to the elements to be food; in other words, they have not fully enough achieved the complexity of a compound. We see evidence of this in the operations of the alchemists, who can easily change one metal into another, in the same way that the simple elements can be transmuted one into the other. Such bodies are dominated by the qualities of earth and water — they do not possess the moist warmth they would need in order to be transformed into the warm moist parts of the animate being. Thus a certain qualitative affinity is necessary before nutrition can occur. A related obstacle to eating stones and metals is that the moist heat of the body is incapable of dissolving them to prepare them for conversion: metals can only be dissolved by dry heat, and stones cannot be dissolved at all. Finally, such substances contain too much sulphur (we see here a shift from Aristotelian to alchemical principles), which is contrary to the natural complexion of living things. In short, in order for the conversion to occur and for the creature to benefit, food must be complex and have particular qualities which can sustain life: "The more similar food is to that which is being nourished, the more effectively it nourishes and the more swiftly it is converted."[29] Thus it is that people with choleric temperaments like dry and bitter foods, while phlegmatics prefer fish (cold and moist). Implied, although not elaborated, is a general dietary theory: things that are close to us make healthy foods; things that are too far removed from the very temperate human complexion are unhealthy. Fruits are bad for us, for they are earthy, and rot in our stomachs or our veins. But for plants, which are themselves earthy, an earthy food, such as manure, is excellent.[30]

---

[28] *De nutr.* tr.1, c.1 (ed. Borgnet 9: 324a).

[29] Ibid. (325b): "Oportet generaliter scire, quod quanto nutrimentum similius est ei quod nutritur, tanto efficacius nutrit et citius convertitur."

[30] Ibid. (323a-326b).

No matter how appropriate a certain food is for the nourishment of a certain individual, it is still only *potentially* similar to that which it is destined to nourish. It is the actualization, the operations by which the living body with its carefully defined characteristics assimilates the food appropriate to it, to which we now turn. In his summary remarks about what takes place during nutrition, Albert declares in his treatise *On Nourishment*: "Thus it is said that nutrition has a triple sense, namely of quadruple digestion, and of the mover and [its] instrument, and of the ways of boiling or baking."[31] He is referring not to a series of separate operations, nor to different types of nutrition that occur under different circumstances (in different orders of animals, for example), but rather to three ways of analyzing a single universal process. Let us examine each of these, taking first the Aristotelian picture of nutrition as a motion involving a mover and its subordinate instruments, turning next to the medical account which focuses on a series of digestions at a series of anatomical sites, and ending with the mineralogical or alchemical model of boiling and baking (elixation and assation — the application of moist and dry heat).

Seen as a motion, nutrition fits the general Aristotelian formula for change: it is an actualization of a potential. The food starts out dissimilar to what it is nourishing, with, however, an incipient similarity (the likeness discussed in the description of food above) which is perfected in the course of the transformation, rendering the product first increasingly similar to what it is becoming (the flesh or bone of a particular animal, for example), and finally completely similar to it.[32] Although in many ways digesting and assimilating food is a feature of matter,[33] Albert, following Aristotle, insists on the formal aspects: especially when he is speaking in terms of motion, he emphasizes that this is a transformation — a change of the food's form, and a change in the nourished creature according to its form. (Freed from the constraints of commentary in his *On Nourishment*, Albert gives a more balanced treatment of the formal and material aspects of nutrition and growth than he does in the *Generation and Corruption*, in which he earnestly applies himself to the elaboration of Aristotle's

---

[31] Ibid., c.3 (331b): "Sic igitur dictum est, quod triplicem modum habet nutritio, scilicet ex digestione quadruplici, et ex motore, et instrumento, et ex modis elixationis vel assationis."

[32] Ibid. (329b).

[33] See, for example, ibid., c.1 (326a).

insistence on form.)[34] If the transformation of food for the fortification and restoration of a plant or animal is a motion that may be considered in formal terms, what kind of motion is it? In the *Generation and Corruption*, where nutrition is secondary to and defined in terms of growth, Albert creates a certain amount of confusion for a reader trying to answer this question. Nutrition, he says, is the same as growth "in essence and subject, but different in being." If we are speaking about food becoming flesh, we are speaking of nutrition; if we are speaking about food becoming a certain amount of flesh, we are speaking about growth.[35] This relation makes general sense: the thing that grows and the thing that is nourished are the same — a living creature — and the event that occurs, the assimilation of food, is also the same. Only the result is different, in that growth involves a change of size. But as Albert characterizes growth in *Generation and Corruption*, the difference is not an insignificant one. In distinguishing between growth and other types of change, he discusses three grounds for distinction: (1) the kind of change, (2) the manner of change, and (3) the subject of change. Growth differs from generation, which is a change from potential substance to actual substance, and from alteration, which is a change from potential quality to actual quality: growth is a change from potential magnitude to actual magnitude.[36] Thus, with regard to the kind (*genus*) of transmutation, the defining characteristic of growth is not shared by nutrition, which is defined without regard to magnitude. Similarly, when Albert discusses the manner (*modus*) of change, the distinguishing feature of growth is not shared by nutrition. Growth, he says, differs from generation and alteration in that it necessarily involves local motion (the expansion of the grower in space).[37] The third ground for distinguishing growth from alteration and coming-to-be allows — if only implicitly in the *Generation and Corruption* — for the inclusion of nutrition. Albert points out that while in generation, something that did not previously exist comes into being, in alteration and growth what changes existed before and persists after

---

[34] *De gen. et corr.* I, tr.3, c.8 (ed. Borgnet 4: 381b-384b) and Aristotle, *De gen. et corr.* I.5 (321b22-35); *De nutr.* tr.1, c.3 (ed. Borgnet 9: 330a-331a).

[35] *De gen. et corr.* I, tr.3, c.13 (ed. Borgnet 4: 386b): ". . . et nutrimentum cum augmentatione ex parte cibantis et augentis id idem est in essentia et subjecto, esse autem est aliud: secundum enim quod est adveniens potentia quanta caro, est augmentum carnis: secundum id autem quod solum potentia caro, est nutrimentum."

[36] Ibid., c.1 (375a) and c.3 (377b-378a).

[37] Ibid., c.1 (375a-b).

the change.[38] Later, he speaks of the conversion of food as an alteration, emphasizing the change of the food and the persistence of the creature which is the subject of the change.[39] In speaking of nutrition and growth in such terms, he momentarily evades the problem that he has created for himself elsewhere in the work: that his definition of nutrition is based on his definition of growth, which in turn relies on a characteristic (quantitative change) not necessarily shared by nutrition. The price he pays for this temporary solution is the loss of the distinction, on which he has previously insisted, between growth and alteration.

Although he does not pose the problem explicitly in his later treatise *On Nourishment*, Albert is clearly unwilling to adopt definitions of nutrition that would either subordinate it to another type of change (indeed, growth is rather subordinated to nutrition here) or attach it inappropriately to the category of quality. When, therefore, he speaks of nutrition as a motion in *On Nourishment*, he turns to and improves the imperfect solution just discussed. He maintains the relation between nutrition and growth established in his *Generation and Corruption* — nutrition is the actualization of potential flesh, growth the actualization of a certain amount of potential flesh[40] — but he softens the distinction by making growth a particular case or result of the more general phenomenon of nutrition. For example, he says that when nutrition occurs, the thing being nourished sometimes does grow and sometimes does not,[41] and that nutrition involves quantity, in the sense that the amount of food consumed is always potentially some indeterminate quantity of the members being nourished.[42] Nutrition without growth thus seems to involve only the quantity necessary to sustain or restore the creature, while nutrition with growth involves some additional quantity. Albert does not, however, rely solely on the introduction of the idea of quantity into the definition of nutrition to solve the problem of distinguishing it from generation and alteration:

> The change of nutriment cannot be generation, although it is substantial change, since in generation a new form is brought out from poten-

---

[38] Ibid., c.5 (379a). He goes on, however, to separate growth from alteration on the grounds that the former involves quantity, the latter quality.

[39] Ibid., c.6 (379b-380a).

[40] *De nutr.* tr.1, c.4 (ed. Borgnet 9: 333a), where he refers to *De gen. et corr.*

[41] *De nutr.* tr.2, c.1 (ed. Borgnet 9: 337a-b).

[42] Ibid. (336b-337b). See also *De an.* II, tr.2, c.6 (ed. Colon. 7/1: 90-91).

cy, and what is generated has new being. Here, [in the case of nutrition,] however, no new form is drawn from potency, since a thing remaining [the same] with respect to form is nourished, and the nutriment receives that form. But neither can it be said to be alteration, since in that the subject remains the same in act, under various accidents with respect to the same substantial being.[43]

Thus nutrition shares with generation substantial change, in that something which was not flesh or bone becomes flesh or bone; it shares with alteration substantial changelessness, in that the dog or the laurel (the thing being nourished) remains the same dog or laurel. In nutrition the form is the same, though the matter is different; in generation, the matter is the same and the form is different; in alteration both the form and the matter are the same.[44] Faced as he was with the problem of working within the general Aristotelian framework, according to which nutrition was regarded as a type of motion or change, Albert radically revised the definitions and distinctions which he had used, however inconsistently, in his commentary on the *Generation and Corruption*. Placing nutrition in a position of primacy with respect to growth, he clarified and modified the role of quantity; taking into consideration both the food and the living thing, he was able to account for the similarities of nutrition to generation and alteration without obscuring its distinctness. The modern reader may be dismayed at the care Albert took to clarify the concept of nutrition, but the breadth of his interests and the standards of thirteenth-century natural philosophy required the integration of philosophic principles and naturalistic explanations.

If nutrition is a kind of motion, then it follows that it is caused by some mover. Here the case is less confused, if no less complicated. The primary mover is the nutritive soul of the living creature.[45] (The heavens have a role in the process in the sense that they preside over the creation of compounds which may eventually come to constitute food, and may, as Plato says, have an important function in the assi-

---

[43] *De nutr.* tr.1, c.3 (ed. Borgnet 9: 331a): "Et ideo generatio motus nutrimenti esse non potest, et tamen est motus substantialis: quia in generatione quidem nova forma de potentia educitur, et esse novum habet id quod generatur. Hic autem nulla nova forma educitur de potentia: quia manens secundum formam nutritur: et illam formam accipit nutrimentum. Sed nec alteratio dici potest: quoniam in illa manet subjectum idem actu sub diversis accidentibus secundum idem esse substantiale."

[44] Ibid.

[45] Ibid. (330a); *De an.* II, tr.2, c.2 (ed. Colon. 7/1: 85 v.52-63); *De gen. et corr.* I, tr.3, c.8 (ed. Borgnet 4: 383a).

milation of food.)[46] The nutritive soul is the changeless cause of change whose substance is unaffected by its actions or their products.[47] It effects the operation by bestowing upon the nutriment the form of the existing creature: the food receives the form from (though not of) this mover, thus becoming assimilated to some part of the living body which is itself the instrument of the soul.[48] The nutritive or vegetative soul has at its disposal a series of secondary movers through which it does its work. Although Albert is eclectic in his choice of instrumental efficient causes, the general pattern of primary and secondary causes of motion is wholly in keeping with the Aristotelian view of change, as is Albert's elaborate abstract description of the process in terms of beginning, middle and end, potentiality, imperfect act, and actuality, "as established in the *Physics*."[49]

Of the three ways of looking at the assimilation of food, the framework of motion and mover just discussed allows Albert the greatest opportunity to relate a specific physiological phenomenon to universal philosophical principles, such as cause, motion, and soul. Yet, as we see in the notion of intermediate or instrumental cause, such abstract analysis does not preclude, but rather invites the examination of more specific features of the process, and it is to the two more concrete ways of regarding nutrition — "quadruple digestion" and "boiling and baking" — we now turn. The first of these, digestion, Albert treats in standard medical terms, though the general idea and some of the specific principles, such as heat, are not, of course, the exclusive property of any school. In different contexts, he emphasizes different aspects of digestion. Thus, in his *Generation and Corruption*, where he is interested in the actions of the nutrimental and radical fluids, he is concerned primarily with the last phase of digestion. In his commentary on Aristotle's *On the Soul* he discusses the natural faculties (*virtutes naturales*) of the nutritive soul — borrowed from Galenic sources — and the spirit (*spiritus*) through which they operate,[50] and he rehearses the standard medical description of the four digestions. The first is initiated in the mouth and completed in the stomach. The juice thus extracted from the food then goes to the liver, where blood is the product of the second digestion. The

---

[46] *De nutr.* tr.1, c.1 (ed. Borgnet 9: 324a) and c.3 (330a-b).
[47] *De gen. et corr.* I, tr.3, c.6 (ed. Borgnet 4: 380a).
[48] *De nutr.* tr.1, c.3 (ed. Borgnet 9: 330a).
[49] Ibid. (329a): ". . . sicut in *Physicis* est determinatum."
[50] *De an.* II, tr.2, c.8 (ed. Colon. 7/1: 92-94).

blood goes then to the veins, where it undergoes the third digestion. Finally, the fourth digestion, which takes place in the various members of the body, transforms the results of the previous digestions into the likeness of the parts.[51] At each stage, residues are drawn off, and it is the successive purifications that Albert later emphasized in the treatise *On Nourishment*: first, in the stomach (or in the earth, in the case of plants), solid impurities are removed; next, in the liver (or in the roots), the liquid superfluities are drawn off; third, in the veins (or in the stalks and branches), there is further refinement; finally, the individual members by their specific powers perfect the digestion by assimilating the nutriment.[52] From this variation on the theme of four digestions, he moves to another, in his still later *On Animals*. He starts, as he did in *On the Soul*, with the picture of the four digestions, but elaborates it so as to strengthen its connections with both the medical and the Aristotelian traditions. By the inclusion of medical terminology, such as "chyle" for the product of the first digestion, and by the integration of the genesis of the four humors, he solidly associates his understanding of digestion with works of Galen and Avicenna.[53] In the same passages, however, he places himself equally squarely in the Aristotelian mold by painstakingly enumerating the formal, final, efficient, and material causes of each of the humors,[54] and by insisting on the primacy of the heart as a source of heat.[55]

To this multiplicity of modes of describing what happens to food in the body, Albert adds (particularly in his later works) notions borrowed from mineralogy. From this area, he derives a striking image of purification by heat:

> In the arts, when gold is dissolved and boiled down, those things which are not of the nature of gold, such as rocks and intermingled metals, are separated from the gold (similarly rusts and the like), and the gold is drawn off and raised up at the same time. In living things, thus we see everything fetid is separated from the nutrimental chyme, and what is gross is separated by withdrawal.[56]

---

[51] Ibid. (93 v.64-86). He refers to Galen and Avicenna on a related topic, ibid., c.6 (91 v.16).

[52] *De nutr.* tr.1, c.3 (ed. Borgnet 9: 329b-330a).

[53] *De animal.* III, tr.2, c.4 (ed. Stadler, pp. 330-334) and XII, tr.1, c.6 (pp. 827-831). Albert refers to Galen (p. 330), and Stadler notes connections with Avicenna's *De animalibus* and *Canon*.

[54] *De animal.* III, tr.1, c.4 (ed. Stadler, p. 332 v. 13-34).

[55] Ibid., XII, tr.1, c.6 (ed. Stadler, p. 828 v. 28-29).

[56] Ibid., XX, tr.1, c.4 (p. 1282 v. 5-10): "In artibus quidem quando aurum digeritur et decoquitur ea quae non sunt de natura auri sicut lapides et immixta metalla, separantur ab auro: similiter autem et rubiginosa et huiusmodi: et aurum attrahitur simul et elevatur. In naturalibus autem sicut videmus omne fetulentum separari a chymo nutrimentali et id quidem quod est grossum separari per secessum. . . ."

In *On Nourishment* he develops the connection more specifically and systematically, elevating it to the rank of one of the three manners of nutrition, along with motion and four-fold digestion. The processes which he calls "elixation" (boiling) and "assation" (baking or roasting) effect the final digestion: the assimilation of the thrice prepared nutrimental fluid to the particular forms of the body's individual members. Having passed through the veins and been divided up in the pores of the various members to facilitate conversion,[57] the nutriment is subjected to the specific heat which varies according to the nature of the part in question:

> All preparation of food is similar either to elixation, which is called *epsesis*, or to assation, which is called *optesis*. . . . And that [heat] which acts in assimilating soft [parts], such as flesh and the like, is similar to *epsesis*, because it works by heat which is in moisture, which is drawing the moisture from what is being cooked. That [heat] which assimilates the nutriment to the hard parts in earthy creatures, such as bones, is similar to roasting, which is *optesis*.[58]

Albert refers us to the fourth book of the *Meteorology* as the source of some of these ideas; he may also have in mind his own or his contemporaries' mineralogical and alchemical studies. Thus the third manner in which Albert examines nutrition has origins and emphases distinct from the other two.

The blending of explanations from various sources is characteristic of Albert's work on living things, especially in his later writings. It is most common at the level of particular phenomena. Uncertain as Albert seems to have been over the years about the exact definition of nutrition and its relation to the definitions of related processes, the terms in which the problems posed themselves and the range of possible answers were limited by the boundaries of Aristotelian ideas about substance, change, and categories. The specific mechanisms of digestion, considered in their anatomical and physiological particulars, seem to have allowed Albert greater freedom in his use of sources, and resulted in the rich, if sometimes frustrating, eclecticism of his studies of nature.

---

[57] *De gen. et corr.* I, tr.3 c.8 (ed. Borgnet 4: 384b).

[58] *De nutr.* tr.1, c.3 (ed. Borgnet 9: 331a-b): ". . . omnis praeparatio nutrimenti est similis aut elixationi quae vocatur epsesis, aut assationi quae optesis appellatur. . . . Et illa quae fit in assimilando nutrimentum mollibus, ut carni, et hujusmodi, similis est epsesis, eo quod illa fit a caliditate quae est in humido, quae est humidum evocans ab eo quod decoquitur. Illa autem quae assimilat nutrimentum partibus duris terrestribus animatis, ut ossibus, similis est assationi quae est optesis."

In some ways, Albertus Magnus gives a full and coherent account of the most fundamental of life's functions, nutrition. The conjunction of form and matter, soul and body, provides a general philosophical context for the specific biological process. The universal principle that the members of the body are suited to the operations they perform as instruments of the soul, seems to guide Albert in matching the physical requirements for nutrition with the ways in which food is made into flesh and bone. Thus the radical and nutrimental fluids make possible the perfection and preservation (as well as reproduction) of the individual, the one by acting as a vehicle of form, the other by embodying and distributing the increasingly perfected food. The qualities of the body likewise serve the purposes of the nutritive soul, the warm as soul's instrument, the humid as the crucial ingredient of the proper kind of primary coction (boiling) and as the basis of the radical and nutrimental fluids. The various anatomical structures also exist for the sake of specific functions, for the stages of digestion and for the phases of purification. In speaking of nutrition, Albert therefore subsumes the multiplicity and variety of nature which he so admires under a single pattern: although, for example, he takes care to elaborate the peculiarities of pores and vessels in plants, he does so in order to show their general equivalence to similar structures in animals.[59] Albert's goal is clearly to order and unify the innumerable aspects of nutrition upon which he touches, yet there is a sense in which, particularly regarding the details of the natural world, his generous and all-encompassing mind preferred inclusiveness to simplicity. When, in the mid-1250s, he looked forward to writing about nutrition and related subjects in books on plants, animals, and nutriment,[60] a goal which he had accomplished within a few years, did he anticipate the multiplication of sources, topics, interpretations which they came to encompass? These works, with their wealth of digressions, their voracious inclusiveness, must — in contrast to the more restrained commentaries on Aristotle's *On the Soul* and *On Generation and Corruption* — be regarded as the mature culmination of Albert's work on nature. It would be improper, therefore, to regard the difficulties which these works present to the modern reader as difficulties in the eyes of Albert and his contemporaries. The citation and explication of many authors, the inclu-

---

[59] Ibid. c.2 (327a).
[60] *De an.* II, tr.2, c.10 (ed. Colon. 7/1: 95 v.63-68).

sion of all available traditions, the juxtapositions of parallel or alternative descriptions and interpretations are all reflections of the universality of Albert's intellect, so admired by his contemporaries and posterity.

13

Albert on the Natural Philosophy of Plant Life

Karen Reeds
University of California at Berkeley

In the seven books of *De vegetabilibus* Albert saw himself as accomplishing three tasks. First, he was paraphrasing and explicating Aristotle's treatise, *De plantis*,[1] as part of his larger endeavor to comment on all of Aristotle's works in natural science. Second, he was reworking the contents of *De plantis* (and other sources of knowledge about plants) into a more coherent, orderly form. Third, he was providing a kind of catalogue of the properties of individual plants and an account of the peculiar features of cultivated plants.

---

[1] For convenience, I will follow Albert's practice, standard in the medieval Latin West, of ascribing *De plantis* to Aristotle. Ernst H. F. Meyer showed that the work was probably written by Nicolaus Damascenus (1st century BC), in his edition of the Latin version of *De plantis* by Alfred of Sareshel: *Nicolai Damasceni De plantis libri duo Aristoteli vulgo ascripti* (Leipzig, 1841). All my references to *De plantis* will be to Meyer's edition. See also Sybil D. Wingate, *The Mediaeval Latin Versions of the Aristotelian Scientific Corpus, with Special Reference to the Biological Works* (London, 1931; reprint, Dubuque, Iowa, n.d.), pp. 55-72, 98-103; E. S. Forster's English translation of Meyer's edition of *De plantis* in *The Works of Aristotle Translated into English*: Vol. 6, *Opuscula*, ed. W. D. Ross (Oxford, 1913). W. S. Hett's English translation of the medieval Greek version of *De plantis* in Aristotle, *Minor Works*, Loeb Classical Library (Cambridge, Mass., 1936) should not be used as a guide to the text Albert knew.

A set of questions by Roger Bacon on *De plantis* refers to Alfred of Sareshel's translation and to an otherwise unknown second version. See *Opera hactenus inedita Rogeri Baconi, Fasc. XI: Quaestiones supra de plantis*, ed. Robert Steele and Ferdinand Delorme (Oxford, 1932) and Wingate (above), pp. 60-68. Professor R. James Long has recently tried to determine which of these two translations Albert used. Like Wingate, he feels that "it cannot be demonstrated conclusively that Albert used the Alfred text." He also notes that there is no sign that Albert knew Alfred's gloss on *De plantis* (personal communication — Professor Long plans an edition of Alfred's gloss).

In the course of carrying out these three tasks, Albert revealed himself to be a very close observer of plant life. We are indebted to the German botanist and historian of botany, Ernst H. F. Meyer, for first drawing attention to the accuracy and originality of Albert's botanical observations.[2] Meyer's two long essays in *Linnaea* in 1835-1837, his discussion in his immense *Geschichte der Botanik*, and above all his critical edition of *De vegetabilibus* (completed by Karl Jessen in 1867) remain the foundations of all subsequent work — including this paper — on Albert's botany.[3] As a botanist, Meyer was especially pleased to find in *De vegetabilibus* descriptions of plants that were so detailed and vivid that he could often identify the particular species Albert had seen. His enthusiasm over Albert's observations of plant morphology and patterns of growth has been shared by all later students, and by far the greater part of work on *De vegetabilibus* has been devoted to pointing out just how keenly Albert had looked at living plants.[4]

However, Albert did not value his observations nearly so highly as his modern readers have done, and he certainly did not value them

---

[2] Conrad Gesner, the one Renaissance botanist and historian of botany who might have been expected to recognize the merits of Albert's *De vegetabilibus*, was apparently only acquainted with the treatises falsely attributed to Albert: *De virtutibus herbarum* and *De mirabilibus mundi*. These Gesner dismissed curtly: "Let him who wishes to spend good hours badly read these." See Gesner's *Praefatio de rei herbariae scriptoribus* in Hieronymus Tragus, *De stirpium. . .commentariorum libri tres* (Strassburg, 1552), sig. c ii r°. For similar eighteenth-century misconceptions about Albert's botanical work, see Meyer, "Albertus Magnus, Ein Beiträg zur Geschichte der Botanik in dreizehnten Jahrhundert," *Linnaea* 10 (1835/6), 641-653.

[3] "Albertus Magnus, Ein Beiträg," *Linnaea* 10 (1835/6), 641-741; "Albertus Magnus. Zweiter Beiträg zur erneuerten Kenntniss seiner botanischen Leistungen," *Linnaea* 11 (1836/7), 545-565; *Geschichte der Botanik* (Königsberg, 1857), 4: 1-83; *Alberti Magni De vegetabilibus libri vii*, ed. E. H. F. Meyer and Karl Jessen (Berlin, 1867); Books i-v take up pages 1-338; Books vi-vii, pages 339-660. All references to *De vegetabilibus* in this paper are to the Meyer and Jessen edition; I have followed their identifications of plants.

[4] See, for example, Heinrich Balss, *Albertus Magnus als Biologe* (Stuttgart, 1947), pp. 79-187; J. Wimmer, *Deutsches Pflanzenleben nach Albertus Magnus (1193-1280)* (Halle a.S., 1908); Stephan Fellner, *Albertus Magnus als Botaniker* (Wien, 1881). For discussions of Albert's theoretical botany, see Adam Paszewski, "Les problèmes physiologiques dans *De vegetabilibus et plantis libri vii* d'Albert von Lauingen," *Actes du XIᵉ congrès international d'histoire des sciences* (Cracow, 1968), 5: 323-330; *A Sourcebook in Medieval Science*, ed. Edward Grant (Cambridge, Mass., 1974), pp. 689-700, selection 87, tr. and annot. Edward Grant, and selection 88, tr. Charles Singer, annot. Edward Grant. (I am indebted to Prof. Frank Egerton for the reference to Paszewski.) Brigitte Hoppe's admirable monograph on ancient and early modern theories of plant growth and nutrition unfortunately devotes only a couple of pages to Albert's ideas: *Biologie, Wissenschaft der belebten Materie von der Antike zur Neuzeit: Biologische Methodologie und Lehren von der stofflichen Zusammensetzung der Organismen* (*Sudhoffs Archiv*, Beiheft 17, 1976), pp. 176-179.

in the same way. To Albert, the knowledge of individual plants was subordinate in importance and interest to the understanding of the phenomena of plant life in general. It takes only a cursory glance through De vegetabilibus to realize that it was intended to be a treatise on the natural philosophy of vegetable life rather than a work of descriptive botany. Although all seven books of De vegetabilibus contain observations which have justifiably aroused the admiration of Meyer and his successors, only Book VI gives a systematic set of discussions of single plants. And Book VI was clearly something of an embarassment to Albert. He knew it was an unphilosophical departure from his plan for the work, and he apologized for it by saying that he had only compiled the information in it (and the equally unphilosophical Book VII) "to satisfy the curiosity of students rather than philosophy."[5]

Since Book VI (especially the first tractate on trees) is where most of Albert's detailed observations are found, it is not surprising that his modern commentators felt none of his misgivings. Meyer, for example, wrote, "To our joy and to his own honor, in the sixth book he came down a bit from these philosophical heights and spread his manifold knowledge of plants before us."[6] T. A. Sprague dismissed the theoretical books of the treatise even more flatly: "Now, after a lapse of nearly seven hundred years, it is his descriptions of species that are important, both for identifying the plants concerned and for ascertaining the precise connotation of the terms employed by him. His general philosophy of the Plant World on the other hand, is now of relatively little interest."[7]

Nevertheless, now, after exactly seven hundred years, I want to redirect our attention to Albert's own stated purposes for De vegetabilibus and to take up once again the long-neglected concerns of the first five books. I do not wish to deny the originality of Albert's observations — I know of nothing else in medieval herbal literature that comes anywhere close to them — but I do believe it is important to know what uses Albert made of them in Books I-V.

---

[5] Book VI, tr.1, c.1, para. 1 (p. 339): "In hoc sexto libro vegetabilium nostrorum magis satisfacimus curiositati studentium quam philosophiae. De particularibus enim philosophiae esse non poterit. Nos autem in hoc sexto libro proprietates quasdam intendimus ponere, quae particularibus plantis convenire videntur." Jerry Stannard discusses Books VI and VII in his essay in this volume, below pp. 355-377.

[6] Geschichte der Botanik, 4: 47.

[7] "Plant Morphology in Albertus Magnus," Kew Bulletin 9 (1933), 431-440. See also Sprague's "Botanical Terms in Albertus Magnus," Kew Bulletin 9 (1933), 440-459.

A. The Organization of Books i-v

Although Albert took *De plantis* to be a genuine treatise of Aristotle's, he must have found it a dismaying piece of work to have come from the philosopher's hand: poorly organized, confusingly written, and too brief to cover the subject. On the whole, Albert was polite about the flaws of the work, speaking of "a certain confusion" and an "imperfect treatment" of the philosophy of plants and ascribing these to the ancients in general rather than to Aristotle in particular.[8] He blamed the unintelligibility of various passages on the translator and went quickly on to remark, "But it is known that Aristotle wanted to say. . . ."[9] To remedy these deficiencies, Albert provided not only his customary close paraphrase of Aristotle's text but also his own formulation of what Aristotle had really meant to say about the natural philosophy of plants. Books i and iv are largely devoted to the exposition of the text; Books ii, iii, and v to Albert's "digressions" which set the subject in order.

Albert introduced the work as a whole by describing its place in his series of commentaries on Aristotle. He had, at this point, just finished discussing the treatises on the universal principles and faculties common to the souls and bodies of all living things. Now it was time to philosophize about the particulars of the bodies of living things, their parts, and the operations peculiar to them. For two reasons it seemed best to begin with those living things which were animated only by the vegetative soul, i.e., plants. They were especially influenced by the movements of the heavens and the qualities of the elements — and thus were closely tied to the subjects of the first books on natural philosophy that Albert had commented upon — and they were simpler than the creatures which possessed both a vegetative and sensitive soul. Albert insisted here, as he would throughout the book, that he intended to treat the bodies of plants as a totality, according to the parts that were common to them all,

---

[8] Book ii, tr.1, c.1, para. 1 (p. 103): "Haec omnia tradita sunt ab antiquis physice de plantis loquentibus, et videntur quandam habere confusionem." Book v, tr.1, c.1, para. 1 (p. 289): "Philosophia de plantis imperfecte tradita ab antiquis multimoda est valde."

[9] Book iv, tr.4, c.1, para. 138 (p. 277): "Et haec est scientia Aristotelis de coloribus lignorum, quae propter malitiam translationis vix est intelligibilis. Sed sciendum est, Aristotelem velle dicere, quod ligna sunt quaedam nigra et quaedam alba." See also Book i, tr.1, c.12, para. 88, 90 (pp. 45-46).

because "as Plato well said, particulars are infinite and there can be no science of them"[10]

After this preface, Albert began to expound Aristotle's first book of *De plantis* which dealt with two issues: the question of the existence of life (and some of its attributes) in plants, and the diversity of kinds of plants. Albert followed his text carefully, but interpolated freely to make antecedents clear, to sharpen the meaning of ambiguous terms, to provide references and transitions, to supply examples, and to fill in the missing steps of arguments. Here is a brief example of Albert's technique of explication applied to part of "Aristotle's refutation of those who said that plants are perfect and sleeping" (Aristotle's text is in italics):

> *For the plant does not crave sleep because* it is *bound* to the earth in which it always rests and thus does not crave the quiet of sleep. *Nor has it movement through which* it is moved *by itself. Nor* has it *the fixed shape* peculiar to each kind of animal in whole and *in parts*, for we see that each kind of animal differs from the others in shape just as it differs from them in kind. However, many plants tend toward one shape both in whole and in parts (as, for instance, in roots and in branches and in leaves) as all pears and all apples, for example, even though they are of different species. *Nor*, we repeat, *does* a plant *have senses*, as was proved above, *nor* has it *voluntary movement, nor* has it *a perfect soul* but it has *only a part of a part of a soul.* And we will show the reason for all of these below.[11]

As even this small example shows, the process of paraphrase and expansion could not completely clarify the difficulties posed by the text. After going through the first part of Book I of *De plantis*, Albert

---

[10] Book I, tr.1, c.1, para. 1-6 (pp. 1-4), especially para. 6: "Propter igitur hanc causam incipiendum est a corporibus plantarum. De quibus in hoc libro intendimus secundum totalitatem et partes ipsarum communia — quaecunque sunt plantis convenientia — prosequentes: eo quod particularia sunt infinita, nec eorum, sicut Plato bene dicit, potest fieri disciplina."

[11] Book I, tr.1, c.8, para. 52 (p. 29): "*Planta enim non indiget somno, eo quod alligata* est terrae in qua semper quiescit; et ideo quiete somni non indiget. *Nec habet motum,* quo per se moveatur; *nec* habet *figuram terminatam,* alicui speciei animalis propriam in toto et *in partibus,* sicut videmus, omnem speciem animalis ab alia differre in figura, sicut differt in specie. Plurimae autem plantae unam figuram praetendunt tam in toto quam in partibus, sicut in radicibus et in ramis et in foliis, sicut omnes piri et omnes mali, cum tamen sint diversarum specierum. *Neque* iterum planta *habet sensum,* sicut superius probatum est, *neque* habet *motum voluntarium, neque* habet *animam perfectam,* sed *tantum habet partem partis animae.* Et horum omnium rationem inferius ostendemus" (Aristotle's text in italics; Meyer and Jessen set off Aristotle's text in *sperrdruck*). Cf. *De plantis,* Book I, c.7, lines 15-18 (p. 12). On the meaning of *species,* see below, n. 28.

declared unhappily: "Everything that has been said since the beginning of this book — excepting what we said on our own account in the first chapter — seems to be quite obscure."[12] The fault lay, he judged, in the translators of the work who either did not understand the philosopher or did not perfectly know the language they were translating. Therefore, Albert decided to recapitulate and summarize all that had been said so far "because then we will teach better what we understand, and the words of the philosopher will be clearer."[13] Immediately he introduced order into the discussion by listing the six problems that were at issue:[14] (1) the souls of plants; (2) the powers of the soul of the plant which were exercised through the body of the plant, such as desire, sensation, and nutrition which some people attributed to plants; (3) phenomena like sleep and waking which were initiated by the body and concluded by the soul (again attributed to plants by some people); (4) sex; (5) the perfection of the plant (which some thought greater than in animals); and (6) the mode of life in plants (a question which the ancients put first).

Albert's chief contribution — beyond clarity of expression — to the discussion of these problems was to provide examples and explanations of the phenomena which could have given rise to the various opinions held by Aristotle and the other ancient philosophers he cited. On the problem of sleeping and waking, for example, Albert noted that some plants seem to go to sleep because their flowers fold up at night. The reason for this was not that they slept, but that the cold around them compressed the humors within them so that the flowers were drawn shut; in the daytime the sun's heat relaxed and expanded the humors within the plant so that the flowers spread open again.[15] To the discussion of sex in plants, he added specific instances of kinds of plants which were commonly called male and female — male and female *pyonia*, male and female olive — and

---

[12] Book I, tr.1, c.9, para. 58 (p. 32): "Omnia autem, quae a principio libri hujus dicta sunt, satis obscura videntur esse, praeter ea sola, quae in primo capitulo ex nostra sententia tradidimus."

[13] Ibid.: "Hanc autem obscuritatem accidisse arbitror ex vitio transferentium librum Aristotelis de plantis, cujus ego sum interpres et relator in capitulis inductis. Aut enim non intellexerunt philosophum, aut forte idioma, ex quo transferre debuerunt, non perfecte cognoverunt. Et ideo summatim, que dicta sunt a principio, recapitulanda sunt et clarius dicenda. Tunc enim et melius docebimus hoc, quod intelligimus, et clariora erunt verba philosophi."

[14] Book I, tr.1, c.9, para. 59 (p. 32): "Omnia autem quae a principio sunt dicta, ad sex reducuntur problemata."

[15] Book I, tr.1, c.11, para. 81 (p. 42).

pointed to the particular features that seemed to justify these names: in the one kind, masculine properties were expressed in the narrowness of the leaves and smallness of the seeds; in the other, female properties appeared in the width of the leaves and the quantity and ripeness of the fruit.[16] He followed Aristotle, however, in regarding such evidence of sexuality in plants as mere accidents.[17]

From general questions about the soul and life in plants in the first part of Book I of De plantis, Aristotle moved abruptly to a long list of the kinds of diversity that could be seen on the bodies and parts of plants. To ease the transition from the one subject to the other, Albert reminded his readers that, without knowledge of the properties of the soul, we could not understand the bodies of plants; to understand the nature of plants fully, we should next inquire into their anatomy.[18] Albert faced two difficulties with this section: first, he had observed more diversity in the plant world than Aristotle apparently had, and second, Aristotle had pretty much dodged the task of explaining the causes of such diversity. Albert's solution was to continue his close paraphrase of the text for the remainder of Book I of De vegetabilibus and then to "digress" at length in Books II and III, following his own ideas and plan rather than Aristotle's about the origins of variation in the parts of plants.[19] The first tractate of Book II described the parts common to and essential to all plants: roots, stems, knots, veins, pith, bark; the second tractate described the parts which assisted in fructification: the leaves and flowers. Because seeds and fruits (and their flavors) reached perfec-

---

[16] Book I, tr.1, c.12, para. 84 (p. 44). The leaves had a reproductive function in that they helped protect the fruit, so it is not surprising to find their qualities being used as a sign of sex differences in plants.

[17] Book I, tr.1, c.7, 12 (pp. 5, 8). Part of c.7 is translated by Grant, Sourcebook in Medieval Science, pp. 691-692, selection 87.

[18] De plantis, Book I, c.8. De vegetabilibus, Book I, tr.1, c.14, para. 109 (p. 55): "Et ex his patere potest satis intellectus omnium eorum, quae de anima plantae dicenda erant; sine quibus corpora plantarum cognosci non poterant, eo quod anima principium est cognitionis corporis animati...." Book I, tr.2, c.1, para 110 (p. 56): "Quaerendo autem de corpore plantarum, oportet nos, via naturae procedere. Secundum autem hanc viam principia composisi sunt partes, ex quibus componitur. Per illas enim habet cognosci, quia compositum, ut eleganter Aristoteles dicit, cognoscimus, quando scimus, ex quot et qualibus compositum est. Unde sicut anatomia, quae divisio vocatur, cognoscuntur animalium corpora; ita per divisionem corporum plantarum cognoscitur natura corporum plantarum. Et ideo de partibus est considerandum primo omnem partium diversitatem."

[19] Book I, tr.2, c.1, para. 110 (pp. 56-57): "Tamen prius assignabimus tantum referendo diversitates has; et postea revertemur, assignando causas omnium diversitatum. Si tamen non Aristotelem, sed nos ipsos sequeremur, pro certo aliter procederemus."

tion only after falling off the plant, the causes of diversity in them were different enough from the causes working in the non-separable parts to warrant devoting Book III wholly to the special questions they raised.[20] The particular problems which interested Albert in these long digressions on plant anatomy were the details of the material composition of the plant parts, their component parts, their shapes, textures, smells, flavors, and colors. (Below we will discuss Albert's use of his observations in explaining the diversity shown in parts of plants.)

In Book IV Albert turned from plant morphology to plant physiology. Adopting the same procedure as before, he gave this book over to the phrase-by-phrase reading of Aristotle's text, Book II of *De plantis*, and then allowed himself to refashion and elaborate the material in the following book (Book V). The contents of Book II of *De plantis* were, if anything, even more of a confusing miscellany than Book I. It successively worried about the expression of the four elements in the material substance of plants, the origins of rivers, earthquakes and sand, the reasons why wood, leaves, oil, and some stones float on water, the places and conditions in which plants grow well or poorly, the growth of trees, the functions of leaves and thorns, the colors of parts of trees, the forms plants take as they grow, the falling of leaves and the ripening of fruits. Albert ran through this jumble of topics rather more briskly than he had done for Book I. Now and then he answered objections which might be raised against Aristotle,[21] or added a significant detail — leaves floated on water only so long as they stayed green[22] — or gave a concrete instance of a general statement — *juncus* and *gladiolus* exemplified the way swamp plants resembled one another in form.[23] He also made one notable observation on the difference between true

---

[20] Book III, tr.1, c.1, para. 1 (pp. 163-164).

[21] For example, to the possible objection that stones, herbs, wood and all such mixed things in which earthiness predominated should sink in water, Albert replied that although earth provided the greater quantity of matter in such things, the admixture of other elements, especially air, bestowed an upwards motion which overruled the earthiness (Book IV, tr.4, c.2, para. 24 [pp. 223-224]).

[22] Book IV, tr.1, c.2, para. 21 (p. 222): "Sed *oleum omne* et *folia* plantarum, quamdiu viridia sunt, *aquae supernatant* nisi aqua intrans per longitudinem in folio omnino ex eis excludat aërem; tunc enim emerguntur."

[23] Book IV, tr.2, c.4, para. 80 (p. 250): "*Si qua* autem *vicositas* sive vaporositas *spiraverit*, proveniet *planta stagnorum*, sicut persicaria aut juncus aut aliquid hujusmodi, et plantae, quae *non multum differunt in figura*, sicut juncus et gladiolus." Cf. *De plantis*, Book II, c.8 (p. 37).

thorns and mere prickles.[24] But, on the whole, Book IV leaves the impression that Albert simply wanted to get through it as quickly as possible.

The second book of De plantis was especially dismaying because it so often touched on fundamental problems of the essential operations of plants and then veered off again. Albert had good reason for declaring at the start of Book V that "the philosophy about plants had been imperfectly treated by the ancients."[25] For, even after going through the Peripatetics' opinions about the diversity of plants and their essential properties, there still remained plenty to say about the things that were common to plants or that distinguished among them and about the marvels of their methods of reproduction, not to mention the effects they have on the bodies of animals.[26] Although Albert did not say so, it must have surprised him that the author of De anima had not used De plantis to consider at length the noteworthy phenomena of plant growth and reproduction.

Although Aristotle had missed the obvious opportunity to discuss these matters, Albert did not. Once again he began by enumerating the possible variations on the general theme.[27] Each plant agreed with every other plant in two respects: a process of generation and a common material substance. The successful generation of any plant required seven things: the vivifying heat of the heavens, the mediating heat of the particular place in which the plant grew, the natural heat within the seed, the intrinsic moisture of the seed, the ministering moisture of the earth around it, the nourishing moisture of the rains and dews, and the tempering air surrounding the plant. The material substance of the plant needed four characteristic qualities: the proper mixture of elements, the presence of the soul, the necessary quantity of matter, and the appropriate form. Once these requirements for generation and material substance were satisfied, plants showed a remarkable diversity. Albert spelled out four different ways one plant could unite with another, two ways in which one plant could be divided into several plants, three ways in which the

---

[24] Book IV, tr.3, c.3, para. 111-117 (pp. 264-267). Sprague, "Plant Morphology," pp. 434-435.

[25] Quoted above, note 8.

[26] Book V, tr.1, c.1, para. 1 (pp. 289-290): "Licet enim jam dictum sit de diversitatibus plantarum et de virtutibus earum essentialibus secundum dicta Peripateticorum, tamen adhuc restat dicere de plantarum convenientia et differentia et unitione et divisione et permanentia et transmutatione, quae valde mirabilia in plantis inveniuntur; adhuc autem de mirabilibus effectibus earum, quos operantur in corporibus animalium diversorum."

[27] Book V, tr.1, c.1 (pp. 289-293).

species of a plant could be maintained, five ways in which one *species* could be changed into another by transmutation.[28] Moreover, each of the four complexions — hot, moist, cold, and dry — also worked in half-a-dozen or more ways to alter the substance, powers, and effects of the plant.

To round off his account of the physiological variability of plants and perhaps also to make the transition from the first five theoretical books to the sixth book on the properties of individual plants, Albert listed very briefly some miscellaneous operations of plants: over twenty different medicinal effects plants could have on other creatures; the rather mysterious power of some plants to mimic others in color, shape, flavor, or effect; and in the very last few lines, the magical properties ascribed to some plants.[29]

B. THE PROBLEM OF DIVERSITY AND THE USES OF OBSERVATIONS

From this outline of Books I-V, it will be clear that Albert's own contributions to the natural philosophy of plants came almost entirely in his long "digressions," i.e., in Books II, III, V. Of all his concerns in these books, the most prominent and important is the problem of diversity in plants. There were a few attributes held in common by all plants — a rather hidden source of life, a material substance in which the qualities of earth predominated, a set of quite simple organs which served the nutritive, augmentative, and generative functions of the vegetative soul — but, beyond these essentials, plants showed all sorts of differences among themselves which needed explaining.[30]

It is important to realize that Albert's concern with understanding the diversity of plants did not lead him into taxonomy. He had no intention of arranging the members of the plant kingdom in hierarchies of *genera* and *species*, although he followed Aristotle in recog-

---

[28] Book V, tr.1, c.3-8. Part of c.7 on the five methods of transmutation of *species* is translated by Grant, *Sourcebook in Medieval Science*, p. 699, selection 87. See also the discussion in Paszewski, pp. 326-328. *Species* and *genus* are not used with our modern technical taxonomic meanings, of course. They both carry the sense of "a group of plants with something in common." Often, Albert's examples of a *species* or *genus* do coincide with a recognized modern taxon, but they may equally well refer to plants which he classed together by categories we do not use in taxonomy today.

[29] Book V, tr.2, c.6, (pp. 336-338).

[30] Book II, tr.1, c.1, para. 2-19 (pp. 104-110).

nizing five major classes of plants which ranged from the most perfect plants, trees, down to the least perfect, small herbs with leaves coming directly from the root, with shrubs, bushes, and large, leafy-stemmed herbs falling in between.[31] These *genera*, Albert readily acknowledged, could be subdivided into smaller *genera* and *species*, but these he declined to discuss:

> If we took each of them individually, it would need more than a book just to put them down by name. But we seek only their causes, according to philosophy, and not to enumerate their diversity one at a time. For there is no way to talk about such individuals in philosophical terms.[32]

To obey this precept, then, Albert could not present a series of unconnected observations of botanical phenomena or individual plants, no matter how interesting they might be in themselves. Every observation and every mention of a particular plant had to be subordinated to the general discussion of the nature and causes of some general feature of plant life. The question of diversity, and especially the diversity of plant organs, however, afforded Albert plenty of chances to use his observations to improve upon *De plantis* and to illustrate the range of possible variations. It was then permissible to cite individual *species* as examples of the variations. Aristotle had provided a limited model of this procedure in the second half of *De plantis*, Book I, and a much more helpful and extensive exemplar in the books on animals.[33]

Albert's chapter on the shapes of leaves is especially rich in origi-

---

[31] Book II, tr.1, c.2, para. 20-28 (pp. 111-113). Fungi were a problematical sixth class of plants. Aristotle often discussed them in *De plantis* as if they were the class of most imperfect plants. Here Albert concluded that they were nothing more than a certain exhalation which evaporated from rotten wood and congealed into mushrooms and the like in the coldness of the air: "nisi quaedam exhalatio humoris, ex putredine ligni vel alicujus alterius commixti et putridi evaporans, et ad frigus aëris constans et coagulata" (para. 21 [p. 111]).

[32] Book II, tr.1, c.1, para. 28 (p. 113): Haec autem genera plantae, quae dicta sunt, plurima sub se habent alia genera et species, quae, si ponantur per singula, modum voluminis excederet, etiamsi nomina solum ponantur plantarum. Nos autem secundum propositam philosophiam non quaerimus nisi causas eorum, quae in plantis apparent, et non enumerationem diversitatis earum per singula. Singula enim talia dicere non est philosophicum. Sufficiant igitur ea, quae dicta sunt de divisione generis plantarum. See also Book I, tr.2, c.5 (pp. 75-79) (parts translated by Grant, *Sourcebook*, selection 87, pp. 694-695).

[33] There are very few explicit references in the first five books to the books on animals — presumably because Albert did not want to refer to works he had not yet commented on — but the discussion of the parts of plants closely resembles Aristotle's treatment of the parts of animals, both in the organization of the whole work and in the use of examples.

nal observations and thus a good place to see how he put his personal knowledge of plants to work.[34] The only comment on shapes of leaves in *De plantis* distinguished briefly between narrow (*stricta*) and broad (*diffusa*) leaves, like grape leaves. In his commentary on this passage in Book I, Albert had supplied additional examples of the very narrow leaves of juniper and pine and the very broad leaves of maple and banana.[35] Now, in Book II, Albert approached the question of leaf shapes from the point of view of the causes of their diversity. He first established that in general leaves were formed from an ill-digested mixture of a watery vapor and a dry earthy excrement which was exuded or exhaled from pores in the stem, propelled by a certain generative, formative power within the plant.[36] He then went on to show how these facts about the material composition and process of formation could explain the variety of figures exhibited by leaves.

He began by describing the simplest and most common form of leaf, found in many perfect plants (i.e. trees). This had the shape of two equal arcs arising from a single straight line at the bottom stem-end of the leaf and converging to a point at the upper tip. This shape was easy to understand. As the substance of the leaf emerged from the pore in the stem, the watery humor made the leaf spread out; but the insufficient quantity of matter in the leaf made it constrict again, and the action of the heat moving the matter outwards and upwards pushed it into a sharp point at the top.[37] The poor mixing of the dry earthy matter with the watery vapor gave rise to the veins in the leaf, and these then served as channels for the watery fluids within the

---

[34] Book II, tr.2, c.2, para. 100-109 (pp. 142-146).

[35] *De plantis*, Book I, c.14, lines 5-6 (p. 20). *De vegetabilibus*, Book I, tr.2, c.4, para. 168 (p. 82). There is some difficulty with the text of this passage. Meyer's edition of *De plantis* speaks of *small* leaves and *split* leaves (*parva; scissa . . . ut folia vitis*). Albert in Book I gives: "*Quarundam* [*plantarum folia*] *sunt stricta valde*, sicut juniperi et abietis, et *quarundam sunt diffusa valde, sicut vitis* et platani, et arboris, quae vocatur arbor paradisi, cujus folium habet latitudinem quasi cubitalem." It is not clear whether Albert was following a different text from Alfred of Sareshel's version or whether he was restating the text to make the two descriptive terms into true opposites. For Albert's description of the banana tree (from hearsay), see Book VI, tr.1, c.4, para. 19 (p. 347). I do not know his source.

[36] Book II, tr.1, c.2, para. 100-103 (pp. 142-143).

[37] Ibid., para. 104 (p. 143): "Et ab humido quidem aqueo facilem habet dilatationem, et a defectu materiae in superioribus habet constrictionem; a calore autem movente ipsum superius constringitur in acumen, quasi in punctum coarctatus. Et haec est causa, quod, ut in pluribus, folia perfectarum plantarum et majorum figuram habent, quae componitur ex duabus aequalibus proportionibus duorum arcuum ex una linea recta linea inferius egredientibus, et superius in puncto convenientibus."

leaf. Variations in venation and the geometrical shapes of leaves derived from the interactions of the qualities of the matter within them. When the veins intersected with the edge of the leaf, the force of the heat in the channeled fluid could push the edge outward a bit and then the fluid would flow back inward. This was the cause of the sharp angular edges which were "plainly seen in the leaves of oak, maple, grapevines, and many other plants."[38] Not all leaves, however, had angular edges or pointed tips. Smooth margins could be the result either of a great excess of the watery fluid (such as in leaves that float on the water, e.g., water lily) or of so good a mixture of earth and water that the dry part could not extend beyond the spread of the watery part (e.g., box and certain other trees). If the watery matter was cold, then the leaf would lack the sharp tip and take on a semicircular shape instead (e.g., mallow).[39]

Albert's other examples in this chapter make it clear that he had carefully examined the veins and leaves of clovers, fig, maple, grape, swamp reeds (probably cyperus and horsetails), cabbage, blite, and plantain.[40] The names, of course, were not necessary to the argument. The formula, "some plants have this shape of leaf, some plants have that one," would have done just as well — and often did. Observations of particular plants were, to Albert, philosophically pointless until they were brought together into a generalization about a characteristic phenomenon of a larger group and coupled to a scientific explanation.

If we take Books I-V on their own terms, not expecting a flora or manual of plant taxonomy or atlas of plant morphology, what can we conclude about Albert's achievement in this half of *De vegetabilibus*? First of all, Albert did accomplish what he had promised: a careful explication of Aristotle's text, and a systematic recapitulation of its contents in a comprehensible form. That in itself was more than any other medieval commentator on *De plantis* managed to do or even saw the need to do. Beyond this, he conscientiously

---

[38] Ibid., para. 105 (p. 144): "Et ideo folium efficitur angulosum et circumpositum quasi angulis acutis, sicut plane videtur in foliis quercus et platani et vitis et multarum aliarum plantarum."

[39] Ibid., para. 106 (pp. 144-145).

[40] Ibid., para. 107-109 (pp. 145-146).

applied the methods Aristotle had used in other, fuller works of natural science to expand and elaborate the unsatisfactory discussions of *De plantis*. He was able to discern the problems of natural philosophy that especially pertained to plants and gave them the attention they deserved but had never before received. Even if Albert had been content with this, he should win our respect. In the end, though, the special novelty and interest of Books i-v lie in Albert's ability to organize his own intimate experience of nature according to the dictates of natural philosophy. He succeeded in giving consistent, convincing explanations of the small details of plant life in the only scientific terms available to him. These books stand as a remarkable reconciliation of the curiosity of a naturalist, the temperament of a philosopher, and the responsibilities of a teacher — unique in the Middle Ages and rare enough in any period.

14

Albertus Magnus and Medieval Herbalism

Jerry Stannard
University of Kansas

At first glance, any connection between Albertus Magnus and herbalism may seem unlikely. Although he is universally recognized as one of the most learned and distinguished scholars of the High Middle Ages for his numerous contributions to philosophy, science, and theology, Albert is not, in any obvious manner, associated today with herbalism or the ancillary disciplines of medicine and pharmacy. Yet, on closer examination, the conjunction of the *Doctor Universalis* and the *herbolarii* may not appear so unusual. He was, even in his own day, famous for the encyclopedic breadth of his knowledge which, in no small measure, was the result of his commentaries on Aristotle.[1] That knowledge covered, inter alia, the biological sciences and mineralogy, portions of which were closely allied to herbalism. Moreover, a substantial segment of one of his major

---

[1] Throughout the Middle Ages, Aristotle was regarded as the author of *De plantis*, now attributed to Nicholas of Damascus. For the complex history of its MS tradition and the several translations, cf. M. Bouyges, "Sur le De plantis d' Aristote-Nicholas à propos d'un manuscript arabe de Constantinople," *Mélanges de l'Université Saint-Joseph, Beyrouth* 9 (1924), 71-89; L. Labowsky, "Aristoteles De plantis and Bessarion," *Medieval and Renaissance Studies* 5 (1961), 132-154; B. Hemmerdinger, "Le De plantis, de Nicholas de Damas à Planude," *Philologus* 111 (1967), 56-65; H. J. Drossaart Lulofs, "Aristoteles Περὶ φυτῶν ," *Journal of Hellenic Studies* 77 (1957), 75-80. For references to the supposed Aristotelian authorship, cf. Thomas of Cantimpré, *De natura rerum*, ed. H. Boese (Berlin: De Gruyter, 1973), p. 317 (Dicit Aristotiles in libro primo De vegetabilibus) and p. 322.

scientific writings covered the same ground, used the same methods and sources, and employed the same terminology, as did contemporary herbals and certain kinds of medical writings.

Book VI of Albert's *De vegetabilibus libri septem* may, in fact, almost be considered to be a herbal.[2] To be sure, there are some differences for, after all, Book VI is but one portion of a much larger work that deals with plant life and plant form in a broad, philosophical sense. Other, more detailed differences will be noted below. But, and this is the major claim here advanced, in form and content alike, Book VI taken independently, may be regarded as a herbal. In partial support of this claim, it may be noted, that later writers treated it as if it were a herbal and sometimes cited Albert by name as a source for various kinds of data traditionally found in herbals.[3]

Before we proceed further, it will be convenient to define the basic terms. A herbal is a series of descriptions of plants (sometimes including animal and mineral substances), usually in alphabetical order, which are regarded as medicinal, accompanied by medical, pharmacological, and scientific data concerning their names, properties, uses, habitat, and related forms of practical information.[4]

A herbalist, hence, is one who practices the art of herbalism and/or who records such findings in a herbal or related document. The medieval herbalist combined the roles of physician, apothecary, plant collector, dietician, and student of natural history. Leaving aside differences in background, training, clientele, and the like, the herbalist's art centered upon a belief, shared by most non-herbalists as well, in the healing properties of plants and plant products.

---

[2] Albertus Magnus, *De vegetabilibus libri septem*, ed. Ernst Meyer and Carl Jessen (Berlin: Georg Reimer, 1867). Book VI (of 492 sections) occupies pp. 339-588. References to *De veg.* are by page and section; the number to the left of the diagonal indicates page, to the right, the section. Similar techniques are used in citing other works, when applicable.

[3] Albert is twice cited on therapeutic matters by Mayster Albrecht: cf. W. L. Wardale, ed., *Albrecht van Borgunnien's Treatise on Medicine*, (Oxford: for St. Andrews University, 1936), 26/47 and 67/237. See also Erich Schmidt, "Die Bedeutung Wilhelms von Brescia als Verfasser von Konsilien," diss., Leipzig, 1922, p. 30: "De hac facit Alberto causam in suo libro De plantis. . . ." "Albre Major" is cited as authority in Clovis Brunel, ed., *Recettes médicales alchimiques et astrologiques* (Toulouse: Privat, 1956), 48/405 and, by A. Feyl, ed., "Das Kochbuch des Eberhard von Landshut," *Ostbairische Grenzmarken* 5 (1961), 352-366, cf.360/81: ". . .als do spricht Albertus." Finally, in the late medieval *Hortus sanitatis germanice* (Mainz: Peter Schöffer, 28 March 1485), Albert is cited in eight chapters: 9 cinoglossa, 204 gagates, 222 Karabe, 242 lapis magnes, 243 lapis margarite, 337 rosa, 410 urtica, 412 verbena. It is possible that the abbreviation "Alber." may also refer to Albertus in T. G. Leporace et alios, ed., *Un inedito erbario farmaceutico medioevale* (Firenze: Olschki, 1952), p. 90.

[4] See Jerry Stannard, "Medieval Herbals and their Development," *Clio Medica* 9 (1974), 23-33.

Most of the claims made by medieval herbalists concerning the therapeutic usefulness of plants cannot be substantiated today, either clinically or pharmacologically. This does not mean, however, that the plants were physiologically inert or that their administration possessed no therapeutic value. It is rather the case that in the absence of other forms of therapy, the majority of persons in the Middle Ages were dependent upon medicaments of plant origin and a supervised diet, mainly of plant substances, as the main lines of defense in times of illness.[5] For this reason, a large body of information accumulated and gradually was systematized in the form of herbals and other kinds of medical writings such as leechbooks, receptaria, and antidotaria.[6] The material found in these writings was organized and explained by the categories available to, and accepted by, their authors or compilers, for example, the theory of the four elements or, alternatively, the theory of the four humors. The writers of these texts were often physicians, although there was nothing to prevent others from compiling their favorite recipes or rearranging traditional material to suit their particular needs.[7] Because much of the information comprising a typical medieval herbal was common knowledge, it was widely diffused and easily modified in accordance with the specific occasion.[8] This has direct bearing upon the present

---

[5] See Gerhard Eis, "Meister Alexanders Monatsregeln," *Lychnos* (1950/51), 104-136; Manfred Koch, "Das Erfurter Kartäuserregimen," diss., Bonn, 1969. For the close connection between plants as foodstuffs and condiments and plants as spices and medicaments, see Hermann Fischer, "Mittelhochdeutsche Receptare aus bayerischen Klöstern und ihre Heilpflanzen," *Mitt d. bayerischen botanischen Gesellsch.* 4 (1925), 69-75 and Jerry Stannard, "The Botanico-Medical Background of Baptista Fiera's *Coena de herbarum virtutibus*" in *Civiltà dell' Umanesimo*, Atti del VIII Convegno del Centro di Studi Umanistici, Montepulciano (Firenze: Olschki, 1972), pp. 327-344.

[6] See Jerry Stannard, "Greco-Roman Materia Medica in Medieval Germany," *Bulletin of the History of Medicine* 46 (1972), 455-468.

[7] This is illustrated by comparing specific entries in medieval herbals. There exist three herbals written within either Albert's lifetime or a decade of his death:
(1) Henrick Harpestraeng (ob. 1244), *Liber de herbarum*, ed. Poul Hauberg (Copehhagen: Bogtrykkeriet Hafnia, 1936).
(2) Rufinus (*fl.* 1280), *De virtutibus herbarum*, ed. Lynn Thorndike (Chicago: Univ. of Chicago Press, 1949).
(3) Hermann of Heiligenhafen (anno 1284), ed. H. Ebel, *Der Herbarius communis des Hermannus de Sancto Portu* (Würzburg: Triltsch, 1940).
All of these writers, like Albert, depend heavily on Matthew Platearius (ob. 1161), *Circa instans.*, ed. Hans Wölfel, "Das Arzneidrogenbuch Circa instans in einer Fassung des XIII. Jahrhunderts aus der Universitätsbibliothek Erlangen," diss., Berlin, 1939.

[8] This is especially clear in the case of two compilers of encyclopedias, both of whose books on plants are indirectly connected with Albert: Thomas of Cantimpré, *Liber de natura rerum*, ed. Boese, (libri X-XII = pp. 312-350) and Konrad of Megenberg, *Das Buch der Natur*, ed. F. Pfeiffer (Hildesheim: Olms, 1962; facsimile reprint of 1861 ed.)

study because, as we shall observe, Albert availed himself of much traditional material, used many of the major sources available in his time, and arranged his material in accordance with his particular purpose.

With this as a brief background of the herbal, the herbalist, and the latter's role, let us return to the herbal, as defined above, and examine the extent to which Book VI fits that pattern.

It is an unprofitable question to ask what is the most important kind of information in a herbal for the simple reason that the various kinds of information are mutually supportive. Despite the difficulty of neatly separating all the components of a medieval herbal, the descriptions of the plants are among the most obvious. This is the case, not only in the few illustrated herbals, but in those which lack any illustrative material.

In herbals, the descriptions have commonly been regarded as subserving a practical function, viz. the identification and recognition of therapeutically useful plants such that they can be collected and used appropriately. To what extent the physician, apothecary, or herbalist actually relied on written description in order to recognize the plants sought, is somewhat unclear. An equally good case can be made for the well-known fact that herbalists, to this day, possess a fund of empirical information on the basis of which they recognize and hence collect the desired plant. The herbal, thus, served other purposes as well, for the description of the plant was only propadeutic to the preparation and administration of medicaments, the composition of which usually included substances of plant origin. Regardless of which way the descriptions be interpreted, it is clear that Albert was not describing plants solely for the purpose of recognizing potentially useful plants for a therapeutic end.

In the opening sentences of Book VI, Albert states, almost apologetically, that he will deal with specific plants and their properties.[9] It

---

[9] "In hoc sexto libro vegetabilium nostrorum magis satisfacimus curiositati studentium quam philosophiae. De particularibus enim philosophia esse non poterit. Nos autem in hoc sexto libro proprietates quasdam intendimus ponere, quae particularibus plantis convenire videntur" (338/1). Concluding his account of *thus* (the oleo-gum resin derived from *Boswellia sacra* Flueck.; frankincense), Albert states, "Est autem stypticum restrictivum, et memoriae confert glutitum, et ad alias multas medicorum praeparatur operationes, de quibus hic non intendimus, nec est per singula dicendum" (457/235).

is significant that his ultimate purpose is not medical, but rather to provide a brief description of those plants whose *formae* and *naturae* were discussed from a philosophical point of view in the preceding five books.[10] It is even more significant that among the *proprietates* chosen to illustrate the specific plants, he includes their medicinal uses and pharmacological action, that is, the *virtutes* and *operationes* whose enumeration is one of the normal features of medieval herbals.

The fact that his descriptions of some four hundred different species were intended to subserve a philosophic or scientific purpose, rather than a medical purpose, permitted Albert to utilize a wide range of descriptive data. The only major limitation was the lack of information concerning exotica, i.e. herbs and trees neither indigenous to nor naturalized in Western Europe in Albert's time. For like the great majority of his contemporaries, Albert had no personal knowledge of the parent plant of Near Eastern exotica. As a consequence, he was restricted to making inferences, based on the dried commercial forms (e.g., cloves, pepper, cinnamon), and repeating, or paraphrasing, descriptions of those plants which he found in his sources, especially Avicenna and Constantine the African, two of the principal literary sources for Book VI.[11]

Taken together, the different kinds of information assembled by Albert in his descriptions are impressive. No doubt it helped the botanically innocent reader to visualize some of the herbs and trees discussed earlier by being able to relate their knowledge of common, everyday plants to Albert's compressed and sometimes technical descriptions.

Leaving aside temporarily the problem of nomenclature — synonyms, varietal names, etymologies and the like — many of Albert's descriptions begin with a simple, yet revealing remark to the effect that the plant in question is well known or common. *Fenugraecum* (*Trigonella foenumgraecum* L.; fenugreek), for example, *est herba*

---

[10] "Sicut enim in animalium scientia non scimus naturam eorum, nisi cognitis cibis et operibus animalium et partibus eorum: ita etiam in scientia plantarum nequaquam cognoscitur natura ipsarum, nisi sciantur et partes earum et qualitates et effectus. Haec igitur in quibusdam plantarum investigata describemus, ut in aliis per similem modum natura plantarum inveniatur" (472/263).

[11] *Liber canonis Avicenne*, revisus et ab omni errore mendaque purgatus summaque cum diligentia impressus (Venice: P. de Paganinis, 1507), (the *simplicia* occupy Lib. II, Tract. ii = folios 88rb-162rb); Constantine the African, "De gradibus simplicium," in his *Opera medica* (Basel: Henr. Petri, 1536), pp. 342-387.

nota, while the description of *diptannus* (*Dictamnus alba* L.; dittany) begins *est herba communis satis*.[12] Like many phrases in herbals, the statement that a plant is *valde nota* or *bene nota* might be regarded as nothing more than a typical herbalistic formula, on a par with the ubiquitous *probatum est*. But in Albert's case it is not a hollow formula; it is a way of setting the stage for subsequent information. By designating a plant as "common" or "known to all," Albert, in effect, is advising his reader to look about or make a simple inquiry if more information is required about the plant or the appropriateness of his analysis.

Immediately following the opening statement that the plant is common or well known, Albert, like most medieval herbalists, turns to a physical description of the plant itself. Ever since the pioneer studies of Ernst Meyer, historians of botany have recognized Albert's attention to details and, equally important, his ability to record them with a degree of precision unmatched in his day.[13] Not only did he observe attentively the plant or some portion thereof, in true scientific fashion he smelled, tasted, or felt some of the plants which he described. Perhaps without completely realizing it, he was engaged in making a differential diagnosis (see below, p. 363). For example, *ruta* (*Ruta graveoleus* L.; rue), described first as *herba nota*, is further characterized as "markedly bitter" (*amara valde*), an apt phrase for one of the most popular of herbs of traditional materia medica.[14] Its dominant characteristics are, in fact, its bluish-grey foliage which, when slightly bruised, emits a pronounced, bitter and pungent odor. Elsewhere, in describing *marmacora* (? *Melissa officinalis* L.; lemon balm), he notes, *est boni odoris, aromaticus*.[15] This too, is appropriate for another old garden favorite whose crinkled leaves when bruised immediately give off an unmistakable but pleasant, lemon-like scent.

In addition to carefully observing, Albert was able to communicate that information in a succinct fashion. His vocabulary was so well chosen, in fact, that the great majority of his plants can be identified readily to the generic level and many even to the specific level.[16]

---

[12] *De veg.* 515/342 (*fenugraecum*); 506/327 (*diptannus*).

[13] Ernst Meyer, "Albertus Magnus. Ein Beitrag zur Geschichte der Botanik im dreizehnten Jahrhundert," *Linnaea* 10 (1836), 641-741; idem, "Albertus Magnus, Zweiter Beitrag zur erneuerten Kenntniss seiner botanischen Leistungen," *Linnaea* 11 (1837), 545-595.

[14] *De veg.* 559/428.

[15] *De veg.* 536/382.

[16] See Jerry Stannard, "Identification of the Plants Described by Albertus Magnus, *De vegetabilibus*, lib. VI," *Res Publica Litterarum* 2 (1979), forthcoming.

Because of the studies by Meyer, Sprague and others, it is unnecessary here to deal at length with the technical aspects of Albert's descriptions or with the various organs denoted by his unusually large botanical vocabulary.[17] Suffice it to say, his descriptions of the various plant organs — root, stem, leaf, flower, fruit and seed — are the sorts of details to which herbalists paid close attention, though for quite a different purpose than Albert. For therapeutic purposes, it is important to know, and to record, which portion of the plant is to be used, as it is well known that some portions of the same plant are more physiologically active than others. It speaks well for Albert's utilization of his sources, written and oral alike,[18] when he remarks that it is the root of *nenufar alba* (*Nymphaea alba* L., white water-lily) which physicians use in the treatment of ulcers.[19] Elsewhere, in concluding his description of *humulus* (*Humulus lupulus* L., hop vine) he states that it is the flower alone which is used.[20] This is an interesting observation and one which may have been based on the use of the dried strobili for flavoring beer. It is possible, moreover, that Albert had seen hops trained along poles during his many travels in southern Germany, for Bavaria is to this day one of the centers for the production of hops in Germany. Like many of the other traditional garden plants, hops may have been grown on lands farmed or owned by monastic houses visited by Albert in the course of his duties as provincial of the Dominican Order *in germania*.[21]

Several other kinds of descriptive information are also included in Book VI, each of which is represented in contemporary herbals. Albert, for example, frequently likens one plant to another though, on occasion, he mentions some character by means of which two plants can be distinguished from one another. In the latter case, the plants may in some respects be similar, but with reference to a stipulated character they are distinguishable.

The use of the former of the two techniques, based on the recogni-

---

[17] T. A. Sprague, "Plant Morphology in Albertus Magnus," *Kew Bulletin* 9 (1933), 431-440; id. "Botanical Terms in Albertus Magnus," *Kew Bulletin* 9 (1933), 440-459. A more detailed study is in preparation and will appear as "Die Botanik des hl. Albert," in *Albertus-Magnus-Festschrift* (Köln, 1980).

[18] In his discussion of the hop vine, Albert states that the dried flower is long lasting and concludes, "ita quod vulgaris opinio est, quod numquam putrescit" (525/361).

[19] *De veg.* 542/395.

[20] *De veg.* 526/361.

[21] It is significant that hops are mentioned in a document (undated) from St. Emeran's in Regensburg, where Albert briefly served as bishop; see C. O. Cech, "Über die geographische Verbreitung des Hopfens in Alterthume," *Bulletin de la Société Imperiale des Naturalistes de Moscou* 57 (1882), 57.

tion of the concept of similarity as an irreducible philosophic category, is an effective device for purposes of emphasizing some noteworthy or conspicuous character. Such a technique, however, assumes that the comparandum be a well-known plant, for otherwise it loses its effectiveness.[22] Albert's use of this technique thus tells us something about the general knowledge of plants in his time. For example, by likening *anetum* (*Anethum graveolens* L.; dill) to *feniculum* (*Foeniculum vulgare* Mill.; fennel) or *cepa* (*Allium cepa* L.; onion) to *porrum* (*Allium porrum* L.; leek) he tells us indirectly that some knowledge of fennel and leek can be expected on the part of nearly everyone.[23] But care must be taken not to misread Albert's use of *similis est* for when describing exotica, any similarity between the parent plant and another plant clearly derives from his sources.

The establishment of certain plants or, as is often the case, certain organs of those plants, as norms for the purpose of comparing and describing less well known plants, is most useful in connection with the description of those plants which lack a conspicuous characteristic. Such plants, often small or weedy, are easily overlooked and, in fact, many such European indigenes are not recorded in medieval herbals. For Albert's purpose, however, the fact that they played no role in medicine is irrelevant. Indeed, with respect to *gauda* (*Reseda luteola* L.; weld), Albert explicitly states that it has no medicinal use or, at least, that it is unknown.[24] Although medieval herbals occasionally devoted a section to a plant of which it is stated that it had no medicinal use, a tell-tale sign of how descriptive botany will eventually outgrow its herbalistic origins, it was not a common practice.[25] It was far more common simply not to mention any medicinal uses, a practice which Albert understandably used frequently.[26]

---

[22] The statement that *gariofilus* (*Syzygium aromaticum* [L.] Merr. et. Perr.; cloves) "est sicut id, quod vocatur *sambacus*" (396/115) is a case in point. Few if any Europeans in Albert's time had seen the clove-bearing tree and second, *sambacus* probably denotes *Jasminum sambac* (L.) Ait.; Arabian jasmine, itself an exoticum, not available as a living plant in Western Europe until the late eighteenth century.

[23] *De veg.* 481/282 (*anetum*); 487/295 (*cepa*).

[24] *De veg.* 521/352.

[25] Rufinus, *teste* Isaac Judaeus, states of *jujube* (*Zizyphus jujuba* Mill.), "nec sanos custodiunt nec sanitatem egris restituunt" (Rufinus, ed. Thorndike, p. 161). Referring to *cinoglossa* (*Cynoglossum offinale* L.; houndstongue) Konrad writes: "daz kraut ist guot für der viertagleichen riten, und sagt daz puoch ze latein niht mêr dâ von. sô vinde ich auch in andern meinen püechern, diu von den kräutern sagent, niht mêr dâ von" (ed. Pfeiffer, 390/20).

[26] For example, of *crocus* (*Carthamus tinctorius* L.; safflower), Albert states, "hic non est bonus, neque bene tingit, neque condit cibaria, et facit nauseam" (488/297). The crocus of commerce or saffron (*Crocus sativus* L.) is termed *crocus hortensis* (488/297) to distinguish it from safflower.

The second of the two methods mentioned above, viz. where two species of plants are distinguished from one another, was readily used by Albert. The basis for this differential diagnosis, as it is termed by some field botanists, is the recognition of a single but typical character by whose presence or absence two species, otherwise regarded as similar, can be differentiated. This method was intuitively employed by herbalists and plant collectors for the practical purpose of distinguishing between two species of similar appearance, one of which was believed to be inert or, alternatively, one of which was believed to be toxic. Referring to *centinodia* (*Polygonum aviculare* L.; knotgrass), Albert first states that its habitat is similar to that of *cicorea* (*Cichorium intybus* L.; chickory) and then proceeds to describe its habit of growth: "it creeps over the ground, but does not entwine itself around those plants which grow nearby. In this, it differs from *volubilis* [*Calystegia sepium* (L.) R. Br.; bindweed] which entwines itself around nearby plants."[27] Another example is interesting for an additional reason. *Esula* (*Euphorbia esula* L.; leafy spurge), he states, has leaves like those of *linaria* (*Linaria vulgaris* Mill.; toadflax) but they differ in that *esula* produces a milky latex but *linaria* does not.[28] The reason why this passage is of double interest is that it indicates an Albertian source not otherwise formally acknowledged. The phrase "esula produces. . .linaria does not" is simply a rewording of a passage from the *Regimen sanitatis salernitanum* though it is possible that Albert derived the passage from a presently unidentified or intermediate source.[29]

It is a tribute to Albert's profound understanding of the diversity of plant life that when he distinguished between two species or compared them, he selected stable characters which were typical and invariant for the species in question. Thus, while he, like all herbalists, often noted the color of the flowers, he recognized that color is not a constant character and that in some species there is a consider-

---

[27] *De veg.* 504/322. Another example concerns the differentiation of *rizum* (*Oryza sativa* L.; rice) and *ordeum* (*Hordeum vulgare* L.; barley), 558/427. The description of the rice grain may have been based on personal observation, but whether Albert saw the rice plant growing is doubtful.

[28] *De veg.* 512/336: "esula lactescit, et non facit linaria."

[29] The Salernitan passage, which later served as a mnemonic aid, appears in the *Collectio Saternitana*, ed. Salvatore de Renzi (Naples: Sebizio, 1859), 5: 33, verse 1173: "esula lactescit, linaria lac dare nescit." The verse does not occur in the *Circa instans*, Hermann of Heiligenhafen, or Henrik Harpestraeng. It appears in Rufinus, ed. Thorndike, p. 173 as "esula lactescit; sine lacte linaria crescit" and, in a still later form as "esula in ir hait milch und linaria keyn milch" (*Hortus sanitatis* [1485 ed.], cap. 235).

able range of normal color forms.[30] The color of the flowers was not, for Albert, the only criterion for either field recognition or taxonomic discrimination. For such purposes, morphological characters were better criteria and these Albert often noted in his descriptions.

In summing up the descriptive data assembled by Albert, it must be emphasized that although he took over some details from herbals and other medical writings, he applied those details to his own ends. This may be illustrated in his descriptions of several different species of trees.

Along with their more obvious features — general appearance, relative size, shape of leaf, nature of their fruit, etc. — several times he described the texture of the wood, insofar as he could without any optical aids, and noted some of the economic uses of the wood. For example, of *platanus* (*Acer pseudoplatonus* L.; plane tree maple), he noted that it is commonly known and that in parts of Germany and elsewhere it is used for domestic wares.[31] Again, he noted that *fagus* (*Fagus silvatica* L.; beech), on the contrary, is useless for construction purposes because of the susceptibility of the wood to insect attack.[32] Both of these statements are the result of personal observation and they remain true even today, as may be determined by walking through provincial farming communities in central Germany.

Closely related to the foregoing kinds of descriptive data is the specification of the habitat of the various plants described in Book VI. For obvious reasons, it was as important for the herbalist to specify as it was for the collector to know the habitat of the desired plant.[33] There is, of course, no guarantee that a written specification of the habitat of a given plant will enable the collector to find it. But Albert, like Theophrastus before him, observed that some species are only found growing in marshes, while other species prefer a dry and rocky soil while still others thrive by the side of heavily travelled

---

[30] For example, the range of colors of *lens* (*Lens culinaris* Medik.; lentil) is described as "flos eius rubeus declinans ad albedinem aliquantulum" (531/372). On the other hand, by describing the flower of *mellilotum* (537/385) as "est aliquantulum flos eius declinans ad albedinem, cum tamen sit croceus," he has probably combined two species under one name, viz. *Melilotus albus* Medik. and *M. officinalis* (L.) Pall., the latter being yellow flowered.

[31] *De veg.* 430/183.

[32] *De veg.* 390/105.

[33] Already by the end of the thirteenth century, this information was assembled in the form of several small, anonymous tracts. Several were edited by Ernest Wickersheimer, "Nouveaux textes médiévaux sur le temps de cueillette des simples," *Archives internationales d'histoire des sciences* series 3, vol. 29 (1950), 342-355.

roadways. Although the early botanists lacked the conceptual apparatus to describe an ecological niche or to isolate the relevant edaphic factors, they empirically recognized the optimum habitat for many species.[34]

Since it was not Albert's object to describe plants in order that they might be collected, it is all the more significant that he included references to their habitat: by the seashore, in moist places, as weeds in grain fields and in domestic gardens, near old walls, even on the roofs of houses.[35] These references to habitat are not only another indication of the care Albert took in recording details, they also illustrate some of the ways he used his sources. As noted above, he relied on herbals and the herbalistic sections of medical writings because they were the major and in some cases the only sources for descriptive data regarding a particular species.[36] Albert often excerpted the accounts of his sources and sometimes supplemented them with data obtained through personal observation.[37] But in other instances, he seems to have recorded the information much as he found it on the grounds, presumably, that it helped to provide a more complete description. This is especially true in the case of plants for which Albert had little or no personal knowledge.[38] But it is also evident when one considers those descriptions in which Albert refers to the actual practice of collecting. In medieval herbals, data regarding the proper place and time to collect are often included for practical purposes, as are instructions on the preparation and administration of medicaments composed of the plants so collected. It is noteworthy that

---

[34] This is well represented by Albert's description of *capparis* (*Capparis spinosa* L.; caper bush): "nascitur autem,. . .in petrosis, et ex loco contrahit caliditatem et siccitatem; in frigido autem et humido dissipatur et perit" (371/68).

[35] "In littore maris" (408/138); "in humidis crescens" (534/376); "nascitur in frumento" (543/396); "visus in viridariis" (524/358); "nascitur. . .in siccis locis iuxta rimas murorum" (510/334); "in tectis plantatur" (484/288).

[36] For example, the discussion of distinct sexes of *palma* (*Phoenix dactilifera* L.; date palm) (425/172) goes back, ultimately to Theophrastus, *Historia plantarum* II, 8, 4; it is unlikely that Albert knew the remarks of Herodotus, *Historia* I, 193.

[37] He notes how the spiny seeds of *lappa* (*Arctium lappa* L.; great burdock) "adhaerent vestibus" (534/376) and how the branches broken from *salix* (*Salix alba* L.; white willow) "cito recrescunt" (449/218). Clearly, his description of the use of torches by the Germans, prepared from the resinous wood of *picea* (*Pinus silvestris* L.; scotch pine), was based on personal observation (432/187).

[38] Of *lignum aloes* (*Aquilaria agallocha* Roxb.; aloes wood), Albert writes: "cuius originem nullus hominum apud nos habitantium sufficienter usque hodie cognovit" (344/11). Elsewhere, in describing *blitus* (*Beta vulgaris* L.; beet) he qualifies the phrase "ut quidam dicunt" by adding, "quia hoc non sum ego expertus" (463/246).

Albert's references to collecting for the most part avoid mention of the various incantations or magical rites which sometimes accompanied the actual collection. But from notices elsewhere in Book VI, it is clear that Albert was aware of such practices (see below pp. 374-375).

Another kind of data that is commonly found in herbals concerns a means of recording the different kinds of relations which exist between different kinds of plants. That is, admittedly, a clumsy way of expressing what taxonomists today would clearly recognize as an early attempt at plant classification. But only at the risk of committing an anachronism can one say that medieval herbalists were studying the relations between different species or different genera; that cannot be said because our modern concepts of species and genus had not been formulated.[39] It was empirically recognized by all herbalists, indeed probably by all farmers and gardeners, that individual plants differ and that some differences are more marked than others. But it was not the herbalists' object to study those empirically recognized differences as ends in themselves. Rather, a recognition of some of the differences was another type of information which was worthy of communicating or recording. In turn, that information was considered to be of practical value for the collection and ultimate utilization of plants for therapeutic purposes.

Throughout Book VI, Albert employed a variety of quasi-taxonomic techniques in order to describe the similarities and differences between plants. Several of these techniques have been noted above. But, like all herbalists, Albert also employed another, more obviously taxonomic device in those cases where the nature of the plants in question required finer discrimination. This technique was essentially the careful application of a linguistic convention to natural objects. That is, by incorporating as part of a plant's biverbal name, some descriptive term that demarcated that plant, not only from many, quite different plants, but from plants which appeared to be similar and thus closely related, a provisional taxonomy was created. By this technique, which was not systematically applied to all plants until much later, the world of plants was, in theory at least, capable of being subdivided into smaller, discrete units about which it was comparatively simple to communicate information. That informa-

---

[39] In medieval nomenclature, *genus, species, varietas,* and *forma* were used interchangeably. Sometimes Albert uses *genus* to denote what we would accept as a *genus* (533/375) but in other places his *genus* is practically equivalent to our *species* (531/372; 547/405; 550/409).

tion, of course, may be practical, as was the intention in herbals, or it may be philosphical, as it was in Albert's case.

A good example of Albert's technique concerns the taxonomy, and inevitably the nomenclature, of *apium*.[40] Normally the term *apium*, without any modifier, denoted *Apium graveoleus* L. or celery and it was used in that fashion from Pliny's time to the Renaissance.[41] But for a variety of reasons, some of which are quite unclear, the term *apium*, plus modifier, was used to designate other species. Hence, Albert's section devoted to *apium* actually refers to four different species, each of which was described by Albert as a kind (*genus*) of *apium*. It is unnecessary here to paraphrase the descriptions of the four taxa recognized by Albert, beyond noting that three of the four taxa represent three different genera of the *Umbelliferae* or carrot family, to which celery belongs; the fourth taxon of *apium*, and the most difficult to identify, is almost certainly not an umbelliferous plant, and its only claim to be called an *apium* is that its many-cleft leaf is typical of many umbellifers.[42]

There is no need to discuss other examples of the taxonomic distinctions made by Albert, e.g., *maior/minor*, *masculus/femina*, *aestivum/hiemale*, etc.[43] In all instances where taxa are so distinguished, his purpose remains the same, to describe the plants and illustrate some of their characteristics for the benefit of his readers.

It is only a short step from taxonomic data to nomenclature. As indicated above, nomenclatural matters cannot easily be separated from taxonomy. But there are other aspects of medieval herbalistic nomenclature which deserve comment. Frequently the individual chapters or sections of medieval herbals begin by listing one or more synonyms of the plant in question. In addition, there are sometimes

---

[40] *De veg.* 480/281.

[41] Pliny, *Historia naturalis* XX, 46, 112 seq.; Isidore, *Etymologiae* XVII, 11, 1; Constantine, "De gradibus," p. 379; Matthew Platearius, *Circa instans*, ed. Wölfel, p. 8; Hermann of Heiligenhafen, ed. Ebel, p. 4; Rufinus, ed. Thorndike, 28/49; Konrad of Megenberg, ed. Pfeiffer, 382/3; Thomas of Cantimpré, ed. Boese, 343/4; *Hortus sanitatis* (1485 ed.), cap. 6 etc.

[42] In summary, Albert's description of *apium* as follows:
  i.   apium domesticum = *Apium graveolens* L.; celery.
  ii. apium montanum = *Athamanta macedonica* (L.) Spr.; no common English name; in French, persil de macedoine.
  iii. apium silvestre = *Smyrnium olusatrum* L.; alexanders
  iv. apium aquaticum = *Ranunculus* sp.; probably including *R. flammula* L. and *R. aquatilis* L.

[43] "Maior/minor": 529/368, 574/461; "masculus/femina": 361/49, 378/81, 396/115 etc.; "aestivum/hiemale": 547/405; "silvestris/hortulana": 507/329, 508/331, 509/332 etc.

etymological notices and differences of opinion concerning the application of the name or synonyms by previous writers. All of these nomenclatural matters are represented in Book VI, though Albert's purpose was certainly not lexicographical nor was the enumeration of synonyms regarded by him as an end in itself.[44] Rather, because of the absence of a universally accepted nomenclature and a precise taxonomic system, the transmission of information regarding a particular plant was difficult and complex.[45]

Many plants were known under a plurality of latin names and there is sufficient evidence to demonstrate that the use of vernacular names and synonyms was widespread outside of academic circles.[46] Moreover, there are instances where one name designated two or more different species.[47] Under these circumstances, and as part of the fund of practical information contained in a herbal, it was essential to describe and name a plant in such a manner that it could be recognized and hence collected or otherwise obtained.[48] Since it is not always possible to recognize a plant solely on the basis of a written description, especially in the absence of any other information regarding that plant, for example, alimentary or economic uses, one or more of the most common synonyms was added. Usually they fol-

---

[44] Over twenty etymologies are provided by Albert, e.g. *agnus castus* 349/24; *ficus* 385/48; *semperviva* 568/447 etc. More interesting are the attempts to explain the origin and meaning of synonyms, e.g. *juniperus* (*Juniperus communis* L.; juniper): "cum cypresso multam habet similitudinem; propter quod etiam cypressus silvestris vocatur" (398/121).

[45] This is well summed up by Rufinus in his discussion of *iacea nigra* (*Centaurea scabiosa* L.; greater knapweed): "item licet non scribantur virtutes istarum herbarum et aliarum multarum, tamen tota die occurrent nomina earum tam in libris medicine quam in libris cyrugie, et ibi videbitur ad quid valent, nam sapientes quos sequor de huiusmodi non locuntur" (ed. Thorndike, p. 155).

[46] Rufinus concludes his discussion of *gratia dei* (*Gratiola officinalis* L.; gratiola) by stating, "alio nomine dicitur portulaca silvatica a laycis" (149/13). Albert uses *carvi* <OHG *garba* to denote *Achillea millefolium* L.; yarrow (503/318) which is thus etymologically unconnected with lat. *carvi*, the normal medieval name of *Carum carvi* L. or caraway; the latter is called *cimunum domesticum* by Albert (493/303). Another vernacular name used by Albert is *fibex* (*Betula pendula* Roth; white birch, 390/107). For the synonymy cf. *Hortus sanitatis* (1485 ed.), cap. 421: "vibex-byrck"; and Leo Jordan, "Ein mittelniederdeutsches Pflanzenglossar," *Ztschr. f. deutsche Wortforschung* 3 (1902), 354.55: "vibex vel pinnosa. berke".

[47] For example, *sponsa solis* designated two different species in *De vegetabilibus*:
(i) "sponsa solis sive solsequium" (570/451) is *Calendula officinalis* L.; calendula.
(ii) "cicorea, quae et sponsa solis vocatur" (504/321) is *Cichorium intybus* L.; chickory.

[48] Plants were obtained not only from the *herbolarii* who collected them (cf. Rufinus, ed. Thorndike, pp. 70, 115, 151 etc.); many simples were also available *in apoteca* or *ex apotecario*. For examples, see Jerry Stannard, "Hans von Gersdorff and some Anonymous Strassburg Apothecaries," *Pharmacy in History* 13 (1971), 55-65.

lowed immediately upon the name chosen as the chapter or sectional heading.

Another reason why synonyms played such an important role is that many of the most commonly used medicinal plants were described under different names in different herbals. For example, *Althaea officinalis* L. or marsh mallow, described by Albert as *altea*[49] was listed in contemporary herbals sometimes as *bismalva*, sometimes as *malvaviscum*, and sometimes as *malva domestica*.[50] But short of knowing that all these, and other vernacular names, denoted the same species, one would be hard pressed to identify the plant solely on the basis of the descriptive data provided by the several herbals. Without such knowledge that *altea*, *bismalva*, and *malvaviscum* denoted the same species, the reader of a herbal would not be able to coordinate the data found in different herbals under different names nor be able to use effectively the many recipes which required a specific ingredient.[51]

In Albert's day, plant nomenclature was further complicated by the introduction of Arabic plant names in herbals and pharmacological writings. Although the transliteration of Greek still exhibited some inconsistencies, at least a pattern was emerging by means of which transliterations and Latin-Greek translations were becoming standardized. This cannot be said, however, of Arabic or for indivi-

---

[49] "Altea — vocatur etiam bismalva, ea quod habet folia sicut malva, sed est major ea, habens crura longa plurima ex radice una. Vocant autem quidam eandem malvaviscum" (*De veg.* 483/285).

[50] A selection of the descriptions and names of *Althaea officinalis* L. in medieval herbals and leechbooks follows: Avicenna (1507 ed.), 96/76 (altea); *Circa instans*, ed. Wolfel, p. 72 (malva domestica); Herman of Heiligenhafen, ed. Ebel, p. 4 (althea); *Hortus sanitatis* (1485 ed.), cap. 12 (altea-ybisch); Konrad of Megenberg, ed. Pfeiffer, 385/10 (alcea-weizpapel); *Regimen sanitatis salernitanum*, ed. De Renzi, p. 24 (altea. . .malvae species). Rufinus has four entries for *Althaea officinalis*, in all of which but the last, the plant is described and its virtues enumerated: "aeviscus" (p. 9); "altea" (p. 18); "bismalva sive herba ungarica alba" (p. 60); "malvaviscus" (p. 178). In addition, see Macer Floridus, De *viribus herbarum*, ed. Lud. Choulant. (Leipzig: Voss, 1832), p. 43 (althaea); Serapion, "De simplicibus ex plantis," in his *Practica* (Lyon: Jac. Myt, 1525), fol. 133 vb (altea; *Boec van Medicinen in Dietsche*, ed. William Daems (Leiden: Brill, 1967), p. 135 (malva. . .groetpappel).

[51] Clearly a knowledge of the names and synonyms underlies the following statement by an anonymous author "Ut perfecte operetur medicus oportet eum cognoscere complexiones cum gradibus, virtutes cum operationibus, ut sciat que herbe vel species et cetera humanum corpus immutantia sint calida, que frigida, que humida, que sicca et in quo gradu" (Léopold Delisle, "Note sur un manuscrit de Tours renfermant des gloses françaises du XIIᵉ siècle," *Bibliothèque de l'Ecole des Chartes* 6ᵉ ser., 5 (1869), p. 322.

dual words which were thought to be Arabic.[52] In the absence of a standardized method for the transcription of Arabic, it was difficult to apply many of the plant names which appeared in herbals and other genres of medical writings.[53] To be certain, Latin-Greek-Arabic synonyms helped, but often the synonymy or the inclusion of the wrong ingredient in a compositum, might make a significant difference. Although this was not Albert's concern, his reliance on Gerard of Cremona's translation of Avicenna's *Liber canonis* occasionally led him onto uncertain ground.[54] In defense, it may be noted, that Albert used the best sources available to him in an effort both to describe and to name, in a meaningful fashion, those plants whose properties were analyzed in Books I-V. He was, moreover, sufficiently critical to query some of the reported claims.[55]

---

[52] Some, but not all, of the Arabic names appearing in *De vegetabilibus*, most of which occur in Book VI, have been collected in Meyer and Jessen's edition (p. 693, "Plantae Albertinae: Dubiae Restant" and pp. 700 seq., "Index Rerum"). A careful study of such names, on the model of F. Schühlein's *Index nominum Arabicorum* (in Stadler's edition of *De animalibus*, 2: 1655-1663) is badly needed.

[53] Inconsistencies in orthography and uncertainties of synonyms can still be dectected much later, e.g., M. Kleemann, "Ein mittelniederdeutsches Pflanzenglossar," *Ztschr. f. deutsche Philologie* 9 (1878), 196-209 (saec. XIV); *Alphita*, ed. J. L. G. Mowat (Oxford: Clarendon Press, 1887), (saec. XV).

[54] A case in point is provided by Albert's discussion of *manna*, which occurs at the end of his description of *zucarum* (*Saccharum officinarum* L.; sugar): "est autem quiddam simile zucaro, quod vocatur etiam zucarum, et cadit in terra, quae vocatur alhusar de aere, et est species mannatis et roris, et est sicut frusta salis" (471/262). Earlier, in his discussion of *ladanum*, a resinous exudation from several related species of *Cistus*, he notes, "non esse aliquid plantae vel arboris, sed potius esse de genere roris, sicut mel et manna et id, quod vocatur tereniabin" (394/113). Albert's account derives, in part, from Avicenna (136/493, s.v. manna) and *Circa instans* (72, s.v. manna), and that another of his sources was of Arabic provenience is indicated by the name *tereniabin* (cf. Pierre Guigues, "Les noms arabes dans Serapion, Liber de simplici medicina," *Journal Asiatique* 6 (1905), 58/360 (*men*) and 84/497 (*tereniabin*). It appears that Albert has confused two substances, similar in appearance and in their properties but distinct in their origin: the exudate of the common manna ash, *Fraxinus ornus* L., which he may easily have known while in Italy or Southern France and *tereniabin*, possibly an exudate produced by an insect; cf. Max Meyerhof, "The earliest mention of a manniparous insect," *Isis* 37 (1947), 32-36.

[55] Cf. the devastating criticism in 430/183 concerning "arbores parvas in insulis Germaniae crescentes," the falsity of which is clear; Albert begins, "quoniam nec Germania insula est. . . ." Elsewhere, in regard to the *arbor peredixion*, Albert queries the description and the association of its fruit with doves (437/198). It is problematic, however, whether Albert realized that this fabulous entity derived from the Φυσιολόγος , c.34: Περὶ δένδρου περιδεξίου , ed. Fr. Lauchert, *Geschichte der Physiologus* (Strassburg: Trübner 1889), pp. 264-265.

At this point, it will be convenient to pause momentarily and to consider the kinds of data discussed so far. One of the functions of the herbal, it will be recalled, was to provide practical information. A large portion of that information, but by no means all of it, has been discussed above as descriptive data. Originally, those data had, as one of their functions, the identification of medicinal plants such that they could be collected or otherwise procured. But there is another kind of practical information contained in medieval herbals which is obviously and thematically connected with the descriptive data. This latter class is generally termed "therapeutic data." It is a fund of practical advice comprised of recipes, dietary counsels, domestic or household recommendations, and explanations regarding the nature of a medicinal plant and the preparation, administration, dosage, and storage of the medicaments made from or containing plant products.

The class of therapeutic data is no less important than the descriptive data, though its application is more restricted and its relations to developments in the history of science less obvious. But in the typical medieval herbal, the therapeutic data supplement the descriptive data in such a fashion that in the absence of either, the value of the herbal is proportionately lessened.

One might expect, *a priori*, to find little therapeutic data in Book VI, especially when it is recalled that *De vegetabilibus* was designed to be a commentary on the Pseudo-Aristotelian *De plantis*. But, as always, Albert was encyclopedic in his coverage and, as if to compensate for the lacunose descriptions of plants and their properties in *De plantis*, he attempted to remedy the situation by bringing together the salient information. Once the decision was made to use herbalistic data, it was difficult to draw a line between the descriptive and the therapeutic data. It may also be noted that in his earlier *De animalibus*, a wide range of therapeutic data is also included, though in no sense could that work be considered a medical text.[56]

Thus, it is not surprising that Book VI contains a selection of therapeutic data. Admittedly, they are neither as systematized nor as complete, as one might expect in a herbal or a leechbook. But whatever the deficiencies, it is more than compensated by Albert's efforts to explain on philosophic grounds the reasons for the therapeutic effec-

---

[56] Extensive portions of Books XXII and XXIII, deal respectively with diseases of horses and falcons, and their remedies. Cf. Albertus Magnus, *De animalibus Libri XXVI*, ed. Hermann Stadler (Münster: Aschendorff, 1916-1920).

tiveness of a given plant. The explanations generally were formulated in humoralistic terms. Both the composition of sound and sick bodies as well as the plant substance itself and its properties were analyzed in terms of qualitative changes. In turn, they were explicated by the theory of the four elemental qualities of heat, coldness, dryness, and moisture. The properties of a plant substance, and hence its physiological action, were thus explained in terms of its elemental nature; for example, *ysopus* (*Hyssopus officinalis* L.; hyssop) is stated to be hot and dry while *portulaca* (*Portulaca oleracea* L.; purslane) is cool and moist.[57] On the basis of its nature, the therapeutic employment of a plant followed as a matter of course. A dry plant dried etc. or a cool plant was beneficial for conditions caused by too much heat, etc.[58]

But this kind of explanation of a plant's therapeutic usefulness did not exhaust the store of therapeutic data found in Book VI. Like the descriptive data, however, the therapeutic data cannot be sharply separated into distinct classes. Because many of the plants and plant products belonging to the traditional materia medica served a plurality of functions, they can be discussed from an equal number of different points of view. Since that is not feasible here, it may merely be noted that Albert's references to the therapeutic uses of plants and their products were, in all essential matters, those of the contemporary herbalist. Depending principally upon the mode of administration, a plant or some portion thereof might serve as part of a normal diet in times of health or in a restorative diet in times of illness.[59] Moreover, those same plants might serve as a relish or condiment,[60]

---

[57] *De veg.* 581/477 (ysopus); 548/406 (portulaca).

[58] Of *lingua arietis* (*Plantago* sp., plantain), Albert states, "et siccitas eius, ut diximus, non est mordicativa, et ideo consolidat ulcera optime, et ad haec nihil est melius ea" (529/369). A cooling and moistening plant such as *orpinum* (*Sedum telephium* L.; orpine) "valet autem calefactioni hepatis, et infrigidat vehementer" (546/402).

[59] Prior to enumerating the *virtutes* of *petroselinum* (*Petroselinum crispum* (Mill.) Nym.; parsley) Albert states, "est autem calidum et siccum, plus medicina quam cibus" (551/413). Nearly the same statement is made in connection with *lupinus* (*Lupinus* spp.; lupine) in 533/375.

[60] Beans (*Vicia faba* L.) are made more acceptable, "cum. . .comedantur cum pipere et asa et sale et oleo et origano et similibus" (513/338). Each of the three plant substances here mentioned, *piper* (*Piper nigrum* L.; black pepper); *asa* (the gum resin from *Ferula assa foetida* L.; asafedita) and *origanum* (*Origanum vulgare* L.; oregano) is described by Albert and its virtues enumerated.

as ingredients in composita[61] or even be taken or applied singly for specified complaints.[62]

On the whole, the rationale underlying the therapeutic data assembled by Albert was humoralistic and was based on the traditional peripatetic theory of the four elements or elemental qualities. But there are traces here and there of a more primitive rationale, though they are partially obscured by the humoralistic interpretation. It is entirely in keeping with his utilization of herbalistic sources that Albert would adopt on occasion references to therapeutic practices based upon the belief in a plant's signature. This belief rests on a supposed likeness between some characteristic of a plant and either some pathological condition or the agent reputed to produce that pathological condition or some other form of bodily harm. For example, *Dracunculus vulgaris*, Schott or dragon arum, known to Albert as *basilicus, dracontea,* or *colubrina* was, somewhat fancifully, likened to a serpent. To what extent one can see any resemblance is irrelevant for once the resemblance is asserted, and bolstered by its set of synonyms, it is no surprise to find that its principal virtue is to cure the bite of a serpent.[63]

As in all medieval herbals and, indeed many later ones as well, there is a further class of data included in Book VI. By its very nature, this class of data is neither easily summarized nor readily organized into sharply defined categories. Yet no account of

---

[61] Many of the sections devoted to the individual plants, conclude with a reference to "praeparationes medicorum" or "operationes in medicorum praeparatum," e.g., 448/216; 457/235; 502/316; 534/376 etc. In addition, Albert sometimes provides simple recipes which include, as one of their ingredients, the plant discussed in that section, e.g., 555/421; 564/436. Under *enula campana (Inula helenium* L.; elecampane), a detailed recipe is included for the preparation of *vinum enulatum* (509/332). A recipe, bearing the following title "Pillole de Alberto: e credesi di Alberto magno: Ma stimiamo sieno di Alberto Bolognese" occurs in *Nuovo Receptario. . .della inclita cipta di Firenze* (Firenze: Ad instantia delli Signori Chonsoli, 21 January 1498), sig. gv^{rb}.

[62] Thus Albert devotes as much space to the *virtutes* as to the alimentary or nutritive properties of such plants as *beta (Beta vulgaris* L.; beet) (485/292); *caulis (Brassica oleracea* L.; cabbage) (493/304); *ordeum (Hordeum vulgare* L.; barley) (544/399); and *rizum (Oryza sativa* L.; rice) (558/427).

[63] "Basilicus [codd. basiliscus] est herba, quae dracontea vel serpentaria dicitur. . . .Primum autem in siliqua profert ea, quae est sicut posterior pars serpentis, in fine sicut cauda serpentis, et habet in stipite suo varietatem serpentis. Et valent contra serpentis morsum succus eius, et etiam dicitur, quod portata tutat a serpentibus omnibus" (484/290).

medieval herbalism would be complete without at least mentioning a stratum of beliefs loosely but popularly termed "magic and folk belief."

Quite apart from the legends associating Albert with magic[64] and the strange assortment of "secret" writings spuriously attributed to him,[65] it cannot be denied that there are over a dozen references to the magical arts in Book VI. In these passages, however, Albert is reporting the uses of plants by magicians and enchanters, not subscribing to those uses. Such references are historically important because they are a reminder that Albert's enlightened, philosophic approach to the therapeutic uses of plants was not representative of all ranks of herbalists and other medical writers. It is no surprise that Albert noticed some of the darker practices since the employment of herbs and portions of trees for a variety of magical purposes was widespread in his time.[66] Not only are these practices well attested in medical literature; they are also mentioned, and sometimes roundly denounced, in various genres of nonmedical literature.[67] The detailed analysis, required to explain the origin of the magical practices reported by Albert, cannot be undertaken here. But a few observations bearing on the relations between herbalism and magic will help to round out an analysis of Book VI.

One very common practice, alluded to in nearly all medieval herbals, concerns the use of plants or plant products as apotropaic devices. A good example is provided by Albert in his discussion of *lingua arietis* (*Plantago* sp.; plantain). "The enchanter says," states Albert, "that its root, suspended from the neck of a child, cures scrophulous sores."[68] This passage, chosen from several similar references, illustrates some of the interrelated aspects of herbalism and

---

[64] For a summary, see Karl Helm, "Albertus Magnus" in *Handwörterbuch des deutschen Aberglaubens*, ed. Hanns Bächtold-Stäubli (Berlin: De Gruyter, 1927), vol. 1, cols. 241-243.

[65] Cf. G. Meersseman, *Introductio in opera omnia B. Alberti Magni O.P.* (Bruges: Beyaert, 1931), pp. 145-148, and Lynn Thorndike, "Further Considerations of the Experimenta, Speculum Astronomiae and De secretis mulierum ascribed to Albertus Magnus," *Speculum* 30 (1955), 413-443.

[66] See the passage pertaining to plants referred to by Berthold of Regensburg, a contemporary of Albert, in Anton Schönbach, "Zeugnisse Bertholds von Regensburg zur Volkskunde," *Stzb. Akad. d. Wissenschaften zu Wein, Phil.-hist. Cl.* 142/7 (1900), esp. pp. 35-50, 138-148.

[67] See Oskar Ebermann, "Zur Aberglaubensliste in Vintlers Pluemen der Tugent (v. 7694-7997)," *Ztschr. f. Volkskunde* 23 (1913), 1-18, 113-136, and Theod. Zachariae, "Abergläubische Meinungen und Gebräuche des Mittelatters in den Predigten Bernardinos von Siena," *Ztschr. f. Volkskunde* 22 (1912) 113-134, 225-244.

[68] *De veg.* 529/369.

magic. By the term "enchanter" (*incantator*), Albert is probably referring to Avicenna whose chapter on *lingua arietis* is almost certainly Albert's source.[69] The passage in question, however, is Albert's paraphrase, not a verbatim quotation.

The use of natural products, whether of plant, animal, or mineral origin, as apotropaia, was not, in itself, regarded as a form of black magic. Not only were various kinds of pendants and medallions tacitly accepted, but various plants, when subjected to ecclesiastically approved rites, were invested with extraordinary properties and hence useful for a wide range of medical and nonmedical purposes alike. Plants which had been consecrated in such a manner can be termed magiferous plants.[70] Unlike imaginary or fabulous plants, they were real plants. But as a result of their exposure to various rituals, their natural properties or *virtutes* were temporarily reinforced with greatly enhanced properties, useful in many forms of healing. Several of the plants described by Albert fall into this category, for example, *betonica, ruta, salvia,* and *verbena,* though the descriptions touch only lightly upon such matters.

There are, finally, other references where the linkage between herbalism and magic is more obvious and potentially more dangerous. Despite the efforts of the authorities to stamp out such practices, references to them in herbals and medical texts seem to suggest that these practices were widely adopted. Albert again provides some examples, not because he advocated them but rather, since they were so well known and commonly accepted parts of the folk scene, they added another dimension to the herbalistic descriptions of plants. For example, the use of oak galls by the *aeromantici* for purposes of divination,[71] the use of *iusquiamus* (*Hyoscyamus niger* L.; henbane) by necromancers intent upon invoking demons,[72] or the use of a drink prepared from the leaves of *populus* (*Populus* sp.; poplar) to be given to women, *non concipiat, sed sterilis efficiatur.*[73] This last example is doubly interesting because Albert attributes to the enchanters

---

[69] Avicenna (1507 ed.), fol. 128va, cap. 433 "De lingua arietis." A different, but no less unusual property of plantain is reported by Henrik Harpestraeng, ed. Hauberg, pp. 104-106.

[70] See Jerry Stannard, "Magiferous Plants and Magic in Medieval Medical Botany," *Maryland Historian* 8 (1977), 33-46.

[71] *De veg.* 441/206. For a summary of the problem, see the fifteenth-century text edited by Gerhard Eis as "Prophezeiung aus dem Gallapfel," in his *Wahrsagetexte des Spätmittelalters* (Berlin/Munich: Erich Schmidt, 1956), p. 69.

[72] *De veg.* 527/363. Cf. the reference to "qui in nigromantia student" (458/235).

[73] *De veg.* 432/185.

(*incantatores*) a typical herbalistic practice. Independent of the question whether there is any valid evidence of its effectiveness, this is a practice which is no more magical than the use of certain plant substances, mentioned by Albert without comment, as abortifacients.[74]

Only a thin, uncertain line separates magical practices from folk beliefs and, for obvious reasons, there is much overlapping. Folk beliefs, as might be expected, are of such a heterogeneous nature, that only a few comments are possible here. Despite the problems involved in defining and separating the two subjects, one point is reasonably clear. The heart and center of medieval herbalism was its empiricism. Albert was far too perceptive not to recognize the force of custom and the pragmatically justified concept of success which together sustained herbalistic practice. Despite the conceptual elegance of scholasticism or the inherent reasonableness of Aristotelian natural philosophy, herbalism was entrenched among the folk. If a certain plant was effective for a specified condition, as measured by the standards that pevailed among the folk, e.g., squill for dropsy, that was important to the patient, significant for the herbalist, and far outweighed the lack of an explanation from the schools.[75]

Thus, to take a simple example, in his description of *sinapis* (*Sinapis alba* L.; mustard), Albert first lists several properties: it clarifies the face of unsightly blotches, opens blockages, dries the tongue, and is effective for baldness. Then, he continues, "And some say, that if drunk while fasting, it produces a good intellect."[76] Short of controlled, laboratory and clinical tests, how could this be gainsaid? Other examples, for which parallels can easily be found today, concern a field test to determine whether fungi are poisonous or not,[77] and the belief that by eating beforehand *allium* (*Allium sativum* L.; garlic) one is protected from harm when collecting *elleborus niger* (*Helleborus niger* L.; black hellebore.)[78]

---

[74] Cf. 374/73; 393/111; 487/294; 499/311, etc.

[75] For Albert's reference to *squilla* (*Urginea maritima* (L.) Bak.; squill) see 561/431. This passage, along with others, has been discussed by Jerry Stannard, "Squill in ancient and medieval materia medica, with special reference to its employment for dropsy," *Bull. New York Academy of Medicine* 50 (1974), 684-713.

[76] *De veg.* 568/446. Another interesting claim is made by Albert in regard to *raphanus* (*Armoracia rusticana* Gaertn., Mey. et Schreb.; horse radish): "apud nos etiam maniacus, raphano contuso cum succo ligato super caput eius rasum, infra triduum recipit beneficium sanitatis" (558/426).

[77] *De veg.* 517/344.

[78] *De veg.* 510/333.

Whether these and the many other similar beliefs found in *De vegetabilibus* and contemporary herbals alike are always capable of being distinguished from the philosophic beliefs of the schools is problematic. But from the point of view of herbalism, the issue is clear cut: whatever offered any help in times of illness or served to avert harm, was deserving of recognition. The collection and compilation of such data, not their validation, was the task of the herbalist. With this in mind, one can, perhaps, better understand why Albert included a range of data over and above what we would term descriptive. Those other data, no less than the descriptions themselves, pertained to plants in all their diversity. It was Albert's task to organize them within the framework available to him.

15

The Medical Learning of Albertus Magnus

Nancy G. Siraisi
Hunter College

Albertus Magnus has an important place in the history of medical learning. Although he was not a physician and produced no medical works as such, his scientific writings treat extensively of human anatomy, physiology, and psychology.[1] Few other Latin authors of the

---

I am grateful to the following scholars for their suggestions and comments: Luke Demaitre, Bert Hansen, Pearl Kibre, James Weisheipl, OP.

[1] This material is contained mainly in the following works: *De generatione et corruptione* (Borgnet 4: 345-476); *De sensu et sensato, De somno et vigilia, De spiritu et respiratione, De aetate sive de juventute et senectute, De nutrimento et nutribili, De morte et vita* (all in Borgnet 9); *De homine* (Part 2 of the *Summa de creaturis*, Borgnet 35); *De anima*, ed. Clemens Stroick (ed. Colon. 7/1); and above all in *De animalibus*, ed. Hermann Stadler, *Beiträge* 15 (1916) and 16 (1920), and *Quaestiones super de animalibus*, ed. Ephrem Filthaut, OP (ed. Colon. 12: 77-351). For discussion of Albert's thought on specific aspects of anatomy, physiology, and psychology, see other contributions to the present volume, in addition to Joan Cadden, "The Medieval Philosophy and Biology of Growth: Albertus Magnus, Thomas Aquinas, Albert of Saxony, and Marsilius of Inghen on Book I, Chapter V of Aristotle's *De generatione et corruptione*," Indiana University dissertation, 1971; James R. Shaw, "Scientific Empiricism in the Middle Ages: Albertus Magnus on Sexual Anatomy and Physiology," *Clio Medica* 10 (1975), 53-64; Nicholas H. Steneck, "Albert the Great on the Classification and Localization of the Internal Senses," *Isis*, 65 (1974), 193-211; Thomas S. Hall, "Life, Death and the Radical Moisture: A Study of Thematic Pattern in Medieval Medical Theory," *Clio Medica*, 6 (1971), 3-23; Michael R. McVaugh, "The '*Humidum Radicale*' in Thirteenth-Century Medicine," *Traditio*, 30 (1974), 259-283; and Pierre Michaud-Quantin, *La Psychologie de l'Activité chez Albert le Grand*, Bibliothèque Thomiste no. 36 (Paris, 1966), especially Part 1, pp. 13-58. Regarding Albert and the medicinal uses of plants, stones, and minerals, see especially the articles by Jerry Stannard, John M. Riddle and James A. Mulholland in this volume.

thirteenth century, whether natural philosophers or physicians, could rival the range and depth of his learning about these subjects. He drew his information not only from the *libri naturales* of Aristotle, but also from Greek, Arabic, and Latin medical authorities. Where he was able, he welded the Aristotelian teachings together with those of the *medici* into a unified system of anatomy, physiology, and psychology; but when the two bodies of doctrine conflicted he did not hesitate boldly to confront the differences between them.

Furthermore, Albert's writings were used by physicians of his own and subsequent generations. While he was not the first author in the Latin West to point out the relevance of the *libri naturales*, and especially the biological works of Aristotle, to learned medicine and vice versa, nor the first to examine the discrepancies between Aristotle and the medical tradition, he was undoubtedly the most distinguished thirteenth-century scholar to turn his attention to those topics. It was perhaps partially due to Albert that Aristotle's books on animals, the works in which the Philosopher's ideas on anatomy and physiology are most comprehensively set forth, came to be eagerly studied in medical schools.[2] It is certain that Albert's commentary on those books, one of the earliest and fullest Latin expositions accorded them, became a valued resource of learned physicians. Thus various medical authors of the later thirteenth and fourteenth centuries, who were normally reluctant to cite by name in their works recent Latin scholars in disciplines other than medicine, made an exception for Albert.[3]

---

[2] As may be seen from the frequency with which they are cited by scholastic medical authors. The Paduan professor of medicine Peter of Abano (d. ca. 1316) endorsed the reading of the works on animals as part of a preparatory study of natural science essential (*necessarissima*) for medical students; see Peter of Abano, *Conciliator differentiarum philosophorum et praecipue medicorum* (Venice, 1496; Klebs 773.5), *differentia* 1, fol. 3r.

[3] For examples of citations of Albert by name by physicians, see Turisanus (Pietro Torrigiano de' Torrigiani, d. ca. 1319), *Turisani monaci plusquam commentum in Microtegni Galieni...* (Venice, 1512), fol. 46v, where Albert is listed among writers on vision; a treatise *De generatione embrionis* attributed to Dino del Garbo but probably actually by Gentile de Foligno (d. 1348), printed with Jacopo da' Forli, *Expositio Jacobi supra capitulum de generatione embrionis...* (Venice, 1502), fol. 17v, where Albert is cited as an authoritative commentator on Aristotle's views on reproduction; and (in a work on natural philosophy written by a physician), Peter of Abano, *Aristotelis Stagirite philosophorum summi problemata atque divi Petri Apponi Patavini eorundem expositiones* 14.15 (Mantua, 1475; Klebs 775.1), where Albert is cited along with Plato, Aristotle, Ptolemy, Galen, Avicenna, and Haly on regional influences on complexion.

The most learned *physician* to comment on the Aristotelian works on animals in the thirteenth century was no doubt Petrus Hispanus (Pope John XXI, d. 1277). The commentary of

And, finally, as is well known, Albert also acquired a reputation as a medical authority at a more popular level during the later Middle Ages and Renaissance. In one sense this was undeserved, since the mostly gynecological tract *De secretis mulierum*, widely circulated in manuscript and later many times printed under his name, is generally considered to be a spurious work (although its compiler seems to have culled some of his material from genuine works of Albert).[4] Yet the attribution of this treatise to Albert may reflect his justly acquired reputation as a scholar learned in human anatomy and physiology who drew upon medical as well as natural philosophical authorities and who wrote extensively concerning human reproduction.

Hence, a general review of the place of medicine in Albert's scientific output may be expected to throw light not only upon a significant aspect of his own achievement as a scientist, but also upon the foundations of the medical learning of the later Middle Ages. The present chapter will therefore briefly survey Albert's conception of medical learning and of its usefulness for the study of natural philosophy, the medical works known to him, and the ways in which he mined those works.

For Albert, as for the learned physicians who were his contemporaries, natural philosophy and medicine shared certain basic principles and overlapped in some of their subject matter, but were nonetheless always clearly distinguished as separate disciplines. Albert's recognition of the autonomy of medicine as a learned discipline is clearly revealed in his well-known remark, often cited as evidence for his belief in the independence of philosophy from theology:

Petrus Hispanus antedated that of Albert; it is however much shorter and apparently had only a limited circulation; see Sybil D. Wingate, *The Mediaeval Latin Versions of the Aristotelian Scientific Corpus, with Special Reference to the Biological Works* (London, 1931), p. 79, and Brian Lawn, *The Salernitan Questions* (Oxford, 1963), p. 77. No attempt has been made in the present study to assess either any possible influence of Petrus Hispanus on the medical learning of Albertus Magnus or the relative importance of their respective influences on subsequent physicians.

[4] On the manuscripts, editions, and authorship of *De secretis mulierum* see Lynn Thorndike, "Further Consideration of the *Experimenta, Speculum astronomiae* and *De secretis mulierum* Ascribed to Albertus Magnus," *Speculum*, 30 (1955), 413-443, and Christoph Ferckel, "Die Secreta mulierum und ihr Verfasser," *Sudhoffs Archiv für Geschichte der Medizin und der Naturwissenschaften*, 38 (1954), 267-274. Thorndike believed the work to be more closely related to Albert's authentic writings than did Ferckel. The present state of knowledge is summed up in Brigitte Kusche, "Zur 'Secreta Mulierum'-Forschung," *Janus*, 62 (1975), 103-123.

Augustine is to be preferred rather than the philosophers in case of disagreement in matters of faith. But if the discussion concerns medicine, I would rather believe Galen or Hippocrates, and if it concerns things of nature, Aristotle or anyone else experienced in natural things.[5]

Yet Aristotle himself had indicated the existence of some common ground between *physica*, the study of the principles of motion and change in the natural world, and medicine; in commenting upon Aristotle's words, Albert revealed his own conception of the interaction of the two disciplines, stating:

It is no part of *physica* to treat of sickness and health, but only of first principles and causes. However, the principles of life are the same as those of health and the principles of death the same as those of sickness...therefore in considering the principles of life one has to consider those of health, and in considering the principles of death one has to consider the first principles of sickness and from them the causes of disease. And therefore many of the *physici* and the more skilled among the *medici* who use the philosophical art to the greatest extent end up at the same point (*terminantur adinvicem*). For the *physici*, coming from first principles to their consequences and from universal considerations to particulars, end their thoughts about living things by reaching matters which have to do with medicine, namely the particular causes of health and sickness. But those physicians (*medici*) who use the art of *physica* ascend from particular diseases to general signs and causes and *accidentia*; for they do not cure the disease unless they remove its causes and draw out the causes of health.[6]

These remarks not only present a striking picture of the overlapping of the interests of natural philosophers and learned physicians as conceived in the thirteenth century, they also constitute a highly positive appraisal of learned medicine as an intellectual discipline. Albert's stand is the more noteworthy, in that respected authors of the twelfth century had been in disagreement as to the place of medicine among the arts and sciences. Although a tradition stretching back to Galen,[7] and indeed as we have just seen to Aristotle, regarded medicine and philosophy as closely akin, some maintained that the ultimately practical purpose of medical learning disqualified

---

[5] In II *Sent.*, distinctio 13, C, art. 2 (ed. Borgnet 27: 247a).

[6] *De sensu et sensato*, I, 1 (ed. Borgnet 9: 2b).

[7] See Owsei Temkin, *Galenism: Rise and Decline of a Medical Philosophy* (Ithaca, 1973), Chapter 1, passim.

it from inclusion among the higher, more speculative, branches of knowledge. For example, Hugh of St. Victor listed medicine among the mechanical arts and declared it to be a suitable occupation for manually adept members of the lower classes.[8] On the other hand, for Gundissalinus, medicine was one of a group of eight sciences, all of practical value, which were "contained under" or were "species of" natural science, itself one of the three divisions of theoretical philosophy.[9] Moreover, while twelfth-century writers on the natural world very commonly drew upon the medical literature known to them, and while the early stages of the reception of Aristotelian natural science in the west owed much to the intellectual activities of physicians,[10] after the full reception of Aristotle the *libri naturales* must have seemed much more impressive in scope and depth than the brief and aphoristic treatises by Hippocrates, Galen, and their followers which formed the core of medical teaching.[11] Thus, Alfred of Sareshel (Alfredus Anglicus), a pioneer in the use of the *libri naturales* writing before 1217, referred with scorn to those who thought they had achieved scientific understanding just because they had read Johannitius' medical *Isagoge* (a brief introduction to Galenic principles) and the *Aphorisms* of Hippocrates while remaining ignorant of the most important parts of *physica*.[12] Yet Albert, a far

---

[8] Hugh of St. Victor, *Didascalicon de studio legendi*, ed. Charles H. Buttimer (Washington, 1939), pp. 37-39, 43-44.

[9] Dominicus Gundissalinus, *De divisione philosophiae*, tr. Marshall Clagett and Edward Grant, in *A Source Book in Medieval Science*, ed. Edward Grant, (Cambridge, Mass., 1974), pp. 61, 63, 68-69.

[10] See Brian Stock, *Myth and Science in the Twelfth Century* (Princeton, 1972), pp. 25-26, and Alexandre Birkenmajer, "Le role joué par les médécins et les naturalists dans la réception d'Aristote au XIIe et XIIIe siècles," *La Pologne au VIe Congrès International des Sciences Historiques, Oslo, 1928* (Warsaw-Lvov, 1930), pp. 1-15.

[11] On the formation of the collection of medical texts known as the *ars medicinae* or *articella*, and its use in medical teaching from the twelfth century, see Paul O. Kristeller, "The School of Salerno: Its Development and Its Contribution to the History of Learning," *Studies in Renaissance Thought and Letters* (Rome, 1956), pp. 514-516; see the same author's "Bartholomaeus, Musandinus, Maurus of Salerno and Other Early Commentators of the 'Articella' with a Tentative List of Texts and Manuscripts," *Italia Medioevale e Umanistica*, 19 (1976), 57-87; the treatises collected in the *ars medicinae* by the thirteenth century usually included the *Aphorisms, Prognostics*, and *De regimine acutorum* of Hippocrates, the *Tegni* (*Microtechne, Ars parva*) of Galen, the medical *Isagoge* of Johannitius, and short tracts on urine and pulse.

[12] "Solo enim phisici nomine elati, quod 'Medicina in duas dividitur partes' aut quod 'Vita brevis' omissis primis praecipuisque Phisicae partibus se legisse gloriantur": Clemens Baümker, ed., *Des Alfred von Sareshel (Alfredus Anglicus) Schrift De motu cordis*, in *Beiträge* 23 (1923), 51-52. The incipits are those of the *Isagoge* and the *Aphorisms*, respectively. On Alfredus, see James K. Otte, "The Life and Writings of Alfredus Anglicus," *Viator*, 3 (1972), 275-291.

greater scholar in Aristotelian natural philosophy than any of his Latin predecessors, evidently regarded medical theory as a valid branch of scientific inquiry, and proclaimed his respect for learned physicians.

Nor was Albert's standpoint merely theoretical: his extensive knowledge of contemporary medical teaching and practice is evident in numerous passages scattered through his scientific works. As will appear from what follows, he was widely read in such learned medical literature as was available to him; but perhaps even more striking are several indications of his personal acquaintance with members of the medical profession and interest in their work. Albert's circle at Cologne included physicians with whom he discussed professional matters; evidently he listened eagerly to what they had to tell him. Sometimes he rejected their ideas, as when he "confounded" a certain physican of Cologne who believed that urine was "simply a simple substance."[13] On at least one other occasion he showed himself perhaps too ready to believe what his medical friends told him; he reported on the authority of one of them the case of a German noblewoman who miscarried of 150 fetuses at one time.[14] And at times he was content merely to note the treatment used, recording, for example, that he had seen skilled physicians cure "*mola*" by the administration of laxatives.[15]

None of the foregoing is, of course, intended to imply that Albert viewed any part of his own writings as primarily medical in purpose. He himself rather clearly indicated the necessary, but subsidiary, role played by medical learning in his own philosophical and scientific enterprise at the beginning of his commentary on *De somno et vigilia*:

> Because we have Aristotle's book about this science, we will follow him in the same way that we have followed him in other works, making digressions from him (*ab ipso*) wherever there seems to be some incorrect or obscure statement, dividing the work into books, and tractates, and chapters, as we did for the others. But, omitting any reference to the works of some of the moderns, we will follow only the opinions of the Peripatetics and especially Avicenna and Averroës and Alfarabi and Algazel, whose works on this subject we perceive as in agreement; however, we will touch from time to time upon the opinion of Galen.[16]

---

[13] *Quaestiones de animalibus* VI, 3 (ed. Colon. 12: 168).

[14] *De animalibus* IX, 1.5 (ed. Stadler, 15: 693).

[15] *Quaestiones de animalibus* X, 5 (ed. Colon. 12: 217).

[16] *De somno et vigilia* I, 1.1 (ed. Borgnet 9: 123a).

It is clear from this passage that Albert perceived Galen as the leading medical authority (as is also indicated by his reference to Galen along with Hippocrates in his remarks on the standard authorities in theology, philosophy, and medicine quoted above); that he regarded some exposition of Galen's views as indispensable; that he was perhaps aware that his knowledge of Galen's ideas was more limited than his knowledge of those of the philosophers named; and that he perceived Galen as at least potentially in opposition to the Peripatetics on certain issues. By and large, further perusal of Albert's scientific works serves only to confirm this assessment of his views regarding the place of medical learning in general, and Galen in particular, in the study of natural philosophy.

Albert also at times proclaimed his unwillingness to discuss particulars better handled by medical writers. For example, in alluding to different factors affecting longevity he remarked that health and sickness were treated to the extent appropriate in natural science (*quantum confert scientiae naturali*) in his work *De natura locorum*.[17] Albert's authentic works are, moreover, almost entirely free of descriptions of symptoms and diseases. Yet the frequency with which Albert indicated the medical effects attributed to the various plant species listed in Book VI of *De vegetabilibus* suggests not merely that he drew upon medical herbals such as that of Dioscorides, but also that he may perhaps have intended this part of his work to be of practical use to physicians. In his own words: "In this sixth book of our work on plants, we are satisfying the curiosity of students rather than philosophy; for concerning particulars philosophy cannot treat."[18] The students he had in mind may well have included students of medicine, since it is presumably they who would be most likely to be interested in the abortifacient, aphrodisiac, or other properties of herbs regularly indicated by Albert and in his occasional remarks naming the diseases for which a particular plant was supposed to be a remedy. It may be noted too that on a few occasions Albert passed on to his readers medical recipes or prescriptions such as the two attributed to Galen in *De animalibus* XXII,2.1.

The question as to where and how Albert acquired his interest in human anatomy and physiology, his respect for at least some learned physicians, and his knowledge of medical authorities, remains open.

---

[17] *De morte et vita* II,1 (ed. Borgnet 9: 351a); see also *De vegetabilibus et plantis* VI, 1.22, under *granatum* (ed. Borgnet 10: 186). Other examples of similar statements could be added.

[18] Ibid., VI, 1.1 (ed. Borgnet 10: 159).

A well-attested and early tradition, supported by references in his own works, holds that as a young man Albert studied at Padua and was there received into the Dominican Order by Jordan of Saxony.[19] Unfortunately, the best efforts both of biographers of Albert and historians of the University of Padua have failed to produce any definite evidence regarding the nature or duration of his studies in that city. The history of the infant Paduan *studium* (the traditional date of the foundation of the University of Padua is 1222, although there were some schools in the city before that date) is extremely obscure.[20] Rhetoric and law were certainly taught there in the 1220s; teaching in the remaining liberal arts and in natural philosophy may have been available, but there are no clues as to the curriculum. It may be noted, however, that twelfth and early thirteenth-century students of arts seem quite frequently to have read medical texts.[21] A document of 1228 suggests that medical teaching may then have been available at Padua, or at any rate desired by the scholars, but this conclusion cannot be drawn with any certainty.[22] Literate medical men and astrologers (the association of medicine with astrology in the thirteenth century was often very close) were certainly present in Padua during the tyranny of Ezzelino da Romano, which began in 1237, although the schools were then either defunct or moribund; but the recorded history of Paduan medical teaching may be said to begin only after the revival of the *studium* in the 1260s. Thus, while it is quite probable that learned — or at least literate — physicians were present and some medical teaching was carried on when Albert was a scholar at Padua in the 1220s, it is impossible to state whether or not he himself ever studied medicine there or, if so, for how long and in what circumstances. That Albert's own interest in natural phi-

---

[19] See pp. 17-19, Chapter 1, above.

[20] For a summary of the available evidence and bibliography concerning the early history of the schools of Padua, see N. Siraisi, *Arts and Sciences at Padua* (Toronto, 1973), Chapter 1.

[21] See Paul O. Kristeller, "Beiträg der Schule von Salerno zur Entwicklung der Scholastischen Wissenschaft im 12. Jahrhundert," in *Artes liberales: von der Antiken Bildung zur Wissenschaft des Mittelalters*, ed. Josef Koch (Leiden and Cologne, 1959), pp. 87-88, and my note 10 above.

[22] A contract negotiated by scholars of Padua with the commune of Vercelli to provide for a migration of scholars to the latter city calls for Vercelli to pay the salaries of two professors of medicine. This provision does not, however, necessarily reflect the contemporary state of affairs at Padua, since the contract also calls for a salaried chair of theology, a subject not taught publicly at Padua until 1363. The document is printed in Andrea Gloria, ed., *Monumenti della Università di Padova (1222-1318)* (Venice, 1884), documents, pp. 5-8, and discussed in Hastings Rashdall, *The Universities of Europe in the Middle Ages*, 2nd ed. (Oxford, 1936), 2: 11-13, and Heinrich Denifle, *Die Entstehung der Universitäten des Mittelalters bis 1400* (Berlin, 1885), pp. 279-281.

losophy was aroused during or before his student days at Padua is indicated by references in his own works to personal observations of natural phenomena made at Padua and Venice.[23] In this period of his life his interests apparently already included human physiology, since he took the trouble to record, and later included in a discussion of the ability of hibernating animals to go with out food, the case of a Paduan woman said to have fasted for forty-five days; Albert suggested that lack of physical exertion facilitated successful fasting.[24]

But the most likely source of much of Albert's knowledge of medical authorities seems not to be his early studies at Padua, whatever they may have encompassed, but private reading pursued over a course of many years. There seems no reason to suppose that his religious vocation would have been a bar to such reading, any more than it proved to be to his various other scientific interests and activities. It is possible, indeed, that those Dominicans who opposed Albert's involvement with Aristotelian natural philosophy[25] may have been more sympathetic to his medical studies; after all, traditional monastic learning up to the twelfth century quite frequently included the study of medical treatises, just as monastic activities frequently included medical practice.[26] That the atmosphere in the Dominican Order in its early years was not hostile to the study of medicine by some friars is perhaps demonstrated by the career of Albert's confrere and approximate contemporary, the surgeon and medical author Theodoric of Lucca, bishop of Bitonto and later of Cervia (d. 1298).[27] The son of the municipal surgeon of Bologna,

---

[23] *Meteorologica*, III, 2.12 (ed. Borgnet 4: 629a); *Mineralium libri quinque* II, 3.1 (Borgnet 5: 48b-49a). Both references are pointed out in Thorndike, *A History of Magic* 2: 523.

[24] *De homine* I, q.10, art.5 (ed. Borgnet 35: 119b).

[25] Albert himself complained of these opponents; see his *In Epist. B. Dionysii Areopagitae*, ep.7, n.2, B (ed. Borgnet 14: 910a). I owe this reference to James Weisheipl, OP, *Friar Thomas D'Aquino, His Life Thought and Work* (Garden City, New York, 1974), pp. 42-43, where discussion is also provided.

[26] Cassiodorus, the author of one of the classics of monastic prescriptive literature, plainly envisaged that monks would both study and practice medicine, which he included under divine, not human, readings; see Cassiodorus Senator, *Institutiones divinarum et humanarum lectionum* I, 31. On several occasions during the twelfth and thirteenth centuries religious were enjoined by the papacy or by conciliar decree from leaving the cloister to study medicine in the schools, but these bans do not appear to refer to the private reading of medical books.

[27] On Theodoric's career and work see Mauro Sarti and Mauro Fattorini, *De claris archigymnasi Bononiensis professoribus*, 2nd ed. (Bologna, 1888-1896), 1: 537-544 and (for his will) 2: 233-237; Mario Tabanelli, *La Chirurgia italiana nell'alto Medioevo* (Florence, 1965), 1: 198-210; E. Campbell and J. Colton, tr., *The Surgery of Theodoric* (2 vols., New York, 1955-1960). The identity of the surgical author Theodoric with the Dominican bishop of Bitonto and Cervia, and of either with the son of the surgeon Ugo of Lucca, was formerly questioned, but now appears to be conclusively established; on the bibliography of this dispute, see Tabanelli, 1: 203-208.

Theodoric had apparently practised medicine before becoming a friar and continued to do so thereafter, yielding to the demands of his numerous and influential patients but returning his fees to the authorities of the Order. His well-known work on surgery was written after he became a bishop. After his elevation to the episcopate he also received papal permission to retain his own fees, to make a will, and, apparently, to reside at Bologna, then already a noted center of medical learning.

In the case of Albert, unlike that of Theodoric, there was, of course, no question of specialization in medicine or medical practice. But there are several indications that Albert's knowledge of medical authorities and teaching grew slowly over a long period, and were by no means solely the fruit of his early studies as a layman at Padua or elsewhere. In the first place, it seems evident that his interest in medicine was largely a by-product of, and developed along with, his general scientific interests and, in particular, his effort to interpret and round out the natural philosophy of Aristotle for the Latins. The latter enterprise, as noted in Fr. Weisheipl's essay above, began at Paris in the 1240s and continued there and in Germany for some twenty years. Moreover, as already noted, at the time when he wrote his *Quaestiones de animalibus* (1258) and *De animalibus* (ca. 1260-1264), Albert was in contact with several learned physicians at Cologne and drew from them information about their practice which he incorporated in those works. In this connection, too, one may note the striking multiplication of allusions to medical works, and particularly of references to Galen by name, in the *De animalibus* as compared to the early *De homine* (ca. 1245-1248). In the latter, a theological work, Albert was content to draw much of his physiological information from the *De natura hominis* of Nemesius of Emesa, which he believed to be a work of Gregory of Nyssa;[28] his reliance upon "Gregory" cannot have been because he wished to quote only theological authorities in a work on the theology, since in *De homine* citations of Aristotle and Avicenna abound. The difference in this regard between *De homine* and *De animalibus* is no doubt in part due to the subject matter; *De homine* is primarily a work on the soul, of which a relatively small portion deals with psychology and sense perception, whereas anatomy and physiology constitute a principal part of the subject matter of *De animalibus*. Moreover, the question of the extent

---

[28] Albertus Magnus, *De fato*, ed. Paul Simon (ed. Colon. 17/1: xxxiv).

of Albert's direct knowledge of Galen at any period of his life is a very vexed one; much of what he knew of Galen's teaching probably came via Avicenna (see below). Nonetheless, even allowing for these factors, it does seem that Albert was unquestionably readier to name medical authorities in the later than the earlier work, a tendency which lends support to the view that his medical interests developed over the years.

A survey of the fairly numerous references by name to medical authors and books in a number of the natural philosophical works of Albert and of the allusions to such sources identified by the editors of his *De animalibus* and *Quaestiones de animalibus*[29] also tends to suggest that Albert's medical learning was acquired by private reading rather than through any formal course of study. He was familiar with some of the standard medical textbooks and encyclopedias, such as the *Pantegni*[30] of Constantinus Africanus and a work, probably the *Continens*[31] of Rasis, but others equally widely used in the medical schools were very infrequently cited by him. For example, he seldom used the *Tegni* of Galen, and I have noticed no references by name in his works to the *Isagoge* of Johannitius.[32]

Furthermore, the paucity of Albert's allusions to Hippocrates is very striking; in all twenty-six books of Albert's *De animalibus* there is apparently only one citation of the *Aphorisms*, which in the thirteenth century not only was the best known and most widely read among the works of Hippocrates, but also served as a basic medical textbook. Albert seldom or never gave the titles of Hippocratic treatises. Hippocrates is called to witness nine times in *De animalibus*; only five of these references have been identified, all except that to the *Aphorisms* being allusions to *De spermate* or *De semine* (that is, *De natura pueri*), and some of them coming via the *Canon* of Avicen-

---

[29] No attempt has been made in the present study to trace medical authors or works used but not named by Albert in the composition of those works contained only in the Borgnet edition; nor has any effort been made to add to the identifications of named and unnamed medical authorities included in Stadler's edition of *De animalibus* and Filthaut's of the *Quaestiones de animalibus*.

[30] *De animalibus* XXII,1.2 (ed. Stadler, 16: 1351).

[31] *De motibus animalium* I, 2.1 (ed. Borgnet 9: 270a), and five citations in a section of *De animalibus* devoted to the anatomy of the arteries and veins (I, 2.20-21 [ed. Stadler, 15: 138, 147, 149]). The cluster of citations in this one part of the work perhaps suggests that Albert had a copy of the *Continens* in his possession while writing this particular section.

[32] The *Tegni* is cited in *Quaestiones de animalibus* I, 16 and XII, 15 (ed. Colon. 12: 91, 233); it is not named by Albert in *De animalibus* nor identified by the editor as among the Galenic sources of that work. On the *Tegni* and *Isagoge* in the medical schools, see note 11 above.

na. In the *Quaestiones de animalibus* Hippocrates is apparently not mentioned by name at all; the editor has identified a single allusion to *De natura pueri*.[33]

Very different is the situation in regard to references to Galen. A perusal of Albert's natural philosophical works leaves the impression that as far as their author was concerned the name of Galen was synonomous with medical wisdom. Albert certainly referred to Galen by name far more frequently than to any other medical author, ancient or medieval, Greek, Latin, or Arabic; in *De animalibus*, for example, Galen is cited as a medical authority over sixty times, and whole chapters are devoted to the exposition of his supposed views. Yet this impression is somewhat misleading. As the careful identification of sources throughout Stadler's edition of *De animalibus* demonstrates, many of Albert's allusions to Galen are in fact derived from Avicenna. It is, as one of Albert's most recent scholarly editors has remarked, very difficult to track down Albert's Galenic references and almost impossible to determine what works of Galen he knew at first hand.[34] Albert rarely gave the titles of works by Galen; Book XXII of *De animalibus* stands out from the rest of that work because in it Albert cited by name treatises of Galen on

---

[33] The study of works attributed to Hippocrates played a major part in medieval medical learning; see Pearl Kibre, "Hippocratic Writings in the Middle Ages," *Bulletin of the History of Medicine*, 18 (1945), 371-412, and the same author's "Hippocrates latinus; Repertorium of Hippocratic Writings in the Latin Middle Ages," part 1, *Traditio*, 31 (1975), 99-126, and subsequent issues.

[34] Ephrem Filthaut, OP, ed. *Quaestiones super de animalibus* (ed. Colon. 12: xlvi-xlvii). A glance at the list of Galenic references identified in Stadler's edition of *De animalibus* confirms the difficulty. In many cases, the editor noted that the citation came via Avicenna, but one suspects that in reality this was the case in an even larger number of instances; among the treatises to which Stadler found allusions are *De usu partium* (Kühn 3: 1-933 and 4: 1-366), *De anhelitu* (presumably either *De causis respirationis*, Kühn 4: 465-469, or *De utilitate respirationis*, Kühn 4: 470-511, or *De difficultate respirationis*, Kühn 7: 753-960), *De theriaca* (either *De theriaca ad Pamphilianum* or *De theriaca ad Pisonem*, Kühn 14: 295-310 and 210-294), *Introductio, seu medicus* (Kühn 14: 674-797), and *De placitis Hippocratis et Platonis* (Kühn 5: 181-805). All of these except the last appear to have been first translated into Latin in the fourteenth century (see TK 1183, 1296, 1353, 1224, 591, and bibliography there cited). *De placitis* is also generally supposed to have been untranslated in the thirteenth century, although there are some indications that Constantinus Africanus may have prepared a version of it (now lost); see Temkin, *Galenism*, p. 100, and Ynez Violé O'Neill, "The Funfbilderserie Reconsidered," *Bulletin of the History of Medicine* 43 (1969), 236-245.

Stadler also identified allusions to the treatises on anatomy, complexion and reproduction named in the following note, and to *De virtutibus naturalibus* (*De facultatibus naturalibus*, Kühn 2: 1-214), *De elementis* (doubtful) (Kühn 1: 413-508), *De regimine sanitatis* (doubtful) (*De sanitate tuenda*, Kühn 6: 1-452), *De simplicibus medicinalibus* (Kühn 11: 379-892 and 12: 1-377), and *De motibus liquidis* (TK 577 and 748).

anatomy, reproduction, disease, and complexion (that is, the balance of humoral qualities in each individual).[35] Since Book XXII also contains the titles of other medical treatises and medical recipes, it is possible that Albert had a collection of medical works at hand while writing it.[36] In addition to the works of Galen just referred to, it seems likely that Albert also knew at first hand the widely distributed introductory *Tegni*, a version of *De simplicibus medicinalibus*, and perhaps *De facultatibus* (or *virtutibus*) *naturalibus*. It must be admitted, however, that most of the information ultimately derived by Albert from Galen could easily have been obtained via Avicenna or other intermediaries.

The uncertainty surrounding Albert's Galenic sources is rather typical of the confused picture of thirteenth-century Galenism. Albert was correct in his assumption, exemplified not only in his use of the name of Galen, along with that of Hippocrates, as that of the standard medical authority, but also in his various references to the views of undifferentiated *medici*, that there existed a more or less unified tradition of medical learning of which Galen was the prototypical representative. The tradition was that of classical Greek and Roman medicine, carried over to the early Middle Ages from late antiquity by means of a variety of brief Latin texts and translations. From the late eleventh century, the often fragmentary and distorted Galenic learning of the Latin West was immeasurably enriched by the translation into Latin of many hitherto unavailable treatises by

---

[35] *De cura membrorum* at XXII, 1.2 (ed. Stadler, 16: 1350); *De spermate*, at XXII, 1.2, 3 (16: 1350, 1351); *De accidenti et morbo*, at XXII, 1.5 (16: 1354); *De complexionibus* at XXII, 2.1 (16: 1367). *De cura membrorum* is presumably the abbreviated Latin version of *De usu partium* commonly known as *De iuvamentis membrorum*; although defective, this translation provided the best account of Galen's anatomical teaching available in Latin in the thirteenth century, and, as Stadler demonstrated, Albert made heavy use of it elsewhere in *De animalibus. De spermate* may be either *De semine* (Kühn 4: 512-651), or another treatise of the same title also attributed to Galen in the Middle Ages (see TK 1520, 1521). *De accidenti et morbo* was, as Filthaut points out (ed. Colon. 12: xlvi-xlvii), a collection made up of *De morborum differentiis* (Kühn 6: 836-880), *De causis morborum* (Kühn 7: 1-41), *De symptomatum differentiis* (Kühn 7: 42-84), and *De symptomatum causis* (Kühn 7: 85-272); there are several thirteenth-century manuscripts of this compilation (TK 684, 745). *De complexionibus* (that is, Books I and II of *De temperamentis*, Kühn 1: 572-694) was twice translated into Latin during the twelfth century, from the Arabic by Gerard of Cremona, and from the Greek by Burgundio of Pisa; see *Galenus Latinus 1: Burgundio of Pisa's Translation of Galen's "De complexionibus,"* ed. Richard J. Durling (Berlin-New York, 1976), p. vii.

[36] The other medical works named are Constantinus Africanus, *Viaticum* at XXII, 2.1 (ed. Stadler, 16: 1368); idem, *De coitu*, at XXII, 1.1 (16: 1349), XXII, 1.3 (16: 1351), XXII, 2.1 (16: 1365); idem, *Pantegni*, at XXII, 1.2 (16: 1351); and Serapion, *De medicinis simplicibus* at XXII, 2.1 (16: 1369).

or attributed to Galen, and by the acquisition in Latin of the works of Arabic medical writers who had a fuller knowledge of Galen's works and understanding of his thought than most physicians in the early medieval west. Nonetheless, in the mid-thirteenth century some of Galen's most important works, and in particular the major anatomical treatises, had not yet been translated into Latin. Moreover, it is probable that access to Galenic teaching was more frequently obtained via the brief treatises enthroned as the standard medical curriculum during the twelfth century, or through the mediation of Arabic authors, rather than through the study of such as Galen's longer and more detailed treatises as were available in Latin.[37] The state of Albert's knowledge of Galen was therefore equal to that of the most learned physicians of his age; it should perhaps be termed a knowledge of "Galenism" rather than of Galen. This thirteenth-century Galenism was a body of traditional medical teaching drawn from a variety of writings more or less Galenic in inspiration. It did not incorporate everything Galen had taught and it was not immune to influence from other sources. Whether his knowledge of it was acquired by the direct study of Galen's treatises, or via the great Arab physicians, or both, Albert was, as will become apparent below, sufficiently master of this body of material to harmonize various Galenic theories with those of Aristotle and to be able to present his readers with coherent and lucid summaries of Galenic doctrines with which he disagreed.

As already suggested, the single medical author most frequently and extensively consulted by Albert was in all probability not Galen but Avicenna. In Book I of his *De animalibus*, indeed, Albert's main purpose seems to have been to supplement Aristotle's teaching on anatomy with material drawn from the *Canon* of Avicenna. Albert may legitimately be regarded as among the pioneers in the Latin West in the use of Avicenna as a medical authority. The *Canon* which was to reign as perhaps the single most consulted textbook in medical schools until the sixteenth century, was translated from Arabic into Latin by Gerard of Cremona (d. 1187),[38] but there is little evidence of any extensive use of it by any Latin writer in the late twelfth or first quarter of the thirteenth century. One of the earliest

---

[37] Temkin, *Galenism*, pp. 95-116, and bibliography there cited.

[38] It is included in a list of Gerard's translations drawn up by his pupils; see Grant, *Source Book in Medieval Science*, p. 38 (no. 63).

Latin works to incorporate whole passages drawn from the *Canon* was, as might be expected, a medical treatise, namely the *Anatomia vivorum* formerly ascribed to Ricardus Anglicus.[39] This treatise, which has been dated between 1210 and 1240 with the greatest likelihood attached to some time about the year 1225, is similar in content and method to parts of the *De animalibus* of Albert. There is evidently a relationship of some kind between the two works; no doubt Albert, the later writer, knew and used his predecessor's treatise. Yet Albert's use of the *Canon* has been shown to be independent of that of the author of the *Anatomia vivorum*.[40] The *Canon* is also frequently cited in Albert's *Quaestiones de animalibus*. Furthermore, Albert may have been familiar with Avicenna's great medical encyclopedia long before he wrote either of his works on the Aristotelian books on animals if one may refer to the *Canon* the allusion to "Avicenna in sua anatomia" found in *De homine*, written in the 1240s.[41] Thus Albert, a theologian and natural philosopher, was familiar with a major medical work that had only recently come into use among learned physicians themselves. Albert was also among the earliest Latin authors to rely extensively upon Avicenna's compendium *De animalibus* (translated between 1220 and 1232).[42] This treatise cannot, of course, be classified as a medical work, but it does contain detailed discussion of discrepancies between Aristotle and Galen and reflects a good many views derived by Avicenna from his medical as well as his philosophical studies.

Averroës, too, was known to Albert as an authority on anatomy and physiology. By the 1240s the latter was already familiar with a brief tractate on the heart, apparently part of an unfinished compendium on the Aristotelian books on animals, in which Averroës, like Avicenna before him, compared the anatomical and physiological teaching of Aristotle with that of Galen. Years later, when Albert came to summarize in his own *De animalibus* the views of major thinkers on some of the points at issue, he incorporated the *De corde* of Averroës into his text "almost word for word in its entirety."[43]

---

[39] See George Corner, *Anatomical Texts of the Earlier Middle Ages* (Washington, 1927), pp. 38-44.

[40] Ibid., p. 40.

[41] *De homine* II, q.23 ad 2 (ed. Borgnet 35: 229b).

[42] *Aristoteles latinus*, ed. G. Lacombe, et al., I (Rome, 1939): 81.

[43] "*Presque textuellement et en entier*" (author's italics): R. de Vaux, "La première entrée d'Averroës chez les Latins," *Revue des Sciences philosophiques et théologiques*, 22 (1933), 193-245, at p. 225; *De homine* II, q.23, a.3, part. 5 (ed. Borgnet 35: 279a-b) and De Vaux, p. 240.

Albert could not, however, have been familiar with Averroës' principal medical work, the comprehensive survey of medicine entitled *Colliget*, as it was not translated into Latin until after Albert's death.[44] The allusions to the *Colliget* pointed out by the editor of the *Quaestiones de animalibus* must therefore be numbered among the later interpolations which that work is known to contain.[45]

Finally, it may be noted that Albert's *Quaestiones de animalibus* both drew upon and helped to shape the tradition of medical question literature. The subjects of a number of the *Quaestiones de animalibus* are drawn from earlier collections of medical questions; other questions discussed by Albert recur in later scholastic medical treatises and commentaries.[46]

Although Albert's references to medical authorities are scattered through works on a variety of different topics, it is possible to discern a pattern in the uses he made of his medical knowledge. Most of his medical material is to be found in one of three contexts: (1) as part of medical anecdotes or admonitions which he retailed apparently because they had caught his attention in personal exchanges with physicians of his acquaintance; (2) in passages supplementing Aristotle, where information drawn from medical authorities is used to fill out areas of anatomy or physiology that the Philosopher had ignored or treated with excessive brevity; (3) in examinations of the differences between Aristotelian teaching and the medical tradition, usually to the detriment of the latter.

The material in the first category is much smaller in quantity than that in either of the other two, but some of it vividly illustrates the immediacy of Albert's own personal interest in the kind of medical detail that he normally sought to exclude from his Aristotelian commentaries as beyond the scope of natural philosophy. Fortunately

---

[44] The *Colliget* was translated into Latin in 1285; see the explicit of the work in Cesena, Biblioteca Malatestiana, MS Pluteo XXV (lato destro) 4, fol. 27v. The dates 1255 or 1289 found in various printed sources spring from misreadings of this explicit. I am grateful to Professor Michael McVaugh for supplying me with a photocopy of the leaf in question.

[45] Filthaut (ed. Colon. 12: xliii-xlv) and his "Um die Quaestiones de animalibus Alberts des Grossen," in ed. Heinrich Ostlender *Studia Albertina: Festschrift für Bernard Geyer zum 70. Geburtstage*, (Münster, Westf., 1952), pp. 112-127.

[46] Lawn, *Salernitan Questions*, p. 85, suggests that Albert's treatment of the question form and much of his question material is modelled upon that incorporated in the commentary on the works on animals by Petrus Hispanus. Examples of questions included in Albert's *Quaestiones de animalibus* which are also to be found in later scholastic medical works include "Utrum solus sanguis inter humores nutriat" (3.20); "Utrum nervi habeant ortum a cerebro vel a corde" (3.7); "Utrum sternutatio sit signum boni vel mali" (1.38); and many others.

for the reader who wishes to understand the nature of Albert's interest in medicine, his curiosity as a student of that subject on occasion got the better of the theoretical limitations he placed on the appropriate uses of medical material in a philosophical work. One may note, for example, his remarks on the importance of excluding wine from the diet of children and restricting its intake by wetnurses;[47] the accounts of *mola* and of a multiple miscarriage already alluded to; and his properly skeptical report of an anecdote supposedly illustrating the relationship of brain tissue and semen recounted to him by master Clemens of Bohemia.[48]

By far the most extensive use of medical authorities by Albert is to be found in the second category, in which medical sources are drawn upon to supplement Aristotle's account of human anatomy, physiology, and psychology.[49] For example, Albert attached importance to the system of humors, fluids, complexions and virtues, which played a central role in medieval medical theory, and derived chiefly from Galenic sources. Thus, in *De animalibus* he inserted lengthy passages in which this system is expounded with approval; one may note that the second tractate of Book III is largely given over to a discussion of the various kinds of *humiditas* in animals and of the humors. Two chapters on these topics were labelled by Albert as digressions from the Aristotelian text, and the citations throughout the tractate are of Galen and Constantinus Africanus; the editor notes that much of the Galenic material comes via the *Canon*. Similarly, *tractatus* 1 of Book XII deals with complexion theory, largely on the basis of Galen's *De complexionibus* via the *Canon* and *De animalibus* of Avicenna.

Even more striking is the way in which Albert systematically expanded Aristotle's treatment of human anatomy by the insertion of material drawn from the *Canon* and other medical sources throughout the second and third tractates of Book I of *De animalibus*. It was, in the first place, Albert's decision to set this

---

[47] *De somno et vigilia* I, 2.8 (ed. Borgnet 9: 151a).

[48] *Quaestiones de animalibus* XV, 14 (ed. Colon. 12: 268). According to this story sexual excess brought about the death of a monk whom an autopsy subsequently revealed to have been drained of brain tissue. If this anecdote really dates from the middle years of the thirteenth century it is remarkable for the casual reference to autopsy to determine the cause of sudden death. Such autopsies were in fact carried out in Bologna by the 1290s, but earlier authentic examples do not seem to be recorded. The editor of the *Quaestiones de animalibus* was unable to identify further Master Clemens of Bohemia, nor have I succeeded in doing so.

[49] See further the secondary sources listed in note 1 above.

material apart in separate tractates dealing with human external and internal anatomy respectively, since no such divisions are found in the Aristotelian text.[50] Aristotle had provided in Book I of the *Historia animalium* only a brief description of the anatomy of man in twelve short chapters (491a19-497b2), explaining its position at the beginning of the work on the grounds that man was the most familiar animal and thus provided a standard of comparison. One may contrast Albert's introduction to his lengthy exposition — 184 closely printed pages in Stadler's edition — of the same material: "It is indeed very necessary to begin by providing an account of the parts of the most perfect animal, which is man, according to the division of his parts that is called anatomy by the Greeks."[51] Elsewhere in *De animalibus* Albert repeatedly referred to the material in Book I as his "anatomy."[52] In *tractatus* 2, on external anatomy, no fewer than fourteen out of twenty-six chapters are labelled as digressions, that is, as including material pertaining to anatomy but not included, even briefly, in Aristotle's text. The subjects of these chapters include the anatomy of the skull, eyes, thorax and arms, the bones of the thighs and feet, the muscles of the chest, arm, back, stomach, and bladder, the nerves, the arteries, and the veins, and the source is, almost exclusively, the *Canon* of Avicenna. Albert justified the inclusion of this material by saying:

> The teaching about the parts of the human body according to the opinion of Aristotle can in no way be transmitted unless we digress to transmit complete teaching about the anatomy of the bones and nerves and muscles and veins and arteries: because these are found in the composition of all the parts into a whole.[53]

The *Canon* is also the source of much additional material in the remaining chapters of the tractate, which expand the brief Aristotelian text. It should perhaps be added that in rounding out Aristotle's anatomy Albert was not content to draw only upon medical authorities; he also richly expanded Aristotle's fairly numerous references to physiognomy, a branch of learning which Albert appears to have regarded as almost as important as anatomy and as inseparably associated with it.

---

[50] See also the passage from Albert's *De somno et vigilia*, p. 384 above.

[51] *De animalibus* I, 1.1 (ed. Stadler, 15: 2).

[52] For example, ibid. XII, 3.2 (15: 870): "De compositione autem nasi satis diximus in anathomia"; ibid. XIII, 1.2 (16: 897): "in nostra quam superius posuimus anathomia."

[53] Ibid., I, 2.11 (15: 90-91).

The length of the anatomical portions of Book I of *De animalibus* precludes any comprehensive analysis here of the way in which Albert made use of his sources in its compilation. Some notion of his method may perhaps be gained, however, from the following summary consideration of his treatment of the anatomy of the head in *De animalibus* I,2.1-10.[54]

In Chapter 1, Albert commented on Aristotle's discussion of the skull, which consisted only of a short definition of sinciput and occiput, of sutures, and of the phenomenon of a double crown in the hair. Albert, according to his editor, confused sinciput and occiput and he repeated Aristotle's mistaken belief that women had only one suture, although providing an elaborate rationalization intended to reconcile this statement with his own knowledge that several bones made up the skull of woman as of man. But he added an explanation of an Arabic medical term and a description of the sutures on a skull he had seen himself, together with an account of the location of the brain, of the skull's function in protecting it, and of the membranes surrounding it. Furthermore, he appended to the chapter a set of general definitions of various parts of the body and their functions, namely the bones, joints, cartilage, nerves, ligament (*ligamenta*, which Albert here, unlike various other ancient and medieval authors, distinguished from the nerves), and membranes.

In his second chapter, Albert took up the subject of the physiognomy of the head, down to the eyes. Aristotle had indicated the physiognomical implications of various facial configurations, but Albert prefaced his much longer and more detailed discussion with a reasoned defence of the science of physiognomy. He was concerned to avoid any suggestion of physiognomical determinism and to indicate the main authorities in this "science." Chapter 3 continues along the same lines with a discussion of the physiognomy of the eyes. While, as we shall see, Albert elsewhere gave detailed descriptions of the anatomy of the eye and of the process of vision, in this chapter he was concerned only with variations in color and shape and their meaning for physiognomy.

Chapter 4 treats of the ears. The main way in which Albert extended Aristotle's description of the outer ear and the ear canal was by the addition of causal explanations: the ear canal is convoluted because if it were straight "abundant heat and excessive cold

---

[54] Ibid. (15: 38-90).

would penetrate along it to the brain and damage it." But he also incorporated a short account of the auditory nerve, with a crossreference to his own subsequent discussion of the anatomy and function of the nerves. Aristotle's statements about the positioning of the ears in various animals Albert supplemented with an explanation of why human ears are placed on the side of the head: the eyes must be in front because light travels along straight lines, and if both eyes and ears were in front the complexion of one pair of organs of sense would be confused with that of the other. The remainder of the chapter deals with the physiognomy of the ears. Thus in the compass of this one brief chapter, which in actuality follows the Aristotelian text fairly closely, Albert called into play his knowledge of medical teaching in regard to the nervous system, the four qualities, and complexion.

Chapters 5 and 6 are "digressions" in which Albert transmitted Avicenna's detailed account of the anatomy of skull, jaw, and teeth, introducing this material as follows:

> We cannot appropriately discuss the composition and physiognomy of the other parts of the head unless we here digress to show the shapes and make-up of the whole head and the anatomy of the bones which constitute it. For then both the material discussed above will be understood to be true, and the more certain facts (*certiora*) which follow will be more plainly understood. And we cannot properly investigate the bones of the head and the way they are put together unless we first point out their shapes and divisions.[55]

Chapter 7 is another *digressio* setting forth in detail the anatomy of the eye, with its three humors and seven tunics, and providing a reference to Albert's earlier discussions of the theory of vision in his expositions of *De anima* and *De sensu et sensato*.[56] In the following three chapters, Albert returned to the exposition of Aristotle's description of the remaining parts of the head, namely, nose, jaws, lips, and tongue, following the Aristotelian text fairly closely, but also drawing upon the *Canon*, Avicenna's *De animalibus*, various physiognomical works and Damascenus.

As the foregoing illustrations from his *De animalibus* show, Albert regarded much of the content of the most advanced contemporary

---

[55] Ibid. (15: 64).

[56] Ibid. (15: 75). For an account of Albert's views on vision, a topic repeatedly discussed by him in various works, see David C. Lindberg, *Theories of Vision from al-Kindi to Kepler* (Chicago, 1976), pp. 104-107.

medical teaching as both true in itself and indispensable for a proper understanding of some parts of natural philosophy. We have seen that he incorporated whole tractates on the system of humors and complexions; that he thought it "very necessary" (*oportet maxime*) to begin the *scientia de animalibus* with an account of human anatomy; and that he believed that a summary he had compiled from Avicenna and other medical authors provided "complete teaching" (*perfectam doctrinam*) on a branch of that subject, teaching without which Aristotle's doctrine could not be understood. Yet no account of Albert's medical learning would be complete which passed over his thoroughgoing hostility to some aspects of the medical tradition. He devoted much attention to the investigation of points of difference between Aristotle and Galen. While he sometimes attempted reconciliation, on major issues he invariably favored Aristotle. On one occasion he did not hesitate to characterize a follower of Galen as lapsing into insanity.[57] We now turn to the background of this third category of Albert's use of his medical sources.

Awareness of the discrepancies between Aristotle and Galen came slowly to the Latin West. These discrepancies are of various kinds, but perhaps the most significant relate to the function of the heart and to the process of conception. Aristotle insisted on the primacy of the heart as the source of heat and life,[58] while the Galenists considered the heart to be only one of several major organs without which life could not continue in the individual or be passed on to the next generation. Aristotle further maintained that the heart was the source of the blood vessels (he did not distinguish between veins and arteries) and of the nervous system. Galen, on the basis of the accumulated anatomical knowledge of his predecessors in medicine, some of it derived from dissection, and of his own dissections and studies, distinguished the veins from the arteries and showed that the brain controlled the nervous system.[59] The Aristotelian-Galenic dis-

---

[57] "Haly autem in tantum insaniae lapsus est, quod Aristotelem qui solus vidit veritatem et necessitatem istam, dicit errare et in errorem inducere medicum, cum ipse adeo errore imbutus sit, quod veritatem errorem appellavit": *De animalibus* III, 1.6 (ed. Stadler, 15: 305).

[58] *De partibus animalium* III, 4 (665b34-666b1); III, 5 (667b15-30), and elsewhere.

[59] A brief statement, no doubt known to all thirteenth-century physicians is found in the medieval Latin version of the *Tegni*: "Principalia igitur sunt cor, cerebrum, epar, et testiculi. Ab illis vero exorta sunt, et illis famulantur, nervi et spinalis medulla cerebro; cordi vero arterie; vene epate; seminalia vasa testiculis" (*Tegni* 2. 28-29). Elaborations of the concept are, of course, worked out throughout Galen's longer treatises. For a recent (although not always wholly satisfactory) account of Galen's teaching see Rudolph E. Siegel, *Galen's System of Physiology and Medicine* (Basel and New York, 1968).

pute over the roles of heart and brain was a living part of medical debate until the seventeenth century; equally long lived were the arguments over conception. Aristotle had held that the power of generation rested solely in the male semen, the female providing only passive matter that was acted upon.[60] Galen, in order to account for the fact that hereditary characteristics could be acquired from both parents, had postulated an active role in conception for the female as well as the male, and hence the existence of female as well as male sperm.[61] Discussion of these differences by Latin authors began only after the full range of Aristotle's anatomical and physiological teaching was revealed by the translation of the books on animals from the Arabic about 1210.[62] Given the peculiar character of the Galenic learning of the period (see above) it is probable that recognition of the specific discrepancies between Aristotle and Galen was greatly aided by the reception of the *Canon* of Avicenna, in which the differences are pointed out. Certainly the first Latin work to lay out some of the differences in detail is the *Anatomia vivorum* (ca. 1225), which, as already noted, is also apparently the first to make extensive use of excerpts from the *Canon*.

Albert's treatment of the discrepancies between Aristotle and Galen is probably the fullest and most thorough of any thirteenth-century scholar. It certainly goes well beyond that in the *Anatomia vivorum*. The author of that work discussed disagreements over the anatomy, complexion, and function of the heart,[63] but seemed unaware of any differences between his authorities on the subject of conception, a topic to which Albert devoted much attention. Moreover, Albert had at his disposal, as we have seen, the *De animalibus* of Avicenna and the *De corde* of Averroës, works which were apparently unknown to the author of the *Anatomia vivorum*. The topic of Aristotle vs. Galen and the *medici* was one to which Albert returned again and again;[64] but his most extended treatment of the subject is to be found in a tractate of seven chapters on the anatomy and origin of

---

[60] *De generatione animalium* I, 19-20 (727b6-729a35) and II, 4 (738b20-739a23).

[61] Siegel, pp. 224-230 and references there cited.

[62] *Aristoteles latinus*, 1: 80; Wingate, pp. 72-75.

[63] Corner, pp. 38-41, 100-101, 108-110.

[64] *De homine* II, q.32, art.3, part.5 (ed. Borgnet 35:279a-b); *De somno et vigilia* I, 1.7, and I, 2.6 (ed. Borgnet 9: 131b-132b, 147a); *De motibus animalium* I, 2.1-3 (ed. Borgnet 9: 269a-273b); *De spiritu et respiratione* I, 2.1 (ed. Borgnet 9: 231a-236a); *Quaestiones de animalibus* I, 55, III, 3, XV, 20 (ed. Colon. 12: 108, 124-125, 272-273); *De animalibus* XIII, 1.7, XV, 1.8, XV, 2.11 (ed. Stadler 16: 916-919, 1013, 1055-1057).

the veins, the blood and the nerves, and one of six chapters on "the dispute between Galen and Aristotle over the origin of the generation of man," both in *De animalibus*.[65] In both of these tractates Albert proceeded by systematically setting out the views of various authorities and concluding with a declaration of his own opinion. He brought to his work a comprehensive knowledge of earlier writing on the subject and a vigorous pro-Aristotelian bias. The authorities whose views on the relationship of the heart to the venous, arterial, and nervous systems he summarized were various "students of the anatomy of the bodies of animals" and natural philosophers among the ancients: Aristotle, Galen, Avicenna, and Averroës. On conception he began with the views of Galen and then proceeded to refute them on the basis of the teaching of Aristotle and "the wisdom of the peripatetics which is contrary to the opinions of the physicians."[66] It may be noted, however, that as regards conception he was prepared to give some weight to Galen's opinions, allowing that female secretions at the time of intercourse, if not possessing *virtus informativa* as the Galenists maintained, at any rate probably contributed to the matter of conception more than did menstrual blood.[67] (In the Aristotelian view, the active principle of conception, residing in the semen, acted upon matter consisting of retained menstrual blood.) But Albert was unyielding in his defence of the Aristotelian doctrine of the primacy of the heart over the whole body, and especially over the arterial, venous, and nervous systems.[68] Nevertheless, he gave a careful and reasonably correct account of Galen's teaching, pointing out that he had distinguished between veins and arteries, as Aristotle had not, that he had depended upon anatomical evidence, and that he had compared the structure of the venous system to that of a plant.[69] Moreover, Albert did not reject what he knew of Galen's system of anatomy outright, but rather asserted that "Galen was deceived" because he had failed to distinguish between the origin of physical branching out and the origin of substance and virtue.[70]

---

[65] Ibid., III, 1.1-7, IX, 2.1-6 (15: 277-308, 706-729).

[66] Ibid., IX, 2.4 (15: 718).

[67] Ibid., IX, 2.3 (15: 717).

[68] Ibid., III, 1.6 (15: 301-305).

[69] Ibid., III, 1.3 (15: 289-291).

[70] "Sapientissime enim dixit Aristoteles, quod idem esset principium venarum quod est principium digestionis completae et sanguinis. Et deceptus est Galienus quia credidit quod idem esset principium venae secundum venae speciem et virtutem digerendi quae est in ipsa, quod est principium ramositatis, et hoc omnino est falsum": ibid. III, 1.6 (15: 303).

The reason for Albert's hostility to the characteristic physiological doctrine of medieval Galenism, the concept of three major organs each giving rise to and dominating its own system (heart/arteries, liver/veins, brain/nerves) becomes apparent in the course of his exposition. In Albert's eyes, rejection of the primacy of the heart and insistence upon the equal importance of the physiological functions of heart, brain, and liver was inseparable from the belief, which he termed Platonic, that the human soul was divided into three parts or that there were three souls, each located in one of three major organs. As Albert put it, "Plato erred (*peccavit*) in placing the appetitive soul in it [the liver], and Galen followed that error,"[71] and, even more explicitly, "According to them [Plato and his followers, specifically including Galen], since the heart, the brain, and the liver differ as to place, subject, and number it follows that their movers differ as to place, subject and number, and hence that there are three souls in man."[72] It was thus his interpretation of Platonic teaching about the human soul that led Albert to reject the Galenic description of the role of the three major organs in physiology.

For Albert, writing in an age when natural science was indeed natural philosophy,[73] it must have seemed entirely proper to choose between rival physiological systems on philosophical (or theological) grounds. Nor indeed were alternative means of choice readily available. Whatever Albert's own empirical bent, the intellectual and material tools for effective empirical verification or disproof of the Aristotelian and Galenic doctrines about the heart and about conception were almost entirely lacking; as was historical information which might have revealed the accumulation of anatomical knowledge in the centuries intervening between the time of Aristotle and that of Galen, and the greater extent of Galen's researches into mammalian anatomy and physiology. The argument had perforce to be grounded chiefly on text interpretation guided by philosophical principles. Aristotle at this point enjoyed over Galen the double

---

[71] "Peccavit Plato ponendo in eo animam concupiscibilem: et peccatum illud secutus est Galienus": ibid. (15: 302).

[72] "Cum igitur, ut inquiunt, cor, cerebrum, et hepar, loco, subjecto, et numero differant, necesse est quod motores eorum loco, subjecto, et numero differant, et inde tres in homine animos, et similiter in animalium corporibus esse concesserunt vel contenderunt," *De spiritu et respiratione* I, 2.1 (ed. Borgnet 9: 231b-232a).

[73] For discussion of the concept of "natural philosophy," see James A. Weisheipl, "The Relationship of Medieval Natural Philosophy to Modern Science: The Contribution of Thomas Aquinas to Its Understanding," *Manuscripta* 20 (1976), 181-196.

advantage of being known in fuller texts and having, to Albert, a more acceptable philosophy.

No doubt, Albert's approval and incorporation into his Aristotelian commentaries of much material drawn from the writings of Galen and Avicenna increased the body of medical information generally available to students of natural philosophy; but his exposition of the differences between Aristotle and Galen may have been the aspect of his thought which most profoundly affected the development of learned medicine itself. It is certainly true that from the later thirteenth century learned physicians habitually devoted some attention in their works to expounding the differences between the "philosophers and physicians" (*philosophi et medici*), that is, usually between Aristotle and Galen and their respective followers.[74] While further research is needed to determine the extent to which Albert's specific arguments on the subjects of heart and brain, conception, etc., were taken up by later medical writers, it seems clear that the practice of explicitly discussing the discrepancies between Aristotle and Galen, and generally reconciling them in such a way as to avoid any repudiation of Aristotle, owes something to Albert's endeavor.

In summing up, it may legitimately be claimed that Albertus Magnus deserves a place in any history of medieval learned medicine, if this is understood, as it should be, to include the development of the sciences of physiology (as yet unnamed) and anatomy. This place was not earned by important original contributions to medical science, nor by extensive empirical investigation, interesting though the evidence of Albert's occasional personal observations may be. Rather, his contribution was to demonstrate more thoroughly than any of his Latin predecessors, with the possible exception of Petrus Hispanus, how the new Aristotelian natural science could be integrated with learned medicine to provide a unified account of the human body and its processes, while at the same time delimiting the areas in which the teachings of Aristotle and those of medical authorities came into conflict. In order to do this, Albert made himself master of the most up-to-date, comprehensive, and authoritative medical works available in his lifetime. Although medicine was for him chiefly an ancillary science to be drawn on in his quest for a truly encyclopedic natural philosophy, his example surely served to endorse the dignity and scientific value of the activities of physicians.

---

[74] Attention may be drawn to the title of the work cited in my note 2 above.

Albert's approach to medicine was for the most part scholastic; but in the thirteenth century scholastic inquiry both served to advance the understanding of ancient authorities on medical science and helped to confirm the status of medicine as a learned profession.

16

Human Embryology and Development in the Works of Albertus Magnus

Luke Demaitre and Anthony A. Travill
New York University; Queen's University

It appears inevitable that St. Albert the Great, the *Doctor Universalis*, should have been deeply interested in human embryology and development. As a theologian he faced questions on human genesis that were either crucial in doctrines on creation and redemption or decisive in moral positions. As an Aristotelian metaphysician he was compelled to apply concepts such as "form," "soul," "cause," "potentiality," and "generation" to man's singular development as unique and different from that of any other living creature.[1] As an encyclopedic natural scientist he marvelled at the biological continuity and variability of creation.[2] As a keen observer of creation's masterpiece he was fascinated by the complex processes of human sexual generation, embryonic growth, and fetal development.[3]

---

[1] The breadth and diversity of thinking on these topics during the time of St. Albert is indicated, for example, in Anton Pegis, *St. Thomas and the Problem of the Soul in the Thirteenth Century* (Toronto, 1934).

[2] The most comprehensive studies to date on Albert as a natural scientist are by Heinrich Balss, *Albertus Magnus als Zoologe*, in *Münchener Beiträge zur Geschichte der Naturwissenschaften und Medizin*, Heft 11-12 (1928); and *Albertus Magnus als Biologe* (Stuttgart, 1947).

[3] There is a paucity of histories of embryology. The most accessible in English are Jane Oppenheimer, *Essays in the History of Embryology and Biology* (Cambridge, Mass., 1967); Joseph Needham, *A History of Embryology* (New York, 1959); A. W. Meyer, *The Rise of Embryology* (Stanford, 1939); F. J. Cole, *Early Theories of Sexual Generation* (Oxford, 1930); E. S. Russell, *Form and Function* (New York, 1917), H. B. Adelmann, *The Embryological Treatises of Hieronymus Fabricius of Aquapendente* (Ithaca, 1942); and idem, *Marcello Malpighi and the Evolution of Embryology*, 5 vols. (Ithaca, 1966).

Therefore Albert's interest in embryology should be treated not as the idiosyncratic expression of an academic genius "out of his time" but, rather, both as a phase in the long history of the discipline and as a facet of scholastic inquiry during the middle two quarters of the thirteenth century. The period was one of intense intellectual ferment in the *studia* and universities, enriched by the growth of the mendicant orders and by the availability of new translations of Greek and Arabic writings.

Historically the study of scientific embryology has fallen into three phases: descriptive, comparative, and experimental.[4] Until very recently it has been the generally accepted principle that human developmental processes were not subject, either morally or technically, to experimentation or *Entwicklungsmechanik*, and thus in its absence speculation on the process of human generation became inescapable. Speculative embryology perennially appealed to students of nature and metaphysicians alike because it provided a basis not only for the treatment of life and growth in the microcosm but also for the contemplation of becoming and being in general.

It has been suggested that the inordinate degree of attention speculative embryology received from medieval scholars retarded its orderly progress as a scientific discipline.[5] A deeper cause of such "retardation," however, lay perhaps in the traditions through which human embryology had been channeled since the time of the pre-Socratics. Based upon observation, description, and — *faute de mieux* — speculation three distinct but closely interwoven traditions sprang from the disciplines of theology, medicine, and natural philosophy. St. Albert was heir to all three and in his turn he contributed, directly or indirectly, to each. We find them represented in his often quoted classification of the major authorities on whom he depended: "Augustine rather than the philosophers in case of disagreement in matters of faith. But if the discussion concerns medicine, I would rather believe Galen or Hippocrates, and if it concerns things of nature Aristotle, or any one else experienced in natural things."[6]

St. Augustine's authority in theological embryology pertained chiefly to three subjects: variability in generation, the creation of the soul, and the moment of animation. His major contribution lay in his

---

[4] W. J. Hamilton and H. W. Mossman, *Human Embryology*, 4th ed. (Baltimore, 1972), p. 2.

[5] Oppenheimer, *Essays*, p. 123. A more sympathetic appraisal is by G. M. Nardi, *Problemi d'embriologia antica e medioevale* (Florence, 1938).

[6] *In II Sent.*, dist.13, art.2 (ed. Borgnet 27: 247).

concept of the *rationes seminales*, derived from the Stoic *logoi spermatikoi*, by which he reconciled the scriptural accounts of instantaneous creation and successive generation during the six days of Genesis. Because of these "seminal reasons" or seeds, indeterminate matter has in itself the principles of all future manifestation and development. This doctrine dovetailed with the traditions of preformationism, which explains generation as the unfolding of preexisting parts, and it was adopted by such thirteenth-century Augustinian scholars as St. Bonaventure, Kilwardby, and Pecham. It could also be associated with the notion of plurality of substantial forms, or of souls in a living organism, which was attributed to Augustine by Pecham in his attacks on Thomist theses.[7] With regard to the creation of the soul, St. Augustine implicitly subscribed to "traducianism," the theory that not only the body but also the soul is derived in generation from the parents (and, with it, original sin from Adam). Paradoxically, Augustine's third noteworthy opinion was that the soul was not created at the time of conception, but that the embryo was animated in the second month.

In the light of his professed deference to Augustine's authority "in matters of faith," Albert's positions on Augustinian doctrines are interesting. In assigning animation to the second month after conception, as we shall see below, he concurred with Augustine whose endorsement, however, he invoked only in the context of the moral question "whether those who cause abortion are murderers."[8] In teaching that the soul infused into each embryo is at that time created by God ("creationism") and thus not derived from the begetter, "*ex traduce*,"[9] Albert unequivocally rejected traducianism, although he argued expressly against Augustine only, and with some ambiguity, on the subject of original sin.[10] Augustine also was not named in relation to embryology throughout Albert's many discussions of generational variability that reflected one of the widest differences between his Aristotelianism and the Augustinian tradi-

---

[7] For a summary and references see James A. Weisheipl, *Friar Thomas D'Aquino* (New York, 1974), pp. 289-290.

[8] *In IV Sent.*, dist.31, art.18 (ed. Borgnet 30: 250).

[9] *In II Sent.*, dist.8, art.8 (ed. Borgnet 27: 324-325).

[10] Ibid., dist.31 (27: 509-517). On the implications of traducianism for the doctrine of original sin see further Albert's *De creaturis*, Pars II: *De homine*, q.17, art.3; (ed. Borgnet 35: 148-161); also *Summa theologica*, q.72, membr.3. For its application to Christology, see the *Problemata determinata XLIII*, probl. 33 (ed. James A. Weisheipl, *Mediaeval Studies* 22 [1960], 45).

tion. In these discussions, Albert adhered to the view of matter as pure potentiality against those who asserted that everything is latently but actually present in something prior, although he did not mention as such the concept of "seminal reasons"; he consistently refuted preformationism in favor of epigenesis or the new development of parts from potentiality to actuality; and he maintained the doctrine of unicity of forms, refining its particular application to human generation. All this not only confirms those features of Albert's thought that have so often been noted, such as his Aristotelianism and his independent spirit of inquiry, but it also suggests that in embryology he assigned a rather limited sphere to St. Augustine's authority in particular and to the theological tradition in general.

The second or medical tradition of embryological study was rooted in the Hippocratic corpus and in the writings of Galen, and much of it reached the Latin West via Arabic translations, commentaries, and compendia. Most prominent among the latter was Avicenna's *Canon*, which also incorporated preclinical doctrines derived from Aristotle. Medical embryology arose primarily from the practical need of the profession to understand the processes of human reproduction and development so that, by regimen and medication, they could be facilitated and maintained through gestation to successful parturition and a live birth. These hygienic and therapeutic concerns, in so far as they went beyond popular lore, did require diagnostic knowledge; but the ancient and medieval physician-obstetrician lacked techniques of *in utero* investigation and was also limited to macroscopic inspection. Therefore, even though he occasionally had access to an inevitable abortion or an accidental maternal death, he could in medical theory contribute little to embryology beyond the knowledge offered by the natural scientist.

Albert's attitude towards medicine and his knowledge of its literature are carefully examined and richly documented elsewhere in this volume. It may be pointed out here that most of the embryological theses which Albert identified with the medical tradition in the "dispute" between Aristotle and Galen, which will be considered later, belonged more properly to the realm of natural philosophy. Hence his repeated challenges of Galenic views and the paucity of his references to Hippocrates were not inconsistent with his earlier cited esteem of these authorities in medical matters but, rather, indicative of his belief that the medical tradition was less significant than natural science in the study of embryology. Nevertheless, Albert's discussions of human genesis and development are enlivened by numerous brief references that illustrate a keen interest in the art of medicine. In addition to citing its academic literature and, more vaguely, alluding to "what physicians say," he occasionally reported what he had learned personally from "a certain physician," from midwives, or from patients with remarkable case histories.

The third embryological tradition, as well as theoretical biology in general, is based on Aristotle's writings on natural philosophy, especially his three works on the history, on the parts, and on the generation of animals. These works, as well as an abridged version by Avicenna, were presented to the thirteenth-century scholarly community through the translations from the Arabic by Michael Scot.[11] They soon became, under the collective title *De animalibus*, part of the arts curriculum, and by their scope, organization, and clarity provided scholars with an unsurpassed canon of philosophical concepts and biological data, particularly on the subject of generation, for the study and teaching of *physica*.

If we tend to translate *physica* somewhat freely as "natural science," we should bear in mind some important qualifications with regard to the ancient and medieval teaching of embryology. This teaching, not only in its philosophical discourses but also in its morphological (or structural) and physiological (or functional) descriptions of human generation in the early stages, was by necessity hypothetical and liable to fallacy. Its conclusions were based on knowledge derived from inadequate sense perception, especially of human development within the first trimester. It is now universally accepted that during this period, following fertilization, the conceptus passes through the stages of blastocyst formation during tubal migration, implantation with concurrent germ layer formation, and primary delineation of the chorion (trophoblastic delineation). Morphological organization and differentiation are completed by about the end of the second month — the termination of embryonic existence. Thereafter until birth fetal maturation occurs by simple growth of the already differentiated embryonic organs. Generally speaking at thirty-two days post conception the fetus is only five mil-

---

[11] William of Moerbeke completed a new translation from the Greek of the five distinct Aristotelian works on animals in 1260, but it seems not to have affected Albert's writing on embryology. In a similar fashion Michael Scot's translations seem to have been just missed by an earlier thirteenth-century author: see M. Kurdzialek, "Anatomische und embryologische Ausserungen Davids von Dinant," *Sudhoffs Archiv* 45 (1961), 1-22.

limeters in length.[12] Thus the order of magnitude of the developing entities on which the authorities, Aristotle, Galen and Avicenna, and their medieval commentators wrote were beyond the range of human perception. Even when the authors argued from avian (chick) analogy, they were forced to base their analysis of human generation upon limited observation until the invention of the optical microscope. Malpighi in the seventeenth century used the microscope to delineate the developmental processes in the chick embryo, but its application became of significance in fully unravelling avian and human morphogenesis only when microscopical optics were improved in the nineteenth century. It was only then, and subsequent to the discovery of the unfertilized human ovum and the definition of the germ-layer theory by von Baer, that one could speculate on the preimplantation and organizational stages of human development.

In addition to inadequate techniques of perception, several deficiencies in "mental techniques" handicapped the advance and accuracy of premodern embryology, as Needham has pointed out.[13] Perhaps the two most serious obstacles were, first, the power of qualitative constructs such as teleology and the theory of humours and complexions; and second, the lack of a scientific terminology, which was compounded by the difficulties inherent in each translation. Only when we are aware of these handicaps can we accurately assess the cogency of theories and fully appreciate the observed data on human generation in the works of Aristotle or any ancient and medieval author.

Albert endeavoured to inform his contemporaries of the entire spectrum of Aristotelian biology as a brilliant though not infallible source of knowledge and guide for further study. In this endeavour he relied quite often on Avicenna's *De animalibus*, and he did not present either a coherent synthesis or even a well-arranged compendium. Nevertheless, it was this pioneering enterprise, together with his attention to fundamental questions and his remarkable powers of observation, that earned for Albert the recognition as "the greatest naturalist of the Latin Middle Ages."[14] His interest in the crucial subject of embryology is evident in the fact that, according to Needham's estimate, he dedicated thirty-one percent of his biological

---

[12] Hamilton and Mossman, *Human Embryology*, pp. 188-191.

[13] Needham, *History of Embryology*, pp. 88-89 and 234-236.

[14] George Sarton, *Introduction to the History of Science*, 2 (Washington, D.C., 1929), 936.

writings to the phenomena of generation;[15] in addition, allusions to human development occur sporadically through most of his other works. His own observations, whether from personal experience or from hearsay, add to his descriptive and comparative embryology a degree of verisimilitude rarely equalled in the analogous scholastic writings, including those of medicine.

More of Albert's own reports, illustrative of particular topics, will be cited later in this study; however, some general comments on the place of *experientia* in his embryology seem in order here. His peripatetic career, in strict keeping with his order's regulations, to make all his journeys on foot, not only provided numerous occasions to observe embryological and teratological phenomena but also gave him time to consider and record them before they were forced out of his mind by the rapid succession of further events. In the process he gained a rich knowledge about, literally, the birds and the bees. Thus, he contradicted Avicenna's description of the mating of partridges, which he had often watched with his "own eyes."[16] He also studied more elusive birds, such as the hawk and the falcon, whose parental behavior he discussed on the basis of "what birdcatchers have told me and what I have observed." Even the most arduous field work did not deter him, as we see in his assertion that the "great eagle"[17] produces two eggs but only one chick because "this is what we have learned by visiting the nest of a certain eagle for six years in a row" — while he admitted that such observation of an eyrie was difficult and could be made "only by lowering someone from the rock on a very long rope."[18] As to the bees, "about whose exact generation even Aristotle and Avicenna confessed to be ignorant," Albert reviewed the prevailing theories at the hand of what he had heard from people "who cultivate swarms" and "who guard beehives;"[19] moreover, he had learned about apian metamorphosis by personal investigation (*"experimento investigavi"*) of a nest.[20] He further enriched his comparative embryology with findings about

---

[15] Compared with Aristotle (37%) and Galen (7%) by Needham, *History of Embryology*, p. 86.

[16] *De animalibus* V, tr.1, c.2 (ed. Stadler, 15: 415-416).

[17] "Aquila magna, quae aput nos est et herodius vocatur." Although *herodius* is listed as "an unknown bird, perhaps the stork" in Lewis and Short's *Latin Dictionary*, the context leaves no doubt that Albert is referring to the eagle.

[18] *De animalibus* VI, tr.1, c.6 (ed. Stadler, 15: 461 and 463).

[19] Ibid., XVII, tr.2, c.2 (16: 1172-1179).

[20] Ibid., XV, tr.1, c.8 (16: 1009).

mammals that ranged from mastiffs to boars; and he questioned ancient lore on fish, for example that females conceive through the mouth, because it conflicted with what he had "seen with his own eyes and heard with his own ears, diligently observing and inquiring from veteran fishermen of sea and rivers."[21]

Whereas for the generation of animals Albert could resort to his own as well as to reported observations, for human embryology itself he depended primarily on hearsay. The sources of his information might range from cases as public as that of the woman in Worms who in spite of her amastia bore several children including triplets,[22] to confidences as intimate as that of the "libidinous woman" who had told Albert "personally" about the pleasure she derived from allowing the wind to enter her womb.[23] In general, Albert felt that "much belief should be given to trustworthy women who have borne many children."[24] At the same time, however, he was conscious of the limited reliability not only of hearsay information, for example when women might miscalculate the duration of their pregnancy,[25] but even of direct observation, especially when it was difficult to discern such characteristics as the anatomy of real hermaphrodites[26] or the resemblance of true identical twins.[27]

Albert's observations on embryology, as those of Aristotle, represent the apogee of his work on practical and theoretical biology. His major writing on animal reproduction was a presentation of Aristotle's treatise *On the Generation of Animals*, along with Avicenna's commentaries and his personal comments. It occurs piecemeal throughout the twenty-six books of his work *De animalibus* and it may be viewed as a sequel to his discussion of plant embryology in the treatise *De vegetabilibus et plantis*, to which he referred regularly. Much of Albert's *De animalibus* was understandably related to developmental processes in submammalian species, but his numerous remarks on avian embryology (particularly in books VI and XVII), based on his careful examination of the developing chick, have broader implications for the entire scope of embryological thought.

---

[21] Ibid., V, tr.1, c.2 (15: 415).
[22] Ibid., IX, tr.1, c.1 (15: 679).
[23] Ibid., VI, tr.3, c.2 (15: 489).
[24] Ibid., IX, tr.2, c.5 (15: 724).
[25] Ibid., IX, tr.1, c.4 (15: 692).
[26] Ibid., XVIII, tr.2, c.3 (16: 1225).
[27] Ibid., IX, tr.1, c.6 (15: 698).

Human conception and development during the first trimester receive special attention in books, I, IX, X, XV, XVI, XVIII, XX and XXII. In several chapters of these books Albert "digressed" from the Aristotelian text to recapitulate or clarify the discussion, but he formulated no synthesis of his own wandering arguments and interjected observations, so that it is difficult to reconstruct a comprehensive picture of his precise position on several major issues.

A similar difficulty is inherent in the *Quaestiones super De animalibus*, the report of lectures given by Albert at Cologne in 1258 but recorded by Friar Conrad of Austria after 1260. Besides coming to us in a student's version not authorized by Albert, these questions and determinations were cast in the didactic mold of the *disputatio* and hence they do not reveal the nuances of the teacher's position. Furthermore, as Brian Lawn has indicated, they were apparently modelled on Peter of Spain's commentary and several of them closely follow the Salernitan Questions. While they may be "less metaphysical and dialectical than the questions of Bacon," as Lawn asserts,[28] the *Quaestiones super De animalibus* are markedly more metaphysical and dialectical — at least on the subject of embryology — than the chapters of Albert's *De animalibus*. Also more metaphysical in thrust are two books on natural philosophy that, according to their modern editor, originally were part of *De animalibus*. One, *De natura et origine animae*, composed about 1260, treats the animation of the embryo in stark ontological terms and without the empirical tone of *De animalibus*; the other, *De principiis motus progressivi*, written between 1256 and 1262, bears only tangentially on embryology when Albert discusses analogies between the movement of the heart and that of the testicles.[29]

Statements on embryology are found not only in Albert's work *De animalibus* and related books, but also throughout his theological and metaphysical writings. These were produced either in conjunction with his teaching in Paris and Cologne or in response to a need for his authoritative opinion. Among the works of the first category, in which embryological material occurred sporadically, we may cite especially his early treatise *De homine*, which was part two of his *Summa de creaturis* (1244-1248); his equally early commentary on Peter Lombard's *Sententiarum Libri IV* (1249); his commentary on

---

[28] Brian Lawn, *The Salernitan Questions* (Oxford, 1963), pp. 85-86.

[29] II, c.9 (ed. Geyer, pp. 68-70).

Aristotle's *De generatione et corruptione*, probably written shortly before *De animalibus*; and his questions *De fato*, disputed at Anagni in 1256 but recorded in the early 1260s. In the second category belongs a tract which Albert wrote to throw his authority into the Averroist controversy, namely *De unitate intellectus contra Averroistas* (1271), and a collection of *determinationes* to questions submitted to him by his order's master general, the *Problemata determinata XLIII* (1271). Adequate answers to questions 34, 35 and 36 of the *Problemata* required embryological expertise if they were to rely on the additional arguments from science, as Albert's did, rather than solely on theological premises, as did those of his colleagues (Kilwardby and Aquinas) who were also consulted.

After examining the place of embryology in Albert's work, we now turn our attention to the content of his teaching on human generation and development. An exhaustive and comprehensive study, as this subject deserves, would require not only an integration of all Albert's scattered statements and a textual collation with each of his sources, but also an exploration of the broader scientific and philosophical implications and a detailed comparison with the writings of his contemporaries. However, since such a study would vastly exceed the confines of this article and should be pursued as a long-range project, as we hope to do in the future, may it suffice here to introduce briefly, and with focus on *De animalibus*, Albert's views on some major embryological questions. These questions pertain, respectively, to the nature, role, and origin of the seed in human conception; the process and moment of animation, and the period of gestation; the form and function of the fetal membranes; and the causes of abnormal embryonic development.

One of the problems on human reproduction to which Albert devoted the greatest attention was that posed by the exact contribution of man and woman to conception. This problem was the most controversial and occupied the central place in embryological thought and writing for two millennia until the advent of modern investigative techniques. Prehellenic generational mythology had treated conception and generation purely as a feminine function, but with the descent of the Hellenes into the eastern Mediterranean Ionian littoral the fertility cult shifted from mother earth and feminine deities.[30] From then onwards the male was considered to play

---

[30] See Wolfgang Jöchle, "Biology and Pathology of Reproduction in Greek Mythology," *Contraception* 4 (1971), 1-13.

an active and, increasingly, a dominant role as cause and determinant of reproduction. Presocratic philosophers gradually diminished the feminine prerogative to that of defining the offspring's sex according to the side of the womb where it developed, into a male if in the right side and into a female if in the left side according to Parmenides.[31] The philosophers' most extreme position would be to reduce woman's role to that of a mere nurse, farm-field, or incubator.[32] Aristotle stopped far short of that position in his views, which were most clearly formulated in the work *On the Generation of Animals*.[33] He held that the female contributed the menstruum or *catamenia* which, however, had no active principle or virtue and provided simply the raw material or *materia prima* and nourishment for the concept; in conception the passive menstruum was informed by the active male seed. Ancient physicians, on the other hand, retained a theory of woman's more active involvement in reproduction. The Hippocratic treatise *On Semen* (Περὶ γονῆς) recognized not only the nourishing menstrual blood but also a female seed, which it identified with the vaginal secretion.[34] Centuries later Galen became the protagonist of the physicians' side with his doctrine, buttressed by anatomical observations and dissection of animals, that the "female testes" or ovaries produced a true albeit imperfect sperm that had the power of generation after mixing with the male seminal fluid.[35]

When thirteenth-century scholars first received these conflicting theories of human generation, they depended on the interpretation of the Arabic commentators. They were also greatly hampered by the lack of a precise embryological terminology, so that they were forced to use terms more appropriate to metaphysics. Moreover, as has been suggested above, the absence of both cellular theory and microscopy still made it impossible to appreciate the discreteness of an individual sperm and ovum or to distinguish them visually from the fluid vehicle in which they are carried. We should not be sur-

---

[31] Owen Kember, "Right and Left in the Sexual Theories of Parmenides," *Journal of Hellenic Studies* 91 (1971), 70-79.

[32] Needham, *History of Embryology*, pp. 43-44.

[33] I, cc.19-20 and II, c.4.

[34] Needham, *History of Embryology*, p. 35.

[35] *On the Usefulness of the Parts of the Body*, XIV, 2; translated with introduction and commentary by Margaret May (Ithaca, 1968), 2: 623-648. For the entire tradition and its context see Erna Lesky, "Die Zeugungs- und Vererbungslehren der Antike und ihr Nachwirken," *Abhandlungen der Akademie der Wiss. und Lit., geistes- und sozialwiss. Klasse* (Mainz, 1950), pp. 1227-1425.

prised, therefore, to find some vagueness and ambiguity in Albert's repeated remarks about the process of conception. A certain confusion in these remarks further results from two practices which Albert shared with contemporary teaching on the causes of conception. First, the complex divergences in the authoritative tradition were reduced to an aspect of the controversy "between philosophers and physicians" or, more simply, to "the disagreement between Aristotle and Galen about the principles of human generation."[36] Secondly, Galen's views were not only cited from the authentic work *De usu partium* but also thought to be represented by the spurious treatise *De spermate*.[37]

In the basic analysis of conception Albert is clear and consistently Aristotelian. Thus, in *De animalibus* I,[38] he recognizes that both distinct male and female contributions are necessary to form "all that is generated out of spermatic moisture" — that is, in modern terminology, all that comes out of the zygote, the immediate product of conception. This moisture, if it is to have the complete faculty or virtue of generation, "must necessarily have in itself both the property of the male's faculty and the property of the female's faculty." Should one infer a Galenic influence from the emphasized necessity of both contributions or from the notion of *virtutes* ($\delta\upsilon\nu\acute{\alpha}\mu\epsilon\iota\varsigma$), the Aristotelian orthodoxy is unmistakable when Albert goes on to define the role of each contribution. He identifies "the first principles of generation" within the zygote: "generating in it actively is the faculty of the male, and generating in it passively is the faculty of the female"; he further illustrates their roles with Aristotle's analogy between cheese and the zygote, in which the male sperm acts like the rennet and the female "sperm" undergoes like the milk, the former giving and the latter receiving "the coagulation, the figuration, and the form." This principal definition of the contributions as active and passive is maintained, and the rennet-milk analogy is repeated not only throughout Albert's *De animalibus* but also in several of his other works. One of his earlier treatises contains a parallel but more suc-

---

[36] *De animalibus* IX, tr.2 (ed. Stadler, 15: 706-727). For the origin, history and complexity of this controversy see Owsei Temkin, *Galenism; Rise and Decline of a Medical Philosophy* (Ithaca, 1973), esp. pp. 69-80. For its application to Albert's writings see the study of Nancy Siraisi in this volume.

[37] Galen's authentic work *De semine* (Kühn 4: 543-678) was apparently not known to Albert. The pseudo-Galenic treatise was translated into Italian with a commentary by Vera Tavone Passalacqua, *Microtegni seu de spermate* (Rome, 1959).

[38] I, tr.1, c.6 (ed. Stadler, 15: 31).

cinct Aristotelian definition: "the semen is twofold, namely effective and material," derived from man and woman respectively.[39]

However, the Aristotelian paradigm and its philosophical clarity grow dimmer with each attempt of Albert to explain conception in greater detail and, presumably as time progressed, with his increasing familiarity with the recently available sources. Most of the confusion stems from Albert's efforts to incorporate into his scheme Galenic ideas, their Avicennan interpretation, and his own observations. This is seen immediately when in the initial discussion in *De animalibus* I, 1.6 he adds, with evident hesitation, that "in some or perhaps in all" animals a third moisture is required which feeds the conceptus and "from which some of its parts are made, not radically but materially."[40] This third component Albert places in the menstrual blood — in modern phraseology we might correct this to the decidual endometrium — for mammals, and in the yolk for oviparous species. It should be noted here that Albert considers the white of an egg as the "*materia radicalis*"[41] from which the embryo develops, and the yolk as only the nutritive component.[42] He also suggests that the menstruum "corresponds to both the white and the yolk in the egg, because from one part of the menstruum the fetus is formed materially and from the other part it is fed."[43]

Thus Albert adopts from Galen, via Avicenna, the division of the female contribution, which Aristotle had left undifferentiated as the *catamenia*, into a nourishing and a material component. This division apparently raises more questions than it answers. One inconsistency can be seen in our paraphrase above, since Albert fails to cor-

---

[39] *De homine* q.17, art.2 (ed. Borgnet, 35: 145).

[40] Ed. Stadler, 15: 31.

[41] Ibid., 15: 30.

[42] In this view Albert followed Aristotle's "backward step" from the Hippocratic description; see Needham, *History of Embryology*, pp. 52-53 and 36. Elsewhere, however, Albert suggested that "the chick is generated in the continuous circumference between the convexity of the yolk and the concavity of the albumen" — *De animalibus* XVII, tr.1, c.2 (ed. Stadler, 16: 1152).

[43] *Quaestiones super De animalibus* XV, q.19 (ed. Filthaut, p. 272). Albert reiterated these descriptions of the egg and the menstruum throughout *De animalibus*, but without an equally explicit statement of the correspondence between both. He did compare, as results of female contributions without male sperm, wind eggs to "the mixture of the female's sperm with the menstrual blood, a matter from which nothing at all is generated" — *De animalibus* VI, tr.1, c.2 (ed. Stadler, 15: 445). He also observed that female birds could be aroused manually and thus secrete "the matter of a conceptus, which is formed into a wind egg, as by a woman's touch in the orifice of her vulva infecund semen is drawn" — ibid. VI, tr.1, c.3 (15: 449).

relate the material cause in the "third moisture" with the passive virtue of the "second moisture" so that here, and at other points in *De animalibus*, it would seem that the female contribution has three components (viz. the passive seed in addition to the menstrual matter and nourishment).[44] If, for the sake of brevity, we treat both the passiveness and the materiality as aspects of the "female sperm," we may be following Albert's intent but we are not resolving all ambiguity in the relationship between the female elements. Their functional distinctiveness as matter and nourishment is blurred, for example, when Albert describes the process of conception as follows: the male sperm in the womb "receives first the female sperm and then also the menstrual blood in which it stamps (*sigillando*) and imprints the creature's form and members."[45] The distinct functions are more clearly stated in a chapter on the three moistures that is largely based on Avicenna's *De animalibus* XIV, 3 and on the pseudo-Galenic *De spermate* I, 3. Albert concludes the chapter with the assertion that, "since according to the opinion of Aristotle generation comes from sperm and nourishment from menstrual blood, we must accept that generation materially is from what is called the woman's sperm and the nourishment is from the menstruum."[46] Here, however, the Aristotelian and Galenic doctrines are linked by a paralogism that not only epitomizes the basic confusion but also suggests the additional difficulties in Albert's analysis of conception.

One of the most pervasive problems lies with terminology, both biological and philosophical. The terms "spermatic moisture" or *humidum spermaticum*, *sperma*, and *semen* are often used interchangeably for the zygote as well as for either the male or the female contribution. Albert does make explicit distinctions between both contributions, but he bases them on potentially confusing criteria

---

[44] Another puzzling fact is that Albert expresses reservations about the universal existence of the "third moisture" or menstruum in his initial analysis paraphrased above, whereas he applies these reservations to the "second moisture" or secretion of the female testes further in *De animalibus*, e.g., in XV, tr.2, c.11 (ed. Stadler, 16: 1055), where he concludes that "it appears more probable to me that this moisture exists in all female animals, though it is not manifest in all."

[45] *De animalibus* XVI, tr.1, c.10 (ed. Stadler, 16: 1090). Compare also ibid. XXII, tr.1, c.1 (16: 1349). A similar indistinctiveness marks Albert's definition of the *materia secundaria* in conception as "the matter of woman's menstrual blood and semen, which is attracted by the [male] semen and is mingled with it so that it is sufficient for the body's quantity" — *De homine* q.17, art.1 (ed. Borgnet, 35: 143).

[46] *De animalibus* XV, tr.2, c.11 (ed. Stadler, 16: 1057).

that are derived from the metaphysical notion of "form." Thus, in line with the Peripatetics and in opposition to Galen, he argues that only the male element can in essence and truth be called "sperm," since this word by definition (for which Albert refers to his own *Physica* II, 2.3) "denotes an efficient and formal (*formantem*) cause." The female contribution, on the other hand, is called "sperm" only equivocally because "we do not give it either a forming or an informing (*formantem vel informantem*) virtue."[47] In a previous chapter, with more precision, Albert also summarizes and rejects as a Galenic tenet the attribution to woman's sperm of an active virtue that is "not formative but informative" (*non formativam sed informativam*).[48] Several chapters later, however, he clarifies that "we do not reject Galen's thesis that woman's sperm has an informative virtue" as long as this virtue is defined as "only preparing and enabling matter to receive the action from the operator, that is, man's sperm."[49] Hence, while making a minor concession to Galen, Albert reaffirms the Aristotelian position that only the male sperm is active and endowed with the formative virtue (*virtus formativa*).[50]

The concept of formative virtue (*virtus formativa*) connects two complex embryological questions, one about the origin of the sperm and the other about the induction of the soul. These questions were not only of paramount concern to the natural philosopher but also of great consequence to the theologian and of practical interest to the physician. It should not surprise us, therefore, either that Albert pursued them throughout his writings or that we cannot do them justice in this brief synopsis. According to Albert the male sperm which carries the formative virtue, as well as the analogous female contribution, was that which is ultimately left after the food has gone through the fourth digestion.[51] This analysis, popular also with thirteenth-century physicians, was based on an Avicennan adaptation of Aris-

---

[47] Ibid., IX, tr.2, c.3 (ed. Stadler, 15: 714). A less metaphysical reason is that female sperm "does not and cannot reach the digestion of sperm" because woman's "natural faculties" are weaker than man's — ibid. XV, tr.2, c.11 (16: 1056).

[48] Ibid., IX, tr.2, c.1 (15: 710).

[49] Ibid., XVI, tr.1, c.16 (16: 1110).

[50] See also ibid., XVI, tr.1, c.7 (16: 1083); XX, tr.2, c.3 (16: 1311); and *De homine* q.17, art.3 (ed. Borgnet, 35: 161).

[51] *De animalibus* XV, tr.2, c.5 (ed. Stadler, 16: 1031-1035). Albert assigns a fifth digestion to the spermatic vessels in ibid. XVI, tr.1, c.7 (16: 1083). For the context of his ideas on digestion see Joan Cadden's contribution to this volume.

totle's doctrine.[52] It was directed against the proponents of pangenesis who held that the seed originates from all parts of the body.[53]

Albert mainly paraphrases Aristotle in presenting the versions of pangenesis, but several of his parenthetical comments on their proponents and arguments may illuminate contemporary thought. He points out that, since the fourth digestion entails a process of general assimilation between its produced moisture and the entire body, the doctrine of pangenesis "is not completely false." In this instance he calls it "the thesis of Stoics and physicians."[54] He agrees with Aristotle that a "rather tolerable" version was the homoeomereity of Anaxagoras, who taught that the seed is derived from the "homogeneous" parts of the body, such as flesh, blood and bones.[55] However, Albert rejects categorically the view of Empedocles that all the body's organs are actually present in the seed, since this view is based on the premise "that the substance of the organs would, without diminishing, emit another substance that invisibly contains all those organs." This false premise, Albert adds, is defended "until today by many men of our time who are full of error."[56] A better known contemporary error is echoed when he attributes the basic doctrine of pangenesis to Plato and interprets it as the negation of a single formative virtue in the sperm. It claims instead, Albert explains, that specific parts of the sperm have the virtues of those organs from which they are derived; and he identifies his target more explicitly as "the position of Plato and his followers who stated that there are many souls in each body according to the division of principal organs."[57] Among

---

[52] *Generation of Animals* I, 17-18. See the important study and further references by Michael McVaugh, "The 'Humidum Radicale' in Thirteenth-Century Medicine," *Traditio* 30 (1974), 259-283.

[53] The development and implications of this theory are examined in C. Zirkle, "The Early History of the Idea of Inheritance of Acquired Characters and Pangenesis," *Transactions of the American Philosophical Society* 25 (1946), 91-151.

[54] *De animalibus* XV, tr.2, c.10 (ed. Stadler, 16: 1054). By "the Stoics" Albert means to include Plato, Socrates, Pythagoras, as well as others, and their followers; see Weisheipl, ed., *Problemata determinata XLIII, Mediaeval Studies* 22 (1960), 335 n. 75. As to the "physicians," it is worth noting that arguments for pangenesis, closely parallel to those refuted by Aristotle, occur in the Hippocratic corpus according to Anthony Preus, "Science and Philosophy in Aristotle's *Generation of Animals*," *Journal of the History of Biology* 3 (1970), 5 n. 3.

[55] *De animalibus* XV, tr.2, c.2 (ed. Stadler, 16: 1022).

[56] Ibid.

[57] Ibid., c.1 (16: 1017). Albert cites this position as erroneous in various embryological contexts, for example, when he supports against "the physicians' crowd" the Aristotelian theory that all the organs develop from the heart; after paraphrasing Avicenna's argument (*De animalibus* IX, c.9) that there is only one formative virtue, Albert adds that "there are not several souls, of which one would be in the brain, another in the heart and a third in the liver, as Plato stated [*Timaeus* 69C]" — *De animalibus* IX, tr.2, c.4 (ed. Stadler, 15: 721).

the "absurd" logical consequences of the doctrine[58] Albert singles out as "patently ridiculous" the implication that the sperm would be a miniature animal or human.[59]

The same implication is raised when Albert adopts Aristotle's refutation of preformationism,[60] and it is also combined with the further *reductio ad absurdum* that the seed would be both "begetting and begotten (*generans et generatum*)" if it contained all the organs in actuality within itself.[61] Albert follows Aristotle quite closely not only in arguing against preformationism but initially also against views that ascribed the animation of the conceptus to outside agents. He considers "destroyed by irrefragable syllogisms" the theses that the zygote receives its formative principle either from the mother, "as some of the physicians' ignorant people say," or from a part that accompanies the father's sperm, as held by Empedocles who treated the soul as a corporeal form.[62] Then, however, Albert supplements Aristotle or unintentionally brings him up to date by devoting special attention to the Avicennan doctrine that the formative virtue and the soul are induced by the "Giver of forms," one of the subsistent intelligences between God and the material world.[63] Believing that this doctrine "originated with Socrates whom Plato followed with the entire school of the Stoics,"[64] Albert does not in this passage include Avicenna among its proponents as he had rightly done earlier.[65] More importantly, even though Albert notes the theological implications in this "thesis of Plato and Pythagoras who thought that several of the immortal gods are in one body,"[66] he bases his rebuttal on premises from natural philosophy rather than on revealed truths not only in *De animalibus* but also in *De homine*. Nevertheless, in the former he admits that details of these "difficult" and "heavy questions" on animation are "to be investigated more subtly in first philosophy, so that we dismiss them now."[67]

---

[58] Ibid., XV, tr.2, c.1 (16: 1018).

[59] Ibid., c.10 (16: 1052).

[60] Ibid., XVI, tr.1, cc.2 and 13 (16: 1064 and 1098).

[61] Ibid., XVI, tr.1, c.8 (16: 1085).

[62] Ibid., cc.2-3 (16: 1064-1072).

[63] Albert defines the *dator formarum* as "forma nuda a materia et mensura materiali" — *De homine* q.17, art.3 (ed. Borgnet 35: 150).

[64] *De animalibus* XVI, tr.1, c.5 (16: 1076).

[65] *De homine* q.17, art.3 (ed. Borgnet 35: 152). Another link in this Neoplatonic tradition, not cited by Albert as a source on this subject, may have been Porphyry, whose views were borrowed by the pseudo-Galenic *De spermate* XI (tr. Passalacqua, p. 39).

[66] *De animalibus* XVI, tr.1, c.2 (16: 1067).

[67] Ibid., c.5 (16: 1077).

We may follow Albert's advice and limit ourselves here to some brief observations on his own theory of animation strictly within the confines of embryology.[68] His theory was a version of the Aristotelian doctrine of recapitulation or gradual development of aliveness within the embryo through the vegetative, sentient and rational stages. He argues from the premise that the sperm carries in its *spiritus* both the potential for the development of all the organs and the formative virtue derived from the man's soul, but not the actual form or soul itself.[69] Hence neither the sperm nor the zygote is truly animated, and the father's soul is in them only as the artisan and his art are in the artifact.[70] There are numerous variations on this analogy throughout Albert's writings but they add up to a cogent theory of animation.[71] Above all, even when it requires him to modify the Aristotelian description of human embryogenesis,[72] he maintains the unicity of substantial form.[73]

The most elusive and the ultimate speculative questions about animation, now as well as seven centuries ago, are whence and when the embryo achieves human essence or, according to the western tradition, receives the rational soul. It is in the latter formulation that each question is addressed directly by Albert. In the language of the Aristotelian scientist and natural philosopher, he argues that since reason is neither a corporeal nor an organic faculty, it must come

---

[68] Albert's theory is placed in the broader context of other medieval doctrines, including those of Averroës, Alfred of Sareshel, Alexander Hales, and Roger Bacon, by Bruno Nardi in the chapter on "L'origine dell'anima umana secondo Dante" in his *Studi di philosophia medioevale* (Rome, 1960). For the soul as *forma corporis* in thirteenth-century thought, especially of Petrus Olivi (d. 1298), see Theodor Schneider, *Die Einheit des Menschen*, in *Beiträge*, Texte, Neue Folge, 8 (Münster, 1972).

[69] See p. 407 above for the related question of traducianism. This question, which was also related to the Averroist controversy, evidently affected Albert's discussion of animation much more in such works as *De anima*, the commentaries on the *Sentences*, the questions *De homine* and the *Summa theologica* than in *De animalibus*.

[70] *De animalibus* XVI, tr.1, c.7 (ed. Stadler, 16: 1082).

[71] The analogy is applied by Albert not only to the presence of man's soul in the sperm but also to such diverse relationships as those between father's soul and zygote, formative virtue and sperm, vegetative soul and embryo, and so on. Some loci, in addition to that cited in the text: *De animalibus* VI, tr.1, c.4; XV, tr.2, cc.3, 5, and 6; XVI, tr.1, cc.2 and 6; XX, tr.2, c.3; *De natura et origine animae* I, 3; *De anima* II, 1-2; *De homine* qq.5-7, and 17; *De unitate intellectus* II; *Summa theologica* II, q.72, membr.3; and *Problemata determinata* XLIII, probl.34.

[72] Augustin Delorme analyzed this aspect of Albert's embryology from a Thomist standpoint in his article "La morphogénèse d'Albert le Grand dans l'embryologie scolastique," *Revue Thomiste* 36 (1931), 352-360.

[73] For a further examination of related questions, see the study by Karen Reeds in this volume.

from a principle that is not only unmixed with the corporeal matter but also extrinsic to the active virtues of the seed. This principle is "the light of the first active intellect" whose work is "the work of nature"; only the "divine intellect, eternal and incorruptible," operates in total freedom from any functions of the corporeal organs, with the *spiritus* of the sperm as its instrument.[74] Here Albert draws the terminus of his philosophical analysis,[75] and while he refers to his other Aristotelian commentaries, especially *De anima* and *De intellectu et intelligibili*, for further metaphysical elaboration, he does not allude to theological implications in these chapters of *De animalibus*.

The autonomy of Albert's natural philosophy is strikingly evident when one compares his straightforward explication in these chapters[76] of a sentence from Aristotle's work *On the Generation of Animals* II, 3 with his later warning that a misinterpretation of that sentence "results in an error that is contrary to faith." The sentence interlocked the Aristotelian concepts of the spermatic *spiritus*, formative virtue, and divine intellect as agents of early embryogenesis;[77] the misinterpretation would consist of a confusion between active and contemplative intellect; and Albert's theological warning concluded his *determinatio* to a question raised by his master general.[78] It may be added that the infusion of the soul is not explicitly characterized as a "creation by God" in Albert's *De animalibus*, as it is in his writings that are more related to theology, for example, *De homine*.[79] In the latter work, on the other hand, due to the different context and perhaps also to the earlier date of composition, a strange lack of precision marks his description of embryogenesis and animation: when the mixture of male seed and female menstruum has cooked until "baked to a crust, then it is formed and the spirit of life is made in it by God's command."[80]

The moment of animation is when "the organs, through which the works of the soul and of life are performed, are completed in shape (*in figura perfectis*), according to Albert's most general definition.[81]

---

[74] *De animalibus* XVI, tr.1, cc.11, 12, and 13 (ed. Stadler, 16: 1094, 1096, and 1098).

[75] "Et hoc est ad quod volumus pervenire in omnibus capitulis quae praemisimus de virtutibus animae et spermatis" — *De animalibus* XVI, tr.1, c.12 (ed. Stadler, 16: 1096).

[76] Ed. Stadler, 16: 1098.

[77] See pp. 429-430.

[78] *Problemata determinata* XLIII, probl.34 (ed. Weisheipl, pp. 45-49).

[79] Q.1, art.2.

[80] *De homine* q.5, art.3 (ed. Borgnet, 35: 80).

[81] *De animalibus* XVI, tr.1, c.11 (ed. Stadler, 16: 1092).

If we depended entirely on *De animalibus* we would have to infer further specifics from his description of morphogenesis and gestation, which we will examine shortly. However, some sharper focus is provided by passages in his commentary on the *Sentences*. There Albert states that the soul, as a spiritual substance "that is the form and perfection of the body, must be made according to the order of nature only when the body has been organized."[82] Later he specifies this further by adding, "the soul needs a body that is such that it has member distinguished from member, and that it is distinguished in shape from the body of another animal."[83] He answers the question, when abortion is murder, primarily by quoting Augustine's assertion that "the unformed child has no soul" and Jerome's view that "the seeds are gradually formed in the womb," and abortion "is not considered murder until the completed elements have received their shapes and members."[84] In his terse *expositio* of the answer Albert refers to *De anima* but not to any other Aristotelian commentaries, and he appends some "useful verses" from the Salernitan collection,[85]

> What is conceived by the semen becomes like milk duly in the first six days; in the next nine it becomes blood; the twelfth added day consolidates it, the eighteenth from then fashions it, and the time that follows leads it to birth.[86]

Albert's descriptions of early embryonic development contain an occasional parallel with this Salernitan doggerel, most notably in

---

[82] *In II Sent.* dist.17, art.2 (ed. Borgnet, 27: 299).

[83] Ibid., art.3 (p. 301).

[84] The text is important enough to warrant quotation in full. "Hic quaeri solet de his qui abortum procurant, quando judicentur homicidae, vel non? Tunc puerperium ad homicidium pertinet, quando formatum est, et animam habet, ut Augustinus super Exodum asserit. Informe autem puerperium ubi non est anima viva, lex ad homicidium pertinere noluit. Dicit etiam Augustinus, quod informe puerperium non habet animam, ideoque mulctatur pecunia, non redditur anima pro anima. Sed jam formato corpori anima datur, non in conceptu corporis nascitur cum semine derivata. Nam si cum semine et anima existit de anima, tunc et multae animae quotidie pereunt, cum semen fluxu non proficit nativitati. Primum oportet domum compaginari, et sic habitatorem induci. Cum ergo lineamenta compacta non fuerint, ubi erit anima? Item, Hieronymus ad Algasiam: Semina paulatim formantur in utero: et tamdiu non reputatur homicidium, donec elementa confecta suas imagines membraque suscipiant. His apparet tunc eos homicidas esse, qui abortum procurant, cum formatum est et animatum puerperium" — *In IV Sent.* dist.31, art.18 (ed. Borgnet, 30: 250-251).

[85] *In IV Sent.* dist.31, art.18 (ed. Borgnet, 30: 251).

[86] Compare *Flos medicinae scholae Salerni* IV, 7; ed. Salvatore de Renzi in *Collectio Salernitana* 5 (Naples, 1859), 51, vv. 1795-1798.

placing the completion of morphogenesis around the fortieth day, but they are much more scientific in tone, intent and detail. It should nevertheless be remembered that until the seventeenth century this topic truly lay in the area of speculation because of the limitations of unaided human vision, as we have indicated earlier. The morphogenetic changes which occur during the preimplantation and organization phases of embryonic development are appreciable to the mother or her physician only through concomitant physiological alteration in general habitus indicated by tenderness and some discomfort in the breasts, morning sickness or nausea, and toward the end of the period slight enlargement of the uterus — to which Albert adds the swelling of the veins in the mother's neck.[87] In addition to accumulated obstetrical observations, analogies taken from avian and mammalian comparative embryology, as well as astrological calculations provided the basis of traditional views on the phasing or staging and length of gestation. These views, available to Albert and his contemporaries through philosophical, medical and encyclopedic sources, have recently been collated by Anthony Hewson.[88] They apparently contained little beyond the description and dating of morphogenesis provided by Avicenna.

The particular sources consulted as well as their presentation by Albert depend upon the context and purpose of each work in which he deals with this feature of human embryology. Though it may be an oversimplification, in general one may recognize his morphological and medically influenced descriptions as derived mainly from Galen and Avicenna, his physiological views as more reflective of Aristotelian natural philosophy, and his metaphysical opinions as susceptible to Neoplatonic and Augustinian inspiration. However, as may have become apparent in the preceding pages of this study, Albert's fields of inquiry retained their independence, and his primary concern was to present all the pertinent data from the authorities and from his own insights, at the risk of redundancy and inconsistency. With this in mind we will neither presume that he

---

[87] *De animalibus* IX, tr.2, c.4 (ed. Stadler, 15: 719).

[88] Anthony Hewson, *Giles of Rome and the Medieval Theory of Conception* (London, 1975), pp. 166-177; for a careful evaluation of Hewson's study in relation to Albert's ideas, see the review by Michael McVaugh in *Speculum* 52 (1977), 987-989. For a comparison between Albert's treatment and that of a contemporary encyclopedist, see Pauline Aiken, "The Animal History of Albertus Magnus and Thomas of Cantimpré," *Speculum* 22 (1947), 205-225.

personally subscribes to every statement he cites, nor expect him to combine the data into an integrated synthesis.

The above observations are borne out by the four chapters which Albert devotes to morphogenesis and gestation in his *De animalibus*. The first of these chapters (IX, tr.1, c.3) is part of his tractate on the principles of human generation and, appropriately, follows Aristotle's *History of Animals* VII, 3. The second (IX, tr.2, c.4) an explication in the context of the dispute between the Peripatetics and the Physicians, is derived from Avicenna's *De animalibus* IX, 9. The fourth and last (X, tr.2, c.3) occurs in Albert's tractate on the causes of sterility and borrows mainly from Avicenna's medical *Canon* III, 21.2. We will concentrate on the third of the four chapters (IX, tr.2, c.5), "On the Time and Order of Formation of the Embryo into a Creature of the Human Species," because it is the most detailed. It is a paraphrase of Avicenna's *De animalibus* IX, 10, enriched with "what is taught unanimously by those more expert in these matters" and with the word of "trustworthy women who have borne many little ones."[89] The stages of development *in utero*, as computed in this chapter, may be summarized as follows:

Stage	*Day*
1. Male sperm and female humor joined together: conception	0
2. Concept becomes foamy and milk-like, with three large vesicles to provide space for the principal organs	6
3. Appearance of blood-like droplet and red thread, in preparation of the vessel of the spirit, that is, the place of the heart	9 (8-10)
4. Coagulation, preparing the substance of the heart	15 (13-17)
5. Formation of the heart and, from it, liver and brain; the whole appears as flesh	27
6. Embryo fully formed, with limbs, hands and feet	36-40

Albert here agrees with Aristotle that the embryo at this time has "only the size of a large ant" and therefore will not be visible in an abortion.

Elsewhere, however, Albert adds that the embryo can be found "through filtering" if aborted into clear cold water, and that in a

---

[89] Ed. Stadler, 15: 723-727.

fresh abortion it is "sometimes seen to have a movement of dilation and contraction when it is pricked with a needle: and thus that creature is known for certain to be animated."[90] He further follows both Aristotle and Avicenna in stating that in the early stages the female embryo develops more slowly than the male, a view consistent with his general explanation of sex differentiation.[91] Faithful to Aristotle and in opposition to Galen, Albert insists that the heart develops before the brain and the liver. It is somewhat puzzling that he cites as coming from "the physicians' camp" the duplication for computing the length of gestation, a method that was in fact Aristotelian.[92] This method of duplication proceeded as follows:

1. Conception until complete embryonic formation .... 40 days
2. Formation to fetal movement .... $40 \times 2 = 80$ days
3. Fetal movement to birth .... $80 \times 2 = 160$ days

Total length of normal gestation .... 280 days.

A close look at the critical apparatus of Stadler's edition of *De animalibus* reveals that Avicenna's formulation and applications of this rule, in themselves inaccurate, are further garbled by Albert who also injects a note of scepticism into his presentation. He replaces Avicenna's guarantee that this computation "is confirmed by all the evidence" by his own reservation that "these things which the physicians assert are not truly proven and not always found to be true." He cautions further that several extraneous factors, such as a "more joyful (*laetius*)" intercourse or a warmer place and time of conception, may accelerate the quickening and thus shorten the gestation.[93]

A premature live birth could be explained by a more rapid devel-

---

[90] *De animalibus* III, t.1, c.3 (ed. Stadler, 15: 607).

[91] About Albert's explanation of sex differentiation, which pertains more properly to the field of genetics, suffice it to note here that he follows Aristotle in asserting that "the female sex is created from an active or passive defect of the principle of generation." The defect, he explains, consists of an "indigestion" suffered by the spermatic moisture and resulting in an imbalance of complexion (too cold) and consistency (too thin). Thus he rejects as primary causes the dominance of male or female seed, as Empedocles and Galen held, and the provenance of sperm or intra-uterine development on the right or left side, as Parmenides taught. Albert admitted only that these might be contributing factors. See *De animalibus* IX, tr.2, c.3 (ed. Stadler, 15: 715); XV, tr.2, c.3 (16: 1023); and especially XVI, tr.1, c.14 (16: 1099-1101) and XVIII, tr.1, c.2 (16: 1192-1201).

[92] See Hewson, *Giles of Rome*, p. 174.

[93] Ed. Stadler, 15: 724.

opment in the early embryonic stages, either with the Aristotelian duplication (for example, 30 + 60 + 120 = 210 days) or with Albert's reference to terrestrial circumstances. However, a commonly held medieval view that premature deliveries were most successful in the seventh month but likely to be stillbirths in the eighth, arose from popular beliefs in the perfection of the number seven and in the malignity of Saturn's influence which controlled the eighth month. In his work on prediction and destiny, *De fato*, Albert explicitly discounts these beliefs of judicial astrology. He counters that the cause of mortality in the eighth month is neither Saturn nor the "celestial cycle" and, more broadly, that stillbirths "occur not on account of the period but because of the corruption of natural principles."[94] Nevertheless in the same treatise, when considering those natural principles that affect embryonic development, Albert's physiological explanation is vague and geared to the premise of astrological influences. He identifies seven distinct physiological changes as influenced by the lunar revolutions. Four powers are attributed to the moon: its own which, according to established tradition, lies in modulating fluid motions such as tides, menses and sperm production; the faculty, in conjunction with Mercury, to promote the mingling of the sperms; the power, derived from the sun, to instill life into the moved moisture; and the influence, drawn from Venus, on the development of the conceptus. Albert then, without specifying respective time periods, describes the functional processes of embryogenesis as a series of necessary changes, summarized below.

Change 1: the semen is converted into the form of the heart.

Change 2: differentiation of the material by the *spiritus*, through the heart, into the vesicles and form of the principal organs. These organs in their turn have secondary creative virtues. The liver creates the natural virtues, the brain the animal virtues, and the seminal vessels the formative virtues of the embryo.

Change 3: further movement of the brain, liver, and seminal vesicles to their definitive positions; their specific differentiation, controlled by the "blowing out" of the cardiac vital spirit.

Change 4: differentiation, by similar cardiac "exsufflation," of the secondary body members which do not possess the creative powers of the principal organs. They perform their functions through power

---

[94] *De fato* art.4, ad 4 and ad 5 (ed. Simon, pp. 74-76).

derived from the heart along channels, veins, and arteries, which follow pathways opened up within undifferentiated tissues (the mesenchyme in modern embryology).

Change 5: final sculpturing and modelling of the embryo's organs in the humid material.

Change 6: desiccation of the superfluous moisture by cardiac warmth, diffused by the *spiritus*, thus condensing and consolidating the organs and reinforcing bones, joints, and ligaments.

Change 7: infusion of movement from the heart into all the bodily organs.

As all movement of the conceptus, as well as each preceding change, is controlled by the moon, Albert concludes, "it must be completed by seven lunar revolutions." Why then is it not until the ninth month that most births occur? Because according to Albert, even though at seven months the embryo is completed and has all things necessary for its being, it may not have a sufficient balance between formative virtue and material. An abundant formative virtue may press undersized material, and then small but agile premature babies will be born in the seventh month. Normally the material is adequate and the formative virtue gains strength by "resting through one revolution of the moon," so that birth comes in the ninth month. If the virtue is hampered by uncooperative material, it is unable either to complete the movement started under lunar impetus in the seventh month or to recover in the eighth month, and then the infant "is born and dies in most cases."[95] Albert's endeavor to weed out judicial astrology while weaving embryology into the texture of astrological influences has resulted in a somewhat hazy description of human development and birth.

Even more nebulous and farther removed from the precision of his *De animalibus* is Albert's synopsis of the early generative and morphogenetic processes in the *Problemata determinata XLIII*. His description there serves merely as a starting point to answer a doctrinal question on the nature of the intellect that animates the conceptus,[96] and it is presented in an idiom tuned to the metaphysical context and in a manner that would not distract his reader from

---

[95] Ibid., art.4 (ed. Simon, p. 76; ed. Colon. 17/1: 76.18-23). In a parallel but much shorter passage Albert insists that reported survivals of eight-month births are spurious and the result of "the pregnant woman's miscalculation" — *De animalibus* IX, tr.1, c.4 (ed. Stadler, 15: 691-692).

[96] See p. 423 above.

the main thesis. In order to demonstrate that not the contemplative or "receptive intellect by which man is man," but only the active intellect is in the seed and directs the formative virtue in the *spiritus* towards the "target" of an animate concept (*"metam animati"*), Albert identifies the following changes in early embryonic development.

1. At conception, the male's seminal liquid is resolved into *spiritus*; this in turn inspires the female's liquid, which it digests and purifies into matter more suitable for its operations.

2. In a first differentiation, the *spiritus* shapes the material into a vessel (*ampulla*) which it fills and from which it then produces two lateral vessels.

3. From the median vessel the *spiritus* extends the material in a thread-like fashion upwards and, thinner but longer, downwards.

4. The *spiritus* forms the median vessel into the heart and develops the lateral ones into the flanks and arms.

5. From its upward extension the *spiritus* becomes the animal spirit in the brain; from its downward extension ("in the shape of a pyramid whose base is near the heart's vessel") it makes the "intestines, abdomen, chest, lungs, liver, back, kidneys, and hips"; and from the thinner part it fashions the legs and feet.[97] This presentation by Albert poignantly documents the need for clearer definitions of morphological and physiological terms, such as *spiritus* and vessel, which lend themselves less to scientific description and analysis than to philosophical speculation and theological embryology.

In this connection we may cite Albert's most succinct synopsis of embryogenesis, in his elucidation of Augustine's teaching on the human development of Christ. Albert claims that Augustine "did not mean to say that [the body of Christ] would have been formed successively like other bodies," and he explains that in Christ all processes, including formation, consolidation and differentiation, took place at one and the same time "on account of the infinite virtue of the operating Holy Spirit, who can do what nature cannot even in succession."[98]

Unlike the questions on animation and embryogenesis, the related subject of fetal membranes was of less concern to the theologian and the natural philosopher than to the physician or obstetrician, for

---

[97] *Problemata determinata XLIII*, probl.34 (ed. Weisheipl, pp. 45-49).
[98] *In III Sent.* dist.11, art.14 (ed. Borgnet, 28: 40-41).

whom the integrity of these appendages including the placenta was of the utmost importance in the frequent cases of postpartum hemorrhage. On this subject, then, Albert tends to quote at length not only from Aristotle but also from Galen and, above all, from Avicenna. For example, his descriptions of the human amnion (*alieas*, Arabic *abgas*) and chorion (*secundina*) are in the main paraphrases from Avicenna's *Canon* and *De animalibus*. It is worth noting that his description of the umbilical cord and its contained blood vessels, one umbilical vein (pulsatile) and two umbilical arteries (non pulsatile veins) is morphologically correct. More significantly, we find Albert making such incisive comments that we must presume that he derived his knowledge from personal observation and oriented his deduction towards his own ideas, unlike many contemporary authors who on this subject merely copied their sources. Thus, a combination of observation and syllogism allows him first to account for and next to reject as erroneous a then current medical theory on the origin of the placenta. Albert observes that the umbilical arteries branch out "into the orifices of the uterine veins" and because these "orifices" are redder and thinner near the uterine wall and become paler and thicker as they approach the umbilical cord, the arteries appear like the continuation of veins that have come out of the uterus. This, he continues, is the reason why some physicians teach that these arteries or cotyledons (the placenta) develop from the uterus as the cotyledons of fruits originate in the tree. He objects that "reason does not accept this" because, according to his basic doctrine, "the formative virtue (*virtus formativa*) of animals is in the seed and not in the uterus — hence the virtue that forms the embryo also forms the cotyledons of the veins."[99]

Somewhat less critically, however, Albert accepts from Galen and Avicenna the existence of a human allantois.[100] Ironically, he may have found support for this error in his own personal observations of domestic animals, such as pigs and dogs, who indeed have extensive and highly vascular allantoid membranes separating the embryos and their amniotic cavities from the surrounding chorio-allantoic placentae.[101] Albert attributes, not unreasonably and to some extent even correctly, a dual purpose to the allantois. He sees it both as a

[99] *De animalibus* IX, tr.2, c.5 (ed. Stadler, 15: 725-726).

[100] For this see A. W. Meyer, "The Elusive Human Allantois in Older Literature," in *Science, Medicine and History; Essays in Honour of Charles Singer* (Oxford, 1953), 1:510-520.

[101] Hamilton and Mossman, *Human Embryology*, pp. 600-602.

reservoir for fetal urine[102] and as a protective shield for the developing embryo. After quoting Avicenna's description of the amniotic fluid and the amnion, he notes that this third membrane "is called the embryo's armor (armatura) by the midwives."[103] This comment is somewhat puzzling, and it is possible that Albert here confuses the allantoid membrane, observed on the farm, with an obstetrical term. On the other hand, he may be referring to the "caul" which was believed to be a good omen; or midwives may have mentioned to him the hard calcareous and granular degeneration of the aging placenta in humans at or close to term. These phenomena, however, would probably have been reported not as common occurrences but as anomalies of childbirth.

Anomalous birth and, even more, the generation of deformed offspring has always held for man a fascination that transcends the bounds of the three embryological traditions and that is usually in inverse relationship to an understanding of the mechanism of teratogenesis.[104] The birth of a deformed child was read as an omen that could even reveal the future of "the Country," as in Mesopotamia four millennia ago.[105] Aberrant individuals were greeted with feelings that ranged from horror or awe to ridicule or marvel. According to Cicero, "they are called manifestations, portents, monsters and prodigies because they show, portend, demonstrate, and predict";[106] Pliny the Elder suggested that they were made by "ingenious Nature as toys for herself and marvels for us."[107] Since abnormal births were usually seen as events outside the orderly process of natural reproduction, they were most often explained by supernatural causes such as divine error, demonic interference, or astrological influences. Even Aristotle conceded that certain malign or astrological influences deflected normal in utero development,[108]

---

[102] Albert states, somewhat differently, that the embryo's secretion through the umbilical cord is received in the chorion, "in vase secundino" — De animalibus I, tr.2, c.24 (ed. Stadler, 15: 159).

[103] De animalibus IX, tr.2, c.5 (ed. Stadler, 15: 726).

[104] The history of teratology has received even less notice than that of embryology. The most comprehensive recent writings on the subject are by Josef Walkany, Congenital Malformations (Chicago, 1971), pp. 6-20; or better still, his article, "History of Teratology" in Handbook of Teratology, ed. J. S. Wilson and F. Clarke Fraser (New York, 1975), pp. 3-45.

[105] See Erle Leichty, The Omen Series Suma Izbu; Texts from Cuneiform Sources (New York, 1970), pp. 7-12.

[106] On Divination I, 8.

[107] Natural History VII, 2.

[108] Anthony Preus, Science and Philosophy in Aristotle's Biological Works (Hildesheim and New York, 1975), pp. 200-204.

although he attempted a natural explanation of teratogenesis within the physiological schema derived from hylomorphism. He attributed aberrations to a divergent movement of the female matter and its imbalance with the formative virtue of the male sperm. Using the analogy of split vortices he taught that, if the maternal matter was equal to the male power, multiple but normal embryos would result and, if it was either overabundant or deficient, an offspring would be formed that either had extra or lacked organs, limbs or digits.[109] Aristotle did not equate congenital abnormality with imperfection because his zoological taxonomy was based on method of reproduction rather than on the external features of the adult of a species; thus, for him "the viviparous animals are the perfect ones, and the first of these is man."[110]

While Aristotle may be said to have provided the only rational comprehensive explanation and scientific characterization of terata until the advent of nineteenth-century embryology,[111] a few authors in the intervening centuries contributed some valuable insights. Among these authors was St. Augustine who observed that "a thing is wonderful only because it is rare" and that "the trouble with a person who does not see the whole is that he is offended by the ugliness of a part because he does not see its context or relation to the whole." In addition to emphasizing that abnormalities are not caused by a divine error, he further elaborated the theological dimension of teratology by raising the question whether monsters have sprung "from one of the sons of Noah or from Adam himself." Augustine made a distinction between the monstrous races of fable, such as the Cyclops and Sciopodes, and the congenitally defective individuals. As to the fabulous races, whose existence he doubted, he declared that, "however abnormal they are in their variation from the bodily shape that all or nearly all men have, if they still fall within the definition of men as being rational and mortal animals, we have to admit that they are of the stock of the first father of all men." As to the numerous individuals "who have been very different from the parents from whom they were certainly born," Augustine asserted that "all these monsters undeniably owe their origin to Adam."[112]

---

[109] *Generation of Animals* IV, 3-5.

[110] Ibid., II, 4 (737b25).

[111] Charles Singer, *A Short History of Scientific Ideas to 1900* (London, 1962), p. 470.

[112] *The City of God*, XVI, 8; tr. Gerald Walsh, et al., ed Vernon Bourke (New York, 1958), pp. 365-367.

It seems safe to claim that Albert was the first since Aristotle to offer a substantial contribution to teratology as a subject of scientific embryology.[113] He followed Aristotle's classification of aberrant offspring by assigning his chapters on teratogenesis in *De animalibus* to book XVIII, "On the Manner of Generation of Perfect Animals." From Albert's theory on animation, examined above,[114] we may infer that like Augustine he believed terata to be human in so far as they have organs that are differentiated, that distinguish them from other animals, and that are necessary for life and reason. A similar belief is implicit in Albert's position that even when an animated offspring has no resemblance either with the individual or with the species of the generant, "it retains at least the generic likeness" because "in all generated beings at least the genus is preserved."[115] Throughout his writings, moreover, runs the idea — perhaps not shared by much of twentieth-century society — that "monsters" are perfect in being the effect of human generation and imperfect only in comparison with Nature's intention to make the best possible.[116]

For the etiology of abnormal births we can identify in Albert's work at least six kinds of causal explanations that pertain, respectively, to demonology, parapsychology, astrology, physiology, morphology, and genetics. The first and most popular kind, which explains the generation of monsters as the work of demons, is peremptorily dismissed by Albert with the verdict that also this generation "is natural."[117] The second explanation, also prominent in popu-

---

[113] Albert's contribution may be contrasted with such mystical views of teratogenesis (and of animation) as those of Hildegard of Bingen, who a century earlier wrote that "often in forgetfulness of God and by the mocking devil, a mistio is made of the man and the woman and the thing born therefrom is deformed, for parents who have sinned against me return to me crucified in their children." Quoted by Charles Singer, "The Scientific Views and Visions of St. Hildegard (1098-1180)," in his *Studies in the History and Method of Science* (Oxford, 1917), pp. 49-51.

[114] See pp. 423-424.

[115] *De animalibus* XVIII, tr.1, c.6 (ed. Stadler, 16: 1214).

[116] This teleological characterization is most explicit in the *Quaestiones super De animalibus*, XVIII, q.2, whether hermaphrodites are natural (ed. Filthaut, p. 297); and q.5, whether monsters are part of nature. The conclusion to the latter question is that "even though a monster is bad when compared with the intended effect in itself, it is good nevertheless in so far as it has the *rationem entis*" (ibid., p. 299). Other pertinent *quaestiones* in the same work, deserving of further analysis but not explored in this study, are IV, 22; VII, 3; X, 5; XV, 13 and 17; and XVIII, 3, 6, and 7.

[117] *In II Sent.* dist.7, art.9 (ed. Borgnet 27: 158). Subsequently, but without much apparent commitment, Albert discusses the question whence devils get the semen for their reproductive role as incubi — *In II Sent.* dist.8, art.5 (ed. Borgnet, 27: 175).

lar accounts, attributes monstrous births to the imprint of the father's or mother's imagination at the moment of conception. This explanation is simply cited by Albert from the sources without further comment,[118] even though such a cause would be perfectly natural in view of his general acknowledgement that embryonic development is influenced by "the parents' regimen, life style, and emotions (accidentibus animae)."[119] The other four explanations, which Albert adopts with varying degrees of emphasis, are anchored in the principle of naturalness and guided by the etiology of Aristotelian natural philosophy.

The astrological explanation of teratogenesis is mentioned only summarily by Albert in De animalibus. In a chapter based on Aristotle's Generation of Animals IV, 3 he states that one of two causes of abnormal birth is "the celestial virtue because, when those heavenly lights that primarily control generation are assembled in certain places, there cannot be a generation of man. However, it will be more fitting to speak of these things in another discipline."[120] The influence of constellations receives slightly more attention in Albert's commentary on the Sentences when he rejects the devil's role in teratogenesis. In one argument he cites Avicenna's view that the only cause of "a birth with a dog's head or a pig's foot" is the aspect of a star at the hour of impregnation. As another argument he refers to "what Ptolemy says in the Quadripartitum," to wit that "when the sun is in a certain place and Aries declining, no human generation takes place; and if the semen then falls into the womb, a monster will be born." Here Albert adds,[121]

and so that it may be believed, I have tested (probavi) this by observation in two honest and good matrons, from whom I have learned that they had borne monsters; asking from them the time and computing the stars, I have found that according to their estimations they had conceived with the sun in that same degree and minute.

---

[118] De animalibus XXII, tr.1, c.3 (ed. Stadler, 16: 1352). Here Albert cites, from the pseudo-Galenic De spermate V and Avicenna's De animalibus IX, 7, the case of a king who during intercourse imagined a "monstruosum nigrum" and mentioned this to the queen who in due course bore such an offspring. In the Quaestiones super De animalibus cases are reported in which according to Avicenna a queen during intercourse imagined and as a result bore a dwarf (VII, q.3 [ed. Filthaut, p. 172]) or "a demon, or a dwarf, or an Ethiopian according to others" (XVIII, q.3 [p. 298]).

[119] De animalibus IX, tr.1, c.6 (ed. Stadler, 15: 699).

[120] Ibid., XVIII, tr.1, c.6 (16: 1214).

[121] In II Sent. dist.7, art.9 (ed. Borgnet, 27: 157-158).

Albert evidently attached special importance to this personal verification, since he referred to it also in the more formal setting of the *Problemata determinata XLIII*.[122]

It is in the *Problemata* that we find Albert's most complete explanation of astrological influences on abnormal births. This explanation occupies half of his *determinatio* to the question "whether anything of the substance of heaven enters into the composition of what is composed out of the four elements and, especially, of what is alive and animate by the effect of its virtue." Albert answers emphatically that "no substance of heaven enters into the generable and animate body."[123] His answer is laconic in comparison with his vehement denunciation in *De animalibus* of any such theses as "fables" and "patent lies" whose absurdity can even be "perceived by barbers."[124] In the latter, however, he rejected the presence of a quintessential heavenly substance in animated bodies and he reasoned from cosmology and natural philosophy, whereas in the *Problemata* he levels his argument at the introduction of celestial substance into generation and he supplements his *determinatio* with astrology. In the process of this argument, then, he affirms the humanity of misshapen offspring.

As first proof of celestial power Albert cites an opinion of "Peripatetic philosophers."[125] This opinion holds that sometimes, even though the generant's virtue aims for an offspring with a similar form and the elemental virtue directs towards that aim, the fetus "will not have a human shape on account of the figurations of the rays and stars that draw it into another shape." The second proof Albert takes from the maxim in the *Quadripartitum* that if conception occurs "when the stars arise in Aries towards Gorgon's Head, if Jupiter does not assist with a strong influence and Venus does not exert influence, what is born will both be man and not have the shape of a human body." By attributing them to the "elemental qualities" which the seminal compound derives from the macrocosm, and to the "irradiation" through what Aristotle called celestial warmth and light, Albert interprets these astrological influences in a "natural" way.[126]

---

[122] Page 49 of the original edition by Prof. Weisheipl, who in n. 59 indicated that he had not found Albert's source of the Ptolemaic idea or a restatement of Albert's observation.

[123] Probl.35 (ed. Weisheipl, p. 49).

[124] *De animalibus* XX, tr.1, cc.5-7 (ed. Stadler, 16: 1284-1294).

[125] Albert names Porphyry and Theophrastus, for whom see n. 58 in'the Weisheipl edition.

[126] *Problemata determinata XLIII*, probl.35 (ed. Weisheipl, pp. 49-50).

"It is more natural and closer to the truth," according to Albert, to identify the "immediate physical causes of monstrous births and offsprings that have abnormal (*occasionata*) organs."[127] He assigns these causes to the physiological and morphogenetic processes of early embryonic development or, with Aristotle, to "the material and manners of impregnation and generation."[128] Albert's etiology is most lucidly summarized in his commentary on the *Sentences*, where he defines a monster as "that which goes outside nature's way (*quod excedit modum naturae*)" on account of one of several factors.

Among the immediate factors, Albert first lists "deficiency" — without further specification — and gives as example "a hand to the shoulder," the phocomelia that became so notorious in the Thalidomide tragedy of 1961. A second cause is "abundance," either of "the virtue to form one member," which results in such offspring as that having "several mouths, noses or ears"; or of "matter in one member, as in the case of six or more fingers."[129] In *De animalibus* he reports having observed two brothers who both had complete polydactylism.[130] A third cause, which Albert attributes (as Aristotle did) to "the movement of the uterus and of the material," leads to multiple embryos that may be inadequately separated from each other. He remembers vividly one personally observed case of incomplete twinning, which he cites three times, namely of a "two-bodied goose, connected only at the back, with two heads, four wings and four feet."[131] About Siamese twins he states that "many trustworthy people have told us that they have seen such a man who was two men connected at the back; one was impetuous and wrathful, the other meek, and they lived more than twenty years; and after one died the other survived until he also died from the putrid decomposition of the dead brother."[132]

Albert devotes special attention to hermaphroditism, which he explains by the imbalance between formative virtue and passive matter in the embryo's genital area. Among related anomalies, he supplements with personally heard cases Aristotle's reports of imperfo-

---

[127] An important discussion of *occasio* in connection with mutation and adaptation follows Albert's description of the eye of the mole (*talpa*). The description is based on personal dissection which leads him to disagree with Aristotle — *De animalibus* I, tr.2, c.3 (ed. Stadler, 15: 51).

[128] *De animalibus* XVIII, tr.1, c.6 (ed. Stadler, 16: 1215-1216).

[129] *In II Sent.* dist.18, art.5 (ed. Borgnet, 27: 319).

[130] XVIII, tr.1, c.6 (ed. Stadler, 16: 1218).

[131] Ibid., (16: 1216); also VI, tr.1, c.5 (15: 457); and XVIII, tr.2, c.3 (16: 1225).

[132] Ibid., XVIII, tr.2, c.3 (16: 1225).

rate anus and deviated urethra. In the latter instance he graphically describes the case of a cryptorchid child whose maleness was discovered by accident but subsequently proven by a prolific marriage.[133] Occasionally Albert may have been misled by reports of abnormal phenomena when he received them from such sources as the "truthful and experienced physician" who told him that "a noblewoman in Germany bore sixty children, five at a time." The same physician claimed that he had been[134]

> called to treat a noblewoman who had aborted 150 at the same time. She thought that she had brought forth worms from her uterus, but when the webs were opened shaped children were found of the size of a human auricular finger; and several of them had a movement of contraction and dilation and many other signs of life; and they were all lying in a basin before his eyes. Their eyes were incomplete, and their fingers and toes were like hairs.

Rather than a multiple abortion, as Albert thought, this was presumably a mole consisting of many cysts (vesicular hydatidiform). Recognized moles, however, he uncompromisingly characterized as inanimate tumors rather than anomalous offsprings.[135] Moreover, he explained that what "midwives think to be a ram's or a goat's head" is nothing but a mole that has hardened after taking such a shape from the cavity of the womb.[136]

A final source for the natural etiology of teratogenesis lies in the realm of genetics and thus could not be explored but only suggested incidentally by Albert. Recognizing the heredity of certain congenital defects, he believed that "it happens very frequently that the abnormalities (*occasiones*) of the parents appear in the offspring with the same or perhaps greater prominence"; furthermore, "the unnatural things of parents are sometimes passed on to the entire generation."[137] To a different branch of genetics belongs Albert's inchoate idea that a monster may be generated "from a deficiency in the mobility of the semen."[138] Another connection with genetics is inherent in his thoughts about the relationship between age and ano-

---

[133] Ibid. (16: 1226).
[134] Ibid., IX, tr.1, c.5 (15: 693).
[135] Ibid., X, tr.1, c.4 and tr.2, c.2.
[136] Ibid., XVIII, tr.2, c.6 (ed. Stadler, 16: 1234).
[137] Ibid., IX, tr.1, c.6 (15: 698).
[138] *In II Sent.* dist.34, art.3 (ed. Borgnet, 27: 550).

maly, which are interesting in the light of modern views of Down's syndrome.[139] Albert observes that when parents beget below or beyond their normal fertile age, the immaturity or debility of the seed may result in "prodigies with regard to birth."[140] There is value for comparative embryology in his understanding of chick anomalies, especially when he realizes that terata may result not only from aberrations in the yolk, albumen or membranes, but also from the age of a developing egg.[141]

While Albert could only hint at the role of genetics in teratology, he discussed at length the normal effects of heredity such as the differentiation of gender and the similarities of the offspring with the generants and ancestors.[142] In addition to the various aspects of heredity, he presented the whole gamut of other subjects related to embryology in his *De animalibus*, at the hand of Aristotle and Avicenna. A further discussion of these subjects, however, pertains more to the study of Albert's and medieval concepts of gynecology, including on the causes of sterility and infertility; of obstetrics, for example on the causes of difficult pregnancy and delivery, on multiple births, and on perinatal care; and of reproductive anatomy and physiology.[143] Similarly, other subjects that could be examined with respect to embryology belong more properly to the area of sexual ethics, in which Albert's views were rather liberal,[144] or to that of sacramental theology, in which Albert may have been "the single prominent author" to explore the relation between the *bonum sacramenti* of marriage and the natural purposes of sexual intercourse.[145] An examination of all these related subjects, as well as a comparison with the concepts of other thirteenth-century authors, would presumably enhance rather than alter our short assessment of Albert's contribution to embryology.

As the Universal Doctor, Albert endeavored to give his contemporaries the most complete possible access to all the data on embryology that were available in the sources, subject to speculation, and

---

[139] Walkany, *Congenital Malformations*, pp. 311-335.

[140] *De animalibus* v, tr.2, c.1 (ed. Stadler, 15: 427).

[141] Ibid., vi, tr.1, c.2 (ed. Stadler, 15: 447).

[142] Ibid., xvi, tr.1, passim; xv, tr.2, c.3; xvi, tr.1, c.14; xviii, tr.1, cc.1-5; and xxii, tr.1, c.3.

[143] See James Shaw, "Scientific Empiricism in the Middle Ages: Albertus Magnus on Sexual Anatomy and Physiology," *Clio Medica* 10 (1975), 53-64.

[144] See Leopold Brandl, *Die Sexualethik des hl. Albertus Magnus* (Regensburg, 1955).

[145] John T. Noonan, Jr., *Contraception: A History of Its Treatment by the Catholic Theologians and Canonists* (Cambridge, Mass., 1966), pp. 286-288.

known from observation. Although he adapted his presentation to the contexts and readers of his various writings, he focused his attention on questions of natural philosophy and science. His theories on human generation and development were not only based on an Aristotelianism that was enriched with more recent insights, but they were also left sufficiently open-ended to allow for further inquiry. To be sure, such inquiry would have to overcome the remaining obstacles of inadequate techniques for observation, confusing terminology for description, and dependence upon analogies for analysis. Because of these handicaps, and perhaps even more on account of shifting concerns in the intellectual climate, little was added to Albert's contributions until the embryological work of Aldovrandus, Coiter, Fabricius and Harvey in the sixteenth and early seventeenth centuries. Even if succeeding generations of scholars did not develop the legacy bequeathed to them by Albert the Great, they nevertheless were indebted to him for the natural and rational orientation that characterized not only his presentation but also his interpretation of embryological and teratological data and doctrines. For modern society, seven centuries after his death, perhaps the most significant aspects of Albert's contribution remain in the idea that embryogenesis is not only an animal process but also a part of the unique human development, and in the reminder that embryology lies at the interface of biology and philosophy.

17

Albertus Magnus on Falcons and Hawks

Robin S. Oggins
State University of New York at Binghamton

Albertus Magnus' work on falcons and hawks comprises roughly half of his book devoted to birds — descriptions of birds listed in roughly alphabetical order according to their Latin names. Most of Albert's 114 descriptions of birds are relatively short, amounting in many cases to only a few lines in Stadler's edition.[1] The section on falcons, on the other hand, covers some 40 pages (out of 84), and another two pages are devoted to entries on the goshawk (*accipiter*), the broad-winged hawk or European common buzzard (*buteo*), and the sparrow hawk (*nisus*).[2] The section on falcons has long interested commentators on Albert's *De animalibus*. Not only is the subject treated at unusual length: it is also treated in unusual detail. What is more, a good deal of the material in this section cannot be traced to

---

[1] Albertus Magnus, *Lib. XXIII animalium*, ed. Stadler, *De animalibus libri XXVI* in *Beiträge* Bd. 16 (1920), 1430-1514. All references to Albert in this paper are to this book in this edition. Of other birds only the eagle, heron, goose, stork, and hen have entries of a page or longer (ibid., pp. 1433-1437, 1440, 1441, 1448-1449, 1497).

[2] Ibid., pp. 1453-1493, 1438-1439, 1445, 1504. A very clear distinction between hawks and falcons was made in medieval falconry (as in modern ornithology). Hawks and falcons have different wing structures, different flight patterns, and different modes of killing their prey; they were trained differently and tended to be flown under different geographical conditions: see R. S. Oggins, "The English Kings and Their Hawks: Falconry in Medieval England to the Time of Edward I," Ph.D. diss. (University of Chicago, 1967), pp. 27-34. It should be noted in passing that "in falconers' phraseology, every falcon is a hawk, although every hawk may not be properly called a falcon" (E. B. Michell, *The Art and Practice of Hawking* [London, 1959], p. 9).

earlier writers. The section therefore becomes important as possibly representing original work by Albert — work which may have been based on first-hand observation.

The section on falcons begins with four chapters on the proper nature of falcons (with reference to shape, color, action, and voice). The twelve chapters which follow describe various "kinds" of falcons. These sixteen chapters make up what Kurt Lindner calls the "ornithological" part of the work. They are followed by the "hunting-veterinary medicine" part:[3] the training and feeding of falcons are discussed in chapter 17, and Albert continues with six chapters on the treatment of avian diseases. The section on falcons concludes with a chapter in which Albert notes two varieties of falcon not previously mentioned.

In compiling the sections on hawks and falcons Albert used a number of written sources, hearsay evidence, and his own observations.[4] A major criterion in his choice of sources, whether written or oral, seems to have been practical experience on the part of the individual cited. Many of the written works Albert used were by falconers, and on a number of occasions he cited information given him at first hand by falconers.[5] As he wrote, at the end of a chapter on treatments for sick falcons, "These . . . are dicta on remedies for falcons determined through the experience of skilled men: nevertheless the wise falconer may add to these or reduce them according to circumstances as it seems advantageous to the constitution of the birds: for experience is the best master in all such things."[6] In his treatment of sources, as Lindner notes, Albert tended to adhere more to the spirit than to the letter of his source.[7] Even when he relied for whole sections of his work on information provided by others, however, he might add material from his own experience: at the end of chapter 18, for instance, he wrote, "We have followed the expert knowledge of William the falconer of King Roger, adding a few words of our own."[8]

---

[3] Kurt Lindner, *Von Falken, Hunden und Pferden: Deutsche Albertus-Magnus-Übersetzungen aus der ersten Hälfte des 15. Jahrhunderts*, Quellen und Studien zur Geschichte der Jagd, vol. 7; 2 parts (Berlin, 1962), 1: 24.

[4] References to personal observation may be found in Albertus Magnus, *De animalibus*, pp. 1457, 1461, 1504; and see pp. 616-617.

[5] Ibid., pp. 1461, 1463-1464.

[6] Ibid., p. 1481.

[7] Lindner, 1: 27.

[8] Albertus Magnus, *De animalibus*, p. 1478.

Among written sources, Albert specifically acknowledged the apocryphal letter from Aquila, Symmachus, and Theodotion to Ptolemy, king of Egypt (hereafter referred to as the "Symmachus letter");[9] a treatise by William the falconer of King Roger of Sicily ("Guillelmus Falconarius");[10] a work by an unnamed falconer of Frederick II (who has been identified as "Gerardus Falconarius");[11] and possibly the work of Frederick II himself.[12] Albert's attributions, however, do not always correspond to extant texts or to texts as they have been edited by modern scholars. For example, while two statements credited by Albert to William the falconer of King Roger are to be found in the work now designated as that of "Guillelmus Falconarius,"[13] a long section also attributed to William does not

[9] Ibid., pp. 1457, 1469, 1471, 1489, 1493. For Catalan and Latin texts of the Symmachus letter see "Epistola Aquilæ Symmachi & Theodotionis ad Ptolemæum regem Ægypti, de re accipitraria, Catalanica lingua," and "Excerpta ex libro incerti auctoris de natura rerum . . . De diversis generibus falconum sive accipitrum, infirmitatibus & medicinis eorum. & hoc secundum Aquilam Symmachum & Theodotionem in epistola directa ad Ptolemæum regem Ægypti," in [Nicholas Rigault], ed., . . . Rei Accipitrariæ scriptores nunc primum editi (Lutetiae, 1612), pp. 183-200 and 201-211. If one compares the two texts printed by Rigault, it is quite clear that the Latin version contains material not present in the Catalan version. Rigault describes his Latin text as being "from a book by an unknown author." As Lindner puts it, "the unknown author is no other than Thomas de Cantimpré, the source his Liber de naturis rerum" (Lindner, 1: 21; and see Luis García Ballester, ed., "Comentarios a la edición facsímil," in Thomas de Cantimpré, De natura rerum (lib. IV-XII) [Granada, Spain, 1974], 2: 34). Thomas seems to have utilized an earlier or original version of the Symmachus letter which appears to be no longer extant (James Edmund Harting, Bibliotheca accipitraria: A Catalogue of Books Ancient & Modern Relating to Falconry [London, 1891; reprint ed., London: Holland Press, 1964], p. 110; and see García Ballester, p. 35). For a discussion of extant texts of the Symmachus letter, see Hermann Werth, "Altfranzösische Jagdlehrbücher nebst Handschriftenbibliographie der abendländischen Jagdlitteratur überhaupt," Zeitschrift für Romanische Philologie, 12 (1888): 160-162; and Lindner, 1: 20-21. For a discussion of the Symmachus letter and its use as a source by Albert and Thomas de Cantimpré, see note 17 below.

[10] Albertus Magnus, De animalibus, pp. 1465, 1468, 1474, 1478, 1484; "Guillelmus Falconarius," in Gunnar Tilander, ed., Dancus Rex, Guillelmus Falconarius, Gerardus Falconarius: Les plus anciens traités de fauconnerie de l'occident, publiés d'après tous les manuscrits connus, Cynegetica, vol. 9 (Lund, 1963), pp. 134-175 (text), 6-9, 118-135 (discussion).

[11] Albertus Magnus, De animalibus, p. 1478; "Gerardus Falconarius," in Tilander, pp. 198-229 (text), 176-199 (discussion). The material in chapter 19, attributed by Albert to "falconers of Frederick the Emperor other than [Guillelmus]," was drawn largely from the text now attributed to Gerardus: see Lindner, 1: 27 and n. 4. Lindner's numbering differs from that of Tilander's edition: Lindner's no. 2 is Tilander's 3, and Tilander gives separate numbers to Lindner's 5^{I}, 5^{II}, and 5^{III}. Tilander also omits parallels between Albertus Magnus, De animalibus, p. 1481, lines 8-17, and "Gerardus Falconarius," Tilander, nos. 20-24.

[12] Albertus Magnus, De animalibus, p. 1465; Frederick II, The Art of Falconry, Being the De Arte Venandi cum Avibus of Frederick II of Hohenstaufen, trans. and ed. Casey A. Wood and F. Marjorie Fyfe (Stanford, Calif., 1943).

[13] Albertus Magnus, De animalibus, pp. 1465, 1468; "Guillelmus Falconarius," in Tilander, pp. 158, 168.

seem to have survived.[14] Albert attributed to William and to Symmachus portions of the treatise now known as "Dancus Rex."[15] And there is a real question as to whether Albert did in fact use Frederick's written work: it seems more likely that Albert was transmitting information passed on to him orally by Frederick's falconers.[16] Finally, it has been shown that (in addition to the sources he specifically acknowledged) Albert drew material (apparently without acknowledgment) from Thomas de Cantimpré's *Liber de natura rerum*.[17] The problem of Albert's indebtedness to Thomas is most complex and will be discussed at length later.

[14] Albertus Magnus, *De animalibus*, pp. 1484-1487. However, the source, described by Albert as "secundum Guilelmi experta . . . dicta," may have been oral rather than documentary.

[15] Albertus Magnus, *De animalibus*, pp. 1471-1474 (chapter 17, citing the Symmachus letter), 1474-1478 (chapter 18, citing Guillelmus); "Dancus Rex," in Tilander, pp. 60-117 (text), 5-59 (discussion). For his chapter 17 Albert used "Dancus Rex," chapter 16, lines 2-8 and 15-26; chapter 17, lines 6-11; and chapter 18, lines 2-4 and 7-10 (Tilander, pp. 80-93). For the parallels between Albert's chapter 18 and "Dancus Rex," see Lindner, 1: 25 n. 1.

[16] Albert cited the emperor twice — first, in his discussion of the black falcon: "Hunc falconem Federicus imperator sequens dicta Guilelmi, regis Rogerii falconarii, dixit primum visum esse in montanis quarti climatis quae Gelboe vocantur . . ." (Albertus Magnus, *De animalibus*, p. 1465). The section appears in "Guillelmus Falconarius" (without an attribution to Frederick) (Tilander, p. 158); but it does not occur in the extant portion of the *De arte venandi*. The second reference to Frederick occurs in chapter 20, where Albert claims to have followed the expert knowledge or techniques worked out in practice (*experta*) of the Emperor (Albertus Magnus, *De animalibus*, pp. 1481-1484). It is possible, of course, that Albert used a lost portion of Frederick's work (see Charles Homer Haskins, *Studies in the History of Mediaeval Science*, 2nd rev. ed. [Cambridge, Mass., 1927; reprint ed., New York: Frederick Ungar, 1960], pp. 307-310; Lindner, however, believes Frederick's work was not finished [Lindner, 1: 29]). But if Albert had used the *De arte venandi* one would expect (a) that it would have been cited more often, and (b) that Albert would have drawn on Frederick's descriptions of various kinds of falcon (see Frederick II, pp. 120-127). That the information was oral might be inferred from Albertus Magnus, *De animalibus*, p. 1478, where he refers to "dicta" of the falconers of Frederick II (see note 14 above).

[17] Pauline Aiken, "The Animal History of Albertus Magnus and Thomas of Cantimpré," *Speculum*, 22 (1947): 205-225. For a history of the scholarly debate over Albert's debt to Thomas, see García Ballester, pp. 24-25. While Albert used the Symmachus letter, he also used sections from Thomas' *Liber de natura rerum* which are independent of the Symmachus letter (those on the *accipiter, aerifylon, buteo* and *nisus*). Albert may have used an original text of Symmachus not transmitted through Thomas; or he may have used Thomas' *Liber de natura rerum*, incorporating the Symmachus letter, without knowing that his source was Thomas' work; or he may have used Thomas' text without acknowledgment, knowing it to be by Thomas. The texts available to us do not make it possible to determine which alternative was in fact the case. As Pearl Kibre has noted, "Both Thorndike and Walstra [indicate] that the [*Liber de natura rerum*] frequently appears in the manuscripts as anonymous or as ascribed to authors other than Thomas of Cantimpré" ("Thomas of Cantimpré," *Dictionary of Scientific Biography*, 14 vols. [New York, 1970-1976], 13: 348). For an extended discussion of the relationship between the Symmachus letter, the material on falcons in Thomas' *Liber de natura rerum*, and Albert's section on falcons, see García Ballester, pp. 34-36. Because Albert used Thomas' work as well as the Symmachus letter, I shall give references to the versions of the Symmachus letter printed by Rigault and to the *Liber de natura rerum* when comparing Albert's work and the Symmachus letter.

Evaluations of Albert's section on falcons vary greatly. Harting calls it a "crude compilation . . . [which] shows the author to have been but imperfectly acquainted with the subject."[18] Killermann, on the other hand, believes that Albert's descriptions correspond to those of modern kinds of birds of prey, and he goes so far as to say that the section on falcons shows the "beginnings of systematization and binary nomenclature" — by inference leading to modern scientific classification.[19] It is true that a good deal of Albert's material — particularly that in the "hunting-veterinary medicine" part of his work — was derived from the writings and experience of others: Albert was quite open in his acknowledgments, as the headings of the "veterinary medicine" chapters show.[20] One can say about these chapters that in them Albert transmitted to his readers methods of training and caring for falcons advocated by contemporary falconers, and that he had himself observed a number of these techniques. To modern readers some of the remedies for sick falcons which Albert drew from his sources seem curious indeed — e.g., ritual incantations, the use of such ingredients as quicksilver mixed with ashes and human saliva, and even, to calm a noisy falcon, the feeding to it of a bat filled with pepper. But these remedies were taken from Albert's sources[21] — sources which, as Hans Epstein points out (with reference to Dancus Rex, Guillelmus, and Gerardus)

> . . . were highly regarded — as a sort of concise, practical vademecum, the falconer's mews-equivalent of Dr. Spock — by the skilled austring-

[18] Harting, p. 162.

[19] Sebastian Killermann, *Die Vogelkunde des Albertus Magnus (1207-1280)* (Regensburg, 1910), p. 26; and see pp. 32-37. See also Lindner, 1: 44-54. Others have made an effort to identify Albert's birds with modern species: e.g., Stadler's editorial identifications; Heinrich Balss, *Albertus Magnus als Biologe: Werk und Ursprung* (Stuttgart, 1947), pp. 240-241; idem, *Albertus Magnus als Zoologe*, Münchener Beiträge zur Geschichte und Literatur der Naturwissenschaften und Medizin, heft 11/12 (Munich, 1928), pp. 132-133.

[20] Chapter 18, for instance, is headed, "Of various cures for the infirmities of falcons according to Guillelmus falconarius;" chapter 19 treats "Of cures . . . according to falconers of Frederick the Emperor other than the preceding" (see p. 443 and note 11 above); the headings of chapters 20 and 21 refer to the "expert knowledge of Frederick the Emperor" and "expert knowledge of Guillelmus," respectively, while that of chapter 23 describes the content of that chapter as being "according to Aquila, Symmachus, and Theodotion" (Albertus Magnus, *De animalibus*, pp. 1474, 1478, 1481, 1484, 1489). Lindner suggests that chapter 22, "Of the regimen for training hawks and [of] the regimen for [keeping] hawks," was probably Albert's own (Albertus Magnus, *De animalibus*, p. 1488; Lindner, 1: 31). As Wallace notes, and as we have already seen, Albert was "usually at pains to distinguish what he had himself seen from what he had read or been told by others" (William A. Wallace, "Albertus Magnus, Saint," *Dictionary of Scientific Biography*, 1: 100).

[21] Albertus Magnus, *De animalibus*, pp. 1481, 1476, 1490. The remedies and rituals cited were drawn, respectively, from "Gerardus Falconarius" (Tilander, pp. 226, 228); from "Dancus Rex" (ibid., p. 72); and from the Symmachus letter (Rigault, p. 207);

er's brotherhood of the late Middle Ages. . . . The limited evidence of modern falconry likewise suggests that the three treatises . . . must be taken seriously. It will be noted that Dancus, William, and Gerard alike rely heavily in their preventives and cures on the efficacy of condiments, drugs, herbs, and natural products: pepper, cinnamon, rue, rock-salt, cardamom, cumin, olive-oil, honey, myrtle, cassia, lye-wash, fresh pigeon's blood, and the flesh of hedgehog, lizard, and chicken (natural food of various species of hawks) are all mentioned. It is at least highly suggestive that many of these are also recommended by a modern falconry authority.[22]

It must be noted, in the context of training and treatment, that Albert recognized that different birds should be treated differently, each according to her natural regimen: of the feeding of falcons he wrote, "Let the falconer feed the falcon food at [such] times and in [such] quantity as she was accustomed to take food [while] in the forest"; and he recommended that perches for falcons be made of stone: "for art should imitate nature," and in their natural habitat, he says, falcons are always found resting on stone or earth.[23]

Albert's recognition that in the practical art of falconry man had to follow nature may have led, in the "ornithological" parts of his work on falcons and hawks, to a more independent and more careful observation of nature than was usual among contemporary writers. The descriptions of falcons and hawks given by Albert are considerably more detailed than those contained in his sources; and it is at these descriptions of the birds themselves that we must look in attempting to evaluate Albert's originality and accuracy in observation. Specifically, two questions must be asked with respect to Albert's descriptions of hawks and falcons: (1) To what extent are the descriptions based on Albert's own observations rather than on existing "authorities"? and (2) How accurate are his observations? The first question involves us in a comparison of Albert's work and his sources; the second requires us to see whether Albert's descriptions can in fact be applied to specific birds of prey.

We may begin with Albert's three sections on hawks — those on

---

[22] Hans J. Epstein, review of "Gunnar Tilander, *Dancus Rex, Guillelmus Falconarius, Gerardus Falconarius* . . .," *Speculum*, 40 (1965): 760. The author he cites, described by Epstein as "an experienced and practising falconer," is Gilbert Blaine, who writes, "For my own part, I feel that great benefit might accrue from the use of many of their [the ancient falconers'] quaint remedies. . ." (*Falconry*, The Sportsman's Library, vol. 15 [London, 1936], pp. 210-211; quoted in Epstein, p. 760).

[23] Albertus Magnus, *De animalibus*, pp. 1473, 1474.

the goshawk (*accipiter*), sparrow hawk (*nisus*), and European common buzzard (*buteo*) — which are found not in a separate section but as entries discussed in their alphabetical order in the body of his book on birds.[24] Some of the material used by Albert in his discussions of the three species of hawk seems to have been derived from Thomas de Cantimpré's *Liber de natura rerum*;[25] and it is particularly illuminating to compare the entries on these hawks in Thomas' work and in that of Albert, as the comparison shows not only the kinds of material the latter added, but demonstrates the essentially different approach to their subject matter taken by the two men.

In both works the longest of the three entries is that on the goshawk. Both authors compare the goshawk with other birds; both note that the goshawk seeks the hearts of the birds it captures; both state that wild hawks hunt domestic fowl; both describe in detail how the goshawk captures and kills hares; and both include a statement from Pliny that a hawk cooked in oil is a good remedy for pains in the limbs. Albert, however, adds a physical description of the goshawk; notes the number of eggs it lays; describes its flight; lists the birds hunted by tamed hawks, noting that tamed hawks will take birds larger than those hunted by untamed hawks; and includes a long description of some of the ailments of hawks (where Thomas has only a few short passages on hawk ailments). Even where Albert's material appears to be derived from Thomas, Albert often includes additional information of his own. While Thomas compares the goshawk and the gerfalcon in respect to speed of flight and caution, Albert compares the goshawk with eagles, falcons, and the sparrowhawk in respect to size, appearance, and manner of flight. Comparing physical structure, for example, he notes that hawks have wings more pointed than those of eagles but less pointed than those of falcons — a distinction still regarded as fundamental by ornithologists.[26] Again, Albert paraphrases Thomas' statement that the goshawk seeks the heart of its prey, but adds that sometimes the hawk seeks the brain as well. And while Thomas merely states that

---

[24] Ibid., pp. 1438-1439, 1504, 1445.

[25] Thomas de Cantimpré, *Liber de natura rerum* [von] *Thomas Cantimpratensis: Editio principes secundum codices manuscriptos*, vol. 1 (Berlin and New York, 1973), pp. 182-183, 217, 185-186 (hereafter cited as *De natura rerum* [Berlin 1973]).

[26] See for example, H. F. Witherby, F. C. R. Jourdain, Norman F. Ticehurst, and Bernard W. Tucker, *The Handbook of British Birds*, vol. 3: (*Hawks to Ducks*) (London, 1939), pp. 2, 38, 72; and see, for particular instances, pp. 9, 50, 73, 79. A drawing showing the differences vividly may be found in Michell opposite p. 11.

the wild hawk preys on domestic birds, Albert notes that it also takes crows and "birds of that kind."

Albert gives the following physical description of the goshawk:

> The whole hawk is mottled in color, but in the first year it has red and black spots. After that, however, this changes to white and black spots, and these are whiter and blacker according to how many times it has moulted; and its feet are yellow, and its claws are great, but not so [great] as those of eagles; and its head is rounder than that of the eagle; and its beak is curved, and according to its own proper analogy, shorter than that of the eagle and longer than those of falcons; and on the back also it has a few white spots and many black ones; and it has wings according to its own proper proportion which are more pointed than [those of] the genus of eagles, and less pointed than [those of] the genus of falcons. It is moreover an irritable bird, and for this reason flies by itself except when raising its young. It lays three or four eggs, or at most five, and its appearance is almost the same as that of the *nisus* which is called the sparrow hawk [*spervarius*], although it [the goshawk] is larger in size. It is smaller than the tree eagle [*aquila truncali*], but is larger than the eagle which catches fish.[27]

This may be compared to H. F. Witherby's description:

> *Resembles a huge Sparrow-Hawk, having same long tail and rounded wings*, though male, as in that species, is considerably smaller than female. Sexes do not differ except in size, and *colouring recalls that of female Sparrow-Hawk*, dark ashy-brown above, with whitish streak from eye over ear-coverts, whitish below closely barred dark brown, and tail strongly barred. Juveniles and first-year birds have upper-parts lighter, more rufous, and less uniform, and *under-parts warm buff streaked, not barred, with broad, drop-like markings* of dark brown.[28]

Albert's observations as to the size and color of the goshawk's feet, its disposition, and the number of its eggs are all borne out by modern authorities, as is his observation that northern goshawks are larger than southern varieties.[29]

Thomas' and Albert's sections on the sparrow hawk (*nisus*) are sub-

---

[27] Albertus Magnus, *De animalibus* p. 1438.

[28] Witherby, et al., 3: 73. All italics his.

[29] Ibid., p. 74; Leslie Brown and Dean Amadon, *Eagles, Hawks and Falcons of the World*, 2 vols. (Feltham, 1968), 2: 452-459; Michell, pp. 32-33, 257-258; Michael Woodford, *A Manual of Falconry* (London, 1960), p. 105. Michell supports Albert's statement that the goshawk takes prey, "not as food, but for the glory of it" (Albertus Magnus, *De animalibus*, p. 1439; and see Michell, p. 157).

stantially shorter than those on the goshawk. Both authors begin their discussion of the sparrow hawk by comparing it with other birds of prey;[30] both discuss the fact that the sparrow hawk hunts alone; and both relate the legend that the sparrow hawk in winter holds a live bird throughout the night for warmth, releasing it, out of gratitude, in the morning — although Albert begins the story of the cold sparrow hawk with the phrase, "It is said that," and concludes it with, "but I have no proof of this." Albert also lists three birds hunted by the sparrow hawk, and adds that he knows "from experience" that if two sparrow hawks are present and the prey escapes, one sparrow hawk will attack the other. Albert alone notes the sparrow hawk's similarity in coloration to the goshawk and its propensity for attacking prey stronger than itself "such as the pigeon, the duck, and the crow or rook."[31] These observations, and that on the solitary nature of the sparrow hawk, once again are supported by modern authors.[32]

Thomas and Albert devote only a few lines to the European common buzzard (*buteo*). Both write of the buzzard's color and sluggish flight. Thomas writes of its method of hunting: "It lives on prey which it can follow by cunning or by making a loud noise, or which is held back by slowness of movement." Albert omits this description of the buzzard's hunting technique, but does describe its claws and beak and lists the prey it takes — including frogs, mice, small slow birds and young and injured birds.[33] Again, Albert's observations are borne out by modern authorities.[34]

Both Thomas and Albert attribute to the Symmachus letter a list of four "kinds" (*genera*) of hawk — the goshawk, the sparrow hawk,

---

[30] Albert compares the coloration of the sparrow hawk to that of the goshawk (Albertus Magnus, *De animalibus*, p. 1504). Thomas compares it to the *herodius* (gerfalcon) (*De natura rerum* [Berlin 1973], p. 217; and see ibid., p. 199); but surely this is an error for the goshawk: elsewhere Thomas describes the *herodius* as being blue (ibid., p. 196); and after the section taken from Symmachus on the four varieties of hawk he notes that the sparrow hawk is similar to the goshawk in disposition and color (ibid., p.199).

[31] Albertus Magnus, *De animalibus*, p. 1504.

[32] Brown and Amadon, 2: 476-482; Witherby, et al., 3: 79-84; David Armitage Bannerman, *The Birds of the British Isles*, 12 vols. (Edinburgh and London, 1953-1963), vol. 5, pt. 2, p. 253; Michell, pp. 32, 35. For the similarity between the goshawk and sparrow hawk, see also Hermann Heinzel, Richard Fitter, and John Parslow, *The Birds of Britain and Europe*, 3d ed. (London, 1974), p. 75.

[33] Thomas de Cantimpré, *De natura rerum* (Berlin 1973), pp. 185-186; Albertus Magnus, *De animalibus*, p. 1445.

[34] Brown and Amadon, 2: 609-616; Witherby, et al., 3: 50-55.

and a smaller but otherwise identical version of each bird. The first, Thomas says, is that already described by him "under the letter A in the chapter on the *accipiter*." The second variety of hawk is "smaller than the first genus, having large wings in relation to the size of its body," and does not hunt well until it has moulted three times. The third variety is the sparrow hawk, and it is discussed "under the letter N in the chapter on the *nisus*." The fourth "genus" is called the *frogellus*, "which in the vulgar tongue we call the musket[:] this bird is much smaller than the sparrow hawk, but very similar to it in color."[35] But where Thomas accepts the list uncritically, Albert objects to it:

> Aquila and Theodotion and Symmachus their associate call all classes [*genera*] of hawks "falcons," and determine them to be of four kinds, placing the larger goshawk [*astur primae quantitatis*] in the first class, and the smaller goshawk which we call the tercel in the second class, and the sparrow hawk [*nisus*] in the third class, and the musket [*muscetus*] in the fourth class; . . . with which we can in no way agree, since the tercel is found in the nest of the goshawk [*accipiter*] and the musket is found only in the nest of the sparrow hawk [*nisus*]: and accordingly the goshawk and the tercel differ only in sex and not in species, because the goshawk is the female and the tercel is the male; and the sparrow hawk and the musket differ in the same way: for the sparrow hawk is the female and the musket is the male.[36]

Albert's main written source for his descriptions of falcons was the letter of Aquila, Symmachus, and Theodotion. From it he seems to have derived the basic idea of grouping falcons together and ranking them, the idea of the "nobility" of different falcons, the names of several of the falcons he listed, and portions of some of the descriptions. Symmachus listed seven or eight falcons — depending on whether the Catalan or Latin version of the letter is followed.[37]

---

[35] Thomas de Cantimpré, *De natura rerum* (Berlin 1973), p. 199. The list does not appear in the Catalan version of the Symmachus letter (see Rigault, pp. 189-190). It should be noted that Thomas also lists the *terciolus* — probably the male peregrine — as a separate "genus" of falcon (*De natura rerum* [Berlin 1973], p. 199; and see Rigault, pp. 190, 205).

[36] Albertus Magnus, *De animalibus*, p. 1493.

[37] Those listed in the Catalan version are the lanner (two kinds), peregrine, "montasi," falcon gentle, "Gathena," a sixth (unnamed) which resembles a white eagle, and "Breton" (Rigault, p. 189). The Latin version lists the lanner (two kinds), peregrine, and "montanum;" a lacuna covers the fourth and fifth varieties; then follow the "spervicum," "Britannicum," and "herodius" or "Giffard" (ibid., pp. 203-205). The first four of Thomas' eight varieties are the lanner (two kinds), peregrine, "montanum," and blue-footed falcon; he does not give a name to the fifth, though he describes it — "Quintum vero genus gracile et longum in dispositione est exertissimum in volatu;" and there follow the "supranicum," "britannicum," and "herodius" or gerfalcon (*De natura rerum* [Berlin 1973], pp. 198-199).

Albert took the list, rearranged the birds in order of excellence, and added several kinds. He also created the formal categories of ignoble and mixed falcons, and expanded the descriptions of the birds. In Albert's finished system there were ten noble varieties of falcon: the saker (*aërifylon* or *britannicus*), gerfalcon, mountain falcon (*montanarius*), peregrine, gibbous falcon (*gybosus*), black falcon, white falcon, red falcon, blue-footed falcon, and the merlin; three ignoble falcons: the black, white, and red lanner; and four mixed falcons resulting from matings of noble and ignoble birds. In addition to these Albert noted the rock and tree falcons, which he did not place in his over-all scheme.[38]

In attempting to identify these birds on the basis of Albert's descriptions, one must make two assumptions. We must assume, first, that Albert knew the difference between a hawk and a falcon, and between these and other families of birds of prey. We have seen that Albert noted the differences in wing shape among eagles, hawks, and falcons. He also commented on differences in the size of the neck and tail among birds of prey — noting that falcons have shorter necks than hawks and eagles, and shorter tails, proportionately, than goshawks or sparrow hawks;[39] and he contrasted hawks and falcons as to their methods of taking prey:

> [The hawk] almost always stays hidden and flies close to the ground, contrary to the manner of falcons, and when it takes a bird, it seizes it from below as if whirling around on itself. . . .
> The proper act of a falcon among raptorial birds is to fall with force on its prey. . . . When it wishes to take game, it is [in the nature] of the falcon to ascend with a swift flight, and with its talons held close to its breast, to fall with force on the bird with so powerful an effort that in descending it raises a sound like the rushing of wind, and it makes this attack not by descending directly or perpendicularly, but at an angle: because striking after such a descent it cuts a long wound with its claws so that sometimes a bird falls divided from head to tail, and sometimes it is found with the whole head torn off.[40]

The second assumption we must make is that Albert knew the difference between male and female birds of prey and did not list males as separate species. We have seen that Albert noted that the goshawk and tercel, and the sparrow hawk and musket, differ in sex

---

[38] Albertus Magnus, *De animalibus*, pp. 1457-1471, 1492-1493.

[39] Ibid., pp. 1453-1454.

[40] Ibid., pp. 1438, 1455. Brown and Amadon, writing on the sparrow hawk, note that "a kill is often made by a quick upward turn from below the prey" (Brown and Amadon, 2: 480).

and not in species. While Albert did not specifically differentiate in *De animalibus* between male and female falcons, it is hard to see how he could have failed to make the same observation about falcons that he did about hawks.

It cannot be assumed, unfortunately, that Albert was always clear as to the true species of young falcons, because of the differences in plumage between birds of the first year and those which have moulted. In his discussion of the color proper to falcons he wrote of the variation in color in the mottling of the breast: "This variation always has black as one color, but in the first year the second color is rufous, of a soft reddish color, and as often as [the bird] moults, it whitens more and more."[41] In practice, as we shall see, he seems in one case to have failed to identify young peregrines correctly.

One other feature of Albert's description of falcons must be noted before we turn to a consideration of individual species. In his section "On the proper color of falcons" Albert provides a visual description of what he regarded as the falcon's basic coloration:

> The color proper to the genus of falcons is (as to the face) to have black spots along the cheeks [the falcon's "moustaches"] and white spots near the sockets of the eyes on both sides from either side of the beak; and to have black eyelids [? *cilia*] and a nearly black ashy color [i.e., a blackish-gray color] on the head and back and on the upper part of the neck and the outer part of the wings and tail; and elsewhere to be full of variation, as if [the color were] falling in stripes, with the stripes broken now and then; and this variation always has black as one color, but in [the falcon's] first year the second color is rufous, of a soft reddish color, and as often as [the bird] moults, it whitens more and more.[42]

This description must be kept in mind when reading Albert's accounts of individual species.

The first of Albert's noble falcons is the saker (*Falco cherrug*). He calls it "regal" and "the noblest falcon of all."[43] He has two entries on the saker, one in the alphabetical section under *aërifylon*, the other in the general section on falcons. Albert's physical description of the *aërifylon* — "It has reddish feathers, a long tail, very long talons and legs, and is a little larger than the eagle" — comes almost

---

[41] Albertus Magnus, *De animalibus*, p. 1454.
[42] Ibid.
[43] Ibid., pp. 1458, 1457.

verbatim from Thomas' entry, "De aeriophilo, qui et aelion [sic]."[44] Albert's description of the saker in the section on falcons follows the Symmachus letter (and Thomas' second description of the saker), though somewhat more loosely: "It has thick and knotty legs, talons more cruel than the eagle's, a terrifying appearance, especially as to the eyes, a large head and very strong beak, and the yoke of the wings [is] large." Albert also records here that Symmachus calls the saker "Britannicus." Many of the details Albert gives on the saker's behavior, however, are drawn from Thomas' account of the aeriophilon.[45]

The gerfalcon (F. rusticolus) has "the perfect nature of the falcon in appearance, color, action, and voice," according to Albert. If we keep in mind what he has described as the proper coloration of falcons, it is clear that he is describing what has been called the "Norway" gerfalcon (formerly differentiated as F. rusticolus rusticolus).[46] Albert says the bird is called gerfalcon from girando because it follows its prey for a long time turning (gyrating) sharply; and he notes that among other falcons it is accustomed to stand erect. Both observations are substantiated to some extent by modern authorities.[47] So far as I can ascertain very little of the detailed material recorded by Albert on the gerfalcon and its behavior and care appears in earlier authorities.

---

[44] Ibid., p. 1444; Thomas de Cantimpré, De natura rerum (Berlin 1973), pp. 184-185; and see ibid., p. 199.

[45] Albertus Magnus, De animalibus, pp. 1457-1458; Rigault, pp. 189-190, 204; Thomas de Cantimpré, De natura rerum (Berlin 1973), pp. 184-185, 199. The saker and the merlin are the only falcons whose physical descriptions by Thomas were used by Albert.

[46] Albertus Magnus, De animalibus, p. 1458. Modern authorities believe there is one species of gerfalcon, though until recently some authorities distinguished a number of forms or subspecies. Three of these subspecies — sometimes called "Greenland," "Iceland," and "Norway" gerfalcons (the last is also simply called "the gerfalcon") — were clearly differentiated in the Middle Ages (see Frederick II, p. 121, and the references in Oggins, pp. 37-45). These three forms of the gerfalcons were distinguished, both in the Middle Ages and later, by color, since the "Greenland" form tends to be whiter than the others, "Iceland" gerfalcons are by and large gray, while "Norway" gerfalcons are the darkest of the three — though the color of each form shades into that of the next (Witherby, et al., 3: 2-9; Heinzel, Fitter, and Parslow, p. 90; Brown and Amadon, 2: 843-844; Michell, pp. 12-14).

[47] Albertus Magnus, De animalibus, p. 1458. According to Michell, "the flight of the ger . . . combines in an extraordinary degree swiftness and the power of turning readily;" and he describes some Norway gerfalcons: "They flew beautifully to the lure, turning more quickly than a peregrine, and stooping with greater dash" (Michell, pp. 15, 14). Harkness and Murdoch note that the gerfalcon holds its head higher than the peregrine (Roger Harkness and Colin Murdoch, Birds of Prey in the Field: A Guide to the British and European Species [London, 1971], p. 114).

Albertus Magnus' third-ranked noble falcon is the "mountain falcon" (*montanarius*):

> It is short and very thick in body and above all has a tail which is short and very thick, a breast which is very round and large, and strong shins which are short in relation to the size of its body, and knotty feet, and on the back and upper surface of the wings it is of an ashy color; and this color, according as it goes through mutations of the feathers over the years becomes clearer and grows lighter compared with the younger and darker variety. . . .
>
> Moreover, this type of falcon is in thickness very like a goshawk, although it may be much shorter; and it has very pale feet and shins which are as it were scaly, with the scales lying close to one another: and its shape when it stands, from the shoulders to the tail, is like a pyramid, if one can imagine a pyramid a little bit compressed toward the back.[48]

Elsewhere, Albert characterizes the *montanarius* as a larger bird than the peregrine.[49]

Now the only falcons which are of goshawk size and larger than the peregrine are the gerfalcon and the saker. The saker, however, is of the wrong color and, as Frederick II wrote of the saker, "the body is proportionately more slender and longer, . . . the breast is less fleshy and thick than in the gerfalcon. . . ,"[50] whereas the *montanarius* has a breast which is notably "round and large." This leaves the gerfalcon as the only falcon large enough and heavy enough in build to be the *montanarius*. Albert discussed the gerfalcon earlier, and it does not seem likely that he would discuss it a second time. However, as we have seen, the bird Albert described earlier is probably the smaller and darker "Norway" gerfalcon (since it has "the perfect color of a falcon"); it is therefore entirely possible that the bird Albert describes as the ashy-colored *montanarius* is the gray "Iceland" variety of the gerfalcon.[51]

---

[48] Albertus Magnus, *De animalibus*, p. 1460. Compare this with Frederick II's description of the best gerfalcons, which he says may be known by the following characteristics: "The body is uniformly proportioned, shapely, and tapering toward the tail, like the figure geometricians call a pyramid. . . . [T]he breast is elevated in front, and is thick and fleshy; the iliac bones are wide. The shin-bones are short and strong . . . " (Frederick II, p. 120).

[49] Albertus Magnus, *De animalibus*, p. 1462.

[50] Frederick II, p. 121.

[51] See note 46 above. For comparative sizes of falcons see Brown and Amadon, 2: 766, 778, 796, 802, 809, 818, 839, 843, 851; Heinzel, Fitter, and Parslow, pp. 90-94. Stadler, Killermann, and Lindner identify the *montanarius* as the peregrine falcon (Albertus Magnus, *De animalibus*, p. 1460 note; Killermann, p. 34; Lindner, 1: 47-48, 2: 188-189). Stadler's identifications are given without explanation, and he queries this particular one; Killermann, too, gives no reason for the identification; Lindner's identification is based largely on what one might call linguistic grounds.

Albert ranks the peregrine (*F. peregrinus peregrinus*) fourth, after the *montanarius*. He claims to have been told by a falconer living in a hermitage high in the Alps between Germany and Italy that the peregrine nested among the highest cliffs and steepest sides of that range; but he also notes that the peregrine is fairly common in all lands. Although her natural prey is usually the wild duck, says Albert, when well trained by a wise falconer the peregrine will capture herons and cranes.[52]

The fifth most noble falcon, according to Albert, is the "gibbous falcon" (*gybosus*). He says he was shown three of these falcons by his hermit friend, who claimed to have sold many others.[53] Albert describes the gibbous falcon as

> ... very small in bodily size, but marvelous in its courage and daring and strength of flight when it follows its prey. Its size is but little superior to that of the *nisus* which the vulgar call the sparrow hawk. . . .
>
> It is called the gibbous [humpbacked] falcon because, due to the shortness of its neck, its head hardly appears in front of the yoke of its wings when it folds [them] over the sides of the back [*super latera dorsi*]; and it has a large head in proportion to the size of its body, and a very short and round beak, and wings [which are] very long and [which] rise very abruptly [*alas . . . valde exortas*] and a short tail . . . and in coloration it is like the other falcons which are called peregrines; and the top of its head is quite flat and the back of the head is not prominent but is like a continuation of the neck; . . . and it makes its nest on inaccessible cliffs like the peregrine. . . . Moreover it is of such daring and strength that it brings down wood geese and herons and cranes; and this kind [of falcon] is extremely fast and climbs so high that it escapes the sight of man: and this falcon is not content to bring down one bird, but wounds many; however, in hunting birds it prefers to have many companions, because of its smallness and the size of the birds which it hunts.[54]

From Albert's description — the very long wings, short tail, and flight characteristics — it would seem that the gibbous falcon is a hobby (*F. subbuteo subbuteo*) or one of its closely related forms. The hobby is a daring, skillful flier which looks like the peregrine but has unusually long wings and a tail shorter in proportion to the body

---

[52] Albertus Magnus, *De animalibus*, pp. 1461-1463. On the peregrine's prey, see Frederick II, pp. 363, 309, 353-354. Frederick says that the peregrine constructs its nest in the far north — i.e., there were then, as now, Scandinavian breeding birds (ibid., pp. 144-145): but perhaps Albert was unaware of the existence of northern peregrines. For modern descriptions see Brown and Amadon, 2: 850-856; Witherby, et al., 3: 9-15.

[53] Albertus Magnus, *De animalibus*, p. 1463.

[54] Ibid.

than that of the peregrine. However, unlike the gibbous falcon, the hobby is somewhat smaller than the sparrow hawk, and it nests in trees.[55]

The light form of Eleonora's falcon (*F. eleonorae*) possesses the characteristics of the hobby described above, but has a longer tail. And in three respects Albert's gibbous falcon resembles Eleonora's falcon still more than it does the hobby. It is slightly larger than the sparrow hawk; it nests on cliffs; and both gibbous falcon and Eleonora's falcon are social birds — birds which fly and hunt in groups.[56]

Albert claims that his gibbous falcon can bring down wood geese, herons, and cranes — quarry far larger than the natural prey of the Eleonora's falcon; but the story he relates to support this assertion is reported at second hand. The fact that the female Eleonora's falcon is roughly the same size as the male peregrine, and that by Albert's account several falcons of this kind were flown together, might possibly justify his statement.[57]

It is possible that in this chapter Albert described Eleonora's falcon but confused it with the smaller hobby and assigned it a shorter tail. The story of the larger prey may have resulted from confusion on his source's part between Eleonora's falcon and *F. peregrinus brookei*. In any case, the identification presents problems, and it is difficult to see how Albert could be correct in all the features he describes.

The names of Albert's next three falcons — the black falcon, white falcon, and red falcon — seem to have been derived from "Guillelmus Falconarius," though Albert credits Guillelmus only in the chapter on the black falcon.[58] Albert, however, only uses the names of the three falcons and a section from Guillelmus on the origins of the black falcon; the rest is his own. He describes the sixth-ranked black falcon as

> ... shorter by a little than the peregrine, but in shape it is similar in every respect; though it is dissimilar in color, since on the back and on the outside of the wings and tail it is wholly of a sort of dusky blackness, and on the breast, belly, and sides it has dusky variations in col-

---

[55] Brown and Amadon, 2: 809-814; Harkness and Murdoch, pp. 121-122.

[56] Brown and Amadon, 2: 818-823; Harkness and Murdoch, pp. 123-124. But see Witherby, who notes that the hobby is "usually seen in pairs or singly, but sometimes several together" (Witherby, et al., 3: 17).

[57] Albertus Magnus, *De animalibus*, pp. 1463-1464.

[58] Ibid., p. 1465; "Guillelmus Falconarius," in Tilander, pp. 158-166.

or; moreover, on the face it has the usual falcon markings, [but] especially black, of a deep blackness, which are surrounded by a sort of dim and dusky pallor. But it has shins and claws and beak just like those of the peregrine. . . .

These falcons are also like the peregrine in their feeding and rearing and in their daring. As the years go by they lighten somewhat through the yearly moult. . . . But since there are two things which we consider with respect to exterior appearance in living things, that is to say, shape and color, *shape indicates conformity to or difference from a species more than does color.* . . . And for this reason, this species of falcon seems to be much like the species of peregrine, although it differs in color.[59]

Wood and Fyfe appear to believe that Albert's black falcon is the dark phase of Eleonora's falcon.[60] Two factors lead me to conclude that this is not the case. While Eleonora's falcon is of the right size, the color of the dark phase of this falcon is more uniformly dark than that of the bird Albert describes. Furthermore, if the black falcon were a variety of Eleonora's falcon, one would expect Albert to have noted the relatively long wings — as he did in the case of the gibbous falcon.

Albert's black falcon is more likely to have been *F. peregrinus minor*, a smaller, dark variety of the peregrine. Today *F. p. minor* is located in Africa south of the Sahara,[61] but as late as the last century its range seems to have been considerably farther north, and Dresser, writing in 1876, notes that it "occasionally wanders into Asia Minor and Europe proper. Its headquarters appear to be Southern and North-western Africa."[62] Archer and Godman describe *F. p. minor* as being (like Albert's black falcon) smaller than the peregrine and as having "blue-grey upperparts banded darker, underparts white with a buff tinge spotted on the chest and barred below. . . . The top of the head and nape . . . is unrelieved dark slate-colour."[63] Other modern varieties of the peregrine would seem to be too light in color to fit

[59] Albertus Magnus, *De animalibus*, pp. 1464-1465. Italics mine.

[60] Frederick II, p. 552: v. "Saker Falcon."

[61] Brown and Amadon, 2: 852; Charles Vaurie, "Systematic Notes on Palearctic Birds. No. 44, Falconidae: The Genus *Falco* (Part 1, *Falco peregrinus* and *Falco pelegrinoides*)," American Museum *Novitates*, no. 2035 (July 7, 1961), p. 4.

[62] Henry Eeles Dresser, *A History of the Birds of Europe: Including All the Species Inhabiting the Western Palæarctic Region*, 9 vols. (London, 1871-1896), 6: 43.

[63] Sir Geoffrey Francis Archer and Eva M. Godman, *The Birds of British Somaliland and the Gulf of Aden: Their Life Histories, Breeding Habits, and Eggs*, 4 vols. (London and Edinburgh, 1937-1961), 1: 156.

Albert's description.[64] Another dark bird, the sooty falcon (*F. concolor*), would appear to be too small and too uniformly dark in color to be Albert's black falcon.[65] Albert himself felt that the black falcon was a species of peregrine and was aware that "shape indicates conformity to or difference from a species more than does color." In the absence of evidence to the contrary, there seems no reason not to accept Albert's view.

The white falcon is seventh in Albert's list:

> [This falcon] comes from the north and the ocean sea, from the regions of Norway and Sweden and Estonia and the neighbouring woods and mountains.
>
> This falcon is in a whitish variety just like that which we have said before is black ... and on the back and wings it is rather whitish, but in other places it has very white spots or drops interposed among other less white [*subpallidis*] spots; and in size it is larger than the peregrine falcon.[66]

The white falcon is clearly the "Greenland" gerfalcon.[67]

Albert places the red falcon eighth in his list of noble falcons. He notes that the red falcon is not entirely red:

> The spots which in some are white, in this kind [of falcon] are red, with black spots interspersed as in the others. This falcon does not appear red on the back or outside of the wing unless the wings are stretched out: then the dark color appears reddish. ...
>
> This falcon is not great (a little smaller than the peregrine) but is strong in claws and feet and beak, and very agile in flight.[68]

Three birds might fit Albert's description. *F. peregrinus brookei* would qualify as far as size and the reddish color underneath are concerned, but it does not seem to have reddish feathers on the wing

---

[64] Brown and Amadon, 2: 851-852. Lindner identifies the black falcon as "probably *F. p. babylonicus*" (Lindner, 1: 50), but Brown and Amadon characterize the latter as "palest of all races" (Brown and Amadon, 2: 851), which would seem to rule it out. Stadler thinks the black falcon may be the hobby, but queries this (Albertus Magnus, *De animalibus*, p. 1464, note).

[65] Brown and Amadon, 2: 823.

[66] Albertus Magnus, *De animalibus*, p. 1465.

[67] See note 46 above. Stadler and Balss agree with this identification (Albertus Magnus, *De animalibus*, ed. Stadler, p. 1465, note; Balss, *Albertus Magnus als Zoologe*, p. 124). Lindner believes the white falcon to be one of two northern varieties of peregrine — *F. p. scandinaviae* Kleinschmidt or *F. p. calidus* (Lindner, 1: 50). The first is not generally recognized as a separate form; and while the second is paler and larger than *F. p. peregrinus* (Brown and Amadon, 2: 851), it is hardly a "whitish" variety (ibid.; and see Vaurie, p. 11).

[68] Albertus Magnus, *De animalibus*, pp. 1466-1467.

or back which show clearly when the wings are extended.[69] The Barbary falcon (*F. peregrinus pelegrinoides*) and the Red-naped Shaheen (*F. peregrinus babylonicus*) are both smaller and more reddish than the peregrine.[70] However, the strength of the red falcon's claws, feet, and beak, as described by Albert, suggest that of the two the Barbary falcon is more likely to have been the red falcon. As Michell points out, "the barbary is even more powerfully armed and feathered than her bigger cousins, having . . . distinctly larger feet and talons [than *F. p. peregrinus*], and a larger beak proportionately to her size." *F. p. babylonicus*, on the other hand, has a foot "smaller proportionately than that of the peregrin."[71] In addition, a number of modern authors comment on the strength and swiftness of the Barbary falcon.[72]

The ninth and next-to-last noble falcon, according to Albert, is the blue-footed falcon. This bird is

> . . . in size and appearance like or equal to the peregrine falcon: but its back and the outside of its wings are not so black, and this kind is even whiter on the breast than the peregrine falcon; and its wings are not so long as the peregrine's, but its tail is a little longer and its voice is sharper . . . and its daring is much less in attacking birds, because that which has blue feet rarely attacks birds larger than magpies or small crows, while peregrines and other large falcons attack whatever birds they please.[73]

Killermann and Wood and Fyfe identify the blue-footed falcon as a young peregrine,[74] and I see no reason not to accept their view. As Frederick pointed out, young peregrines before moulting exhibit a range of coloration from brown to reddish to fawn. He also noted a correspondence between the color of a young peregrine's plumage and the color of its feet: ". . . the browner the peregrine the greener are her 'feet', and the redder she is the more citron yellow her feet become."[75] Albert seems to have identified properly the more red-

---

[69] Brown and Amadon, 2: 851. Vaurie states that "some adults of *brookei* from the Western Mediterranean . . .[show] a narrow and vague band of rufous on the nape" (Vaurie, p. 14); and see Witherby, et al., 3: 15.

[70] Vaurie, pp. 2, 3, 16; Brown and Amadon, 2: 851; Archer and Godman, 1: 152-153.

[71] Michell, p. 21; and see Vaurie, p. 3; and Dresser, 6: 49.

[72] Vaurie, p. 3; Archer and Godman, 1: 152, 155; Dresser, 6: 49.

[73] Albertus Magnus, *De animalibus*, pp. 1467-1468.

[74] Frederick II, p. 552: v. "Saker Falcon."

[75] Ibid., p. 123, and see p. 122.

dish young peregrines,[76] but he appears to have classified the browner, "blue-footed" young peregrines as a separate species rather than merely a color variant.

The last of Albert's noble falcons is the merlin (*F. columbarius*), which he describes in two different places. In the chapter in the section on falcons Albert says that despite the merlin's small size it is not deficient in audacity, and he relates Guillelmus Falconarius' claim that the merlin can be trained to take cranes,[77] although Albert adds that the prey appropriate to the strength of the merlin is the lark, or at most the partridge and dove.[78] Albert's other description of the merlin, listed alphabetically, is essentially an abridged paraphrase of Thomas de Cantimpré's account.[79]

Albert considers the lanner (*F. biarmicus*) to be an "ignoble" falcon.[80] He says there are three varieties of lanner: white and black lanners "of the same size as falcons," and the red lanner, which is smaller and like the merlin. He says the lanner is timid, but he describes how over a three-year period she can be taught to overcome her fear and hunt ducks and geese.[81] Albert's black and white lanners may represent extreme color variations — the color of older birds lightens somewhat.[82] His red bird may be a kestrel (*F. tinnunculus*).[83]

At the very end of his section on falcons Albert mentions briefly two birds which he has not discussed previously — the rock falcon

---

[76] Albertus Magnus, *De animalibus* (p. 1454; and see above, p. 452.

[77] Albertus Magnus, *De animalibus*, pp. 1468-1469; "Guillelmus Falconarius," in Tilander, p. 168.

[78] Albertus Magnus, *De animalibus*, p. 1468. Compare Brown and Amadon, 2: 802-806; Witherby, et al., 3: 21-25.

[79] Albertus Magnus, *De animalibus*, pp. 1502-1503; Thomas de Cantimpré, *De natura rerum* (Berlin 1973), p. 215.

[80] Albertus Magnus, *De animalibus*, p. 1469. This is in contrast to Frederick, who considers the lanner to be a noble falcon (Frederick II, p. 110).

[81] Albertus Magnus, *De animalibus* p. 1469. See Brown and Amadon, 2: 831-834; Harkness and Murdoch, pp. 116-118; and, on the timidity of one variety of lanner, Frederick II, p. 127. It is interesting to note how Albert's technique for training the lanner parallels Frederick's description of how to train the gerfalcon to catch cranes (ibid., pp. 257-266).

[82] H. Kirke Swann, *A Monograph of the Birds of Prey (Order Accipitres)*, ed. Alexander Wetmore, 2 vols. (London, 1930-1945), 2: 404. Another possibility is that Albert could be differentiating between darker and paler races of lanner: see Brown and Amadon, 2: 831. Lindner thinks the birds are two kinds of buzzard — *Buteo buteo* and *Buteo lagopus* (Lindner, 1: 51-52): but (a) Albert described the *buteo* recognizably (see above, p. 449); and (b) as we have seen, Albert distinguished clearly between the wing structure of hawks and that of falcons.

[83] Brown and Amadon, 2: 776-784. Lindner also thinks the red lanner is a kestrel (Lindner, 1: 51).

and the tree falcon. The rock falcon is "halfway in strength and size between the peregrine and the gibbous falcon, and is found in the cliffs of the Alps, and has the same diet and way of life as the peregrine." The tree falcon is "halfway in size and strength between the gibbous falcon and the merlin and has the same way of life as the merlin."[84] There is relatively little information here to help identify either bird. The rock falcon may have been *F. peregrinus brookei*. The similarity of habits and nesting site with those of the peregrine would support this, as would Albert's characterization of the rock falcon as "halfway in size and strength between the peregrine and the gibbous falcon [Eleonora's falcon]."[85] The tree falcon may well have been the hobby (*F. subbuteo*). The size, strength, habits, and nesting site of the tree falcon seem to correspond to those of the hobby.[86] In both cases, however, the identification depends, to some degree, on accepting Eleonora's falcon as Albert's gibbous falcon.

It is difficult to sum up Albert's writings on hawks and falcons. Much of his material and organization he derived quite openly from others, and a number of what appear to be Albert's own additions may have come from oral accounts, rather than from his own direct experience. Some of Albert's chapters (e.g., that on the gerfalcon) provide information not available from other contemporary sources; other chapters are less valuable. Nevertheless, Albert's writings on hawks and falcons represent an important account of the birds used in falconry in thirteenth-century Europe. To some extent Albert's work on hawks and falcons stands in the shadow of the work of his kinsman Frederick II. But Frederick was able to devote far more of his time both to the practical and to the theoretical study of falconry than could Albert, to whom hawks and falcons were only a small, albeit important, part of the universe he surveyed. Frederick's work, moreover, exceptional as it was, did not have the influence that Albert's writings did. But perhaps the more meaningful comparison is that between Albert and his pupil (and source) Thomas de Cantimpré. Both were encyclopedists, and both wrote at roughly the

---

[84] Albertus Magnus, *De animalibus*, p. 1492.

[85] Brown and Amadon, 2: 850-851; Vaurie, p. 9. Stadler, Killermann, and Lindner identify the rock falcon as a variety of peregrine (Albertus Magnus, *De animalibus*, p. 1491, note; Killermann, p. 37; Lindner, 1: 50).

[86] Brown and Amadon, 2: 809-814; Witherby, et al., 3: 17-21. Stadler, Killermann, Balss, and Lindner concur (Albertus Magnus, *De animalibus*, p. 1492, note; Killermann, p. 37; Balss, *Albertus Magnus als Zoologe*, p. 132; and Lindner, 1: 52).

same time. The significant difference between them lies in the fact that Thomas looked backward to an older tradition of writing about natural phenomena, while Albert looked about him at the natural world. Thomas cited traditional authorities — "the Experimenter," Ambrose, Pliny, Aristotle;[87] and as García Ballester points out, Thomas followed the medieval approach of seeking symbolical meanings in natural phenomena:[88] his stories draw morals. (The sparrow hawk story, for instance, is "an example of compassion to be remembered."[89]) García Ballester characterizes Albert's approach, on the other hand, as "wholly scientific"[90] — with some justice. Albert's authorities tended to be practitioners, and he was not afraid to question his sources. He, too, drew conclusions, but they were of a totally different kind: e.g., "shape indicates conformity to or difference from a species more than does color."[91] Albert was concerned with telling the reader what the birds he was describing looked like, how they behaved, what their prey was. In consequence, many of the birds he described can be identified from the descriptions he gave, though he gave them names we do not recognize; and many of his incidental observations on the birds' behavior are borne out by modern authorities. Albert's work on hawks and falcons was not only important in itself, therefore, but more important as an example for his time of how scientific enquiry into the natural world might be carried out.

---

[87] Thomas de Cantimpré, *De natura rerum* (Berlin 1973), pp. 182-183, 217.
[88] García Ballester, p. 24.
[89] Thomas de Cantimpré, *De natura rerum* (Berlin 1973), p. 217; and see above p. 449.
[90] García Ballester, p. 24.
[91] Albertus Magnus, *De animalibus*, p. 1465; and see above, pp. 457-458.

18

Mathematics in the Thought of Albertus Magnus

A. G. Molland
University of Aberdeen

A. Introduction

In a letter to Marin Mersenne in 1638 René Descartes claimed that his physics was nothing but geometry.[1] The exact interpretation of this statement may be difficult, but similar sentiments abounded in the seventeenth century, and, with justice, are taken to reflect one of the most important characteristics of the science of the time. Thus for Galileo the book of the universe was written in the language of mathematics, and for Kepler geometry had provided God with the exemplars for the creation of the world. Newton produced the significant title *Mathematical Principles of Natural Philosophy*, and Leibniz saw infinitesimal analysis as linking geometry and physics. These writers had many differences, but they all shared a faith in the power of mathematics in natural philosophy.

There were many ancient precedents for this, particularly among Pythagoreans and Platonists. The complete tradition was not available to the Middle Ages, but there was quite sufficient evidence to show how much some ancient thinkers had valued mathematics. The part of Plato's *Timaeus* that appeared in Latin with Chalcidius'

---

[1] *Oeuvres de Descartes*, ed. Charles Adam and Paul Tannery (Paris, 1897-1913), 2: 268.

commentary[2] discussed the mathematical structure of the world and of the World Soul. Boethius regarded arithmetic as the exemplar for the creation, and dwelt on the harmonic structure of the world.[3] (We should remember that music in its theoretical aspect was regarded as a mathematical science.) And great authority had to be attached to the assertion to God in the *Wisdom of Solomon* that "Thou hast ordered all things in number, weight and measure."[4] Passages such as these received much attention in the twelfth century, but the situation became more complicated after the advent of the huge quantity of new learning in translation from Greek and Arabic. Not only was more knowledge of Greek mathematics available, but there was the massive achievement of Aristotle to be tangled with. And Aristotle was ambivalent towards mathematics.

This was largely a result of his historical situation. Mathematics was advancing rapidly in Aristotle's time, and he himself spent many years in Plato's Academy. In his *Posterior Analytics* geometry clearly provided the model for demonstrative science, and mathematical examples abound throughout the corpus. He regarded mathematics as providing one of the three branches of theoretical knowledge, along with physics and metaphysics, and he was also very conscious of what may be called separate branches of applied mathematics, such as optics, astronomy and musical theory. Nevertheless he was concerned to counter what he saw as an excessive exaltation of mathematics in the work of some of his predecessors, and in his writings there were many points of tension between what mathematics could seem to say the world was like and what he held it actually to be. These tensions were not adequately resolved, and so there was room for much divergence of opinion in the commentatorial tradition. As always this was abetted by the very laconic form of Aristotle's extant writings. In this paper my central aim will be to glean some understanding of Albertus Magnus' attitude towards mathematics mainly on the basis of what he says in the so-called Aristotelian paraphrases. I shall make little allusion to the commentary on Euclid ascribed to Albert, which is discussed by Paul Tummers in the next essay in this volume.

---

[2] Plato, *Timaeus a Calcidio translatus commentarioque instructus*, ed. J. H. Waszink (London and Leiden, 1962).

[3] Boethius, *De institutione arithmetica* I.1, *De institutione musica* I.2, ed. Gottfried Friedlein (Leipzig, 1867; repr. Frankfurt, 1966), pp. 10, 187-188.

[4] Wisd. 11: 21.

Albert is more renowned as a biologist than as a mathematician. This is just. His interests and abilities were far more in that direction, and he could be in danger of being classified by a mathematician as woolly minded. If we adopted Pierre Duhem's notorious contrast between the French and the English,[5] we should have to say that Albert's mind was ample and weak rather than deep and narrow. This incidentally means that the method of close textual analysis of particular passages can often lead us into a morass. To re-create effectively his vision of mathematics we need to adopt a more impressionistic approach and always search for the thought behind the words, for often this does not shine clearly through them. As we shall see, Albert did not think that mathematics was very important to natural philosophy, and at times it could be a snare and a delusion. Factors such as this have led to the comparative neglect of his mathematical thought. Nevertheless he had frequently to discuss mathematics, and it is important to assess his attitude towards the subject, both as part of his own intellectual make-up and as a foil to those scholastic writers who had a far higher opinion of the value of mathematics.

An important exception to the neglect of this aspect of Albert's thought is a valuable article by J. A. Weisheipl.[6] Weisheipl held that Albert was particularly concerned to attack "Plato's error" as it appeared in the work of Robert Grosseteste, Roger Bacon and Robert Kilwardby (particularly the last of these). These Weisheipl referred to as the "Oxford Platonists." This provides the important reminder that Albert was not operating in an intellectual vacuum, but I am not so confident as Weisheipl in identifying precise contemporary targets for Albert's attacks. In any case the onslaught must be regarded as oblique, for the explicit targets are almost always the views of ancient Pythagoreans and Platonists as presented by Aristotle and others. For this and other reasons I shall at this stage make little reference to Albert's contemporaries, although I shall occasionally bring in Grosseteste for purposes of comparison.

In Albert's time pure mathematics comprised arithmetic and geometry. Arithmetic was akin to what we should now call number theory. Its principal source was the work on the subject by Boethius,

---

[5] Pierre Duhem, *The Aim and Structure of Physical Theory*, tr. Philip P. Wiener (New York, 1962), pp. 55-104.

[6] James A. Weisheipl, "Albertus Magnus and the Oxford Platonists," *Proceedings of the American Catholic Philosophical Association*, 32 (1958), 124-139.

and its subject matter was number conceived as a collection of units, so that unity was in some sense the principle of number. Geometry derived mainly from Euclid's *Elements* and was concerned with continuous quantity. Albert made particular use of the commentary on this work by al-Nairīzī, the Anaritius of the Latins, and this encouraged him to regard the point as the principle of magnitude. By its motion the point generated a line, and from the motion of a line there arose a surface, and from that of a surface a body.[7] Important questions were the degree of independence of these disciplines and the ways in which they related to other branches of knowledge.

B. The Division of the Sciences

Albert followed Aristotle's division of theoretical science into metaphysics, mathematics and physics, although with some difference of nuance.

> The first in the real order (*secundum ordinem rei*) is that which is generally about being (*ens*) as being and not conceived with motion and sensible matter in itself or in its principles, neither according to being (*esse*) nor according to reason (*ratio*). And this is first philosophy, which is called metaphysics or theology. The second in the same real order is mathematics, which is conceived with motion and sensible matter according to being but not according to reason. The last is physics, which is totally conceived with motion and sensible matter, according to being and reason.[8]

Mathematics abstracts from motion and sensible qualities and considers quantity as pictured in the imagination, although its objects have real existence only in sensible bodies. A further act of abstraction removes quantity also, and we are in the realm of metaphysics.

This abstractive ascent can suggest that we are by steps approaching the real causes of things, and that we may reverse the process and see how all things proceed from their principles. And indeed in his *Physics* Albert hinted that this was the case.

---

[7] *Anaritii in decem libros priores Euclidis commentarii*, ed. Maximilian Curtze, in *Euclidis opera omnia*, ed. J. L. Heiberg and H. Menge, *Supplementum* (Leipzig, 1899), p. 1. On Albert's use of Anaritius see Paul M. J. E. Tummers, "The 'Commentary' of Albertus (Magnus?) on Euclid's 'Elements of Geometry'; Anaritius as his Source," xvth *International Congress of the History of Science. Abstracts of Scientific Section Papers* (Edinburgh, 1977), p. 51.

[8] Albert, *Physica* I, tr.1, c.1 (ed. Borgnet, 3: 2a). On Albert's classification of theoretical science, see Joseph Mariétan, *Problème de la classification des sciences d'Aristote à St-Thomas* (St. Maurice and Paris, 1901), pp. 166-171.

Since the quiddity which is unqualifiedly (*simpliciter*) first gives first being, from which flows the being of this [particular] quantity in what is measured by quantity, from which further flows forth the being of this sensible [thing], distinguished by quantity and distinguished by active and passive forms, the first will without a doubt be the cause of the second and third. Wherefore both mathematicals and naturals are caused by metaphysicals and take their principles from them.[9]

But in the later *Metaphysics* he made abundantly clear that he intended no such simple priority in the order of things.

Here there is need to beware of the error of Plato, who said that naturals were founded in mathematicals and mathematicals in divines, just as the third cause is founded in the second, and the second is founded in the primary, and therefore he said that mathematicals were principles of naturals, which is completely false.[10]

Sensible qualities do not inhere in bodies by reason of their spatial extension, but because the bodies have the aptitude of being extended.[11] Thus, as it were, quantity and sensible qualities arise simultaneously, and mathematics and physics are twin births from metaphysics.

C. MATHEMATICS AND THE UNDERSTANDING OF NATURE

This complex view of the interrelations of physics, mathematics, and metaphysics, with its distinction between what can be thought and what actually is the case, makes particularly difficult the question of how mathematics relates to physics. I have elsewhere spoken of realist and conceptualist poles in scholastic attitudes to this question.[12] The realist places his focus on the actual existence of mathematical objects in the outside world, and expects mathematics to tell him quite a lot about the world; he often hints at mathematical design in nature. The conceptualist on the other hand pays particular attention to the fact that the mathematician operates on objects pictured in the imagination, and he often seems to lose sight of their anchorage in external bodies. In this matter Albert veers very much towards the conceptualist pole.

---

[9] Albert, *Physica* I, tr.1, c.1 (ed. Borgnet 3: 3b).

[10] Albert, *Metaphysica* I, tr.1, c.1 (ed. Colon. 16/1: 2, vv. 31-35).

[11] Albert, *Metaph.* I, tr.1, c.1 and tr.4, c.1 (ed. Colon. 16/1: 2, vv. 62-67; 47, vv. 58-64).

[12] Andrew George Molland, "An Examination of Bradwardine's Geometry," *Archive for History of Exact Sciences*, 19 (1978): 113-175.

One passage where Aristotle seemed to assign a particularly important role to mathematics for the understanding of nature was in *Posterior Analytics* I, 13.

> The reason why [*propter quid* in medieval discussions] differs from the fact [*quia* in medieval discussions] in another fashion, when each is considered by means of a different science. And such are those which are related to each other in such a way that the one is under the other, e.g., optics to geometry, and mechanics to solid geometry, and harmonics to arithmetic, and star-gazing to astronomy. . . . For here it is for the empirical [scientist] to know the fact and for the mathematical [to know] the reason why; for the latter have the demonstrations of the explanations, and often they do not know the fact, just as those who consider the universal often do not know some of the particulars through lack of observation.[13]

Albert did not disagree, but he was careful to circumscribe the power of mathematics. Its proper concern was only with quantity as quantity.

> An example of this is a ray, which is a line. A line as line has [the property of] being straight or curved, and, by meeting another, of making an angle and the quantity of an angle, in that it meets it perpendicularly or obliquely. Because all these things are [properties] of quantity as such, the geometer considers them as to cause. But in as much as a ray has [the property of] being bent (*reflecti*)[14] at a clean and polished [surface], and in a concave mirror of being bent towards the middle, and in a round pervious [object] of being bent (*reflecti*) towards the opposite point, because these properties are not caused by a line in that it is quantity, therefore lines as such are not appropriate, nor can [the properties] be produced by the geometer from the proper principles of quantity, and so he cannot pronounce the *propter quid* in them. But they are properties of the visual line in that it is visual, and so the *propter quid* in such is pronounced by the *perspectivus*.[15]

For Albert mathematical properties were very much on the surface of things. Robert Grosseteste on the other hand had a vision which emphasised the penetration of mathematics into physics, and we may think of the basic mathematical structure being decked out by the addition of other qualities.

---

[13] Aristotle, *An. post.* I.13 (78b35-79a6). Translation from *Aristotle's Posterior Analytics*, trans., ann. Jonathan Barnes (Oxford, 1975), pp. 22-23.

[14] On the vagueness of Albert's terminology see Carl B. Boyer, *The Rainbow: From Myth to Mathematics* (New York and London, 1959), pp. 95-96.

[15] Albert, *Posteriora analytica* I, tr.3, c.7 (ed. Borgnet 2: 86a).

But one must know that an inferior science always adds a condition by which it appropriates to itself the subject and properties (*passiones*) of the superior science, and in the conclusion of the subordinated science they are like two natures, namely the nature which it receives from the superior and its own nature which it superadds of itself. And so the superior science does not pronounce the causes of what is superadded, and sometimes the inferior science pronounces these causes and sometimes not, but the superior science pronounces the causes of what the inferior science receives from the superior.[16]

With this type of view Grosseteste can well maintain the importance of mathematics. But Albert, although he agrees with Aristotle that there is no falsehood in abstraction,[17] gives a minimal assessment of what mathematics can say about the physical world. Significantly he does not attempt to reproduce the sophisticated mathematical account of the rainbow that Aristotle gave in *Meteorologica* III, 5.[18]

i. Geometrical Exactness

Albert may explicitly affirm that there is no falsehood in abstraction, but there are many passages in his writings where he seems strongly inclined to the opposite view. For instance, there is the problem of geometrical exactness, which provides a central difficulty for an abstractionist view of mathematics. How can the exact nature of geometrical objects arise from a mere stripping away of qualities in thought from the more chaotic sensible world? An example of the problem (as given by Albert) is the following: "A sensible circle, which has a bent line [as circumference] does not touch a ruler, which is a sensible straight line, in a point, while yet, as is demonstrated in the fifteenth and sixteenth [propositions] of the third [book] of our geometry, a line touching a circle only touches it in a point."[19] As a biologist too Albert was very conscious that the limited range of shapes with which the geometer operated did not fit well with the forms of living creatures. "Many of the geometers' figures are in no

---

[16] Robert Grosseteste, *In Aristotelis Posteriorum analyticorum libros* I.12 (78b35-79a16), ed. Pamphilus de monte Bononiensis (Venice, 1514; repr. Frankfurt, 1966), f. 14v. Cf. Alistair C. Crombie, *Robert Grosseteste and the Origins of Experimental Science*, 2nd imp. (Oxford, 1961), pp. 91-98, and William A. Wallace, *Causality and Scientific Explanation* (Ann Arbor, 1972-74), 1: 27-47.

[17] E.g., Albert, *Physica* II, tr.1, c.8 (ed. Borgnet 3: 108b).

[18] Aristotle, *Meteor.* III.5 (375b16-377a27). Albert, *Meteora* III, tr.4 (ed. Borgnet 4: 666-700).

[19] Albert, *Metaph.* III, tr.2, c.3 (ed. Colon. 16/1: 118, vv. 35-40).

way found in natural bodies, and many natural figures, and particularly those of animals and plants, are not determinable by the art of geometry."[20] Neither Aristotle nor Albert satisfactorily solved problems of this kind, but Albert's general strategy was to make more tenuous than had Aristotle the link between the geometer's mind and the outside world.

> All abstract and mathematical [objects] are received according to the understanding (*intellectus*), for, according to being (*esse*) in nature, this one or that one is not found, unless it be in light alone, as some say, although their view is not in accord with the philosophers. Although they are received according to the understanding, this understanding is caused by particular things, and refers to being, because it expresses the nature and being of the thing as it is what it is.[21]

Interestingly enough this passage may contain a reference to the "Oxford Platonists," for, when facing the problem of exactness, Roger Bacon had grounded geometry primarily in the multiplication of species and in celestial things.[22]

ii. Infinity

If the problem of exactness should seem rather pernickety, Albert could turn to questions of infinity for more dramatic instances of the power of mathematics to mislead, for the impulse of mathematics towards the infinite chafed at Aristotelian restrictions. Let us consider the question of infinite spatial extension. The Aristotelian world was finite. Moreover space did not exist apart from body, and so it was nonsensical to posit an infinite space beyond the outermost heaven. On the other hand geometry seemed to demand an infinite space for its constructions, and indeed Euclid's postulates asserted that a straight line could always be produced further, and that a circle could be described on any centre with radius of any length. Aristotle considered the apparent conflict between physics and geometry in a not altogether satisfactory passage of his *Physics*. He said that

---

[20] Albert, *Physica* III, tr.2, c.17 (ed. Borgnet 3: 235b).

[21] Albert, *De praedicamentis*, tr.3, c.3 (ed. Borgnet 1: 199b): "Omnia enim abstracta et mathematica sunt accepta secundum intellectum; secundum enim esse in natura hoc vel hoc, non invenitur nisi in sola luce, ut quidam dicunt, quamvis dictum eorum cum Philosophis non concordet. Quamvis ergo ista accepta sint secundum intellectum, iste tamen intellectus a specialibus rebus causatus est et ad esse refertur."

[22] Oxford, Bodleian Library, MS Digby 76, f. 78ra; Molland, "An Examination of Bradwardine's Geometry," *Archive for History of Exact Sciences*, 19 (1978), 113-175.

the geometers only postulated that a line be produced so far as was wished rather than to infinity, and he also suggested a scaling down argument. If a configuration was too large to allow the constructions necessary for proving a theorem, then the same theorem could be proved on a similar but smaller configuration. Albert refers to these points, but the main emphasis of his approach was to say that the geometer was dealing with imagined objects rather than real ones. "The mathematicians do not need in their science an infinite magnitude according to act, because they do not receive quantity according to being (*esse*), but according to imagination, and they proceed according to the power of the imagination to compose figures and angles, and not according to the power of the thing imagined."[23] In another part of his discussion of infinity Albert suggests an even more tenuous link between mathematics and the world when he accuses mathematicians of begging the question. "Because the mathematicians posit the principles, they prove something to be infinite from their positing it to be so (*ex illo probant aliquid esse infinitum hi qui ponunt ipsum esse*)."[24] This can suggest a purely formal view of mathematics in which the concern is not with the truth of categorical statements but only with validity of inference.

iii. The Nature of Space

The size of space was not its only feature that made for an uneasy relationship between Euclid and Aristotle; there was also the question of its structure. Alexandre Koyré proposed as one of the leading characteristics of the new science of the seventeenth century the following: "the geometrization of space — that is, the substitution of the homogeneous and abstract space of Euclidean geometry for the qualitatively differentiated and concrete world-space of the pre-Galilean physics."[25] Whatever terminology one uses, the Aristotelian world was certainly structured in a way different from that of the mechanical philosophy. Different properties followed from mere difference of position. Heavy bodies moved towards the centre of the world because it was the centre, and not because the earth was located there. If the whole earth were displaced it would naturally return to the centre. Moreover, Aristotle wished to ascribe an above

---

[23] Albert, *Physica* III, tr.2, c.17 (ed. Borgnet 3: 235b).

[24] Ibid., c.3 (p. 210a).

[25] Alexandre Koyré, *Metaphysics and Measurement* (London, 1968), pp. 19-20.

and a below, a right and a left, a front and a back to the world as a whole, just as a human being unambiguously possessed these features. Aristotle even hinted that these characteristics were also in geometrical objects: "Above is the principle of length, right of breadth, front of depth."[26] But elsewhere he modified this: "Though [the objects of mathematics] have no real place, they nevertheless, in respect of their position relatively to us, have a right and a left as attributes ascribed to them only in consequence of their relative position, not having by nature these various characteristics."[27] Albert developed Aristotle's references to animals and plants in this regard, but was concerned to emphasise that right and left, etc. were not intrinsic characteristics of mathematical objects.[28]

> Their difference of position is only in the intellect, just as they are received abstractly by the intellect alone, and they do not have by nature any difference among these six positions, because if they did have them by nature they would move to them by nature, and this is false since they are separated from motion.

Once again the burden is to maximise the gap between mathematics and the natural world.

iv. The Continuum

The term "mechanical philosophy" is applied to a varying cluster of ideas and images about the structure of the world, and how one is to give a scientific account of it. One image, by analogy with machines, is that one may understand the world by taking it to pieces — in thought if not in fact. In common with many biologists Albert has a far more holistic approach, and denies that enlightenment is to be gained by such metaphorical butchery. In this instance he could draw some support from mathematics, for even the continuum did not seem to be properly resolvable into parts, and continuous quantity was the subject-matter of geometry. In antiquity contradictions had appeared to arise within mathematics itself. This produced a divorce between arithmetic and geometry, and Albert

---

[26] Aristotle, De coelo II.2 (284b24-25). Cf. Leo Elders, Aristotle's Cosmology: A Commentary on the De caelo (Assen, 1966), p. 185, and Geoffrey E. R. Lloyd, "Right and Left in Greek Philosophy," Journal of Hellenic Studies, 82 (1962), 56-66.

[27] Aristotle, Physica IV.1 (208b22-24).

[28] Albert, De caelo et mundo II, tr.1, c.4 (ed. Colon. 5/1: 109, v. 47–113, v. 92). Quotation from Albert, Physica IV, tr.1, c.2 (ed. Borgnet 3: 243a).

was glad to welcome the weakening of the upper link in the "Platonic" hierarchy that descended from arithmetic to geometry to the world.

Let us see where some of the problems of continuity lay. It seemed quite innocuous, and indeed necessary to geometry, to hold that any straight line may be divided (at least mentally) into two equal parts, which are themselves straight lines. But this has the effect of asserting that a straight line may be divided into smaller and smaller parts without limit. There is then a temptation to take a leap, and ask what would be the result of an infinite process of division. Surely the line would have been decomposed into points or other indivisibles. But then how many points are necessary to form a line? If two suffice, the line they form is not divisible into divisible lines, contrary to our initial assumption, and the same difficulty applies to any finite number. But if an infinite number is necessary, we may ask, as Zeno did, whether or not the points have magnitude, and we seemed to be faced with the dilemma of having either a line of no length or one of infinite length.

By the time of Aristotle it was realised that one had to tread warily concerning such matters. Aristotle's basic strategy was to distinguish between potential infinite division and actual infinite division. Lines were potentially infinitely divisible, but this did not license one to speak of the state of affairs when they were actually infinitely divided. Lines were not made up of points, nor could one speak of one point of a line being immediately adjacent to another of its points. All this entailed a certain lack of correspondence between thought and its objects. Thought, or its verbal expressions, takes place in a finite number of discrete units, but the continuum seemed inextricably bound up with what F. Solmsen has called the "ocean of infinity,"[29] and as such could not be fully controlled. It was similar with the discovery of incommensurability. This meant, for example, that it was impossible to find any line, however small, that would exactly fit a whole number of times into both the side of a square and its diagonal. The Pythagorean dream of the dominance over geometry by number was severely shaken.

These matters were not simply the concern of mathematicians, for continuity emphatically entered into the physical world. Aristotle was firm in maintaining that physical bodies were essentially conti-

---

[29] Friedrich Solmsen, *Aristotle's System of the Physical World* (Ithaca, 1960), p. 201.

nuous, and not composed of indivisibles, and in this Albert enthusiastically followed him. Albert was concerned to maintain the autonomy of physics, but in this instance he was prepared to draw arguments from mathematics, for although his reasoning did not reach the sophistication of later writers, geometry seemed definitely to be on his side. "We shall first draw reasons from those things that are said in mathematics. The sayings of those sciences must either remain firmly supposed, or, if they are to be removed, they must be removed by stronger and more credible reasons than they are in themselves."[30] This is notably different from Albert's attitude in the case of infinite extension, but there still remained here differences between mathematics and physics. Two natural bodies could touch without being united into one, but this was not the case with mathematical objects. If two lines touched end to end, their end-points became one.[31] Moreover division in thought was not the same as actual physical division. In the division of a natural body "a minimum is received, and it is [the smallest part] that can perfect the operation of the natural body, because if it were divided it would be corrupted from operation and essence, because it could not resist alteration."[32] Nevertheless even mathematical quantity cannot be actually divided at every possible point, although at any stage it may be divided at any point.

Albert saw Plato (even more than Leucippus and Democritus) as the chief villain in the matter of composing continua from points. In this he may not have been altogether just, but the view could seem a plausible extension of the doctrine of the *Timaeus*, encouraged by Aristotle, who perhaps interpreted the "likely story" there presented in too literal a fashion. In the *Timaeus* the small particles of the four elements were assigned the shapes of four of the five regular solids, and in a somewhat obscure fashion these solids were said to be derived from triangles.[33] Like Aristotle, Albert regarded this sort of

---

[30] Albert, *De indivisibilibus lineis*, c.3 (ed. Borgnet 3: 469b). Pseudo-Aristotle expressed a similar sentiment in the corresponding place; *De lineis insecabilibus* 2 (969b29-970a17). See also Aristotle, *De coelo* III.1 (299a4-6).

[31] Albert, *Physica* V, tr.2, c.1, 3 (ed. Borgnet 3: 379a-b, 383a). Cf. John E. Murdoch, "Superposition, Congruence and Continuity in the Middle Ages," in *Mélanges Alexandre Koyré* (Paris, 1964), 1: 416-441.

[32] Albert, *De generatione et corruptione* I, tr.1, c.14 (ed. Borgnet 4: 357a-b).

[33] Plato, *Timaeus* 53c-57D. Cf. Aristotle, *De coelo* III.1 (299a1-300a19); III.7-8 (305b29-307b24); *De gen. et corr.* I.2 (315b25-317a17); and A. T. Nicol, "Indivisible Lines," *Classical Quarterly*, 30 (1936), 120-126.

thing as undesirable mathematicism, taking eternal geometrical principles to be the principles of corruptible things.

> But the Platonists and some other ancients, on account of love of their teachers who exalted mathematicals too much on account of the incorruptibility and necessity that they found in them, and so said that they were the principles of natural things, did what those were wont to do who posit impossible sayings. For in the beginning and without much consideration they had used an impossible supposition, and since they posited impossibles as principles it was necessary that they take pains to justify and prove those things that they had said, which were often contrary to the truth, lest they seem to surrender, and lest the doctrines of their teachers with which they were imbued be annihilated.[34]

An objection to Plato's doctrine was that it involved constituting bodies from non-bodies (namely surfaces), and we see the affinity with the doctrine of the composition of lines from points. Elsewhere Aristotle had reported that Plato regarded indivisible lines (rather than points) as the principle of the line.[35]

In such questions Albert was usually a firm partisan of Aristotle, but there is one significant point of divergence which somewhat muddies the waters. This is Albert's frequent use of the idea of the generation of a line from the motion of a point, of a surface from that of a line, and of a body from that of a surface. At times he insists that the motion is merely imaginary, and indeed he agreed with Aristotle that it was impossible in the nature of things for an indivisible to move, except *per accidens*.[36] But on other occasions Albert seemed to slip into a rather more realist interpretation, and this allowed him to make an apparent concession to "Plato," while still preserving intact the mystery of the continuum.

> It is further to be noted that every quantity flows from an indivisible, as we have already remarked, for, if the essential flux of a point be taken, it will without doubt constitute a line, and the line a surface, and the surface a body. And so potentially there is a point everywhere in the line, and a line everywhere in the surface, and a surface everywhere in the body potentially. And so a point is in two ways related to a line. For if the line be considered as the essential flux of a point, the point is its material part or matter, and similarly the line of the surface, and the

---

[34] Albert, *De caelo* III, tr.2, c.7 (ed. Colon. 5/1: 236, vv. 27-39).

[35] Aristotle, *Metaph.* I.9 (992a20-23).

[36] Albert, *Physica* VI, tr.3, c.4 (ed. Borgnet 3: 456a-459a).

surface of the body. But if formal line be taken in the way said by Euclid, namely as longitude terminated at two points, the two points are formally in the line according to the act of the form, which is to terminate and bound. And so the point is in some way the form of the line.[37]

The concept of flow may seem an odd way to preserve the continuity of body, but Albert's use of a flowing now to express the continuity of time appears more natural.[38]

v. Measure and Number

After Albert's time several medieval writers disagreed with Aristotle, and held that the continuum was composed from indivisibles, and earlier Grosseteste had at least veered towards this view. Grosseteste raised the question[39] of how, if there were only one line in the universe, it was to be measured, when there was nothing else to compare it with. His answer was that it was properly measured by the number of points that it contained. This was infinite, but for God an infinite number was finite, and so he could know measure in this way. This entailed there being different infinite numbers, and one number could even have to another the irrational ratio of the diagonal of a square to its side. This position may be grounded in Grosseteste's cosmogony, in which extension derived from the infinite self-multiplication of a point of light. Grossesteste showed an awareness of the tradition. "This, as I believe, was the understanding of the philosophers who posited all things to be composed from atoms, and said that bodies were composed from surfaces, and surfaces from lines, and lines from points."[40] The overall effect of such a view was to place a quasi-numerical structure at the very heart of physical reality — something that Albert was vehemently opposed to.

Albert's own account of measure, like Aristotle's, was on a more superficial level. Measurement was basically the expression of the quantity of something in numerical terms. Numbers were collections

[37] Albert, *De gen. et corr.* I, tr.1, c.15 (ed. Borgnet 4: 358a). Cf. Wolfgang Breidert, *Das aristotelische Kontinuum in der Scholastik*, in *Beiträge*, NF 1 (1970), 23-32.

[38] Albert, *Physica* VI, tr.1, c.7 (ed. Borgnet 3: 420a-421b); *Metaph.* V, tr.3, c.2 (ed. Colon. 16/1: 260, vv. 47-51).

[39] Robert Grosseteste, *Commentarius in VIII libros physicorum Aristotelis* IV (219b5-8), ed. R. C. Dales (Boulder, 1963), pp. 90-95.

[40] Robert Grosseteste, *De luce seu de inchoatione formarum*, in *Die Philosophischen Werke des Robert Grosseteste*, ed. Ludwig Baur, *Beiträge* 9 (1912), 53-54.

of units, and these units were indivisibles. Their indivisibility was ultimately rooted in the indivisibility of substantial form.[41] (What, for example, could be meant by half a substantial form?) Thus numbers seemed firmly anchored in the physical world. They were consequences of the existence of things, and not causes. As Albert frequently put it, they arose from the division or separation or discreteness that there was between things. In reality they were inseparable from the things numbered, but there were also abstract units in the soul from which were formed the numbers by which we number. Thus Albert often spoke of a twofold aspect of number: materially it was in the numbered or numerable things, while formally it was primarily located in the soul. By means of number quantity was known. For discrete collections this involved simple counting, but in the case of continuous objects there was no true minimum to be reached by division. Nevertheless certain quantities could be conventionally treated as atoms or units, as, for example, the foot in measuring length.[42] These units could then be counted, as was done in measuring discrete multitudes.

This conventional imposition of units could appear very arbitrary, and Anneliese Maier has suggested this as one of the main reasons why the Schoolmen did not develop an exact natural science.[43] In the next century the apparent arbitrariness was mitigated by the development of an elaborate language of ratios, which was also used to discuss intensive quantities in abundance. We do not know how Albert would have reacted to the new mathematicism, but we may suspect that it would have incurred his displeasure, if only because its empirical grounding was weak.

D. Conclusion

With hindsight it is tempting to say that Albert's influence was inimical to the growth of mathematical physics. But this is with hindsight, and is in any case too negative. More positively we may see Albert as warning against the dangers of fitting the variety of nature into an ill-fitting mathematical strait-jacket. Such warnings have

---

[41] See principally Albert, *Metaph.* v, tr.1, c.8, 10 (ed. Colon. 16/1: 227, v. 41 - 229, v. 26; 231 v. 61 - 233, v. 52).

[42] Albert, *Metaph.* x, tr.1, c.3 (ed. Colon. 16/2: 434, v. 22 - 435, v. 68).

[43] Anneliese Maier, *Metaphysische Hintergründe der spätscholastischen Naturphilosophie* (Rome, 1955), pp. 398-402.

continued in various contexts until the present. A. N. Whitehead remarked that, "It often happens. . . that in criticising a learned book of applied mathematics, or a memoir, one's whole trouble is with the first chapter, or even with the first page. For it is there, at the very outset, where the author will probably be found to slip in his assumptions."[44] Quantification may be applied with an inappropriate conceptual basis, although this may only be discoverable by trial and error. Moreover examples such as astrology and some modern social science show that mathematisation may be a smokescreen: it frightens off the uninitiated and conceals what may be shaky assumptions. Mathematics may be a very great aid to the natural and social sciences, but neither of them may be reduced to it.

Modern mathematics is very different from ancient and medieval mathematics. In particular the description of mathematics as the science of quantity has increasingly been seen to be too restrictive. Nevertheless there are numerous features which make them recognisably the same subject. Partly this is because of the perennial nature of the philosophical problems thrown up by mathematics. One such is the extent to which the objects of mathematics are simply given as opposed to being creations of the human intellect. In practice mathematicians talk of both existence and construction. Philosophers often try to interpret one in terms of the other, and the polar positions are sometimes dubbed platonism and constructivism. Aristotle rejected the existence of a separate realm of mathematicals, and instead grounded mathematical objects in physical objects. But he also spoke of constructions in terms of the geometer's thought bringing to actuality what had only been present potentially. Albert, as we have seen, emphasized the extent to which mathematics was a mental activity, and referred constantly to the generation of geometrical objects from imaginary motions. All this may be seen as a freeing of mathematics from the shackles of the physical world, and opening the way for a creative flowering of pure mathematics. But, for one reason or another, the times were not ripe for such a development. Perhaps the mathematician feels that it is solipsistic to talk of things that are only in his own mind.

---

[44] Alfred North Whitehead, *Science and the Modern World* (New York: Mentor Books, 1948), p. 29.

19

The Commentary of Albert on Euclid's Elements of Geometry

Paul M. J. E. Tummers
Filosofisch Instituut, Katholieke Universiteit

Most readers of Albertus Magnus are well aware of his vigorous opposition to the "error of Plato" and his contemporary "Oxford Platonists" who would make numbers and mathematical structures the immediate and proper causes of natural phenomena.[1] His insistence on the impossibility of discovering the "real causes" of natural phenomena *qua* natural by way of mathematics immediately raises the question of St. Albert's own competence in mathematics, including geometry. His passing references to mathematics in general and to geometry in particular in the course of his Aristotelian paraphrases shed some light on the question.[2] However, it would be much more illuminating were we to have a purely mathematical treatise written by Albertus Magnus.[3] There are numerous indications in the authentic writings of Albertus Magnus to show that he actually did write a commentary on Euclid's *Elements of Geometry*. If in fact he did write such a commentary, the first question that comes

---

I am greatly indebted to J. A. Weisheipl for his help, and I wish to thank H. A. G. Braakhuis for his comments and other colleagues for their help with the English translation.

[1] J. A. Weisheipl, "Albertus Magnus and the Oxford Platonists," *Proceedings of the American Catholic Philosophical Association*, 32 (1958), 124-139.

[2] See the preceding article by George Molland in this volume.

[3] The paraphrase of Albertus Magnus *De lineis indivisibilibus* can also be regarded as a mathematical treatise (ed. Borgnet 3: 463-481).

to mind is: Where is it? That is, can this commentary be produced? Is it known to exist?

These questions were first discussed in this century by Bernhard Geyer in an article completed in 1944 and published in 1958.[4] In it Geyer collected quotations from various authentic writings, especially from the *Metaphysics* paraphrase, and the testimony of ancient catalogues to prove that Albertus Magnus wrote a commentary on Euclid's *Elements*.[5] Geyer further argued that this commentary must have been written after the *De animalibus* and before the *Metaphysics*, which can be dated in the early 1260s.[6] Finally, Geyer claimed to have found this commentary in a thirteenth-century manuscript preserved in the Dominikanerkloster in Vienna as MS 80/45, which contains, among other things, a commentary on the first four books of Euclid's *Elements*.[7] On the top folio of the beginning of this text (fol. 105r) there is inscribed: *Primus Euclidis cum commento Alberti.* Geyer believed that this manuscript was written in Albert's own hand, although at first sight it does not resemble the other known autographs of Albertus Magnus.

In 1960 J. E. Hofmann published another article on the Vienna manuscript.[8] Accepting Geyer's conclusions concerning the authenticity and dating of this text, Hofmann analyzed its mathematical contents and identified its main sources as Alfarabi and Anaritius.[9] Moreover, he announced the eventual publication of the full text.[10]

---

[4] B. Geyer, "Die mathematische Schriften des Albertus Magnus," *Angelicum*, 35 (1958), 159-175. In this article Geyer published two parts of the text, namely the prologue and the question "Utrum angulus sit quantitas." Henceforth this article will be cited as "Geyer."

[5] Geyer, p. 163: the catalogue of Henry of Herford lists "[scripsit] expositionem Euclidis. . ."; the *Tabula* of Stams: "item exposuit Euclidem."

[6] Concerning the dating of *De animalibus*, see B. Geyer, *Prolegomena* to *Metaphysica* (ed. Colon. [1951] 16/1: vii-viii). Geyer dates *De animalibus* around 1261 and the *Metaphysics*, 1262-1263.

[7] For a description of the MS, see Geyer, p. 167. The MS has two parts: (fol. 1-104) Petrus de Alvernia, *Super libros metheorum*; (fol. 105r-145r) Commentary on the first Four Books of Euclid's *Elements*, inc.: "Sicut triplex est philosophia, ut dicit Aristoteles in sexto Philosophie Prime. . ."; expl.: ". . .Hec autem est figura utriusque demonstrationis."

[8] J. E. Hofmann, "Ueber eine Euklid-bearbeitung die dem Albertus Magnus zugeschrieben wird," *Proceedings of the International Congress of Mathematicians*, 14-21 Aug. 1958, ed. J. A. Todd (Cambridge, 1960), pp. 554-566. This article will be cited as "Hofmann."

[9] Alfarabius, al-Farābī (880-950); Anaritius, al-Naīrizī (fl. Bagdad, ca. 897-ca. 922). See A. I. Sabra, *Dictionary of Scientific Biography*, s.v., 10: 5-7.

[10] Since Hofmann has not been able to publish this text, I have agreed to take on this task for the Albertus-Magnus-Institut, and I hope to have everything ready for the Cologne edition of the *Opera Omnia* in the near future.

As we shall see below, the three passages quoted by Geyer from the paraphrase of the *Metaphysics* are sufficient proof that Albertus Magnus did write a commentary on Euclid's *Elements*. Therefore our main concern in this article is whether or not the commentary found in Vienna, Dominikanerkloster MS 80/45 is in fact by him.[11] First, we will give a general picture of the contents of this commentary, looking particularly at Book I and the influence of Anaritius. Second, we will consider some examples of the mathematical level of the commentary with particular emphasis on its sources, notably Alfarabi. Finally, we will reexamine the date of this work and its authorship by Albertus Magnus.

Some preliminary remarks, however, should be made about the various Latin translations of Euclid's *Elements* in order to put the Albert commentary in its proper perspective. The reception of Euclid in the Latin Middle Ages is a highly complex problem. But after the pioneering work of M. Clagett,[12] J. E. Murdoch,[13] and M. Folkerts[14] the foundations have been laid for further studies.[15] For our purposes it is sufficient to note that there were two main traditions to which the various translations of Euclid belong. The first, is the Greek-Latin tradition with which the name of Boethius is particularly associated. This tradition contained at least most of the definitions, postulates, and axioms of the first five books, most of the propositions enunciated in the first four books, and the demonstrations of Book I, prop. 1-3.[16] To this tradition belong the two works attributed to Boethius himself, namely a *Geometria* in five books (eighth century) and one in two books (first half of the eleventh century). Second, there is the Arabic-Latin tradition which was far bet-

---

[11] The following is an elaboration of a paper originally presented at the Fifteenth International Congress of the History of Science at Edinburgh in August 1977.

[12] M. Clagett, "The Medieval Latin Translations from the Arabic of the *Elements* of Euclid, with Special Emphasis on the Versions of Adelard of Bath," *Isis*, 44 (1953), 16-42.

[13] J. E. Murdoch, "The Medieval Euclid: Salient Aspects of the Translation of the *Elements* by Adelard of Bath and Campanus of Novara," *Revue de Synthèse*, 3e ser., nrs. 49-52 (1968), 68-94.

[14] M. Folkerts, *Boethius' Geometrie II: Ein mathematisches Lehrbuch des Mittelalters* (Wiesbaden, 1970).

[15] Dissertations of students of M. Clagett and others: G.D. Goldat, "The Early Medieval Traditions of Euclid's Elements," Ph.D. thesis (Madison, Wisconsin, 1956) (in this work Goldat edited a mélange version found in Paris, Bibl. Nat., MS lat. 10257, to be cited in this article as "mélange Goldat"); Sister Mary St. Martin van Ryzin, O.S.F., "The Arabic-Latin Tradition of Euclid's Elements in the Twelfth Century," Ph.D. thesis (Madison, Wisconsin, 1960).

[16] See Folkerts, *Boethius' Geometrie II*.

ter known. To this tradition belong the three versions attributed to Adelard of Bath (twelfth century), the second being the most famous.[17] But within this tradition the most influential was the translation by Campanus of Novara (d. 1296) which relegated all others[18] to the background. Then, finally, there are the "mixed" or mélange versions of the twelfth or thirteenth centuries which combined elements of both "Boethius" and Adelard versions.[19] As far as is known at present, there existed in Latin only two "commentaries" properly so-called on Euclid's *Elements* prior to the more popular *quaestiones*. The first of these is the commentary by al-Nairīzī, known in Latin as Anaritius, on the first ten books of Euclid, translated from the Arabic by Gerard of Cremona in the twelfth century. This was published by Maximilian Curtze in 1899 from a single manuscript.[20] The second is the Albert commentary we are discussing from the thirteenth century, presumably the first original Latin commentary on Euclid. As we will see, it uses Anaritius as its main source, but it also incorporates much other material. For these reasons alone, Albert's commentary deserves the attention of historians of mathematics. But for this article, we will limit our consideration to the three points mentioned above.

A. GENERAL STRUCTURE OF THE ALBERT COMMENTARY[21]

i. Introduction

Albert begins his commentary with an introduction, more philo-

---

[17] Among the Adelard versions are *Adelard I* (a literal trans. dating from around 1125), *Adelard II* (an abridged version, which became the most influential), and *Adelard III* (an "editio specialis," ca. 1200, to be considered also as one of the first commentaries). References to Adelard in this article are, if not otherwise stated to *Adelard II*, Bk. I, as edited by Mary Van Ryzin (see above note 15).

[18] Other versions in this tradition are a translation by Gerard of Cremona (1114-1187) and one attributed to Hermann of Carinthia (fl. 1140-1150).

[19] For further information on the reception of Euclid in the Latin Middle Ages, see J. E. Murdoch, *Dictionary of Scientific Biography*, 4: 437.

[20] *Anaritii in decem libros priores Elementorum Euclidis commentarii, ex interpretatione Gherardi Cremonensis in codice Cracoviense 569 servata*, ed. Maximilian Curtze, in *Euclidis Opera Omnia*, I. L. Heiberg and H. Menge, *Supplementum* (Leipzig, 1899). Today more MSS of this version are known, and a new edition is badly needed, since Curtze based his edition on only one MS, which has many lacunae, and he himself often misread the text and skipped many lines. I am currently preparing a new edition of Anaritius (Books I-IV), which I hope will soon be published.

[21] As a convenient device for this article, the author of our commentary will simply be called *Albert* until the identification with Albertus Magnus is made in the last section.

sophical than mathematical, which was published accurately by Geyer.[22] This introduction, or prologue, opens with the well-known Aristotelian[23] division of philosophy into physics, mathematics, and the study of "divine things" (divina separata). Albert's estimation of these three branches seems to be more in keeping with the Platonic-Pythagorean tradition than with the Aristotelian. The first branch (physics), quoting Ptolemy,[24] cannot give any certain knowledge because of the instability and variability of its subject-matter; an indication of this is the great diversity of opinions, which cannot be harmonized. The third branch (divina) is elevated so far beyond us "that our intellect gazes upward as the eyes of a bat into the light of the sun."[25] Only the middle branch (mathematics) properly deserves the name of "science."[26] Albert then divides this science into two parts, following, as he says, the tradition of the Pythagoreans: one part deals with discrete quantity, the other with continuous.[27] From this follows the well-known subdivision into arithmetic, music, geometry, and "astrology." "After treating of arithmetic and music," the author says,[28] "one comes to the treatise on geometry." Albert gives the usual etymology of the word, giving two explanations for it without deciding in favour of either one, namely that in Egypt, where mathematics originated, the lands and fields were divided in a geometrical way (ratione geometrica) or that man and animals differ in the way in which they possess their respective portions of land.[29]

---

[22] Geyer, pp. 170-173.

[23] Aristotle, Metaph. VI (1026a18-19). See incipit in note 7 above.

[24] "Testatur autem magnus in disciplinalibus Ptolemeus quod prima harum partium hominem ad certitudinem sui propter subiecti sui mobilitatem ac varietatem perducere nequit, cuius etiam signum est diversa opinantium in ea diversitas que usque ad hodie ad concordiam revocari . . . non potuit" (fol. 105r). See Albertus Magnus, Physica I, tr.1, c.2 (ed. Borgnet 3: 5).

[25] Arist., Metaph. II.1 (993b9-10). See below note 67.

[26] For this view, which is close to the Platonic-Pythagorean tradition, see parallel passages in Albertus Magnus.

[27] This division of mathematics into discrete and continuous quantity is based on Boethius, Arithmetica I.1 (ed. Friedlein, p. 8), but the wording used by Albert is found only in Hugh of St. Victor (d. 1141), Didascalicon II.6 (ed. B. H. Buttimer [Washington, 1939], p. 30) and in Vincent of Beauvais (1192-1265), Speculum doctrinale, XVI, c.1 (ed. Duaci [1624], p. 1503), where Michael Scot (d. 1235) is quoted.

[28] "Post arismetice igitur musiceque tractatum geometricum ordinatur negotium" (fol. 105r). This does not necessarily mean that Albert in fact wrote treatises on arithmetic or music. This passage could be a "Topos"; cf. Gerbert of Aurillac (972-1003), Geometria, ed. Bubnov (Berlin, 1899), p. 48.

[29] ". . . a mensura terre sic vocatur vel eo quod in egipto ubi primo mathematice exstiterunt scientie, terra et agri ratione geometrica partita sunt, aut quia aliter hominis est accipere terre portionem et aliter aliorum animalium" (fol. 105r). Cf. Gundissalinus, De divisione philosophiae (ed. Baur [Münster, 1903], p. 110) and Isidore of Seville, Etymologie III.10.3 (ed. Lindsay).

The fact that geometry is proper to human beings, Albert supports with a quotation from Aristotle[30] and the story of Aristippus, who considered geometrical figures as being typical of human beings.[31]

Since geometry deals with quantity without motion, the question arises as to what its principle might be and the number of species that might be derived from it. Alfarabi[32] is quoted as saying, "There are only three primary kinds of continuous quantity: line, surface, and solid." Albert explains how one kind can be imagined to derive from another by an imaginary motion, the solid being last, since, according to Aristotle,[33] a four-dimensional figure cannot exist. Further, place, movement, and time do not belong to this category of "continuous quantity."[34] In discussing the relationship between point, line, surface, and solid, Albert notes that of all these the point alone is the ultimate principle. For this reason the work begins with the definition of point. For the rest of the text Albert follows Euclid closely and presents, generally with some comment, the definitions, postulates, axioms, and propositions of Euclid's *Geometry*.

ii. The Definitions

Albert gives nineteen definitions. He includes all twenty-three Euclidean definitions, sometimes combining two into one, and he introduces a new one (no. 15) on the portion of a circle (*portio circuli*), which Euclid gives in Book III as the sixth definition. In this arrangement, including the additional definition, Albert follows the tradition of Adelard, whereas Anaritius follows Euclid.[35] Albert's formulation of the definitions, with several minor variants, is the same as in the Adelard versions.

---

[30] The quotation is from Aristotle, *Metaph.* I.1 (980b27-28): "Hominum genus, ut dicit Aristoteles, arte utitur et ratione propter quod omnia sua redigit ad normam rationis qui homo est" (fol. 105r). Cf. *Auctoritates Aristotelis* 1, 3 (ed. Hamesse [Louvain-Paris, 1974], p. 115) and Albertus Magnus, *Metaphysica* I, 1, 6 (ed. Colon. 16/1: 10).

[31] The story of Aristippus, "socraticus philosophus" (435-360 BC), is mentioned by Vitruvius, *De architectura* VI, 1.

[32] See below sect. B.

[33] Arist., *De caelo* I (268a20-b10): "Corpori autem, ut in primo celi et mundi probat Aristoteles, impossibile quartam addi dimensionem quocumque moveatur" (fol. 105r). Cf. Albertus Magnus, *De caelo* I, tr.1, c.2 (ed. Colon. 5/1: 7).

[34] Cf. Anaritius, ed. Curtze, p. 3. Albertus Magnus says the same thing in his *De caelo* I, 1, 2 (ed. Colon. 5/1: 3).

[35] I quote Anaritius from my own collation of the MSS, but I will give the page of Curtze's edition. See note 19 above.

On examining two definitions,[36] one can see that Albert considers the version he is using to be a "translation from the Greek." Thus he believed the Adelard version to be a translation from the Greek. In four cases[37] Albert gave an additional definition, which he labelled "Arab" or "in the translation from the Arabic." An examination of these additional definitions reveals that they are taken from the translation of Euclid as given by Anaritius in his commentary, translated from the Arabic. After each definition, Albert makes a short comment and when he presents an alternative definition, he always mentions its source. In general, he takes all this from Anaritius, but there are several exceptions. On seven occasions Albert attributes an alternative definition to an individual who is not mentioned in Anaritius.[38] One would like to know Albert's source for this additional information. In several cases Albert gives very significant information, which is not found in Anaritius, but indeed in other authentic works of Albertus Magnus.[39]

iii. Postulates

Albert gives five postulates. For the first four, he gives all five Euclidean postulates, combining the first two of Euclid into one, and as a fifth, he gives "Two straight lines do not contain a surface," which is axiom 9 in the Greek Euclid. In this Albert is simply following the Adelard tradition, saying that the fifth postulate always appears in the Greek translation, but is not found in the Arabic tradition. In fact, this postulate is not found in the versions attributed to Boethius, but only in the Adelard versions. This confirms our earlier conclusion that for Albert, "the Greek translation" means an Adelard version. Each of these postulates is followed by some comment, and for postulates 1, 3, and 5, Albert gives demonstrations. Most of this material he took from Anaritius.[40]

---

[36] Def. 6 (= Euclid 8) and 18 (= Euclid 22).

[37] Def. 3 (= Euclid 4), def. 4 (= Euclid 7), def. 6 (= Euclid 8), and def. 18 (= Euclid 22).

[38] Albert cites Plato once, Sambelichius (= Simplicius) three times, Hermides (= Heron) twice, Yrinus (= Heron) once in cases where Anaritius simply says *alii* or *aliquis.*

[39] For example, a definition of a straight line, and the names which Pythagoreans and Platonists give to a surface. See below section C.

[40] Except, for example, the statement that it is proper to philosophers to indicate an angle with three letters.

iv. Axioms or Common Notions

The situation is more complicated when we examine the number and contents of the axioms, which Albert calls *communes animi conceptiones*. Albert gives no less than thirteen axioms: all of the usual ones attributed to Euclid, except the fifth, and additional ones from the versions of Adelard and Anaritius. In grouping his axioms, Albert says that the first three are to be found "in the old books," the next are "presupposed by the later people," axioms 8 and 9 are "added by the moderns," and the last four Albert adds, saying that "one can add others *ad infinitum*." One would like to know the source for this grouping. Thus, here in the number and sequence of the axioms, one finds a rather personal touch in combining his material from Adelard, Anaritius, and other unknown sources. However, Albert makes no comment on the axioms, except on the first, which is taken from Anaritius.

Regarding the formulation of the axioms, one realizes at first glance that Albert is following the Adelard tradition and not that of Boethius.

After the axioms are presented, Albert divides geometry into *theorica* and *practica*, which is not original. He gives three meanings for *practica* with texts from Aristotle and Avicenna.[41] Albert ends this part of the work with additional material,[42] inspired by Anaritius.

v. Theorems

The Albert text gives 47 theorems for the first book, including all forty-eight Euclidean propositions, except no. 45, as do all the versions of Adelard, although Boethius does not.

At first glance the formulation of the theorems is similar to that of Adelard II. But upon careful examination, a remarkable difference can be seen: sometimes there is an altogether different reading[43] and in a few cases Albert gives a mixture of Adelard and probably a

---

[41] "Et cum practica tripliciter dicitur, scil. [1] accipiens formam prout est principium operis, sicut ars secundum Aristotelem est factivum principium cum ratione [cf. *Ethica Nicom.* IV, 4 (1140a21)], et [2] docens sententiam de opere modo, quemadmodum pars medicine est practica, ut dicit Avicenna in primo Canone sue medicine; [3] accipitur in disciplinalibus praxis tertio modo, scil. pro constitutione figure que fieri docetur sicut in prima figura. . ." (fol. 108r).

[42] "His habitis sumende sunt figure, nos autem in omnibus locis addemus ubi alii philosophi quedam addiderunt Euclidi, ut plenior habetur scientia" (fol. 108r).

[43] E.g., Bk. I, prop. 5, 24, 25, 30, and 40.

mélange version.[44] Nevertheless, the overall impression is that Albert's version of Euclid belongs to the Adelard tradition or is to be situated very near to it. However, it is important to note that the formulation of the theorems which Albert gives is not identical to that of Adelard I, II, or III. Quite remarkably the version of Euclid in MS Vat. Reginensis 1268, fol. 1r-69r (versio a) has the very same reading as in our Albert, with only minor variants.[45]

In addition to the "normal" formulation of the theorems, Albert gives alternatives for three propositions.[46] These formulations are identical with those of version b in Vat. Reg. 1268, mélange Goldat,[47] and Boethius. Our tentative conclusion can only be that these alternative formulations came from Boethius or a mélange version following the Boethian tradition.

vi. Books II-IV

Briefly the remaining books present the same picture that we have seen for Book I: the wording of Albert's version is not entirely the same as any of the Adelard versions; sometimes they are completely different. As for the commentary, Albert used mainly Anaritius, but he also added material from other sources.

B. CONTENTS OF THE COMMENTARY

Although a fuller comparison of the demonstrations given by Albert and those found in the Adelard versions has yet to be done,

---

[44] E.g., Bk. I, prop. 4.

[45] This MS (prob. late 13th cent.) is described by Björnbo in *Abhandlungen zur Geschichte der mathematischen Wissenschaften*, 14 (1902), 138-142. This version of Euclid (fol. 1r-69r) has two sets of readings for the theorems: (1) For *all* propositions heading each demonstration and written in larger script (to be called Vat. Reg. versio a). One may assume that these formulations were inserted into the text before or after the demonstrations had been written. These are identical with the usual rendering of the theorems by Albert; they are not the same as the Adelard versions, but they are very close to that tradition. (2) For *six* propositions only, namely 2, 3, 6, 7, 9, and 10, written in smaller script, similar to that of the demonstrations, appended to the demonstration of the *preceding* theorem, and thus not written at the same time as version a (to be called Vat. Reg. versio b). One may assume that this version of the six propositions was in the same exemplar from which the scribe or author took the demonstrations. These six are identical with Boethius, mélange Goldat, and Albert's *alternative* readings. In my opinion there is evidence that Vat. Reg. versio a was borrowed from Albert, and not vice versa. Thus the work of Albert must have been known to others, who used it.

[46] Bk. I, prop. 6, 9, and 10.

[47] See note 15 above.

we perhaps already have some idea of Albert's abilities as a mathematician. Frankly, in my opinion, Albert does not reveal as much insight into mathematics in this commentary as one might have expected. He does not always understand his sources,[48] and he makes mistakes. As an illustration, we might consider only two demonstrations taken from Book I.

First, consider Euclid's proposition 11 of Book I: *To draw a straight line perpendicular to a given straight line from a given point on it.* The demonstration for this given by Albert is, in fact, more suitable for the following proposition 12: *To draw a straight line from a given point perpendicular to a given straight line.* The procedure Albert gives is as follows:[49] "Let AB be a given straight line and C a point on it. Let two equal parts be taken at both sides of C (by theorem 2) and construct an equilateral triangle on that part of the line on which is the assigned point C (by theorem 1)." Then Albert curiously adds: "or if too long (*vel si maius*), divide the straight line (by the previous theorem) into equal parts at C, then construct on the whole an equilateral triangle (by theorem 1), as was said." If this is an intermediate step for the practical purpose of construction, one can understand it, and in this sense it may be correct. However, this remains a strange remark, because point C is given, and that is exactly the starting point of the construction asked for.

Then Albert continues, "Then by dividing the angle of the triangle which is ABD, I mean the angle ADB, (by theorem 9) produce line DC,

---

[48] As an example, I would refer to a passage in Bk. III, prop. 13, where Albert gives a demonstration of Yrinus, taken from Anaritius. He, however, deletes this, saying, "Figura autem hec, et est non multum valens" (fol. 135r). He appears not to have understood Anaritius well.

[49] Fol. 111v: "*Data recta linea a puncto in ea assignato perpendicularem extrahere.* Sit enim data linea AB, sitque punctus in ea assignatus C. Deinde in data linea partes equales ex utraque parte assignati puncti resecentur per secundum theorema. Postea super partem illam linee in qua est assignatus punctus, equilaterus triangulus per primum theorema statuatur; vel si maius, rectam lineam per antecedens theorema per equalia divide ad C, deinde super utramque partem simul equilaterum triangulum per primum theorema, ut dictum est, statue. Deinde dividendo angulum trianguli qui ABD, angulum inquam ADB, per nonum theorema, produc lineam DC et affirma ipsam esse perpendicularem. Hoc igitur facto, patebunt duo trianguli sc. DCB et DCA et duo latera unius equalia sunt duobus lateribus alterius, quia DC <a mistake. It must be DB> est sicut DA per hoc quod totus triangulus fuit equilateralis. Latus autem DC est commune. Item angulus equis lateribus contentus, sc. CDB, est equalis angulo equis lateribus contento, sc. angulo CDA. Ergo basis basi (!) et reliqui anguli reliquis angulis per quartum theorema. Ergo angulus DCB est sicut angulus DCA. Ergo lineam rectam DC directe stantem super lineam rectam AB circumstant duo anguli equales, sc. DCB et DCA. Ergo per diffinitionem perpendicularis ipsa est perpendicularis, et hoc est quod demonstrare voluimus."

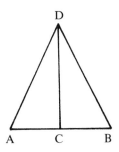

and state that it is perpendicular." Albert does not say explicitly, "Divide the angle in two equal parts," but it is clear that he means such for he refers to theorem 9 and he uses in the remainder of the demonstration the fact that the two parts of the angle are equal. In fact it is here where he makes his mistake. Why should the bisector of angle ADB arrive at point C? Albert continues: "Draw the line DC (thus the bisector!) and state that this line is perpendicular." This he proves by stating that $\triangle$ ACD and $\triangle$ BCD are congruent, while in both triangles two sides and the contained angle are equal. In this he uses explicitly the datum he should not have used, namely that DC would be a bisector. In fact he argues correctly that in an equilateral triangle the bisector is also a perpendicular, a conclusion which is correct, but which is not asked for. He should have demonstrated that in such a triangle the *median* is also a perpendicular, arguing that the triangles ACD and BCD are congruent because all three sides of one triangle are equal to the corresponding sides of the other.

The second example, which I want to mention only briefly, is proposition 16 of Book I, or rather, a conclusion Albert takes from this theorem. After proving proposition 16 (*that in a triangle an exterior angle is greater than each of the two interior and opposite angles*), he states that he can now give immediately a demonstration for proposition 32 of Book I: *That in a triangle an exterior angle is precisely equal to the sum of the two interior and opposite angles, and consequently the three interior angles of the triangle are equal to two right angles.*[50] This "demonstration" Albert gives without using the parallel postulate — which is impossible. The special figure he uses (an equilateral triangle) is in my opinion the source of his error.

Many more examples could be offered, including in some cases a *petitio principii.*

i. Albert's Use of Earlier Euclid Versions

As has already become clear, Albert's most important source is the

---

[50] *Albert* (fol. 112v): "Mirum autem videtur quod nec Yrinus neque Aganiz neque aliquis aliorum aliquid inveniuntur addidisse huic theoremati cum facile theorema XXXII statim per istud possit concludi" follows the demonstration.

commentary of Anaritius. However, he mentions him only four times as "Anarizus": once in Bk. I, prop. 21 and later on in Bk. IV, prop. 5, three times.[51] In four other cases he refers to him as "Arabs" or "translatio ex Arabico." Besides Anaritius, Albert used several editions of Euclid. At least two of them are discernable, as we have seen above, namely one translation, which in its definitions, postulates, and common notions, is the same as the Adelard versions, and even though its formulation of the propositions is somewhat different, this too is to be situated close to the Adelard tradition. However, this translation which Albert generally followed, he considered to be "ex Graeco." The other is one type of the mixed or mélange versions, which seems to be used only three times by Albert.[52]

In proposition 5 of Book I, Albert explicitly mentions the "commentaries" of Adelard and Boethius.[53] Does this mean that Albert had at hand an edition of Boethius? We have already seen that Albert almost never follows Boethius, with the possible exception of the three alternative wordings of theorems in Book I. The remark in Bk. I, prop. 5, however, gives sufficient evidence that Albert did *not* use Boethius, for he attributes to Boethius a "commentum," and indicates a way in which Boethius would have demonstrated this theorem. Folkerts' reconstruction of Boethius, based on all the available material, shows that Boethius had demonstrations only for the first three theorems of Book I. Thus it is not likely that Albert actually had a translation by Boethius at hand. All we can conclude from Albert's remark is that he knew that Boethius had made a translation, thus indicating that Albert had a fair knowledge of the relevant literature. But the basic question remains, which "commentum" did Albert consider to be the Boethian one and which the Adelardian? One might presume that the version Albert saw as the Greek translation (and which, in fact, is an Adelardian one) is what he meant by the *commentum Boethii*, while the mélange version is what he meant by the *commentum Adelardi*, although one might also argue the other way around.

---

[51] Hofmann says, incorrectly (p. 556): "Die Besprechung von IV 4 taucht schliesslich auch der Name Anarizus auf."

[52] It is not impossible that this could be the version of Boethius, but Boethius does not have most of the demonstrations, but only those for prop. 1-3! It is more probable that it is a mélange version.

[53] Bk. I, prop. 5: "Est autem sciendum hoc theorema ab Euclide non esse demonstratum eo ordine quo partes eius proponuntur. Nam ipse primo probat ultimam partem et postea ex illa probat primam, et simul faciunt commenta Boethii et Adelardi" (fol. 109v).

ii. Other Sources

In this commentary Albert quotes many authors, most of the quotations being taken from Anaritius, except, of course, in the Prologue. However, there are several names mentioned by Albert which are wholly absent in Anaritius, such as the Pythagoreans and Platonists. Aristotle is quoted ten times. These quotations are both of a general nature (especially in Albert's prologue) and specifically mathematical. In five of them Albert refers to specific works of Aristotle, namely *Metaphysics* VI, *De caelo* I, *De anima* I and II, and the *Posterior Analytics*. Four of the other quotations can be traced to Aristotle's *Metaphysics* I and II, *De anima* I, and *Ethics* VI.[54] The definition of a straight line, attributed by Albert to Aristotle, will be considered below.

The author who needs special attention is Alfarabi (880-950),[55] who is quoted four times by Albert:

1. In his introduction, Albert notes that Alfarabi states in his commentary on the theorems of Euclid that there are only three primary species of continuous quantity.[56]
2. Albert notes Alfarabi as the author of a definition of a point.[57]
3. and as the author of a definition of a plane surface.[58]
4. Finally, Albert says that Alfarabi represents Euclid as giving a definition of a trapezoid in his *Liber divisionum*.[59]

We know of only one mathematical work of Alfarabi that must be intended here, namely his "Commentary on the difficulties in the introductions of the First and Fifth Book of Euclid." This work is preserved in a Hebrew translation, of which we have at present only

---

[54] These quotations are found in the works of Aristotle mentioned, but they could have been taken from some *florilegium*, which were plentiful. See one such list, dating from the fourteenth century, edited by J. Hamesse, *Auctoritates Aristotelis* (Louvain-Paris, 1974).

[55] Hofmann, p. 556: "Erwähnt wird Alfarābī, auf den sich Albertus (Magnus) auch in philosophischen Fragen so häufig bezieht."

[56] "Sicut ergo tradit Alfarabius in commento theorematum Euclidis, non sunt nisi tres continue quantitatis species prime, sc. linea, superficies et corpus" (fol. 105r).

[57] "Alfarabius vero sic: punctum est quod non habet dimensionem quantitatis continue habentis situm" (fol. 105v).

[58] "Alfarabius vero planam superficiem generaliter diffiniens dixit: superficies plana est cuius spatium est equale spatio linee que ipsum comprehendit, aut spatio linearum ipsam comprehendentium" (fol. 106r).

[59] "Sed tamen Alfarabius inducit Euclidem in Libro divisionum dicentem quod non nominamus figuras quadrilateras trapezias nisi illas quarum duo latera que sibi oponuntur, fuerunt equidistantia, et alia sicuti eveniunt . . . et alias vocamus similes trapeziis" (fol. 107r).

a Russian translation.[60] Recently an Arabic manuscript of this work was discovered in Madrid, but, as far as I know, this text has not yet been studied.[61] However, none of the four passages mentioned by Albert are to be found in the Hebrew translation. It would seem that if Alfarabi were indeed the author of these passages, they should be found in this work. In this commentary Alfarabi discusses the relation of physical to mathematical reality and gives the definitions of point, line, and the rest, but all of these definitions are the "usual" ones, and not those Albert attributes to Alfarabi.

Let us examine these items more carefully, reserving the first to the end:

2. The definition of a point that Albert attributes to Alfarabi can be found in precisely the same words in *Anaritius*, who gives the definition as his own and as more in conformity with the meaning of Euclid.[62] Paris, Bibl. Nat., MS lat. 7215, a kind of commentary based largely on Anaritius, gives this very definition and attributes it to Anaritius.[63]

3. The definition of a plane surface is also found in Anaritius, given as his own definition.[64]

4. The last remark of Albert concerning Alfarabi claims that "Alfarabi represents Euclid as giving a definition of a trapezoid in his *Liber divisionum* as that quadrilateral figure which. . . ." But, in fact, it is Anaritius who does this very thing, for we read in his commentary: "Euclid is found to be saying in his *Liber divisionum*. . . ."[65]

---

[60] Munich, Staatsbibliothek, Codices Hebr. 290, fol. 2-6, and Cod. Hebr. 36, fol. 17-18. For the Russian trans., see M. F. Bokstein and B. A. Rozenfeld, *Akademiia Nauk, S.S.S.R. Problemy Vostokovedeniya*, 4 (1959), 93-104.

[61] Madrid, Escorial MS Arab. 612, fol. 109-111. Cf. J. E. Murdoch, *Dictionary of Scientific Biography*, 4: 454.

[62] Anaritius: "Sed definitio magis propinqua intentioni est ut dicatur quod punctum est quod non habet dimensionem quantitatis continue habentis situm." See ed. Curtze, p. 31, and above note 57.

[63] Paris, Bibl. Nat., MS lat. 7215, fol. 4r: "Anaricius dicit: punctum est. . . ."

[64] Anaritius: "Quod si quis voluerit reducere hanc definitionem ad hoc ut non solum sit superficierum quas recte comprehendunt linee, sed etiam superficierum rotundarum et mediarum, minuat ex eo parum. Dicat ergo: superficies plana est cuius spatium est equale spatio linee que ipsum comprehendit, aut spatio linearum que ipsum comprehendunt." See ed. Curtze, p. 1, and note 58 above.

[65] "Euclides tamen in Libro divisionum invenitur dixisse quod non nominavimus figuras quadrilateras trapezias nisi illas quarum duo latera que sibi opponuntur, sunt equidistantia et alia duo sicuti eveniunt; alias vero vocamus similes trapeziis" (ed. Curtze, p. 24). See note 59 above. Hofmann (p. 563, n. 26) concludes, incorrectly in my opinion, "Anscheinend gibt es also eine Einleitung Alfarābī zum Euklidschen liber Divisionum."

1. The first item attributed to Alfarabi cannot be found in Anaritius, nor, as I have said, in the Hebrew translation. We find this item in the Prologue of Albert's work, which is not borrowed by him from Anaritius, but compiled by himself, using other sources. It is only here that Albert indicates the title of Alfarabi's work. Further, the item is a rather general remark. That is to say, this item occupies a special place in relation to the other three items. It could be from Alfarabi. But then one would like to know which work it might be. It should be noted that a similar remark is to be found in Aristotle's *Categories* and in the *De caelo* of Albertus Magnus.[66]

It is not improbable as an hypothesis to assume that in Albert's commentary on Euclid, with the possible exception of the Prologue, the name Alfarabius is really to be taken as Anaritius and not Alfarabi. One can only conjecture why Albert should have made such a "mistake." If the manuscript is a copy, it could easily be a misreading of the scribe. If, however, the manuscript is an autograph or if the "mistake" is in the original, then one might suppose that at the beginning of his commentary Albert did not realize that Anaritius was the author of the commentary which he used as his principal source, and that he thought it was a work by Alfarabi.

The latter hypothesis seems likely in view of the following considerations. Although Anaritius is Albert's most important source, the name of "Anarizus" is mentioned only three times (Bk. I, prop. 21 and Bk. IV, prop. 5), as we have said. Further, in the beginning of Albert's commentary and before Bk. I, prop. 21 he mentions "Arabs" twice, where "translation from the Arabic" has been scratched out, and twice more "translation form the Arabic." In these four passages Albert clearly meant Anaritius, as has already been shown above.

My conclusion, then, is that name Alfarabius signifies Anaritius in at least three of the four cases in which the name Alfarabius appears in the text of this commentary on Euclid.

C. DATE AND AUTHORSHIP

It is precarious to say anything definite about the date of this commentary. However, there are certain points that can be stated as a basis for a tentative dating. First, the author used Anaritius' com-

---

[66] Aristotle, *Categ.* 4b20; Albertus Magnus, *De caelo* I, tr.1, c.2 (ed. Colon. 5/1: 3): "Huius autem continui sic in genere accepti sunt species: corpus, superficies, et linea."

mentary, which was translated by Gerard of Cremona in the last half of the twelfth century. Second, Albert used a version of Euclid which is near to the Adelard tradition and one which is a mélange version. But from these two certainties nothing more can be gathered than that our text was written no earlier than the end of the twelfth century. However, in my opinion, there is a passage in the text which offers a more precise date *post quem*, namely a quotation from Aristotle's *Metaphysics*.[67] In this quotation Albert used the word *vespertilio* for "bat." All the Greek-Latin translations of this passage of the *Metaphysics* used *noctua* or *nicticorax* (the little owl or screech owl) in the singular or plural. Only the *translatio nova*, made from the Arabic, has *vespertilio* here. This *translatio nova* is attributed to Michael Scot and became known in the Latin West sometime before 1236.[68] Quotations from other Aristotelian works are of no help in narrowing the date of Albert's commentary on Euclid's *Elements*, not even the single reference to Aristotle's *Ethics*.[69] From these considerations, all one can say is that Albert's commentary seems to have been written after 1235.

It is more difficult to indicate a date "ante quem," since we must argue from the weakest of all testimonies, that of silence. Albert in no way refers or uses the new edition of Euclid made by Campanus of Novara before 1259. In a very short time this edition became the most widely used in the Middle Ages. It must therefore be concluded that it is most unlikely that Albert wrote his commentary much later than 1260. The situation, however, is more complicated, for he seems not to know or use the Euclid versions made by Gerard of Cremona or Hermann of Carinthia.[70] One reason for this may be that these

---

[67] "Tertia vero sic super nos est elevata ut ipse princeps philosophorum dicat quod ad illam <dispositio nost>ri intellectus quemadmodum dispositio oculorum *vespertilionis* ad lumen solis" (fol. 105r). The reference is to Arist., *Metaph.* II, 1 (993b9).

[68] Cf. G. Diem, "Les traductions gréco-latines de la Métaphysique au Moyen Age," *Archiv für Geschichte der Philosophie*, 49 (1967), 22. The fact that the word *vespertilio* also appears in the *Auctoritates Aristotelis* (ed. Hamesse, I.63, p. 120) strengthens the argument, since there it is attributed to Averroës.

[69] In the *Lib. introd.*, written by Michael Scot (d. 1236) there is a quotation from a version of the *Ethics*: "Ars est habitus quidam cum experta ratione veritatis factivus que non fallit" (see AL 26: 1-3 [*Ethica*, fasc. 2: 105] and introduction to fasc. 1: cxxxvii). Thus there is no need to assume that Albert quotes from the translation of Robert Grosseteste (ca. 1243). Cf. *Auctoritates Aristotelis*: "Ars est recta ratio factibilium" (ed. Hamesse, XII.111, p. 240).

[70] E.g., Bk. I, prop. 35 has six "cases" in the version of Hermann and of Gerard. Albert does not mention them. Moreover, Albert's formulation of the theorems in no way resembles that of Hermann or of Gerard.

versions were not widely known toward the middle of the thirteenth century or in Albert's part of the world.

Our tentative conclusion, therefore, is that internal evidence seems to suggest that the commentary of Albert on Euclid's *Geometry* was written between 1235 and 1260. This agrees with the opinion of Geyer, who, starting from the authorship of Albertus Magnus, dated the commentary around 1260.

Up to this point we have simply used the name "Albert" for the author of the text we have been discussing because of the explicit ascription on fol. 105r of our manuscript: *Primus Euclidis cum commento Alberti.* Assuming that Albert is the author and that the work was written sometime between 1235 and 1260, we must try to identify this author. The first Albert that comes to mind is Albertus Magnus. Which other Albert could it be? If another Albert had been intended, some further specifications would normally be expected, as in most other cases. A thirteenth-century attribution of a work to Albert without further qualification most strongly suggests that Albert to be Albertus Magnus. While this argument is not conclusive, it certainly is most probable.

Geyer and Hofmann have both identified our Albert with Albertus Magnus, as we have already said. Geyer regarded the Vienna manuscript to be an autograph, but, in my opinion, he has not given convincing proof. Geyer stated that this manuscript presents the general features of an autograph, but in his article he did not give specific indications.[71] Even the passage in Bk. III, prop. 13, as mentioned above,[72] is not conclusive. Moreover, if the manuscript is an actual autograph of Albertus Magnus, and not some copy, then many paleographical difficulties arise when it is compared to other known autographs, as Geyer himself admitted.[73]

Passing over the question of autograph, let us turn to the identity of the author of the text. Surprisingly, Geyer himself had some doubts about this as well, as one can see in his article[74] and in his

---

[71] Geyer, p. 169: "Der Schreiber des Werkes ist auch sein Verfasser. Es finden sich nämlich alle jene Eigentümlichkeiten, durch die sich die Autographe also solche ausweisen, Verbesserungen die nur während des Schreibens vorgenommen sein können, nachträgliche Veränderungen des Textes, die eine Verbesserung darstellen."

[72] See note 48 above.

[73] Geyer, p. 168: "Prima vista ist nämlich unser Autograph denen Alberts keineswegs ähnlich."

[74] Geyer, p. 169: "Diese Ausführungen . . . sind aber nicht so dass sie deutlich die Autorschaft Alberts verrieten. Auch die Stileightümlichkeiten sprechen nicht eindeutig für Albert."

introduction to the *Metaphysics* paraphrase of Albertus Magnus.[75]

To resolve this question, we should first consider the three passages where Albertus Magnus himself speaks about his *Geometria*:

1. *Metaphysica* V, tr.3, c.1: Albertus Magnus states that two straight lines do not contain a surface, "as we have proved in Book I of *our geometry*."[76] This refers to the fifth postulate of which our Albert does indeed give a demonstration, analogous to that of Anaritius.

2. *Metaphysica* III, tr.2, c.3: Here Albertus Magnus speaks about physical and mathematical objects, and he refers to Bk. III, prop. 15 and 16 *"of our geometry,"* in which he demonstrated that a line touches a circle only at a point.[77] This passage proves without a doubt that Albertus Magnus wrote a treatise in the form of a commentary on Euclid's *Elements*. Our Albert treats this question explicitly in the second definition of Book III,[78] but refers to proposition 15 of the same book and to Aristotle's *De anima*,[79] which is the "classical" reference for this question. Since Albertus Magnus often quotes from memory, there is no contradiction between the *Metaphysics* passage and our text.

3. *Metaphysica* I, tr.2, c.10: Here Albertus Magnus makes a remark about the incommensurability of the diagonal and the side of a square, as "has already been *demonstrated by us* in the books of *geometry*."[80] Immediately one should note that

---

[75] *Prolegomena*, ed. Colon. 16/1: xix (1960): "Quae sit haec Geometria Alberti, adhuc non constat, specialiter utrum opus autographum quod est in Cod. Vind. Dom. 80/45, fol. 105-145 re vera sit Alberti Magni. In codice laudato Alberto cuidam adscribitur et multa cum iis quae in Metaphysica dicuntur, congruunt. . . . Quare adhuc sub iudice lis est."

[76] "Licet superficies nulla intra duas lineas contineatur *sicut nos in I nostrae geometriae* ostendimus" (ed. Colon. 16/1: 256).

[77] ". . . cum tamen sicut in XV et XVI tertii *geometriae nostrae* demonstratum est, linea contingens circulum non nisi secundum punctum contingat ipsum" (ed. Colon. 16/1: 118).

[78] "Cum enim rectum mathematice sumptum tangat tactu mathematico circulum vel speram absque dubio in puncto tangit ut in primo De anima dixit Aristoteles, tactu punctuali non dividitur peryferia circuli" (fol. 130v). Note that Albertus Magnus refers to Bk. III in his *De caelo* (ed. Colon. 5/1: 244), but the context and purpose are different.

[79] Arist., *De anima* I.1 (403a12-16). In his commentary on this passage of *De anima* (ed. Colon. 7/1: 13) St. Albert does not refer to his work on geometry, but see the passage in *De lineis indivisibilibus*, 5 (ed. Borgnet 3: 477): "Cum linea recta circulum non tangat nisi in puncto uni secundum rei veritatem."

[80] "Vir enim geometricus sciens causam non mirabitur si diametrum non fiat commensurabile lateri quadrati. . . . Hoc autem *iam a nobis in geometricis* est demonstratum" (ed. Colon. 16/1: 27).

Albert does not say "in *our* geometry," and that he does not —
in contrast to the above passages — give an exact reference.
This passage, of course, is not found in our text, since Euclid
discusses this incommensurability only in Book X, and our text
goes no further than Book IV. Does this mean, then, that
Albertus Magnus commented on Book X of Euclid? Not necessarily. As already noted, Albertus Magnus here gives no
exact reference to book or proposition. It could refer to other
geometrical works of Albertus Magnus. In fact, he does discuss Book X of Euclid and incommensurability in his authentic *De lineis indivisibilibus*.[81] But the possibility remains that
Albertus Magnus did in fact comment on the whole of
Euclid's *Elements*.

From these considerations we must conclude that there is no reason why the author of our text cannot be identified with Albertus
Magnus. Even if we accept the suggestion implied in the third passage quoted above, namely that Albertus Magnus wrote a commentary on the whole of Euclid's *Elements*, and not merely on four
books, our own text itself suggests that it is incomplete.[82]

A more positive identification can be made from the first passage
from the *Metaphysics* quoted above. As far as I am aware, only
Anaritius and our Albert give a "demonstration" for postulate 5 of
Book I. And Albertus Magnus, in the above reference, claims to have
done so.

For further evidence, one must examine the doctrinal contents of
Albert's commentary. In this respect, the prologue to the commentary is most important, because it is more philosophical than technically mathematical. First, one can point to the many parallel passages in the works of Albertus Magnus, especially the *Metaphysics*
paraphrase, to our prologue, which Geyer has already done in his
article.[83] Second, many points of view expressed in our prologue are
similar to those found in Albertus Magnus, e.g., the relation between
mathematical and physical objects, the way in which a point generates a line and a line a surface, the notion of point as the principle of
a line, and so forth. Further, both the prologue and Albertus Magnus
agree that mathematics concerns forms existing in moveable matter,

---

[81] Ed. Borgnet 3: 465b ff.

[82] Cf. fol. 123v (Book II): "Et de hac agetur infra in quinto."

[83] E.g., *Metaphysica* I.1.1; I.1.6; I.1.10; II.1.2; VI.1.2, and *De caelo* I.1.2.

but conceived without matter in its "ratio diffinitiva"; that imagination plays a crucial role in the constitution of mathematical objects; that mathematics deserves the real name of "science" because it belongs to human imagination, which conceives the "ratio diffinitiva" of mathematical being without the variation of sensible matter.[84] Although major research still needs to be done, there seem to be no major differences in content between our prologue and the known authentic works of Albertus Magnus. In fact, there is every indication of identity.

Finally, there are two very important points found in the definitions of Book I that would seem to confirm our view that the commentary on Euclid's *Elements*, which we have examined, is by Albertus Magnus:

1. The definition of a straight line which our Albert attributes to Aristotle is not found in Aristotle or in Anaritius. In this passage, Albert gives two similar definitions of a straight line. One he attributes to Plato: "a straight line is one whose middle covers its two extremities."[85] The other he attributes to Aristotle: "a straight line is one whose middle does not extend above or below the extremities."[86] Only the first definition, which Albert rightly attributes to Plato, can be found in Anaritius, Aristotle, or pseudo-Aristotle.[87] Remarkably, Albertus Magnus himself gives both definitions. In one work, *De lineis indivisibilibus*, he attributes both definitions to Plato in the exact words quoted above but in his *De predicamentis* he gives only the last definition twice.[88] In *De caelo*, he gives the last definition again in almost the same words.[89]

---

[84] See the article by G. Molland in this volume. Cf. Albertus Magnus, *Physica* VI.1.1 (ed. Borgnet 3: 408) *et passim* here as well as in the *Metaphysics*, and *De lineis indivisibilibus*.

[85] "Plato autem diffinit eam a signo recti dicens linea recta est cuius medium duas ipsius extremitates cooperit" (fol. 105v).

[86] "Quod quare sit, Aristoteles in sua tangit diffinitione dicens: rectam esse cuius medium non exit ab extremis" (fol. 105v).

[87] Anaritius (see ed. Curtze, p. 6): "Plato vero diffinivit rectam lineam dicens: linea recta est cuius medium duas ipsius extremitates cooperit." Cf. Arist., *Topica* VI, 11 (148b29): ". . .cuius medium superadditur finibus." See also T. L. Heath, *The Thirteen Books of Euclid's Elements* (New York: Dover, 1956), 1: 165.

[88] Albertus Magnus, *De lineis indivisibilibus*, 3 (ed. Borgnet 3: 469): "Recta autem linea secundum Platonem est cuius medium non exit ab extremis, vel cuius medium extrema cooperiunt." *De predicamentis*, tr.3, c.7 (ed. Borgnet 1: 207): "linea est longitudo cuius medium non exit ab extremis." Ibid., tr.4, c.5 (ed. Borgnet 1: 230): "Rectum ut diffinit Plato, cuius medium ab extremis non exit."

[89] Albertus Magnus, *De caelo* I, tr.1, c.10 (ed. Colon. 5/1: 25): "quia rectum non exit ab extremo."

2. Albert says that the Platonists call a surface sometimes *rasura* and sometimes *lamina*, claiming that all bodies are composed of these and resolved into them.[90] Only in Albertus Magnus' *De caelo* do we find both names and the same remark.[91]

3. Perhaps a third point is indicative. In definition 17, Albert mentions the names of three species of triangle, the Greek as well as the Latin. He confuses, however, the meaning of "ysocheles" (isosceles) and "ysopleurus" (equilateral). This same mistake is found in *De lineis indivisibilibus* by Albertus Magnus.[92]

To sum up our study we can state our conclusions briefly. Internal evidence indicates that the commentary on Euclid in the Vienna manuscript was originally composed between 1235 and 1260. That manuscript gives "Albert" as the author. Given the parallel passages and important correspondences between our commentary and the authentic works of Albertus Magnus, we cannot but take as a well-grounded hypothesis that Albertus Magnus is the author of that work. Evaluating the commentary itself, we can say that it throws new light upon the role of mathematics in the view of Albertus Magnus and on his estimation of that science. While it is true that the mathematical level of the commentary is not as high as one would expect, it is also true that its author had a fair knowledge of the relevant literature, and that he tried to rework many of his sources critically, and to make Euclid "intelligible" to his readers.

In the last analysis, even if one rejects our hypothesis that Albertus Magnus is the author of this commentary, it is still an important work. It is one of the earliest, if not the earliest original Latin commentary on Euclid's *Elements*, contemporary with Albertus Magnus. It at least gives us a fuller picture of the state of mathematics in that period.

---

[90] "In omni autem corpore sensibiliter apparet superficies, et ipsa est hoc quod de corpore videtur. Propter quod Greci Pythagorici eam epyfaniam vocaverunt, Platonici autem quandoque rasuram, quandoque autem laminam, ex superficiebus dicentes corpora esse composita et in hec resolvi, sed de his determinare est philosophi primi" (fol. 106r). The term "lamina" is to be found in Aristotle's *De caelo* III, 1 (299a1-3), as translated by Gerard of Cremona. For the text, see Albertus Magnus, *De caelo* (ed. Colon. 5/1: 203 line 77 and 204 line 66).

[91] Albertus Magnus, *De caelo* III, tr.1, c.2 (ed. Colon. 5/1: 204): "Hi autem qui omnia corpora generata esse dixerunt, corpus quidem universaliter ex superficiebus per modum laminarum et rasurarum corpus componentium componi dixerunt et ad superficies resolvi."

[92] See *De lineis indivisibilibus* (ed. Borgnet 3: 471): "Isopleurus, hoc est qui duo aequa habet latera." Albert has the same mistake again in Bk. II, prop. 9 (fol. 127v).

20

Albert's Influence on Late Medieval Psychology

Katharine Park
Harvard University

Although the word *psychologia* seems to have been a creation of the sixteenth century,[1] the discipline itself — the systematic study of the soul — was an established part of classical and medieval philosophy. Albert's description of it is both typical and influential for scholastic writers. He identifies the soul as that which differentiates living from non-living beings and accounts for the functions which distinguish the former from the latter and among each other. An immaterial principle, it is nonetheless closely related to the body, since all of its operations depend ultimately on the body's members — perception on the sense organs, for example, and reproduction on the genitals. Psychology, along with disciplines like physics and biology, belongs to "natural science," the study of *mobilia* or movable things. Because part of its object is incorporeal, human psychology occupies a special place among these sciences, as both the noblest and the most certain.[2] Situated on the intersection of the material and spiritual realms, it touches physiology and medicine on one side and metaphysics, ethics, and theology on the other. As a result, it is an excellent window through which to view general developments in medieval science and philosophy.

---

[1] François H. Lapointe, "Who originated the term 'Psychology'?" *Journal of the History of the Behavioral Sciences*, 8 (1972), 328-333.

[2] Albert the Great, *De anima* I.1.1-2 (ed. Colon. 7/1: 1a-3b).

Albert laid the foundations of much of later medieval psychology. As Martin Grabmann writes, Albert's great contribution was to bring Aristotelian thought into scholastic philosophy, establish it within university education, and disseminate it through commentaries and paraphrases of Aristotle, as well as through his own monographs.[3] Albert's Aristotelianism was based on the writings of Aristotle and his Islamic commentators, all of which had recently been translated from the Arabic. This is particularly true in psychology; the effect of Albert's work was to detach it from the previous Augustinian and encyclopedic tradition — largely theological in orientation — and to establish it as an Aristotelian discipline with Arabic sources.[4]

By 1252 Aristotle's De anima was required for the licentiate at Paris, and the systematic study of the soul had entered the arts curriculum.[5] The sources for the history of medieval academic psychology are the commentaries, questions, and occasional monographs which grew out of this study of De anima. Most are still unpublished and unstudied; our knowledge of their contents, and in many cases of the philosophical environment they reflect and express, is sketchy at best. For the purposes of this paper, I have chosen to concentrate on published sources and on the works of a relatively small number of well-known and representative medieval philosophers associated with the universities of northern Europe in the fourteenth and fifteenth centuries.[6] My conclusions are therefore tentative. Nonetheless, I think it is possible to identify the most impor-

---

[3] Grabmann's studies, although only preliminary, are still the fundamental guide to Albert's later influence. See "Der Einfluss Alberts des Grossen auf das mitterlalterliche Geistesleben," in his Mittelalterliches Geistesleben (Munich, 1936), 2: 324-412; and "Die Aristoteleskommentar des Heinrich von Brüssel und der Einfluss Alberts des Grossen auf die mittelalterliche Aristoteleserklärung," Sitzungsberichte der bayerische Akademie der Wissenschaften, phil.-hist. Abteilung, 10 (Munich, 1944).

[4] Albert was not the only Latin philosopher to whom this applies. A number of his slightly earlier contemporaries, like Jean de la Rochelle, John Blund, Alexander of Hales, and William of Auvergne, worked along these lines; see Ernest A. Moody, "William of Auvergne and his Treatise De anima," in his Studies in Mediaeval Philosophy, Science and Logic (Berkeley, 1975), p. 17. Albert continued the efforts of these writers and in many cases drew on their work. His theory received much wider dissemination, however, and had an incomparably greater influence on later psychology.

[5] Lynn Thorndike, University Records and Life in the Middle Ages (New York, 1944), pp. 54 and 65.

[6] The psychological literature produced in Italian universities in the later Middle Ages is equally rich and reflects many of the same developments; see the contribution by Edward P. Mahoney, next in this volume.

tant developments in later medieval psychology and relate them to changes in the nature and extent of Albert's influence.

From this point of view, I hope to show that there are two stages in the study of the soul after 1300. In the first, which occupies the first half of the fourteenth century, Albert's ideas were used in the critical interpretation of Aristotle at Paris by some of the best minds in medieval philosophy. In the second, which extends to ca. 1520, psychological theory fragmented into a number of rival approaches grouped under the two main rubrics of *via antiqua* and *via moderna*. This tendency is especially marked in Germany. For the self-styled "Albertists," who in several universities represented a sizeable fraction of the *antiqui*, Albert's teaching on the soul became a dogma of absolute authority. At the same time, while defending their master against all criticism, the Albertists developed his psychology in ways that he would certainly have rejected.

A. Albert's Psychological Theory

Albert wrote four major works on the soul: a commentary on *De anima; Summa de homine* (Book ii of his *Summa de creaturis*); *De natura et origine animae*; and *De intellectu et intelligibili*. The first two were the most important for later psychology.[7] In these, as in Albert's other writings on the subject, the most frequently cited philosopher, apart from Aristotle, was Avicenna, whose *De anima seu Sextus de naturalibus* was Albert's principal source. Albert depends on Avicenna for many of his particular doctrines and for much of his method.[8]

Concerning his method, Albert distinguishes two approaches to the study of the soul.

On this subject, Avicenna says in *Sextus de naturalibus* that there are

---

[7] While the *Summa de homine* was written somewhat earlier than *De anima*, later medieval writers were unconcerned with the evolution of Albert's thought and treated it as a consistent whole. On the evolution of Albert's psychology, see Bruno de Solages, "La cohérence de la métaphysique de l'âme d'Albert le Grand," in *Mélanges offerts au R. P. Ferdinand Cavallera* (Toulouse, 1948), pp. 367-400. See especially the contribution of L. Dewan to this volume, above pp. 291ff.

[8] See Arthur Schneider, *Die Psychologie Alberts des Grossen* in *Beiträge* 4/5-6 (Münster, 1903-1906). Schneider's book, which is still the standard and most comprehensive survey of Albert's teaching on the soul, is organized according to Albert's sources. Also helpful is Etienne Gilson, "Les sources gréco-arabes de l'augustinisme avicennisant," *Archives d'histoire doctrinale et littéraire du moyen âge* 4 (1929), 5-158.

two ways of defining a sailor: in one he is considered in himself and is called a worker governing a boat by skill; in the other he executes his functions through the instruments of the boat, namely the yard, mast, sail, and oars. In the same way, the soul has two definitions: one according to which it performs the operations of life in the body and its organs. The other is given of the soul in itself and as it is separable from the body.[9]

Thus, says Albert, one may study the soul *a priori*, in itself, or *a posteriori*, as it performs its various operations in and through the body.[10] Considered in the first sense, the human soul is in its essence a separable spiritual substance which differs from the angels only in its affinity for the human body. Because it is, in principle, independent of matter, it should be thought of as the body's motor, perfection, or act, not as its form.[11] It is joined to the body only incidentally by the lower functions of nutrition and sensation.

From this point of view, the human soul is not simple: it contains a lower part which performs the corporeal functions, and a higher which performs these and the intellectual ones as well. The two enter the foetus at different times and in different ways. The lower vegetative and sensitive faculties are drawn out from the matter of the embryo by the formative power (*virtus formativa*) in the seed of the parents. They remain incomplete, however, until God, the First Cause, illuminates the composite by infusing into it from outside the higher part, individually created.[12] As soon as this happens, the "intrinsic" lower parts are transformed by their association with the "extrinsic" higher ones, and the product is a substantial whole —

---

[9] Albert, *Summa theologia* II.12.69.2 (ed. Borgnet 33: 15b-16a): "Quod tractans Avicenna in VI *de naturalibus* dicit, quod sicut nauta duplicem habet definitionem: unam secundum quam consideratur in seipso, secundum quam dicitur artifex arte regens navim: aliam secundum quam operationes nauticas operatur instrumentis navis, antenna (text: artemone) scilicet, malo, velo, remis: ita anima duplicem habere definitionem: unam secundum quod operatur opera vite in corpore et in organis eius . . . . Alia definitio est, quae datur de anima secundum se, et secundum quod separabilis est a corpore . . . ." Cf. Avicenna, *De anima seu Sextus de naturalibus* I.1, ed. by S. Van Riet (Louvain and Leiden, 1968-72), 1: 27-37. For a survey of Avicenna's teaching on the soul, see G. Verbeke, "Introduction sur la doctrine psychologique d'Avicenne," in the same edition, 1: 1*-90* and 2: 1*-72*.

[10] *Summa de homine* I.1.1 (ed. Borgnet 35: 3b).

[11] Ibid., I.4.1 (ed. Borgnet 35: 34a-35a). The clearest treatment of this matter is in Anton Charles Pegis, *Saint Thomas and the Problem of the Soul in the Thirteenth Century* (Toronto, 1934), ch. 3.

[12] Albert, *De natura et origine animae* I.5 (ed. Colon. 12: 14a). See Bruno Nardi, "L'origine dell'anima umana secondo Dante," in his *Studi di filosofia medievale* (Rome, 1960), pp. 24-33; and "La dottrina d'Alberto Magno sull'"inchoatio formae'," in the same, pp. 75-95.

entirely human, rather than vegetable or animal, in nature. In more technical terms, "the powers are faculties which follow the constitutive species."[13] For example, although the faculty of vision is the same in a man and an ass with respect to object and operation, nonetheless in the man it belongs to a different species — the human — because it is completed and "denominated" by the intellectual soul.[14]

As a result, the intellectual soul becomes united to the body, depending on the lower parts, such as sensation, for all its operations. This is why, in Aristotle's maxim, "the soul never thinks without an image."[15] Despite its association with the body, however, the intellectual soul remains fundamentally separable from it and closely allied with the other spiritual substances: with the angels and God himself. Therefore, while relying on sense images for its initial stimulation, the intellect may proceed beyond them to contemplate first itself, then the celestial intelligences, and finally God. Albert calls this intellectual state "assimilative":

> the assimilative intellect is that in which man, as much as is possible or permitted, springs up analogically toward the divine intellect, which is the light and cause of all things. . . . Therefore from the light of its own agent intellect, it reaches the light of the intelligence, and from that extends itself toward the intellect of God.[16]

Thus the soul ascends to virtue, wisdom, and prophetic powers.

In his *a priori* discussion of the soul as substance, Albert depends heavily, as many scholars have shown, on the neoplatonic elements in Avicenna's version of Aristotle; using them, he can integrate Aristotle's psychology into the Augustinian tradition of the separable and immortal soul. In his *a posteriori* inquiry into the soul as revealed through its various operations, he takes as the center of his analysis Avicenna's transformation of the Aristotelian notion of faculty.

Aristotle considered the faculties as potentialities of the soul for

---

[13] Albert, *De homine* I.6 (ed. Borgnet 35: 88a): "vires vero sunt potentiae quae species constitutas sequuntur . . . ."

[14] Albert, *De natura* I.6 (ed. Colon. 12: 14b-15a); see also *De anima* II.2.15 (ed. Colon. 7/1: 59b).

[15] Aristotle, *De anima* III.7 (431a16-431a17).

[16] Albert, *De intellectu et intelligibili* II.9 (ed. Borgnet 9: 516a). "Est autem intellectus assimilativus, in quo homo quantum possibile sive fas est proportionabiliter surgit ad intellectum divinum, qui est lumen et causa omnium . . . . Devenit ergo ex lumine sui agentis in lumen intelligentiae, et ex illo extendit se ad intellectum Dei."

different kinds of action and used them mainly as convenient categories to classify different levels of living things.[17] Albert, following Avicenna, visualizes them as really existing and distinct powers which possess, in some sense, continuous actuality. For all practical purposes, the study of the soul *a posteriori* reduces to the study of the faculties as powers of the soul. In Albert's image, the soul is divided into diverse powers as a potestative whole. Like the organs of the body, the faculties are separate from each other, but mutually dependent and arranged in a hierarchy of nobility and command which resembles the chain of authority in a well-ordered monarchy.[18]

The element of order is central for Albert. He takes the division of the soul into faculties as the organizing principle of his psychological works. In *De anima* and *Summa de homine*, after introductory sections which define the soul, each tractate or question is devoted to a different power or group of powers.[19] These are of great importance in the history of medieval psychology. At the time Albert wrote the *Summa de homine*, his first extended discussion of the soul, Latin psychological theory was in chaos. The powers were acknowledged as central to any account of the soul, but there was no consensus as to what they were and how they should be divided, since the various Greek, Arabic, and Christian authorities had all proposed different models. Most earlier Latin writers, like Jean de la Rochelle, were content to give several different classifications without attempting to reconcile them.[20]

Albert transforms the situation with his *Summa de homine*. He replaces the chaos of authorities and opinions with a coherent system based on Avicenna and incorporates elements from earlier Latin writers. In the first place, he rationalizes the enumeration of the powers of the soul by establishing a single system derived from Avicenna's interpretation of Aristotle and by relegating the faculties according to Augustine and Lombard to an appendix as motive pow-

---

[17] See David Hamlyn, *Sensation and Perception* (London, 1961), pp. 17-18.

[18] Albert, *De anima* II.1.11 (ed. Colon. 7/1: 80a); Albert's simile is taken from Boethius' *De divisione* (PL 64: 887-888), but it appears in other writers of the twelfth and thirteenth centuries.

[19] See the scheme of the faculties in Schneider, pp. 547-548; note the similarities to Avicenna's scheme as diagrammed by Gilson, p. 62. The organization is of course based on Aristotle, but Aristotle uses the powers far less explicitly as an organizing principle, and his divisions are not nearly as clear.

[20] Jean de la Rochelle, *Tractatus de divisione multiplici potentiarum animae*, ed. by Pierre Michaud-Quantin (Paris, 1964), pp. 69-136.

ers "according to the Platonists and theologians."[21] The result is a complex but coherent scheme of faculties divided into four main groups — vegetative, sensitive, motive, and intellectual. These correspond to three types of soul: vegetative, sensitive (or animal), and intellectual (or human). Albert's divisions and subdivisions, with minor variations, remain standard in Latin psychological theory through the end of the fifteenth century, and they persist in many authors well past 1600.

In the second place, Albert establishes the general philosophical terms in which the powers will be discussed throughout the later Middle Ages. As actual operative principles, the powers demanded a much more rigorous and systematic discussion than was found either in earlier literature or in Aristotle himself. Albert is the first adequately to provide such a discussion, and he does so by drawing on both Avicenna and earlier Latin sources, notably Boethius. Applying Aristotelian logical and philosophical principles, he develops a coherent explanation of the faculties which addresses the issue of their ontological status.

For Albert, the central questions are the following: In what sense can the soul be said to be composed of parts? What are the logical and ontological relations between the soul as a single substance and its multiple powers and sub-powers? He takes his answer, as he takes his image of the soul as hierarchy of authority, from Boethius' *De divisione*. Although physical objects may be divided into essential or integral parts, a spiritual entity like the soul has only "potestative" parts — natural powers — which flow from it; it must be considered as a whole composed of powers, or what Boethius and Albert call a "potential whole" (*totum potestativum, potentiale*, or *virtuale*).[22] If the faculties of the soul are natural powers, then they lie, according to Aristotle's *Categories*, in the category of quality.[23]

Albert uses Boethius not only to establish the logical status of the faculties, but also to answer a question which had plagued psychological theory for more than a century: Is the soul identical with its faculties? The Augustinian tradition, dominant through the end of

---

[21] The problem of the faculties has been treated in two articles by Pierre Michaud-Quantin, "La classification des puissances de l'âme au douzième siècle," *Revue du moyen âge latin* 5 (1949), 15-34; and "Albert le Grand et les puissances de l'âme," in the same, 11 (1955), 59-86. For Albert's special contribution see "Albert le Grand," especially pp. 79-85.

[22] Albert, *De anima* II.1.11 (ed. Colon. 7/1: 80a). Cf. Boethius, *De divisione* (PL 64: 887-888).

[23] Albert, *De homine* I.8.2 (ed. Borgnet 35: 106a). Aristotle, *Categories* VIII (9a19-9a28).

the twelfth century, had argued that the distinction between the powers was only verbal; the different powers were in fact various names given to the soul as it performed various actions, but in essence identical to the soul. With the new translations, it became clear that both Avicenna and Averroës accepted a real distinction between the soul and its powers and that Latin psychologists would have somehow to accommodate this position.[24] Albert is the first to do so in a satisfactory manner. Rejecting the strained compromises of his earlier contemporaries, he demonstrates that, if the soul is truly a *totum potestativum*, it is a substance, while the powers are qualities which function as its powers. Logic thus demands a real distinction between them.[25]

Albert's significance for the history of medieval Latin psychology mirrors his significance for medieval science in general. His influence extended beyond that of a generalized Aristotelianism, however; many specific aspects of his thought on the soul entered the tradition of medieval psychology. He took over two particular strains of Avicenna's theory: the Platonic strain which emphasized that nature of the soul as a separable substance — the "perfection" rather than the form of the body — and the scholastic strain which manifested itself in the elaborate hierarchical subdivisions of the faculties. Using concepts from the logical writings of Aristotle and Boethius, he developed a clear and reasonably consistent explanation of the way in which the faculties could be really distinct from the soul and from each other, but still of the same essence and substance.[26] Later Latin writers often reject or alter Albert's conclusion, but they remain interested in Albert's questions, asked in his own terms.

The psychological theory of Thomas Aquinas, once a student of Albert, is both an index and a vehicle of his master's influence on late medieval psychology. Thomas' principal philosophical concerns are different from Albert's: for him metaphysics and theology replace natural philosophy and physiology as the center of attention. Nonetheless, his thought on the soul and its faculties clearly reflects

---

[24] For an exhaustive examination of the history of this problem, see Pius Künzle, *Das Verhältnis der Seele zu ihren Potenzen: Problemgeschichtliche Untersuchungen von Augustin bis und mit Thomas von Aquin*, Studia Friburgensia, n.s., 12 (Freiburg, 1956). Künzle treats the twelfth-century authors on pp. 43-96, and Albert's immediate predecessors on pp. 97-144.

[25] Albert, *De homine* I.73.2.2.2 (ed. Borgnet 35: 616b). See Künzle, pp. 144-158.

[26] From this point of view, Schneider's characterization of Albert's psychology as ill-digested eclecticism is clearly overstated.

that of his early teacher. On the one hand, he is very similar to Albert in his account of the faculties. He describes them as composing a hierarchy of authority like that in a monarchy,[27] and his list of them in the *Summa theologiae* follows Albert's quite closely.[28] The soul is a virtual whole and its powers are its natural properties — accidents flowing from its essence — and therefore really different from the soul itself. Thomas elaborates on Albert's conclusions by noting that the faculties, as natural powers, must lie in the second species of quality.[29]

On the other hand, Thomas explicitly rejects the *a priori* discussion of the soul as separable substance that Albert took over from Avicenna and other Arabic sources. For Thomas, the soul is first and foremost the substantial form of the body. This leads him to reject a number of Albert's other claims. In the first place, he denies that the developing foetus derives its vegetative and sensitive powers from the formative power of the semen and is only later perfected by the infusion of a rational soul. This would mean, he argues, that

> the substantial form would be continuously perfected. It would further follow that the substantial form would be drawn not all at once, but progressively from potency into act, and further that generation would be a continuous motion, like alteration. All of these things are naturally impossible.[30]

What really happens, according to Thomas, is that the embryo receives a succession of increasingly perfect forms: it is first animated by a purely vegetative soul. At a certain point, this is wholly corrupted and replaced by a sensitive soul and, later, through God's direct creation, by a rational soul.

By the same token, Thomas rejects Albert's apparent contention that in this life the human intellect, as a separable spiritual substance, can know itself directly or the other separable substances —

---

[27] Thomas Aquinas, *Questio disputata de spiritualibus creaturis* 11, ad 19, in *Opera omnia* (Parma, 1852-1873), 8: 464b.

[28] Thomas, *Summa theologiae* 1.78-79, in *Opera* 1: 297b-308a; the only major differences are that Thomas reduces the internal senses from five to four and eliminates several motive and intellectual powers as redundant.

[29] Ibid., 1.77.1,5 and 6, in *Opera* 1: 297b-302a; and I *Sent.* 3.4.2, in *Opera* 6: 40a-41a.

[30] Thomas, *Summa contra gentiles* II.89, in *Opera* 5: 147b: "forma substantialis continue magis ac magis perficeretur; et ulterius sequeretur quod non simul, sed successive, educeretur forma substantialis de potentia in actum, et ulterius quod generatio esset motus continuus, sicut et alteratio; quae omnia sunt impossibilia in natura." See Nardi, "Alberto Magno e San Tommaso," in *Filosofia medievale*, especially p. 107.

God and the celestial intelligences. As the substantial form of the body, the human soul is bound inextricably to corporeal modes of cognition. To know immaterial reality it must rely on what it can abstract from sense images, and as a result may understand this reality only reflectively and by analogy. Albert's "assimilative" intellect does not exist: "according to the state of present life, neither by the possible nor by the agent intellect can we understand the separate immaterial substances in themselves."[31]

While in actual fact both Albert and Thomas insist that the proper object of the human intellect is the essence of material things and that everything above man can be known only *by analogy* to what is proper to man, Albert seems to stress the self-sufficiency of the human intellect in self-knowledge more than does Thomas. But even Thomas admitted with Albert that the human intellect can through discourse know itself as an intellectual substance.

These differences were later exaggerated by opposing camps of Albertists and Thomists, and they became central to the debates over Albert's authority in fifteenth-century discussions of the soul.

B. Albert's Influence in the Fourteenth Century: the University of Paris

As Grabmann has shown, Albert's later influence in northern Europe was not confined to the universities. In many of the German town schools, for example, the principal textbook was the *Philosophia pauperum*, a compilation of his natural philosophy probably made by one of his students.[32] Although Albert's psychology was widely disseminated through this book and the associated commentaries and epitomes, it was not developed in ways that were particularly interesting or sophisticated. For that we must turn to the arts faculties, particularly at the university of Paris.

---

[31] Thomas, *Summa theologiae* I.88.1, in *Opera* 1: 351a: "Secundum statum praesentis vitae, neque per intellectum possibilem neque per intellectum agentem possumus intelligere substantias separatas immateriales secundum seipsas." See also Albert, *Liber de causis* II, tr.1, cc.24-25 (ed. Borgnet 10: 474b-476b), and many other places.

[32] Grabmann, *Die Philosophia pauperum und ihr Verfasser Albert von Orlamünde*, in *Beiträge* 20/1 (Münster, 1918), pp. 29-46. On the problem of authorship, see Bernhard Geyer, *Die Albert dem Grossen zugeschriebene Summa naturalium (Philosophia pauperum)*, in *Beiträge* 35/1 (Münster, 1938), pp. 1-3 and 42-47; Geyer includes an edition of both recensions of the fifth book on the soul. Other titles of the work include *Compendium totius philosophiae naturalis* and *Isagogae in libros Aristotelis Physicorum, De caelo et mundo, De generatione et corruptione, Meteororum, et De anima.*

The chief concern of fourteenth-century natural philosophy was the explanation and interpretation of Aristotle's works on the subject. In this endeavor, the three principal guides were Avicenna, Averroës, and Albert. Albert thus appears as the single most important Latin authority — a position Roger Bacon complains he occupied even during his lifetime.[33] The nature and extent of his influence do not remain constant, however. In order to understand the various ways in which his teachings were used by Parisian philosophers in the fourteenth century, it is necessary to consider two major philosophical developments of the period: the ascendance of the general authority of Averroës and the emergence of the cluster of ideas and methods which came to be known as the *via moderna*.

As Van Steenberghen has shown, there was no generalized Averroism in later thirteenth-century philosophy. Rather, certain writers, of whom the most familiar is Siger de Brabant, adopted certain positions held by Averroës without accepting him as their only or their principal source for the interpretation of Aristotle.[34] In psychology, Thomas and Bonaventure identified as *Averroistae* those who, like Siger, argued that the human intellect was unique, separate and central. Albert had attacked this position;[35] nonetheless, as Nardi has argued, his teaching concerning the soul is in many respects compatible with it — particularly in the general neoplatonic character of his notion of the intellectual soul as separable substance. Tacitly or explictly, his arguments concerning the introduction of the human soul into the embryo and the ability of the human soul to know the other spiritual substances seem to have influenced Siger and others.[36]

The years around 1400 saw the emergence of a strain of thought which can be more accurately described as "Averroist" and which is embodied in the writings of men like Jean de Jandun, who described himself as "the most faithful imitator of Aristotle and Averroës."[37]

---

[33] William Wallace, "Saint Albertus Magnus," DSB 1: 99.

[34] See Fernand Van Steenberghen, *La philosophie au treizième siècle*, Philosophes médiévaux 9 (Louvain/Paris, 1966), pp. 391-400.

[35] For some of the literature on this topic, see D. Salman, "Albert le Grand et l'averroisme latin," *Revue des sciences philosophiques et théologiques* 24 (1935), 38-59.

[36] Nardi, "La posizione di Alberto Magno di fronte all'Averroismo," in *Filosofia medievale*, pp. 119-150; and "L'anima umana secondo Sigieri," in the same, pp. 151-161.

[37] Jean de Jandun, *Questiones super tres libros de anima* (Venice, 1519), proemium, f. 2ra: "fidelissimi Aristotelis et Averrois imitatoris." See Zdzisław Kuksewicz, *De Siger de Brabant à Jacques de Plaisance: La théorie de l'intellect chez les averroïstes latins des treizième et quatorzième siècles* (Wrocław, 1968), pp. 12-15.

The psychological theory associated with this school of thought, as analyzed by MacClintock in the works of Jandun, seems to have been conciliatory and philosophically conservative. It drew heavily on the idea of the intellectual soul as separable substance as it was developed on the one hand by Siger, and on the other by thirteenth-century Franciscans like Bonaventure, Bacon, Pecham, and Olivi, who, refusing to interpret the intellectual soul as substantial form, were forced to maintain some kind of plurality of forms.[38]

Two of the most typical and influential members of this school were Jandun and Walter Burley, both active at Paris during the second and third decades of the fourteenth century. Both acknowledged Averroës as their principal authority in psychology while drawing on Albert's works and on the doctrine of Siger in his *Tractatus de anima intellectiva*.[39] In this work, Siger had developed his psychological theory in a way strongly reminiscent of Albert's teaching on the different origins of the higher and lower souls. According to Siger, only the intellectual soul comes *ab extrinseco*, and even then it functions not as a substantial form of the body but as an intrinsic operating principle (*intrinsecum operans*) united to the composite of human body and lower souls only in its operations and not in its being.[40] It may therefore, as Averroës claimed in *De anima* III, be one in all

---

[38] Stuart MacClintock, *Perversity and Error: Studies on the "Averroist" John of Jandun*, Indiana University Publications, Humanities Series 37 (Bloomington, 1956), pp. 51-55. MacClintock's description of the various elements — Franciscan, Averroist, Avicennian — in this synthesis is extremely useful. His designation of it as "Augustinian" is misleading, however, since most of the authors he discusses make no reference to Augustine and identify themselves strongly with Averroës. See Arrigo Pacchi, "Note sul commento al *De anima* di Giovanni di Jandun I," *Rivista critica di storia della filosofia* 13 (1958), 373-374 n. 2; and "Note sul commento al *De anima* di Giovanni di Jandun II, III," in the same, 14 (1959), 456-457.

[39] Siger's doctrine of the human soul changed in response to criticisms by Aquinas; the *Tractatus de anima intellectiva* represents an intermediate stage in his philosophical development. See Van Steenberghen, pp. 430-456; and Edward P. Mahoney, "Saint Thomas and Siger of Brabant Revisited." *The Review of Metaphysics* 27 (1973-1974), 531-553. Siger's ideas in the *Tractatus* may have been transmitted through Thomas Wylton's *Questio de anima intellectiva*. See Anneliese Maier, "Ein unbeachteter 'Averroist' des 14. Jahrhunderts: Walter Burley," in her *Ausgehendes Mittelalter*, 1 (Rome, 1964): 119-120; Konstanty Michalski, "La lutte pour l'âme à Oxford et à Paris au quatorzième siècle," *Proceedings of the Seventh International Congress of Philosophy*, ed. by Gilbert Ryle (London, 1931), p. 508; and Władysław Senko, "Jean de Jandun et Thomas Wilton," *Bulletin de la Société internationale pour l'étude de la philosophie médiévale* 5 (1963), 139-143.

[40] Siger de Brabant, *Tractatus de anima intellectiva* 3, in Bernardo Bazan, ed., *Siger de Brabant: Quaestiones in tertium de anima, De anima intellectiva, De aeternitate mundi*, Philosophes médiévaux 13 (Louvain/Paris, 1972), pp. 81-85. See Mahoney, pp. 540-544. Cf. Albert, *De natura* I.5 (ed. Colon. 12: 14a).

men; human beings are individualized only by their bodies and lower souls, which are corporeal and therefore mortal.[41]

In his *Questiones super tres libros de anima*, probably written between 1315 and 1318, Jandun adopts Siger's solution, which stems ultimately from Albert.[42] Early in the second book, he notes that "there are two types of form. Some confer both being and operation. The vegetative and sensitive soul is of this kind, and in truth such a form is corrupted [with its body]. But other forms, like the intellectual soul, confer only operation, as the Commentator says in *De anima* III."[43] This kind of form, of course, is immortal and eternal. Later, citing Albert and Averroës, Jandun denies that the intellectual soul is the substantial form of the body in the first sense — a position he ascribes to Alexander of Aphrodisias and elsewhere to Aquinas. Citing Siger, he identifies it as an *operans intrinsecum* like the celestial intelligences.[44] But by appealing to the Franciscan doctrine of the plurality of forms, he can retain the sensitive soul as the individuating material form of the body and still keep the intellectual soul as form in the second sense — a compromise that allows him to avoid one of the errors condemned by the Council of Vienne.[45]

Although Jandun develops this position with reference to Albert and adopts some of the ramifications, like the ability of the soul to comprehend the separate substances,[46] it is important to note that his position on the intellectual soul differs fundamentally from Albert's. A true "Averroist" in Thomas' sense, he maintains against Albert's objections that Aristotle and reason demand the intellect to be one for all men.[47] There are other areas in which he follows Albert much more closely and explicitly. One of these is his account of the powers of the soul, which appears in Questions 7 through 10 of Book II.

Like Albert, Jandun teaches that the soul is composed of a hierarchy of powers and sub-powers, the higher "nobler," "more per-

---

[41] Averroës, *Commentarium magnum in Aristotelis de anima libros* III. com.5 (429a21-429a24), ed. by F. Stuart Crawford (Cambridge, Mass., 1953), especially p. 403.

[42] MacClintock gives a survey of Jean's life on pp. 4-7 and discusses his works and their dates on pp. 103-129. For a longer biographical study, see Ludwig Schmugge, *Johannes von Jandun (1285/89-1328)* (Stuttgart, 1966), pp. 1-38.

[43] Jean de Jandun, f. 16ra: "Duplex est forma, quia quedam est forma dans esse et dans operari, cuiusmodi est anima vegetativa et sensitiva. Et de tali verum est quod corrumpitur. Sed alia est forma dans solum operari, sicut anima intellectiva, ut dicit commentator 3° huius."

[44] Ibid., III.5; f. 51va-51vb. See Pacchi, "Note II, III," pp. 436-441.

[45] See MacClintock, p. 159 n. 36; and Van Steenberghen, p. 497.

[46] Jean de Jandun, III.37; ff. 85rb-88vb.

[47] Ibid., III.7; ff. 54rb-56vb. See Pacchi, "Note II, III," pp. 441-451.

fect," and "prior in dignity." The order of the powers and their names are clearly taken from Albert.[48] Similarly, Jandun defends Albert's position when he argues that the powers of the soul, taken as "immediate principles of operation," are essentially different from the soul itself, since the latter is in the category of substance, while the former, as natural potencies, are in the second species of quality. It would be ridiculous and absurd, Jandun notes, if vision were essentially identical with a lower, non-cognitive power like nutrition.[49] In fact, he argues, the soul must be considered as a "potestative whole" (totalitas potestativa).[50] He later claims that "the powers of the soul flow from its essence; Albert especially used this way of speaking."[51]

There are a number of other passages in which Jandun follows Albert on specific points concerning the faculties. Discussing nutrition, for example, he copies the list of differences between generation and growth given in Albert's De nutrimento et nutribili.[52] Again, in his questions on the internal senses, he notes that Albert distinguished the internal apprehensive powers from the external and located the former in the brain. "We commonly call internal those senses which are not affected by external things except through the other senses. They are five according to Albert, namely common sense, imagination, cogitation, phantasy, and memory."[53] Jandun's treatment of this issue is particularly interesting because it is one on which Albert clearly disagrees with Averroës, who acknowledges only four internal senses. Jandun concludes that his two authorities may be reconciled if cogitation and phantasy are counted as a single operation. Averroës' placement of common sense in the heart is tacitly ignored.[54]

---

[48] Ibid., II.10; f. 24ra.

[49] Ibid., II.9; f. 23ra.

[50] Ibid., II.7; f. 21rb.

[51] Ibid., II.10; f. 23va: "potentie anime fluunt ab anima, et hoc modo loquendi precipue utitur Albertus."

[52] Ibid., II.13; ff. 25vb-26ra. Cf. Albert, De nutrimento et nutribili II.1 (ed. Borgnet 9: 336a-338b).

[53] Ibid., II.37; f. 45rb: "hec communiter vocantur sensus interiores qui non immutantur a rebus exterioribus nisi mediantibus aliis sensibus, et isti sunt quinque secundum Albertum, scilicet sensus communis, imaginativa, cogitativa, phantasia et memorativa." See Nicholas H. Steneck, "Albert the Great on the Classification and Localization of the Internal Senses," Isis 65 (1974), 193-211; and "The Problem of the Internal Senses in the Fourteenth Century," (University of Wisconsin, 1970), ch. 6.

[54] Averroës, De anima III. com.6 (429a24-429a29); pp. 415-416.

This kind of solution is characteristic of Jandun's use of Albert. Throughout the *De anima*, his instincts are entirely conciliatory; rather than emphasizing the differences between Albert's and Averroës' opinions on various psychological issues, he relies on logical distinctions to reconcile them. Thus he notes that Albert is correct in saying that the powers flow from the essence of the soul, if "flow" is taken *improprie*, in a causal sense. In the same way, the soul is not whole in every part of the body, if "whole" is construed in the sense of a *totum potestativum*, not a *totum essentiale*. Again, the faculties, when taken as genera, are as Albert enumerated them, while they are infinite in number from the point of view of Averroës, who considered them as *species specialissime*.[55] If Jandun must reject one of Albert's propositions, he does so anonymously. By arguing in this manner, he can gloss over contradictions and emphasize opinions held in common. The result is an eclectic and synthetic psychological theory; in it the basic vision is of the intellectual soul not as the substantial form of the body, but rather as a spiritual substance which is eternal, not to mention unique, because fundamentally separate and external to the body.[56]

Much the same can be said of the psychology of Walter Burley, an English contemporary of Jandun's who began his studies at Oxford, but who was in Paris by 1310 and remained there until 1327.[57] Like Jandun, Burley was an Averroist, in that he took the Commentator as his principal guide in interpreting the text of Aristotle; again, Albert is his main Latin authority on the soul, even though less often referred to. In his commentary on *De anima*, probably written after 1315,[58] Burley adopts a theory of the intellect as a unique *operans intrinsecum* very similar to Jandun's. Albert's influence can be seen much more clearly in his most widely circulated psychological work,

---

[55] Jean de Jandun, III.38; f. 89rb.

[56] As one would expect, Jandun rejects Aquinas' doctrine of the human soul as "irrational" and contrary to the Commentator and philosophy: this is only one of a number of passages in which he writes denigratingly of Thomas' opinions. See Pacchi, "Note II, III," pp. 451-456; and Grabmann, "Heinrich von Brüssel," pp. 52-54.

[57] On Burley's life and works, see Weisheipl, "Ockham and Some Mertonians," *Medieval Studies* 30 (1968), 174-188; and "Repertorium Mertonense," *Medieval Studies* 31 (1969), 185-208. On his general significance for medieval logic and science, see John E. Murdoch and Edith Sylla, "Walter Burley," DSB 2: 608-612.

[58] Maier, pp. 119-120, bases this date on a conjecture that Burley makes use of Thomas Wylton. In fact, the dating of many of Burley's works is problematic; see Weisheipl, "Ockham," p. 183. Burley's *De anima* is as yet unpublished; my remarks are based on the list of questions and long extracts published by Maier in her "Walter Burley."

the *De potentiis animae*. Here, after explaining that the faculties, as natural powers, are in the second species of quality,[59] Burley proceeds to a schematic enumeration of the faculties, taken as much from Albert as from Aristotle or Averroës. His list of the five internal senses, for example, comes from Albert's *Summa de homine*, even though he cites Avicenna as its source.[60] Similarly, he quotes Albert's *De sensu* repeatedly when discussing light, color, and other aspects of sensation.[61]

Both Burley's *De anima* and his *De potentiis animae* illustrate the approach characteristic of fourteenth-century Averroism which we have seen embodied in the work of Jean de Jandun. Conservative and conciliatory, Burley integrated ideas from thirteenth-century psychology and philosophy into an interpretation of Aristotle based largely on Averroës. He echoed Albert's questions — whether the powers of the soul flow from its essence, for example, and whether the soul is a separable substance or a material and substantial form — and answered them in terms which would have been familiar to Albert.

The school of thought that gained the upper hand at Paris during the third and fourth decades of the fourteenth century was entirely different. Known later as the *via moderna*, it derived much of its force from the thought of William of Ockham, many of whose opinions were attacked by Burley. This school exercised its widest influence in psychology, and in natural philosophy in general, through the works of Jean Buridan. While Averroës and Albert, like Burley and Jean de Jandun, were primarily interested in using a variety of sources to understand and interpret Aristotle's ideas, Buridan and his followers engaged in a critical reformulation of them. Inspired partly by nominalism and partly by analytical "languages" and methods recently available to logic and philosophy, Buridan and his followers developed new questions and answered the old ones in new ways.[62] It was

---

[59] Burley, *De potentiis animae* 1, ed. by M. Jean Kitchel in "The 'De potentiis animae' of Walter Burley," *Medieval Studies* 33 (1971), 88.

[60] Ibid., 22; Kitchel, p. 104. See Steneck, "Fourteenth Century," pp. 170-187.

[61] Ibid., 11 and 17; Kitchel, pp. 96-97 and 102. See also Burley's *De sensibus*, ed. by Herman Shapiro and Frederick Scott, Mitteilungen des Grabmann-Instituts der Universität München 13 (Munich, 1966).

[62] On these new languages and methods, see Murdoch, "Philosophy and the Enterprise of Science in the Middle Ages," in *The Interaction between Science and Philosophy*, ed. by Y. Elkana (Atlantic Highlands, 1974), pp. 51-74; and "From Social into Intellectual Factors: An Aspect of the Unitary Character of Late Medieval Learning," in *The Cultural Context of Medieval Learning*, ed. by Murdoch and Sylla (Dordrecht and Boston, 1975), pp. 271-339.

above all in the area of philosophical method and argument that they broke with the psychological tradition linking Albert and the fourteenth-century synthetic Averroists. For the *modernistae*, new kinds of argument led to new conclusions.

The main aim of these new kinds of argument was to eliminate explanatory principles and distinctions based on imprecise thought or misleading use of language rather than on logic. They were used most characteristically by Ockham, who treated a number of important psychological questions in his commentary on the *Sentences* (before 1320) and his *Quodlibeta* (between 1320 and 1323).[63] In both works, for example, he appeals frequently to the principle of economy — the maxim that "it is unnecessary for that which can be accomplished by fewer to be accomplished by more." On these grounds he rejects the tradition that treated the intellectual soul as an independent and eternal substance separate from matter and connected to the body only temporarily through its operations — a view common to Averroës as well as to Burley and Jandun. Arguing from economy, he notes that it is "superfluous" for natural philosophers to posit an intellectual soul which is either an immaterial and incorruptible form or a detached motor of the body; pure reason and experience can "demonstrate" no more than an extended material form which dies with the body, the position traditionally identified with Alexander of Aphrodisias. Therefore

if we mean by "intellectual soul" an immaterial and incorruptible form which is entire in the whole body and in every part, *it cannot be evidently known by reason or experience* that such a form is in us, or that the understanding of such a substance is proper to us, or that such a soul is the form of the body. I do not care what Aristotle thought on this matter, since he seems everywhere to have spoken dubitatively. *We hold these three opinions by faith alone.*[64]

In the same way, Ockham denies by the principle of economy the

---

[63] Weisheipl gives a sketch of Ockham's life and works in "Ockham," pp. 164-174. For a survey of Ockham's thought, see Moody, "William of Ockham," in his *Studies*, pp. 409-439.

[64] William of Ockham *Quotlibeta* 1.10, in *Quotlibeta septem una cum tractatu de sacramento altaris* (Strasbourg, 1491; repr. Louvain, 1962), sig. b1ra: "Intelligendo per animam intellectivam formam immaterialem incorruptibilem que tota est in toto et tota in qualibet parte, *non potest sciri evidenter per rationem vel experientiam* quod talis forma sit in nobis, nec quod intelligere talis substantie proprium sit in nobis, nec quod talis anima sit forma corporis. Quicquid de hoc senserit Arestotiles non curo, quia ubique dubitative videtur loqui. *Sed ista tria sola fide tenemus*" (emphasis added). See also *Quotlibeta* 1.12.

real distinction between the soul and its powers that had been championed by Albert and his followers, as well as the formal distinction proposed by Scotus.[65] Again, he argues that although the sensitive differs really from the intellectual soul because its objects are distinct, there is no real difference between the various sub-powers of the intellectual soul.[66] He makes this last point by analyzing the concepts of agent and passive intellect and concluding that they signify the same entity (the intellectual soul) while merely connoting its different actions.[67]

Jean Buridan, master in the Parisian arts faculty from shortly after 1320 until at least 1358, adopts very similar logical methods in his *Questiones in tres libros Aristotelis de anima*.[68] Like Ockham, he uses the principle of economy and the analysis of how and what terms signify to eliminate what were for him purely verbal distinctions in psychology: the intellectual soul is the material form of the body;[69] the soul is its powers.[70] But where Ockham acknowledged a real distinction between the intellective and lower souls, even this is denied by Buridan. Using a favorite logical technique of the *via moderna*, he argues that if this were true God, by his absolute power, could remove the intellectual soul from the human composite, leaving what amounts to an animal — a conclusion which must be rejected for its absurd implications.[71]

By this argument and the preceding ones, Buridan not only rejects all the characteristic doctrines concerning the nature of the soul and its powers which had been developed by Albert and adopted by nearly all Latin philosophers during the century after the *Summa de homine*, but he also enters a new realm of philosophical discourse,

---

[65] Ockham, II *Sent.* 24, in *Opera plurima* (Lyon, 1494-1496; repr. London, 1962), 4: sig. h7rb.

[66] Ibid., sig. h7ra-h7rb. See also *Quotlibeta* II.10, sig. c6ra-c6va.

[67] Ibid., sig. h7vb.

[68] Maria Elena Reina, in *Note sulla psicologia de Buridano* (Milan, 1959), emphasizes that Buridan also uses more classical modes of inquiry. For a more general introduction to Buridan's life, works, and thought, see Moody, "Jean Buridan," in *Studies*, pp. 441-453; and Edmond Faral, *Jean Buridan, Maître ès arts de l'université de Paris*, in *Histoire littéraire de la France*, 38 (Paris, 1949), 462-605. Faral discusses the *De anima* on pp. 494-495.

[69] Jean Buridan, *Questiones in libros Aristotelis de anima* II.7 and III.3, in *Questiones et decisiones Alberti de Saxonia, et al.*, ed. by George Lokert (Paris, 1518), ff. 8rb-8vb and 23va-24rb. Michalski argues that this "Alexandrinism" is characteristic of the entire tradition of psychology influenced by Buridan; see "La lutte autour de l'âme au quatorzième et au quinzième siècles," *Résumés des communications présentées au VIe congrès international des sciences historiques* (Oslo, 1928), pp. 116-117.

[70] Ibid., II.6 and II.9; Lokert, ff. 8ra and 10ra. See Reina, pp. 3-7.

[71] Ibid., II.5; Lokert, f. 6vb.

where the radical contingency of creatures becomes the basis for logical argument and where the most proper form of demonstration is the linguistic analysis of propositions. It is not surprising that Albert figures as an authority in only one question in Buridan's *De anima* — that concerning the internal senses. And here it is significant that the opinion which is ascribed to Albert among others is, in fact, that of Averroës:[72] Ockham, Buridan, and other philosophers of the *via moderna* rely much less often on arguments from authority than their Parisian predecessors, but when they do it is the Commentator who bears the most weight.[73]

Under Buridan's influence the *via moderna* became the standard moderate position in the Parisian arts faculty by the middle of the fourteenth century;[74] Buridan's *De anima*, like his other commentaries on Aristotle, set the tone and content of psychological theory at Paris and in the German universities. Toward the end of the century, however, there was a marked decline in the vigor and originality of philosophical thought at Paris. With it came a retreat from critical trends and a tendency toward a new strain of eclecticism and conciliation similar in many respects to what we saw in Burley and Jandun. This development is evident in psychology. In his *Tractatus brevis de anima* (1372), for example, Pierre d'Ailly presents a synthetic theory of the soul which is based on the works of Ockham and Buridan, but into which he has introduced elements directly traceable to the influence of Albert.[75]

D'Ailly acknowledges that his book is a collection of others' opinions. In the prologue he writes, "With the help of God I will compile this brief treatise on the soul and the things in it from the more probable sayings of wise men, not to teach advanced students, but to instruct myself."[76] Buridan is clearly the principal source for his theories concerning the nature of the soul and its relations to its faculties. D'Ailly follows him word for word in support of the Alex-

---

[72] Ibid., ii.22; Lokert, f. 21ra-21rb. See Steneck, "Fourteenth Century," pp. 187-205.

[73] Thus Kuksewicz seems to be in error when he claims that Averroës' influence had disappeared at Paris by 1327.

[74] Moody, "Ockham, Buridan, and Nicholas of Autrecourt," in *Studies*, p. 160.

[75] For a survey of d'Ailly's life and writings, see Claudia Kren, "Pierre d'Ailly," DSB 1: 84; and Francis Oakley, *The Political Thought of Pierre d'Ailly* (New Haven, 1964), pp. 8-14.

[76] Pierre d'Ailly, *Tractatus brevis de anima* (Paris, 1505), sig. alv: "Quapropter, de anima et his quae sunt in ea tractatum hunc brevem non ad provectorum eruditionem, sed ad meipsius instructionem, ex probabilioribus sapientum sententiis cum Dei auxilio compilabo."

andrine description of the rational soul as corruptible and material form,[77] and in denying the real distinction of the powers.[78]

Because d'Ailly rejects so many of the opinions first codified and established by Albert, it is all the more interesting to see that he relies on Albert in his description and enumeration of the faculties. He does not use Albert's works directly, however; his source, as a careful comparison of wording and content shows, is the spurious *Philosophia pauperum* compiled from Albert's writings in the thirteenth century and attributed to him.[79] It was widely used as a trot or introduction to Aristotelian philosophy and generated in its turn a large number of commentaries and epitomes. Most of these are associated with German schools and universities, but it appears from d'Ailly's *Tractatus* that the *Philosophia pauperum* also circulated in Parisian circles in the later fourteenth century.

The fifth book of the *Philosophia pauperum*, which appears in two recensions, is a summary of Albert's writing on the soul. It seems to have had a wide circulation, both independently and as part of the longer work. The psychological theory of the book is highly schematic and unoriginal. The author lays great emphasis on the number and division of the faculties; Albert's arguments concerning the relation of the soul to its powers are given in shortened form. One of its most notable features is a strong Augustinian element, which had been removed by Albert, but is here reintroduced into the sections on intellection.

D'Ailly uses all these elements. He organizes the first half of his *Tractatus* according to powers, on the model of the *Philosophia pauperum*; he adopts the order and names of the powers in his German source. Even more telling are the verbal parallelisms between the *Tractatus* and Recension A of the *Philosophia pauperum*. In his treatment of the vegetative and sensitive powers, d'Ailly repeatedly

---

[77] Ibid., 6.1; sig. b3r. Cf. Buridan, III.3; Lokert, f. 23vb.

[78] Ibid., 1.QQ2-3; sig. a4r-v. Cf. Buridan, II.6 and III.3; Lokert, ff. 8va and 23vb.

[79] See note 32 above.

draws on the latter for the text of his explanations and definitions.[80] He also retains references to the Augustinian theological faculties, such as synderesis and higher and lower reason, which Albert had relegated to an appendix.[81]

Pierre d'Ailly thus in some sense completes a process which had begun during the first decades of the fourteenth century at Paris. The eclectic psychological theory in the *De anima* commentaries of Jean de Jandun and Walter Burley, although strongly influenced by Averroist ideas, remained closely within the structural framework established by Albert in the middle of the previous century. With the appearance of the new logical and philosophical methods of the *via moderna* in the works of William of Ockham and Jean Buridan, a break with Albert's psychology was inevitable; the questions he had first posed concerning the nature of the soul and its faculties remained central, but Buridan and Ockham, arguing in different ways, gave entirely different answers. For them, Albert's opinions became irrelevant.

With the *Tractatus* of Pierre d'Ailly, we witness the emergence of a new eclecticism which draws on the general model of the soul and its powers developed by the writers of the *via moderna* and on the details of these powers and their operations in the *Philosophia pauperum*, a popularization of Albert's ideas. This process makes a

---

[80] There are many parallel passages. Compare, for example, the introduction to the external senses:

Tractatus brevis 3.1 (sig. a6r)
Et dicitur apprehensiva deforis que
apprehendit per organum
quod est extra in corpore, vel quia hoc
non est verum de sensu tactus
potest dici apprehensiva deforis
quia apprehendit rem deforis presentem.

Philosophia pauperum VA.5 (Geyer, pp. 47*-48*)
Apprehensiva deforis dicitur, quae
apprehendit per organum vel in organo
quod est extra in corpore. Sed forte hoc
non invenitur in sensu tactus . . .
Unde apprehensiva deforis dicitur,
quia apprehendit rem deforis in materia
subiecto praesente.
(Note that no such passages appears in any of Albert's original works.) Other parallel passages are sig. a6v and p. 49*; sig. blr and p. 54.*

[81] Pierre d'Ailly, 6.4; sig. b4r. Cf. *Philosophia pauperum* VA.8; Geyer, pp. 58*-59*.

good deal of historical sense. In particular it provides an appropriate context for the major development in the Parisian arts faculty of the 1380s and 1390s — the rise of the *via antiqua*, an opposition movement which rejected nominalist methods and advocated a return to the ideas of the great philosophers of the thirteenth century. One of the most important subdivisions of the *via antiqua* in the next century was the philosophical school kown as Albertism, which originated at Paris but flourished above all in Germany. It is to this school that we now turn.

C. ALBERTISM IN THE FIFTEENTH CENTURY: THE GERMAN UNIVERSITIES

The splintering of the arts faculties into rival schools came to dominate philosophy in the northern universities during the late medieval period.[82] The first stage in this development was marked by the appearance of the *via antiqua* at Paris during the last decades of the fourteenth century. This movement was above all an attempt to reform teaching methods: the *antiqui* advocated a return to a simpler and more eclectic approach than that of the *via moderna*, with more emphasis on the text of Aristotle and the opinions of Latin and Arabic authorities.[83] Around 1400, the *antiqui* seem to have split into several opposing schools. The most prominent were the *Albertistae*, followers of Albert, and the *Thomistae*, whose master Aquinas had only recently been rescued from comparative oblivion.[84]

As a result of a rather complicated set of political circumstances, the Albertists, under their leader and founder Jean de Maisonneuve

---

[82] The two classic works on this development are Gerhard Ritter, *Studien zur Spätscholstick II: Via antiqua und via moderna auf den deutschen Universitäten des 15. Jahrhunderts,* Sitzungsberichte der Heidelberger Akademie der Wissenschaften, Phil.-hist. Klasse 13 (Heidelberg, 1922); and Franz Ehrle, *Der Sentenzkommentar Peters von Candia des pisaner papstes Alexanders V: Ein Beitrag zur Scheidung der Schulen in der Scholastik des 14. Jahrhunderts und zur Geschichte des Wegestreites* (Münster, 1925); see especially pp. 114-140 on Paris.

[83] Ritter, *Via antiqua,* pp. 100-104. Gerson attempted to introduce this kind of reform in the faculty of theology; see Palémon Glorieux, "Le chancelier Gerson et la réforme de l'enseignement," in *Mélanges offerts à Etienne Gilson* (Toronto, 1959), pp. 285-298.

[84] On the status of Thomas at Paris at the end of the fourteenth century, see Marie-Dominique Chenu, " 'Maître' Thomas est-il une 'authorité'?" *Revue thomiste,* 30 (1925), 187-194; and Grabmann, "Die Kanonisation des heiligen Thomas von Aquin in ihrer Bedeutung für die Ausbreitung und Verteidigung seiner Lehre im 14. Jahrhundert," *Divus Thomas* ser. 3, 1 (1923), 233-249. A broader view is given by Weisheipl in "Thomism — Introduction and General Survey," *New Catholic Encyclopedia* 14: 128-132.

(Joannes de Nova Domo), gained the upper hand over both the *moderni* and the other *antiqui* in 1407, and kept it until 1437.[85] During this time, Paris exported large numbers of masters and students to the German universities, which were still predominantly nominalist.[86] The Parisians brought with them their school allegiances and animosities, and these flourished in Germany, generating conflicts even more acrimonious than those at Paris.[87] The German focus of these controversies seems to have been the university of Cologne, which also served as the center of German Albertism. Albert had had close personal ties with the city, and the strength of his influence there, in psychology and philosophy in general, is not surprising.

When the university of Cologne was founded in 1388, its orientation, like that of the other early German foundations, had been nominalist; during the following thirty years, realism came slowly to dominate the curriculum, although the *via moderna* remained strong.[88] The situation crystallized at the end of the second decade of the fifteenth century with the arrival of two masters trained in Paris. The first, Heinrich von Gorkum (d. 1431), came to Cologne in 1419 and soon after founded a *bursa* — a college where students in arts lived and received instruction — later called the Bursa Montana.[89] Its doctrinal orientation, like its founder's, was Thomist. Then in 1422 or 1423, Heymerich van den Velde (Heimericus de Campo, d. 1460), an ex-student of the Parisian Albertist Jean de Maisonneuve, arrived in Cologne to teach in the arts faculty, apparently at the invi-

---

[85] Gilles Meersseman, *Geschichte des Albertismus I: Die Pariser Anfänge des Kölner Albertismus*, Dissertationes historicae, Institutum historicum FF. praedicatorum Romae 3 (Rome, 1933), pp. 11-12. Little is known of Jean de Maisonneuve; see Meersseman, *Albertismus I*, pp. 21-191 (includes an edition of his *De esse et essentia*); and A. G. Weiler, "Un traité de Jean de Nova Domo sur les universaux," *Vivarium*, 6 (1968), 108-154 (includes an edition of his *De universali reali*.)

[86] Astrik L. Gabriel, " 'Via antiqua' and 'Via moderna' and the Migration of Paris Students and Masters to the German Universities in the Fifteenth Century," in *Antiqui und Moderni*, ed. by Albert Zimmermann, Miscellanea medievalia 9 (Berlin and New York, 1974), pp. 439-483.

[87] For a vivid illustration of the importance of these divisions in German student life of the later fifteenth century, see *The Manuale Scholarium*, tr. by Robert Seybolt (Cambridge, Mass., 1921), pp. 43-45, 54-55, and 102-103.

[88] On the relations between the various schools during the early history of the university of Cologne, see Ehrle, pp. 146-157; Weiler, *Heinrich von Gorkum (d. 1431): Seine Stellung in der Philosophie und der Theologie des Spätmittelalters* (Hilversum, 1962), pp. 42-45 and 56-81; Meersseman, *Geschichte des Albertismus II: Die erster Kölner Kontroversen*, Dissertationes historicae, Institutum historicum FF. praedicatorum Romae 5 (Rome, 1935), pp. 7-10.

[89] On Gorkum's life and works, see Weiler, *Heinrich von Gorkum*.

tation of Gorkum, who had known him at Paris.[90] Several years later he fell out with Gorkum and founded his own college, the Bursa Laurentiana, which was avowedly Albertist.[91]

According to Heymerich, the doctrinal lines had been drawn before he appeared on the scene. On his arrival in Cologne, he claimed, he found "the same tripartite controversy which he left at Paris, between the terminists, who were then called *moderni*, and the Thomists and Albertists, who were then called *antiqui*."[92] Whatever the case, Heymerich certainly exacerbated whatever animosity existed previously.[93] In 1428, he circulated a work called *Problemata inter Albertum Magnum et Sanctum Thomam*, attacking both the "epicurean nominalists" and the Thomists, and elevating the authority of Albert. This treatise effectively split the masters of the *via antiqua* into two camps, corresponding roughly to the two major *bursae*, and created a tension between them which was to persist into the sixteenth century.

The Thomist response did not appear until 1456, in the form of a moderate and conciliatory treatise, usually referred to as the *Tractatus concordiae*, by Gerhard ter Stegen s'Herrenberg (Gerardus de Monte, d. 1480), rector of the Bursa Montana after Gorkum.[94] Heymerich replied from Louvain, where he had been teaching since 1426, with a letter to the Cologne masters which became known as the *Invectiva*; Gerhard answered with an *Apologia*.

The first part of Heymerich's *Problemata* concerns universals and is directed against the *moderni*. The second, which provoked far more controversy, consists of eighteen questions of interest to the

---

[90] On Heymerich, see Rudolf Haubst, "Zum Fortleben Alberts des Grossen bei Heymerich von Kamp und Nikolaus von Kues," *Studia Albertina*, ed. by Heinrich Ostlender (Münster, 1952), especially pp. 221-235.

[91] On teaching at Cologne and the *bursae*, see Sophronius Clasen, "Der Studiengang an der Kölner Artistenfakultät," in *Artes liberales*, ed. by J. Koch (Leiden and Cologne, 1959), especially p. 132. There were two other lesser fifteenth-century *bursae* at Cologne: the Kuck (Albertist) and the Corneliana (Thomist).

[92] Heymerich van den Velde, *Invectiva*, ed. by Meersseman in *Albertismus II*, p. 113: "cum reperiret ibi similem cum ea, quam reliquit parisius, inter terministas, qui dicebantur tunc moderni, thomistasque et albertistas, qui dicebantur antiqui, controversiam . . . tripartitam."

[93] The most complete account of this controversy, including a partial edition of the *Invectiva* and *Apologia*, is in Meersseman, *Albertismus II*, pp. 11-106. See also Weiler, *Heinrich von Gorkum*, pp. 79-82.

[94] On Gerhard, see Meersseman, "Ergänzungen zur Kenntnis des literarischen Nachlasses des Kölner Professors Gerhard ter Steghen de Monte," *Jahrbuch des Kölnischen Geschichtsvereins* 17 (1935), 264-268.

antiqui, classified by subject, on which Heymerich perceives Albert and Thomas to differ; in each question, after analyzing the issue, he decides in favor of the Albertist position. Four of these "problems" (11-14) belong to psychology.

Problem 11 deals with the generation of the rational soul in the embryo. In it, Heymerich rejects Thomas' position that the vegetative, sensitive, and intellectual souls are consecutively present in the foetus, and sides with Albert's position that the lower functions are performed by the *virtus formativa* in the semen until the entry of the intellectual soul, from which "all life formally proceeds." Problem 12 is related. Here, Heymerich notes that whereas Thomas and his followers hold that the species of the faculties depend on the species of their objects, Albert maintains that they depend on the species of soul from which they flow, so that an ass's faculty of vision differs specifically from a man's. Similarly, Problems 13 and 14 go together. The former concerns the ability of the embodied human intellect to think without images (*phantasmata*) — something Albert admits, says Heymerich, and Thomas erroneously denies. The latter concerns its ability to know itself and the other separable substances; again, Heymerich claims to stand by Albert and against Thomas in holding this as possible.[95]

Heymerich's arguments are at best confused and parochial, and at worst inflammatory. Gerhard's reply in the *Tractatus concordiae*, on the other hand, is moderate and conciliatory. He goes through Heymerich's psychological questions one by one, citing and comparing passages from Albert and Thomas in an effort to show that the two fundamentally agree, usually on the position identified as Thomist.[96] His basic exegetical principle is that "it is not certain that the things the venerable Albert writes in his philosophical works are his own opinion, unless the same things are found in his theological works"[97] — works with which Heymerich, as a mere master of arts, was probably unfamiliar. Heymerich's *Invectiva* and Gerhard's *Apologia* add little of substance to the debate.

From the point of view of style, tone and sophistication, Gerhard's

---

[95] Heymerich, *Problemata inter Albertum Magnum et Sanctum Thomam* (Cologne, 1496), ff. 37r-46r.

[96] Gerhard von Herrenberg, *Ad favorabilem dirigens concordiam quedam problemata* (Cologne, 1497), ff. 33r-34v.

[97] Ibid., f. 27v: "Non est certum quod ea que venerabilis Albertus scribit in philosophicis sint sue proprie opinionis, nisi eadem reperiantur in operibus suis theologicis."

work is of much higher quality. His careful collection and analysis of the relevant texts in Albert's and Thomas' works put Heymerich's abstract and often confused arguments to shame. Many modern scholars, in fact, repeat his assessment of the relative authority of Albert's philosophical and theological works. Nonetheless, his reconciliation of the views of Thomas and Albert on the soul require further investigation. On the other hand, Heymerich, for all his faults, has effectively isolated a number of significant points of disagreement between Albert and Thomas, although other of his dichotomies are specious or exaggerated.

The first point has to do with the generation of the vegetative and sensitive soul in man prior to infusion of the rational soul. For Albert, the male seed contains within itself a *virtus formativa* through which celestial intelligences operate to produce first the vegetative, then the sentient soul, after the manner of a "developing" process. Thomas, on the other hand, rejects this continuous process and insists on a succession of souls, each one of which is the substantial form of the foetus at that particular stage, until the rational soul is finally created and infused by God. While both Thomas and Albert insist on the immediate creation and infusion of the human soul by God for each individual person, the event seems to happen more "naturally" for Thomas, whereas for Albert the whole "process" is more the work of intelligence (*opus intelligentiae*).

From this, a second point seems to follow. For Albert, although the rational soul is the unique form of the human person, it has much in common with separated intelligences. The act of human thinking (*ratiocinare* and *intelligere*) shares in the activity of separated intelligences, namely knowledge of self (*se intelligere*), more perhaps than Thomas would allow. Nevertheless, both insist that all human knowledge comes through the senses and that all of our knowledge of supra-sensible things is strictly by analogy and through sensible effects. For both, neither God nor intelligences can be known quidditatively in this life, and the need for sensible phantasms in this life is essential for human thought. However, although Thomas admits that the human intellect (even in this life) can directly reflect upon its own activity, it would seem that Albert emphasizes more than Thomas this ability of the human intellect.

In any case, Heymerich spells out fully what he sees to be the clear-cut conflict between Thomists and Albertists. Whether or not this conflict really existed before his *Problemata* is difficult to ascertain. But it was of indisputable importance during the following hundred years. A look at the literature on psychology generated in the arts faculty during the last half of the fifteenth century confirms this: it divides neatly into two schools — Albertists and Thomists — corresponding to the two Bursae for and in which this literature was produced. It consists largely of longer commentaries and shorter epitomes or *reparationes* on Aristotle's *De anima*, which was taught "cursorily" through commentaries rather than through the Aristotelian text.[98] Each begins with a declaration of allegiance to the master (Thomas or Albert) it followed and a reference to the college (Montana or Laurentiana) it was intended for; each is further introduced by a woodcut showing either Thomas or Albert flanked by admiring students. The sameness of the format reflects the monotony of the contents. These later products of the *via antiqua* are generally unoriginal and repetitive. The arguments are extremely formalized — sometimes each demonstration is presented as a syllogism — and adhere slavishly to the chosen authority. Within a given school, the authors tend to repeat each other literally.

The Albertists at the Bursa Laurentiana are not free of these faults. Johannes Hulstadt von Mechlin (d. 1475) begins his *De anima* with the boast that he has throughout followed Albert, "from whose commentaries everything that follows is excerpted with few changes."[99] He punctuates his commentary by questions to which he gives the expected answers. Against Thomas, he argues that there are five rather than four internal senses and that the soul is not located whole in every portion of the body.[100] He adopts Heymerich's conclusions concerning the four disputed questions and supports them by long passages lifted word-for-word from the *Problemata*.[101] The "revised and corrected" version of his commentary put out by another illustrious Albertist of the next generation, Gerhard von Harderwijck (d. 1503), is almost identical. It is printed with Harderwijck's epitome of the *De anima*, consisting almost entirely of Albert's powers of the soul arranged into neat mnemonic diagrams.[102]

---

[98] Classen, pp. 126-127.

[99] Johannes Hulstadt von Mechlin, *Textus trium librorum de anima Aristotelis cum commentario secundum doctrinam venerabilis domini Alberti Magni* 1.1 (Cologne, 1491), sig. b3r: "ex cuius commentariis paucis mutatis totum sequens excerptum est."

[100] Ibid., II.1.Q7 and III.1.Q1.

[101] E.g. ibid., II.2.dubium 7 and III.2.Q4. Mechlin's arguments on the separable substances seem to show some originality.

[102] Ibid., revised and corrected by Gerhard von Harderwijck (Cologne, 1496).

Under masters like Johannes von Nürtigen (d. 1515) and Arnold von Tongern (d. 1540), Albertism continued into the sixteenth century at Cologne. The division between Albertists and Thomists at Cologne was satirized in the *Epistolae obscurorum virorum* (1515-1517)[103] By the end of the fifteenth century, it had also been exported into the theological faculty and Dominican studium at Cologne,[104] as well as into the arts faculties of other universities, like Krakow and Heidelberg, although its force was somewhat diminished in the process. The Cologne controversies are themselves a fascinating episode in the history of later scholasticism and in the history of Albert's influence. Grounded in legitimate and important differences in the interpretation of Aristotle, they were clearly fueled by institutional rivalries which had little to do with the philosophical issues. The two principal Bursae of the Cologne arts faculty were apparently competing for students, textbook sales, and prestige. One of the products of that competition was the emergence of an explicit and dogmatic Albertism in psychology.

Perhaps the most interesting element of the Cologne controversies was Heymerich's thirteenth problem — whether intellection can take place without mental images or *phantasmata*. In his conclusion Heymerich, taking what he identifies as an Albertist position, argues that it can. The divine part (*particula divina*) of the human intellect — also known as the synderesis — may rise above sense and imagination to contemplate God in His purity. For this reason,

> Dionysius said that everything created by the highest good returns to it according to its proper nature. . . . But the rational soul has been created immediately by the highest good. Therefore it returns to it on the intelligible plane without using the mirror of imagination or sense, first through the inclination of desire and finally through the habit of god-like intelligence, . . . the knowledge of man's super-essential divine good.[105]

---

[103] Epistolae obscurorum virorum II.45, ed. by Francis Griffin Stokes (New Haven, 1925), pp. 224-225; English translation, pp. 482-483.

[104] Gabriel M. Löhr, *Die Kölner Dominikanerschule vom 14. bis zum 16. Jahrhundert* (Freiburg, 1946), pp. 64-70; *Die theologische Disputationen und Promotionen an der Universität Köln im ausgehenden 15. Jahrhundert*, Quellen und Forschungen zur Geschichte des Dominikanerordens in Deutschland 21 (Leipzig, 1926), pp. 15-32.

[105] Heymerich, (f. 46r): "Dicit Dyonisius quod omne creatum a summo bono secundum modum nature sue convertitur ad ipsum. . . . Sed anima rationalis est immediate a summo bono creata, ergo sine medio enigmatis fantastici vel sensibilis convertitur ad ipsum intelligibiliter. Primo quidem per inclinationem desiderii, et tandem per habitum intelligentie deiformis . . . divinum et super essentiale bonum hominis."

It is significant that Heymerich confirms this conclusion by citing Pseudo-Dionysius rather than Albert's *De anima, Summa de homine*, or even his commentaries on the Pseudo-Dionysian corpus. In fact, Albert explicitly denies that even the assimilative intellect can function entirely without images. Why then does Heymerich insist? The answer apparently lies in the strong Pseudo-Dionysian character of fifteenth-century Albertism.

The association between Albert and Pseudo-Dionysius, like most other elements in German philosophy of this period, was established at Paris by masters like Gerson and Maisonneuve.[106] It was then exported to Germany with the migration of Parisian Albertists like Heymerich. Throughout his work, Heymerich heavily emphasizes the Neoplatonic aspects of Albert's philosophy and frequently cites Albert's commentaries on the works that formed the backbone of the medieval Neoplatonic tradition — the *Liber de causis* and Pseudo-Dionysius' *De divinis nominibus, De caelesti hierarchia*, and *Theologia mystica*. It seems, in fact, that the Pseudo-Dionysian strain in fifteenth-century philosophy is based as much on Albert's commentaries as on the Dionysian corpus itself.[107]

Intellection without images looms large in Pseudo-Dionysius' psychology. One of his central prescriptions is that "we must contemplate things divine in a manner becoming God."[108] Thus, since God is above name and likeness, "it is appropriate that, by ascending from obscure images to the cause of all, we contemplate with otherworldly eyes all things in the cause of all. . . ."[109] Pseudo-Dionysius identifies this ascent beyond image as ignorance (*agnosia*) and darkness — the only state in which God may be truly known. As a result of these ideas, the problem of intellection without images haunts German literature in both philosophy and mystical theology, separating the Albertists from the other adherents of the *via antiqua*. Ironically, one of the most influential elements of Albertist psychology is based on a position Albert never held.

The most illustrious exponent of the Albertist position on intellec-

---

[106] André Combes, in his *Jean Gerson, commentateur Dionysien*, Etudes de philosophie médiévale 30 (Paris, 1940), claims that Albert's thought informs Gerson's interpretation of Dionysius; see p. 445. Jean de Maisonneuve cites Dionysius in his Albertist tract *De esse et essentia*, in Meersseman, *Albertismus I*, e.g., p. 110.

[107] Eusebio Colomer, "Nikolaus von Kues und Heimeric van den Velde," *Mitteilungen und Forschungsbeiträge der Cusanus-Gesellschaft* 4 (1964), p. 201.

[108] Pseudo-Dionysius the Areopagite, *De divinis nominibus* VII.1.

[109] Ibid., V.7.

tion without images was Nicholas of Cusa, who attended lectures at the university of Cologne between 1423 and 1426, when the Albertist controversies, according to Heymerich, were at their height. Cusanus probably came into contact with Heymerich at this time. He was certainly familiar with Heymerich's Pseudo-Dionysian brand of Albertism, for he acquired a manuscript (Codex Cusanus 106) containing a number of Heymerich's works.[110] He supplemented this years later, in 1453, with a copy of Albert's commentaries on Pseudo-Dionysius (Cod. Cus. 96). Both codices were read with care: they are the most heavily annotated of Cusanus' surviving manuscripts, and their influence appears clearly in his own writing.[111]

Cusanus takes an Albertist position on images in intellection, and in a certain sense he incorporates Heymerich's conclusions into the core of his philosophy and theology — the notion of "learned ignorance." The premise of *De docta ignorantia*, his first and most influential work, is that man is ordinarily in a state of ignorance because his reason is dependent on sensible objects and their images; it is only in the state of "learned ignorance" that he may finally understand the essence of all things in God:

> one must necessarily cast out those things which are attained through sense, imagination or reason, to arrive at the simplest and most abstract intelligence, where all things are one — where the line is the triangle, circle and sphere; where unity is trinity and vice versa; where accident is substance; where body is spirit, motion is rest. . . .[112]

Later, in *De coniecturis*, Cusanus is even more explicit: when the soul "looks on things in their simple intellectual nature, it grasps them without phantasmata, in the clarity of truth."[113]

This element of Cusanus' thought did not go unchallenged. It generated a controversy like that in Cologne, although somewhat more

---

[110] Colomer, p. 199. Haubst gives details on the contents of the codex in "Fortleben," pp. 423-435.

[111] Haubst, "Fortleben," pp. 436-437.

[112] Nicholas of Cusa, *De docta ignorantia* I.10 ed. by E. Hoffmann and Raymond Klibansky, in *Opera omnia*, 1 (Leipzig, 1932): 20: "Illa, quae aut per sensum aut imaginationem aut rationem cum materialibus appendiciis attinguntur, necessario evomere oporteat, ut ad simplicissimam et abstractissimam intelligentiam perveniamus, ubi omnia sunt unum; ubi linea sit triangulus, circulus et sphaera; ubi unitas sit trinitas et e converso; ubi accidens sit substantia; ubi corpus sit spiritus, motus sit quies et cetera huiusmodi."

[113] Ibid., *De coniecturis* II.16, ed. by J. Koch and C. Bormann, in *Opera omnia*, 3 (Hamburg, 1972): "Dum autem res . . . in sua simplici intellectuali natura intuetur, eas extra ipsa phantasmata in claritate veritatis amplectitur."

limited in scope, which further illuminates the nature of Albert's influence in psychological theory. The locus of this controversy was the university of Heidelberg, which, like Cologne, had been dominated by the *moderni* after its foundation. The *via antiqua* was introduced by Johannes Wenck von Herrenberg, a Parisian master who arrived in 1426 or 1427 and who later founded the first realist college, the Bursa Parisiensium.[114] Wenck lay in the same intellectual tradition as Heymerich. An Albertist from Paris, he emphasized the Dionysian and Neoplatonic aspects of Albert's work and of Aristotelian philosophy in general. His sources included Albert's *De natura et origine animae, Summa de homine*, and commentaries on *De caelesti hierarchia* and the *Liber de causis*, as well as Jean de Maisonneuve's *De esse et essentia*.[115] In his early commentary on *De anima*, probably written at Paris before 1427, Wenck apparently even maintained a qualified Albertist position concerning intellection without images.[116] Because he had so much in common with Cusanus, his controversy with him is all the more interesting.

In 1422, just after Cusanus had written his *De docta ignorantia*, Wenck composed *De ignota litteratura*, a bitter attack on it in the form of a letter to Johannes von Gelnhausen, who had sent him the book.[117] Cusanus responded with his *Apologia doctae ignorantiae*, and Wenck replied in turn with *De facie scolae doctae ignorantiae*. In the first, Wenck accuses Cusanus of holding certain un-Aristotelian opinions, one of which is that it is possible to think without images. Appealing to Aristotle and Boethius, he argues that according to our corporeal nature, "it does not happen that we understand without a *phantasma*."[118] God is only comprehensible "through His footprint and image, appearing under the likeness of creatures; He is described to us by Scripture in the similitudes of creatures adapted to

---

[114] Ritter, *Die Heidelberger Universität I: Das Mittelalter (1386-1508)* (Heidelberg, 1936), pp. 382-394; *Via antiqua*, pp. 50-54. The colleges of the *moderni* were the Bursa suevorum and, after 1456, the Bursa nova.

[115] Haubst, "Johannes Wenck aus Herrenberg als Albertist," *Recherches de théologie ancienne et médiévale* 18 (1951), pp. 308-323; and *Studien zu Nikolaus von Kues und Johannes Wenck*, in *Beiträge*, 38/1 (Münster, 1955), pp. 87-92. Most of my conclusions are dependent on the accuracy of Haubst's research.

[116] Ibid., pp. 318-319; and *Studien*, pp. 87-88. Wenck's position on the issue as it appears in the passages of the manuscript quoted by Haubst, is not as unambiguous as Haubst seems to indicate.

[117] Haubst argues for this dating in *Studien*, p. 99.

[118] Johannes Wenck, *Le "De ignota litteratura" de Jean Wenck de Herrenberg contre Nicolas de Cuse*, in *Beiträge*, 8/6 (Münster, 1910), p. 21: "nec sine fantasmate contingit nos intelligere."

our understanding, so that He can be understood here, in this life."[119]

Thus in *De ignota litteratura* Wenck contradicts a doctrine which he may have supported earlier in his *De anima* and which was dear to the hearts and minds of his fellow Albertists at Cologne. Why then does he take so vehement a stand on the issue? One possible answer is political: Wenck, a confirmed conciliarist, had been one of Cusanus' antagonists at the Council of Basel in 1441, and he may have attacked *De docta ignorantia* in retaliation.[120] But this does not explain why Wenck wrote as he did concerning images and intellection. On this point one of the other passages in *De ignota litteratura* is more illuminating. In it, Wenck compares Cusanus' position to the doctrine of Eckhart that man must "despoil and denude himself of the image of himself and of any creature; then . . . his whole being, living, knowing, and loving will be from God, in God, and God."[121] Wenck thus equates Cusanus' philosophical error with the heresy of Eckhart and the Beghards, who also pursue the life of detachment — the "*abgescheiden leben.*"[122]

The remarks make sense in the context of other pieces of information we possess. We know from Trithemius that the theological faculty of Heidelberg had condemned seventeen articles of Eckhart in 1430.[123] We further know from Johannes of Gelnhausen that Wenck had argued in his sermons and lectures against errors attributed to the Waldensians and Beghards.[124] In addition we have another letter from him to Johannes of Gelnhausen, written in 1443, in which he attacks seventeen propositions put forth by one of his colleagues in the Heidelberg theology faculty; he deplores this man's intention to take up the "*abgescheiden leben*" and adds that he fears Lollard influence is involved.[125]

These events relate directly to Wenck's attack on Cusanus' notion

---

[119] Ibid., p. 27: "Deus in vestigo et in ymagine est cognoscibilis sub nocione similitudinis creaturarum innotescens, quia per scripturam sub similitudinibus creaturarum nobis descriptus ydonee ad nostram comprehensionem eo modo quo hic in via comprehendi potest."

[120] Haubst, *Studien*, pp. 110-113.

[121] Wenck, pp. 24-25: "Homo deberet esse multum diligens ut spoliaret et denudaret se ipsum a propria ymagine et cuiuscumque creature . . .; tunc . . . totum suum esse, vivere et nosse, scire, amare est ex Deo, in Deo, et Deus." (The quotation is from Eckhart's *Buch der göttlichen Tröstung*, 1.)

[122] Wenck, p. 31.

[123] Johannes Trithemius, *De scriptoribus ecclesiasticis* (Paris, 1512), f. 118r.

[124] Ritter, *Heidelberger Universität*, p. 433 n. 2.

[125] Haubst, "Johannes von Franckfurt als der mutmassliche Verfasser von *Eyn deutsch Theologia*," *Scholastik* 33 (1958), 375-398.

of learned ignorance and his rejection of Cusanus' contention that intellection is possible without images. Wenck perceived the Beghards and Lollards as a clear and present danger — one that had infiltrated the Heidelberg faculty itself. He identified their ideas with the mystical theology of Eckhart, particularly that aspect which emphasized the synderesis, a divine element in man through which he could cast out the images of all creatures and become one with God. Given the close ties between Heidelberg and Cologne, the two centers of the *via antiqua* in Germany, Wenck must have known of the Albertist-Thomist controversies at Cologne and of Heymerich's teaching concerning the synderesis and the *particula divina* in the human intellect. When he saw these ideas adopted and transformed by Cusanus, Wenck may have thought it necessary to demonstrate their similarity to the teachings of Eckhart, who had recently been condemned at Heidelberg.[126] Thus his attack was probably motivated by a combination of doctrinal and political reasons, as well as by a sincere fear of resurgent heresy. It is unclear how real the danger was; Wenck's "Beghards" may have been real Hussites, or they may simply have been adherents to the tradition of the Modern Devotion or Rhineland mysticism. In any case, Wenck identified them as true heretics, and it was for these reasons that, while fundamentally an Albertist in his own psychology, he rejected that element of Albertist theories of the soul which seemed to him closest to their errors.

It is important to note that the controversies at Cologne and Heidelberg mark a very late stage in the history of Albert's influence in psychology. They were also fairly localized. Elsewhere the lines between Thomists and Albertists were not so clearly drawn as at the university of Cologne. Even at Heidelberg, there was only one realist Bursa; Albertism formed part of the general philosophical and pedagogical orientation called the *via antiqua*. Its main distinguishing characteristic seems to have been a strong interest in the medieval Neoplatonic tradition and in the ideas of Pseudo-Dionysius as interpreted by Albert. By the end of the century at Heidelberg, according to the student interlocutors in the *Manuale scholarium*, even this was on the wane: "those who follow Albert are few, merely three or four masters graduated at Cologne, and probably just as many follow

---

[126] The question of the extent of Albert's influence on Eckhart and Eckhart's on Heymerich is still open.

Scotus, but their audience is small and they receive little."[127] Nearly all the masters owed their primary intellectual allegiance to Thomas.

By 1500 this was generally true of the German arts faculties, except Cologne. At the university of Krakow, for example, the *via antiqua* had been introduced from Paris during the 1460s, and Albertism had had a definite influence on the work of masters like Jan z Głogowa (Johannes Glogoviensis, d. 1507), who incorporated Dionysian elements into his physchology.[128] But even Glogoviensis, in his revision of Le Tourneur's questions on *De anima*, strikes a balance between Albertist and Thomist positions and tends to compromise on sensitive subjects like intellection without images.[129] At Krakow, as throughout the universities of northern Europe, the impetus of the *via antiqua* was toward Thomism, and this definitively established its ascendancy in the next century.[130]

Thus the explicit influence of Albert on late medieval psychology, like his influence in natural philosophy in general, had an extended history and a reasonably abrupt end. During the first decades of the fourteenth century at Paris, Albert was accepted as the chief Latin authority in philosophy, to be cited along with Aristotle, Averroës, and Avicenna. Eclectic thinkers like Burley and Jean de Jandun integrated his teaching on the soul into a synthesis of Averroës and thirteenth-century Latin philosophy. The rise of the *via moderna* transformed the tone and nature of psychological thought; Albert's methods and conclusions became peripheral to writers like Ockham and Buridan. With the emergence of the *via antiqua* at the end of the fourteenth century in Paris and later in the German universities, Albert's influence reached its highest mark among Albertists like Heymerich and, at least indirectly, Cusanus, who were anxious to defend the psychological theories they attributed to their master.

---

[127] *Manuale scholarium*, p. 103.

[128] See Kuksewicz, " 'Via antiqua' and 'via modernorum' in der Krakauer Psychologie im 15. Jahrhundert," in *Antiqui und Moderni*, pp. 509-514.

[129] See Michalski, "La philosophie thomistique en Pologne à la fin du quinzième et au début du seizième siècle," *Bulletin international de l'académie des sciences de Cracovie, Classe de philologie* (Jan.-July, 1916), p. 69; and Kuksewicz, "Le prolongement des polémiques entre les albertistes et les thomistes vu à travers le *Commentaire du De anima* de Jean de Głogow," *Archiv für Geschichte de Philosophie* 44 (1962), 151-171. Note that in this article, Kuksewicz attempts to make Glogoviensis more of an Albertist than he actually is by grossly misreading certain questions, like the one concerning intellection without images; cf. Jean Le Tourneur (Johannes Versor), *Questiones librorum de anima . . . per magistrum Joannem Glogoviensem . . . emendatum* III.13 (Krakow, 1514), f. 174r-174v.

[130] Weisheipl, "Thomism," pp. 132-134.

Albert's authority gradually ceded to Thomas' during the second half of the fifteenth century, and by the beginning of the sixteenth century his eclipse was assured. Except at Cologne, which had a special allegiance to him, his philosophical influence was minimal after 1530. He retained his reputation and respect through the sixteenth and into the seventeenth centuries, but less as a philosopher than as, on the one hand, a symbol of the greatness of German thought for humanists and nationalists like Celtes and Aventinus and, on the other, a magus wise in the occult properties of natural objects and author of a number of enormously popular *spuria* like the *Liber secretorum*.[131]

In another sense, however, Albert's influence on theories of the soul was more persistent. As the writer who integrated Aristotle and Arabic Aristotelianism into Christian natural philosophy, he was for all practical purposes the inventor of systematic Latin psychology. Even after Albert's works were no longer printed and read, writing on the soul remained fundamentally Aristotelian, and many of its most basic elements — the lists of the faculties, the questions asked concerning the nature of the soul and its relation to the faculties and the body — can be traced back to Albert's *De anima* and *Summa de homine*. It is not until the seventeenth century, when philosophers like Descartes and later Locke transformed and redirected psychological theory, that Albert's influence can be truly said to be eclipsed.

---

[131] Grabmann, "Einfluss," p. 393; Frances Yates, *The Art of Memory* (London, 1966), pp. 201-202.

21

Albert the Great and the *Studio Patavino* in the Late Fifteenth and Early Sixteenth Centuries

Edward P. Mahoney
Duke University

Albert the Great came as a youth to Northern Italy, and to Padua in particular, presumably in order to engage in studies in the liberal arts.[1] Although there has been a long scholarly debate over the last

---

I am indebted to Father James A. Weisheipl, OP, for various suggestions regarding the focus and scope of this essay, for his encouragement over the last few years, and above all for his remarkable persistence in bringing me to finish. I hope that my presentation of an over-looked chapter in the history of Albertism will repay him for his exemplary patience and untir-ing efforts in assembling a volume in honour of his Dominican confrere. I must also express my gratitude to the Duke University Research Council for fellowships and grants over the years which helped make possible the research on which my essay is based. The Latin of the early printed books cited in the essay was standardized.

[1] Joachim Sighart, *Albert the Great*, trans. T. A. Dixon (London, 1876), p. 16; Franz Pelster, *Kritische Studien zum Leben und zu den Schriften Alberts des Grossen* (Freiburg im Breisgau, 1920), pp. 58-60; P. Mandonnet, "Albert le Grand," *Dictionnaire de théologie catholique*. 1 (Paris, 1930), col. 666; Heribert C. Scheeben, "Zur Chronologie des Lebens Alberts des Grossen," in *St. Albertus-Magnus-Festschrift* (Fribourg, Switzerland, 1932), pp. 231-241; Ludo-vico De Simone, "Il B. Alberto Magno in Italia," *Memorie Domenicane*, 48 (1931), 366-367; M. H. Laurent, "Les grandes lignes de la vie du bienheureux Albert," *Revue thomiste*, 36 (1931), 257. For a useful survey of the secondary literature regarding Albert's life, see Pietro Castagno-li, "La vita e gli scritti di Sant'Alberto Magno (Rassegna bibliografica)," *Divus Thomas* (Piacenza), 37 (1934), 129-137. For a good general introduction to his life and thought, see Wil-liam A. Wallace, "Albertus Magnus, Saint," in *Dictionary of Scientific Biography*, 1 (New York, 1970), pp. 99-103. See now the life of Albert in this volume written by James A. Weisheipl, especially the discussion regarding the age at which Albert entered the Dominicans (pp. 17-19).

several decades regarding Albert's stay in Padua, it appears to be generally held today that he studied there during the school year of 1222-1223 and entered the Order of Preachers after hearing the preaching of its second general, Jordan of Saxony.[2] Whether he also completed his novitiate year at Padua seems doubtful, as do the claims that he studied either philosophy or theology and later taught the same subjects in a Dominican convent in the same city.[3] Weish-

---

[2] Andrea Gloria, "Quot annos et in quibus Italiae urbibus Albertus Magnus moratus sit," Atti del Reale Instituto Veneto, ser. 5, vol. 6 (1879-1880), 1025-1050; Heinrich Denifle, *Die Entstehung der Universitäten des Mittelalters bis 1400* (Berlin, 1885), pp. 280-281, n. 231 and n. 232; J. A. Endres, "Das Geburtsjahr und die Chronologie in der ersten Lebenshälfte Alberts des Grossen," *Historisches Jahrbuch*, 31 (1910), 295-298; Emil Michael, "Wann ist Albert der Grosse geboren?" *Zeitschrift für Katholische Theologie*, 35 (1911), 562-563; Franz Pelster, *Kritische Studien zum Leben und zu den Schriften Alberts des Grossen* (Freiburg im Breisgau, 1920), pp. 52-53; P. Mandonnet, "La date de naissance d'Albert le Grand," *Revue thomists*, 36 (1931), 236; Heribert C. Scheeben, *Albert der Grosse zur Chronologie seines Lebens* (Leipzig, 1931), pp. 8-14; idem, *Albertus Magnus* (Cologne, 1955), pp. 36-39.There seems no good evidence for the claim of Albert Garreau, *Saint Albert le Grand* (Paris, 1932), pp. 35-36, that Albert first went with his uncle in 1222 to study at the University of Bologna. Also dubious is his further claim (pp. 43-44) that he entered the Dominicans in that city and only later went to Padua.

[3] Scheeben (*Albert der Grosse*, pp. 8-11 and 14) argues that since Jordan of Saxony had won thirty students for the order and they were foreigners in Italy, it is out of the question that they would have done their novitiate there and not in one of the Dominican houses in their native countries. He therefore proposes that while Albert took the habit in Padua, he did his novitiate in a Dominican convent in Germany. For an attack on Scheeben and a reassertion that Albert was a novice in Italy, see Alberto Zucchi, "Sant'Alberto Magno a Padova," *Memorie Domenicane*, 49 (1932), 393-394. J. Quétif and J. Echard, *Scriptores ordinis praedicatorum*, 1 (Paris, 1719), p. 162, state that Albert studied philosophy, mathematics and medicine at Padua. In like fashion, Angiolo Puccetti, *San Alberto Magno dell'Ordine dei Predicatori, vescovo e dottore della chiesa: profilo biografico* (Rome, 1932), p. 10, claims not only that Albert studied the arts and medicine at Padua, but also that he gained there his first knowledge of the physical and ethical works of Aristotle. Georg von Hertling argues that it was highly likely that during this period of his life Albert made his first, basic acquaintance with the writings of Aristotle and his Arab commentators. See von Hertling's *Albertus Magnus: Beiträge zu seiner Würdigung*, in Beiträge, 14/5-6 (Münster, 1914), p. 5. Paul Simon, in his article, "Alberto Magno (s.)," in *Enciclopedia filosofica*, 2nd ed., 1 (Florence, 1967), col. 152, also believes that Albert's knowledge of Aristotelian philosophy had its origins in Italy. Sighart (pp. 24-25) assumes that Albert studied philosophy at Padua unti the age of thirty. Quétif-Echard (p. 162) believe that after his entry into the Dominican Order Albert studied theology for some years either at Padua or Bologna. Scheeben (*Albertus Magnus*, p. 30) rules out Albert studying theology at Padua. On the other hand, Zucchi argues that Albert studied for twenty years in Italy, since he there became a physicist, geologist and moralist (p. 397). He thus concludes (p. 399) that it is in Italy, above all at Padua, that we find the place of Albert's intellectual and religious formation. On the basis of a tradition of the Dominican convent of Saint Augustine at Padua, Zucchi argues (pp. 402-406) that Albert not only studied philosophy and theology at Padua but also taught those subjects there. Pelster (*Kritische Studien*, pp. 59-60) rejects the view that Albert had studied philosophy and medicine at Padua before he entered the Dominicans, just as he rejects the claim that Albert taught at Padua.

eipl has argued that Albert could not have become acquainted with Aristotelianism at Padua since Aristotelian teaching had not yet been introduced into the curriculum, and he adds that Roger Bacon testifies that Albert was self-taught in philosophy.[4] Albert refers to Padua and Venice on a few occasions in his writings, and modern scholars have frequently alluded to these texts.[5] It does not seem too much of an exaggeration to suggest that his relatively brief stay at Padua had a life-long effect on his intellectual and spiritual life, since it was here that he first entered a university environment and it was also here that he found his vocation as a son of Saint Dominic.[6] Consequently, since there are already several studies about the influence of Thomas Aquinas and John Duns Scotus on the philosophers at Padua toward the end of the fifteenth century and the beginning of the sixteenth century, it seems appropriate to chronicle and evaluate

---

[4] James A. Weisheipl, "Albertus Magnus and the Oxford Platonists," *Proceedings of the American Catholic Philosophical Association*, (1958), 124 n. 2. See also Nancy G. Siraisi, *Arts and Sciences at Padua: The "Studium" of Padua before 1350* (Toronto, 1973), pp. 112-113. However, in his article for this volume, above 000, Weisheipl now considers it likely that Albert had studied some of Aristotle's works at Padua, but that it is unlikely that he absorbed much of Aristotle at that time.

[5] See Albert, *Meteora* III, tr.2, c.12 (ed. Borgnet 4: 629a); *De nat. loc.*, tr.3, c.2 (ed. Borgnet 9: 570b-571a); *De mineral.*, II, tr.3, c.1 (ed. Borgnet 5: 48b-49a). The first two references are to Padua and the third is to Venice. There is also a reference to a woman at Padua who fasted for forty days in *De hom.*, q.10, a.5 (ed. Borgnet 35: 119b). While the first three references are to be found in a variety of secondary sources, the fourth is only rarely cited. For references to the first three texts, see Quétif and Echard, *Scriptores*, p. 163 n. 4; Paul de Loe, "De vita et scriptis D. Alberti Magni," *Analecta Bollandiana*, 20 (1901), 277; Emil Michael, *Geschichte des deutschen Volkes seit dem 13. Jahrhundert bis zum Ausgang des Mittelalters*, 3 (Freiburg im Breisgau, 1903), p. 71; von Hertling, *Albertus Magnus*, p. 5; Pelster, *Kritische Studien*, pp. 52-53; "Alberts des Grossen Jugendaufenthalt in Italien," *Historisches Jahrbuch*, 42 (1922), 102-105; Lynn Thorndike, *A History of Magic and Experimental Science*, 2 (New York, 1929), p. 523; Laurent, "Les grandes lignes," p. 257; Scheeben, *Albertus Magnus*, pp. 29-30; Siraisi, *Arts and Sciences*, p. 17 n. 8. The sole scholar I have found referred to the fourth text cited above is Zucchi, "Sant'Alberto Magno," p. 395.

[6] "Wichtig ist jedenfalls, dass italienische Umgebung und italienisches Geistesleben auf den schon damals recht aufgeschlossenen Sinn des jungen Deutschen ihren Einfluss ausübten und dass höchst wahrscheinlich auf dem Boden der Lombardei die jugendfrische Stiftung des hl. Dominikus, wie sie in der liebenswürdigen Gestalt seines ersten Nachfolgers Jordanis sich verkörperte, werbend an Albert herantrat" (Pelster, "Alberts des Grossen Jugendaufenthalt in Italien," p. 106).

Albert's fortunes at his *alma mater* during the same period.[7] In this

---

[7] For the general influence of Thomas in the Italian Reniassance see Paul Oskar Kristeller's essay, "Thomism and the Italian Thought of the Renaissance," in his *Medieval Aspects of Renaissance Learning*, ed. and trans. E. P. Mahoney (Durham, N.C., 1974), pp. 27-91, which originally appeared in his *Le Thomisme et la pensée italienne de la Renaissance* (Montreal, 1967). For Thomas' influence at Padua there is the older work of Pietro Ragnisco, *Della for tuna di S. Tommaso d'Aquino nella Università di Padova durante il Rinascimento* (Padua, 1892), as well as my own essay, "Saint Thomas and the School of Padua at the End of the Fifteenth Century," in *Thomas and Bonaventure: A Septicentenary Commemoration, Proceedings of the American Catholic Philosophical Association*, 48 (1974), 277-285. Related studies on Thomas' reputation during the Italian Renaissance are John W. O'Malley, "Some Renaissance Panegyrics of Aquinas," *Renaissance Quarterly*, 27 (1974), 174-192, and Glori Cappello, "Umanesimo e scolastica: Il Valla, gli umanisti et Tommaso d'Aquino," *Rivista di filosofia neo-scolastica* 69 (1977), 423-442. See also F. Edward Cranz, "The Publishing History of the Aristotle Commentaries of Thomas Aquinas," *Traditio*, 34 (1978), 157-192.

The role of Duns Scotus' thought in the philosophical life of the University of Padua at the end of the fifteenth and the beginning of the sixteenth centuries has been illuminated by various studies of Antonino Poppi. See his "Lo scotista patavino Antonio Trombetta (1436-1517)," *Il Santo*, 2 (1962), 349-367; "L'antiaverroismo nella scolastica padovana alla fine del secolo XV," *Studia Patavina*, 11 (1964), 102-124; "Il contributo dei *formalisti* padovani al problema delle distinzioni," in *Problemi e figure della Scuola scotista del Santo* (Padua, 1966), pp. 671-702; *Causalità e infinità nella scuola padovana dal 1480 al 1513* (Padua, 1966), pp. 273-348; "Padova, Scuola di," in *Enciclopedia filosofica*, 2nd ed., 4 (Florence, 1967), col. 1263-1270; "Trombetta, Antonio," in *New Catholic Encyclopedia*, 14 (New York, 1967), p. 314; "Per una storia della cultura nel Convento del Santo dal XIII al XIX secolo," *Quaderni per la storia dell'Università di Padova*, 3 (1970), 1-29. See also the studies of Pietro Scapin, "Maurizio O'Fihely editore e commentatore di Duns Scoto" and "La metafisica scotista a Padova dal XV al XVII," in *Storia e cultura al Santo*, ed. Antonino Poppi (Vicenza, 1976), pp. 303-308 and 485-538. I have discussed Scotus' influence in my essay, "Duns Scotus and the School of Padua around 1500," in *Regnum Hominis et Regnum Dei: Acta Quarti Congressus Scotistici Internationalis*. ed. Camille Bérubé, 2 (Rome, 1978), pp. 215-227.

The existence of an Albertistic tradition at Padua, exemplified in the writings of Nicoletto Vernia, Agostino Nifo and Marcantonio Zimara appears to have escaped the notice of some of the most important scholars in the field of Renaissance Aristotelianism, namely Bruno Nardi, Eugenio Garin and Antonino Poppi. However, both Nardi and Poppi do note references to Albert in some of the Paduans. See for example Nardi, *Saggi sull'aristotelismo padovano dal secolo XIV al XVI* (Florence, 1958), pp. 104-105 and 106. It remains, nonetheless, that the presence of Albertism at Padua is not brought out either in Garin's magisterial *Storia della filosofia italiana* (Turin, 1966) or in Poppi's "Padova, Scuola di" and *Introduzione all'aristotelismo padovano* (Padua, 1970), all of which are important contributions to the scholarly literature. I have found no mention of the Paduans in Paul Simon, "Albertisti," *Enciclopedia filosofica*, 2nd ed., 1 (Florence, 1967), col. 151, nor does there appear to be any allusion to them in any of the literature on Albertism. For this reason, the present study represents a new contribution both to our understanding of the nature of the philosophical community at Padua around 1500 and also to the history of Albertism. Although limitations of space forbid adequate discussion of why Vernia, Nifo and Zimara were attracted to Albert, I would like to offer some tentative suggestions. In the first place, all three were attracted to Averroës as an interpreter of Aristotle, just as all three were acquainted with the writings of John of Jandun, one of the most important exponents of so-called "Averroism" in the Middle Ages. Let us recall that Jandun himself mentions Albert frequently and with respect. It should also be underscored that at various points in his writings Albert shows regard for Averroës, says that he differs but little from him, and reveals a strong interest in the philosophical doctrine of the human intellect's union with a

survey we shall first indicate briefly Albert's initial impact on Italian culture in the late Middle Ages and the early Renaissance. Then, after presenting some of the references to Albert which are to be found in the writings of Paul of Venice, we shall concentrate our attention on several of the major philosophers who dominate the Paduan intellectual milieu of the late fifteenth and early sixteenth centuries. These philosophers are Gaetano di Thiene, Nicoletto Vernia, Agostino Nifo, Pietro Pomponazzi and Marcantonio Zimara. By reason of limits of space, references to other philosophers will be regrettably sparse.

In his studies on Albert's influence on later medieval thought, Martin Grabmann has documented references to Albert in the writings of a variety of Italian medieval figures. These include both Dominicans like Ptolemy of Lucca, Ranieri da Pisa, Giovanni Balbi da Genova, Remigio di Girolami and Savonarola, and also others outside the Order, such as Taddeo da Parma, Pietro d'Abano, Antonio da Parma, Angelo d'Arezzo, Apollinare Offredi, Guido Cavalcanti, Dante, Fernando de Cordoba, Cardinal Bessarion and Giovanni Pico della Mirandola.[8] The thesis of Albert's influence on Dante was repeatedly defended by Bruno Nardi in various of his scholarly publications.[9] Recent scholars have demonstrated the heavy use that the fourteenth-century Jewish philosopher, Jehudàh Romano, makes of Albert's *De natura et origine animae* in his own doctrine regarding the intellect and the strong influence of Albert on

---

separate intellect. We may suggest, then, that the shift of allegiance from Averroës to Albert was not as dramatic a shift as would have been involved in a shift from Averroës to Thomas. While Vernia, Nifo and Zimara also show respect for Thomas, it is noteworthy that they, all laymen, advance Albert's cause, while the Dominicans at Padua seem to promote Thomas almost exclusively. In a word, "Paduan Albertism" exists solely outside the Dominican Order and primarily among lay philosophers who started by accepting Averroës as the "Commentator" on Aristotle, though two of them (that is, Vernia and Nifo) later abandoned him as an accurate guide to Aristotle. Cajetan may approach Thomas' doctrine of finite being from an Albertist perspective (see n. 77 below), but he does not consciously put Albert before Thomas as a guide to Aristotle or as a philosopher.

[8] Martin Grabmann, *Mittelalterliches Geistesleben*, 2 (Munich, 1936), pp. 290, 395-400 and 407-408. See also pp. 242, 245 and 254-255 for Taddeo's references to Albert. Grabmann also discusses Albert's influence on Taddeo and Apollinare Offredi in his "Die Aristoteleskommentar des Heinrich von Brüssel und der Einfluss Alberts des Grossen auf die mittelalterliche Aristoteleserklärung," *Sitzungsberichte der Bayerischen Akademie der Wissenschaften*, Philosophisch-historische Abteilung Jahrgang 1943, Heft 10 (Munich, 1944), pp. 55-57.

[9] See Bruno Nardi, *Dante e la cultura medievale* (Bari, 1942); *Nel mondo di Dante* (Rome, 1944); *Studi di filosofia medievale* (Rome, 1960).

the Mariology of Sant'Antonino, archbishop of Florence.[10] In an early letter of 1454, Marsilio Ficino cites Albert's commentary on the *Physics* and states that he is not afraid to put him in second place among the Latin philosophers (*Latinorum philosophorum secundo loco*), apparently intending to rank him just after Thomas Aquinas.[11] There are also references to Albert's commentary on the Pseudo-Dionysius and his *Speculum astronomiae* in Ficino's *Theologia platonica* and *Liber de vita coelitus comparanda*.[12] If we turn to Padua itself, we find that Albert's writings were not unknown to professors there during the fourteenth century. In her study on the curriculum of the university before 1350, Siraisi has brought out the use of Albert's writings, especially his *De natura locorum*, by Pietro d'Abano in his own *Expositio Problematum Aristotelis*, and also Jacopo Dondi's acquaintance with Albert's views on tides.[13] Moreover, she has suggested that some of the commentaries on Aristotle written at Paris in the thirteenth century — presumably including those of Albert — may have been introduced at Padua from about 1260 to 1315 by such people as Zambonino da Gaza, Pietro d'Abano or the Augustinian Hermits, who had a convent at Padua and contacts with Paris.[14] Albert was not unknown to philosophers at Padua in the latter part of the fourteenth century. In questions on the *De anima* which he disputed at Padua in 1385, Biagio Pelacani da Parma explicitly refers to Albert's *De natura elementorum*.[15] While these allusions to Albert in Pietro d'Abano, Jacopo Dondi and Biagio Pelacani da Parma demonstrate that he was studied and cited during the fourteenth century, only in the next century would the wide range of his writings be well known to philosophers at Padua and have a striking influence on discussions of major philosophical issues.

---

[10] On Jehudàh and Albert, see Josef Barùkh Sermonetta, "La dottrina dell'intelletto e la 'fede filosofica' di Jehudàh e Immanuel Romano," *Studi medievali*, 3rd ser., 6 (1965), 41-48. For Sant'Antonino's use of Albert, see Eberhard Brand, *Die Mitwirkung der seligsten Jungfrau zur Erlösung nach dem hl. Antonin von Florenz mit besonderer Berücksichtigung des Verhältnisses zur Lehre Alberts des Grossen* (Rome, 1945), pp. 60-62. On Albert's own Mariology, see Robert J. Buschmiller, *The Maternity of Mary in the Mariology of St. Albert the Great* (Carthagena, Ohio, 1959), and Albert Fries, *Die Gedanken des heiligen Alberts Magnus über die Gottesmutter*, Thomistische Studien 7 (Freiburg, 1959).

[11] For the text, see Paul Oskar Kristeller, *Studies in Renaissance Thought and Letters* (Rome, 1969), p. 149. See also Kristeller's remark on p. 143.

[12] Ibid., p. 39.

[13] Siraisi, pp. 117-125.

[14] Ibid., pp. 141-142.

[15] Biagio Pelacani da Parma, *Le Quaestiones de anima*, ed. Graziella Federici Vescovini (Florence, 1974), p. 78.

Paul of Venice (1372-1429) shows so little interest in Albert that it would be impossible to characterize him as one of Paul's major authorities. For example, in the *Summa naturalium* there are rarely any explicit references to Albert,[16] while there appear to be no references at all to Albert in Paul's commentary on *De generatione*.[17] In Book III of his commentary on the *De anima*, Paul does list Albert, along with Egidio Romano and Thomas Aquinas, among *nostri antiqui* who held that the universal is the first object of the intellect.[18] However, Albert does not enjoy the central role played by Egidio in the work. There thus appears to be no other mention of Albert in Book III and none whatever in Book I, not even in the discussion on the nature of universals.[19] All the other references to Albert appear in Book II, though not all are to Albert's commentary on the *De anima*. Paul does cite Albert to show that matter is not of itself an individual something (*hoc aliquid*), since it does not provide the specific definition of a thing.[20] He presents in detail Albert's views when discussing the nature of nutrition, its relationship to the reproductive structure of animals, and how both individual and species are benefited.[21] There are also passages in which Paul alludes to Albert's explanations of how things composed of water and how brass bodies make sound, the relation between nerves and the brain in head wounds, how a decapitated cock can continue to sing, how a spider can sense a fly at a great distance, and the mode of vision of aquatic animals.[22] In a word, Paul seems to cite Albert primarily for odd and

---

[16] See *Summa naturalium* (Venice, 1476), especially *Liber de anima*, sig. o3ra and *Liber metaphysicae*, sig. r2ra, sig. v1rb, sig. v9va and sig. x6vab. There may of course be other unacknowledged borrowings from Albert.

[17] Paul of Venice, *Expositio super libros de generatione et corruptione Aristotelis* (Venice, 1498).

[18] Paul of Venice, *In libros de anima explanatio* (Venice, 1504), III, t.c.11, f. 136va. Their view is contrasted to *nostri antiqui*, such as Walter Burley, William of Ockham, Gregory of Rimini and John Buridan (f. 137ra). On the different positions, see Camille Bérubé, *La connaissance de l'individuel au moyen âge* (Montreal and Paris, 1964), especially pp. 27-31 and 259-277.

[19] Paul of Venice, *In . . . de anima*, I, t.c.8, f. 7rab.

[20] Ibid., II, t.c.2, f. 36vb. See Albert, *De anima*, II, tr.1, c.1 (ed. Colon. 7/1: 64 v. 32-36).

[21] Paul of Venice, *In . . . de anima*, II, t.c.46, f. 61rb-61va, and t.c.47, f. 62ra. See Albert, *De anima*, II, tr.2, c.4 and c.7 (ed. Colon. 7/1: 87-88 and 91-92).

[22] Paul of Venice, *In . . . de anima*, II, t.c.82, f. 81ra and t.c.85, f. 83ra; t.c.82, f. 81rb; t.c.87, ff. 84va-85ra; t.c.94, f. 88ra t.c.92-93, f. 87ra. I have been unable to find the passages Paul attributes to Albert's *Mineralia* regarding sound. For the headless cock who sings, see Albert, *De anima*, II, tr.3, c.22 (ed. Colon. 7/1: 130 vv. 24-32). Albert's remarks on the spider and the mode of sight of aquatic beings can be found in his *De anima*, II, tr.3, c.23 (ed. Colon. 7/1: 132 vv. 58-67; 133 vv. 55-69).

interesting facts and theories. He does not cite him in his own discussions on the central questions of the nature of the soul, the intellect and human cognition. Albert also plays only a rather minor role in Paul's commentary on the *Physics*. In this entire large work there seems to be only one reference to Albert in Book I and about six in Book V. On the question of whether forms preexist in some fashion before generation, Albert's views are listed along with those of Plato, Grosseteste (*Lincolniensis*) and Egidio.[23] The major authorities from the Latins used in the commentary are Egidio and Walter Burley, though there are occasional references to William of Ockham, John Buridan, and Gregory of Rimini.

If Paul of Venice made very little use of Albert himself as an authority, the same cannot be said of his student, Gaetano di Thiene (1387-1465).[24] Albert's commentaries on *De animalibus, Physics, De anima* and *De generatione* are among the books mentioned in his will, and Albert must be listed, along with Averroës, Egidio Romano, John of Jandun and Walter Burley, among the authors he most frequently cites.[25] For example, in his commentary on the *Physics*, finished around 1439, Thiene cites and quotes Albert with great frequency. He gives Albert's views that mobile body is the subject of natural science, presents an argument against this *via*, but then shows how to reply.[26] When discussing the metaphysical status of universals, he considers Albert to be a member of the *via media*, along with Averroës, Thomas and Egidio, and he presents Albert's distinction of universals *ante rem, in re*, and *post rem*.[27] In the following question, on the relation of our knowledge of singulars to that of universals, he lists Albert's as one of two basic positions.[28] In his discussion regarding the preexistence of form in matter, where Albert is again listed among the holders of the middle view, Thiene presents Albert's ideas on form, privation and *incohatio*, but he then argues

---

[23] Paul of Venice, *Expositio super octo libros physicorum Aristotelis necnon super commento Averrois cum dubiis eiusdem* (Venice, 1499), I, t.c.34, sig. d3vb-d4ra (which concerns preexistence of forms); V, sig. A7ra, A8rb, B3va, B3va, B5rab and B5vb. See Albert, *Physica*, V, tr.1, cc.5, 6, 8 and 9.

[24] On Thiene's life and works, see Silvestro da Valsanzibo, *Vita e dottrina di Gaetano di Thiene*, 2nd ed. (Padua, 1949), pp. 1-2 and 21-39.

[25] Ibid., pp. 18 and 221.

[26] Gaetano di Thiene, *Recollectae super octo libros physicorum cum annotationibus textuum* (Venice, 1496), I, q.1, f. 2vab. See Albert, *Physica*, I, tr.1, c.3 (ed. Borgnet 3: 6b-7b).

[27] Thiene, *Physicorum*, I, q.4, f. 5ra. See Albert, *Physica*, I, tr.1, c.6 (ed. Borgnet 3: 13b).

[28] Thiene, *Physicorum*, I, q.5, f. 6rb. See Albert, *Physica*, I, tr.1, c.6 (ed. Borgnet 3: 14b-15a).

against them.[29] Thiene occasionally tries to show how Albert follows Averroës or disagrees only slightly with him,[30] but elsewhere he characterizes Albert as the best defender of Avicenna (*optimus defensator Avicennae*) against the criticisms of Averroës.[31] The striking number of explicit quotations from Albert in this early commentary surely indicates Thiene's knowledge of and high regard for Albert. Albert.

In his commentary on the *De anima*, which was composed around 1443, Thiene again reveals a strong interest in Albert. He speaks on occasion of following the *via Alberti*, and he makes close and constant use of Albert's commentary, often giving word for word quotations. On the question of universals, Thiene again presents Albert as holding the middle position in agreement with Averroës, Thomas and Egidio.[32] He appears to have great respect for Albert's conception of the soul, carefully citing Albert's views on how the soul and its powers are present in the heart and the rest of the body, on the relation of the powers to the soul, and on how there is only one soul in each animate thing, though a soul is more perfect the more powers it has.[33] Albert's influence on Thiene is especially noticeable in regard to the question whether the senses are passive or active powers, a classic topic for late medieval and Renaissance Aristotelians. The four arguments which he presents for the agent sense, along with the replies to them, are taken from Albert. Thiene astutely observes that Averroës had touched on another opinion, one which he did not assert in his own name, namely, that an Intelligence is one of the causes required for sensation to occur. However, Thiene immediately rejects this opinion, basing himself on Albert and following him

---

[29] Thiene, *Physicorum*, I, q.13, f. 11rb-11vb. See Albert, *Physica*, I, tr.3, c.15 (ed. Borgnet 3: 83b-84b) for Thiene's quotation from Albert. Albert is also cited along with Burley as a major position on privation in I, q.18, f. 17va.

[30] Thiene, *Physicorum*, I, q.17, f. 17ra; V, q.6, f. 37vb; and VIII, q.1, f. 44va. In the first of these passages Thiene is quoting from Albert, *Physica*, I, tr.3, c.4 (ed. Borgnet 3: 56a).

[31] Thiene, *Physicorum*, I, t.c.61, f. 19rb and II, q.3, f. 21vb. See Albert, *Physica*, I, tr.3, c.18, and II, tr.1, c.10 (ed. Borgnet 3: 19a and 112b).

[32] Gaetano di Thiene, *Super libros de anima* (Venice, 1493), I, comm. 8, f. 4rb-4va.

[33] Ibid., II, comm. 22, f. 18rb; comm. 29, f. 20ra; comm. 32, f. 21vab. In regard to the first and third of these passages, see Albert, *De anima*, II, tr.1, c.7 and c. 11 (ed. Colon. 7/1: 74b-75b and 79b-81a). Thiene appears to accept in II, comm. 50, f. 25vb Albert's fourfold difference between the nutritive and augmentative powers. See Albert, *De anima*, II, tr.2, c.6 (ed. Colon. 7/1: 90a-91b).

almost word for word.[34] And when he takes up the number and operations of the internal senses, he announces that he will limit himself to three distinguished men (*clarissimi viri*), namely, Averroës, Avicenna and Albert. In his discussion, he sets forth Albert's four levels of abstraction and also shows how Albert does not wholly agree with Avicenna, though each says there are five internal senses.[35] Especially noteworthy is Thiene's emphasis on Albert's doctrine regarding the human intellect's union with the separate intellect and how Albert agrees and disagrees with Averroës on this topic.[36] Subsequent Paduan philosophers would be much intrigued with Averroës' teaching regarding such a union and the happiness that it would bring human beings. Moreover, by drawing attention to Albert's benign adoption of such a view as true to Aristotelianism and by showing such respect for Albert's discussions regarding the nature of the soul Thiene doubtlessly prepared the way for the Albertistic tendencies of both Vernia and Nifo.

Nicoletto Vernia (ca. 1420-1499), who was himself a student of Thiene's, also owned copies of various works of Albert. There is a manuscript at Venice in the Biblioteca Nazionale San Marco (Marciana, Cod. Lat. VI, 214 [= 2566]) containing Albert's commentaries on the *Prior Analytics* and the *De interpretatione* which was once owned by Vernia.[37] Among the incunabula presently in the Biblioteca Universitaria at Padua are two which had belonged to Vernia. They are copies of Albert's *De animalibus* (Rome: Simon Chardella, 1478; GKW # 587) and his *De anima* printed together with his *De intellectu et intelligibili* (Raynaldus de Novimagio, 1481; GKW #

---

[34] Thiene, *De anima*, II, comm. 62, f. 28rab. See Albert, *De anima*, II, tr.3, cc.1, 3 and 6 (ed. Colon. 7/1: 97b-98a, 101a, 104ab and 105b-106a). For further discussion, see Ermenegildo Bertola, "La questione del 'Senso agente' in Gaetano di Thiene," in his *Saggi e studi di filosofia medievale* (Padua, 1951), pp. 53-69; Edward P. Mahoney, "Agostino Nifo's *De sensu agente*," *Archiv für Geschichte der Philosophie*, 53 (1971), 121 and 133-134; Adriaan Pattin, "Pour l'histoire du sens agent au moyen âge," *Bulletin de philosophie médiévale*, 16-17 (1974-1975), 109. In II, comm. 88, f. 35vb, Thiene cites Albert's discussion to explain how *grilli* continue to sing after they have been decapitated. See n. 22 above.

[35] Thiene, *De anima*, II, comm. !62, f. 53rb-53vb. See also Albert, *De anima*, II, tr.3, c.4, and tr. 4, c.7 (ed. Colon. 7/1: 101b-102a and 156b-158a).

[36] Thiene, *De anima*, III, comm. 36, f. 70vb-71ra and 72vb; comm. 39, f. 74rb. There is also heavy use of Albert in Thiene's *In quattuor Aristotelis metheororum libros expositio* (Venice, 1491), which was completed in 1460-1461.

[37] For a description of the manuscript, see Giuseppe Valentinelli, *Bibliotheca manuscripta ad S. Marci Venetiarum*, 4 (Venice, 1871), pp. 25-26. See f. 41v (second foliation) for proof that this manuscript belonged to Vernia. I examined the manuscript in 1972 and again in 1974.

585).[38] Moreover, works of Albert are included among the books mentioned in Vernia's will. They are the commentaries on the *Physics, De coelo, De generatione, Meteora* and *Metaphysics*.[39] We shall see from our analysis of his writings that he referred to a wide range of Albert's works.

In his early *Quaestio de gravibus et levibus*, composed by 1474, Vernia rejects the impetus theory of such thinkers as Walter Burley, Albert of Saxony, and Cajetan of Thiene, claiming instead that only Averroës grasped the thought of Aristotle on the motion of bodies. However, he also cites Albert's commentaries on the *Physics* and *De coelo*, identifies Albert with Averroës' position, and questions who in his right mind could doubt, when two such excellent philosophers agree, that this is undoubtedly the mind of Aristotle.[40] Interest in Albert is also evident in the later *Quaestio an ens mobile sit totius naturalis philosophiae subiectum*, which was finished in 1480. Vernia first presents Thomas as opposed to Albert, his teacher, since he thought that mobile being and not mobile body is the subject of natural philosophy. Thomas' position deviates from the principles of Aristotle and his Commentator, whose mind is rather that mobile body is the subject of natural science. However, in Vernia's judgment Albert himself is not wholly in agreement with their view, since he would understand by body the composite of matter and substantial form, while Averroës would say that the heaven is a mobile body and yet is not composed of matter and form. Toward the end of the question, Vernia cites Albert on the subalternation of the sciences in order to correct some remarks of Egidio.[41] On the other hand, Egidio is frequently cited in the *Quaestio an coelum sit ex materia et forma*

---

[38] These volumes are respectively Incunabula #186 and #360. The existence of the former was drawn to my attention by Pietro Ragnisco, *Nicoletto Vernia: Studi storici sulla filosofia padovana nella 2ᵃ metà del secolo decimoquarto* (Venice, 1891), p. 35 (= p. 625) n. 1, whereas the latter volume I discovered and identified on my own. Both volumes, which were examined in 1972, contain a statement that the book was left by Vernia to the monastery of San Bartolomeo in Vicenza. The first of these two volumes has also been identified recently by Barbara Marx, "Handschriften Paduaner Universitätsdozenten und Studenten aus San Bartolomeo di Vicenza," *Quaderni per la storia dell'Università di Padova*, 9-10 (1976-1977), 143.

[39] See Paolo Sambin, "Intorno a Nicoletto Vernia," *Rinascimento*, 3 (1952), 261-268. In the 1499 list (p. 267), the entry "Albertus in libro Phisicorum . . ." should be corrected to "libros."

[40] See Nicoletto Vernia, *De gravibus et levibus*, in Gaetano di Thiene, *De coelo et mundo* (Padua, 1474?; British Library copy: IC. 29956), ff. 174vb-175rb and 176vb-177rb. Cf. Anneliese Maier, *Zwei Grundprobleme der scholastischen Naturphilosophie*, 2nd ed. (Rome, 1951), p. 295.

[41] Nicoletto Vernia, *Quaestio an ens mobile sit totius naturalis philosophiae subjectum*, in *Marsilius de generatione et corruptione cum expositione Egidii* (Venice, 1500), ff. 226ra-228rb.

constitutum (1481), whereas Albert is never mentioned.[42] But when Vernia again treats of the heavens in his *Quaestio an coelum sit animatum* (1491), he cites Albert as a major source. Vernia seems especially concerned to find points of agreement between Averroës and Albert regarding the heavens, claiming once again that when these two are in agreement the true mind of Aristotle has been reached.[43] In his *Quaestio an dentur universalia realia* (1492), Vernia shows a close knowledge of a wide variety of Albert's writings. He carefully sets forth Albert's distinction of universals *ante rem, post rem*, and *in re*, and he claims, after presenting at length Albert's views on *incohatio* of form, that Averroës differs only slightly with Albert on that topic. These two distinguished philosophers (*clarissimi philosophi*) also agree that matter does not belong to the quiddity of a thing. Vernia sets it down as axiomatic that whenever these two greatest peripatetics are in agreement (*hii duo summi peripatetici concordent*), only someone inept at philosophy could say that this was not the mind of Aristotle.[44]

Vernia's preference for Averroës as the true commentator of Aristotle is strikingly evident in an early treatise on the intellective soul, certainly written before Pietro Barozzi's decree of 1489, which forbade further public discussion regarding the unity of the intellect, and probably written before 1483. Making no attempt to show any agreement between Albert and Averroës, Vernia lists the difficulties regarding individuation of the soul which result from the interpretation of Aristotle adopted by Albert and others, namely, that the intellective soul is the substantial form of the body, united to it in

---

[42] For Vernia's *Quaestio an coelum sit ex materia et forma constitutum* (1482), see Walter Burley, *Super octo libros phisicorum* (Venice, 1501).

[43] "Quando enim hi duo, Albertus et Averroës, concordant in aliquo, illud certe est Aristotelis intentio" (*Quaestio an caelum sit animatum*, Oxford, Bodleian Library, Canonici Latini Codex 506, f. 326v). Vernia also tries to conciliate them in his *An detur equale ad pondus* (1490), which is contained in the same manuscript. See f. 325r.

[44] Nicoletto Vernia, *Quaestio an dentur universalia realia*, in *Urbanus Averoysta philosophus summus ex Almifico Servorum Dive Marie Virginis ordine Commentorum omnium Averoys super librum Aristotelis de physico auditu expositor clarissimus* (Venice, 1492), unnumbered folios 1ra-2vb. "Averroes vero Cordubensis in paucis discrepat ab Alberto. . ." (f. 1rb); ". . .quia materia ad quidditatem non spectat secundum hos clarissimos philosophos, Averroem scilicet et Albertum. . ." (f. 2va); "Et sic est concludendum indubitanter istam fuisse Aristotelis intentionem, cum enim hii duo summi peripatetici concordent, Averroes et Albertus, in hoc quaesito, quis nisi ineptus ad philosophandum poterit dicere istam non fuisse Aristotelis intentionem" (f. 2vb). Such passages led Ragnisco, p. 38 (= 628), to speak of Vernia's "Albertistic Averroism." See also Ragnisco, pp. 51 and 64.

existence and numbered according to the number of human bodies.[45] However, Vernia dramatically rejects Averroës' interpretation of Aristotle on the soul in his later treatise against the Commentator, supposedly finished in 1492, but only published in 1504 after Vernia's death. There is heavy and constant use of Albert's works throughout this treatise, including his *De natura et origine animae, De homine, De intellectu et intelligibili,* and also his commentaries on the *Metaphysics, De anima,* and *Physics.* The sketch that Vernia presents of Plato's psychology is based not on the works of Plato himself, which Vernia knew in Ficino's translation, but rather on Albert. The essential features of this account are that for Plato the intellective souls were created from all eternity by God and contain in themselves the intelligible forms. The latter are forgotten once the soul is poured into the body, but the soul can be excited to recollect them if the body is purged through study. Plato also thought, we are told, that if souls lived correctly during their incarnation here, they would return to their respective stars.[46] Vernia then goes on to ascribe the doctrine of preexistence to Aristotle as well, claiming support from the Greek commentators. While he presents considerations from Albert against the preexistence and transmigration of the soul, he gives careful replies to these arguments. On the other hand, he accepts the authority of Albert, Thomas and Scotus to argue that the

---

[45] Nicoletto Vernia, "Utrum anima intellectiva humano corpore unita tanquam vera forma substantialis dans ei esse specificum substantiale, aeterna atque unica sit in omnibus hominibus" (Venice, Biblioteca Nazionale Marciana, Cod. Lat. vi, 105, ff. 156r1-160v1, at f. 157bisv2). See my article, "Nicoletto Vernia on the Soul and Immortality," in *Philosophy and Humanism: Renaissance Essays in Honor of Paul Oskar Kristeller,* ed. E. P. Mahoney (New York and Leiden, 1976), pp. 145-148. This question has also been studied by Giulio F. Pagallo, "Sull'autore (Nicoletto Vernia?) di un'anonima e inedita quaestio sull'anima del secolo xv," in *La filosofia della natura nel medioevo* (Milan, 1966), pp. 670-682.

[46] Nicoletto Vernia, *Contra perversam Averrois opinionem de unitate intellectus et de animae felicitate* (Venice, 1505), f. 3rb-4rb. The first edition was published at Venice in 1504 with Albert of Saxony's *Acutissimae quaestiones super libros de physica auscultatione.* Another printing with Albert of Saxony's work appeared at Venice in 1516. For the relevant references in Albert the Great, see my "Nicoletto Vernia on the Soul," p. 151 n. 19-20. One philosopher who may have inclined Vernia toward a greater interest in Albert's psychology was Apollinare Offredi, whose *De anima* was a major source in Vernia's own early treatise on the intellective soul. In his *Expositio in libros de anima* (Venice, 1496), Offredi cites Albert regarding intelligible species (f. 38vb and f. 43va) and considers Albert to be close to Averroës except where the *opinio fidei* is involved (f. 40ra). In his *Quaestiones,* which were published with the *Expositio,* Offredi cites Averroës and Albert together regarding the meaning of such terms as *imago* and *intentio* (ii, q.38, f. 69vb), shows special interest in Albert's conception of the internal sense powers (f. 70ra), and carefully examines Albert's arguments against Averroës on the manner in which the intellective soul is the form of the human body (iii, q.2, ff. 74vb-75va).

faith, the truth, and Aristotle all maintain, contrary to Averroës, that there is only one soul in a living thing and it is united in existence to the body.[47] Consequently, although he does at one point cite Albert's remark that he differs in few things with Averroës, he now rejects Averroës as the true interpreter of Aristotle.[48] When Vernia raises the question whether the soul undergoes an alteration in the process of knowledge because of its receiving intelligible species, he argues on the authority of Albert and the Greek commentators, namely, Simplicius and Themistius, that there is no need to postulate such species, and he further uses Albert as an authority to argue that Averroës himself did not maintain intelligible species.[49] This assimilation of Albert to the Greek commentators, who served as the major interpreters of Aristotle in Vernia's late philosophical development, is also evident in his *Proemium in libros de anima*. In that work, he seeks both to determine whether speculative or practical science has priority in the order of *doctrina* and also to investigate the nature of the science regarding the soul, that is, its relation to natural science, mathematics, and metaphysics. Themistius, Simplicius, and Albert appear to be treated as the major authorities on these issues. For example, after deciding how the science regarding the soul exceeds all other sciences in the certitude of its demonstrations, he remarks that this is the mind of Aristotle, gathered from "the entrails of the fathers," that is, from Themistius, Simplicius, and Albert.[50]

---

[47] Ibid., f. 8ra and f. 10rb-10vb. See my "Nicoletto Vernia on the Soul," p. 155 n. 32 and p. 158 nn. 43-44. It is also to be noted that the arguments for immortality ascribed to Avicenna (f. 8vab) are actually borrowed from Albert, *Summa de creaturis, II: De homine*, q. 61, a.1 and 2 (ed. Borgnet 35: 518-519 and 523-528). I shall not discuss here all references to Albert in this treatise, since I hope to discuss its Albertism in a separate essay on another occasion.

[48] Offredi, f. 5ra. See my "Nicoletto Vernia on the Soul," pp. 162-163.

[49] Offredi, ff. 4vb, 5vb and 6vab. The need for intelligible species in cognition and the question whether Averroës postulated such species were hotly debated at Padua toward the close of the fifteenth century. See my article, "Antonio Trombetta and Agostino Nifo on Averroës and Intelligible Species: A Philosophical Dispute at the University of Padua," in *Storia e cultura nel Convento del Santo a Padova*, ed. Antonino Poppi (Vicenza, 1976), 289-301. For a fine discussion on this debate, see also Father Poppi's essay, "La discussione sulla 'species intelligibilis' nella Scuola Padovana del cinquecento," in his *Saggi sul pensiero inedito di Pietro Pomponazzi* (Padua, 1970), 139-194. It should be noted that in his *Conclusiones*, ed. Bohdan Kieszkowski (Geneva, 1973), Giovanni Pico della Mirandola presents sixteen theses regarding Albert, the first of which is the following: "Species intelligibiles non sunt necessariae, et eos ponere non est bonis Peripateticis consentaneum" (p. 27).

[50] Nicoletto Vernia, *Proemium de libro de anima*, Oxford, Bodleian Library, Canonici Latini Codex 506, ff. 319-327. See especially f. 321v: "Is ergo est intellectus Aristotelis collectus ex visceribus patrum, Themistii, Simplicii et Alberti." Vernia rejects (f. 319v) Thiene's account of Albert on a supposed triple distinction regarding matter. For Vernia's interest in and explication of another Greek commentator on Aristotle, see my article, "Nicoletto Vernia and Agostino Nifo on Alexander of Aphrodisias: An Unnoticed Dispute," *Rivista critica di storia della filosofia*, 23 (1956), 268-296.

Agostino Nifo (ca. 1470-1538) was one of Vernia's students at Padua, and he also began his teaching career there. His first work was an edition with commentary of the medieval Latin version of Averroës' *Destructio destructionum* published together with his treatise on the agent sense. In the *De sensu agente*, written during the summer of 1495, Nifo at one point cites with approval Albert's commentary on the *De anima*, though he elsewhere questions Albert's analysis of Averroës' doctrine regarding the role of the sensible in sensation and decides to determine the mind of Aristotle and Averroës himself. But even while disagreeing with Albert, he maintains an attitude of respect for him.[51] In the commentary on the *Destructio*, which was composed between 1494 and 1497, Nifo frequently refers to the *latini* or the *expositores latini*, a group composed of Albert, Thomas, and Egidio Romano, but he consistently rejects their views and appears to prefer the Greek commentators, namely, Themistius, and Simplicius. Nonetheless, he shows both acquaintance with Albert's commentaries on the *Posterior Analytics, Physics, De coelo, De anima*, and *Liber de causis* and also interest in determining when Albert and Averroës agree. However, in one passage he accuses Albert of having contradicted Averroës without showing any willingness to understand him, and in another he simply says that Albert misunderstood Avicenna.[52] On the other hand, he defends both Albert and Thomas from attacks of hostile Franciscans.[53]

In his early commentary on the *De anima*, Nifo again shows far more respect for Averroës and the Greek commentators than he does for Albert, Thomas, and Egidio. Those works of Albert that are cited by title are the commentaries on the *Posterior Analytics, Physics, De anima*, and *Metaphysics*, as well as the *De homine*. The commentary contains one hint of the Albertism that would emerge in the *De intellectu*. It is Nifo's adoption of Albert's conception of the soul as a potestative whole. When commenting on the *De anima*, I, c. 1 (402b1-402b5), Nifo says that Aristotle believes that the soul is one in subject but divided according to is powers and Aristotle therefore

---

[51] Agostino Nifo, *De sensu agente*, in *Destructiones destructionum Averroys cum Augustini Niphi de Suessa expositione* (Venice, 1497), f. 126ra and f. 128ra. In these passages Nifo cites Albert, *De anima*, II, tr.3, cc.1 and 6. For further discussion, see my article, "Agostino Nifo's *De sensu agente*," *Archiv für Geschichte der Philosophie*, 53 (1971), 119-142, especially 141 n. 74.

[52] Agostino Nifo, *Destructiones destructionum*, I, dub. 11, f. 15ra, and XIV, dub. 1, f. 118r2-118v1.

[53] Ibid., V, dub. 3, f. 69vb. The hostile Franciscan is in fact Antonius Andreae in his *Quaestiones super duodecim libros metaphysicae*. See my article, "Antonio Trombetta and Agostino Nifo," p. 290.

thinks that it is a certain potestative whole (*totum quoddam potestativum et virtuale*).[54] Nifo again ascribes this concept to Aristotle when commenting on 402b9-11, and he uses it elsewhere in the commentary.[55]

The strong Albertistic orientation that we discovered in Vernia's treatise against Averroës is more than matched by Nifo's straightforward adoption of an Albertist psychology in his own *De intellectu*.[56] Nifo presents without acknowledgment long excerpts from Albert's works when setting forth the "true position" on most of the major topics discussed. It is true that Nifo calls Saint Thomas "the first expositor of the Latins" (*latinorum primus expositor*), but it would be a serious mistake to take this as an indication either that Thomas is the prime authority in the *De intellectu* or that Nifo has here become a Thomist.[57] What must be noted is that he calls Albert "the first of the Latins" (*Albertus latinorum primus*), thus giving him primacy even over Thomas.[58] The works of Albert to which he refers by title are the *De natura et origine animae, De intellectu et intelligibili, Liber de homine*, and the commentaries on the *De anima*, and *Metaphysics*, but it is the first of these that will have the most important influence.

---

[54] Agostino Nifo, *Collectanea ac commentaria in libros de anima* (Venice, 1522), I, comm. 7, f. 12vb. I have used this edition for the sake of convenience. For the first edition, see *Augustini Niphi super tres libros de anima* (Venice, 1503). I have found no reference to Albert in all of Book III of Nifo's commentary.

[55] Ibid., I, text. comm. 9, f. 16rb. On Albert's conception of the soul as a "potestative whole," see Odon Lottin, "L'identité de l'âme et de ses facultés pendant la première moitié du XIIIe siècle," *Revue néoscolastique de philosophie*, 36 (1934), 205-209; idem, *Psychologie et morale aux XIIe et XIIIe siècles*, 1 (Louvain, 1942), pp. 497-501; A.-M. Ethier, "La double définition de l'âme humaine chez Saint Albert le Grand," *Études et recherches publiées par le Collège Dominicain d'Ottawa: I Philosophie*, cahier 1 (Ottawa, 1936), 107-109; Bruno de Solages, "La cohérence de la métaphysique de l'âme d'Albert le Grand," in *Mélanges offerts au R. P. Ferdinand Cavallera* . . . (Toulouse, 1948), p. 385; Pius Künzle, *Das Verhältnis der Seele zu ihren Potenzen, Problemgeschichtliche Untersuchungen von Augustinus bis und mit Thomas von Aquin* (Fribourg, Switzerland, 1956), pp. 145-158.

[56] Agostino Nifo, *Liber de intellectu cum gratia et privilegio* (Venice, 1503). The commentary on the *De anima*, which was also first printed in 1503, would appear to be an earlier work than the *De intellectu*. I have discussed this problem in my article, "Agostino Nifo's Early Views on Immortality," *Journal of the History of Philosophy*, 8 (1970), 451-460, at 454-458.

[57] Nifo, *De intellectu*, I, tr.1, c.16, f. 9va; tr.2, c.17, f. 23va.

[58] Ibid., I tr.4, c.14, f. 41va. Two prominent historians have mistakenly believed that Nifo's *De intellectu* is Thomistic. See Giovanni di Napoli, *L'immortalità dell'anima nel Rinascimento* (Turin, 1963), pp. 203-214; Etienne Gilson, "Autour de Pomponazzi, Problematique de l'immortalité de l'âme en Italie au debut du XVIe siècle," in *Archives d'histoire doctrinale et litteraire du moyen âge*, 36 (1961), 237-238. It is interesting to note that Nifo (*De intellectu*, I, tr.3, c.26, f. 35vb) refers to Siger of Brabant as a "student" (better "follower"?) of Albert: "Ad secundam quaestionem Suggerius, vir gravis, sectae Averroisticae fautor aetate expositoris [i.e., Saint Thomas], discipulus Alberti, persolvit in suo de intellectu tractatu. . . ." For discussion, see Bruno Nardi, *Sigieri di Brabante nel pensiero del Rinascimento Italiano* (Rome, 1945), pp. 20, 34, and 145-146; Fernand van Steenberghen, *Maître Siger de Brabant*, Philosophes médiévaux, 21 (Louvain and Paris, 1977), pp. 395-396.

The basic structure of the *De intellectu* is a division into a Book I, which treats of the nature of the soul and intellect, and a Book II, which examines the nature of beatitude. The first book is itself divided into treatises (*tractatus*) dealing with the following topics: (1) the origin and immortality of the soul; (2) the nature of the soul's separability from the body; (3) the question of whether there is only one intellective soul for all men; (4) the nature of the agent and possible intellect; and (5) the nature of the speculative and practical intellects.[59] Toward the end of the first, third and fourth treatises of Book I there are chapters in which Nifo presents as the true position unacknowledged excerpts from Albert's writings, especially the *De natura et origine animae*. When he attempts in Book II to explain both Aristotle's position on human happiness and also the place of the soul after death, he borrows heavily and frequently word for word from that same work of Albert.[60] Besides these striking indications of

---

[59] For further details on the structure of this work, see my article, "Agostino Nifo and Saint Thomas Aquinas," *Memorie Domenicane*, 7 (1976), 195-226. In the article (pp. 202-203), I show that Nifo gave preference to Albert over Thomas in his *Die intellectu*.

[60] The following chart indicates Nifo's "borrowings" from Albert's *De natura et origine animae* for the "true position":

NIFO, *De intellectu*	ALBERT, *De nat. et or. an:*
I, tr.1, c.28, f. 15vab	I, c.5 (ed. Colon. 12: 13 vv. 23, 34-41, 45-94; 14 vv. 1-7 and 16-40)
I, tr.3, c.31, f. 37va	I, c.6 (ed. Colon. 12: 14 vv. 51-54, 15 vv. 2-25 and 45-53)
I, tr.4, c.23, f. 45rb-45va	I, c.6 and 7 (ed. Colon. 12: 14 vv. 53-54; 16 vv. 3-33)
I, tr.4, c.24, f. 45va	I, c.7 (ed. Colon. 12: 16 vv. 34-76)
II, tr.3, c.2, f. 74rb	II, c.11 and 12 (ed. Colon. 12: 35 vv. 24-26; 36 vv. 44-52 and 59-62)
II, tr.3, c.3, f. 74va	II, c.8 (ed. Colon. 12: 32 vv. 46-49)
II, tr.3, c.4, f. 74vb-75ra	II, c.13 (ed. Colon. 12: 37b-39a)
II, tr.3, c.6, f. 75vab	II, c.11 and 14 (ed. Colon. 12: 35 vv. 23-26 and 41-70; 36 vv. 22-35; 41 vv. 40-74)
II, tr.3, c.8, f. 76, 1-2	II, c.7 (ed. Colon. 12: 30 vv. 35-36 and 80-94; 31 vv. 1-11, 21-24 and 30-33)

On Nifo's use of Thomas Aquinas in his *De intellectu*, see my article, "Agostino Nifo and

Nifo's indebtedness to Albert for the "true" position, other examples of his reliance on Albert's writings can be found scattered through the *De intellectu*. For example, when presenting some arguments for immortality which are identified as arguments of Avicenna and Plato, he in fact borrows from Albert, just as he borrows arguments against the preexistence of the soul from him.[61] In sum, Nifo's *De intellectu* must be recognized as essentially an Albertistic work. However, one important qualification must be made. It is that while Nifo denies the need for intelligible species in cognition and claims that Averroës himself denied such species, he does not, like his teacher Vernia, ascribe such views to Albert.[62]

Although they cannot be classified as Albertistic, Pietro Pomponazzi's (1462-1525) early Paduan works contain frequent respectful references to Albert. He is cited together with Thomas in regard to the composition of the heavens, and his agreement with both Thomas and Scotus, all outstanding peripatetics (*praecipui peripatetici*), is used to argue that Aristotle and Averroës both held that all created nature, the heaven included, depends on God not only as its final and exemplar cause but also as its efficient cause.[63] Pomponazzi correctly remarks that Albert put forth various arguments for immortality in his *De homine* and *De natura et origine animae* which go beyond Aristotle's own arguments, but he also makes the somewhat dubious claim that Albert attempted to sustain in his *De homine, De natura et origine animae, De coevis*, and *De intellectu et intelligibili* the belief in the preexistence of the rational soul, an article condemned by the faith.[64] On the other hand, Albert is one of the authorities Pomponazzi uses to denounce those like Nifo and Achillini who

---

Saint Thomas Aquinas," pp. 204-211. Tullio Gregory has translated Nifo's *De intellectu*, I, tr.4, c.24, but apparently without realizing that it is derived from Albert. See *Grande antologia filosofica*, vol. 6: *Il pensiero della Rinascenza e della Riforma* (Milan, 1964), pp. 738-739.

[61] Nifo, *De intellectu*, I tr.1, c.8, f. 6rab. See Albert, *De natura et origine animae*, II, cc.2 and 6 (ed. Colon. 12: 22 and 26-27). See also Nifo, I, c.13, ff. 8va-9ra, where he summarizes Albert's arguments against Alexander in the *De natura et origine animae*, II, c.5. In this case, Nifo himself gives the reference in Albert.

[62] See for example *De intellectu*, I, tr.5, c.14, f. 50rb-50va. For discussion, see my article cited in n. 49 above.

[63] Pietro Pomponazzi, *Expositio libelli De substantia orbis*, c.2, in *Corsi inediti dell'insegnamento padovano*, ed. Antonino Poppi, 1 (Padua, 1966), p. 130. See also *Quaestiones super libello De substantia orbis*, q.4, p. 302.

[64] Pietro Pomponazzi, *Utrum anima rationalis sit immaterialis* and *An anima intellectiva sit unica vel numerata*, in *Corsi inediti dell'insegnamento padovano*, II, ed. Antonino Poppi (Padua, 1970), pp. 5 and 81.

deny the need for intelligible species in cognition and who claim Averroës did not hold to them. He hurls against this "stupid" and "bestial" view the authority of the whole university of Paris (*totum gymnasium parisiensium*), especially Thomas, Albert, Egidio, and Scotus, all of whom hold the true view that our intellect knows through species and so expound the text of Aristotle.[65]

Another professor at Padua, Marcantonio Zimara (1460-ca. 1532), shows critical respect for Albert in the various treatises he wrote during his first Paduan period. In his early *De principio individuationis*, first published in 1505, Zimara rejects the Scotist doctrine of *haecceitas* as an accurate gloss on either Aristotle or Averroës, whose common doctrine he is trying to establish, and he turns instead to Albert. He explains that demarcated individuals (*individua signata*) are not properly and per se in any category, since they are infinite in number and therefore unknown, though an individual can be considered to be in a category reductively, that is, when it is treated as a vague individual (*individuum vagum*). Zimara claims to find verification for this stand in the fact that when Aristotle speaks of first substance in the *Categories* his examples are those of vague individuals, namely, "some man" (*aliquis homo*) and "some cow" (*aliquis bos*). What is noteworthy for our purposes, however, is that Zimara then admits that long before him Albert had already set forth (*propalavit*) this position in his commentaries on Aristotle's *Categories* and Porphyry's *Isagoge*.[66] He also valiantly tries to reconcile Albert with his own reading of Averroës, whom he takes to say that matter and form diversify one another.[67]

Albert plays an important role in Zimara's annotations on John of Jandun's *Metaphysics*. For example, when Zimara rejects Jandun's statement that logic is not absolutely necessary for learning other sciences and says that Jandun's position is false according to the mind

---

[65] Pietro Pomponazzi, *Quaestio de speciebus intelligibilibus et intellectu speculativo*, in *Corsi inediti*, II, pp. 186-187. These same four medieval philosophers are cited again on pp. 200 and 209. In a later work, Pomponazzi claims to find in Albert justification for a sharp distinction between philosophy and theology, and he goes on to suggest that Dominicans should want to burn Albert. See Bruno Nardi, *Studi su Pietro Pomponazzi* (Forence, 1965), p. 27 n. 2.

[66] Marcantonio Zimara, *De principio individuationis ad intentionem Averrois et Aristotelis*, in John of Jandun, *Quaestiones in duodecim libros metaphysicae ad intentionem Aristotelis et magni commentatoris Averrois subtilissimae disputatae* (Venice, 1505), ff. 153rb and 154rab. See Albert, *De praedicabilibus*, I, tr.2, c.2 (ed. Borgnet 1: 169a).

[67] Zimara, *De principio*, f. 155vb. See Albert, *De intellectu et intelligibili*, I, tr.1, c.5 (ed. Borgnet 9: 484a).

of Aristotle and Averroës, he immediately cites Albert with approval for having said that if one who is ignorant of logic seems to know something, he does not, since he no more knows the object he is thinking about than does the fire which is capable of enkindling wood but does not know its own nature. Albert stands at center stage in Zimara's determination of Averroës' position on universals. He cites Albert's remark that Avicenna, Algazel, Averroës, Abubacher and almost all peripatetics agree that the universal does not exist in reality, but he chides Albert for having explained only in confused language (*sub confusis verbis*) the manner in which universality comes from the intellect and has a foundation in reality. Nevertheless, after claiming that for Averroës universals have a certain existence in the thought of God and the Intelligences, he desperately tries to conciliate Albert with such a view by suggesting that this is what Albert meant by the universal *ante rem*.[68] Moreover, Zimara attacks those who try to separate Albert from the opinion of Avicenna, Algazel, and Averroës by saying that while they held the universal to exist only in the mind Albert meant to distinguish his view from theirs as the "middle way," that is to say, he intended to posit universals as real. They thus take Albert to be in agreement with Scotus. After curtly rejecting such an interpretation of Albert, Zimara goes on to argue that universality is actually a second intention of the mind, that is, a certain quality that is an accident of the first intention but not of its essence. However, he claims no originality for this position, since he admits that it was put forth long before him by Albert, though in very obscure language (*sub obscurissimis verbis*).[69]

Zimara's question *De triplici causalitate intelligentiae* is in effect his annotation on Jandun's *Metaphysics*, Book XII, q. 12, namely, whether the heaven is animated. Zimara only makes one reference to Albert. He remarks, first of all, that Averroës predicates the definition of the soul analogically and not univocally or equivocally of the Intelligences and the souls of animals and plants. The reason is because the Intelligence is the soul of the heaven only as it provides

---

[68] Marcantonio Zimara, *Annotationes in Joannem Gandavensem super questionibus metaphysicae*, in John of Jandun, *Quaestiones*, f. 157b. Cf. Albert, *De praedicabilibus*, I, tr.1, c.3 (ed. Borgnet 1: 5b-6a). See also *Annotationes*, f. 157vab for use of Albert's *De homine* as an authority.

[69] Ibid., ff. 166ra; 166va; 168rab; 168va. See also f. 169va. See Albert, *De intellectu et intelligibili*, I, tr.2, c.2 (ed. Borgnet 9: 492a-493a) and *Metaphysica*, III, tr.3, c.18 and V, tr.6, c.5-7 (ed. Colon. 16/1: 157b and 285a-288a). He also cites Albert's *De coelo* in these annotations (f. 166vb and f. 167rb).

the heaven with its activity, and not by being united to it in existence. Zimara then claims that Albert says the same thing in his *Metaphysics*. This, however, is a serious misreading, since Albert in fact says that the Intelligences are in no way the actualities of bodies, and that soul is therefore predicated only equivocally of them and the lower souls.[70] What is important to note is not Zimara's inacuracy here, but his expectation to find in Albert some confirmation of his own interpretation of Averroës. This inclination is also evident in Zimara's "solutions" of the contradictions between statements which Averroës makes in his various works. When he has to reconcile Averroës listing the senses under the passive powers and yet speaking of the soul as the moving or efficient cause of all motions in the living thing, he turns for help to Albert's commentary on the *De anima*. Zimara accepts Albert's explanation that sense is passive insofar as it is in potency to receiving the sensible species, but it becomes active after it has received that species. He also uses Albert in his argument against those who claim Averroës made a separate Intelligence the active cause of sensation comparable to the separate agent intellect which is a cause of intellectual cognition.[71] The great respect that Zimara had for Albert as a guide to the thought of Averroës is set in sharp focus when he announces that "all our Latins" (*omnes latini nostri*), namely, Albert, Saint Thomas, Duns Scotus, and Egidio Romano, as well as various "distinguished Averroists" (*praeclari Averroistae*), think that Averroës' doctrine was that the intellect was not the substantial form of the human being. Leading the list of names is Albert the Great, "who was the most faithful interpreter of the doctrines of the Arabs, especially of Averroës" (*Albertus cognomento magnus, qui opinionum Arabum, et praecipue Averrois, fuit fidelissimus interpres*).[72]

---

[70] Ibid., f. 173ra.

[71] Marcantonio Zimara, *Solutiones contradictionum in dictis Averroys*, in his *Questio de primo cognito* (Venice, 1508), f. 22vab. See Averroës, *Commentarium magnum in Aristotelis De anima libros*, ed. F. Stuart Crawford, Corpus Commentariorum Averrois in Aristotelem v, 1 (Cambridge, Mass., 1953), II, comm. 37, p. 188; comm. 52, p. 212; comm. 60, p. 221. For the passages from Albert, see Albert, *De anima*, II, tr.3, c.1 and 6 (ed. Colon. 7/1: 96-97 and 104-107). It is almost a certainty that Zimara is here attacking Nifo, who proposed precisely this interpretation of Averroës in his own *De sensu agente*. For further details, see my article, "Agostino Nifo's *De sensu agente*," cited above in n. 51. What is ironic is that Zimara is using Albert as an authority against Nifo, who is himself an Albertist in his general psychology. We shall see below that Zimara will also make much use of Albert to argue against Nifo's belief that Averroës denied the need for intelligible species in cognition.

[72] Zimara, *Solutiones*, f. 35rb. See Jean Rohmer, "L'intentionnalité des sensations de Platon à Ockham," *Revue des sciences religieuses*, 25 (1951), 32.

Among the topics most heatedly debated by the philosophers at Padua toward the end of the fifteenth century and the beginning of the sixteenth century were whether intelligible species are required in cognition and whether Aristotle or Averroës maintain the need for them. We have already noted the interest of Vernia, Nifo, and Pomponazzi in these questions.[73] Zimara himself wrote a separate treatise on the question whether Averroës thought that the possible intellect received intelligible species as in a subject (subiective) during cognition. Although other philosophers, such as Jandun, Scotus, and Thomas, are occasionally cited, both the frequency of reference to Albert and also the manner in which the citations from Albert are used leave no doubt that Zimara considers him to be the best source for discovering the mind of Averroës on this question. The works of Albert cited are the De homine, the De intellectu et intelligibili, and the commentaries on the Posterior Analytics, Physics, and De anima. Zimara presents the views and arguments of various self-proclaimed Averroists (Averroistae) who deny that Averroës maintained intelligible species. At one point in his reply to them, he cites Averroës and Albert jointly as having condemned Theophrastus and Themistius for denying intelligible species and for thereby having contradicted both Aristotle and truth.[74] Zimara does not even hesitate to call Albert an "Averroist," when he insists that all the ancient Averroists, such as Albert the Great (omnes antiqui Averroistae ut Albertus

---

[73] See Poppi, Saggi sul pensiero inedito pp. 139-194 and also my article, "Antonio Trombetta and Agostino Nifo," both of which are cited in n. 49 above.

[74] "Ecce igitur auctoritate Averrois et Alberti magni quod omnes negantes species intelligibiles, sicut negaverunt Theophrastus et Themistius, contradicunt verbis Aristotelis et contradicunt veritati in se" (Marcantonio Zimara, Quaestio de speciebus intelligibilibus ad mentem Averrois [s. l., s. a.], sig. blr, which was later reprinted as Quaestio qua species intelligibiles ad mentem Averrois defenduntur in Asclepii ex voce Amonii Hermeae in Metaphysicam Aristotelis praefatio, ed. F. Storella [Naples, 1575-1576], sig. B2r). See also ibid., sig. b3v (1575-1576 edition: sig. B5v). See Averroës, Commentarium magnum, III, comm. 5, pp. 389-392, and Albert, De anima, III, tr.2, cc.3 and 5 (ed. Colon. 7/1: 180 vv. 30-44; 184 vv. 38-72). One of Zimara's targets is surely Agostino Nifo, who denies the need for intelligible species in his early works and who attacks those (like Jandun) who said that Averroës had held to them. Nicoletto Vernia may also be a target. I have used a copy of the rare undated edition of Zimara's question on intelligible species which is presently in the Beinecke Library at Yale University. I have also examined a copy in the Biblioteca Civica Bertoliana at Vicenza (G. 4. 4. 14bis). This early edition apparently escaped the notice of Antonio Antonaci, Ricerche sull'aristotelismo del Rinascimento: Marcantonio Zimara, vol. 1: Dal primo periodo padovano al periodo presalernitano (Lecce and Galatina, 1971), pp. 47-48 and 74-78. On the later printing, see Antonio Antonaci, Francesco Storella, filosofo Salentino del cinquecento (Bari and Galatina, 1966), pp. 9, 134-136 and 204-205; Nardi, Saggi, p. 233 n. 28 and p. 328, who also refer to printings in 1554 and 1561. I have examined a copy of the former in the Biblioteca Apostolica Vaticana.

Magnus), ascribed intelligible species to Averroës.[75] Moreover, he gathers together passages from Albert's writings to argue that Albert not only attributed intelligible species to Averroës but also maintained them himself. He adds the refinement that for Albert, as also for Scotus, the quiddity or essence shining forth in the species is not present in the intellect as an accident in a subject but is present rather in the objective existence (*esse obiectivum*) or the abstracted existence (*abstractum esse*) provided by the intellect.[76]

Besides Vernia, Nifo, and Zimara, all of whom were in varying degrees "Albertistic," other Paduan philosophers of the late fifteenth and early sixteenth centuries refer to Albert in their writings. The famed Dominican Thomas de Vio, better known as Cardinal Cajetan (1469-1534), cites Albert several times in the commentary on Thomas' *De ente et essentia* which he wrote in 1493-1494 while he was at Padua, and it has been argued that his approach on occasion reveals an Albertist perspective.[77] Albert is referred to by Maurice O'Fihely (d. 1513),[78] and he also is mentioned by O'Fihely's better known Franciscan confrère, Antonio Trombetta (1436-1517), who taught *in via Scoti* against Cajetan.[79] On the other hand, references to Albert in

---

[75] Zimara, *De speciebus*, sig. blv (sig. B3r). By speaking of "ancient Averroists" Zimara presumably means to distinguish the older, medieval philosophers who explicated Averroës from his own contemporaries.

[76] Ibid., sig. b3r (sig. B5r-B5v).

[77] Thomas de Vio, *In De ente et essentia D. Thomae Aquinatis commentaria*, ed. M.-H. Laurent (Turin, 1934), ch. 2, q.5, § 39, p. 60; ch. 5, q.11, § 94, p. 149; and q. 12, § 100-101, pp. 222-223; ch. 7, q. 16, § 136, p. 222. See Norman J. Wells, "On Last Looking into Cajetan's Metaphysics: A Rejoinder," *The New Scholasticism*, 42 (1968), 112-117, especially p. 116: "In short, Cajetan reads Aquinas with Albertian spectacles on." Wells argues that this Albertistic turn results in part from Cajetan's use of Capreolus: "However, one must not overlook the fact that Capreolus ground the lenses for these spectacles" (p. 116 n. 17). Cajetan also refers to Albert in his Paduan lectures on the *Sentences*. See Armand Maurer, "Cajetan's Notion of Being in His Commentary on the *Sentences*," *Mediaeval Studies*, 28 (1966), 268-278, especially 277, where Cajetan cites Albert's *De causis*, I, tr.1, c.8.

[78] See O'Fihely's comments in his edition of *Quaestiones subtilissimae Scoti in Metaphysicam Aristotelis, eiusdem De primo rerum principio tractatus atque Theoremata* (Venice, 1497), ff. 101ra, 102rab, 103ra, 103va, 105rab, 105va, 106ra, 107rab, 108ra, 108va, and 111va. Albert is one of the *antiqui*. Noteworthy is his reference to Albert in the letter to Antonio Trombetta (f. 99r). O'Fihely mentions by name only Albert among the Latins, along with Alexander of Aphrodisias, Themistius and Simplicius, as having expounded the obscure and concise language of Aristotle more clearly and more fully. This passage can also be found in Duns Scotus, *Opera*, 4 (Lyons, 1639; repr. Hildesheim, 1968), p. 509a.

[79] See Antonio Trombetta, *Opus in Metaphysicam Aristotelis Padue in thomistas discussum* (Venice, 1502), ff. 31vb, 34va, 80va, 83vab, 84va and 88va. However, Albert does not appear to be mentioned in Trombetta's *Tractatus singularis contra Averroistas de humanarum animarum plurificatione* (Venice, 1498).

the printed Latin works of Elia del Medigo are very rare.[80] In contrast, there are many references to Albert in Gieronimo Taiapietra, who lists Albert along with Thomas and Egidio Romano as from Italy (*ex regione nostra*).[81] He also includes Albert among *moderniores nostri*, the *latinorum maiores*, and *clarissimi illi latinorum philosophi*.[82] Among the works of Albert which Taiapietra cites are the commentaries on the *De coelo, Physics, Metaphysics*, and *Liber de causis*, as well as the *De natura et origine animae*. Like Nifo, whom he is probably following here, Taiapietra calls Siger a "follower of Albert" (*discipulus Alberti*).[83] In his still unpublished work on the *De anima*, Pietro Trapolin appears to consider Albert, along with Thomas and Egidio Romano, as one of the most important of the *latini* — in doing this he is following a view common among the Paduans.[84] He makes frequent use of Albert's commentary on the *De anima* and his *De homine*, and he occasionally cites Albert's commentaries on the *De sensu et sensato* and *Metaphysics*. Cristoforo Marcello cites various works of Albert in marginal annotations to be found in his own work on the soul. These include Albert's commentaries on the *Metaphysics, De generatione et corruptione, De anima* and *De nutrimento*, as well as his *De homine*. Marcello appears to pay special attention to Albert when discussing sensation, sight, sound, *vox*, vapor, and the common sense, but it is noteworthy that there appear to be no references to Albert in the fifth and sixth books of his work, namely, those which especially concern the nature of the rational soul.[85]

One other aspect of Albert's influence at the University of Padua at the end of the fifteenth and the beginning of the sixteenth century must be mentioned at least in passing. It is that exhibited by the vari-

---

[80] The reference that I have found is in Elia's *De primo motore*, in *Ioannis de Ianduno philosophi acutissimi super octo libros Aristotelis De physico auditu subtilissimae quaestiones* (Venice 1551), f. 139rb. Averroës, Jandun and Burley are the favored philosophers cited. Elia's treatise was first published in 1488.

[81] Geronimo Taiapietra, *Summa divinarum ac naturalium difficilium quaestionum* (Venice, 1506), proemium, sig. a4v.

[82] Ibid., I, tr.1, c.1, sig. blv; tr.6, c.2, sig. o2rv; tr.6, c.3, sig. o2v. See also II, tr.1, c.10, sig. y3r.

[83] "Et ista videtur esse plana sententia Averrois in hoc quaesito, ut de mente eius tenent praeclarissimi viri, et maxime inter alios Subgerius. Et iste fuit discipulus Alberti et contemporaneus Thomae." Ibid., II, tr.1, c.19, sig. &3v.

[84] Pietro Trapolino, *Collecta in libro de anima Aristotelis*, Perugia, Biblioteca Comunale Augusta, Codex F82. He takes Albert, Thomas and Egidio to hold to intelligible species (f. 88-88v).

[85] Cristoforo Marcello, *Universalis de anima traditionis opus* (Venice, 1508).

ous printings of Albert's works during those decades.[86] Although they were not the first to print Albert at Venice or Padua, the de Gregoriis brothers' attempt to print in the 1490s all the works of Albert is especially interesting, since they make revealing statements regarding their project on the different title pages of their edition. For example, on the title page of the *Metaphysics* volume they state that they have promised to aid all students of philosophy by printing the works of Albert.[87] This statement also appears on the title page of the *Logica*.[88] We are told on the title page of the *Physics* volume that some of Albert's most learned followers (*doctissimi sequaces*) were responsible for the emendations of the text, though no one is identified.[89] The most striking claim made in behalf of Albert is that found on the title page of the *De anima* volume, which was issued in 1494. It is that one could hardly deserve the name of philosopher at that time without studying Albert's works.[90] This remark may be a publisher's exaggerated blurb, but it also may indicate the existence of an Albertist trend at Padua, one that was cultivated by Vernia.[91] Zimara subsequently saw to the press four volumes of various works of Albert. The first is an edition with annotations of Albert's *Parva naturalia*, which also includes the *De natura et origine animae* and the

---

[86] For a listing of fifteenth-century printings of Albert's various works, see *Gesamtkatalog der Wiegendrucke*, 1 (Leipzig, 1925), #581-783, col. 264-385.

[87] *Aureus liber Metaphysicae Divi Alberti Magni* (Venice, 1494; GKW #683). I have examined a copy of this volume and copies of the volumes cited in notes 88-90 at the Beinecke Library of Yale University. This is the first printed edition of the work.

[88] *Opera Alberti Magni ad logicam pertinentia* (Venice, 1494; GKW #677). There appears to have been no earlier edition printed at Venice or Padua. See n. 91 below.

[89] *Divi Alberti Magni physicorum sive de phisico auditu libri octo* (Venice, 1494; GKW #717). An earlier edition had been published by the de Gregoriis brothers in 1488-89 (GKW #716).

[90] *Divi Alberti Magni De anima libri tres, De intellectu et intelligibili libri duo* (Venice, 1494; GKW #586). An edition of these two works had already been published at Venice in 1481 (GKW #585).

I have also examined at Yale the de Gregoriis printing of *Prima pars summae Alberti Magni De quattuor coequevis una cum secunda eius quae est De homine* (Venice, 1498-1499; GKW #779), which does not contain a letter to the reader regarding the edition.

[91] There is no hard evidence, however, that Vernia was responsible for the edition of Albert published by the de Gregoriis brothers in the 1490s. Thomas Accurti does claim in his *Editiones saeculi XV pleraeque bibliographis ignotae* (Florence, 1930), p. 111, to have seen a copy of the 1494 edition of Albert's *Logica* which contains a dedicatory epistle of Vernia to Grimani and his *Quaestio de medio potissimae demonstrationis*. This information is repeated by Mario E. Cosenza, *Biographical and Bibliographical Dictionary of the Italian Humanists*, 4 (Boston, 1962), p. 3629, col. 3, and Pio Paschini, *Domenico Grimani, Cardinale di S. Marco (1523)* (Rome, 1943), p. 127. I have not found this letter or question in the copies of the edition (GKW #677) which I have examined in Florence, Biblioteca Nazionale Centrale, and at Yale and Harvard.

De unitate intellectus.[92] The second is a joint edition with annotations of Albert's *Physics, De coelo, De generatione et corruptione, De meteoris, De mineralibus, De anima, De intellectu et intelligibili,* and *Metaphysics.*[93] The third is an edition of Albert's *De animalibus,*[94] while the fourth is an edition of his *De creaturis.*[95]

In the above survey, we first noted the widespread interest for Albert that existed in Italy during the late Middle Ages and the Renaissance, from Dante to Ficino, and both within and without the Dominican order. Then, after chronicling the few references to Albert which we have from such fourteenth-century philosophers at Padua as Pietro d'Abano and Biagio Pelacani da Parmi, we turned to two of the most important philosophers at Padua during the fifteenth century, namely, Paul of Venice and Gaetano di Thiene. While Paul of Venice had only spotty interest in Albert's writings and did not treat him as a major authority in philosophy, Gaetano di Thiene regularly cited Albert and obviously considered him to be a major figure in the medieval Christian tradition. Although Thiene's student, Nicoletto Vernia, began his philosophical career with a decided prejudice in favor of Averroës, he very early linked together Albert and Averroës, arguing that when they agreed on the interpretation of Aristotle one surely had achieved the true mind of the Philosopher. And even when Vernia abandoned Averroës for the Greek Commentators he maintained his attachment to Albert. This allegi-

---

[92] *Tabula tractatuum parvorum naturalium Alberti Magni* (Venice, 1517). I have examined a copy of this work in Paris, Bibliothèque Nationale. Cf. Antonaci, *Ricerche,* p. 34; Nardi, *Saggi,* pp. 334-335. There is also reference made to this edition by M. Pereira in the recent edition of Albert's *Speculum astronomiae,* ed. Paola Zambelli, with S. Caroti, M. Pereira, S. Zamponi (Pisa, 1977), pp. 185-186.

[93] *Divi Alberti Magni summi in via peripatetica philosophi theologique profundissimi naturalia ac supernaturalia opera per Marcum Antonium Zimaram philosophum excellentissimum nuper castigata erroribusque purgata, necnon cum marginibus optimis annotationibus ornatis doctrinaque excultis atque fideliter impressis feliciter incipiunt . . .* (Venice, 1517-1518). Cf. Antonaci, *Ricerche,* p 34; Nardi, *Saggi,* p. 334. I have examined a copy in the British Library.

[94] *Divi Alberti Magni De animalibus libri vigintisex novissime impressi* (Venice, 1519). Cf. Antonaci, *Ricerche,* p. 35 I have examined copies in the British Library, the Wellcome Library, and the Library of the University of California at Berkeley. I also verified its existence in the Library of the University of Amsterdam.

[95] *Divi Alberti Magni Ratisponensis episcopi summi peripatetici due partes summae, quarum prima De quattuor coequevis, secunda De homine inscribitur, una cum pulcherrimis additionibus editis ab excellente artium et medicinae doctore Marco Antonio Zimara sanctipetrinate nuperrime castigatae ac pristinae integritati restitutae* (Venice, 1519). I have examined a copy in the Biblioteca Apostolica Vaticana, and I have verified the existence of copies in the Bodleian Library and Florence, Biblioteca Nazionale Centrale. Cf. Antonaci, *Ricerche,* p. 35; Nardi, *Saggi,* p. 335.

ance to Albert reappeared in the *De intellectu* of Vernia's student, Agostino Nifo, who also had had leanings toward Averroës. Indeed, Albert is there called "the first of the Latins" and given priority even over Saint Thomas. A strong interest in Albert was noted in the early Paduan writings of Marcantonio Zimara, whose attachment to Averroës as the true interpreter of Aristotle is well known. Albert appears to be given special attention precisely because he is himself a faithful interpreter of Averroës. In fact, Zimara goes so far as to call Albert an "Averroist."

What appears to emerge from the study which we have attempted here is the fact that Albert's devotees at Padua were those who had a special interest in Averroës.[96] It is noteworthy that it is not the Dominicans there who serve as his most forceful proponents but rather lay philosophers. The latter appeared to find in Albert an orthodox medieval philosopher who had himself professed his interest in Averroës and openly declared that on some points he was not far from "the Commentator." Both Vernia and Nifo had accepted Averroës as the true guide to Aristotle before they made their shift to Albertism, and Zimara always gave preference to Averroës, despite his great interest in Albert. The further reputation and influence that Albert would have at Padua and in Italy in general during the sixteenth century should be the subject of further investigation. What is incontestable is that Albert's writings and ideas had great impact on the course of philosophy at his *alma mater* toward the end of the fifteenth century and the beginning of the sixteenth century. Indeed, in some ways it was a more striking and significant influence than that of his celebrated student.[97]

---

[96] One of the most ambitious monographs on Albert and Thomas and their influence is M. M. Gorce, *L'essor de la pensée au moyen âge: Albert le Grand — Thomas d'Aquin* (Paris, 1933). In his allusions to Vernia and Nifo (pp. 195-199), he shows no realization of the role that Albert played in their thought. See also n. 7 above.

[97] On Albert's stay at Padua, see now Paolo Marangon, *Alle origini dell'Aristotelismo padovano (sec. XII-XIII)* (Padua, 1977), pp. 35-37. This work was unavailable to me until the present article was in proofs.

Appendix 1

Albert's Works on Natural Science
(*libri naturales*)
in Probable Chronological Order

James A. Weisheipl, OP

1. *Physica* libri VIII (ed. Borgnet 3: 1-632; partial autograph).
   Undoubtedly this is the first of the Aristotelian paraphrases composed by Albert, probably in Cologne not much before 1250. His intention, as he declares at the outset, is to explain the whole of human knowledge so that his Dominican confreres could competently understand the text of Aristotle, thereby making the "new learning" intelligible to the Latins (*Latinis intelligibiles*). The procedure was deliberate, systematic, and fundamentally consecutive with the Aristotelian corpus (logic, *libri naturales*, moral philosophy, and first philosophy) as it was known in the Latin West from both Greek and Arabic. For the *Physica* Albert used at least two Latin versions: (1) the *vetustior*, translated from Greek by James of Venice ca. 1170, and (2) the *Physica cum commentario magno Averrois*, translated by Michael Scot from Arabic ca. 1230.

2. *De lineis indivisibilibus* (ed. Borgnet 3: 363-481; inserted after *Physica* VI).
   This paraphrase is based on Grosseteste's translation from the Greek of Aristotle's authentic work on this title (cf. Milan, MS Ambros. E. 71 sup., fol. 156ra-157rb), which S. Harrison Thomson dates ca. 1245 for no firm reason (*Writings of Robert Grosseteste*, Cambridge, 1940, p. 67). Albert's paraphrase seems to have been written not only after the six Books of the *Physica*

(passim), but also after *Physica* VIII ("sicut in VIII *Physicorum* declaravimus. Sed quia ibi diximus . . ." [ed. Borgnet 3: 475b]) and after his *Geometry* ("sicut diximus in geometricis" [ibid. 3: 465b]). It would seem, however, that the paraphrase was finished and inserted into its proper place by the time Albert wrote *De caelo* III, tr.1, c.2: "Et nos etiam hoc sufficienter improbavimus in eis quae prius tractavimus in 6° *Physicorum* qui est de motus divisione, et etiam in libro *De indivisibilibus lineis*" (ed. Colon. 5/1: 205.2-5). The treatise was certainly in place by the time Albert wrote *De anima* I, tr.2, c.10: "quia iam probavimus in VI *Physicorum* et in libello *De lineis indivisibilibus*, quod . . ." (ed. Colon. 7/1: 46.17-18). It is clear, however, that Albert intended to insert Aristotle's *De lineis indivisibilibus* in the present place when he wrote *Physica* III, tr.2 c.3: ". . .quem nos inferius in sexto huius scientiae adiungemus" (ed. Borgnet 3: 209b).

3. *De caelo et mundo* libri IV (ed. Borgnet 4: 1-321; ed. Colon. 5/1 [n.9, 1971]; autograph in Vienna, Oesterreichische National-bibliothek, Cod. misc. lat. 273, fol. 72v-142r).

   Written at Cologne after the *Physica*, to which he frequently refers, and immediately before *De natura loci*, as shown in the autograph. Therefore it must have been composed around 1250. The paraphrase is based mainly on the *vetus translatio* from the Arabic made by Gerard of Cremona (d. 1187) at Toledo (printed in the Cologne edition), with constant reference to the version of Michael Scot, translated with the *Commentarium magnum* of Averroës between 1231 and 1235. Albert does not seem to be aware of the new translation of Grosseteste from the Greek for Bks. I-III.1, made between 1247 and 1253. In this paraphrase Albert opposes mainly Averroës and the "new astronomy" of Alpetrugi (al-Bitrūjī) in his *De motibus caelorum*, which was translated by Michael Scot at Toledo in 1217.

4. *De natura loci* (or *De natura locorum quam habent ex longitudine et latitudine eorum*) (ed. Borgnet 9: 527-585; autograph in Vienna, Oesterreichische Nationalbibliothek, Cod. misc. lat. 273, fol. 142r-156r).

   Definitely written in Cologne after *De caelo*, as in autograph: "Agrippinam, quae nunc Colonia vocatur, in qua istud volumen compilatum est" (autog. fol. 152v; ed. Borgnet 9: 570b), and before *De causis et proprietatibus elementorum*. Moreover,

Albert refers to his own paraphrase of *De caelo*: "sicut in libro *De caelo et mundo* diximus" (tr.1, c.4 [ed. Borgnet 9: 534b]). This seems to be Albert's personal contribution to the science of geography, although he knew the tradition of Ptolemy and Avicenna in a work entitled *De divisione locorum habitabilium* (tr.1, c.6 [ed. Borgnet 9: 541]) and fragments of a geography attributed to Aristotle (tr.3, c.1 [ed. Borgnet 9: 566a]). Grosseteste also seems to have written a work entitled *De natura locorum*, known to Roger Bacon, (cf. S. H. Thomson, p. 110).

5. *De causis et proprietatibus elementorum* (ed. Borgnet 9: 585-653; autograph missing last chapter in Vienna, Oesterreichische Nationalbibliothek, Cod. misc. lat. 273, fol. 156r-168v).
A paraphrase of the pseudo-Aristotelian treatise of this name, translated from Arabic at Toledo by Gerard of Cremona as *Tractatus primus de causis proprietatum et elementorum quatuor* (minus *Tractatus secundus*, due to incomplete Arabic exemplar). Albert however, supplied from his own ingenuity what he thought was lacking (cf. I, tr.2, c.2 [ed. Borgnet 9: 601a]). This work deals with the effects of planets on the elements and is often referred to by Albert as *De effectibus planetarum in elementis* (*Meteor*. II, tr.3, c.1 [ed. Borgnet 4: 563]; see also ibid., tr.2, c.15 [ed. Borgnet 4: 560]; and tr.3, c.17 [ed. Borgnet 4: 581]). Like *De natura loci*, this science was considered an adjunct to *De caelo et mundo*. It was completed prior to his paraphrase of *Meteor*. II, in which the work is frequently cited as already composed by him.

6. *De generatione et corruptione* (or *Peri geneseos*) libri II (ed. Borgnet 4: 345-457).
A paraphrase based mainly on the Graeco-Latin translation of Henricus Aristippus (d. 1162), although Albert apparently knew the version made by Gerard of Cremona before 1187 from the Arabic. Composed prior to the *Meteora* and after *De causis et proprietatibus elementorum*, which is frequently referred to as already completed.

7. *Meteora* libri IV (ed. Borgnet 4: 477-832).
Composed after *De gen. et corrup.*, to which it refers several times, but before *De mineralibus*. The first three books of the paraphrase are based on the Latin translation from the Arabic by Gerard of Cremona, but the last book is clearly based on the Latin version from the Greek made by Henricus Aristippus (d. 1162); this fact is conspicuous from the use of corrupted Arabic

and Greek words in the two respectve parts. One of the main sources of Albert's paraphrase is Seneca's *Quaestiones naturales*, mainly because Albert was particularly conscious of the obscurity of the Aristotelian text and the imperfections of the manuscript. Although Aristotle is often taken to task for errors of observation, it is mainly Averroës, Avicebron, and Michael Scot who are severely criticized. Three additional chapters were attached to the Latin version from the Arabic by Alfred of Sareshel, the Englishman, around 1210. These chapters were a translation of Avicenna's *De congelatis* or *De mineralibus*. Albert omitted these from his *Meteora*, since they deal not with simple bodies and their motions, but with mixed bodies; for these Albert composed an entirely separate work.

8. *De mineralibus et lapidibus* (or *Mineralia*) libri v (ed. Borgnet 5: 1-116).

As the opening words indicate, this work follows immediately upon the *Meteora*. It was composed at Cologne, as we shall see, before becoming provincial. Albert complained that he had seen only excerpts from Aristotle's work on this subject and Avicenna's third chapter — both of which were insufficient to understand the subject matter, which includes stones, metals, and intermediary chemical substances (*De mineral.* I, tr.1, c.1). While using Alfred of Sareshel's translation of Avicenna's *De congelatis* as a framework, he expanded many sections and added a typical medieval lapidary from A to Z in II, tr.2, a mineralogical list in Bk. IV, and a list of intermediary chemicals in Bk. v. Kraków, Bibl. Jagiellońska MS 6392 (s. xv), fol. 46va-vb, has the following colophon after the five books of *Mineralia*: "Explicit liber mineralium editus a fratre Alberto quo<n>dam Ratisponense nacione theutonico professore de ordine fratrum predicatorum precipuo philosopho, editus anno domini M°CC°L in civitate Colonia Agrippina, presidente dicto Cunrado archiepiscopo civitatis memorate. Amen, etc." As previously suggested (above, p. 35 n. 75), this reading might have originally been "M°CC°L iv," which could have been shortly after 25 March 1254.

9. *De anima* libri III. (ed. Borgnet 5: 117-443; ed. Colon. 7/1 [n.8, 1969]).

One of Albert's most important works, it was written when he was prior provincial of *Teutonia*, i.e., between June 1254 and June 1257: "fratris alberti provincialis fratrum predicatorum per theutoniam liber de anima" (Basel, Universitätsbibliothek MS, F IV 34, fol. 50ra). Throughout the paraphrase Albert carefully compared and utilized two Latin versions of Aristotle's *De anima*: (1) the *vetustior* from the Greek, perhaps by James of Venice, ca. 1160 or at least by 1175, and (2) the version from the Arabic with the *Commentarium magnum* of Averroës made by Michael Scot ca. 1220. Albert did not know of the new translation from Greek made by William of Moerbeke, completed in 1268. In Bk. III Albert rejected the Averroist doctrine of one intellect for all men with arguments that are perhaps earlier than those later used at the papal curia in Rome (1256-57), arguments that were still later (ca. 1264-67) turned into the *Libellus de unitate intellectus contra Averroistas* (see ed. Colon. 17/1: 1-30). Certainly the present form of the *Libellus* is later than his *Metaphysica*, that is, after 1267.

a. This paraphrase of *De anima* seems to have the earliest indication that Albert had begun to work on Aristotle's *Organon* in systematic exposition for he refers to his *De praedicamentis* in II, tr.3, c.1: "dictum est in 1° *De generatione et corruptione*, ubi posuimus naturales rationes de ipsa agere et pati, in *Praedicamentis* vero logicas quasdam et communes posuimus rationes, et ideo ibi non inquisivimus . . ." (ed. Colon. 7/1: 96. 30-34).

b. He announced his intention of discussing *De motibus animalium* later, I, tr.2, c.16 (ed. Colon. 7/1: 63. 2-3); also in III, tr.4, c.8 (ibid., 237. 60).

10. *Parva naturalia* (a generic title) (ed. Borgnet 9: 1-525).
All of the following were written immediately after *De anima* and before *De vegetabilibus*. The Latin version used of the seven authentic works of Aristotle (minus *De longitudine et brevitate vitae* as a separate work) was the anonymous *corpus veteris translationis* from the Greek, ca. 1175. Having combined *De longitudine et brevitate vitae* with *De morte et vita*, Albert added two important works of his own and included Aristotle's *De motibus animalium* ("out of his own ingenuity" as he did not yet have the text) in the following order:

a. *De nutrimento et nutribili* (ed. Borgnet 9: 323-343).
An original work by Albert, although he seems to have thought that Aristotle wrote such a work. No such treatise, however, is known to have existed in the medieval Aristotelian corpus. Albert explicitly refers to his own paraphrase of *De anima* in tr.1, c.1: "Diximus autem in libro *De anima*" (ed. Borgnet 9: 323b).

b. *De sensu et sensato* (ed. Borgnet 9: 1-96).

Usually the first of the Aristotelian *parva naturalia* in the medieval corpus, Albert refers to his own previous work *De nutrimento et nutribili*: "sicut in libro *De nutrimento* dictum est" (tr.1, c.2, ed. Borgnet 9: 3).

c. *De memoria et reminiscentia* (ed. Borgnet 9: 97-119).
Refers to his earlier *De sensu* in tr.1, c.1: "dictum sit qualiter sensibilia veniunt ad animam, relinquitur considerandum . . ." (ed. Borgnet 9: 97b), and projects his next work in tr.1, c.3: "partim autem dicetur in libro *De intellectu et intelligibili*" (ed. Borgnet 9: 102a).

d. *De intellectu et intelligibili* libri II (ed. Borgnet 9: 477-525).
A very important and original work, Albert interpolated it in the series because the subject matter is "essential to all disciples of Aristotle" (Alkindi, Alfarabi, Alexander of Aphrodisias, etc.). But he notes in I, tr.1, c.1: "Interdum etiam Platonis recordabimur in his quibus Peripateticorum sententiis in nullo contradixit" (ed. Borgnet 9: 478a). All the material seems to be taken "ex epistola quadam Aristotelis, quam scripsit de universitatis principio, cuius mentionem in *Metaphysica* facit Avicenna" (I, tr.1, c.1, ed. Borgnet 9: 479b). Among other works, Albert refers to his own *De anima, De nutrimento et nutribili*, and *De sensu sensato* (cf. ed. Borgnet 9: 477b).

   i. Book I seems originally to have been entitled *De natura intellectus* (cf. Borgnet 9: 478a and 502a).

   ii. Book II, which is sometimes cited as a separate treatise, is frequently called *De perfectione animae intellectualis*, as in *Metaph.* V, tr.6, c.7: "De his autem alibi et praecipue in libro *De perfectione animae* determinatum est" (ed. Colon. 16/1: 288. 8-9). Note also the statement in *De unit. intel.* P.3, § 1: "Disputavimus de hoc latius in libro *De perfectione animae*, qui secundus est in libro *De intellectu et intelligibili* quem scripsimus" (ed. Colon. 17/1: 23. 46-48).

The intention to write this special work is announced in *De memoria* (cf. above), and it is prior to *De nat. et orig. animae.*

e. *De somno et vigilia.* (ed. Borgnet 9: 121-212).
Projected in *De intellectu* (I, tr.1, c.1, ed. Borgnet 9: 478a) with other *parva naturalia*, this paraphrase combines the three Aristotelian or pseudo-Aristotelian treatises *De somno, De somnio*, and *De divinatione per somnium*.

f. *De spiritu et respiratione* libri II. (ed. Borgnet 9: 213-255).
   Announced as prior to *De motibus animalium*: "prius *De inspiratione et respiratione* dicendum est quam de motu secundum locum . . . propter hoc etiam de motu spirationis sermo, sermoni *De motibus animalium* est anteponendus" (ed. Borgnet 9: 213b). Although Albert here follows Aristotle's principles, he is in fact explaining the *De differentia spiritus et animae* of Costa-ben-Luca, translated from Arabic at Toledo by John of Seville (fl. 1133-42). Three times he refers to his preceding work: "sicut in *Somno et Vigilia* dictum est" in I, tr.1, c.4 (ed. Borgnet 9: 220a), in I, tr.2, c.2 (ibid., p. 235a), and in II, tr.1, c.2 (ibid. p. 243a).

g. *De motibus animalium* (ed. Borgnet 9: 257-303).
   Announced at end of *De intellectu* above (ed. Borgnet 9: 519b) as *De principiis motuum animalium* shortly to follow. This work was composed by Albert "out of his own ingenuity" before he discovered Aristotle's text *De principiis motus processivi* in Italy (see below). As previously noted, this work was already projected in *De spiritu* (see above).

h. *De iuventute et senectute* (or *De aetate*) (ed. Borgnet 9: 305-321).
   In his opening words Albert announced the position of this work in the ensemble, and notes that it is to be followed by *De causis longioris et brevioris vitae* (ed. Borgnet 9: 305b), better known as *De morte et vita*. In this paraphrase Albert frequently refers to his previous compositions: *Mineralia, Peri geneseos, Meteororum* IV, *De somno et vigilia*, and *Physica*.

i. *De morte et vita* (or *De causis longioris et brevioris vitae*) (ed. Borgnet 9: 345-373).
   This work was announced at the beginning of the previous work, and is sometimes called *De longitudine et brevitate vitae*, because it combines two Aristotelian titles, and combines the subject matter of both. It is composed of two tractates. The first, however, is not a commentary on the Aristotelian work of either title, but Albert's own compilation utilizing the Aristotelian doctrine of aging and relevance of environment to the aging process. Throughout he frequently refers the reader to his earlier *De natura locorum* and *Meteororum* IV. In tr.2, c.11 (ed. Borgnet 9: 369a) Albert announces his forthcoming discussions of *De plantis* and *De animalibus*.

11. *Quaestiones super De animalibus* (not to be included among Albert's "Aristotelian paraphrases", even if authentic) (ed. Colon. 12 [n.3, 1955]: 281-309).

These questions, disputed at Cologne in 1258, exist as a *reportatio* and a compilation by Friar Conrad of Austria, as noted in Milan, Bibl. Ambros. MS, H 44 inf. (s. xiv), fol. 87vb: "Expliciunt questiones super de animalibus, quas disputavit frater albertus repetendo librum animalium fratribus colonie, quas reportavit quidam frater et collegit ab eo audiens dictum librum nomine cunradus de austria. Hoc actum est anno domini 1258." This would have been when Albert was *lector* at Cologne, following his resignation as provincial, and prior to his appointment to head the special commission on studies in the Order for the General Chapter of Valenciennes in June 1259. Albert was not appointed bishop of Regensburg until January 1260.

12. *De vegetabilibus* (or *De plantis*) libri VII (ed. Borgnet 10: 1-320; ed. E. Meyer and C. Jessen [Berlin, 1867]).

Only Books I and IV are a paraphrase of the pseudo-Aristotelian (Nicholas of Damascus) book of this title, translated from Arabic by Alfred of Sareshel around 1200. There seems to have been another translation of this pseudo-Aristotelian work, but it is still impossible to determine which version Albert used. Books II, III, and V are Albert's personal digressions, while Books VI and VII together constitute a typical medieval herbal in alphabetical order. The sources for this herbal are disputed, although it is commonly claimed that the main source was *De naturis rerum* by Friar Thomas of Cantimpré, one of Albert's students. However, the purpose of each of these works is fundamentally different: Thomas' being for the sake of preachers, Albert's for the sake of students "to satisfy curiosity rather than philosophy" (VI, c.1). *De vegetabilibus* is clearly prior to the vast collection *De animalibus*, as he states in I, tr.1, c.4: "de sensu enim et qualiter constituit animam in libro *De animalibus* habet determinari" (ed. Borgnet 10: 6b). And it explicitly refers to his earlier work *De intellectu*, noted above. Apparently it was written before Albert became bishop and occupied the See of Regensburg in March 1260.

13. *De animalibus* libri XXVI. (ed. Borgnet 11-12; ed. H. Stadler, *Beiträge*, 15-16 [1916, 1920]; autograph in Cologne, Historisches Archiv W 258a).

The first nineteen books are a paraphrase of the authentic Aristotelian corpus translated from Arabic by Michael Scot at Toledo around 1230. The others are Albert's own additions, mainly a bestiary in five books:

a. Bks. I-X are a paraphrase of Aristotle's *De historia animalium* in the Latin version from Arabic. Avicenna's brief paraphrase of *De animalibus* which Albert found useful, had also been translated by Michael Scot in the 1220s in Italy and dedicated to the Emperor Frederick II (1198-1250). At least part of this work was composed while Albert was bishop of Regensberg, for in Bk. VII, tr.2, c.6 (ed. Stadler, p. 523, v.1) Albert explicitly refers to his observations "in my villa above the Danube," which can only refer to the episcopal castle of Donaustauff, about three miles from the city, on the Danube. This can only be between April 1260 and December 1261.

b. Bks. XI-XIV are a paraphrase of Aristotle's four books *De partibus animalium*, in the Scot translation. In this section Albert refers to almost all of his previous compositions, notably the *Physica, De gen. et corrup., De anima*, various *parva naturalia*, and *De plantis*.

c. Bks. XV-XIX are a paraphrase of the five books of Aristotle's *De generatione animalium* in the Scot translation. Clearly, Bk. XVI (= *De gen. animal.* II) contained the key to much of Albert's understanding of Aristotle's doctrine concerning God's creation of the human soul, the operations of intelligences in nature, and the development of the human embryo; he often refers to it in connection with *Meteora* IV to explain how God and celestial bodies influence terrestrial generation and physical composition.

d. Bks. XX-XXI, in their present numbering, constitute Albert's unique contribution to comparative anatomy of perfect and imperfect animals. They are not a paraphrase of any known text. Therein he quotes among his previous compositions: *De caelo* IV, *Mineralia, Meteora, De vegetabilibus*, and various *parva naturalia*.

e. Bks. XXII-XXVI, in their present numbering, are a typical medieval bestiary, the sources of which are still much disputed, although many of his statements and descriptions are the result of his personal observation:

XXII man and quadrupeds

XXIII birds
XXIV aquatic animals
XXV serpents
XXVI vermin.

14. *De natura et origine animae* (ed. Borgnet 9: 375-436; ed. Colon. 12 [n.3, 1955]; autograph in Cologne, Historisches Archiv W 258a, fol. 308-326v).

This was originally intended as Book XX of *De animalibus* to follow the nineteen authentic works of Aristotle, but soon excerpted to form a separate "letter" (*epistula*) to discuss certain problems concerning the immortality of the human soul. Consequently Albert had to alter earlier passages in the *De animalibus*, wherein he had promised to discuss the problem, notably in XVI, tr.1, c.7, n.44 (ed. Stadler, p. 1083, v.5). It is clear that Albert changed his original plans and intended this treatise to be a separate work, explicitly a "letter" (*epistula nostra*) destined for certain confreres who urgently requested some such reply, probably in connection with the Averroist controversy beginning at Paris in the early 1260s. It is a magnificent culmination of his paraphrase of Aristotle's *De gen. animal.* II. By the time Albert wrote tr.1, c.4, he had already completed the second book of his *Metaphysica*; "De hoc tamen causa est a nobis dicta in secundo Primae Philosophiae" (ed. Colon. 12: 11. 41-42), added in the margin, but see editors, ibid. n.42. Albert refers to this work under various titles in his *Metaphysica* (cf. ed. Colon. 12/1, Proleg. § 1, pp. viii-ix), frequently by the title *De immortalitate animae*.

15. *De principiis motus processivi* (ed. Borgnet 10: 321-360; ed. H. Stadler [Munich, 1909]; ed. Colon. 12 [n.3, 1955]; autograph in Cologne, Historisches Archiv W 258a, fol. 339v-350v).

After *De natura et origine animae* had been extracted, this work was intended as Bk. XXII of *De animalibus* and as a continuation of Bk. XXI prior to the bestiary. After numerous changes in the autograph, this became a separate book even before the present ordering of *De animal.* XXI, tr.1, c.8, n.46, and *De animal.* XXIII, tr.un., n.2. It was clearly composed before Albert's paraphrase of *Metaph.* V, tr.1, c.7 (ed. Colon. 16/1: 224. 5-7). This separate work is a paraphrase based on a new version from the Greek, which Albert discovered "in campania iuxta Graeciam" during his second visit to Italy (August 1261 to February 1263), which apparently was *not* that of Moerbeke (cf. ed. Colon. 12/1, Proleg, § 2, pp. xxiv-xxvii). In this work Albert states that he had earlier composed a work on the movement of animals (*De motibus animalium*, above) "out of his own ingenuity" and now wished to see how closely he had come to the authentic teaching of Aristotole.

De principiis motus processivi presupposes all of Albert's paraphrases at least prior to *De animal.* XIII, inclusive. Therefore it must have been written after 1261/62 and before most of the *Metaphysica*, i.e., before ca. 1265.

* * * * *

With this we come to the end of Albert's paraphrases of Aristotle's *libri naturales*, including some pseudo-Aristotle and some of his own elaborations and additions. However, to place the *libri naturales* in the fuller perspective of the entire Aristotelian philosophy that Albert intended "to make intelligible to the Latins", the following *addendum*, even though brief, should be kept in mind:

16. *Ethica* libri x:

   a. *per modum commenti*, that is, commentary-with-questions, composed at Cologne on the basis of Grosseteste's translation from the Greek (ca. 1246/47) of all ten Books, delivered as lectures and reported by Thomas Aquinas ca. 1250-52 (ed. Colon. 14/1 [n.7, 1968-72]).

   b. *per modum scripti* (ed. Borgnet 7).
       This is the commonly known paraphrase in all the printed editions, and intended to be part of the "Aristotelian paraphrases." It was completed before the *Metaphysica* was begun, I, tr.1, c.5: "sicut ostendimus in VII *Physicorum* et in *Ethicorum* libro x" (ed. Colon. 16/1: 7. 78-79). Perhaps around 1262/63.

17. *Politica* libri VIII. (ed. Borgnet 8)
     Unlike his other commentaries on Aristotle, this is not a paraphrase but a literal commentary based on the Latin translation of Moerbeke from the Greek, which was finished at Orvieto in 1260. Albert probably obtained this version during his second visit to Italy (August 1261 to February 1263). Toward the end of this work, as a kind of epilogue, Albert not only protests the common complaint against his supposed adherence to Peripatetic philosophy and the *ipsissima verba* of Aristotle, but he begs all actually to read this work together with the other books of

natural and moral philosophy, which he had explained for the benefit of students: "Ecce hunc librum cum aliis physicis et moralibus exposui ad utilitatem studentium, et rogo omnes legentes . . ." (ed. Borgnet 8: 803). It was apparently composed before the *Metaphysica*, perhaps before he was assigned to preach another crusade to the Holy Land in February 1263.

18. *Metaphysica* libri XIII (ed. Borgnet 6; ed. Colon. 16 [n.5, 1960; n.6, 1964].

The text Albert utilized was the *Metaphysica media*, which was used only between 1250 and 1270, when William of Moerbeke revised the whole and introduced a translation of Book *Kappa*, making Book *Lambda* Book XII. It was composed not only after the entire *De animalibus*, but also after *Ethica* X, the *Poetica* (lost), and the very important two Books of the *Analytica posteriora* ("*per modum scripti*"). The paraphrase of the latter presupposes and refers to almost all of the earlier books of Aristotle's *Organon*. Albert's paraphrases of the entire *Organon*, together with some of Boethius' treatises and *Liber sex principiorum*, constitute an independent series that is difficult to date, but we have suggested above that he might have begun with the early books when he was prior provincial of *Teutonia* (1254-57), while composing his *De anima*. Although Albert may have begun his paraphrase of the *Metaphysica* earlier, the most likely date for the composition of the whole is between 1264 and 1267 while he was living as a retired bishop with his brother Henry at the Dominican Kloster in Würzburg. Even this late dating is prior to St. Thomas' own *Sententia super Metaphysicam* and Moerbeke's revision of all fourteen books of Aristotle's *Metaphysics*.

19. *De causis et processu universitatis* libri II. (ed. Borgnet 10: 361-628).

This seems to be the last of Albert's Aristotelian paraphrases; it is the last major work in the Aristotelian corpus known to the Middle Ages. Albert himself considered the pseudo-Aristotelian *De causis* to be "the natural complement" to *Metaph.*, Lambda: "Et haec quidem quando adiuncta fuerint undecimo Primae Philosophiae opus perfectum erit" (II, tr.5, c.24, ed. Borgnet 10: 619a). Bk. I is really an introduction to the whole of metaphysics as the study of the First Principle of all Being (*esse*) with particular emphasis on the absolute freedom of the First Being, who created all things with intelligence and freedom of choice. With this Albert undermines the foundation of all Moslem and Stoic determinism, while still allowing for a hierarchy of intelligences that influence the "formation" of natural things and phenomena, always "in virtue of the first cause." Bk. II is a systematic paraphrase of the pseudo-Aristotelian *De causis*, translated from Arabic by Gerard of Cremona (d. 1187), often known as *Liber Aristotelis de bonitate pura*. Although Albert knew that this was not an authentic work by Aristotle, he considered it a thematic composition with propositions and comment (after the manner of Euclid) by David, a certain Jew, who drew his material "ex quadam Aristotelis epistola quam de principio universi esse composuit, multa adiungens de dictis Avicennae et Alfarabi" (ed. Borgnet 10: 435b). It was not until William of Moerbeke completed his translation of Proclus' *Elementatio theologica* at Viterbo on 18 May 1268, that the Arabic *Liber de causis* was seen to be drawn from Proclus' propositions with commentary. This Albert did not know.

All of these philosophical works were completed by April 1271, at the latest, since in his *Problemata determinata* XLIII, of that date, Albert refers to his explanations already given in *De causis, Ethica, De animalibus*, and *Philosophia Prima* (see ed. Colon. 17/1 [n.12, 1975], xxvii-xxix, 45-64). One might conclude that all of Albert's Aristotelian paraphrases, that is, what he thought to be the "whole of Peripatetic philosophy" and human knowledge was "made intelligible to the Latins" by 1270 or by April 1271, at the latest. Within this period between ca. 1250 and ca. 1270, not only must the rest of the *Organon* be included, particularly the *Topics* and *Elenchi* (which could have been simultaneous with the *Metaphysica*), but also whatever other authentic philosophical commentaries of Albert be admitted by scholars, e.g., *Geometry, Poetics*, etc. We have said nothing about the vast number of theological works, notably exegetical and systematic, written by Albert the Great in his long life, nor anything about his sermons, letters, pious prayers, or poetry. The authenticity of some works, currently the *Speculum astronomiae*, is still being disputed.

Appendix 2

Apostolic Letter of Pope Piux XII
Ad Deum (16 December 1941)

Saint Albert the Great, Bishop, Confessor, and Doctor of the Church, is Declared Patron before God of Students of the Natural Sciences

For the perpetual remembrance of this subject. — To praise Almighty God, the source of all wisdom, the creator of nature, its master and its ruler (*Physica*, I, tr.1, c.1), St. Albert the Great, bishop, confessor and doctor of the Church, endeavored to mount to God through the knowledge of the natural creation; and to this end applied his genius to master the whole body of scientific knowledge known to his age. His grasp on the sciences was astonishing enough to earn for him, even among contemporary writers in their amazement at the depth and extent of his learning, the characterization: the wonder of the world and the universal doctor. And, in truth, apart from theology, philosophy and the elucidation of Holy Scriptures, to which he devoted himself with such zeal and skill that he had scarcely an equal in his knowledege of them, the saintly doctor, bent on banishing the conflict between faith and reason which a group of philosophers were introducing into the universities in the guise of the counterfeit principle of the double truth, busied himself from the early days of his youth to the end of his long life with the diligent, painstaking study of nature: "For from the creation of the world God's invisible attributes are plainly observable, being perceived through created things — His eternal power, namely, and divinity." (Rom. 1: 20.)

The fruits of his research he passed on to posterity in a copious literature composed with the utmost care in which he undertook to expound, in all its branches, nearly every natural science which was known in his time by the experimental method or induction; although not all the fruits were gathered which might have been expected, even in those days, from the example and the industry of so brilliant a teacher, owing chiefly to the conditions of the age and the lack of the necessary instruments. For, had the principles established by the great bishop of Regensburg on the necessity of experimentation and keen observation, on the importance of induction to find the truths of nature, been rightly understood and effectively exploited in his day, the marvellous advances made in the sciences, the proud boast of our own and the generations of the recent past, might have been attained centuries ago and have been fixed upon firm foundations to the best advantage of human society.

It is no wonder, then, that the universities and the more important Catholic colleges, not only in Italy, but in Germany, France, Hungary, Belgium, Holland, as well as in Spain, America and the Philippine Islands, besides numbers of professors of physics and other natural sciences, at the present time, look upon Albert the Great, as a beacon shining in a world engulfed in gloom. To make sure of the help of Almighty God in their exacting researches into the world of nature, they eagerly desire to have for their guide and heavenly intercessor him who, even in his own day, when many, puffed up with a hollow science of words, were turning their eyes away from the things of the spirit, has taught us by his example how we should rather mount from the things of earth to the things above.

It is, therefore, with sentiments of deepest pleasure that we accede to the wish expressed by the Catholic Academicians at their recent convention in Triers, by universities and by other international gatherings of scientists, and brought to our notice by the master general of the Order of Friar Preachers, who, on behalf of himself and of the order over which he presides, adds a fervent plea that We may deign to constitute St. Albert the Great the heavenly patron of students of the natural sciences. Accordingly, on this tenth anniversary of the Decree of December 16, 1931, which our predecessor of late memory, Pope Pius XI, issued, enjoining upon the universal Church the veneration of St. Albert the Great, bishop and confessor, with the additional title of doctor, it is altogether fitting that, as our supreme spiritual office requires, We foster a devotion so timely begun: moved also by the sad state of affairs of our day when the latest advances of science are employed, unhappily, not for God's praise and man's salvation, but to visit the calamities of war even upon civilian centers and cities. May St. Albert, who in his own very difficult times, proved by his wonderful work that science and Faith can flourish harmoniously in men, through his powerful intercession with God arouse the hearts and minds of those who devote themselves to the sciences to a peaceful and orderly use of the natural forces, the laws of which, divinely established, they investigate and seek after.

After consultation on this subject with our venerable brother, the bishop of Palestrina, Prefect of the Sacred Congregation of Rites, with due consideration of all the circumstances and regardless of anything to the contrary, by this letter and out of the fullness of our Apostolic authority We declare and constitute St. Albert the Great, bishop, confessor and Doctor of the church, forever the PATRON before God of Students of the Natural Sciences with the supplemental privileges and honors which belong, of its nature, to this heavenly patronage.

We decree that these presents shall ever be and remain firm, valid and effective; and shall have and hold their effects whole and entire; that they shall now and hereafter be upheld to the full by those whom they reach or shall teach; and that they shall be duly adjudged and defined in such wise that should any attempt be made upon them, wittingly or unwittingly, by anyone whomsoever, by whatsoever authority it shall be null and void from this time forward.

Given at Rome, at St. Peter's, under the ring of the Fisherman, the 16th day of the month of December in the year 1941, the third of our pontificate.

A. CARD. MAGLIONE,
Secretary of State.

AAS 34 (1942): 89-91.

Contributors

Benedict M. ASHLEY, OP, is professor of philosophy as well as professor of moral theology at the Aquinas Institute of Theology in Dubuque, Iowa, and former Regent of Studies for the Dominican Province of St. Albert the Great.

Joan CADDEN, a graduate of Harvard University, teaches the history of medieval science in the Department of History at Kenyon College in Gambier, Ohio.

Luke DEMAITRE, a disciple of Pearl Kibre at CUNY, writes on medieval medicine and teaches history at New York University; his current book is on Bernard de Gordon.

Lawrence DEWAN, OP, a graduate of the University of Toronto, is professor of philosophy at the Collège dominicain de philosophie et de théologie in Ottawa, Ontario.

Nadine F. GEORGE teaches history of science in the Department of Classics at Hamilton College in Clinton, New York, although most of her researches have been comparative studies of chemical theory and process in the later Middle Ages.

Jeremiah M. G. HACKETT, of the National University of Ireland, is a junior associate of the Pontifical Institute of Mediaeval Studies, and a graduate student of the Centre for Medieval Studies in the University of Toronto.

Pearl KIBRE is Professor Emeritus of History at The Graduate School of The City University of New York, where she is still in residence, and currently working on the Hippocratic Corpus of the Latin Middle Ages.

Edward P. MAHONEY, a disciple of Paul O. Kristeller, is professor of philosophy at Duke University in Durham, North Carolina, and an authority on Renaissance philosophy.

Ernest J. McCULLOUGH, a graduate of the University of Toronto, is professor of philosophy at St. Thomas College in the University of Saskatchewan, Saskatoon.

A. George MOLLAND, a graduate of Cambridge University, is lecturer in history and philosophy of science at the University of Aberdeen, Scotland.

James A. MULHOLLAND is an assistant professor in the history of science at North Carolina State University, and member of the American Society for Medals.

Robin S. OGGINS is an associate professor of history at the State University of New York at Binghamton, teaching medieval history there, and is a fellow of SUNY-Binghamton's Center for Medieval and Early Renaissance Studies.

Betsey Barker PRICE is a graduate student of the history of medieval science at the Centre for Medieval Studies in the University of Toronto.

John M. RIDDLE is a professor of ancient and medieval history in the Department of History at North Carolina State University at Raleigh; his research interests are in the history of early medicine.

Nancy G. SIRAISI, a disciple of Pearl Kibre, is associate professor of history at Hunter College of The City University of New York; her current research concerns the medical circle around Taddeo Alderati.

Jerry STANNARD, one of the outstanding authorities on medieval botany and herbals, is professor of history of science and medicine at the University of Kansas in Lawrence, Kansas.

Nicholas H. STENECK, a graduate of the University of Wisconsin, is associate professor of history, and fellow of the Collegiate Institute for Values and Science at the University of Michigan in Ann Arbor.

Edward A. SYNAN is professor of philosophy in the University of Toronto, senior fellow and past-president of the Pontifical Institute of Mediaeval Studies, Toronto.

Anthony A. TRAVILL is professor of anatomy and former chairman in the Faculty of Medicine at Queen's University in Kingston, Ontario.

Paul M. J. E. TUMMERS teaches the history of medieval philosophy in the Filosofisch Instituut at the Katholieke Universiteit of Nijmegen in the Netherlands; he is currently editing the commentary on Euclid's *Elements* for the Cologne edition of Albert's *Opera Omnia*.

William A. WALLACE, OP, is professor of philosophy and history of science at The Catholic University of America in Washington,

DC, and Director General of the Leonine Commission preparing the critical edition of the *Opera Omnia* of St. Thomas Aquinas.

James A. WEISHEIPL, OP, a graduate of Oxford University, is professor of the history and philosophy of medieval science at the University of Toronto, and senior fellow of the Pontifical Institute of Mediaeval Studies in Toronto.

Bibliography

Abu-l'Abbâs al-Fadhl ibn Ḥatim al-Narizi. *Anavitii in decem libros priores Elementorum Euclidis commentavii, ex interpretatione Gherardi Cremonensis in codice Cracoviense 569 vervata.* Ed. Ernst Maximilian Curtze. In *Supplementium* of *Euclidis opera omnia,* ed. J. L. Heiberg and H. Menge. Leipzig, 1899.

Accurti, Thomas. *Editiones saeculi xv pleraeque bibliographis ignotae.* Florence, 1930.

Adelardus of Bath. *Quaestiones naturales.* Eng. tr. Hermann Gollancz. In Berachya Hanakdan, *Dodi Ve-Nechdi [Uncle and Nephew],* pp. 85-161. London, 1920.

Adelmann, Howard Bernhardt. *The Embryological Treatises of Hieronymus Fabricius of Aquapendente.* Ithaca, 1942.

———. *Marcello Malpighi and the Evolution of Embryology.* 5 vols. Ithaca, 1966.

Adler, Mortimer J. "Sense Cognition: Aristotle vs. Aquinas." *New Scholasticism* 42 (1968): 578-591.

Agricola, Georg. *De natura fossilium.* Tr. Mark C. Bandy and Jean A. Bandy. New York: Geological Society of America, 1955.

———. *Georgius Agricola De re metalica.* Tr. Herbert C. Hoover and Lou H. Hoover. London, 1912; rpt. New York, 1950.

Aiken, Pauline. "The Animal History of Albertus Magnus and Thomas of Cantimpré." *Speculum* 22 (1947): 205-225.

Albertus Magnus. *Aureus liber Metaphysicae Divi Alberti Magni.* Venice, 1494.

———. *Book of Minerals.* Tr. Dorothy Wyckoff. Oxford: Clarendon Press, 1967.

———. *De anima et De intellectu et intelligibili.* Venice, 1481.

———. *De anima libri tres, De intellectu et intelligibili libri duo.* Venice, 1494.

———. *De anima.* See Aristotle, *Textus trium librorum De anima.* . . .

———. *De animalibus.* Rome, 1478.

———. *De animalibus libri vigintisex novissime impressi.* Venice, 1519.

———. *De animalibus libri xxvi.* Ed. Hermann Stadler. *Beiträge* 15 (1916) — Bk. i-xii; 16 (1920) — Bk. xiii-xxvi.

———. "De occultis naturis." Ed. Pearl Kibre. *Osiris* 13 (1958): 157-183.

———. *De vegetabilibus libri vii.* Ed. Ernst Meyer and Carl Jessen. Berlin: Georg Reimer, 1867.

———. *Euclid Commentary.* In Bernhard Geyer, "Die mathematischen Schriften des Albertus Magnus." *Angelicum* 35 (1958): 159-175.

———. *Opera . . . ad logicam pertinentia.* Venice, 1494.

———. *Opera Omnia.* Ed. Institutum Alberti Magni Coloniense. Münster i. Westf.: Aschendorff, 1951—.

_____. *Opera Omnia*. Ed. Auguste Borgnet. 38 vols. Paris: Vives, 1890-99.

_____. *Physicorum sive de phisico auditu libri octo*. Venice, 1494.

_____. *Prima pars summae Alberti Magni De quattuor coequevis una cum secunda eius quae est De homine*. Venice, 1498-1499.

_____. "The Problemata determinata XLIII. Ascribed to Albertus Magnus (1271)." Ed. James A. Weisheipl. *Mediaeval Studies* 22 (1960): 303-354.

_____. *Speculum astronomiae*. Ed. Paola Zambelli, S. Caroti, M. Pereira, S. Zamponi. Pisa: Domus Galilaeana, 1977.

_____. *Summi in via peripatetica philosophi theologique profundissimi naturalia ac supernaturalia opera per Marcum Antonium Zirnaram . . . purgata. . . .* Venice, 1517-1518.

_____. *Summi peripatetici due partes summae, quarum prima De quattuor coequevis, secunda De homine inscribitur, una cum pucherrimis additionibus editis ab . . . Marco Antonio Zirnara. . . .* Venice, 1519.

_____. *Tabula tractatuum parvorum naturalium Alberti Magni*. Venice, 1517.

Pseudo-Albertus Magnus. *Libellus de Alchimia ascribed to Albertus Magnus*. Tr. Sr. Virginia Heines. Berkeley: University of California Press, 1958.

Albrecht van Borgunnien. *Albrect van Borgunnien's Treatise on Medicine*. Ed. Walter Lawrence Wardale. Oxford, 1936.

Alfred von Sarashel (Alfredus Anglicus). *Des Alfred von Sarashel Schrift De motu cordis*. Ed. Clemens Baumker. *Beiträge* 23 (1923).

Andreae, Antonius. *Quaestiones super XII libros metaphysicae*. Venice, 1523.

Annales Basileenses. Ed. Phillippe Jaffé. MGH, Scriptorum 17: 193-202.

Antonaci, Antonio. *Francesco Storella, filosofo salentino del cinquecento*. Bari-Galatina, 1966.

_____. *Ricerche sull'aristotelismo del Rinascimento: Marcantonio Zimara*. Vol. 1. Lecce-Galatina, 1971.

Archer, Geoffrey [Francis], and Eva M. Godman. *The Birds of British Somaliland and the Gulf of Aden: Their Life Histories, Breeding Habits and Eggs*. 4 vols. London, 1937-1961.

Aristotle. *Aristoteles Latinus*. Ed. George Lacombe, et al. Rome, 1939—.

_____. *The Basic Works of Aristotle*. Ed. Richard McKeon. New York: Random House, 1941.

_____. *Textus trium librorum De anima . . . cum commentario secundum doctrinam venerabilis domini Alberti Magni*. Commentary by Johannes de Mechlinia, ed. Garardus de Harderwyck. Cologne, 1491 and Cologne, 1497.

_____. *De Caelo*. Tr. J. L. Stocks. In *The Works of Aristotle*, ed. W. D. Ross, 2: 268a1-313b24. Oxford, 1930.

_____. *De generatione animalium*. Tr. Arthur Platt. In *The Works of Aristotle*, ed. J. A. Smith and W. D. Ross, 5: 715a1-789b21. Oxford, 1912.

_____. *De partibus animalium*. Tr. William Ogle. In *The Works of Aristotle*, ed. J. A. Smith and W. D. Ross, 5: 639a1-679b30. Oxford, 1912.

_____. *De sensu et sensibili*. Tr. J. I. Beare. In *The Works of Aristotle*, ed W. D. Ross, 3: 436a1-449a33. Oxford, 1931.

_____. *De somno et vigilia*. Tr. J. I. Beare. In *The Works of Aristotle*, ed. W. D. Ross. 3: 453b11-458a33. Oxford, 1931.

_____. *Historia animalium*. Tr. D'Arcy Wentworth Thompson. In *The Works of Aristotle*, ed. J. A. Smith and W. D. Ross, 4: 486a5-633b9. Oxford, 1910.

_____. *Meteorologica*. Tr. H. D. P. Lee. Cambridge, Mass., 1962.

_____. *Meteorologicorum libri IV*. Ed. I. L. Ideler. Lipsiae, 1836.

_____. *Physica*. With commentary of Averroës. [Padua, about 1472-1475.]

_____. *Aristotle's Physics*. Tr. Hippocrates G. Apostle. Bloomington, Indiana, 1969.

_____. *Aristotle's Physics*. Tr. Richard Hope. Lincoln, Nebraska, 1961.

_____. *Aristotle's Physics I and II*. Tr. Waltar Charlton. Clarendon Aristotle Series. Oxford, 1970.

_____. *The Physics*. Tr. Philip H. Wicksteed and Francis M. Cornford. 2 vols. Cambridge, Mass., 1957.

_____. *Posterior Analytics*. In Thomas Aquinas, *Commentary on the Posterior Analytics of Aristotle*, tr. F. R. Larcher. Albany, 1970.

_____. *Aristotle's Posterior Analytics*. Tr. and ann. Jonathan Barnes. Oxford, 1975.

_____. *Posteriorum opus cum duplici traductione: antiqua scilicet et Argiropyli; ac eius luculentissimum interpretem Lincolniensem Burleumque*. Venice, 1521.

_____. *Topica*. Tr. W. A. Pickard-Cambridge. In *The Works of Aristotle*, ed. W. D. Ross, 1: 100a18-164b19. Oxford, 1928.

Pseudo-Aristotle. *De indivisibilibus lineis*. See *De lineis insecabilibus*.

_____. *De lineis insecabilibus*. Tr. Harold H. Joachim. In *The Works of Aristotle*, ed. W. D. Ross, 6: 968a1-972b31. Oxford, 1913.

_____. *De mirabilibus auscultationibus*. Tr. Launcelot D. Dowdall. In *The Works of Aristotle*, ed. W. D. Ross, 6: 830a4-847b10. Oxford, 1913.

_____. *De plantis*. Tr. E. S. Forster. In *The Works of Aristotle*, ed. W. D. Ross, 6: 815a10-830b4. Oxford, 1913.

_____. *De plantis*. Tr. W. S. Heth. In Aristotle, *Minor Works*, pp. 141-233. Loeb Classical Library. Cambridge, Mass., 1936.

Arnaldus de Villanova. *Aphorimi de gradibus*. In *Opera medica omnia*, ed. Michael R. McVaugh, vol. 2. Granada-Barcelona, 1975.

Arnold of Saxony. *De coelo et mundo*. Ed. Emil Stange. Erfurt, 1905.

Ashley, Benedict M. and Pierre Conway. "The Liberal Arts in St. Thomas Aquinas." *Thomist* 22 (1959): 460-532.

Augustine of Hippo, St. *The City of God*. Tr. Gerald Walsh et al, ed. Vernon Burke. New York, 1958.

_____. *De Genesi ad litteram libri duodecim*. PL 34: 245-486.

_____. *De quantitate aminae liber unus*. PL 32: 1033-1080.

_____. *De Trinitate libri XV*. Ed. W. J. Mountain and F. Glorie. Corpus Christianorum, Series Latina, vols. 50-50A. Turnhout, 1968.

Averroës. *Aristotelis Metaphysicorum libri XIII cum Averrois Cordubensis in eosdem commentariis, et Epitome*. Venice: Juntae, 1574; rpt. Frankfurt am Main, 1962.

_____. *Averrois Cordubensis commentarium magnum in Aristotelis De anima libros*. Ed. F. Stuart Crawford. Cambridge, Mass., 1953.

——. *Commentaria in libros Physicorum*. In *Aristotelis De Physico audito libri octo cum Averrois Cordubensis variis in eosdem commentariis*. Venice, 1562.

——. *Commentarium medium in Aristotelis De generatione et corruptione libros*. Ed. Francis H. Forbes. Corpus Commentariorum Averrois in Aristotelem, versionum Latinarum, 4, 1. Cambridge, Mass., 1956.

——. *Compendium libri Aristotelis De sensu et sensate*. Ed. A. L. Shields. Cambridge, Mass., 1949.

Avicenna. *De congelatione et conglutinatione lapidum*. Ed. and tr. E. J. Holmyard and D. C. Mandeville. Paris, 1927.

——. *De philosophia prima*. In his *Opera philosophica*. Tr. B. Cecilius Fabranensis. Venice, 1508; rpt. Louvain: Bibliothèque S. J., 1961.

——. *The Letter to King Hasen*. In *Theatrum chemicum*, ed. L. Zetzner, vol. 4. Strasbourg, 1622.

——. *Liber canonis Avicenne*. Venice: Paganinis, 1507.

——. *Liber de anima seu sextus de naturalibus*. Ed. Simone Van Riet. 2 vols. Avicenna Latinus. Louvain-Leiden, 1968-1972.

——. *Liber de animalibus Avicennae super librum De animalibus Aristotelis*. Tr. Michael Scot. Lyons, 1515.

——. *Sufficientia*. Venice, 1508.

Pseudo-Avicenna. *De anima in arte alchemiae*. In *Bibliotheca chemica curiosa*, ed. Jean Jacques Manget, 1: 633-636. Geneva, 1702.

Bach, Josef. *Des Albertus Magnus Verhältniss zu der Erkenntnislehre der Griechen, Lateiner, Araber und Juden*. Vienna, 1881.

Bacon, Roger. *Compendium studii philosophiae*. In *Opera quaedam hactenus inedita*, ed. John Sherren Brewer, 1: 391-519. Rolls Series, 15. London, 1859.

——. *Liber primus Communium naturalium fratis Rogeri*. Ed. Robert Steele. Oxford, 1909.

——. *Opus majus*. Ed. John Henry Bridges. 3 vols. Oxford, 1897-1900.

——. *Opus minus*. In *Opera quaedam hactenus inedita*, ed. John Sherren Brewer, 1: 311-389. Rolls Series, 15. London, 1859.

——. *Opus tertium*. In *Opera quaedam hactenus inedita*, ed. John Sherren Brewer, 1: 3-310. London, 1859.

——. *Quaestiones supra De plantis*. In *Opera hactenus inedita Rogeri Baconi*, ed. Robert Steele and Ferdinand Delorme, 11: 171-252. Oxford, 1932.

——. *Questiones supra libros octo Physicorum Aristotelis*. In *Opera hactenus inedita*, eds. Ferdinand M. Delorme and Robert Steele, vol. 13. Oxford, 1935.

——. *Rogeri Baconis moralis philosophia*. Ed. Eugenio Massa. Turin, 1953.

——. *Secretum secretorum*. In *Opera hactenus inedita Rogeri Bacon*, ed. Robert Steele, vol. 5. Oxford, 1920.

Balss, Heinrich. *Albertus Magnus als Biologe: Werk und Ursprung*. Stuttgart, 1947.

——. *Albertus Magnus als Zoologe*. In *Münchener Beiträge zur Geschichte und Literature der Naturwissenschaften und Medizin*, vols. 11-12. 1928.

Balthasar, Hans Urs von. *Herrlichkeit: Eine theologische Ästhetik*. 3 vols. Einsiedeln, 1961-1969.

Bannerman, David Armitage. *The Birds of the British Isles.* 12 vols. Edinburgh, 1953-1963.

Bayschlag, Franz et al. *The Deposits of Useful Minerals and Rocks: Their Origin, Form and Content.* Tr. S. J. Truscott. 2 vols. London, 1914-1916.

Beaujouan, Guy. "Motives and opportunities for science in the medieval universities." In *Scientific Change: Historical Studies in the Intellectual, Social and Technical Conditions for Scientific Discovery and Technical Invention, From Antiquity to the Present,* ed. Alistair Cameron Crombie, pp. 219-236. London, 1963.

Berthelot, Marcelin. *Archéologie et histoire des sciences avec publication nouvelle du papyrus grec chimique de Leyde et impression originale du Liber de septuaginta de Geber.* Paris, 1906.

―――. *La chimie au moyen âge.* paris, 1893.

Bertola, Ermenigildo. *"La questione del 'Senso agente' in Gaetano di Thiene."* In his *Saggi e studi di filosofia medievale,* pp. 53-69. Padua, 1951.

Bérubé, Camille. *La connaissance de l'individuel au moyen âge.* Montréal-Paris, 1964.

Binchy, Daniel A. *Church and State in Fascist Italy.* London, New York, 1941; rpt. with new preface, London, 1970.

Biringuccio, Vannoccio. *Pirotechnia.* Tr. Cyril S. Smith and Martha T. Gnudi. New York: American Institute of Mining and Mineralogical Engineering, 1941; rpt. Cambridge, Mass., 1966.

Birkenmajer, Alexandre. "Le role joué par les médécins et les naturalistes dans la réception d'Aristote au xxie et xiiie siècles." In *La Pologneau au vi^e Congrès International de Sciences Historiques, Oslo, 1928,* pp. 1-15. Warsaw-Ivov, 1930.

al-Biṭrūjī. *De motibus celorum.* Latin trans. by Michael Scot, ed. Francis J. Carmody. Berkeley and Los Angeles, 1952.

―――. *On the Principles of Astronomy.* Ed. and tr. Bernard R. Goldstein. New Haven, Conn., 1971.

Björnbo, Axel Anthon. "Studien über Menelaos, Sphärik: Beiträge zur Geschichte der Sphärik und Trigonometrie der Griechen." *Abhandlungen zur Geschichte der mathematischen Wissenschaften mit Einschluss ihrer Anwendungen.* 14 (1902): iii-viii, 1-154.

Blaine, Gilbert. *Falconry.* London, 1936.

Bochenski, I. M. *A History of Formal Logic.* South Bend, Ind.: University of Notre Dame Press, 1961.

Bock, Hieronymus. *De stirpium . . . nomenclaturis . . . commentariorum libri tres.* Strassburg, 1552; rpt. Louisville, 1960.

Boethius. *De divisione.* PL 64: 875-892.

―――. *De institutione arithmetica, De institutione musica.* Ed. Gottfried Friedlein. Leipzig, 1867; rpt. Frankfurt, 1966.

―――. *Philosophiae consolatio.* Ed. Ludwig Bieler. Corpus Christianorum, Series Latina, 94. Turnhout, 1957.

―――. *The Theological Tractates and the Consolation of Philosophy.* Tr. H. F. Stewart, E. K. Rand and S. J. Tester. New, rev. ed. Cambridge, Mass. 1973.

Boll, Franz. *Studien über Claudius Ptolomäus.* Leipzig: Teubner, 1894.

Bonnaud, R. "L'education scientifique de Beoce." *Speculum* 4 (1929): 198-206.

Bonné, Jacob. *Die Erkenntnislehre Alberts des Grossen mit besonderer Berücksichtigung des arabischen Neoplatinismus.* Bonn, 1935.

Borgognoni, Teodorico. *The Surgery of Theodoric.* Tr. Eldridge Campbell and James Colton. 2 vols. New York, 1955-1960.

Borman, F. "Thomas de Cantimpré indiqué comme une des sources où Albert le Grand et surtout Maerlant ont puisé les materiaux de leurs écrites sur l'histoire naturelle." *Bulletin de l'Academie Royales des Sciences . . . de Belgique,* 19/1 (1852): 132-159.

Bouyges, M. "Sur de *De plantis* d'Aristote-Nicholas à propos d'un manuscript arabe de Constantinople." *Mélanges de l'Université Saint-Joseph, Beyrouth* 9 (1924): 71-89.

Boyer, Carl B. *The Rainbow: From Myth to Mathematics.* New York, 1959.

Brand, Eberhard. *Die Mitwirkung der seligsten Jungfrau zur Erlösung nach dem hl. Antonin von Florenz mit besonderer Berücksichtigung des Verhältnisses zur Lehre Alberts des Grossen.* Rome, 1945.

Brandl, Leopold. *Die sexualethic des hl. Albertus Magnus.* Regensburg, 1955.

Breidert, Wolfgang. *Das aristotelische Kontinuum in der Scholastik.* Beiträge, new series, vol. 1 Münster, 1970.

Brennan, Sheila O'F. "Sense and the Sensitive Mean in Aristotle." *New Scholasticism* 47 (1973): 279-310.

Brockelman, Carl. *Geschichte der arabischen Literatur.* 2 vols. Leiden, 1943.

Brown, Leslie and Dean Amadon. *Eagles, Hawks and Falcons of the World.* 2 vols. Feltham, 1968.

Brunel, Clovis, ed. *Recettes médicales alchimiques et astrologiques.* Toulouse (privately published), 1956.

Burghoff, Henry L. "Corrosion of Copper Alloys." In *Corrosion of Metals,* ed. Carl William Borgmann. Cleveland, 1946.

Buridan, Jean. *Quaestiones in tres libros De anima.* Paris, 1516.

———. *Questiones in libros Aristotelis De anima.* In *Questiones et decisiones insignium vivorum,* ed. George Lokert. Paris, 1518.

Burley, Walter, "The 'De potentiis animae' of Walter Burley." Ed. M. Jean Kitchel. *Mediaeval Studies* 33 (1971): 85-113.

———. *De sensibus.* Ed. Herman Shapiro and Frederick Scott. Munich, 1966.

———. *Super octo libros physicorum.* Venice, 1501.

Burtt, Edwin Arthur. *The Metaphysical Foundations of Modern Physical Science.* New York, 1954.

Buschmiller, Robert J. *The Maternity of Mary in the Mariology of St. Albert The Great.* Carthagena, Ohio, 1959.

Cadden, Joan. "The Medieval Philosophy and Biology of Growth: Albertus Magnus, Thomas Aquinas, Albert of Saxony, and Marsilius of Inghen on Book I Chapter v of Aristotle's *De generatione et corruptione.*" Indiana University diss., 1971.

Callisthenes. See Khālid ibn Yazīd.

Callus, Daniel Angelo. "Introduction of Aristotelian Learning to Oxford." *Proceedings of the British Academy* 29 (1943): 229-281.

_____. "Robert Grosseteste as a Scholar." In *Robert Grosseteste: Scholar and Bishop*, ed. D. A. Callus, pp. 1-69. Oxford, 1955.

Cappello, Glori. "Umanesimo e scolastica: Il Valla, gli umanisti et Tommaso d'Aquino." *Rivista de filosofia neo-scolastica* 69 (1977): 423-442.

Cassiodorus Senator, Flavius Magnus Aurelius. *Cassiodori Senatoris Institutiones*. Ed. by Roger A. B. Mynors. Oxford, 1937.

Castagnoli, Pietro. "La vita e gli scritti di Sant'Alberto Magno (Rassegna bibliografica)." *Divus Thomas* (Piacenza) 37 (1934): 129-137.

Caws, Peter. *The Philosophy of Science*. Princeton, 1965.

Cech, C. O. "Über die geographische Verbreitung des Hopfens in Alterhume." *Bulletin de la Société Impériale des Naturalistes de Moscou* 57 (1882): 54-83.

Chenu, Marie-Dominique. "The Platonists of the Twelfth Century." In his *Nature, Man and Society in the Twelfth Century*, pp. 49-98. Oxford, 1968.

_____. " 'Maître' Thomas est-il un 'authorité'?" *Revue Thomiste* 30 (1925): 187-194.

Chevalier, Cyr Ulysse Joseph. *Répertoire des sources historiques du moyen âge*. 2nd ed. 2 vols. Paris, 1905-1907.

Clagett, Marshall. "The Medieval Latin Translations from the Arabic of the *Elements* of Euclid, with Special Emphasis on the Versions of Adelard of Bath." *Isis* 44 (1953): 16-42.

Classen, Sophronius. "Der Studiengang an der Kölner Artistenfacultät." In *Artes liberales von der antiken Bildung zur Wissenschaft des Mittelalters*, ed. Josef Koch, pp. 124-136. Leiden-Cologn, 1959.

Cole, Francis Joseph. *Early Theories of Sexual Generation*. Oxford, 1930.

Colomer, Eusebio. "Nikolaus von Kues und Heimeric van den Welde." *Mitteilungen und Forschungsbeiträge der Cusanus-Gesellschaft* 4 (1964): 198-213.

Combes, André. *Jean Gerson, commentateur Dionysien*. Paris, 1940.

Constantinus Africanus. . . .*Opera*. Basel: Petrus, 1536.

_____. . . .*Operum relique*. Basel: Petrus, 1539.

Corner, George. *Anatomical Texts of the Earlier Middle Ages*. Washington, D. C., 1927.

Cosenza, Mario E. *Biographical and Bibliographical Dictionary of the Italian Humanists and the World of Classical Scholarship in Italy*. 6 vols. Boston, 1962-1967

Costa ben Luca. See Qusṭā ibn Lūqā.

Cranz, F. Edward. "The Publishing History of the Aristotle Commentaries of Thomas Aquinas." *Traditio* 34 (1978): 157-192.

Crombie, Alistair Cameron. *Augustine to Galileo: The History of Science A.D. 400-1650*. London, 1952.

_____. "Avicenna's Influence on the Mediaeval Scientific Tradition." In *Avicenna: Scientist and Philosopher. A Millenary Symposium*, ed. G. M. Wickens, pp. 84-107. London: Luzac, 1952.

_____. *Robert Grosseteste and the Origin of Experimental Science, 1100-1700*. Oxford, 1953.

Crowley, Theodore. *Roger Bacon: The Problem of the Soul in his Philosophical Commentaries.* Louvain-Dublin, 1950.

Daems, Willem F. *Boec van Medicinen in Dietsche.* Leiden: Brill, 1967.

Daguillon, Jean, ed. *Ulrich de Strasbourg, la "Summa de bono," livre I.* Paris: Vrin, 1930.

Dähnert, Ulrich. *Die Erkenntnislehre des Albertus Magnus gemessen an den Stufen der 'abstractio'.* Leipzig, 1933.

Darmstaedter, Ernst. *Die Alchemie des Geber.* Berlin, 1922.

Deely, John N. "The Immateriality of the Intentional as Such." *New Scholasticism* 42 (1968): 293-306.

Delisle, Léopold. "Note sur un manuscrit de Tours renfermant des gloses françaises du XIIe siècle." *Bibliothèque de l'École des Chartes,* 6e sér., 30 (1869): 320-333.

Delmedigo, Elijah ben Moses Abba. *De primo motore.* In Joannes de Janduno, *Super octo libros Aristotelis De physico auditu subtilissimae quaestiones.* Venice, 1551.

Delorme, Augustin. "La morphogénèse d'Albert le Grand dans l'embryologie Scolastique." *Revue Thomiste* 36 (1931): 352-360.

Denifle, Heinrich. "Die Constitutionen des Prediger-Ordens vom Jahre 1228." *Archiv für Litterature- und Kirchen-geschichte des Mittelalters* 1 (1885): 165-227.

———. *Die Entstehung der Universitäten des Mittelalters bis 1400.* Berlin, 1885.

———, and Emile Chatelain. *Chartularium Universitatis Parisiensis.* 4 vols. Paris, 1889-1897.

Descartes, René. *Oeuvres de Descartes.* Ed. Charles Adam and Paul Tannery. 12 vols. Paris, 1897-1910.

Dewey, John. "A Recovery of Philosophy." In *Creative Intelligence,* pp. 3-69. New York, 1917.

Diem, Gudrun. "Les traductions gréco-latines de la Métaphysique au moyen âge." *Archiv für Geschichte der Philosophie* 49 (1967): 7-71.

Dietrich, Albert. *Medicinalia Arabica.* Gottingen, 1966.

Dijksterhuis, Eduard Jan. *The Mechanization of the World Picture.* Tr. C. Dikshoorn. Oxford: Clarendon Press, 1961.

Pseudo-Dionysius the Areopagite. *De divinis nominibus.* PG 3: 586-996.

Doesser, Henry Eeles. *A History of the Birds of Europe: Including All the Species Inhabiting the Western Palaearctic Region.* 9 vols. London, 1871-1896.

Drossaart Lulofs, H. J. "Aristoteles Περὶ φυτῶν." *Journal of Hellenic Studies* 77 (1957): 75-80.

Duhem, Pierre. *The Aim and Structure of Physical Theory.* Tr. Philip P. Wiener. New York, 1962.

———. *Le système du monde: Histoire de doctrines cosmologiques de Platon à Copernic.* 10 vols. Paris, 1913-1959.

Duns Scotus, Johannes. *Opera omnia.* 12 vols. Lyons, 1639; rpt. Hildesheim, 1968—.

———. *Quaestiones . . . in Metaphysicam Aristotelis.* Ed. Maurice O'Fihely. Venice, 1497.

Easton, Stewart C. *Roger Bacon and His Search for a Universal Science.* Oxford-New York, 1952.

Ebermann, Oskar. "Zur Aberglaubernsliste in Vintlers Pluemen der Tugent (v. 7694-7997)." *Zeitschrift für Volkskunde* 23 (1913): 1-18, 113-136.

Eckhart, Meister. *Das Buch der göttlichen Tröstung.* Ed. Joseph Quint, pp. 101-139. München, 1955.

Ehrle, Franz. *Der Sentenzkommentar Peters von Candia des Pisaner papstes Alexanders V: Ein Beitrag zur Scheidung der Schulen in der Scholastik des 14. Jahrhunderts und zur Geschichte des Wegestreites.* Münster, 1925.

Eis, Gerhard. "Meister Alexanders Monatsregeln." *Lychnos* (1950-1951): 104-136.

_____. *Wahrsagetexte des Spätmittelalters.* Berlin-Munich: Schmidt, 1956.

Elders, Leo. *Aristotle's Cosmology: A Commentary on the De caelo.* Assen, 1966.

Endres, J. A. "Das Geburtsjahr und die Chronologie in der ersten Lebenshälfte Alberts des Grossen." *Historishes Jahrbuch* 31 (1910): 293-304.

Ethier, Albert-Marie. "La double définition de l'âme humaine chez Saint Albert le Grand." *Études et recherches publiées par le Collège Dominicain d'Ottawa: I Philosophie,* Cahier I, pp. 79-110. Ottawa, 1936.

Euclid. *Elementa geometriae.* Trans. into Latin by Adelard of Bath, ed. Campanus de Novara. Venice, 1482.

_____. *Opera a Campano interprete fidissimo translata.* Venice, 1509.

_____. *The Thirteen Books of Euclid's Elements.* Tr. Thomas Little Heath. Cambridge, 1908.

_____. *The Translation of the Elements of Euclid from the Arabic into Latin by Herman of Carinthia.* Ed. H. L. L. Busard. Leiden, 1968.

Evans, Joan. *Magical Jewels of the Middle Ages and the Renaissance.* Oxford, 1922.

Evax, King of the Arabians. *Letters.* In *Marbode of Rennes' (1035-1123) De lapidibus,* ed. John M. Riddle, pp. 28-31. Sudhoffs Archiv, Beiheft 20. Wiesbaden, 1977.

Extensionis seu concessionis Officii et Missae addito Doctoris titulo ad universam Ecclesiam in honorem B. Alberti Magni. Rome: Sacred Congregation of Rites, 1931.

al-Fārābī, Abū Naṣr. "Kommentarii Abu Nasra al-Farabi k trudnostyam vo vvedeniakh k pervoy i pyatoy knigam Evklida" ("The Commentary of Abū Naṣr al-Fārābī on the Difficulties in the Introduction to Books I and V of Euclid"). Ed. M. F. Bokshteyn and B. A. Rozenfeld. *Problemy Vostokovedeniia* [= *Narody Azii I Afriki*], Akademiia Nauk S. S. S. R., no. 4 (1959): 93-104.

_____. *Alfarabi's Philosophy of Plato and Aristotle.* Tr. Muhsin Mahdi. Glencoe, Ill., 1962.

_____. *Über den Ursprung der Wissenschaften (De ortu scientiarum).* [Interpres Dominicus Gundissalinus.] Ed. Clemens Baeumker. Beiträge, vol. 19, part 3. Münster, 1916.

Faral, Edmond. "Jean Buridan, Maître ès arts de l'université de Paris." In *Histoire littéraire de la France* 38 (Paris, 1949): 462-605.

Fellner, Stephan. *Albertus Magnus als Botaniker.* Wien, 1881.

Ferckel, Christoph. "Die *Secreta mulierum* und ihr Verfasser." *Sudhoffs Archiv für Geschichte der Medizin und der Naturwissenschaften* 38 (1954): 267-274.

Ferrua, Angelico, ed. *S[ancti] Thomae Aquinatis vitae fontes praecipuae.* Alba: Ed. Domenicane, 1968.

Festugière, André Marie Jean. "La place du *De anima* dans le système Aristotélicienne d'après S. Thomas." *Archives d'histoire doctrinale et littéraire du moyen âge* 6 (1931): 25-47.

Feyl, A., ed. "Das Kochbuch des Eberhard von Landshut." *Ostbairische Grenzmarken* 5 (1961): 352-366.

Filthaut, Ephrem. "Um die Quaestiones de animalibus Alberts des Grossen." In *Studia Albertina, Festschrift für Bernard Geyer zum 70. Geburtstage*, ed. Heinrich Ostlender, pp. 112-127. Münster, Westf., 1952.

Firmicus Maternus, Julius. *Iulii Firmici Materni Matheseos libri VIII.* Ed. Wilhelm Kroll, Franz Skutsch and Konoat Julius Fürchtegott. 2 vols. Leipzig, 1897-1913.

Fischer, Hermann. "Mittelhochdeutsche Receptare aus bayerischen Klostern und ihre Heilpflanzen." *Mitteilungen der bayerischen botanischen Gesellschaft* 4 (1925): 69-75.

Folkerts, Menso. *Boethius' Geometrie II: Ein mathematische Lehrbuch des Mittelalters.* Wiesbaden, 1970.

Forbes, Robert James. *Studies in Ancient Technology.* vol. 1. Leiden, 1955.

Foster, Kenelm. *The Life of Saint Thomas Aquinas: Biographical Documents.* London: Longmans, 1959.

Frank, Jerome D. *Persuasion and Healing: A Comparative Study of Psychotherapy.* Baltimore, 1961.

Frederick II. *The Art of Falconry, Being the* De arte venandi cum avibus *of Frederick II of Hohenstaufen.* Ed. and tr. C. A. Wood and F. M. Fyfe. Stanford, Ca., 1943.

Freed, John B. *The Friars and German Society in the Thirteenth Century.* Cambridge, Mass.: Mediaeval Academy of America, 1977.

Fries, Albert. "Albertus Magnus." In *Die deutsche Literatur des Mittelalters Verfasserlexicon*, 1: 124-139. Berlin, 1977.

———. *Die Gedanken des heiligen Alberts Magnus über die Gottesmutter.* Thomistische Studien, 7. Freiburg, 1959.

———. *Marienkult bei Albertus Magnus.* (Forthcoming.)

Gabriel, Astrik L. " 'Via antiqua' and 'Via moderna' and the Migration of Paris Students and Masters to the German Universities in the Fifteenth Century." In *Antiqui und Moderni*, ed. Albert Zimmermann, pp. 439-483. Berlin-New York, 1974.

Galen. Ἅπαντον. *Opera omnia.* Ed. Karl Gottlob Kühn. 20 vols. Leipzig, 1821-1833.

———. *De complexionibus.* Tr. Burgundio of Pisa, ed. Richard J. Durling. Berlin-New York, 1976.

———. *Galenus Latinus I: Burgundio of Pisa's Translation of Galen's "De complexionibus."* Ed. Richard J. Durling. Berlin, 1976.

———. *On the Usefulness of the Parts of the Body.* Tr. Margaret Tallmadge May. 2 vols. Ithaca, 1968.

Pseudo-Galen. *Microtegni seu De spermate.* Trans. into Italian by Vera T. Passalacqua. Rome, 1959.

García Villoslada, Ricardo. *Storia del Collegio Romano dal suo inizio (1551) alla soppressioni della Compagnia di Gesu (1773).* Analecta Gregoriana 66. Rome, 1954.

Garin, Eugenio. *Storia della filosofia italiana.* Turin, 1966.

Garreau, Albert. *Saint Albert le Grand.* Paris, 1932.

Gasparrini Leporace, Tullia, et al., edd. *Un inedito erbario farmaceutico medioevale.* Florence: Alschki, 1952.

Gaul, Leopold. *Alberts des Grossen Verhältnis zu Plato.* Beiträge vol. 12, part 1. Münster, 1913.

Gentile da Foligno. *De generatione embrionis.* In Jacopo da Forli, *Expositio Jacobi supra capitulum de generatione embrionis.* . . .Venice, 1502.

Gerard de Frachet. *Vitae fratrum Ordinis Paedicatorum.* Ed. Benedictus M. Reichert. Monumenta Ordinis Fratrum Praedicatorum Historica, 1. Louvain, 1896.

Gerard of Cremona. *Theorica planetarum Gerardi.* Berkeley (privately published), 1942.

Gerhard von Herrenberg. *Ad favorabilem dirigens concordiam quedam problemata.* [Cologne], 1497.

Gesamtkatalog der Wiegendrucke. 7 vols. Leipzig, 1925-1938; rpt. Stuttgart-New York, 1968.

Geyer, Bernhard. "De mathematische Schriften des Albertus Magnus." *Angelicum* 35 (1958): 159-175.

_____. *Die Albert dem Grossen zugeschriebene* Summa naturalium (Philosophia pauperum). Beiträge vol. 35, part 1. Münster, 1938.

al-Ghazzālī. *Metaphysica.* Ed. Joseph Thomas Muckle. Toronto, 1933.

Gilson, Etienne. "L'âme raisonnable chez Albert le Grand." *Archives d'histoire doctrinale et littéraire du moyen âge* 14 (1943-1945): 5-72.

_____. "Autour de Pomponazzi, Problematique de l'immortalité de l'âme en Italie au debut du XVIe siècle." *Archives d'histoire doctrinale et littéraire du moyen âge* 36 (1961): 163-279.

_____. *History of Christian Philosophy in the Middle Ages.* New York: Random House, 1955.

_____. "Les sources gréco-arabes de l'augustinisme avicennisant." *Archives d'histoire doctrinale et littéraire du moyen âge* 36 (1961): 163-279.

Gliedman, Lester H., Earl H. Nash Jr., Stanley D. Imber, et al. "Reduction of Symptoms by Pharmacologically Inert Substances and by Short-term Psychotherapy." *A. M. A. Archives of Neurology and Psychiatry* 79 (1958): 345-351.

Gloria, Andrea, ed. *Monumenti della Università di Padova (1222-1318).* Venice, 1884.

_____. "Quot annos et in quibus Italiae urbibus Albertus Magnus moratus sit." *Atti del Reale Instituto Veneto* ser. 5, 6 (1879-1880): 1025-1050.

Glorieux, Palémon. "Le chancelier Gerson et la réforme de l'enseignement." In *Mélanges offerts à Etienne Gilson*, pp. 285-298. Toronto-Paris, 1959.

_____. *Répertoire des Maîtres en Théologie de Paris.* 2 vols. Paris: Vrin, 1933-1934.

Goldat, George David. "The Early Medieval Traditions of Euclid's Elements." University of Wisconsin, Madison: Unpublished Ph.D. thesis, 1956.

Goldstein, Bernard R. "Some Medieval Reports of Venus and Mercury Transits." *Centaurus* 14 (1969): 49-59.

Goldwater, Leonard J. *Mercury: A History of Quicksilver.* Baltimore, 1972.

Gorce, M.M. *L'essor de la pensée au moyen âge: Albert le Grand–Thomas d'Aquin.* Paris, 1933.

Grabmann, Martin. *Die Aristoteleskommentar des Heinrich von Brüssel und der Einfluss Alberts des Grossen auf die mittelalterliche Aristoteleserklärung.* Sitzungsberichte der Bayerischen Akademie der Wissenshaften, philosophisch-historische Abteilung, 1943, part 10.

——. "Die Kanonisation des heiligen Thomas von Aquin in ihrer Bedeutung für die Ausbreitung und Verteidigung seiner Lehre in 14. Jahrhunderts." *Divus Thomas* ser. 3, 1 (1923): 233-249.

——. *Mittelalterliches Geistesleben: Abhandlungen zur Geschichte der Scholastik und Mystik.* 3 vols. Munich, 1926-1956.

——. *Die Philosophia pauperum und ihr Verfasser Albert von Orlamünde.* Beiträge vol. 20, part 2, Münster, 1918.

——. *Die theologische Erkenntnis- und Einleitsungslehre des heiligen Thomas von Aquin und Grund seiner Schrift 'In Boethium de Trinitate'.* Freiburg-im-Schweiz, 1948.

Grant, Edward, ed. *A Sourcebook in Medieval Science.* Cambridge, Mass., 1974.

Grataroli, Guglielmo. *Turba philosophorum.* Tr., ed. Arthur E. Waite. London, 1896; rpt. London-New York, 1970.

Gregory of Nyssa, St. *De natura hominis.* PG 40: 503-818.

Gregory, Tullio. *Il pensiero della Rinascenza e della Riforma. Grande antologia filosofica,* vol. 6. Milan, 1964.

Grosseteste, Robert. *Commentarius in VIII libros Physicorum Aristotelis.* Ed. Richard C. Dales. Boulder, Colorado, 1963.

——. *In Aristotelis Posteriorum analyticorum libros.* Ed. Pamphilus de Monte Bononiensis. Venice, 1514.

——. *Die philosophischen Werke des Robert Grosseteste.* Ed. Ludwig Baur. Beiträge, vol. 9. Münster, 1912.

Guigues, Pierre. "Les noms arabes dans Serapion, 'Liber de simplici medicina': Essai de restitution et d'identification de noms arabes de médicaments usités au moyen âge." *Journal Asiatique,* ser. 10, 5 (1905): 473-546; 6 (1905): 49-112.

Gundissalinus, Dominicus. *De divisione philosophiae.* Ed. Ludgwig Baur. Beiträge vol. 14, parts 2-3. Münster, 1903.

Hall, Thomas S. "Life, Death and the Radical Moisture: A Study of Thematic Pattern in Medieval Medical Theory." *Clio Medica* 6 (1971): 3-23.

Hamesse, Jacqueline. *Les Auctoritates Aristoteles.* Louvain, 1974.

Hamilton, William James, and Harland Winfield Mossman. *Hamilton, Boyd and Mossman's Human Embryology: Prenatal Development of Form and Function.* 4th ed. Cambridge, 1972.

Hamlyn, David. *Sensation and Perception.* London, 1961.

Harkness, Roger, and Colin Murdoch. *Birds of Prey in the Field: A Guide to the British and European Species.* London, 1971.

Harpestraeng, Henrik. *Liber de herbarum.* Ed. Poul Hauberg. Copenhagen: Bogtrykkeriet Hafnia, 1936.

Harting, James Edmund. *Bibliotheca acciptravia: A Catalogue of Books Ancient and Modern Relating to Falconry.* London, 1891; rpt. London: Holland Press, 1964.

Haskins, Charles Homer. *Studies in the History of Mediaeval Science.* 2nd rev. ed. Cambridge, Mass., 1927; rpt. New York: Frederick Ungar, 1960.

Haubst, Rudolf. "Zum Fortleben Alberts des Grossen bei Heymerich von Kamp und Nikolaus von Kues." In *Studia Albertina,* ed. Heinrich Ostlender, pp. 420-447. Münster, 1952.

———. "Johannes von Franckfurt als der mutmassliche Verfasser von *Eyn deutsh Theologia.*" *Scholastik* 33 (1958): 375-398.

———. "Johannes Wenck aus Herrenberg als Albertist." *Recherches de théologie ancienne et médiévale* 18 (1951): 308-323.

———. *Studien zu Nikolaus von Kues und Johannes Wenck.* Beiträge vol. 38, part 1. Münster, 1955.

Heath, Thomas L. *Aristarchos of Samos.* Oxford University Press, 1913.

Hegel, Georg Wilhelm Friedrich. *The Logic of Hegel.* Tr. William Wallace. 2nd. ed. Oxford, 1892.

Heinzel, Hermann, Richard Fritter and John Parslow. *The Birds of Britain and Europe.* 3rd. ed. London, 1974.

Helm, Karl. "Albertus Magnus." In *Handwörterbuch des deutschen Aberglaubens,* ed. Hanns Bächtold- Stäubli, 1: 241-243. Berlin-Leipzig, 1927.

Hemmerdinger, Bertrand. "Le *De plantis,* de Nicholas de Damas a Planude." *Philologus* 111 (1967): 56-65.

Hempel, Carl G. *Philosophy of Natural Science.* Englewood Cliffs, N.J., 1966.

Henry of Hereford. *Chronica seu Liber de rebus moralibus.* Ed. A. Potthast. Göttingen, 1859.

Hermann of Heiligenhafen. *Der Herbarius communis des Hermannus de Sancto Portu.* Ed. H. Ebel. Würzburg: Triltsch, 1940.

Hermes. *Tabula smaragdena.* In Pseudo-Aristotle, *Secreta secretorum,* trans. from the Arabic into Latin by Philip of Tripoli. In Roger Bacon, *Opera hactenus inedita,* ed. Robert Steele, 5: 173-175. Oxford, 1920.

Herodotus. *Historiae.* Ed. Charles Hude. 3rd. ed. 2 vols. Oxford, 1951.

Hertling, Georg von. *Albertus Magnus: Beiträge zu seiner Würdigung.* Beiträge, vol. 14, pts. 5-6. Münster, 1914.

Hesse, Mary B. *Models and Analogies in Science.* South Bend, Ind., 1966.

———. *Science and the Human Imagination.* New York, 1953.

Hewson, Anthony. *Giles of Rome and the Medieval Theory of Conception.* London, 1975.

Hinnebusch, William A. *The History of the Dominican Order.* 2 vols. New York, 1966-1973.

Hippocrates. *The Medical Works of Hippocrates.* Tr. John Chadwick and William Neville Mann. Oxford, 1950.

——. *Du régime des maladies aiguës, De l'aliment de l'usage des liquides*. Texte établi et traduit par Robert Joly. Paris, 1972.

Hofmann, J. E. "Ueber eine Euklid-bearbeitung die dem Albertus Magnus zugeschrieben wird." In *Proceedings of the International Congress of Mathematicians, 14-21 Aug. 1958*, ed. J. A. Todd. Cambridge, 1960.

Hoppe, Brigitte. *Biologie, Wissenschaft von der belebten Materie von der Antike zur Neuzeit: Biologische Methodologie und Lehren von der stofflichen Zusammensetzung der Organismen*. Sudhoffs Archiv, Beiheft 17. Wiesbaden, 1976.

Horne, R. A. "Aristotelian chemistry." *Chymia* 11 (1961): 21-27.

Hortus sanitatis germanice. Mainz, 1485.

Hufnagel, Alfons. "Zur Echtheitsfrage der *Summa Theologiae* Alberts des Grossen." *Theologische Quartalschrift* (Tübingen) 146 (1966): 8-39.

Hugo of St. Victor. *Didascalicon de studio legendi*. Ed. Charles H. Buttimer. Washington, D.C., 1939.

Ḥunayn ibn Isḥāq al-ʿIbādī. *L'introduction de Hunain Ibn Ishāq, Joannitius, à la Mikrotechne de Galien; d'après un manuscrit, Codex 32, du XIIe siècle, conservé à la Bibliothèque d'Einsiedeln*. Ed., tr. Werner Wasserfallen. Berne, 1951.

Isidorus, Saint, bp. of Seville. *De scriptoribus ecclesiasticis*. Ed. Johannes Albertus Fabricius. In *Bibliotheca ecclestiastica*, ed. Johannes Albertus Fabricius, pp. 47-72. Hamburg, 1718; rpt. Farnborough, Hants., Eng., 1967.

——. *Isidori Hispalensis episcopi Etymologiarum sive Originum libri XX*. Ed. Wallace Martin Lindsay. 2 vols. Oxford, 1911.

Jābir ibn Ḥaiyān. *The works of Geber, Englished by Richard Russell, 1678*. London, 1928.

Jaeger, Werner Wilhelm. *Aristotle, Fundamentals of the History of His Development*. Tr. Richard Robinson. 2nd. ed. Oxford, 1955.

Jaki, Stanley L. *Science and Creation: From Eternal Cycles to an Oscillating Universe*. Edinburgh, 1974.

James, William. *Pragmatism*. Cambridge, Mass., 1975.

Jean de la Rochelle. *Tractatus de divisione multiplici potentiarum animae*. Ed. Pierre Michaud-Quantin. Paris, 1964.

Joannes de Janduno. *Quaestiones in duodecim libros metaphysicae ad intentionem Aristotelis et magni commentatoris Averrois subtilissimae disputatae*. Venice, 1505.

——. *Quaestiones super tres libros de anima*. Venice, 1519.

——. *Super libros Aristotelis De anima subtilissimae quaestiones*. Venice, 1589.

——. *Super octo libros Aristotelis De physico auditu subtilissimae quaestiones*. Venice, 1551.

Jöchle, Wolfgang. "Biology and Pathology of Reproduction in Greek Mythology." *Contraception* 4 (1971): 1-13.

Johannes de Mechlinia. See Aristotle. *Textus trium librorum De anima*. . . .

John Damascene, St. *De fide orthodoxa*. PG 94: 781-1228.

Jordan, Leo. "Ein mittelniederdeutsches Pflanzenglossar." *Zeitschrift für deutsche Wortforschung* 3 (1902): 353-356.

Jordan of Saxony. *Beati Iordani de Saxonia Epistulae.* Ed. Angelus M. Walz. Monumenta Ordinis Fratrum Praedicatorum Historica, 23. Rome, 1951.

Kearney, Hugh. *Science and Change: 1500-1700.* New York, 1971.

Kember, Owen. "Right and Left in the Sexual Theories of Parmenides." *Journal of Hellenic Studies* 91 (1971): 70-79.

Khālid ibn Yazīd [Callisthenes]. *Liber trium verborum.* In *Bibliotheca chemica curiosa,* ed. Jacques Jean Marget, 2: 189-191. Geneva, 1702.

Kibre, Pearl. "An Alchemical Tract Attributed to Albertus Magnus." *Isis* 35 (1944): 303-316.

_____. "Alchemical Writings Attributed to Albertus Magnus." *Speculum* 17 (1942): 499-518.

_____. "The *Alkimia Minor* ascribed to Albertus Magnus." *Isis* 32 (1940 [1949]): 267-300.

_____. "The *De occultis naturae* attributed to Albertus Magnus." *Osiris* 11 (1954): 23-39.

_____. "Further Manuscripts Containing Alchemical Tracts Attributed to Albertus Magnus." *Speculum* 34 (1959): 238-247.

_____. "Hippocratic Writings in the Middle Ages." *Bulletin of the History of Medicine* 18 (1945): 371-412.

_____. "Hippocratus latinus: Repertorium of Hippocratic Writings in the Latin Middle Ages." *Traditio* 31 (1975): 99-126, 32 (1976): 257-292, 33 (1977): 253-295, 34 (1978): 193-226.

_____. "Thomas of Cantimpré." In *Dictionary of Scientific Biography,* ed. C. C. Gillispie, 13: 347-349. New York, 1976.

Killermann, Sebastian. *Die Vogelkunde des Albertus Magnus (1207-1280).* Regensburg, 1910.

Kilwardby, Robert. *De ortu scientiarum.* Ed. Albert G. Judy. Toronto-London, 1976.

Kleemann, M. "Ein mittelniederdeutsches Pflanzenglossar." *Zeitschrift für deutsche Philologie* 9 (1878): 196-209.

Klubertanz, George. *The Discursive Power: Sources and Doctrine of the 'Vis cogitativa' according to St. Thomas Aquinas.* St. Louis, 1952.

Koch, Manfred. *Das Erfurter Kartäuserregimen.* Bonn: Unpublished dissertation, 1969.

Konrad of Megenberg. *Das Buch der Natur.* Ed. Franz Pfeiffer. Stuttgart, 1861; rpt. Hildesheim: Olms, 1962.

Koudelka, Vladimír J. "Zur Geschichte der böhmischen Dominikanerprovinz im Mittelalter." *Archivum Fratrum Praedicatorum* 25 (1955): 75-99, 26 (1956): 127-160.

Koyré, Alexandre. *Metaphysics and Measurement.* London, 1968.

Kren, Claudia. "Ailly, Pierre d'." In *Dictionary of Scientific Biography,* ed. C. C. Gillispie, 1: 84. New York, 1970.

Kristeller, Paul Oskar. "Beiträg der Schule von Salerno zur Entioicklung der Scholastischen Wissenschaft in 12. Jahrhundert." In *Artes liberales: von der Antiken Bildung zur Wissenschaft des Mittelalters*, ed. Josef Koch, pp. 84-90. Leiden, 1959.

———. "The School of Salerno: Its Development and Its Contribution to the History of Learning." In his *Studies in Renaissance Thought and Letters*, pp. 495-551. Rome, 1956.

———. *Studies in Renaissance Thought and Letters*. Rome, 1956.

———. "Thomism and the Italian Thought of the Renaissance." In his *Medieval Aspects of Renaissance Learning*, ed., tr. E. P. Mahoney, pp. 27-91. Durham, N. C., 1974. Originally published in his *Le Thomisme et la pensée italienne de la Renaissance*. Montreal, 1967.

Kroll, Jerome. "A Reappraisal of Psychiatry in the Middle Ages." *Archives of General Psychiatry* 29 (1973): 276-283.

Kuksewicz, Zdzistaw. *De Siger de Brabant à Jacques de Plaisance: La théorie de l'intellect chez les averroïstes latin des treizième et quatorzième siècles*. Wrocław, 1968.

———. "Le prolongement des polémiques entre les albertistes et les thomistes vu à travers le *Commentaire du De anima* de Jean de Glogow." *Archiv für Geschichte der Philosphie* 44 (1962): 151-171.

———. " 'Via antiqua' and 'via moderna' in der Krakauer Psychologie im 15. Jahrhundert." In *Antiqui und Moderni*, ed. Albert Zimmermann, pp. 509-514. Berlin-New York, 1974.

Künzle, Pius. *Das Verhältnis der Seele zu ihren Potenzen: Problemgeschichtliche Untersuchungen von Augustinus bis und mit Thomas von Aquin*. Freiburg, 1956.

Kurdzialek, M. "Anatomische und embryologishe Ausserungen Davids von Dinant." *Sudhoffs Archiv* 45 (1961): 1-22.

Kusche, Brigitte. "Zur 'Secreta mulierum'-Forschung." *Janus* 62 (1975): 103-123.

Labowsky, Lotte. "Aristoteles *De plantis* and Bessarion." *Medieval and Renaissance Studies* 5 (1961): 132-154.

Lapointe, François H. "Who Originated the Term 'Psychology'?" *Journal of the History of the Behavioural Sciences* 8 (1972): 328-333.

Lauchert, Friedrich, ed. *Geschichte des Physiologus*. Strassburg: Trübner, 1889.

Laurent, Marie Hyacinthe, and Yves Congar, edd. "Essai de bibliographie Albertinienne." *Revue Thomiste* 36 (1931): 422-468.

———. "Les grandes lignes de la vie du bienheureux Albert." *Revue Thomiste* 36 (1931): 257-259.

Lawn, Brian. *The Salernitan Questions*. Oxford, 1963.

Lazzarini, Vittorio. "I libri, gli argenti, le vesti di Giovanni Dondi dall'Orologio." *Bolletino del Museo Civico di Padova*, Series 1, vol. 1 (1925).

Leichty, Erle. *The Omen Series Summa Izbu: Texts from Cuneiform Sources*. New York, 1970.

Lesky, Erna. *Die Zeugungs- und Vererbungslehren der Antike und ihr Nachwirken*. Mainz, 1951.

Le Tourneur, Jean (Johannes Versor). *Questiones librorum de anima. . .per magistrum Joannem Glogoviensem. . .emendatum.* Cracow, 1514.

Lieser, Ludwig. *Vincenz von Beauvais als Kompilator und Philosoph. Eine Untersuchung Seiner Seelenlehre im Speculum Maius.* Leipzig, 1928.

Lindberg, David C. *Theories of Vision from Al-Kindi to Kepler.* Chicago, 1976.

———, and Nicholas Steneck. "The Sense of Vision and the Origins of Modern Science." In *Science, Medicine and Society in the Renaissance,* ed. Allen Debus, 1: 29-45. New York, 1972.

Lindner, Kurt. *Von Falken, Hunden und Pferden: Deutsche Albertus-Magnus-Übersetzungen aus der ersten Hälfte des 15. Jahrhunderts.* 2 vols. Berlin, 1962.

Little, Andrew George. "Roger Bacon." In *Franciscan Papers, Lists and Documents,* pp. 72-97. Manchester, 1943.

———. "Roger Bacon." *Proceedings of the British Academy* 14 (1928): 265-296.

———, ed. *Roger Bacon Essays.* Oxford, 1914.

Lloyd, Geoffrey E. R. "Right and Left in Greek Philosophy." *Journal of Hellenic Studies* 82 (1962): 56-66.

Loë, Ludwig Dietrich, *in religion* Paulus von. "De vita et scriptis B. Alberti Magni." *Analecta Bollandiana* 19 (1900): 257-284; 20 (1901): 273-316; 21 (1902): 361-371.

Löhr, Gabriel M. *Die Kölner Dominikanerschule vom 14. bis zum 16. Jahrhundert.* Freiburg, 1946.

———. *Die theologische Disputationen und Promotionen an der Universität Köln im ausgehenden 15. Jahrhundert.* Leipzig, 1926.

Lottin, Odon. "L'identité de l'âme et de ses facultés pendant la première moitié du XIIIe siècle." *Revue néoscolastique de philosophie* 36 (1934): 205-209.

———. "Problèmes concernant la 'Summa de creaturis' et le Commentaire des Sentences de Saint Albert le Grand." *Recherches de théologie ancienne et médiévale* 17 (1950): 321.

———. *Psychologie et morale aux XIIe et XIIIe siècles.* 6 vols. Louvain, 1942-1960.

[Luís of Valladolid.] *Brevis historia de vita et doctrina Alberti Magni.* In *Catalogus codicum hagiographicorum bibliothecae regiae Bruxellensis,* ediderunt Hagiographi Bollandiani, part 1, vol. 2, pp. 95-105. Bruxelles, 1889.

———. See also Scheeben, Heribert C.

Macer Floridus. *De viribus herbarum.* Ed. L. Choulant. Leipzig: Voss, 1832.

Maddin, Robert, James D. Muhly and Tamara S. Wheeler. "How the Iron Age Began." *Scientific American* 237 (Oct., 1977): 122-131.

Mahoney, Edward P. "Agostino Nifo and Saint Thomas Aquinas." *Memorie Domenicane* 7 (1976): 195-226.

———. "Agostino Nifo's *De sensu agente.*" *Archiv für Geschichte der Philosophie* 53 (1971): 119-142.

———. "Agostino Nifo's Early Views on Immortality." *Journal of the History of Philosophy* 8 (1970): 451-460.

———. "Antonio Trombetta and Agostino Nifo on Averroës and Intelligible Species: A Philosophical Dispute at the University of Padua." In *Storia e cultura al santo*, ed. Antonio Poppi, pp. 289-301. Venice, 1976.

———. "Duns Scotus and the School of Padua around 1500." In *Regnum hominis et regnum Dei: Acta quarti congressus Scotistici internationalis*, ed. Camille Bérubé, 2: 215-227. Rome, 1978.

———. "Nicoletto Vernia and Agostino Nifo on Alexander of Aphrodisias: An Unnoticed Dispute." *Rivista critica di storia della filosofia* 23 (1956): 268-296.

———. "Nicoletto Vernia on the Soul and Immortality." In *Philosophy and Humanism: Renaissance Essays in Honor of Paul Oskar Kristeller*, ed. Edward P. Mahoney, pp. 144-163. Leiden, 1976.

———. "Saint Thomas and Siger of Brabant Revisited." *The Review of Metaphysics* 27 (1973-1974): 531-553.

———. "Saint Thomas and the School of Padua at the End of the Fifteenth Century." In *Thomas and Bonaventure: A Septicentenary Commemoration, Proceedings of the American Catholic Philosophical Association* 48 (1974): 277-285.

Maier, Anneliese. *Ausgehendes Mittelalter: Gesammelte Aufsätze zur Geistesgeschichte des 14. Jahrhunderts*. 3 vols. Rome, 1964-1977.

———. *Metaphysische Hintergründe der spätscholastischen Naturphilosophie*. Rome, 1955.

———. "Die scholastische Wesenbestimmung der Bewegung als forma fluens oder fluxus formae und ihre Beziehung zu Albertus Magnus." *Angelicum* 21 (1944): 97-111.

———. *Die Vorläufer Galileis in 14. Jahrhundert*. Rome, 1949.

———. *Zwei Grundprobleme der scholastischen Naturphilosophie*. 2nd, ed. Rome, 1951.

———. *Zwischen Philosophie und Mechanik*. Rome, 1958.

de Maisonneuve, Jean. *De esse et essentia*. In *Geschichte des Albertismus I: Die Pariser Anfänge des Kölner Albertismus*, ed. Gilles Meersseman. Paris, 1933.

Mandonnet, Pierre. "Albert le Grand." In *Dictionnaire d'histoire et de géographie ecclésiastiques*, 1: 1515-1524. Paris, 1912.

———. "Albert le Grand." In *Dictionnaire de théologie catholique*, 1: 666-674. Paris, 1902; rpt. 1923.

———. "La date de naissance d'Albert le Grand." *Revue Thomiste* 36 (1931): 233-256.

———. "Polémique Averroiste de Siger de Brabant." *Revue Thomiste* 5 (1897): 95-105.

———. "Roger Bacon et le *Speculum astronomiae*." *Revue néo-scolastique* 17 (1910): 313-335.

Mantell, C. L. *Tin, Its Mining, Production, Technology and Application*. 2nd. ed. New York: American Chemical Society, 1949.

Mappae clavicula. Ed., tr. Cyril S. Smith and John G. Hawthorne. Philadelphia. 1974.

Marbode of Rennes. *Marbode of Rennes' (1035-1123) De lapidibus*. Ed. John M. Riddle. Sudhoffs Archiv, Beiheft 20. Wiesbaden, 1977.

Marcello, Cristoforo. *Universalis de anima traditionis opus*. Venice, 1508.

Mariétan, Joseph. *Problème de la classification des sciences d'Aristote à St. Thomas*. Paris, 1901.

Marx, Barbara. "Handschriften Paduaner Universitätsdozenten und Studenten aus San Bartolomeo di Vicenza." *Quaderni per la storia dell'Università di Padova* 9-10 (1976-1977): 143.

Matthew of Paris. See Paris, Matthew.

Matthew Platearius. *Das Arzneidrogenbuch* Circa instans *in einer Fassung des XIII. Jahrhunderts aus der Universitatsbibliothek Erlangen.* Ed. Hans Wölfel. Berlin: Unpublished dissertation, 1939.

Maurer, Armand. "Cajetan's Notion of Being in His Commentary on the *Sentences.*" *Mediaeval Studies* 28 (1966): 268-278.

Mazzantini, C. "La teoria della conoscenze in Alberto Magno." *Rivista di filosofia neo-scholastica* 29 (1937): 329-335.

McClintock, Stuart. *Perversity and Error: Studies on the "Averroist" John of Jandun.* Bloomington: Indiana University Press, 1956.

McVaugh, Michael R. "The 'Humidum radicale' in Thirteenth-Century Medicine." *Traditio* 30 (1974): 259-283.

Meadows, A. J. *The High Firmament.* Leicester: Leicester University Press, 1969.

Meersseman, Gilles. "Ergänzungen zur Kenntnis des literarischen Nachlasses des Kölner Professors Gerhard ter Steghen de Monte." *Jahrbuch des Kölnischen Geschichtvereins* 17 (1935): 264-268.

_____. *Geschichte des Albertismus I: Die Pariser Anfänge des Kölner Albertismus.* Paris, 1933.

_____. *Geschichte des Albertismus II: Die erster Kölner Kontroversen.* Rome, 1935.

_____. *Introductio in Opera Omnia B. Alberti Magni O. P.* Bruges: Beyaert, 1931.

Meyer, A. W. "The Elusive Human Allantois in Older Literature." In *Science, Medicine and History: Essays. . .in Honour of Charles Singer,* ed. E. Ashworth Underwood, 1: 510-520. Oxford, 1953.

_____. *The Rise of Embryology.* Stanford, 1939.

Meyer, Claudius Franz. "Die Personallehre in der Naturphilosophie von Albertus Magnus." *Kyklos* 2 (1929): 201.

Meyer, Ernst H. F. "Albertus Magnus, Ein Beiträg sur Geschichte der Botanik in dreizehnten Jahrhundert." *Linnaea* 10 (1836): 641-741.

_____. "Albertus Magnus. Zweiter Beiträg zur erneuerten Kenntniss seiner botanschen Leistungen." *Linnaea* 11 (1837): 545-595.

_____. *Geschichte der Botanik.* 4 vols. Köningsberg, 1854-1857.

Meyerhof, Max. "The Earliest Mention of a Manniparous Insect." *Isis* 37 (1947): 32-36.

Michael, Emil. *Geschichte des deutschen Volkes seit dem dreizehnten Jahrhundert bis zum Ausgang des Mittelalters.* 6 vols. Freiburg im Breisgau, 1897-1915.

_____. "Wann ist Albert der Grosse geboren?" *Zeitschrift für Katholishe Theologie* 35 (1911): 562-563.

Michalski, Konstanty. "La lutte autour de l'âme au quatuorzième et au quinzième siècles." In *Résumés des communications présentées au VIe congrès international des sciences historiques.* Oslo, 1928.

_____. "La lutte pour l'âme à Oxford et à Paris au XIVe siècle et sa répercussion à l'époque de la Renaissance." In *Seventh International Congress of Philosophy*

(Oxford, 1931), Proceedings, ed. Gilbert Ryle, pp. 508-515. London, 1931.

———. "La philosophie thomistique en Pologne à la fin du quinzième et au début du seizième siècle." *Bulletin international de l'académie des sciences de Cracovie, Classe de philologie* (Jan-July 1916).

Michaud-Quantin, Pierre. "Albert le Grand et les puissances de l'âme." *Revue du Moyen Âge Latin* 11 (1955): 59-86.

———. "La classification des puissances de l'âme au douzième siècle." *Revue du Moyen Âge Latin* 5 (1949): 15-34.

———. *La psychologie de l'activité chez Albert le Grand*. Bibliothèque Thomiste, 36. Paris, 1966.

Michell, E. B. *The Art and Practice of Hawking*. London, 1959.

Miller, Walter M. *A Canticle for Leibowitz*. Philadelphia, 1959.

Molland, Andrew George. "An Examination of Bradwardine's Geometry." *Archive for History of Exact Sciences* 19 (1978): 113-175.

Moody, Ernest A. *Studies in Mediaeval Philosophy, Science and Logic*. Berkeley, 1975.

Morienus. *A Testament of Alchemy*. Ed., tr. Lee Stavenhagen. Hanover, N. H., 1974.

Mowat, John Lancaster Gough, ed. *Alphita: A Medico-Botanical Glossary from the Bodleian Manuscript, Selden B. 35*. Oxford, 1887.

Mueller, Robert F., and Surendra K. Saxena. *Chemical Petrology*. New York, 1977.

Murdoch, John E. "Euclid: Transmission of the Elements." In *Dictionary of Scientific Biography*, ed. C. C. Gillispie, 4: 437-459. New York, 1971.

———. "From Social into Intellectual Factors: An Aspect of the Unitary Character of Late Medieval Learning." In *The Cultural Context of Medieval Learning*, ed. John E. Murdoch and Edith Sylla, pp. 271-339. Dordrecht-Boston, 1975.

———. "The Medieval Euclid: Salient Aspects of the Translation of the Elements by Adelard of Bath and Campanus of Novarra." *Revue de Sythèse*, 3e ser., nos. 49-52 (1968): 68-94.

———. "Philosophy and the Enterprise of Science in the Middle Ages." In *The Interaction between Science and Philosophy*, ed. Y. Elkana, pp. 51-74. Atlantic Heights, 1974.

———. "Superposition, Congruence and Continuity in the Middle Ages." In *Mélanges Alexandre Koyré*, 1: 416-441. Paris, 1964.

———, and Edith Sylla. "Burley, Walter." In *Dictionary of Scientific Biography*, ed. C. C. Gillispie, 2: 608-612. New York, 1970.

Napoli, Giovanni di. *L'immortalità dell'anima nel Rinascimento*. Turin, 1963.

Nardi, Bruno. *Dante e la cultura medievale*. Bari, 1942.

———. *Nel mondo di Dante*. Rome, 1944.

———. *Saggi sull'aristotelismo padovano dal secolo XIV al XVI*. Florence, 1958.

———. *Sigieri di Brabante nel pensiero del Rinascimento Italiano*. Rome, 1945.

———. *Studi di filosofia medievale*. Rome, 1960.

———. *Studi su Pietro Pomponazzi*. Florence, 1965.

Nardi, G. M. *Problemi d'embriologia antica e medioevale*. Florence, 1938.

Naudé, G. *Apologie pour tous les grands personnages qui ont esté faussement soupçonnes de magie*. Paris, 1625.

Neckam, Alexander. *De naturis rerum libri duo*. Ed. Thomas Wright. London: Longman, Green and Roberts, 1863.

Needham, Joseph. *A History of Embryology*. New York, 1959.

Nef, John U. "Mining and Metallurgy in Medieval Civilization." In *The Cambridge Economic History of Europe*, 2: 429-492. Cambridge, 1952.

Neugebauer, O. *A History of Ancient Mathematical Astronomy*. Studies in the History of Mathematics and Physical Sciences, 1. New York: Springer-Verlag, 1975.

Newton, Isaac. *Mathematical Principles of Natural Philosophy*. Ed. Florian Cajori. Rev. 2nd trans. by Andrew Motte. 2 vols. Berkeley: University of California Press, 1962.

Nicol, A. T. "Indivisible Lines." *Classical Quarterly* 30 (1936): 120-126.

Nicolaus Cusanus. *De coniecturis*. Ed. J. Koch and C. Bormann. In *Opera omnia*, vol. 3. Hamburg, 1972.

———. *De docta ignorantia*. Ed. E. Hoffman and R. Klibansky. In *Opera omnia*, vol. 1. Leipzig, 1932.

Nicolaus Damascenus. *Nicolae Damasceni De plantis libri duo Aristoteli vulgo ascripto*. Ed. Ernst H. F. Meyer. Leipzig, 1841.

Niebyl, Peter H. "Old Age, Fever, and the Lamp Metaphor." *Journal of the History of Medicine* 26 (1971): 351-368.

Nifo, Agostino. *Collectanea commentariaque in libros. . .de anima*. Venice, 1522.

———. *De sensu agente*. In Averroës, *Destructiones destructionum Averroys cum Augustini Niphi de Suessa expositione*. Venice, 1497.

———. *Liber de intellectu cum gratia et privilegio*. Venice, 1503.

———. *Super tres libros de anima*. Venice, 1503.

Noble, Henri-D. "Note pour l'étude de la psychophysiologie d'Albert le Grand et de S. Thomas: le cerveau et les facultés sensibles." *Revue Thomist* 13 (1905): 91-101.

Noonan, John T., Jr. *Contraception: A History of Its Treatment by the Catholic Theologians and Canonists*. Cambridge, Mass., 1966.

Nuovo receptario . . . della inclita cipta di Firenze. Florence: Ad instantia delli Signori Chonsoli, 1498.

O'Brien, Thomas. *Metaphysics and the Existence of God*. Washington, D. C., 1960.

Oakley, Francis. *The Political Thought of Pierre d'Ailly*. New Haven, 1964.

Ockham, William. *Quotlibeta septem una cum tractatu de sacramento altaris*. Strasbourg, 1491; rpt. Louvain, 1962.

———. *Super quattuor libros Sententiarum*. In his *Opera plurima*, ed. Jadocus Badius Ascensius, vols 3 and 4. Lyons, 1495-1496; rpt. London, 1962.

———. *The* Tractatus de succesivis *attributed to William of Ockham*. Ed. Philotheus Bochner. St. Bonaventure, New York: Franciscan Institute, 1944.

Oesterle, John A. "The Significance of the Universal *Ut Nunc*." *The Thomist* 24 (1961): 163-174.

Offredi, Appolinare. *Expositio in libros de anima et Quaestiones*. Venice, 1496.

Ogarek, S. *Die Sinneserkenntnis Alberts des Grossen verglichen mit derjenigen des Thomas von Aquin*. Fribourg, 1931.

Olsen, Richard. *Science as Metaphor*. Belmont, California, 1971.

O'Malley, John W. "Some Renaissance Panegyrics of Aquinas." *Renaissance Quarterly* 27 (1974): 174-192.

O'Neill, Ynez Viole. "The Fünfbilderserie Reconsidered." *Bulletin of the History of Medicine* 43 (1969): 236-245.

Oppenheimer, Jane. *Essays in the History of Embryology and Biology*. Cambridge, Mass., 1967.

Ostlender, Heinrich, ed. *Studia Albertina: Festschrift für Bernard Geyer zum 70. Geburtstage*. Münster, Westf., 1952.

Otte, James K. "The Life and Writings of Alfredus Anglicus." *Viator* 3 (1972): 275-291.

Pacchi, Arrigo. "Note sul commento al *De anima* di Giovanni di Jandun I, II, and III." *Rivista critica di storia della filosofia* 13 (1958): 373-374, 14 (1959): 456-457.

Pagallo, Giulio F. "Sull'autore (Nicoletto Vernia?) di un'anonima e inedita quaestio sull'anima del seclo xv." In *La filosofia della natura nel medioevo*, Congrès international de philosophie médiévale, 3 (Mendola, 1964), pp. 670-682. Milan, 1966.

Paneth, Fritz. "Über eine alchemistische Handschriften des 14. Jahrhunderts und ihr Verhältnis zu Albertus Magnus' Buch 'De mineralibus'." *Archiv für Geschichte der Mathematik, der Naturwissenschaften und der Technik*, ser. 3, 12 (1929): 35-45; 13 (1930): 408-413.

Parent, J. M. *La doctrine de la création dans l'école de Chartres: études et textes*. Ottawa, 1938.

Paris, Matthew. *Matthew Paris's English History from the Year 1235 to 1273*. Tr. J. A. Giles. 3 vols. London, 1852-1854.

Park, Charles F., Jr., and Ray A. MacDiarmen. *Ore Deposits*. 3rd. ed. San Francisco: Freeman, 1975.

Partington, J. R. "Albertus Magnus on Alchemy. *Ambix* 1 (1937): 3-20.

———. *A History of Greek Fire and Gunpowder*. Cambridge, 1960.

Paschini, Pio. *Domenico Grimani, Cardinale di S. Marco (†1523)*. Rome, 1943.

Paszewski, Adam. "Les problèmes physiologiques dans *De vegetabilibus et plantis libri VII* d'Albert von Lauingen." In *Actes du XIe congrès international d'histoire des sciences* (1965), 5: 323-330. Cracow, 1968.

Pattin, Adrian. "Pour l'histoire du sens agent au moyen âge." *Bulletin de philosophie médiévale* 16-17 (1974-1975).

Paul of Venice. *Expositio super libros de generatione et corruptione Aristotelis*. Venice, 1498.

———. *Expositio super octo libros physicorum Aristotelis necnon super commento Averrois cum dubiis eiusdem*. Venice, 1499.

———. *In libros de anima explanatio*. Venice, 1504.

———. *Summa naturalium*. Venice, 1476.

Peckham, John. *Registrum epistolarum fratris Johannis Peckham*. Ed. Charles Trice Martin. London, 1885.

Pedersen, Olaf. "The *Corpus astronomicum* and the Tradition of Medieval Latin Astronomy." In *Colloquia Copernica III*, ed. Jerzy Dobrzycki, pp. 57-96. Studia Copernica 13. Wrocław: Polska Akademia Nauk, 1975.

Pegis, Anton. *St. Thomas and the Problem of the Soul in the Thirteenth Century.* Toronto, 1934.

Pelacani da Parma, Biagio. *Le Quaestiones de anima.* Ed. Graziella Federici Vescovini. Florence, 1974.

Pelster, Franz. "Alberts des Grossen Jugendaufenthalt in Italien." *Historisches Jahrbuch* 42 (1922): 102-105.

_____. "Der beiden ersten Kapitel der Erklärung Alberts des Grossen zu De animalibus in ihrer ursprünglichen Fassung." *Scholastik* 28 (1953): 229-240.

_____. *Kritische Studien zum Leben und zu den Schriften Alberts des Grossen.* Freiburg-im-Breisgau: Herder, 1920.

_____. "Zur Datierung der Aristotelesparaphrase des hl. Albert des Grossen." *Zeitschrift für katholische Theologie* 56 (1932): 423-436.

Peter of Abano. *Aristotelis Stagirite philosophorum summi problemata atque divi Petri Apponi Patavini eorundem expositiones.* Mantua, 1475.

_____. *Conciliator differentiarum philosophorum et praecipue medicorum.* Venice, 1496.

Peter of Spain. *Tractatus.* Ed. L. M. De Rijk. Assen, 1972.

Pico della Mirandola, Giovanni. *Conclusiones.* Ed. Bohdan Kieszkowski. Geneva, 1973.

Pierre d'Ailly. *Tractatus brevis de anima.* Paris, 1505.

Pius XII, pope. Litterae apostolicae [16 Dec. 1941]: "Ad Deum." *Acta Apostolicae Sedis* 34 (1942): 89-91.

Plato. *Timaeus a Calcidio translatus commentarioque instructus.* Ed. J. H. Waszink. In *Plato Latinus*, ed R. Klibansky. Leiden, 1962.

Plessner, Martin. *Vorsokratische Philosophie und griechishe Alchemie in arabish-lateinischer Überbeferung: Studien zu Text und Inhalt der Turba philosophorum.* Nach dem Manuskript ediert von Felix Klein-Franke. Wiesbaden, 1975.

Pliny the Younger. *Natural History.* Ed., tr. H. Rackham, et al. 10 vols. Loeb Classical Library. Cambridge, Mass.-London, 1938-1963.

Pomponazzi, Pietro. *Corsi inediti dell'insegnamento padovano.* Ed. Antonino Poppi. 2 vols. Padua, 1966-1970.

Poppi, Antonino. "L'antiaverroismo nella scolastica padovana alla fine del secolo XV." *Studia Patavina* 11 (1964): 102-124.

_____. *Causalità e infinità nella scuola padovana dal 1480 al 1513.* Padua, 1966.

_____. "Il contributo dei *formalisti* padovani al problema delle distinzioni." In *Problemi e figure della Scuola scotista del Santo*, pp. 671-702. Padua, 1966.

_____. *Introduzione all'aristotelismo padovano.* Padua, 1970.

_____. "Padova, Scuola di." *Enciclopedia filosofica*, 2nd. ed., 4: 1263-1270. Florence, 1967.

_____. "Per una storia della cultura nel convento del Santo dal XIII al XIX secolo." *Quaderni per la storia dell'Università di Padova* 3 (1970): 1-29.

_____. *Saggi sul pensiero inedito di Pietro Pomponazzi.* Padua, 1970.

——. "Lo scotista patavino Antonio Trombetta (1436-1517)." *Il Santo* 2 (1962): 349-367.

——, ed. *Storia e cultura al Santo*. Fonti e Studi per la Storia del Santo a Padova, Studi 1. Venice, 1976.

——. "Trombetta, Antonio." *New Catholic Encyclopedia* 14: 314. New York, 1967.

Porta, Giovanni Battista della. *Natural Magick*. 1658; rpt. New York, 1958.

Preus, Anthony. *Science and Philosophy in Aristotle's Biological Works*. New York, 1975.

——. "Science and Philosophy in Aristotle's *Generation of Animals*." *Journal of the History of Biology* 3 (1970).

Ptolemy. *The Almagest*. Tr. R. Catesby Taliaferro. Great Books of the Western World, 16. Chicago: Encyclopaedia Britannica, 1952.

——. *Karpos*. Ed. Ae. Boer. In *Opera quae extant omnia*, vol. 3.2. Leipzig: Teubner, 1952.

——. *Tetrabiblos*. Ed., tr. F. E. Robbins. Loeb Classical Library. Cambridge, Mass.-London, 1940.

Puccetti Angiolo. *San Alberto Magno dell'Ordine dei Predicatori, vescovo e dottore della chiesa: profilo biografico*. Rome, 1932.

Quétif, Jacques and Jacques Échard. *Scriptores Ordinis Praedicatorum*. 2 vols. Paris, 1719-1721.

Quṣṭā ibn Lūqā. *De Physicis ligaturis*. In Arnaldus de Villanova, *Hec sunt opera . . .*, pp. 344-345. Venice, 1505.

——. *De physicis ligaturis*. In *Constantinus Afficanus, . . .Opera*, pp. 317-320. Basle, 1536.

Ragnisco, Pietro. *Della fortuna di S. Tommaso d'Aquino nella Università di Padova durante il Rinascimento*. Padua, 1892.

——. *Nicoletto Vernia: Studi storici sulla filosofia padovana nella 2ª metà del secolo decimoquarto*. Venice, 1891.

Randall, J. H. *Aristotle*. New York: Columbia University Press, 1960.

Rashdall, Hastings. *The Universities of Europe in the Middle Ages*. 2nd. ed. 3 vols. Oxford, 1936.

al-Rāzī, Abū Bakr Muḥammad ibn Zakarīyā. *Kitāb al-ḥāwī fī al-ṭibb* [*Continens of Rhazes*]. 23 vols. Hyderabad-Deccan (India), 1955-1971.

——. *De aluminibus et salibus*. In Robert Steele, "Practical Chemistry in the Twelfth Century." *Isis* 12 (1929): 10-46.

Reichert, Benedikt Maria, ed. *Acta capitulorum generalium Ordini praedicatorum*. 9 vols. Monumenta Ordinis Fratrum Praedicatorum Historia, 3-4, 8-14. Rome 1898-1904.

Reina, Maria Elena. *Note sulla posicologia de Buridano*. Milano, 1959.

Renzi, E. Salvatore de, ed. *Regimen sanitatis salernitanum (Flos medicinae scholae Salerni)*. In *Collectio Salernitana*, 5: 1-104. Naples, 1859.

Rhomer, Jean. "L'intentionnalité des sensations de Platon à Ockham." *Revue des sciences religieuses* 25 (1951): 5-39.

Rigault, Nicholas, ed. . . .*Rei accipitrariae scriptores nunc primum editi*. Lyons, 1612.

Ritter, Gerhard. *Die Heidelberger Universität I: Das Mittelalter (1386-1508)*. Heidelberg, 1936.

―――. *Studien zur Spätscholastik II: Via antiqua und via moderna auf den deutschen Universitäten des 15. Jahrhunderts*. Heidelberg, 1922.

Rose, Valentin. "Aristoteles *De lapidibus* und Arnoldus Saxo." *Zeitschrift für deutsches Alterthum* 18 (1875): 321-455.

Ross, William David. *Aristotle*. London, 1956.

Rudolph de Novimagio. *Legenda beati Alberti*. Ed. H. Chr. Scheeben. Ed. altera. Cologne, 1928.

Rufinus (fl. 1280). *De virtutibus herbarum*. Ed. Lynn Thorndike. Chicago: University of Chicago Press, 1949.

Ruska, Julius. *Turba philosophorum: ein Beitrag zur Geschichte der Alchemie*. Berlin, 1931.

Russell, Edward Stuart. *Form and Function: A Contribution to the history of animal morphology*. New York, 1917.

Sabra, A. I. "al-Nayrīzī." In *Dictionary of Scientific Biography*, ed. C. C. Gillispie, 10: 5-7. New York, 1974.

Salman, D. "Albert le Grand et l'averrïsme latin." *Revue de sciences philosophiques et théologiques* 24 (1935): 38-59.

Sambin, Paolo. "Intorno a Nicoletto Vernia." *Rinascimento* 3 (1952): 261-268.

Sarti, Mauro, and Mauro Faltorini. *De claris archigynasii Bononiensis professoribus*. 2nd. ed. 2 vols. Bologna, 1888-1896.

Sarton, George. *Introduction to the History of Science*. 3 vols. Baltimore: Williams and Wilkins, 1927-1950.

Scapin, Pietro. "Maurizio O'Fihely, editore e commentatore di Duns Scoto." In *Storia e cultura al Santo*, ed. Antonino Poppi, pp. 303-308. Venice, 1976.

―――. "La metafisica scotista a Padova dal XV al XVII secolo." In *Storia e cultura al Santo*, ed. Antonino Poppi, pp. 485-538. Venice, 1976.

Scheeben, Heribert C. *Albert der Grosse; zur Chronologie seines Lebens*. Quellen und Forschungen zur Geschichte des Dominikanerordens in Deutschland, 27. Vechta: Albertus-Magnus Verlag, 1931.

―――. *Albertus Magnus*. Cologne, 1955.

―――. "Handschriften I." *Archiv der deutschen Dominikaner* 1 (1937): 149-202.

―――. "Die Tabulae Ludwigs von Valladolid im Chor der Praedigerbrüder von St. Jakob in Paris." *Archivum Fratrum Praedicatorum* 1 (1930): 223-263.

―――. "Zur Chronologie des Lebens Alberts des Grossen." *Divus Thomas* (Fribourg) 10 (1932): 363-377. Also printed in *St. Albertus-Magnus-Festschrift* (Fribourg, 1932), pp. 231-245.

Schipperges, Heinrich. *Die Assimilation der arabischen Medizin durch das Lateinische Mittelalter*. Sudhoffs Archiv, Beiheft 3. Wiesbaden, 1964.

Schmidt, Erich. *Die Bedeutung Wilhelms von Brescia als Verfasser von Konsilien*. Leipzig: Unpublished dissertation, 1922.

Schmugge, Ludwig. *Johannes von Jandun (1285/89-1328)*. Stuttgart, 1966.

Schneider, Arthur. *Die Psychologie Alberts des Grossen: nach den Quellen dargestellt*. Beiträge vol. 4, parts 5-6. Münster, 1903-1906.

Schneider, Theodor. *Die Einheit des Meschen*. Beiträge, new series, vol. 8, Münster, 1972.

Schönbach, Anton. "Zeugnisse Bertholds von Regensburg zur Volkskunde." *Sitzungsberichte der Philosophisch-Historischen Classe der Kaiserlichen Akademie der Wissenschaften* (Wien.) 142/7 (1900): 1-156.

Schwertner, T. M. *St. Albert the Great*. Milwaukee: Bruce, 1932.

Senko, Wtadystaw. "Jean de Jandun et Thomas Wilton." *Bulletin de la Société international pour l'étude de la philosophie médiévale* 5 (1963): 139-143.

Serapion Junoir. *De medicinis simplicibus*. Milan, 1473.

Serapion Maior. *Practica*. Lyon, 1525.

Sermonetta, Josef Barùkh. "La dottrina dell'intelletto e la 'fede filosofica' di Jehudàh e Immanuel Romano." *Studi medievali*, 3rd. ser., vol. 6, part 2 (1965): 3-78.

Seybolt, Robert, tr. *The Manuale scholarium: An Original Account of Life in the Mediaeval University*. Cambridge, Mass., 1921.

Sezgin, Fuat. *Geschichte des arabischen Schrifttums*. 3 vols. Leiden, 1970.

Shapiro, Arthur K. "The Placebo Effect in the History of Medical Treatment: Implications for Psychiatry." *American Journal of Psychiatry* 116 (1959): 298-304.

Sharp, Dorothea Elizabeth. *Franciscan Philosophy at Oxford*. London, 1930.

Shaw, James R. "Scientific Empiricism in the Middle Ages: Albertus Magnus on Sexual Anatomy and Physiology." *Clio Medica* 10 (1975): 53-64.

Siegal, Rudolph E. *Galen's System of Physiology and Medicine*. Basel-New York, 1968.

Siger of Brabant, *Tractatus de anima intellectiva*. In *Siger de Brabant: Quaestiones in tertium De anima, De anima intellectiva, De aeternitate mundi*, ed. Bernardo Bazán, pp. 70-112. Louvain, 1972.

Sighart, Joachim. *Albert the Great*. Tr. A. Dixon. London, 1876.

Silvestro da Valsanzibio. *Vita e dottrina di Gaetano di Thiene*. 2nd. ed. Padua, 1949.

Simon, Paul. "Albertisti." In *Enciclopedia filosofica*, 2nd. ed., 1: 151. Florence, 1967.

――――. "Alberto Magno (s.)." In *Enciclopedia filosofica*, 2nd. ed., 1: 151-158. Florence, 1967.

Simone, Ludovico de. "Il B. Alberto Magno in Italia." *Memorie Domenicane* 48 (1931): 366-370. Reprinted as "Il Beato Alberto Magno in Italia, Influsso di lui nella Cultura Italiana." In *Alberto Magno: Atti della Settimana Albertina*, pp. 271-279. Rome, 1931.

Singer, Charles. *From Magic to Science*. New York, 1958.

――――. *A Short History of Scientific Ideas to 1900*. London, 1962.

――――. *Studies in the History and Method of Science*. Oxford, 1917.

Singer, Dorothy Waley. *Catalogue of Latin and Vernacular Alchemical Manuscripts in Great Britain and Ireland*. 3 vols. Brussels, 1928-1931.

Siraisi, Nancy G. *Arts and Sciences at Padua: The "Studium" of Padua before 1350*. Toronto: Pontifical Institute of Mediaeval Studies, 1973.

Solages, Bruno de. "La Cohérence de la métaphysique de l'âme d'Albert le Grand." In *Mélanges offerts au R. P. Ferdinand Cavallera*, pp. 367-400. Toulouse, 1948.

Solmsen, Friedrich. *Aristotle's System of the Physical World.* Ithaca, 1960.

Sprague, T. A. "Botanical Terms in Albertus Magnus." *Kew Bulletin* 9 (1933): 440-459.

———. "Plant Morphology in Albertus Magnus." *Kew Bulletin* 9 (1933): 431-440.

Stadler, H. "Albertus Magnus, Thomas von Cantimpré und Vinzenz von Beauvais." *Natur und Kultur* 4 (1906/07): 86-90.

Stannard, Jerry. "The Botanico-Medical Background of Baptista Fiera's Coena de herbarum virtutibus." In *Civilità dell'Umanesimo*, Atti del VIII Convegno del Centro di Studi Umanistici, Montepulciano, pp. 327-344. Florence: Olschki, 1972.

———. "Greco-Roman *Materia Medica* in Medieval Germany." *Bulletin of the History of Medicine* 46 (1972): 455-468.

———. "Hans von Gersdorff and Some Anonymous Strassburg Apothecaries." *Pharmacy in History* 13 (1971): 55-65.

———. "Identification of the Plants Described by Albertus Magnus, De vegetabilibus, lib. VI." *Res Publica Litterarum* 2 (1979). (Forthcoming).

———. "Magiferous Plants and Magic in Medieval Medical Botany." *Maryland Historian* 8 (1977): 33-46.

———. "Medieval Herbals and their Development." *Clio Medica* 9 (1974): 23-33.

———. "Squill in Ancient and Medieval Materia Medica, with Special Reference to Its Employment for Dropsy." *Bulletin of the New York Academy of Medicine* 50 (1974): 684-713.

Stapleton, Henry Ernest, R. F. Azo, M. Hidayat Husain and G. L. Lewis. "Two Alchemical Treatises Attributed to Avicenna." *Ambix* 10 (1962): 41-82.

Steele, Robert. "Practical Chemistry in the Twelfth Century." *Isis* 12 (1929): 10-46.

Steenberghen, Fernand van. *Aristotle in the West: The Origins of Latin Aristotelianism.* Louvain. 1955.

———. *Maître Siger de Brabant.* Louvain-Paris, 1977.

———. *La philosophie au treizième siècle.* Louvain, 1966.

———. *Siger de Brabant d'après ses oeuvres inédites.* 2 vols. Les philosophes belges, textes et études, 12-13. Louvain, 1931-1942.

Steneck, Nicholas. "Albert the Great on the Classification and Localization of the Internal Senses." *Isis* 65 (1974): 193-211.

———. "A Late Mediaeval Debate Concerning the Primary Organ of Perception." *Proceedings of the XIIIth International Congress of the History of Science*, 3.4: 198-204. Moscow, 1974.

———. "The Problem of the Internal Senses in the Fourteenth Century." Unpublished Ph.D. thesis, University of Wisconsin, 1970.

Stephanus de Salaniaca [and Bernardus Guidonis]. *De quatuor in quibus Deus Praedicatorum ordinem insignivit.* Ed. Thomas Kaeppeli. Monumenta Ordinis Fratrum Praedicatorum Historica, 22. Rome, 1949.

Stillman, John Maxson. *The Story of Early Chemistry.* New York-London, 1924. Rpt. as *The Story of Alchemy and Early Chemistry.* New York, 1960.

Stock, Brian. *Myth and Science in the Twelfth Century*. Princeton, N. J., 1972.

Stokes, Francis Griffin, ed. *Epistolae obscurorum virorum*. New Haven, 1925.

Stratton, George M. "Brain Localization by Albertus Magnus and Some Earlier Writers." *The American Journal of Psychology* 43 (1931): 128-131.

Sudhoff, Karl. "Codex Fritz Paneth, Eine Untersuchung." *Archiv für Geschichte der Mathematik, der Naturwissenschaften und der Technik* series 3, 12 (1929): 2-32.

Swann, Harry Kirke. *A Monograph of the Birds of Prey (Order Accipitres)*. Ed. Alexander Wetmore. 2 vols. London, 1930-1945.

Sylvester II, pope (Gerbert von Aurillac). *Opera mathematica*. Ed. Nicolaus Bubnov. Berlin, 1899.

Tabanelli, Mario. *La chirurgia italiana nell'alto Medioevo*. 2 vols. Florence, 1965.

Taiapietra, Geronimo. *Summa divinarum ac naturalium difficilium quaestionum*. Venice, 1506.

al-Tamīmī, Abū 'Abd Allāh Muhammad ibn Ahmad. *Über die Steine: das 14. Kapital aus dem Kitāb al-Muršid des Muhammad ibn Ahmad at-Tamīmī*. Ed., tr. Jutta Schönfeld. Freiburg, 1976.

Taton, René. *Histoire générale des sciences*. Paris, 1951.

Teilhard de Chardin, Pierre. *The Phenomenon of Man*. New York, 1961.

Temkin, Owsei. *Galenism: Rise and Decline of Medical Philosophy*. Ithaca, New York, 1973.

Themistius. *De anima*. Tr. William of Moerbeke. Ed. Gérand Verbeke. Louvain-Paris, 1957.

Theodoric. See Borgognoni, Teodorico.

Theophilus. *On Divers Acts*. Tr. J. G. Hawthorne and C. S. Smith. Chicago, 1963.

Theophrastus. *Enquiry into Plants and Minor Works on Odours and Weather Signs*. Tr. Arthur Hort. Loeb Classical Library. London-New York, 1916.

Thomas, Antonius Hendrik, ed. *Constitutiones antique Ordinis Fratrum Praedicatorum (1215-1237)*. Louvain, 1965.

Thomas Aquinas. *Commentaria in octo libros Physicorum Aristotelis*. In *Opera Omnia*, Leonine edition, vol. 2. Rome, 1884.

———. *De veritate Catholicae fidei contra Gentiles*. In *Opera Omnia*, vol. 5. Parma, 1855.

———. *In Aristotelis librum De anima commentarium*. Ed. Angelo Maria Pirotta. Turin, 1948.

———. *In libros Peri hermeneias expositio*. In *Opera Omnia*, Leonine edition, 1: 1-128. Rome, 1882.

———. *Quaestio disputata De spiritualibus creaturis*. In *Opera Omnia*, 8: 425-464. Parma, 1856.

———. *Scriptum super libros Sententiarum magistri Petri Lombardi episcopi Parisiensis*. Ed. Pierre Felix Mandonnet, and Marie Fabien Moos. 4 vols. Paris, 1929-1947.

———. *Summa theologiae*. Ed. Pietro Caramello. 4 vols. Turin-Rome, 1948-1956.

Thomas de Cantimpré (Cantimpratanus). *Bonum universale de apibus*. Douai: Baltazar Bellerus, 1627.

———. *De natura rerum* (lib. IV-XII). Transcribed and tr. by Luís García Ballester, Wolfram Schmitt, José Manuel Pita Andrade, et al. 2 vols. Granada, 1974.

———. *Liber de natura rerum.* vol. 1: *Text.* Ed. H. Boese. Berlin-New York, 1973.

Thomson, S. Harrison. *The Writings of Robert Grosseteste.* New York, 1940.

Thorndike, Lynn. "Further Consideration of the *Experimenta, Speculum astronomiae,* and *De secretis mulierum* ascribed to Albertus Magnus." *Speculum* 30 (1955): 413-443.

———. *A History of Magic and Experimental Science.* 8 vols. New York, 1923-1958.

———. *The Sphere of Sacrobosco and Its Commentators.* Chicago: University of Chicago Press, 1949.

———. *University Records and Life in the Middle Ages.* New York, 1944.

———, and Pearl Kibre. *A Catalogue of Incipits of Mediaeval Scientific Writings in Latin.* New and augmented ed. Cambridge, Mass., 1963.

Tiene, Gaetano. *Commentaria super quatuor libros metheororum Aristotelis.* Venice, 1491.

———. *Expositio in Aristotelem De Anima / Super libros de anima.* Venice, 1493.

———. *Expositio supra libro Aristotelis De caelo et mundo* Padua, 1475.

———. *Recollectae super octo libros physicorum cum annotationibus textuum.* Venice, 1496.

Tilander, Gunnar, ed. *Dancus Rex, Guillelmus Falconarius, Gerardus Falconarius: Les plus anciens traités de fauconnerie de l'occident, publiés d'après tous les manuscrits connus.* Cynegetica, vol. 9. Lund. 1963.

Tolomeo da Lucca. *Historia ecclesiastica.* In *Rerum Italicarum Scriptores,* ed. Lodovico Antonio Muratori, 11: 740-1249. Milan, 1727.

Torrigiano di Torrigiani, Pietro (Turisanus). *Turisani monaci plusquam commentum in Microtegni Galieni Cum questione ejusdem de ypostasi.* Venice, 1512.

Trithemius, Johannes. *De scriptoribus ecclesiasticis.* Paris, 1512.

Trombetta, Antonio. *Opus in Metaphysicam Aristotelis Padue in thomistas discussum.* Venice, 1502.

———. *Tractatus singularis contra Averroistas de humanarum animarum plurificatione.* Venice, 1498.

Trum, Beda. "La dottrina di S. Alberto Magno sui sensi interni." *Angelicum* 21 (1944): 279-298.

Tummers, Paul M. J. E. "The 'Commentary' of Albertus (Magnus?) on Euclid's 'Elements of Geometry'; Anaritius as his Source." *XVth International Congress of the History of Science. Abstracts of Scientific Section Papers,* p. 51. Edinburgh, 1977.

Ulrich of Strassburg. *Summa de bono, livre 1.* Ed. Jeanne Daguillon. Paris, 1930.

Valentinelli, Guiseppe. *Bibliotheca manuscripta ad S. Marci Venetiarum.* 6 vols. Venice, 1868-1873.

van den Welde, Heymerich. *Invectiva.* In *Geschichte des Albertismus II,* ed. Gilles Meersseman, pp. 109-121. Rome, 1935.

———. *Problemata inter Albertum Magnum et Sanctum Thomam.* Cologne, 1496.

van Ryzin, Sr. M. St. Martin. "The Arabic-Latin Tradition of Euclid's Elements in the Twelfth Century." Madison, Wis.: Unpublished Ph. D. thesis, 1960.

Vaurie, Charles. "Systematic Notes on Palearctic Birds, No. 44, Falconidae: The Genus Falco (Part 1, Falco peregrinus and Falco pelegrinoides)." *American Museum Novitates*, no. 2035.

Vaux, Roland de. "La première entrée d'Averroës chez les Latins." *Revue des sciences philosophiques et théologiques* 22 (1933): 193-245.

Verbeke, Gérard. "Introduction sur la doctrine psychologique d'Avicenne." In Avicenna, *Liber de anima seu sextus de naturalibus*, ed. Simone Van Riet, 1: 1*-90*, 2: 1*-73*. Louvain-Leiden, 1968-1972.

Vernia, Nicoletto. *Contra perversam Averrois opinionem de unitate intellectus et de animae felicitate*. Venice, 1505. Also in Albertus de Saxonia, *Acutissimae quaestiones super libros de Physica auscultatione*. Venice 1504, and Venice 1516.

———. *De gravibus et levibus*. In Gaetano Tiene, *Liber Aristotelis De celo et mundo*. Padua, 1474?

———. *Quaestio an coelum sit ex materia et forma constitutum*. 1482. Also in Walter Burley, *Super octo libros phisicorum*. Venice, 1501.

———. *Quaestio an dentur universalia realia*. In Urbanus Averroista, *Commentorum omnium Averroys super librum Aristotelis de Physico auditu expositor clarissimus*. Venice, 1492.

———. *Quaestio an ens mobile sit totius naturalis philosophiae subjectum*. In Marsilius van Inghen, *Quaestiones super libris Aristotelis de generatione et corruptione cum expositione Egidii*. Venice, 1500.

Vincent de Beauvais. *Bibliotheca mundi/Speculum quadruplex*. 4 vols. Douai, 1624; rpt. Graz, 1964-1965.

Vio, Thomas de. *In De ente et essentia D. Thomae Aquinatis commentaria*. Ed. Marie-Hyacinthe Laurent. Turin. 1934.

Vitruvius Pollio. *De architectura libri decem*. Ed. F. Krohn. Leipzig, 1912.

———. *The Ten Books on Architecture*. Tr. Morrio Hickey Morgan. Cambridge, Mass., 1914.

Walkany, Josef. *Congenital Malformations*. Chicago, 1971.

———. "History of Teratology." In *Handbook of Teratology*, ed. J. S. Wilson and F. C. Fraser, pp. 3-45. New York, 1975.

Wallace, William A. "Albertus Magnus, Saint." In *Dictionary of Scientific Biography*, ed. C. C. Gillispie, 1: 99-103. New York, 1970.

———. "Aquinas on the Temporal Relation between Cause and Effect." *The Review of Metaphysics* 27 (1974): 569-584.

———. *Causality and Scientific Explanation*. 2 vols. Ann Arbor: University of Michigan Press, 1972-1974.

———. "Galileo and Reasoning *Ex Suppositione*: The Methodology of the Two New Sciences." In *Proceedings of the 1974 Biennial Meeting of the Philosophy of Science Association*, ed. R. S. Cohen, et al., pp. 79-104. Dordrecht-Boston, 1976.

———. *Galileo's Early Notebooks: The Physical Questions*. Notre Dame, Ind., 1977.

Weiler, Antonius Gerardus. *Heinrich von Gorkum (d. 1431): Seine Stellung in der Philosophie und der Theologie des Spätmittelalters.* Hilversum, 1962.

_____. "Un traité de Jean de Nova Domo sur les universaux." *Vivarium* 6 (1968): 108-154.

Weisheipl, James Athanasius. "Albertus Magnus and the Oxford Platonists." *Proceedings of the American Catholic Philosophical Association* 32 (1958): 124-139.

_____. "Albert the Great, St." In *New Catholic Encyclopedia*; 1: 255-258. New York, 1967.

_____. "Classification of the Sciences in Medieval Thought." *Mediaeval Studies* 27 (1965): 54-90.

_____. *Friar Thomas d'Aquino: His Life, Thought and Works.* Garden City: Doubleday, 1974.

_____. "The Nature, Scope and Classification of the Sciences." *Studia Mediewistyczne* 18 (1977): 85-101.

_____. *The Development of Physical Theory in the Middle Ages.* New York, 1959; rpt. Ann Arbor, 1971.

_____. "Ockham and Some Mertonians." *Mediaeval Studies* 30 (1968): 163-213.

_____. "The Parisian Faculty of Arts in Mid-Thirteenth Century: 1240-1270." *American Benedictine Review* 25 (1974): 200-217.

_____. "The *Problemata determinata* XLIII Ascribed to Albertus Magnus (1271)." *Mediaeval Studies* 22 (1960): 303-354.

_____. "The Relationship of Medieval Natural Philosophy to Modern Science: The Contribution of Thomas Aquinas to Its Understanding." *Manuscripta* 20 (1976): 181-196.

_____. "Repertorium Mertonense." *Mediaeval Studies* 31 (1969): 185-208.

_____. "Thomism—Introduction and General Survey." In *New Catholic Encyclopedia*, 14: 126-135. New York, 1967.

Wells, Norman J. "On Last Looking into Cajetan's Metaphysics: A Rejoinder." *The New Scholasticism* 42 (1968): 112-117.

Wenck, Jean. *Le "De ignota litteratura" de Jean Wenck de Herrenberg contre Nicolas de Cuse.* Ed. E. Vansteenberghe. Beiträge vol. 8, part 6. Münster, 1910.

Werner, Karl. *Der Entwicklungsgang der mittelalterlichen Psychologie von Alcuin bis Albertus Magnus.* Vienna, 1876.

Werth, Hermann. "Altfranzösische Jagdlehrbücher nebst Handschriftenbibliographie der abendländischen Jagdlitteratur überhaupt." *Zeitschrift für Romanishe Philologie* 12 (1888): 146-191.

Whitehead, Alfred North. *Science and the Modern World.* New York: Mentor Books, 1948.

Wickersheimer, Ernest. "Nouveaux textes médiévaux sur le temps de cueillette des simples." *Archives internationales d'histoire des sciences* 29 (1950): 342-355.

Wimmer, Josef. *Deutsches Planzenleben nach Albertus Magnus (1193-1280).* Halle a. S., 1908.

Wingate, Sybil Douglas. *The Mediaeval Latin Versions of the Aristotelian Scientific*

Corpus, with Special Reference to the Biological Works. London, 1931.

Witherby, Harry Forbes, Francis C. R. Jourdain, Norman F. Ticehurst, and Bernard W. Tucker. *The Handbook of British Birds.* 5 vols. London, 1938-1943.

Wolfson, Harry. "The Internal Senses in Latin, Arabic, and Hebrew Philosophic Texts." *Harvard Theological Review* 28 (1935): 69-133.

Woodford, Michael. *A Manual of Falconry.* London, 1960.

Wyckoff, Dorothy. "Albertus Magnus on Ore Deposits." *Isis* 49 (1958): 109-122.

[Wylton, Thomas.] "Tomasza Wiltona: Quaestio disputata De anima intellectiva." Ed. W. Seńko. *Studia Mediewistyczne* 5 (1964): 3-190.

Yates, Frances. *The Art of Memory.* London, 1966.

Zabrzykowska, Anna, Zerzy Zathey, et al. *Inwentarz Rekopisów Biblíoteki Jagiellónskie.* 7 vols. Krakow, 1962.

Zachariae, Theod. "Abergläubische Meinungen und Gebräuche des Mittelalters in den Predigten Bernardinos von Siena." *Zeitschrift für Volkskunde* 22 (1912): 113-134, 225-244.

Zambelli, Paola. "Da Aristotele a Abū Ma'Shar, da Richard de Fournival a Guglielmo da Pastrengo: Un'Opera Astrologica Contoversa di Alberto Magno." *Physis* 15 (1973): 375-400.

Zimara, Marco Antonio. *Annotationes in Joannem Gandavensem super quaestionibus metaphysicae.* In Joannes de Janduno, *Quaestiones in duodecim libros metaphysicae.* Venice, 1505.

——. *De principio individuationis ad intentionem Averrois et Aristotelis.* In Joannes de Janduno, *Quaestiones in duodecim libros metaphysicae.* Venice, 1505.

——. *Quaestio de speciebus intelligilibus ad mentem Averrois.* N.p., n.d. Reprinted as *Quaestio qua species intelligilibiles ad mentem Averrois defenduntur.* In [Asclepius], *Hoc volumine contenta: Asclepii ex voce Ammonii Hermeae in Metaphysicam Aristotelis praefatio . . .,* ed. Franciscus Maria Storella. Naples, 1575 and 1576.

——. *Solutiones contradictionum in dictis Averroys.* In his *Questio de primo cognito.* Venice, 1508.

Zimmermann, Albert, ed. *Antiqui und Moderni.* Berlin-New York, 1974.

Zirkle, Conway. "The Early History of the Idea of Inheritance of Acquired Characters and of Pangenesis." *Transactions of the American Philosophical Society* 35 (1945): 91-151.

Zucchi, Alberto. "Sant'Alberto Magno a Padova." *Memorie Domenicane* 49 (1932): 339-406.

Index of Persons and Places

Note: anonymous works have been listed by title in the subject index.

283 n68; *Parva naturalia, De somno*, 283 n70; *Physics*, 81, 82 n44, 84, 86 n67, 88, 98, 98 n103, 104 n6, 106, 106 n10, 107 n11, 109, 111, 113, 114, 116, 126, 130-132, 134, 135, 136 n22, 140, 140 n37, 287 n84, 470, 472 n27, 538 n3; *Posterior Analytics*, 6 n10, 103 n3, 104, 109, 136 n21, 305 n25, 464, 468, 468 n13, 491, 491 n54; *Politics*, 101; *Topics*, 498 n7

Aristotle, pseudo, 473, 498; works of: *De causis proprietatum et elementorum quatuor*, 567; *De lineis insecabilibus*, 85, 565, 566; *De mirabilibus auscultationibus*, 226 n98; *De natura deorum*, 87 n67; *De natura locorum*, 566; *De plantis*, 27, 341, 344, 345 n11, 346 n13, 347, 348, 351 n31, 572 (see also Nicholas of Damascus); lapidary, 228, 230, 230 n119, 231, 231 n121, 233, 371, 572; *Liber de causis*, 30, 576; *Liber de indivisibilibus lineis see De lineis insecabilibus*; *Secreta secretorum*, 190 n23

Arnaldus de Villanova, 206 n15, 206 n16, 207 n18, 211 n46

Arndelhousen *see* cloister of Arndelhousen

Arnold of Saxony, 224 n91, 230, 230 n119, 231, 231 n121, 232, 233, 234, 234 n133, 234 n134

Arnold von Tongern, 528

Ashley, Benedict M., 78 n29, 95 n94, 101 n119

Austria, 15

Augsburg, 15, 50

Augustine of Hippo, 32, 99, 182, 274, 274 n35, 382, 406, 407, 424, 425, 430, 433, 434, 506, 508 n24; *City of God*, 433 n112; *De Genesi ad litteram*, 274 n35; *De quantitate animae*, 306 n28; *De trinitate*, 288, 288 n87

Averroës, 14, 44-45, 63, 64, 75, 79 n31, 93, 130, 132, 143 n50, 145, 148, 149, 168, 265 n4, 274 n35, 275, 295, 296, 313, 328, 328 n23, 384, 393, 393 n43, 401, 422 n68, 494 n68, 508, 511, 512, 512 n38, 513 n41, 514, 515, 516, 517, 519, 519 n73, 534, 540-541, 541 n7, 544, 549 n46, 550 n49, 552, 554-559, 562-563, 566, 568; on spiritual being *see* spiritual being; theory of sensation, 302; theory of vision, 307; works of, 134 n15; *Colliget*, 394, 394 n44; *De anima*, 294, 294 n8, 295, 311 n36, 319 n55, 514 n54, 557 n71, 558 n74; *De corde*, 393, 400; *De generatione et corruptione*, 328 n22; *De sensu et sensato*, 272, 272 n28, 276 n42, 302, 302 n19; *Meteorologicorum*, 93 n89; *Physics*,

107, 107 n12, 116, 117, 130, 132, 134-135, 134 n16, 135 n17, 135 n18, 135 n19, 137 n27, 140, 140 n38, 141 n39, 142 n43, 144 n55, 148 n63, 148 n65

Avicebron, 568

Avicenna, 14, 32, 63, 64, 75, 79 n31, 130, 132, 143, 151, 170, 183, 188, 188 n6, 188 n8, 189, 190, 195, 196 n49, 201, 203, 205, 205 n8, 206, 265 n4, 270, 274 n35, 275, 281, 282, 288, 299, 312, 313, 316 n48, 336, 336 n51, 359, 369 n50, 370 n54, 375, 375 n69, 380 n3, 384, 388, 389, 399, 401, 403, 409, 410, 411, 417, 419, 420 n57, 421, 425, 427, 431, 432, 435, 439, 486, 486 n41, 503, 505, 506, 506 n19, 507, 508, 511, 516, 534, 545, 546, 550 n47, 554, 567, 570; theory of sensation, 299, 300; theory of vision, 307; works of: *Book of the Healing of the Soul*, 189; *Canon of Medicine*, 206 n15, 227 n103, 232, 232 n127, 336 n53, 359 n11, 370, 389, 392, 393, 395, 396, 398, 400, 408, 426, 431; *De anima*, 268 n11, 268 n12, 268 n13, 269 n17, 270 n19, 270 n20, 272, 272 n28, 273 n32, 273 n33, 274 n36, 276 n41, 277 n48, 279 n51, 281 n62, 282 n65, 282 n66, 293 n6, 299 n14, 316 n48, 503, 504, 504 n9; *De anima in arte alchemiae see* Pseudo-Avicenna; *De animalibus*, 336 n53, 393, 395, 398, 400, 410, 411, 418, 426, 431, 435 n118, 573; *De congelatione et conglutinatione lapidum*, 189, 189 n17, 198, 568; *De mineralibus see De congelatione*; *Kitab al-shifa' see Book of Healing*; *Liber canonis see Canon of Medicine*; *Metaphysica*, 101 n117; *Sextus de naturalibus see De anima*; *Sufficientia*, 130, 132-134, 133 n8, 133n 9, 133 n10, 133 n11, 133 n12, 134 n13, 134 n14, 137 n26, 141, 141 n40, 143 n46, 143 n47, 144 n56, 150 n67, 151 n69, 151 n70, 151 n71, 151 n74

Pseudo-Avicenna, *The Physical Stone (De anima in arte alchemiae)*, 188

Azo, R. F., 188 n8

Bach, Josof, 266 n5

Bacon, Roger, 14, 19, 25, 55-72, 76, 76 n13, 76 n15, 78 n29, 79, 87, 94, 95, 99 n107, 100, 179 n49, 101, 196, 207, 341 n1, 413, 422 n68, 465, 470, 511, 512, 539, 567; career, 53-55, 77; on *experimentum*, 76, 76 n17, 77 n23; reputation, 53; scholarship on, 55-57; works of, 53, 57-58, 79 n31, 172; *Communia naturalium*, 58, 66, 70-71, 70 n32, 71 n33, 72, 94 n93, 101 n117; *Compendium studii*

497, 510 n32, 521 n81

al-Ghazzālī (Algazel), 556; *Metaphysica*, 270 n20, 274 n36, 288, 302 n18, 384

Giambattista della Porta *see* Porta, Giovanni Battista della

Gigil (Abu Da'ud ibn Juljul), 188, 191, 191 n26, 192, 201

Giles, J. A., 221 n79

Giles of Rome, 425 n88, 427 n92, 543, 545, 547, 551, 555, 557, 560, 560 n84

Gilson, Etienne, 16 n8, 27, 45, 45 n106, 99 n107, 103 n2, 129, 129 n1, 306 n28, 503 n8, 506 n19, 552 n58

Gliedman, Lester H., 213 n57

Gloria, Andrea, 386 n22, 538 n2

Glorieux, Palémon, 16, 16 n8, 22 n38, 522 n83

Gnudi, Martha T., 241 n12

Godfrey of Bleneau, 24

Godman, Eva M., 457, 457 n63, 459 n70, 459 n72

Goldat, George David, 481 n15, 487 n45

Goldstein, Bernard R., 170 n26, 171 n29, 172 n31

Goldwater, Leonard J., 227 n103

Gollancz, Hermann, 263 n2

Gorce, M. M., 563 n96

Goslar, 20, 217, 219

Gottfried of Duisburg, 46

Grabmann, Martin, 101 n118, 502, 502 n3, 510 n32, 515 n56, 522 n84, 535 n131, 541, 541 n8

Grant, Edward, 158 n4, 342 n4, 347 n17, 350 n28, 351 n32, 383 n9, 392 n38

Graupen, Bohemia, 220

Greece, 28 n51

de Gregoriis brothers, 561, 561 nn89-91

Gregory, Tullio, 554 n60

Gregory X, pope, 58

Gregory of Nyssa, 274, 274 n35, 275, 388; *De natura hominis*, 269 n18, 273 n32, 274 n35, 276 n41

Grosseteste, Robert, 58, 60, 64, 71, 75, 79 n31, 306 n28, 319, 319 n56, 465, 468, 469, 476, 476 n40, 494 n69, 544, 565-567, 575; *De motu supercaelestium*, 172; *Physics*, 75, 476 n39; *Posterior Analytics*, 75, 113, 469 n16

Guéric of St.-Quentin, 23, 24

Guglielmo da Pastrengo, 179 n50

Guibert de Tournai, 24

Guillaume d'Etempes, 24

Guillemus Falconarius (William the Falconer of King Roger of Sicily), 442, 443, 443 n10, 443 n13, 444, 444 n14, 444 n16, 445, 445

n20, 446, 446 n22, 456, 456 n58, 460

Gundissalinus, Dominicus (Gundissalvi), 94, 383; *De divisione philosophiae*, 94 n91, 383 n9, 483, n29

Gundissalvi, Dominic *see* Gundissalinus, Dominicus

Halberstadt, 48

Hales, Alexander of *see* Alexander of Hales

Hall, Thomas S., 379 n1

Haly ibn Rodan, 184, 380 n1, 399 n57

Hamburg, 48

Hamesse, Jacqueline, 491 n54

Hamilton, William James, 406 n4, 410 n12, 431 n101

Hamlyn, David, 506 n17

Hammarskjöld, Dag, 2

Hansen, Bert, 379

Hardie, R. P., 108, 108 n14, 119

Harkness, Robert, 453 n47, 456 n55, 456 n56, 460 n81

Harting, James E., 443 n9, 445, 445 n18

Harvard University, 561 n91

Harvey, 440

Harz Mountains, Lower Saxony, 20

Haskins, Charles Homer, 74 n6, 444 n16

Hauberg, Paul, 357 n7

Haubst, Rudolf, 524 n90, 530 n110, 530 n111, 531 n115, 531 n116, 531 n117, 532 n120, 532 n125

Hawthorne, John G., 224 n89, 237, 237 n7

Heath, Thomas L., 164 n15, 498 n87

Hegel, George Wilhelm Friedrich, 5 n8

Heiberg, J. L., 466 n7, 482 n20

Heidelberg, 528, 531, 532 n124, 533

Heimericus de Campo *see* van den Welde, Heymerich

Heines, Sr. Virginia, 197 n53

Heinrich von Brussel, 502 n3, 515 n56

Heinrich von Gorkum, 523, 523 n88, 523 n89, 524, 524 n93

Heinzel, Hermann, 449 n32

Helm, Karl, 374 n64

Hemmerdinger, B., 355 n1

Hempel, Carl G., 104 n5

Henrick Harpestraeng, 357 n7, 363 n29, 375 n69

Henricus Aristippus, 567

Henry of Cologne, Friar, 26

Henry of Cologne, Friar (Jordan's classmate), 19, 28

Henry of Ghent, 30

Henry of Herford, 16, 16 n8, 18, 20, 20 n28,

Subject Index

comparative
chickory, 363, 368 n47
chiromancy, 89
chlorine, 248
chorion *see* fetal membranes
Christ, 430
Church, 66
chryselectrum see amber
cinnabar *see* mercury sulphides
classification of plants *see* plants
classification of sciences *see* science
cloister of Arndelhousen, 221 n82
clovers, 353
cloves, 362 n22
coagulation, 196, 213, 246, 247, 249, 259
Collegio Romano, 127, 127 n64
colour, 193, 206, 208, 268, 271, 272, 280, 291
n2, 292, 296, 298, 299, 300, 301, 303, 304,
305, 307, 314, 319, 348, 350, 363, 364, 397;
nature of, 295, 296, 307, 312 n41; spiritual
being of, 296, 306, 307, 309 (*see also* air;
contraries; light; medium; *species*; water)
comets, 20, 159, 180
common sense *see* senses, internal
complexion theory in medicine, 390 n34, 391,
395, 398, 399, 400; in plant life, 350
conception, 180, 181, 183, 209, 390 n34, 399,
400, 401, 402, 403, 407, 412, 414, 417, 425,
427, 430, 435
condemnation of 1277, 43, 45
contingency, 106
continuum, the 472-76
contraries, coincidence of, 182, 294, 295, 307,
309, 310 (*see also* celestial bodies; colour;
sound)
copper, 191, 200, 216, 218, 219, 224, 225, 225
n94, 227, 242, 247, 248; alloys, 216, 225,
225 n95, 227, 227 n105; pyrites, 220 n76
(*see also* poling)
coral, 213
cornelius, 210, 210 n41, 212
corpus mobile see body
creation, *ex nihilo*, 87; of human soul *see* animation
creationism, 407
crows, 128
crystallus, 210
cursor biblicus, 22
cyperus, 353

dator formarum see Giver of forms
De cura membrorum see Galen, *De iuvamentis
membrorum*

De divisione locorum habitabilium (Anon.),
567
death, 322, 323, 324, 382
deduction, 80
definition, 79, 80, 80 n36, 111, 112, 113, 139;
geometric *see* geometry; material, 131, 136,
139
demons, 375, 434
demonstration, 84, 103, 111, 112, 113, 114,
121, 126, 156; *ex suppositione finis* (per suppositionem), 105, 107, 110, 111, 113, 121,
125, 126 (*see also* necessity, conditional);
mathematical, 119, 120; *propter quid*, 125;
quia, 125; *see also* supposition
desire, 209, 346
determinism, 157, 175, 182-185, 576-577
diagnosis, differential, 360, 363
dialectics, 97, 120, 126
diet, 213, 330, 357, 371, 372, 395
digestion, 195, 326, 327, 331-337, 338, 419
dill, 362, 362 n23
disputation, academic, 23, 413
distillation, 192, 198, 200, 253 (*see also* alcohol)
dittany, 360, 360 n12
divination, 182, 375
doctor expertus see Albertus Magnus, designa-
tions
doctor of the Church *see* Albertus Magnus,
Doctor Ecclesiae
doctor universalis see Albertus Magnus, desig-
nations
Dominican chair, 21; for "externs", 23; for
foreigners, 24; for France, 24
Dominican General Chapter, 1248 in Paris,
27-28; 1255 in Milan, 33; 1256 in Paris, 34;
1257 in Florence, 36; 1259 in Valenciennes, 37, 572; May 1277, 50; 1303, 51
Dominican Order, 1, 3, 9, 13, 17, 28, 45, 69,
157, 158, 231, 386, 387, 388, 538, 538 n3,
540-541 n7, 555 n65, 562, 563; chapter
house of St.-Michael, 221 n82; constitutions, 18; houses in Teutonia, 48-51; *ratio
studiorum*, 30, 37; *studium generale*, 14, 25,
28, 78; *see also* Albertus Magnus, entry
into Dominican Order; *studium generale*
Dominican provinces; Saxony, 33, 51; Teutonia, 14, 33, 51
Dominican Provincial Chapter, 1225, of Magdeburg, 48; 1254, of Worms, 33, 48; 1255,
of Regensburg, 33-34; 1256, of Erfurt, 34;
1257, of Augsburg, 35; 1257, of Strassburg,
36

essence; sensible, the
forma fluens, 129, 130, 141, 142, 143, 145, 147
n61, 147 n62, 149, 149 n66, 150, 151 n70,
152, 153
formative virtue *see virtutes*
fortuitous events *see* chance
foundry practice, 243
Franciscan chair, 24
Franciscan Order, 45, 54, 61, 68, 512, 551
freedom, 83, 89
fumes *see* evaporation
fungi, 351 n31, 376

gagates see jet
Galenism, 286, 391-292, 402
gall *see* humour(s)
garden, 365
garlic, 376
gecolitus, 211
generation, 106, 122, 123, 128, 175, 177, 180,
317, 332, 333, 334, 400, 406 (*see also* conception; embryonic development; plants)
genetics, 438, 439
genus, genera, 206, 350, 350 n28, 351, 360,
366, 366 n39, 367
geomancy, 89
geometry, 6, 56, 477-479; axioms, 486; definitions, 484-485, 491, 492, 498; etymology,
483; generation of objects, 466, 475; postulates, 485; *practica*, 486; problem of exactness, 469-470; *theorica*, 486
German Dominican Province *see* Dominican
provinces
Gesamtkatalog der Wiegendrucke, 561 n86
gestation *see* embryonic development; pregnancy
Giver of forms, 421, 421 n63
gladiolus, 348
glass, 200, 204, 242, 255, 256, 257, 258, 294
God, 12, 61, 86 n67, 99, 100, 174, 184, 214,
215, 556
gold, 191, 192, 193, 194, 195, 198, 199, 216,
217, 218, 219, 219 n73, 222, 223, 225, 225
n94, 228, 235, 242, 255, 256, 257, 260, 336
grape vines, 352, 353
Great Convent of the Cordeliers, 23
Greek(s), 64, 106, 107, 108, 159, 369, 396
growth, 269, 322, 324, 326, 328, 332, 333, 333
n38, 334
gynecology, 384, 395

haeccietas, 555
hair *see also* anatomy, 277, 397
happiness, 553

head, 397-398 (*see also* anatomy)
hearing *see* senses, external
heart, 209, 264, 283, 284, 285, 286, 300, 336,
393, 399, 400, 401, 402, 403, 425, 427, 428,
545 (*see also* anatomy; organs of the body)
heat, 176, 177, 180, 193, 205, 218, 219, 238,
239, 240, 248, 250, 254, 255, 256, 271, 284,
322-323, 326-327, 330, 335, 336, 337, 346,
349, 352, 353, 372, 397, 399
heavenly bodies *see* celestial bodies
heavens, 155, 547-548, 554, 556 (*see also*
celestial bodies)
Heilige Kreuz, Dominikanerkloster, 28
hellebore, 376
hematite, 212
hemorrhoids, 209, 210, 212
henbane, 375
herb, herbal(s), 4, 213, 214, 343, 348 n21, 356,
357, 357 n7, 358, 359, 361, 362, 363, 365,
366, 367, 368, 369, 369 n50, 370, 371, 373,
377, 385, 572; contents of, 358; definition
of, 356; rubrics used in, 359-360; therapeutic data in, 371-373
herbalism, 355, 356, 376, 377
herbalist, 356, 357, 358, 360, 361, 363, 364,
366, 372, 376, 377
heredity *see* genetics
hermaphroditism, 412, 437-438
hermetic *see* Hermes
hiena see opal
hippopede, 163
hop vine, 361, 361 n18, 361 n21
horsetails, 353
humidum nutrimentale, 325, 327, 335, 338
humidum radicale, 322, 323 n8, 325, 327, 335,
338
humors, 9, 181, 275, 296, 327, 330, 336, 346,
357, 372, 373, 395, 398, 399, 410 (*see also*
digestion; semen)
hydragyrum *see* mercury
hydromancy, 89
hypotheses, 88, 96, 106, 107, 116, 174, 223,
230
hypothetics, 77 n23, 79-80, 126, 127 (*see also*
deduction; scientific method)
hyssop, 372, 372 n57

ideas, innate, 79
images *see* intellection; senses, internal,
phantasy
imagination, 179, 303, 435 (*see also* senses,
internal)
immateriality, 98, 100, 291, 291 n1, 297, 304,
305, 309 n32

zation of, 277, 278 n49, 282; memory, 273, 274, 275, 275 n39, 282, 288; phantasy, 273, 278, 279, 280, 282, 288, 505-506, 514, 525, 526, 528, 530-532

sensible, the, 271, 291, 292, 293, 297, 298, 300, 301, 302, 304, 305, 306, 307, 311, 312, 313, 318, 319, 320, 551; as pure, 302 n19, 312; formally or materially, 122, 316; having spiritual being, 307, 314; intrinsically sensible, 305, 319; its primacy in natural science, 298; mover of the sense, 293 n6, 312; *per se*, 305; proper, 268, 271, 272, 273, 280

Sentences, Bachelor of, 21

separable substances *see* intellection

serpent, 373

sex, differentiation of, 415, 427, 437-438; in plants, 346, 347, 347 n16, 365 n36; of stones, 211

sight, the sense of *see* senses, external

sigils, 212, 231

signatures, doctrine of, 373

silver, 191, 192, 193, 194, 198, 199, 216, 217, 221, 222, 223, 228, 240, 242, 243

skin, 277, 300 (*see also* anatomy)

sin, original, 407

sleep, 284; in plants, 345-346

smell *see* senses, external

sodium chloride *see* salt

solution, chemical, 195, 198

soul, human, 86, 182, 183, 267, 270, 271, 283, 286, 296, 302 n18, 318, 327, 388, 402, 546, 553, 556-557, 560; definition of, 501, 504; immortality of, 550 n47, 553, 554; individuation of, 548-549, 549 n46, 555; origin of, 406, 407, 422-423, 504, 509, 512-513, 525-526, 549-550, 554 (*see also* animation; creation); powers of, 266, 267, 271, 279, 344, 505-508, 509, 513-515, 516, 518, 520-521, 525, 545, 551-552 (*see also virtutes*); union with body, 407, 419, 504, 508, 509, 513, 515-516, 517-518, 527, 548-549

soul, vegetative, 322, 324-325, 326, 328, 329, 334, 335, 338, 344, 345, 346, 347, 349, 350, 545 n33, 556-557

sound, 268, 271, 272, 276, 281, 297-298, 301, 303, 305, 314, 318, 319, 560; medium of, 304, 317; nature of, 312 n41; spiritual being of, 305, 307, 309, 310

space, 162, 470-472

species, 112, 181, 206, 273, 280, 293 n6, 302, 303, 304, 311, 315, 316, 345 n11, 350, 350 n28, 351, 360, 363, 366, 366 n39, 367, 368, 369, 515; intelligible, 549 n46, 550, 550 n49,

554, 558-559; multiplication of, 306 n28, 307 n29, 311; of colour, 272, 275-276, 298; of flavour, 314; of sensibiles, 270, 271, 276, 280, 297, 298; of smell, 273; of sound, 272, 298; of visibles, 293, 295, 296

specularis, 210

sperm, 181; origin of, 419

spirit, 86, 276, 277, 280, 282, 284, 295, 306, 335; animal, vital, 271, 284, 286, 301, 302; its role in touch and taste, 300, 301-302, 302 n18, 316

spiritual being, 296, 298, 299, 300, 301, 302, 304, 304 n22, 309 n32, 311-312, 313, 316, 318, 319, 319 n55; in the medium, 294, 297, 298, 303-307, 308, 310, 318, 319, 320; *see also* colour; immaterial entities

spiritus, 423, 428, 430

squill, 376, 376 n75

star(s), 4, 162, 163, 167, 173, 177, 178, 180, 181, 549

steel, 236, 237, 239, 240, 244, 245

Stoics (Platonists), 32, 93, 95, 96, 175, 407, 420, 420 n54, 421, 577

stomach, 209, 325, 328, 330, 335, 336, 396 (*see also* anatomy)

stones, 4, 189, 194, 203, 204-206, 208-215, 217, 218, 222, 227, 233, 237, 254, 324, 330, 348, 348 n21, 379 n1; *see also* bladder stone; kidney stone

storms, 208, 213

studium generale, 28, 59, 69, 158, 386 (*see also* Dominican Order)

subalternation of sciences *see* science(s), subalternation of

sublimation, 189, 192, 193, 195, 198, 200, 217, 223, 247, 253

substance, 133, 134, 137, 139, 194, 266, 332, 337, 350, 401

substantial change *see* change

substantial form *see* form

succinus see amber

sugar, 252

sulphur, alchemical, 188, 189, 192, 193, 194, 195, 199, 217-219, 220 n76, 222, 238, 242, 246, 247, 248, 250, 251, 252, 253, 254, 330

sun, 159, 283-284, 308, 327

superstition, 5, 89

supposition, 110, 111, 116, 117, 119, 119 n43, 122, 125, 126, 127, 128 (*see also* demonstration; necessity, conditional)

surgery, 388

syllogism, 6, 7

synonyms, 359, 367, 368, 368 n44, 368 n46,

Index of Citations of Works Attributed to Albert

Index of Citations of the Works of Albertus Magnus

I, tr.1, c.7: 284 n71, 284 n72, 284 n74, 400 n64
I, tr.2, c.6: 400 n46
I, tr.2, c.8: 395 n47
II, tr.2, c.1: 284 n76
II, tr.2, c.2: 284 n75
III, tr.2, c.12: 10 n28, 10 n29
III, tr.2, c.5: 182 n64

—. *De spiritu et respiratione, libri II*: 93, 93 n85, 379 n1, 571
I, tr.1, c.4: 571
I, tr.2, c.1: 400 n64, 402 n72
I, tr.2, c.2: 571
II, tr.1, c.2: 571

—. *De unitate intellectus*: 34, 414, 562
Liber II: 422 n71
III, §1: 570

—. *De vegetabilibus et plantis, libri VII*: 31, 93, 93 n87, 207, 321, 321n, 324, 341, 342, 342 n2, 342 n3, 343, 356 n2, 370 n52, 412, 569, 571, 572, 573
Libri I-V: 342 n3, 343, 344-350, 353, 354, 370
Liber I: 344, 345, 347, 351
I, tr.1, c.1: 345 n10, 346
I, tr.1, c.4: 572
I, tr.1, c.7: 347 n17
I, tr.1, c.8: 345 n11
I, tr.1, c.9: 346 n12, 346 n13, 346 n14
I, tr.1, c.11: 346 n15
I, tr.1, c.12: 344 n9, 347 n16, 347 n17
I, tr.2, c.1: 347 n18, 347 n19
I, tr.2, c.4: 352 n35
I, tr.2, c.5: 351 n32
Liber II: 344, 347, 350, 352
II, tr.1: 347
II, tr.1, c.1: 327 n15, 328 n21, 344 n8, 350 n30, 351 n32
II, tr.1, c.2: 351 n31, 352 n36, 352 n37, 353 n38, 353 n39, 353 n40
II, tr.2: 347
II, tr.2, c.2: 352 n34
Liber III: 344, 347, 348, 350
III, tr.1, c.1: 348 n20
III, tr.1, c.6: 207 n19
Liber IV: 344, 348, 349
IV, tr.1, c.1: 326 n14, 327 n18
IV, tr.1, c.2: 348 n22
IV, tr.2, c.4: 348 n23
IV, tr.3, c.3: 349 n24

IV, tr.4, c.1: 344 n9
Liber V: 344, 348, 349, 350
V, tr.1, c.1: 344 n8, 349 n26, 349 n27
V, tr.1, cc.3-8: 350 n28
V, tr.1, c.7: 350 n29
V, tr.2, c.6: 350 n29
Libri VI-VII: 342 n3, 345 n5
Liber VI: 343, 350, 356, 356 n2, 358, 358 n9, 359, 361, 364, 366, 368, 368 n44, 370 n52, 371, 372, 373, 374, 385, 572
VI, tr.1, c.1: 6 n13, 6 n14, 27, 87 n68, 385 n18
VI, tr.1, c.2: 207 n19
VI, tr.1, c.4: 352 n35
VI, tr.1, c.22: 385 n17

—. *Liber de aetate sive de juventute et senectute*: 93, 93 n86, 379 n1, 571

—. *Liber de memoria et reminiscentia*: 93, 93 n83, 275 n39, 283, 283 n69, 570
I, tr.1, c.1: 570

—. *Liber de morte et vita*: 93, 93 n86, 379 n1, 569, 571
tr.2, c.1: 385 n17
tr.2, c.11: 571

—. *Liber de natura et originae animae*: 32, 92-93, 93 n82, 413 531, 541, 549, 552, 553, 554, 560, 561, 570, 574
tr.1, c.3: 422 n71
tr.1, c.4: 574
tr.1, c.5: 504 n12, 553 n60
tr.1, c.6: 505 n14, 553 n60
tr.1, c.7: 553 n60
tr.2, c.2: 554 n61
tr.2, c.6: 554 n61
tr.2, c.7: 553 n60
tr.2, c.8: 553 n60
tr.2, c.11: 553 n60
tr.2, c.12: 553 n60
tr.2, c.13: 553 n60
tr.2, c.14: 553 n60
tr.2, c.15: 554 n60

—. *Liber de natura locorum*: 31, Plate 3, 92, 228, 385, 542, 566-567, 571
tr.1, c.1: 87 n68
tr.3, c.2: 17 n14, 539 n5

—. *Liber de nutrimento et nutribili*: 92, 92 n81, 321, 321 n1, 324, 331, 333, 336, 337, 379

Index of Manuscripts